Thermodynamics of Solutions

Eli Ruckenstein and Ivan L. Shulgin

Thermodynamics of Solutions

From Gases to Pharmaceutics to Proteins

Eli Ruckenstein
SUNY, Buffalo
Department of Chemical &
  Biological Engineering
303 Furnas Hall
Buffalo NY 14260-4200
USA
feaeliru@buffalo.edu

Ivan L. Shulgin
SUNY, Buffalo
Department of Chemical &
  Biological Engineering
303 Furnas Hall
Buffalo NY 14260-4200
USA
ishulgin@hotmail.com

ISBN 978-1-4419-0439-3   e-ISBN 978-1-4419-0440-9
DOI 10.1007/978-1-4419-0440-9
Springer Dordrecht Heidelberg London New York

Library of Congress Control Number: 2009929313

© Springer Science+Business Media, LLC 2009
All rights reserved. This work may not be translated or copied in whole or in part without the written permission of the publisher (Springer Science+Business Media, LLC, 233 Spring Street, New York, NY 10013, USA), except for brief excerpts in connection with reviews or scholarly analysis. Use in connection with any form of information storage and retrieval, electronic adaptation, computer software, or by similar or dissimilar methodology now known or hereafter developed is forbidden.
The use in this publication of trade names, trademarks, service marks, and similar terms, even if they are not identified as such, is not to be taken as an expression of opinion as to whether or not they are subject to proprietary rights.

Printed on acid-free paper

Springer is part of Springer Science+Business Media (www.springer.com)

Preface

This book consists of a number of papers regarding the thermodynamics and structure of multicomponent systems that we have published during the last decade. Even though they involve different topics and different systems, they have something in common which can be considered as the "signature" of the present book. First, these papers are concerned with "difficult" or very nonideal systems, i.e. systems with very strong interactions (e.g., hydrogen bonding) between components or systems with large differences in the partial molar volumes of the components (e.g., the aqueous solutions of proteins), or systems that are far from "normal" conditions (e.g., critical or near-critical mixtures). Second, the conventional thermodynamic methods are not sufficient for the accurate treatment of these mixtures. Last but not least, these systems are of interest for the pharmaceutical, biomedical, and related industries.

In order to meet the thermodynamic challenges involved in these complex mixtures, we employed a variety of traditional methods but also new methods, such as the fluctuation theory of Kirkwood and Buff and ab initio quantum mechanical techniques.

The Kirkwood-Buff (KB) theory is a rigorous formalism which is free of any of the approximations usually used in the thermodynamic treatment of multicomponent systems. This theory appears to be very fruitful when applied to the above mentioned "difficult" systems. Indeed, in some cases (see Chapters 3–5) this theory allows one to obtain results that are scarcely attainable by traditional thermodynamic methods. For example, in mixtures of three or more components, thermodynamics can not provide any rigorous relations connecting the thermodynamic properties, such as the activity coefficients of components in multicomponent mixtures to those in binary mixtures of the constituents, whereas the KB theory of solutions provides such relations. It allows one, for example, to develop a method for predicting or correlating the solubility of drugs in multicomponent solvents. Let us emphasize again that the above results are possible because the KB theory provides a rigorous method for the treatment of multicomponent (not only binary!) mixtures.

We have employed ab initio quantum mechanical techniques to obtain information about the intermolecular interactions and the geometry of large molecular clusters. Such information was very helpful in the understanding of the properties of the "difficult" systems and is not readily obtained by other computational methods.

The results presented in this book have both theoretical and practical importance. Among the most important theoretical results one can list: I) the explanation of density augmentation in dilute supercritical mixtures (chapter 2); II) a new method for computing the excess (or deficit) around a central molecule in multicomponent mixtures (chapter 1); III) new criteria for the salting-in or salting-out effect on gas solubility (Chapter 3); and IV) new methods for analyzing protein-solvent and protein-protein interactions in aqueous protein solutions (Chapter 5).

On the other hand, the results regarding: A) the Henry constant in binary and in multicomponent solutions (Chapter 3); B) the entrainer effect in supercritical mixtures

(Chapter 2); C) methods for correlating the solubility of drugs and environmentally important substances in multicomponent solutions (Chapter 4); and D) protein solubility and its correlation with the preferential binding parameter (Chapter 5) can have practical applications.

We also hope that the book will be helpful to undergraduate and graduate students who are interested in the applications of modern methods of thermodynamics and statistical mechanics to complex systems.

Contents

1 The Kirkwood–Buff integrals and their applications to binary and ternary solutions 1

2 Supercritical mixtures 75

3 Solubility of gases in mixed solvents 153

4 Solubility of pharmaceuticals and environmentally important compounds 197

5 Aqueous solutions of biomolecules 251

6 Water and dilute aqueous solutions 317

Chapter 1

The Kirkwood–Buff integrals and their applications to binary and ternary solutions

1.1 Kirkwood–Buff integrals in aqueous alcohol systems: comparison between thermodynamic calculations and X - ray scattering experiments.
1.2 Kirkwood–Buff integrals in aqueous alcohol systems: aggregation, correlation volume and local composition.
1.3 Range and energy of interaction at infinite dilution in aqueous solutions of alcohols and hydrocarbons.
1.4 Aggregation in binary solutions containing hexafluorobenzene.
1.5 Hydrophobic self-assembling in dilute aqueous solutions of alcohols and hydrocarbons.
1.6 Effect of a third component on the interactions in a binary mixture determined from the fluctuation theory of solutions.
1.7 The Kirkwood-Buff theory of solutions and the local composition of liquid mixtures.
1.8 Excess around a central molecule with application to binary mixtures.
1.9 Reply to Comment on The Kirkwood–Buff theory of solutions and the local composition of liquid mixtures.
1.10 An improved local composition expression and its implications for phase equilibrium models.

Introduction to Chapter 1

Chapter 1 is devoted to the application of the Kirkwood–Buff theory of solutions to the investigation of the structures of binary and multicomponent mixtures. The analysis involves the quantity Δn_{ij}, which represents the excess (or deficit) number of molecules of species i around a central molecule of species j compared with a hypothetical mixture in which molecules of species i are distributed randomly around a central molecule of species j.

The KBIs can be accurately determined from measurable macroscopic properties, such as the derivatives of the chemical potentials with respect to concentrations, the isothermal compressibility and the partial molar volumes. To date, the KBIs have been calculated for numerous binary (see 1.1, 1.4–1.5, 1.7–1.8) and ternary (see 1.6) mixtures. The KBI could also be determined experimentally by small-angle X-ray scattering, small-angle neutron scattering, light scattering, and other methods. Comparison of these experimentally determined KBIs with those obtained from thermodynamic calculations demonstrated good agreement between them (see 1.1).

The KB theory provides a unique opportunity to obtain information about the structure of liquid mixtures at a nanometer level from the excesses $\left(\Delta n_{ij}\right)$. However, it took a long time to find the correct procedure to calculate the above excess (or deficit). For several decades the calculations of Δn_{ij} were based on the expression suggested by Ben-Naim [1]:

$$\Delta n_{ij} = c_i G_{ij} \qquad (1)$$

where c_i is the molar concentration of species i in the mixture and G_{ij} are the Kirkwood - Buff integrals (see 1.7–1.9 for details). However, the excesses (or deficits) calculated with (Eq. 1) provide nonzero values for ideal binary mixtures and the opinion [2] was expressed that they should be equal to zero in that case. For this reason it was proposed in [Ref. 2], and also by us (see 1.1), that the excess (or deficit) number of molecules around a central molecule should be calculated with respect to a reference state (for example, the ideal binary mixture [2])

$$\Delta n_{ij} = c_i (G_{ij} - G_{ij}^R) \qquad (2)$$

where G_{ij}^R are the Kirkwood–Buff integrals of the reference state. More recently we suggested a different reference state, rather than the ideal mixture. However, this yielded results comparable to those obtained with the ideal mixture reference state (see 1.1 for details). The excesses (or deficits) calculated with Eq. (2) for various reference states were dramatically different (in many cases) from those provided by Eq. (1).

There is a basic issue regarding the suggested reference states:

Is, indeed, in an ideal mixture, the excess (or deficit) number of molecules around a central molecule equal to zero? This is true if the distribution of the components in an ideal mixture is random. However, because in an ideal mixture the volumes of the components are different, there is no absolute randomness in them.

Such considerations led us to a new treatment, which took into account the existence of a volume which is not accessible to the molecules surrounding a central molecule and which revealed that, for ideal mixtures, the excesses (or deficits) are not zero (1.7-1.10). This observation allowed us to apply the KB theory to the investigation of the microstructure of a wide range of mixtures starting from binary Lennard-Jones fluids to mixtures with complex intermolecular interactions including hydrogen bonding (see 1.7–1.9).

In addition, this new insight suggested an improved method for the treatment of phase equilibria (1.10).

[1] Ben-Naim, A. J. Chem. Phys. 1977, 67, 4884-4890.
[2] Matteoli, E.; Lepori, L. J. Chem. Soc., Faraday Trans. 1995, 91, 431-436 and Matteoli, E. J. Phys. Chem. B 1997, 101, 9800-9810.

Kirkwood−Buff Integrals in Aqueous Alcohol Systems: Comparison between Thermodynamic Calculations and X-Ray Scattering Experiments

I. Shulgin[†] and E. Ruckenstein*

Department of Chemical Engineering, State University of New York at Buffalo, Amherst, New York 14260

Received: August 18, 1998; In Final Form: January 19, 1999

Thermodynamic data were used to calculate the Kirkwood−Buff integrals for the aqueous solutions of methanol, ethanol, propanols, and butanols. The calculated values have been compared to those obtained from small-angle X-ray scattering (SAXS) experiments, and satisfactory agreement was found. Improved expressions have been suggested and used to calculate the excess number of molecules around central ones. On this basis, information about the structure of the solutions was gathered, the main conclusion being that the interactions among similar molecules prevail; this leads to clustering dominated by the same kind of molecules.

Introduction

Many models are available for describing the thermodynamic behavior of solutions.[1−2] However, so far no one could satisfactorily simulate the solution behavior over the whole concentration range and provide the correct pressure and temperature dependencies. This generated interest in the thermodynamically rigorous theories of Kirkwood−Buff[3] and McMillan−Mayer.[4] In the present paper, the emphasis is on the application of the Kirkwood−Buff theory to the aqueous solutions of alcohols, because it is the only one which can describe the thermodynamic properties of a solution over the entire concentration range.[5] The key quantities in the Kirkwood−Buff theory of solution are the so-called Kirkwood−Buff integrals (KBIs) defined as

$$G_{ij} = \int_0^\infty (g_{ij} - 1) 4\pi r^2 \, dr \qquad i,j = 1, 2 \qquad (1)$$

where g_{ij} is the radial distribution function between species i and j and r is the distance between the centers of molecules i and j. From the KBI's valuable information regarding the structural and energetic features of the binary, ternary and multicomponent solutions could be obtained.[3] Ben-Naim indicated how to calculate the KBIs from measured thermodynamic properties.[6] Since then, the KBIs have been calculated for numerous systems[7−14] and the results used to examine the solution behavior with regard to (1) local composition, (2) various models for phase equilibrium, and (3) preferential solvation and others.[15−18]

The KBIs could also be determined experimentally from small-angle X-ray scattering, small-angle neutron scattering (SANS) and light-scattering[5,19−20] experiments. It is worth mentioning that SAXS and SANS experiments allow one to obtain not only the KBIs, but also to gather information about the formation of complexes and clusters in solutions.

Numerous SAXS determinations for aqueous alcohol solutions are available.[20−30] The aqueous solutions of tertiary butanol (*tert*-butyl alcohol) were examined in refs 20−24, the methanol−water system in ref 25, and the aqueous solutions of ethanol in ref 26. The aqueous solutions of 1-propanol were studied in a number of papers,[27−29] and the 2-propanol + water system was also investigated.[27−28] It is worth mentioning that these investigations were performed over a wide range of concentrations and that measurements at different temperatures were conducted.[24,28] All the systems investigated in refs 19−29 were homogeneous. SAXS measurements were also carried out for binary aqueous solutions of 1-butanol, 2-butanol, and iso-butanol for a few compositions near the transition to the two-phase region.[30] The aggregation or cluster formation was the subject of SAXS research as well: aggregates were identified in the aqueous systems of any of the propanols or butanols and the dependence of the aggregate size on composition obtained. SANS was much rarely employed; it was used to study the aqueous alcohol solutions at a few concentrations.[19]

The aqueous alcohol systems were chosen for the current research because (1) aggregation in these systems was found by several independent experimental methods; (2) while there are several investigations regarding their KBI's, there is no agreement between the reported data. This becomes clear if we compare the peak G_{11} values in the system 1-propanol (1)−water (2). (Throughout this paper, component 1 represents the alcohol and component 2 the water.) (3) For all the selected systems, accurate thermodynamic data are available. (4) Last but not least, SAXS data are available for most of the systems chosen (except the methanol−water system), but a systematic comparison between the KBI's extracted from the SAXS data and those obtained thermodynamically was not yet made.

Therefore the aim of this paper is to calculate the KBIs for aqueous solutions of alcohols from thermodynamic data, to compare the results to those obtained from SAXS measurements, and to examine some specific features regarding the structure of aqueous alcohol solutions.

Theory and Formulas. *1. KBI's Calculation.* The main formulas for calculations[7,13−14] are

$$G_{12} = G_{21} = RTk_T - \frac{V_1 V_2}{VD} \qquad (2)$$

$$G_{ii} = G_{12} + \frac{1}{x_i}\left(\frac{V_j}{D} - V\right) \qquad i \neq j \qquad (3)$$

where

$$D = \left(\frac{\partial \ln \gamma_i}{\partial x_i}\right)_{P,T} x_i + 1 \qquad (4)$$

In eqs 2−4, k_T is the isothermal compressibility, V_i is the partial

---

* To whom the correspondence should be addressed. E-mail: feairu@acsu.buffalo.edu. Fax: (716) 645-3822.
[†] Current e-mail address: ishulgin@eng.buffalo.edu.

molar volume of component i, x_i is the molar fraction of component i, V is the molar volume of the mixture, T is the absolute temperature and γ_i is the activity coefficient of component i. Because the dependence of k_T on composition is not known for all the systems investigated and because of the small contribution of $RT\, k_T$ to the KBI's,[7] the dependence of k_T on composition will be approximated by

$$k_T = k_{T,1}^0 \varphi_1 + k_{T,2}^0 \varphi_2 \tag{5}$$

where φ_i is the volume fraction of component i in solution and $k_{T,i}^0$ is the isothermal compressibility of the pure component i.

The analysis of the possible errors in the calculation of KBI's clearly indicated that the main error is introduced through the D value.[7,31] The usual way for calculating D is from isothermal vapor−liquid equilibrium data, by assuming that

$$(\partial \ln \gamma_i / \partial x_i)_{P,T} = (\partial \ln \gamma_i / \partial x_i)_T \tag{6}$$

The main uncertainty in the derivative of the activity coefficient or partial pressure is caused by the vapor composition which is needed to calculate the activity coefficient or the partial pressure. Almost 35 years ago, Van Ness[32] suggested to measure the dependence of the total pressure on liquid composition at constant temperature and to use the data to calculate the vapor composition and the activity coefficient. This suggestion has the advantage that the vapor pressure can be measured much more precisely than the vapor composition. As shown in Appendix 1, D can be related to the pressure via the expression

$$D = \frac{\left(\frac{\partial \ln P}{\partial x_1}\right)_T}{P_1^0 \gamma_1 - P_2^0 \gamma_2} \tag{7}$$

where P_1^0 and P_2^0 are the saturated vapor pressures of the pure components 1 and 2 at a given temperature T, and γ_i can be taken from experiment or calculated through any of the usual models, such as the Wilson, NRTL, or UNIQUAC model.[2]

At infinite dilution, the following limiting expressions are valid for KBIs[7,13−14]

$$\lim_{x_i \to 0} G_{12} = RTk_{T,j}^0 - V_i^\infty \tag{8}$$

$$\lim_{x_i \to 1} G_{12} = RTk_{T,i}^0 - V_j^\infty \tag{9}$$

$$\lim_{x_i \to 0} G_{ii} = RTk_{T,j}^0 + V_j^0 - 2V_i^\infty - V_j^0 \left(\frac{\partial \ln \gamma_i}{\partial x_i}\right)_{P,T,x_i=0} \tag{10}$$

and

$$\lim_{x_i \to 1} G_{ii} = RTk_{T,i}^0 - V_i^0 \tag{11}$$

where V_i^0 is the molar volume of the pure component i and V_i^∞ is the partial molar volume of component i at infinite dilution. The limiting value $(\partial \ln \gamma_i / \partial x_i)_{P,T,x_i=0}$ was calculated using for the dilute region the following expression:[33]

$$\left(\frac{\partial \ln \gamma_i}{\partial x_i}\right)_{P,T} = k_i(P,T) \tag{12}$$

where $k_i(P,T)$ can be obtained directly from isothermal vapor−liquid equilibrium data in the dilute region.

2. Excess Number of Molecules around a Central Molecule. Almost all the considerations regarding the KBI's are based on the quantity

$$\Delta n_{ij} = c_i G_{ij} \tag{13}$$

(where c_i is the molar concentration of species i in the mixture) which is usually interpreted as the excess (or deficit) number of molecules i around a central molecule j.[6]

Matteoli and Lepori[13,14] noted that Δn_{ij} calculated with eq 13 have nonzero values for ideal systems, even though they are expected to vanish. In addition, there are many systems for which all KBIs (G_{11}, G_{12}, and G_{22}) are negative in certain ranges of composition.[17,34] As a result, in such cases all Δn_{ij} would be negative, and this is not possible.

Because the KBIs have nonzero instead of zero values for ideal systems, Matteoli and Lepori suggested to replace G_{ij} in eq 13 by $(G_{ij} - G_{ij}^{id})$[13,14]

$$\Delta n_{ij}' = c_i(G_{ij} - G_{ij}^{id}) = c_i \Delta G_{ij} \tag{14}$$

with the KBIs for ideal systems (G_{ij}^{id}) given by the expressions[13,14]

$$G_{12}^{id} = RTk_T^{id} - \frac{V_1^0 V_2^0}{V^{id}} = RTk_{T,2}^0 - V_1^0 - \varphi_1(V_2^0 - V_1^0 + RTk_{T,2}^0 - RTk_{T,1}^0) \tag{15}$$

$$G_{11}^{id} = G_{12}^{id} + V_2^0 - V_1^0 \tag{16}$$

and

$$G_{22}^{id} = G_{12}^{id} - (V_2^0 - V_1^0) \tag{17}$$

where k_T^{id} and V^{id} are the isothermal compressibility and the molar volume of an ideal solution, respectively.

However, because "the volume occupied by the excess j molecules around an i molecule must be equal to the volume left free by the i molecules around the same i molecule",[13]

$$V_j \Delta n_{ji}' = -V_i \Delta n_{ii}' \tag{18}$$

Equation 14 does not satisfy identically eq 18, because its insertion in the latter equation leads to

$$RTV(k_T - k_T^{id}) = V_i V + V_i(V_j^0 - V^{id}) - \frac{V_i^0 V_j^0 V}{V^{id}} \tag{19}$$

Equation 19 indicates that only if G_{ij}^{id} is replaced by G_{ij}^V (eqs 20 and 21), which is obtained from the former by substituting k_T^{id}, V_i^0, and V^{id} with k_T, V_i, and V, respectively, can eq 18 be satisfied identically.

$$G_{12}^V = G_{21}^V = RTk_T - \frac{V_1 V_2}{V} \tag{20}$$

$$G_{ii}^V = G_{12}^V + V_j - V_i \qquad i \neq j \tag{21}$$

Figure 1. Comparison between G_{ij}^{id} and G_{ij}^{V} for methanol–water at 298.15 K. G_{ij}^{id} is given by the solid line and G_{ij}^{V} by the broken line.

TABLE 1: Peak G_{11} for the 1-Propanol–Water System

x_1 (max)	G_{11} [cm³/mol]	ref
~0.25	~390	7
0.06	876	11
~0.15	~1900	12

TABLE 2: Original Data Used for Calculating the Kirkwood–Buff Integrals

system	D data	ref	V, V_1, V_2 data	ref
methanol–water	$P-x-y$, $T = 298.14$ K	35	V^E ^{a}, $T = 298.15$ K	36
ethanol–water	$P-x$, $T = 298.15$ K	37	V^E, $T = 298.15$ K	36
1-propanol–water	$P-x-y$, $T = 303.15$ K	38	V^E, $T = 298.15$ K	39
2-propanol–water	$P-x-y$, $T = 298.15$ K	40	V^E, $T = 298.15$ K	41
1-butanol–water	$P-x$, $T = 323.23$ K	42	V^E, $T = 308.15$ K	43
2-butanol–water	$P-x$, $T = 323.18$ K	42	V^E, $T = 293.15$ K	44
iso-butanol–water	$P-x$, $T = 323.15$ K	42	V^E, $T = 293.15$ K	45
tert-butanol–water	$P-x$, $T = 323.13$ K	42	V^E, $T = 303.15$ K	46

^{a} V^E is the excess volume.

Consequently

$$\Delta n_{12}' = c_1(G_{12} - G_{12}^V) = c_1 \Delta G_{12} = c_1 \Delta G_{21} = -\frac{c_1 V_1 V_2}{V}\left(\frac{1-D}{D}\right) \quad (22)$$

$$\Delta n_{21}' = c_2(G_{12} - G_{12}^V) = c_2 \Delta G_{12} = c_2 \Delta G_{21} = -\frac{c_2 V_1 V_2}{V}\left(\frac{1-D}{D}\right) \quad (23)$$

and

$$\Delta n_{ii}' = c_i \Delta G_{ii} = c_i(G_{ii} - G_{ii}^V) = \frac{c_i x_j V_j^2}{x_i V}\left(\frac{1-D}{D}\right) \quad i \neq j \quad (24)$$

A comparison between G_{ij}^V and G_{ij}^{id} for the methanol–water system is presented in Figure 1. It is worth mentioning that for ideal mixtures $\Delta n_{ij}' = 0$.

Data Sources and Treatment Procedure. The calculation of the KBIs from thermodynamic data requires information about the dependence on composition of the following variables: D, molar volume V, partial molar volumes V_1 and V_2, and the isothermal compressibility k_T. The sources of these data are listed in Table 2.

The vapor–liquid equilibrium data used in the calculations have been selected on the basis of the following two criteria: (1) the thermodynamic consistency tests (the integral test and the point test[47]) should be fulfilled by the systems chosen, (2) the sets chosen should contain at least 10–15 experimental

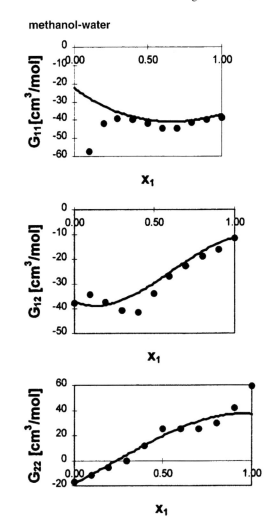

Figure 2. Comparison between G_{ij} calculated by us (solid lines) and ref 8 (●).

TABLE 3: The Sources for the Kirkwood–Buff Integrals from SAXS Measurements

system	data	ref
ethanol–water	G_{ij}, $T = 298.15$ K	26
1-propanol–water	G_{ij}, $T = 293.15$ K	27
	G_{ij}, $T = 298.15$ K	29
2-propanol–water	$I(0)$, $T = 293.15$ K	27
1-butanol–water	G_{ij}, $T = 298.15$ K	30
2-butanol–water	G_{ij}, $T = 298.15$ K	30
iso-butanol–water	G_{ij}, $T = 298.15$ K	30
tert-butanol–water	G_{ij}, $T = 293.15$ K	23

points. For several systems the vapor–liquid equilibrium data and those for excess volume are available at different temperatures. This has no significant effect on the results.[48] The partial molar volumes at infinite dilution used for calculating the limiting values of KBIs have been taken from refs 49–52 or calculated from excess volume data.

The isothermal compressibilities have been calculated with eq 5, using for the isothermal compressibilities of the pure substances the data from refs 53–56 (only the value for 2-butanol was taken as that for isobutanol). The V^E data have been fitted using the Redlich–Kister equation.[57] The values of D have been obtained from the activity coefficients or total pressure data by the sliding polynomials method.[58] To check the accuracy of our calculations, the D values have been

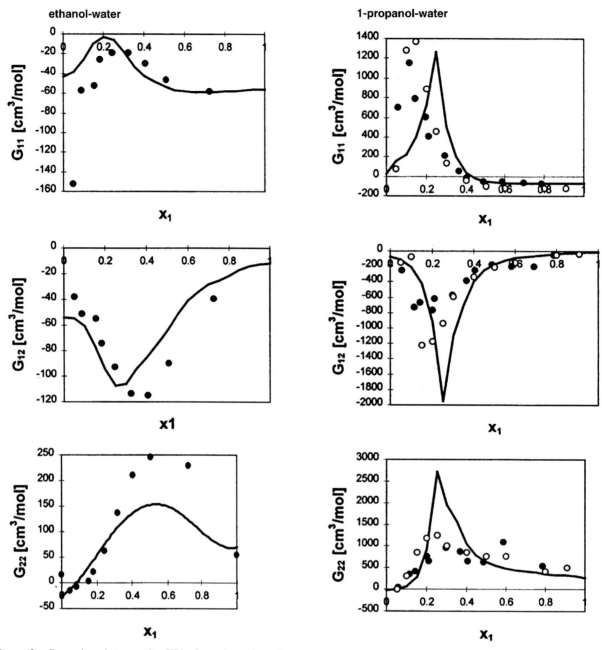

Figure 3. Comparison between the KBIs from thermodynamic calculations (solid line) and those from SAXS (●, ref 26).

Figure 4. Comparison between the KBIs from thermodynamic calculations (solid line) and those from SAXS (●, ref 29; ○, ref 27).

additionally evaluated using the NRTL equation. The KBIs obtained from SAXS measurements were taken from the original publications,[24–30] and the sources of the data are listed in Table 3.

The KBI's data for the 2-propanol−water system have been obtained from the so-called X-ray zero-angle intensity $I(0)$.[27] The equations used to calculate the KBIs are summarized in Appendix 2.

Results

A comparison between the KBIs obtained thermodynamically and those obtained from SAXS measurements is made in Figures 2−9. For the methanol−water system no SAXS data are available, and we compared our calculations to those of Donkersloot[8] and found satisfactory agreement between the two sets (Figure 2). Only for G_{11} in the dilute range of methanol there are differences between the compared sets. However, by calculating the limiting values of KBIs with eqs 8−11 a value of G_{11}^0 equal to -25 ± 5 [cm³/mol] was found for $T = 298$ K, which is in good agreement with our results. Our calculations for the ethanol−water mixture (Figure 3) are in agreement with the G_{ij} obtained from the SAXS measurements,[26] with the exception of G_{11} at low concentrations of ethanol and the peak value of G_{22}. The limiting value of G_{11} (G_{11}^0), calculated using eq 10, and that obtained by extrapolating the results of the calculation for the nondilute region nearly coincide. The comparison of our results with those from SAXS measurements for 1- or 2-propanol−water systems (Figures 4 and 5) show agreement, except for the dilute region of the 2-propanol−water system.

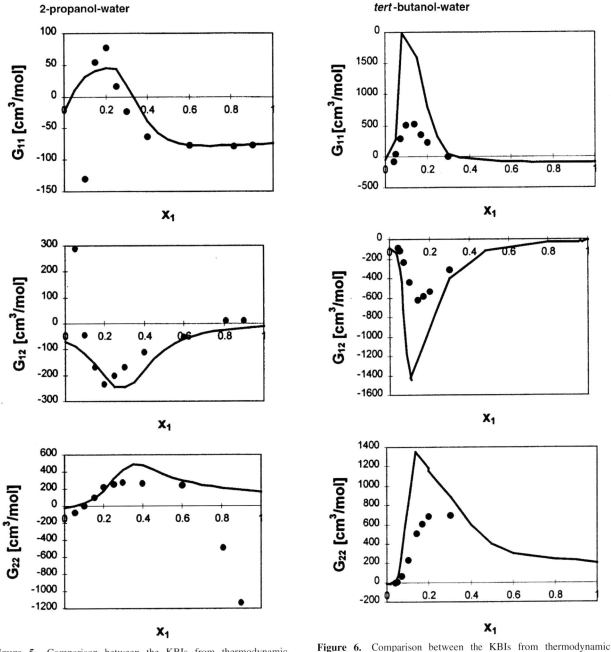

Figure 5. Comparison between the KBIs from thermodynamic calculations (solid line) and those from SAXS (●, ref 27).

Figure 6. Comparison between the KBIs from thermodynamic calculations (solid line) and those from SAXS (●, ref 23).

The experiments provide somewhat lower values for G_{11} ($x_1 < 0.2$) and for G_{22} ($x_2 < 0.2$), but the limiting values G_{ij}^0 are in good agreement with the calculated extrapolated results. There are some differences in the peak values of the KBI's for the 1-propanol−water system. Table 1 shows that for this system there is a large scattering of the KBI's peaks. This scattering is caused by the difficulty to obtain reliable D values in the composition range $0 \leq x_1 \leq 0.3$. More precise vapor−liquid equilibrium data are needed to obtain more exact peak values for the KBI's. As well-known,[42,51] the mixture tert-butyl alcohol−water (Figure 6) is in normal conditions the only homogeneous one among the butanol−water systems. The vapor−liquid equilibrium in this system was thoroughly investigated in ref 42, where it was also pointed out that it is difficult to describe the vapor−liquid equilibrium for this system with the conventional equations (e.g., NRTL and UNIQUAC equa-

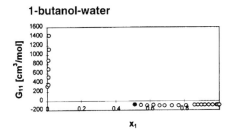

Figure 7. Comparison between the G_{11} from thermodynamic calculations (○, this work) and those from SAXS (●, ref 30).

tions wrongly indicate phase separation). This generates some deviations between the calculated KBI's and those obtained from SAXS measurements, in the composition range $0.06 \leq x_1 \leq 0.25$.

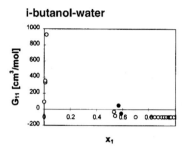

Figure 8. Comparison between the G_{11} from thermodynamic calculations (○, this work) and those from SAXS (●, ref 30).

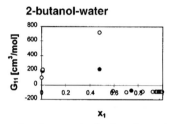

Figure 9. Comparison between the G_{11} from thermodynamic calculations (○, this work) and those from SAXS (●, ref 30).

Limited information is available concerning the KBI's for systems with phase separation. The KBI's for the 1-butanol–water system were calculated in ref 7. The behavior of the KBI's in all butanol (1-, 2-, and iso-)–water systems is similar (Figures 7–9). In these systems, the G_{ij}s change rapidly and become infinite at the phase separation point

$$\lim_{x_1 \to x_1^f} G_{ij} = \pm\infty \qquad (25)$$

where x_1^f is the butanol concentration at the point where the mixture becomes partially miscible. In the butanol-rich region, G_{ij} tends to infinity when the concentration approaches the boundary of the two phase region. Only few SAXS data are available for the systems 1-, 2-, and iso-butanol + water,[30] and we found satisfactory agreement between the results of our calculations and the SAXS measurements.

Generally Figures 2–9 show satisfactory agreement between the KBI's calculated and those obtained from SAXS measurements.

Discussion

Using the expressions suggested in this paper, we calculated the excess number of molecules $\Delta n'_{ij}$ around a central one (eqs 22–24). Figure 10 provides the excess (or deficit) number of molecules in the vicinity of an alcohol molecule and Figure 11—in the vicinity of a water molecule as the central molecule.

The present calculations are in agreement with the conclusion of ref 59 (which employed both a lattice and the McMillan–Mayer theories of solution[4]) that the solute–solute interactions in the systems investigated increase in the sequence MeOH < EtOH < 2-PrOH < 1-PrOH ≅ t-BuOH. There are, however, essential differences between the lower alcohols (MeOH and EtOH) and the higher ones.

Figures 10 and 11 reveal that, for the methanol–water and ethanol–water systems, the local compositions are close to the bulk ones over the entire concentration range. This means that,

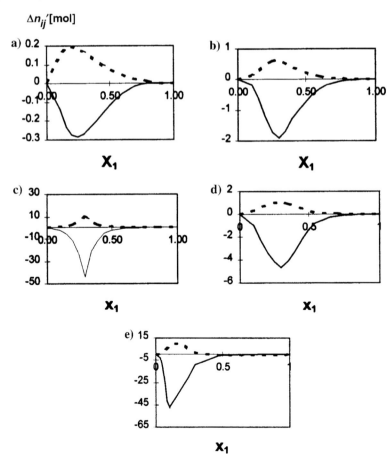

Figure 10. $\Delta n_{ij}'$ in the vicinity of an alcohol molecule. $\Delta n_{11}'$ is given by broken line, $\Delta n_{21}'$ by the solid line. (a) methanol–water, (b) ethanol–water, (c) 1-propanol–water, (d) 2-propanol–water, (e) *tert*-butanol–water.

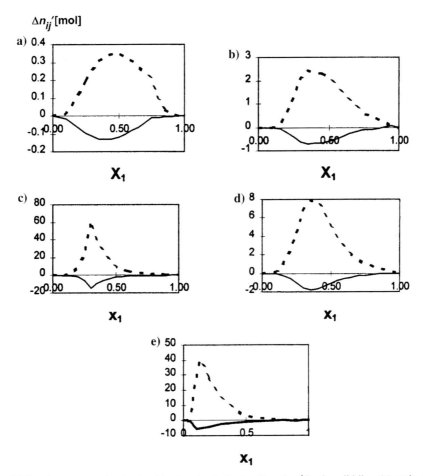

Figure 11. $\Delta n_{ij}'$ in the vicinity of an water molecule. $\Delta n_{22}'$ is given by the broken line, $\Delta n_{12}'$ by the solid line. (a) methanol−water, (b) ethanol−water, (c) 1-propanol−water, (d) 2-propanol−water, (e) *tert*-butanol−water.

in these systems, little clustering occurs, a conclusion in agreement with the SAXS measurements.[25,26]

For the ethanol−water system, there are somewhat larger changes in the vicinity of a water molecule. Indeed, in the concentration range $0.25 < x_1 < 0.65$, $2.5 > \Delta n_{22}' > 1$ and small clusters, dominated by water molecules, seems to be generated. Figures 10 and 11 bring evidence for clustering in aqueous binary systems of 1- and 2-propanols and *tert*-butyl alcohol; SAXS experiments also indicated the presence of clusters in these systems.[23−24,27−29]

One may note that the values of $\Delta n'_{ij}$ for the 2-propanol−water system are much smaller than those for 1-propanol−water and *tert*-butyl alcohol−water; they are, however, about twice as large as those in the ethanol−water system.

The calculation of the $\Delta n'_{ij}$ in aqueous systems of propanols and butanols indicates that clustering in these systems is caused by the accumulation of alcohol molecules near a central alcohol molecule and water molecule near a central molecules of water. This means that the hydrophobic interactions between the alcohol molecules and the interactions between water molecules are dominant in these systems.

Conclusion

The aqueous systems of methanol, ethanol, propanols, and butanols were examined in the framework of the Kirkwood−Buff theory of solution. The Kirkwood−Buff integrals were calculated using thermodynamic equations, in which the derivatives $(\partial \ln \gamma_i/\partial x_i)_{P,T}$ were expressed in terms of $(\partial \ln P/\partial x_i)_T$, which can be more accurately determined from isothermal $P-x$ data. The calculated KBI's were compared to those obtained from SAXS measurements and satisfactory agreement found.

New expressions for the excess number of molecules near a central molecule were suggested and used to calculate the distribution of species near the water or alcohol molecules as central ones. The main conclusion is that separate clusters of water and alcohol are formed and that the clustering increases with the length of the alcohol chain.

Acknowledgment. We are indebted to Dr. K. Fischer (University of Oldenburg, Germany) for providing information regarding the phase equilibria and excess volume.

Appendix 1

For vapor−liquid equilibrium in binary system with ideal behavior of the vapor phase one can write:[1−2]

$$P = P_1^0 x_1 \gamma_1 + P_2^0 x_2 \gamma_2 \qquad (A1\text{-}1)$$

Differentiating with respect to x_1 for isothermal condition, one obtains

$$\left(\frac{\partial P}{\partial x_1}\right)_T = P_1^0 \gamma_1 - P_2^0 \gamma_2 + P_1^0 x_1 \left(\frac{\partial \gamma_1}{\partial x_1}\right)_T - P_2^0 x_2 \left(\frac{\partial \gamma_2}{\partial x_2}\right)_T$$

$$(A1\text{-}2)$$

Kirkwood−Buff Integrals in Alcohol Systems

According to eq 4

$$x_1\left(\frac{\partial \gamma_1}{\partial x_1}\right)_T = \gamma_1(D-1) \quad (A1\text{-}3)$$

and

$$x_2\left(\frac{\partial \gamma_2}{\partial x_2}\right)_T = \gamma_2(D-1) \quad (A1\text{-}4)$$

Inserting the last two equations in eq A1-2 yields

$$D = \frac{\left(\frac{\partial P}{\partial x_1}\right)_T}{P_1^0 \gamma_1 - P_2^0 \gamma_2} \quad (A1\text{-}5)$$

For nonideal vapor phase, corrections for nonideality should be included.[1−2]

Appendix 2

To calculate the KBIs from the SAXS data for the 2-propanol−water mixture, the following equations were used:[23,27,60]

$$\frac{I(0)}{\bar{N}} = \frac{\bar{N}}{v} k_B T z^2 k_T + [z\delta - (z_1 - z_2)]^2 [\bar{N}\langle(\Delta x_1)^2\rangle] \quad (A2\text{-}1)$$

$$\langle \Delta N \Delta x_1 \rangle = -\delta \bar{N} \langle (\Delta x_1)^2 \rangle \quad (A2\text{-}2)$$

$$\langle(\Delta N)^2\rangle/\bar{N} = \frac{\bar{N}}{v} k_B T k_T + \delta^2 \bar{N} \langle(\Delta x_1)^2\rangle \quad (A2\text{-}3)$$

where $\bar{N}$ is the mean total number of molecules in the considered volume v of mixture, z_1, z_2, and z are average numbers of electrons per mole of alcohol, water, and mixture, respectively, k_B is the Boltzman constant, $\delta = (\bar{N}/v)(V_1 - V_2)$ is the dilatation factor, $\bar{N}\langle(\Delta x_1)^2\rangle$, $\langle(\Delta N)^2\rangle/\bar{N}$, and $\langle(\Delta x_1)(\Delta N)\rangle$ are the mean-square fluctuation in concentration, the mean-square fluctuation in particle number, and their correlation, respectively.

The three kinds of fluctuations have been obtained from eqs A2-1−A2-3, and the KBIs have been calculated using the expressions:

$$G_{11} = \frac{v}{\bar{N}}\left\{\frac{1}{x_1^2}[x_1^2\langle(\Delta N)^2\rangle/\bar{N} + 2x_1\langle\Delta N \Delta x_1\rangle + \bar{N}\langle(\Delta x_1)^2\rangle - x_1 x_2] - 1\right\} \quad (A2\text{-}4)$$

$$G_{22} = \frac{v}{\bar{N}}\left\{\frac{1}{x_2^2}[x_2^2\langle(\Delta N)^2\rangle/\bar{N} - 2x_2\langle\Delta N \Delta x_1\rangle + \bar{N}\langle(\Delta x_1)^2\rangle - x_1 x_2] - 1\right\} \quad (A2\text{-}5)$$

$$G_{12} = \frac{v}{\bar{N}}\left\{\frac{1}{x_1 x_2}[x_1 x_2\langle(\Delta N)^2\rangle/\bar{N} + (x_2 - x_1)\langle\Delta N \Delta x_1\rangle - \bar{N}\langle(\Delta x_1)^2\rangle]\right\} \quad (A2\text{-}6)$$

References and Notes

(1) Reid, R. C.; Prausnitz, J. M.; Poling, B. E. *The Properties of Gases and Liquids*, 4th ed.; McGraw-Hill: New York, 1987.
(2) Prausnitz, J. M.; Lichtenhaler, R. N.; Gomes de Azevedo, E. *Molecular Thermodynamics of Fluid−Phase Equilibria*, 2nd ed.; Prentice−Hill: Englewood Cliffs, NJ, 1986.
(3) Kirkwood, J. G.; Buff, F. P. *J. Chem. Phys.* **1951**, *19*, 774.
(4) McMillan, W.; Mayer, J. *J. Chem. Phys.* **1945**, *13*, 276.
(5) Kato, T.; Fujiyama, T.; Nomura, H. *Bull. Chem. Soc. Jpn.* **1982**, *55*, 3368.
(6) Ben-Naim, A. *J. Chem. Phys.* **1977**, *67*, 4884.
(7) Matteoli, E.; Lepori, L. *J. Chem. Phys.* **1984**, *80*, 2856.
(8) Donkersloot, M. C. A. *J. Solution Chem.* **1979**, *8*, 293.
(9) Patil, K. J. *J. Solution Chem.* **1981**, *10*, 315.
(10) Zaitsev, A. L.; Kessler, Y. M.; Petrenko, V. E. *Zh. Fiz. Khim.* **1985**, *59*, 2728.
(11) Cheng, Y.; Page, M.; Jolicoeur, C. *J. Phys. Chem.* **1993**, *93*, 7359.
(12) Patil, K. J.; Mehra, G. P.; Dhondge, S. S. *Indian J. Chem.* **1994**, *33A*, 1069.
(13) Matteoli, E.; Lepori, L. *J. Chem. Soc., Faraday Trans.* **1995**, *91*, 431.
(14) Matteoli, E. *J. Phys. Chem. B* **1997**, *101*, 9800.
(15) Ben-Naim, A. *Pure Appl. Chem.* **1990**, *62*, 25.
(16) Pfund, D. M.; Lee, L. L.; Cochran, H. D. *Fluid Phase Equil.* **1988**, *39*, 161.
(17) Rubio, R. G.; et al. *J. Phys. Chem.* **1987**, *91*, 1177.
(18) Ben-Naim, A. *Cell Biophys.* **1988**, *12*, 3694.
(19) D'Arrigo, G.; Teixeira, J. *J. Chem. Soc., Faraday Trans.* **1990**, *86*, 1503.
(20) Bale, H. D.; Sherpler, R. E.; Sorgen, D. K. *Phys. Chem. Liq.* **1968**, *1*, 181.
(21) Koga, Y. *Chem. Phys. Lett.* **1984**, *111*, 176.
(22) Nishikawa, K. *Chem. Phys. Lett.* **1986**, *132*, 50.
(23) Nishikawa, K.; Kodera, Y.; Iijima, T. *J. Phys. Chem.* **1987**, *91*, 3694.
(24) Nishikawa, K.; Hayashi, H.; Iijima, T. *J. Phys. Chem.* **1989**, *93*, 6559.
(25) Donkersloot, M. C. A. *Chem. Phys. Lett.* **1979**, *60*, 435.
(26) Nishikawa, K.; Iijima, T. *J. Phys. Chem.* **1993**, *97*, 10824.
(27) Hayashi, H.; Nishikawa, K.; Iijima, T. *J. Phys. Chem.* **1990**, *94*, 8334.
(28) Hayashi, H.; Udagawa, Y. *Bull. Chem. Soc. Jpn.* **1992**, *65*, 155.
(29) Shulgin, I.; Serimaa, R.; Torkkeli, M. Report Series in Physics HU-P-256; Helsinki, 1991.
(30) Shulgin, I.; Serimaa, R.; Torkkeli, M. Helsinki, 1991. Unpublished Data.
(31) Zaitsev, A. L.; Petrenko, V. E.; Kessler, Y. M. *J. Solution Chem.* **1989**, *18*, 115.
(32) Van Ness, H. *Classical Thermodynamics of Non-Electrolyte Solution*; Pergamon Press: New York, 1964.
(33) Debenedetti, P. G.; Kumar, S. K. *AIChE J.* **1986**, *32*, 1253.
(34) Marcus, Y. *J. Chem. Soc., Faraday Trans.* **1995**, *91*, 427.
(35) Kooner, Z. S.; Phutela, R. C.; Fenby, D. V. *Aust. J. Chem.* **1980**, *33*, 9.
(36) Benson, G. C.; Kiyohara, O. *J. Solution Chem.* **1980**, *9*, 791.
(37) Rarey, J. R.; Gmehling, J. *Fluid Phase Equil.* **1993**, *65*, 308.
(38) Udovenko, V. V.; Mazanko, T. F. *Izv. Vyssh. Uchebn. Zaved, Khim. Khim. Tekhnol.* **1972**, *15*, 1654.
(39) Dethlefsen, C.; Sorensen, P. G.; Hvidt, A. *J. Solution Chem.* **1984**, *13*, 191.
(40) Sazonov, V. P. *Zh. Prikl. Khim.* **1986**, *59*, 1451.
(41) Davis, M. I.; Ham, E. S. *Thermochim. Acta* **1991**, *190*, 251.
(42) Fischer, K.; Gmehling, J. *J. Chem. Eng. Data* **1994**, *39*, 309.
(43) Singh, P. P.; Sharma, P. K.; Maken, S. *Indian J. Technol.* **1993**, *31*, 17.
(44) Altsybeeva, A. L.; Belousov, V. P.; Ovtrakht, N. V. *Zh. Fiz. Khim.* **1964**, *38*, 1243.
(45) Nakanishi, K. *Bull. Chem. Soc. Jpn.* **1960**, *33*, 793.
(46) Kim, E. S.; Marsh, K. N. *J. Chem. Eng. Data* **1988**, *33*, 288.
(47) Gmehling, J.; et al. *Vapor−Liquid Equilibrium Data Collection*; DECHEMA Chemistry Data Series I; DECHEMA: Frankfurt, 1977−1996 (in 19 parts).
(48) Zielkiewicz, J. *J. Phys. Chem.* **1995**, *99*, 3357.
(49) Friedman, M. E.; Scheraga, H. A. *J. Phys. Chem.* **1965**, *69*, 3795.
(50) Franks, F.; Smith, H. T. *J. Chem. Eng. Data* **1968**, *13*, 538.
(51) Franks, F.; Desnoyers, J. E. *Water Sci. Rev.* **1985**, *1*, 171.
(52) Sakurai, M.; Nakagawa, T. *J. Chem. Thermodyn.* **1984**, *16*, 171.
(53) Lide, D. R., Ed. *Handbook of Chemistry and Physics*, 77th ed.; CRC Press: Boca Raton, 1996−1997.
(54) Hellwege, K.-H., Ed. *Landolt−Boernstein: Numerical Data and Functional Relationship in Science and Technology: New Series*; Springer-Verlag: Berlin, 1979.
(55) Diaz Pena, M.; Tardajos, G. *J. Chem. Thermodyn.* **1979**, *11*, 441.
(56) Moriyoshi, T.; Inubushi, H. *J. Chem. Thermodyn.* **1977**, *9*, 587.
(57) Redlich, O.; Kister, A. T. *Ind. Eng. Chem.* **1948**, *40*, 345.
(58) Savitsky, A.; Gulay, M. J. F. *Anal. Chem.* **1964**, *36*, 1627.
(59) Kozak, J. J.; Knight, W. S.; Kauzmann, W. *J. Chem. Phys.* **1968**, *48*, 675.
(60) Bhatia, A. B.; Thornton, D. E. *Phys. Rev. B* **1970**, *2*, 3004.

Kirkwood−Buff Integrals in Aqueous Alcohol Systems: Aggregation, Correlation Volume, and Local Composition

I. Shulgin[†] and E. Ruckenstein*

Department of Chemical Engineering, State University of New York at Buffalo, Amherst, New York 14260

Received: September 11, 1998; In Final Form: December 2, 1998

The Kirkwood−Buff theory of solution was used to investigate the formation of clusters in aqueous alcohol solutions. The correlation volume (volume in which the composition differs from the bulk one) was calculated for the systems 1-propanol−water and *tert*-butyl alcohol−water and compared with the sizes of clusters determined by various physical techniques. The calculations indicated that two types of clusters, alcohol- and water-rich clusters, are present in the solutions. Their sizes, which depend on composition in a similar way, exhibit maxima in the water-rich region. The calculated values are in a satisfactory agreement with experiment. The composition inside the clusters (the local composition) was calculated as a function of the correlation volume for dilute aqueous methanol, ethanol, propanols, and *tert*-butyl alcohol solutions. The results were compared with the local compositions provided by the Wilson and NRTL equations.

Introduction

The alcohol−water systems have attracted attention[1−5] for a number of reasons:

They are used as industrial solvents for small- and large-scale separation processes,[2] and they have unusual thermodynamic properties, which depend in a complicated manner on composition, pressure, and temperature; for example, the excess molar enthalpy (H^E) of ethanol + water mixture against concentration exhibits three extrema in its dependence on composition at 333.15 K and 0.4 MPa.[6] The thermodynamic behavior of these systems is particularly intricate in the water-rich region, as illustrated by the dependencies of the molar heat capacity and partial molar volume on composition.[7−9] This sensitivity of the partial molar properties indicates that structural changes occur in the water-rich region of these mixtures.[2,3] Of course, the unique structural properties of water are responsible for this behavior.[5,10]

One of the peculiar features of the alcohol−water systems is the clustering that takes place in the solutions of the higher alcohol members (PrOH, BuOH, and higher). Direct experimental evidence regarding clustering was provided by small-angle X-ray scattering (SAXS),[11−18] small-angle neutron scattering (SANS),[19] light scattering (LS),[20−23] fluorescence emission spectroscopy,[24] microwave dielectric analysis,[25] and adsorption.[26] No clustering could be found, however, in the aqueous solutions of methanol and ethanol.[27,28] The results obtained by SAXS, SANS, and LS are summarized in Table 1.

The data in Table 1 show that clustering occurs in the water-rich region of solutions of propanols and *tert*-butyl alcohol, for alcohol molar fractions <0.3−0.4. Numerous models have been suggested to explain the properties of water-alcohol mixtures. They can be roughly subdivided in the following groups: (a) Chemical models,[5,29−33] based on chemical equilibrium between clusters and the constituent components, which can explain some thermodynamic properties of these solutions, but involve oversimplified descriptions of the structure of the alcohol−water mixtures;[2] (b) clathrate-like models for dilute aqueous solutions of several alcohols;[13,15,34] (c) micellar-like models in which the alcohol molecules aggregate like surfactants in water, with the alkyl chain inside and the polar OH group outside;[19] (d) models that combine any of the above models with the representation of water as ice-like domains.[29−30,35] In the latter cases, the solution behavior in the water-rich region was attributed to a second-order phase transition involving the disordering of an ice-like network;[36,37] (e) the Kirkwood−Buff theory[38] provides a useful tool for the investigation of the structural features of solutions. The latter approach became popular particularly after Ben-Naim,[39] using the Kirkwood−Buff equations,[38] calculated the integrals G_{ij} (KBI)

$$G_{ij} = \int_0^\infty (g_{ij} - 1) 4\pi r^2 \, dr \qquad i,j = 1, 2 \qquad (1)$$

from measurable macroscopic thermodynamic properties. In the relation of eq 1, g_{ij} is the radial distribution function between species i and j and r is the distance between the centers of molecules i and j. The KBIs have been calculated for numerous binary and ternary systems including the water−alcohol mixtures[40−48] and used to examine the solution behavior regarding the local composition, preferential solvation, and various models for phase equilibria.[49−52] In this paper, the KBI combined with the NRTL expression[53,54] for the local composition will be utilized for the calculation of the correlation volume (the volume in which the composition differs from the bulk composition). The correlation volumes for several alcohol−water systems will be calculated and compared with the sizes of clusters determined by various physical techniques (Table 1).

Theory and Formulas

1. Excess Number of Molecules near a Central One. The key quantities in the Kirkwood−Buff approach are the integrals

$$c_i G_{ij} = c_i \int_0^\infty (g_{ij} - 1) 4\pi r^2 \, dr \qquad (2)$$

---

* To whom the correspondence should be addressed. E-mail: fealiru@acsu.buffalo.edu; fax: (716) 645-3822.
† Current e-mail address: ishulgin@eng.buffalo.edu.

TABLE 1: Clustering in Aqueous Solutions of Alcohols

system	method	clustering	size (Å)^a	composition range (x_1)	temperature (K)	reference
MeOH/H$_2$O	SAXS	no			298.15	27
EtOH/H$_2$O	SAXS	no			293.15	28
1-PrOH/H$_2$O	SAXS	yes	3–13.5	0.05–0.3	293.15	15
	SAXS	yes	–	–	278.15	16
	SAXS	yes	2–12	0.056–0.365	298.15	17
	LS	yes	8	0.05–0.4	298.15	20
1-PrOH/D$_2$O	SANS	yes	14.4	0.114	298.15	19
	SANS	yes	18.7	0.114	278.15	19
2-PrOH/H$_2$O	SAXS	yes	1.5–6.5	0.1–0.3	293.15	15
2-PrOH/D$_2$O	SANS	yes	7.4	0.234	298.15	19
1-BuOH/H$_2$O	SAXS	yes	8.2	0.018	298.15	18
2-BuOH/H$_2$O	SAXS	yes	9	0.041	298.15	18
	SAXS	yes	12	0.48	298.15	18
i-BuOH/H$_2$O	SAXS	yes	13	0.019	298.15	18
	SAXS	yes	13	0.568	298.15	18
t-BuOH/H$_2$O	SAXS	yes	–	–	278.15	11
	SAXS	yes	–	–	300.15	11
	SAXS	yes	–	–	329.15	11
	SAXS	yes	0–17.1	0.05–0.4	300.15	12
	SAXS	yes	–	–	293.15	13
	SAXS	yes	3–19	0.05–0.3	293.15–348.15	14
t-BuOH/D$_2$O	SANS	yes	29.1	0.107	278.15	19
	SANS	yes	18.4	0.107	298.15	19
	SANS	yes	30.4	0.107	310.15	19
t-BuOH/H$_2$O	LS	yes	9.8–22	0.0725–0.26	257.15–345.15	21
	LS	yes	9.3–19	0.04–0.27	298.15–318.15	23

In Table 1, x_1 is the molar fraction of alcohol (throughout this paper, component 1 represents alcohol and component 2, water). ^{a} The size is given by the Debye correlation length (l_D), which is related to the radius of the cluster (R_C) through the expression $l_D = 1.1 R_C$.[13,14]

where c_i is the molar concentration of species i in the mixture and $c_i G_{ij}$ represents the excess (or deficit) number of molecules i around a central molecule j.[39] The integrals G_{ij} (see *Appendix* for their expressions in terms of macroscopic thermodynamic quantities[40]) provide information about the tendency of molecules to stay away or to aggregate. However, these integrals have finite values for ideal solutions, which should be considered nonaggregated. Consequently, a better measure of the above tendency of the molecules can be obtained by introducing new quantities, Δn_{ij}, defined with respect to a reference state. Matteoli and Lepori[46,47] suggested to use the ideal mixture as the reference state and hence considered that the excess (or deficit) number of molecules i around a central molecule j is given by

$$\Delta n_{ij} = c_i(G_{ij} - G_{ij}^{id}) = c_i \Delta G_{ij} \qquad (3)$$

where G_{ij}^{id} are the KBIs of an ideal system (See *Appendix* for their expressions in terms of macroscopic thermodynamic quantities[40,46,47]). However, Δn_{ij} are not independent quantities, because the volume occupied by the excess i molecules around an i molecule must be equal to the volume left free by the j molecules around the same i molecule.[46] This volume conservation condition leads to

$$V_j \Delta n_{ji} = -V_i \Delta n_{ii} \qquad (4)$$

where V_i is the partial molar volume of component i. Eq 3 does not satisfy identically eq 4, because its insertion in the latter equation leads to

$$RTV(k_T - k_T^{id}) = V_i V + V_i(V_j^0 - V^{id}) - \frac{V_i^0 V_j^0 V}{V^{id}} \qquad (5)$$

where k_T and V are the isothermal compressibility and the molar volume of the liquid mixture, respectively, the superscript "id" indicates ideal mixture, the superscript 0 refers to the pure

component, R is the universal gas constant, and T is the temperature in Kelvin. Eq 4 is satisfied identically only if k_T^{id}, V_i^0, and V^{id} are replaced in the expressions of G_{ij}^{id} with k_T, V_i, and V, respectively. Consequently, G_{ij}^{id} in eq 3 has to be replaced with G_{ij}^V, which is given by the following expressions[48]

$$G_{12}^V = G_{21}^V = RT k_T - \frac{V_1 V_2}{V} \qquad (6)$$

$$G_{ii}^V = G_{12}^V + V_j - V_i \quad i \neq j \qquad (7)$$

Consequently Δn_{ij} can be written as

$$\Delta n_{12} = c_1(G_{12} - G_{12}^V) = c_1 \Delta G_{12} = c_1 \Delta G_{21} = -\frac{c_1 V_1 V_2}{V}\left(\frac{1-D}{D}\right) \qquad (8)$$

$$\Delta n_{21} = c_2(G_{12} - G_{12}^V) = c_2 \Delta G_{12} = c_2 \Delta G_{21} = -\frac{c_2 V_1 V_2}{V}\left(\frac{1-D}{D}\right) \qquad (9)$$

and

$$\Delta n_{ii} = c_i \Delta G_{ii} = c_i(G_{ii} - G_{ii}^V) = \frac{c_i x_j V_j^2}{x_i V}\left(\frac{1-D}{D}\right) \quad i \neq j \qquad (10)$$

where D is given by

$$D = \left(\frac{\partial \ln \gamma_i}{\partial x_i}\right)_{P,T} x_i + 1 \qquad (11)$$

In eq 11, γ_i is the activity coefficient of component i, P is the pressure, and x_i is the molar fraction of component i in the mixture. Eqs 8–10 have been used[48] to calculate the Δn_{ij} for the aqueous solutions of the following alcohols: MeOH, EtOH,

1-PrOH, 2-PrOH, and t-BuOH. It should be noted that for ideal mixtures $\Delta n_{ij} = 0$.

2. Local Composition and Correlation Volume. One of the most attractive features of the Kirkwood–Buff approach is its capability to provide values for the local composition.[49,52] Let us consider a molecule i whose correlation volume is V_{cor}^i. The total number of molecules i and j in this volume is given by the expressions

$$n_{ii} = \Delta n_{ii} + c_i V_{cor}^i \qquad (12)$$

and

$$n_{ji} = \Delta n_{ji} + c_j V_{cor}^i \qquad (13)$$

Consequently the local molar fractions, x_{ii} and x_{ji}, can be expressed as

$$x_{ii} = \frac{n_{ii}}{n_{ii} + n_{ji}} = \frac{\Delta n_{ii} + c_i V_{cor}^i}{\Delta n_{ii} + \Delta n_{ji} + (c_i + c_j)V_{cor}^i} \qquad (14)$$

and

$$x_{ji} = \frac{n_{ji}}{n_{ii} + n_{ji}} = \frac{\Delta n_{ji} + c_j V_{cor}^i}{\Delta n_{ii} + \Delta n_{ji} + (c_i + c_j)V_{cor}^i} \qquad (15)$$

Combining with eqs 8–10 and taking into account that $c_j = n_j/V$ and $x_j = n_j/\Sigma n_i$, where n_j is the number of moles of component j in solution and V is the solution volume, eqs 14 and 15 become

$$x_{ii} = \frac{x_i(G_{ii} - G_{ii}^V) + x_i V_{cor}^i}{x_i(G_{ii} - G_{ii}^V) + x_j(G_{ji} - G_{ji}^V) + V_{cor}^i} \qquad (16)$$

and

$$x_{ji} = \frac{x_j(G_{ji} - G_{ji}^V) + x_j V_{cor}^i}{x_i(G_{ii} - G_{ii}^V) + x_j(G_{ji} - G_{ji}^V) + V_{cor}^i} \qquad (17)$$

Similar equations can be written for the local composition near a central molecule j. To calculate the correlation volumes with eqs 16 and 17, it is necessary to express x_{ii}, x_{ji}, G_{ij}, G_{ij}^V as a function of composition. The values of G_{ij}, G_{ij}^V for the systems investigated were calculated in our previous paper.[48] For x_{ii} and x_{ji}, the Wilson[55] and NRTL[53] equations can be used. Although not always satisfactory,[56] the NRTL expression represents the local composition better than the Wilson equation.[56] The NRTL provides the following expressions for the local compositions:[53]

$$x_{ii} = \frac{x_i}{x_i + x_j \exp(-\alpha_{12}\tau_{ji})} \qquad (18)$$

$$x_{ji} = \frac{x_j \exp(-\alpha_{12}\tau_{ji})}{x_i + x_j \exp(-\alpha_{12}\tau_{ji})} \qquad (19)$$

where α_{12}, τ_{ji} are parameters. The Wilson model leads to the following expressions for the local compositions:[55]

$$x_{ii} = \frac{x_i}{x_i + x_j \exp\left(-\dfrac{\lambda_{ij} - \lambda_{ii}}{RT}\right)} \qquad (20)$$

and

$$x_{ji} = \frac{x_j \exp\left(-\dfrac{\lambda_{ij} - \lambda_{ii}}{RT}\right)}{x_i + x_j \exp\left(-\dfrac{\lambda_{ij} - \lambda_{ii}}{RT}\right)} \qquad (21)$$

where $(\lambda_{ij} - \lambda_{ii})$ are parameters in the Wilson equation.

3. Local Composition in the Dilute Region. Let us consider the alcohol molecule as the central one ($i = 1$). Consequently, for the dilute region of alcohol one can write that

$$V_1 = V_1^\infty \qquad (22)$$

$$V_2 = V_2^0 \qquad (23)$$

$$V = x_1 V_1^\infty + x_2 V_2^0 \qquad (24)$$

$$V_{cor}^1 = V_{cor}^1(P,T) \qquad (25)$$

where V_i^0 and V_i^∞ are the molar volume of the pure component i and the partial molar volume of component i at infinite dilution, respectively. In addition, in the dilute region[57]

$$D = K(P,T)x_1 + 1 \qquad (26)$$

where[57]

$$K(P,T) = \left(\frac{\partial \ln \gamma_1}{\partial x_1}\right)_{P,T,x_1 \to 0}$$

Inserting expressions 22–26 into eq 16 yields

$$x_{11} = \frac{\alpha_1(P,T)x_1 + \alpha_2(P,T)x_1^2 + \alpha_3(P,T)x_1^3}{\beta_0(P,T) + \beta_1(P,T)x_1 + \beta_2(P,T)x_1^2} \qquad (27)$$

and

$$x_{21} = 1 - x_{11} \qquad (28)$$

where

$$\alpha_1 = -(V_2^0)^2 K + V_{cor}^1 V_2^0$$

$$\alpha_2 = (V_2^0)^2 K + V_{cor}^1 V_2^0 K + V_{cor}^1 V_1^\infty - V_{cor}^1 V_2^0$$

$$\alpha_3 = V_{cor}^1 V_1^\infty K - V_{cor}^1 V_2^0 K$$

$$\beta_0 = V_{cor}^1 V_2^0$$

$$\beta_1 = -(V_2^0)^2 K + V_1^\infty V_2^0 K + V_{cor}^1 V_1^\infty - V_{cor}^1 V_2^0 + V_{cor}^1 V_2^0 K$$

$$\beta_2 = (V_2^0)^2 K - V_2^0 V_1^\infty K + V_{cor}^1 V_1^\infty K - V_{cor}^1 V_2^0 K$$

It should be noted that all the above quantities have a linear dependence on V_{cor}^1.

Calculations and Results

The correlation volumes were calculated using eqs 16 and 17 for the aqueous systems 1-PrOH ($T = 303.15$ K) and t-BuOH ($T = 323.15$ K). These systems were chosen because reliable data about their clustering are available (Table 1). The values of the KBIs were taken from our previous paper,[48] and the

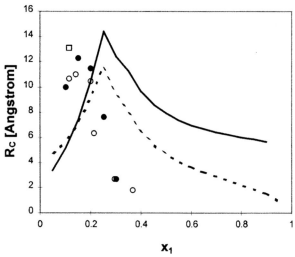

Figure 1. Correlation radius R_C against x_1 for the 1-propanol − water system. The broken line is for the alcohol-rich cluster and the solid line is for the water-rich cluster; experimental data from: ●, ref 15; ○, ref 17; □, ref 19.

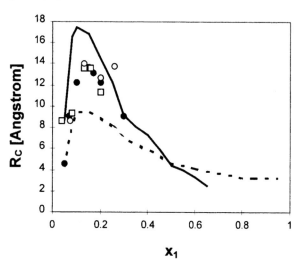

Figure 2. Correlation radius R_C against x_1 for the *tert*-butanol−water system. The broken line is for the alcohol-rich cluster and the solid line is for the water-rich cluster; experimental data from: ●, ref 14 ($T = 301.15$ K); ○, ref 21 ($T = 293.15$ K); □, ref 23 ($T = 293.15$ K).

NRTL equation was used to express x_{ij} in terms of the overall composition. The parameters of the NRTL equation were taken from the Gmehlings vapor−liquid equilibrium (VLE) data compilation.[58] The results of the calculations are compared with experimental data in Figures 1 and 2.

The local composition x_{11} was calculated with eq 27 for dilute aqueous solutions of MeOH, EtOH, 1-PrOH, 2-PrOH, and *t*-BuOH at 298.15 K in the composition range $0 \leq x_1 \leq 0.06$. The partial molar volumes of alcohols at infinite dilution were taken from literature.[2] The parameter $K(P, T)$ was calculated using data from the Gmehlings VLE compilation.[58] The results are compared with those calculated with the Wilson and the NRTL equations in Figures 3−7. The parameters of the Wilson and NRTL equations were taken from the above-mentioned compilation.[58]

Discussion

A satisfactory agreement was found between the sizes of the clusters determined experimentally and calculated. For the

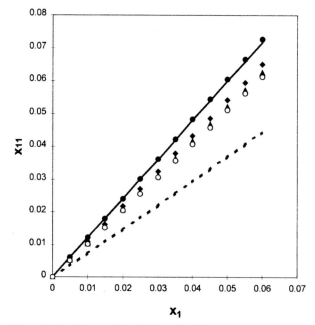

Figure 3. Local composition in the methanol−water system. The solid line is calculated with the NRTL and the broken line with the Wilson equation; ●, eq 27 with $V^l_{cor} = 100$ cm^3/mol; ◆, eq 27 with $V^l_{cor} = 250$ cm^3/mol; ▲, eq 27 with $V^l_{cor} = 500$ cm^3/mol; ○, eq 27 with $V^l_{cor} = 1000$ cm^3/mol.

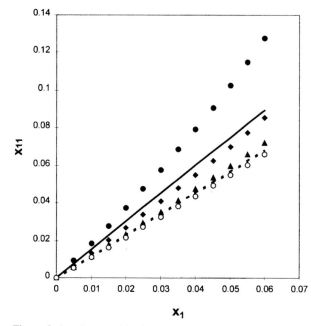

Figure 4. Local composition in the ethanol−water system. The solid line is calculated with the NRTL and the broken line with the Wilson equation; ●, eq 27 with $V^l_{cor} = 100$ cm^3/mol; ◆, eq 27 with $V^l_{cor} = 250$ cm^3/mol; ▲, eq 27 with $V^l_{cor} = 500$ cm^3/mol; ○, eq 27 with $V^l_{cor} = 1000$ cm^3/mol.

1-PrOH + H$_2$O solutions, the calculations indicate (Figure 1) that two types of clusters, namely alcohol- and water-rich, are present in the solution and that the dependencies of their sizes on composition are similar. They reach a maximum at an alcohol molar fraction $x_1 \approx 0.2-0.3$, after which they decrease. A similar behavior was found for Δn_{ii}.[48] Even though the systems investigated are homogeneous at all compositions (in normal conditions), a "phase separation" at molecular scale, in water-

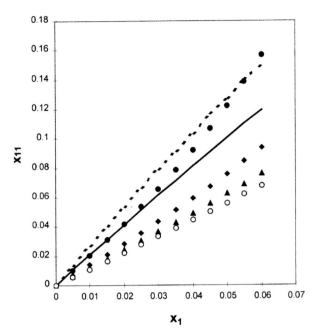

Figure 5. Local composition in the 1-propanol–water system. The solid line is calculated with the NRTL and the broken line with the Wilson equation; ●, eq 27 with $V_{cor}^l = 100$ cm³/mol; ◆, eq 27 with $V_{cor}^l = 250$ cm³/mol; ▲, eq 27 with $V_{cor}^l = 500$ cm³/mol; ○, eq 27 with $V_{cor}^l = 1000$ cm³/mol.

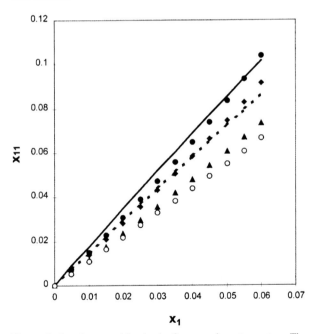

Figure 6. Local composition in the 2-propanol–water system. The solid line is calculated with the NRTL and the broken line with the Wilson equation; ●, eq 27 with $V_{cor}^l = 100$ cm³/mol; ◆, eq 27 with $V_{cor}^l = 225$ cm³/mol; ▲, eq 27 with $V_{cor}^l = 500$ cm³/mol; ○, eq 27 with $V_{cor}^l = 1000$ cm³/mol.

and alcohol-rich clusters, takes place. Although there is qualitative agreement with the experimental results provided by the SAXS and SANS, the calculated values do not decay for $x_1 > 0.4$ as rapidly as the experimental ones. A similar conclusion can be noted for the system t-BuOH + H_2O (Figure 2). The cluster size combined with the values of Δn_{ij} allow one to estimate the cluster composition. At the molar fraction $x_1 = 0.1$, the alcohol composition in the alcohol-rich cluster is 0.145

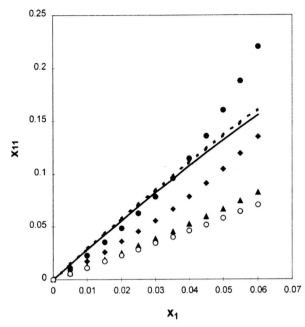

Figure 7. Local composition in the *tert*-butanol–water system. The solid line is calculated with the NRTL and the broken line with the Wilson equation; ●, eq 27 with $V_{cor}^l = 100$ cm³/mol; ◆, eq 27 with $V_{cor}^l = 175$ cm³/mol; ▲, eq 27 with $V_{cor}^l = 500$ cm³/mol; ○, eq 27 with $V_{cor}^l = 1000$ cm³/mol.

and in the water-rich cluster is 0.096; for $x_1 > 0.35$, the alcohol-rich cluster is essentially free of water molecules and the alcohol molar fraction in the water-rich clusters is about 0.14–0.15.

Because the cluster composition is not constant, a clathrate-like structure cannot be used for its representation. Nevertheless, the clathrate-like structure $[(PrOH)_8(H_2O)_{40}]$ suggested in ref 20 is compatible with the water-rich clusters for $x_1 > 0.35$. Indeed, the alcohol molar fraction of about 0.15, calculated for the water-rich clusters when $x_1 > 0.35$, corresponds to the clathrate $(PrOH)_8(H_2O)_{40}$.

The Kirkwood–Buff model can be used for the calculation of the local composition in the dilute region. Our calculation of the local composition in the aqueous systems of MeOH, EtOH, 1-PrOH, 2-PrOH, or t-BuOH is compared with the local compositions provided by the Wilson and NRTL equations (Figures 3–7). The local composition of an aqueous solution of MeOH, calculated with eq 27 for $V_{cor}^l = 100$ cm³/mol (Figure 3), is in agreement with the local composition obtained from the NRTL equation; this means that even if clusters are present in this solution, they are small. (For comparison, the molar volume of pure MeOH in normal condition is 40.74 cm³/mol.) The local composition of EtOH + H_2O system plotted in Figure 4 shows that there is agreement between eq 27 and the NRTL equation for $V_{cor}^l \approx 250$ cm³/mol; consequently the clusters in this system are also small. There is some experimental evidence that small clusters are present in the EtOH + H_2O system.[19,28,59] It is worth mentioning that for the systems MeOH + H_2O and EtOH + H_2O there is no agreement between the Wilson and NRTL equations. For the 1-propanol-water system and $x_1 < 0.03$ (Figure 5), there is agreement between eq 27 and the Wilson and NRTL equations for $V_{cor}^l \approx 100-250$ cm³/mol. The size of the clusters determined experimentally (Table 1) is in agreement with the above values of V_{cor}^l. For the 2-PrOH + H_2O system (Figure 6), small clusters are present in the dilute region. Figure 7 shows that for the system t-BuOH + H_2O, there is quantitative agreement between eq 27 and the

NRTL equation for $V_{cor}^1 \approx 100-175$ cm³/mol and $x_1 < 0.04$. The latter values of V_{cor}^1 are in agreement with the SAXS results (Table 1).

Conclusion

The clustering in aqueous solutions of alcohols was examined by combining the Kirkwood−Buff theory of solution with the Wilson and the NRTL equations. The correlation volumes were calculated for the aqueous systems of 1-PrOH and t-BuOH. Two type of clusters, alcohol- and water-rich, were found with similar dependencies of size on composition. Satisfactory agreement was found between the calculated cluster sizes and those provided by the SAXS, SANS, and LS experiments.

An analytical expression was derived for the local composition in the dilute region, which was used for the dilute aqueous solutions of MeOH, EtOH, 1-PrOH, 2-PrOH, and t-BuOH. The results were compared with those obtained with the Wilson and NRTL equations and on this basis the correlation volume in the dilute region evaluated. It was found that small clusters (such as dimers and trimers) can be present in the dilute region of alcohols.

Appendix

The main formulas for the calculation of the KBIs[40,46−47] are

$$G_{12} = G_{21} = RTk_T - \frac{V_1 V_2}{VD} \quad (A1)$$

$$G_{ii} = G_{12} + \frac{1}{x_i}\left(\frac{V_j}{D} - V\right) \quad i \neq j \quad (A2)$$

where

$$D = \left(\frac{\partial \ln \gamma_i}{\partial x_i}\right)_{P,T} x_i + 1 \quad (A3)$$

In eqs A1−A3, k_T is the isothermal compressibility, V_i is the partial molar volume of component i, x_i is the molar fraction of component i, V is the molar volume of the mixture, and γ_i is the activity coefficient of component i.

Consequently, the KBIs of an ideal binary mixture, G_{ij}^{id}, are given by the expressions:[40,46,47]

$$G_{12}^{id} = RTk_T^{id} - \frac{V_1^0 V_2^0}{V^{id}} \quad (A4)$$

$$G_{11}^{id} = G_{12}^{id} + V_2^0 - V_1^0 \quad (A5)$$

and

$$G_{22}^{id} = G_{12}^{id} - (V_2^0 - V_1^0) \quad (A6)$$

where k_T^{id} and V^{id} are the isothermal compressibility and the molar volume of an ideal mixture, respectively, and V_i^0 is the molar volume of the pure component i.

References and Notes

(1) Franks, F.; Ives, D. J. G. *Q. Rev. Chem. Soc.* **1966**, *20*, 1.
(2) Franks, F.; Desnoyers, J. E. *Water Sci. Rev.* **1985**, *1*, 171.
(3) Belousov, V. P.; Panov, M. Y. *Thermodynamics of Aqueous Solutions of Non-Electrolytes* (in Russian); Khimiya: Leningrad, 1983.
(4) Kozak, J. J.; Knight, W. S.; Kauzmann, W. *J. Chem. Phys.* **1968**, *48*, 675.
(5) Shealy, G. S.; Sandler, S. I. *AIChE J.* **1988**, *34*, 1065.
(6) Ott, J. B. *J. Chem. Thermodyn.* **1990**, *22*, 1129.
(7) Benson, G. C.; Kiyohara, O. *J. Solution Chem.* **1980**, *9*, 791.
(8) Benson, G. C.; Kiyohara, O. *J. Solution Chem.* **1980**, *9*, 931.
(9) Ogawa, H.; Murakami, S. *Thermochim. Acta* **1986**, *109*, 145.
(10) Narten, A. H.; Levy, H. A. In *Water—A Comprehensive Treatise*; Franks, F., Ed.; Plenum Press: New York, 1972; Vol. 1, Chapter 8.
(11) Bale, H. D.; Sherpler, R. E.; Sorgen, D. K. *Phys. Chem. Liq.* **1968**, *1*, 181.
(12) Koga, Y. *Chem. Phys. Lett.* **1984**, *111*, 176.
(13) Nishikawa, K.; Kodera, Y.; Iijima, T. *J. Phys. Chem.* **1987**, *91*, 3694.
(14) Nishikawa, K.; Hayashi, H.; Iijima, T. *J. Phys. Chem.* **1989**, *93*, 6559.
(15) Hayashi, H.; Nishikawa, K.; Iijima, T. *J. Phys. Chem.* **1990**, *94*, 8334.
(16) Hayashi, H.; Udagawa, Y. *Bull. Chem. Soc. Jpn.* **1992**, *65*, 155.
(17) Shulgin, I.; Serimaa, R.; Torkkeli, M. Report Series in Physics, HU-P-256: Helsinki, 1991.
(18) Shulgin, I.; Serimaa, R.; Torkkeli, M. Unpublished data, 1991.
(19) D'Arrigo, G.; Teixeira, J. *J. Chem. Soc. Faraday Trans.* **1990**, *86*, 1503.
(20) Grossmann, G. H.; Ebert, K. H. *Ber. Bunsen−Ges. Phys. Chem.* **1981**, *85*, 1026.
(21) Euliss, G. W.; Sorensen, C. M. *J. Chem. Phys.* **1984**, *80*, 4767.
(22) Ito, N.; Kato, T.; Fujiyma, T. *Bull. Chem. Soc. Jpn.* **1981**, *54*, 2573.
(23) Bender, T. M.; Pekora, R. *J. Phys. Chem.* **1986**, *90*, 1700.
(24) Zana, R.; Eljebari, M. J. *J. Phys. Chem.* **1993**, *97*, 11134.
(25) Mashimoto, S.; Umehara, T.; Redin, H. *J. Chem. Phys.* **1991**, *95*, 6257.
(26) Marosi, T.; Dekany, I.; Lagaly, G. *Colloid Polym. Sci.* **1994**, *272*, 1136.
(27) Donkersloot, M. C. A. *Chem. Phys. Lett.* **1979**, *60*, 435.
(28) Nishikawa, K.; Iijima, T. *J. Phys. Chem.* **1993**, *97*, 10824.
(29) Mikhailov, V. A.; Ponomareva, L. I. *J. Struct. Chem.* **1968**, *9*, 8.
(30) Mikhailov, V. A. *J. Struct. Chem.* **1968**, *9*, 332.
(31) Iwasaki, K.; Fujiyma, T. *J. Chem. Phys.* **1977**, *81*, 1908.
(32) Iwasaki, K.; Fujiyma, T. *J. Chem. Phys.* **1979**, *83*, 463.
(33) Roux, A. H.; Desnoyers, J. E. *Proc. Indian Acad. Sci. Chem. Sci.* **1987**, *98*, 435.
(34) Onori, G. *J. Chem. Phys.* **1988**, *89*, 4325.
(35) Anisimov, M. A., et al. *J. Struct. Chem.* **1977**, *18*, 663.
(36) Vuks, M. F. *Opt. Spectrosc.* **1976**, *40*, 86.
(37) Vuks, M. F. *Light Scattering by Liquids and Solutions* (in Russian); Izdatelstvo Leningradskogo Universiteta: Leningrad, 1977.
(38) Kirkwood, J. G.; Buff, F. P. *J. Chem. Phys.* **1951**, *19*, 774.
(39) Ben-Naim, A. *J. Chem. Phys.* **1977**, *67*, 4884.
(40) Matteoli, E.; Lepori, L. *J. Chem. Phys.* **1984**, *80*, 2856.
(41) Donkersloot, M. C. A. *J. Solution Chem.* **1979**, *8*, 293.
(42) Patil, K. J. *J. Solution Chem.* **1981**, *10*, 315.
(43) Zaitsev, A. L.; Kessler, Y. M.; Petrenko, V. E. *Zh. Fiz. Khim.* **1985**, *59*, 2728.
(44) Cheng, Y.; Page, M.; Jolicoeur, C. *J. Phys. Chem.* **1993**, *93*, 7359.
(45) Patil, K. J.; Mehra, G. P.; Dhondge, S. S. *Indian J. Chem.* **1994**, *33A*, 1069.
(46) Matteoli, E.; Lepori, L. *J. Chem. Soc. Faraday Trans.* **1995**, *91*, 431.
(47) Matteoli, E. *J. Phys. Chem. B* **1997**, *101*, 9800.
(48) Shulgin, I.; Ruckenstein, E. *J. Phys. Chem. B* **1999**, *103*, 2496.
(49) Ben-Naim, A. *Pure Appl. Chem.* **1990**, *62*, 25.
(50) Pfund, D. M.; Lee, L. L.; Cochran, H. D. *Fluid Phase Equilib.* **1988**, *39*, 161.
(51) Rubio, R. G., et al. *J. Phys. Chem.* **1987**, *91*, 1177.
(52) Ben-Naim, A. *Cell Biophys.* **1988**, *12*, 3694.
(53) Renon, H.; Prausnitz, J. M. *AIChE J.* **1968**, *14*, 135.
(54) Prausnitz, J. M.; Lichtenhaler, R. N.; Gomes de Azevedo, E. *Molecular Thermodynamics of Fluid-Phase Equilibria*, 2nd ed.; Prentice Hill: Englewood Cliffs, NJ, 1986.
(55) Wilson, G. M. *J. Am. Chem. Soc.* **1964**, *86*, 127.
(56) Gierycz, P.; Nakanishi, K. *Fluid Phase Equilib.* **1984**, *16*, 255.
(57) Debenedetti, P. G.; Kumar, S. K. *AIChE J.* **1986**, *32*, 1253.
(58) Gmehling, J., et al. *Vapor−Liquid Equilibrium Data Collection*; DECHEMA Chemistry Data Series, Vol. 1; 19 Parts, DECHEMA, Frankfurt, 1977−1996.
(59) Nishi, N. Z. *Z. Phys. D: At. Mol. Clusters* **1990**, *15*, 239.

Range and Energy of Interaction at Infinite Dilution in Aqueous Solutions of Alcohols and Hydrocarbons

I. Shulgin[†] and E. Ruckenstein*

Department of Chemical Engineering, State University of New York at Buffalo, Amherst, New York 14260

Received: February 16, 1999; In Final Form: April 14, 1999

Infinitely dilute hydrocarbon/water and alcohol/water systems were examined, with the objective to gather information about the size of the volume which is affected by the presence of a solute molecule and the interaction energy parameter between solute and solvent. First, an expression for the local composition at infinite dilution was obtained on the basis of the Kirkwood–Buff theory of solution. Second, equations for the activity coefficients at infinite dilution were derived using a modified Flory–Huggins equation for the excess free energy. In this modified expression, the molar fractions in the volume fractions were replaced by the local compositions provided by the first step. Finally, an additional expression for the local molar fraction was selected, which was coupled with that obtained on the basis of the Kirkwood–Buff approach. Experimental data regarding the activity coefficient at infinite dilution combined with the above equations allowed to obtain the values of the correlation volume (volume which is affected by a single solute molecule) and the interaction energy parameter between solute and solvent. The thickness of the layer of solvent influenced by the presence of a solute molecule was found to be equal to several molecular shells of water molecules (from 4 for propane to 7–8 for dodecane).

Introduction

The infinite dilution state is very suitable for the investigation of intermolecular interactions between solute and solvent molecules, because in that state a single solute molecule is completely surrounded by solvent molecules and thus information regarding the solute–solvent interaction in the absence of the solute–solute interactions can be obtained. This is particularly relevant for systems with complicated intermolecular interactions such as the aqueous systems.

To characterize the process of dissolution in water, Butler suggested that "the process of bringing a solute molecule into a solvent may be supposed to consist of two steps: (1) making a cavity in the solvent large enough to hold the solute molecule; (2) introducing the solute molecule into the cavity".[1] In the calculation of the contribution of the second step to the enthalpy of hydration, Butler considered only the interactions between the solute molecule and the nearest-neighboring water molecules. He also assumed that the water molecules are distributed around a solute molecule as randomly as in its absence. Butler's assumption of random distribution in solution was replaced by Frank and Evans[2] by a nonrandom distribution. The latter authors suggested that around a molecule of a nonpolar solute, there is a layer of more ordered water molecules ("iceberg"). The approach can account for the negative enthalpy and entropy of solution (see discussion for additional comments). No evaluation of the thickness of this layer was, however, provided. Nemethy and Scheraga[3] extended their model for water to aqueous solutions of hydrocarbons, but considered that only a monolayer of solvent molecules is affected by a solute molecule. The aim of this paper is to evaluate the thickness of the layer of water, which is affected by a solute molecule. In addition, information about the interaction energy parameter between the solute molecule and a solvent molecule will be obtained. In order to evaluate those quantities, expressions for the correlation volume (volume which is affected by a single solute molecule), the local compositions, and the activity coefficients at infinite dilution are needed, and they are presented below.

Theory and Formulas

1. Correlation Volume. We consider a molecule of species i and its surrounding correlation volume V^i_{cor} in which the composition and/or the structure differ from the bulk one. The total number of molecules i and j in this volume is given by the expressions

$$n_{ii} = \Delta n_{ii} + c_i V^i_{cor} \tag{1}$$

$$n_{ji} = \Delta n_{ji} + c_j V^i_{cor} \tag{2}$$

where Δn_{ii} and Δn_{ji} are excess (or deficit) number of molecules i and j around a central molecule i. Using the Kirkwood–Buff theory (KB)[4] of solution and a suitable reference state for the excess quantities, Δn_{ii} and Δn_{ji} can be expressed as follows:[5]

$$\Delta n_{12} = -\frac{c_1 V_1 V_2}{V}\left(\frac{1-D}{D}\right) \tag{3}$$

$$\Delta n_{21} = -\frac{c_2 V_1 V_2}{V}\left(\frac{1-D}{D}\right) \tag{4}$$

and

$$\Delta n_{ii} = \frac{c_i x_j V_j^2}{x_i V}\left(\frac{1-D}{D}\right) \quad i \neq j \tag{5}$$

---

* Author to whom the correspondence should be addressed. E-mail: fealiru@acsu.buffalo.edu. Fax: (716) 645-3822.
[†] Current e-mail address: ishulgin@eng.buffalo.edu.

Infinitely Diluted Aqueous Solutions

where c_i is the overall molar concentration of species i in the mixture, x_i is the overall molar fraction of component i in solution, V is the molar volume of the solution, V_i and V_j are the partial molar volumes of components i and j, respectively, $D = (\partial \ln \gamma_i/\partial x_i)_{P,T} x_i + 1$, γ_i is the activity coefficient of component i, P is the pressure, and T is the absolute temperature. Because the local molar fractions are given by

$$x_{ii} = \frac{n_{ii}}{n_{ii} + n_{ji}} = \frac{\Delta n_{ii} + c_i V_{cor}^i}{\Delta n_{ii} + \Delta n_{ji} + (c_i + c_j)V_{cor}^i} \quad (6)$$

$$x_{ji} = \frac{n_{ji}}{n_{ii} + n_{ji}} = \frac{\Delta n_{ji} + c_j V_{cor}^i}{\Delta n_{ii} + \Delta n_{ji} + (c_i + c_j)V_{cor}^i} \quad (7)$$

one obtains

$$x_{ii} = \frac{x_j V_j^2(1-D) + x_i V_{cor}^i VD}{x_j V_j^2(1-D) - x_j V_i V_j(1-D) + V_{cor}^i VD} \quad (8)$$

$$x_{ji} = \frac{-x_j V_i V_j(1-D) + x_j V_{cor}^i VD}{x_j V_j^2(1-D) - x_j V_i V_j(1-D) + V_{cor}^i VD} \quad (9)$$

where x_{ji} is the local molar fraction of component j in the vicinity of a central molecule i. Because for the dilute region one can consider that[6] $(\partial \ln \gamma_i/\partial x_i)_{P,T}$ is independent of x_i, one can write that

$$D = K_i(P,T)x_i + 1 \quad (10)$$

where

$$K_i(P,T) = \left(\frac{\partial \ln \gamma_i}{\partial x_i}\right)_{P,T,x_i \to 0}$$

For infinite dilution, eq 8 leads to

$$\lim_{x_i \to 0} \frac{x_{ii}}{x_i} = 1 - \frac{V_j^\infty K_i(P,T)}{V_{cor}^{i,0}} \quad (11)$$

where $V_j^\infty = \lim_{x_j \to 0} V_j$ is the partial molar volume of component j at infinite dilution and $V_{cor}^{i,0} = \lim_{x_i \to 0} V_{cor}^i$. One may note that $\lim_{x_i \to 0} (x_{ii}/x_i) > 1$ when $K_i(P,T) < 0$ (or $(\partial \ln \gamma_i/\partial x_i)_{P,T,x_i \to 0} < 0$). To calculate the correlation volume $V_{cor}^{i,0}$, hence the size of the region affected by a solute molecule, an additional expression for $\lim_{x_i \to 0} (x_{ii}/x_i)$ is needed. Such an expression, proposed in refs 7–9, will be used in what follows. Since that expression contains the interaction energy parameter between the solute and solvent molecules and this quantity is not known, an equation for the activity coefficients at infinite dilution will be also derived. This equation coupled with the two expressions for the local composition and experimental values for the activity coefficients at infinite dilution will allow to obtain both the correlation volume and the interaction energy parameter.

2. Activity Coefficient at Infinite Dilution. A procedure similar to that employed by Wilson[10] will be used here to obtain an expression for the excess Gibbs energy. Wilson started from the Flory and Huggins expression[11,12] for the excess free energy of athermal solutions, but expressed the volume fractions in terms of local molar fractions. We selected Wilson's approach from a number of approaches,[13] because it provided a better description of phase equilibria and because the interactions that count the most are the local one, but started from the more complete Flory–Huggins equation:

$$g^E = x_1 \ln\frac{\varphi_1}{x_1} + x_2 \ln\frac{\varphi_2}{x_2} + \chi\varphi_1\varphi_2 \quad (12)$$

where φ_i are volume fractions and χ is the energetic parameter, given by

$$\chi = w/kT = z\frac{\Gamma_{12} - 0.5(\Gamma_{11} + \Gamma_{22})}{RT} \quad (13)$$

In eq 13, w is the interchange energy, k is the Boltzmann constant, R is the universal gas constant, Γ_{ij} is the energetic parameter for the interaction between molecules of species i and j and z is the coordination number.

We write for the volume fractions the expressions

$$\varphi_1 = \frac{V_1 x_{11}}{V_1 x_{11} + V_2 x_{21}} \quad (14)$$

$$\varphi_2 = \frac{V_2 x_{22}}{V_1 x_{12} + V_2 x_{22}} \quad (15)$$

which differ from those used by Wilson because they contain the partial molar volumes V_i instead of the molar volumes V_i^0 of the pure components. Eliminating the local molar fractions between eqs 8–9 and 14–15 yields

$$\varphi_1 = x_2 \frac{V_1 V_2^2}{V_{cor}^1 V^2}\left(\frac{1-D}{D}\right) + \frac{x_1 V_1}{V} \quad (16)$$

$$\varphi_2 = x_1 \frac{V_2 V_1^2}{V_{cor}^2 V^2}\left(\frac{1-D}{D}\right) + \frac{x_2 V_2}{V} \quad (17)$$

Assuming that $\chi = \chi(x_1)$, the activity coefficients are given by the expressions

$$\ln \gamma_1 = \ln\frac{\varphi_1}{x_1} + x_2\left[x_1\left(\frac{\partial\left(\ln\frac{\varphi_1}{x_1}\right)}{\partial x_1}\right) + x_2\left(\frac{\partial\left(\ln\frac{\varphi_2}{x_2}\right)}{\partial x_1}\right)\right] +$$
$$\chi\left[\varphi_1\varphi_2 + x_2\varphi_2\frac{\partial\varphi_1}{\partial x_1} + x_2\varphi_1\frac{\partial\varphi_2}{\partial x_1}\right] + x_2\varphi_1\varphi_2\frac{\partial\chi}{\partial x_1} \quad (18)$$

$$\ln \gamma_2 = \ln\frac{\varphi_2}{x_2} - x_1\left[x_1\left(\frac{\partial\left(\ln\frac{\varphi_1}{x_1}\right)}{\partial x_1}\right) + x_2\left(\frac{\partial\left(\ln\frac{\varphi_2}{x_2}\right)}{\partial x_1}\right)\right] +$$
$$\chi\left[\varphi_1\varphi_2 - x_1\varphi_2\frac{\partial\varphi_1}{\partial x_1} - x_1\varphi_1\frac{\partial\varphi_2}{\partial x_1}\right] - x_1\varphi_1\varphi_2\frac{\partial\chi}{\partial x_1} \quad (19)$$

Inserting eqs 16–17 into eqs 18–19, one obtains at infinite dilution, by assuming that the derivative of the correlation volume with respect to the molar fraction is negligible

$$\ln \gamma_1^\infty = \ln\left(\frac{-K_1(P,T)V_1^\infty}{V_{cor}^{1,0}} + \frac{V_1^\infty}{V_2^0}\right) + 1 - \frac{V_1^\infty}{V_2^0} + \frac{1}{V_2^0}\left(\frac{\partial V_2}{\partial x_1}\right)_{x_1=0} +$$
$$\chi(x_1 = 0)\left(\frac{-K_1(P,T)V_1^\infty}{V_{cor}^{1,0}} + \frac{V_1^\infty}{V_2^0}\right) \quad (20)$$

$$\ln \gamma_2^\infty = \ln\left(\frac{-K_2(P,T)V_2^\infty}{V_{\text{cor}}^{2,0}} + \frac{V_2^\infty}{V_1^0}\right) + 1 - \frac{V_2^\infty}{V_1^0} + \frac{1}{V_1^0}\left(\frac{\partial V_1}{\partial x_2}\right)_{x_2=0} +$$

$$\chi(x_2=0)\left(\frac{-K_2(P,T)V_2^\infty}{V_{\text{cor}}^{2,0}} + \frac{V_2^\infty}{V_1^0}\right) \quad (21)$$

Each of the above two equations contains two unknown, namely $V_{\text{cor}}^{i,0}$ and $\chi(x_i=0)$. The additional equation needed to calculate them can be obtained by equating eq 11 with another expression for the local compositions.

3. Local Composition. There have been many attempts to express the local composition in terms of the bulk composition and intermolecular interactions.[7–10,13–17] While Guggenheim was the first to introduce the concept of local composition,[17] this idea became extensively used after it was applied by Wilson to phase equilibria.[10] A number of authors proposed various expressions for the local composition,[7–10,14–17] among which we selected the following equations proposed in refs 7–9, because they were derived on the basis of some plausible theoretical considerations:

$$x_{ii} = \frac{x_i \exp\left(x_j \frac{\Delta}{RT}\right)}{x_j + x_i \exp\left(x_j \frac{\Delta}{RT}\right)} \quad (22)$$

$$x_{ji} = \frac{x_j}{x_j + x_i \exp\left(x_j \frac{\Delta}{RT}\right)} \quad (23)$$

where $\Delta = 2e_{12} - e_{11} - e_{22}$ and e_{ij} is the interaction energy parameter between species i and j. Equations 22 and 23 have been obtained by Lee et al.[7] on the basis of Monte Carlo simulations for the nonrandom behavior of off-lattice square-well molecules, and derived theoretically by Aranovich and Donohue.[8–9] The latter authors extended the Ono–Kondo lattice model for the density profile near a surface to the concentration profile around a solute molecule, by assuming that only the first shell has a composition different from that in the bulk. While approximate, their expression will be extended in this paper to the entire correlation volume. At infinite dilution, eq 22 becomes

$$\lim_{x_i \to 0} \frac{x_{ii}}{x_i} = \exp\left[\frac{2e_{12} - e_{11} - e_{22}}{RT}\right] \quad (24)$$

Data Sources and Numerical Calculations

The calculations were conducted for aqueous solutions of alcohols (methanol, ethanol, propanols, butanols, and tert-pentanol) and hydrocarbons (normal saturated aliphatic hydrocarbon from propane through dodecane, isobutane, cyclopentane, cyclohexane, cycloheptane, benzene, toluene).

The parameters Γ_{ii} (in eq 13) for pure substances can be identified with e_{ii} in eq 24 and evaluated from the heats of vaporization ΔH_{vap}^i using the expression

$$\Gamma_{ii} = e_{ii} = -\frac{2}{z}(\Delta H_{\text{vap}}^i - RT) \quad (25)$$

where ΔH_{vap}^i is the heat of vaporization of component i. The latter quantity was obtained from eq 26, which is based on the Antoine equation for the vapor pressure.[18]

$$\Delta H_{\text{vap}}^i = \frac{RT^2 B_i}{C_i + T} \quad (26)$$

Equations 11, 13, 20, 21, 24–26 were employed to calculate $K_i(P,T)/V_{\text{cor}}^{i,0}$ and Γ_{ij}. The correlation volume was finally obtained by extracting $K_i(P,T) = (\partial \ln \gamma_i / \partial x_i)_{P,T,x_i\to 0}$ from vapor–liquid equilibrium data. The Antoine parameters B_i and C_i were taken from ref 18 and the activity coefficients at infinite dilution of the alcohols in water and water in alcohols from refs 19–22. The molar volume of alcohols was taken from ref 23, and the partial molar volumes at infinite dilution of alcohols in water and water in alcohol were taken from refs 23–27 and for tert-pentanol–water was calculated from the excess molar volume.[28] The values of $K_i(P,T)$ were extracted from vapor–liquid equilibria data[20,21,29] in the dilute region. The values of γ_i^∞, V_i^∞, $K_i(P,T)$ used for the calculation of the correlation volume in hydrocarbon + water systems were taken from ref 30. The derivatives $(\partial V_k/\partial x_k)_{x_k=0}$ were estimated by $(V_k^0 - V_k^\infty)$ and were taken zero for all hydrocarbon/water systems. The coordination number was taken to be 4 for water and infinitely dilute aqueous solutions[31] and 6 for alcohols and infinitely dilute alcohol solutions.[32] All the data used in the calculations are listed in Tables 1–3. Throughout this paper, component 1 is an alcohol or a hydrocarbon and component 2 is water.

Results, Discussion, and Conclusion

The calculated correlation volumes and energetic parameters for the alcohol–water and hydrocarbon–water systems are listed in Tables 4 and 5. The calculated volumes are compared with the sizes of clusters in several alcohol/water systems determined by small-angle X-ray scattering or light scattering[33–37] at low concentrations (Table 6). Table 6 shows that there is reasonable agreement between them and the calculated correlation volumes at infinite dilution.

For the systems investigated, the free energies, enthalpies and entropies of hydration are known.[38–40] They exhibit linear dependencies on the number of carbon atoms for different homologous series.[38–40] One may note that, for the systems investigated, ΔH_{hyd} is negative and much smaller in absolute value than $T\Delta S_{\text{hyd}}$, which is also negative. Frank and Evans[2] concluded that the decrease in entropy caused by the organization of the water molecules as an iceberg is responsible for the low solubility of hydrocarbons in water. In reality,[41–44] the change in entropy due to the ordering is compensated by the change in the enthalpy caused by the interactions between the hydrocarbon molecule and water, and the free energy associated with the formation of a cavity is mainly responsible for hydrophobic bonding.[41–44] Shinoda[41,42] concluded that the formation of a cavity constitutes the main effect, while Ruckenstein[43,44] has shown on the basis of a simple thermodynamic approach that while the formation of a cavity provides the largest contribution, the iceberg formation plays also a role. The emphasis in this paper is, however, only on the hydrophobic layer.

One can see from Tables 4 and 5 that the correlation volume at infinite dilution increases for both normal hydrocarbons and normal alcohols with the number of carbon atoms. A comparison between the two shows that they are several times larger for hydrocarbons than for the corresponding alcohols, but that the difference between them decreases as the number of carbon atoms increases (Figure 1).

The smaller values for alcohols are due to the presence of the hydroxyl group, which, because of its favorable interaction with the solvent, does not disturb, as much as the hydrocarbon

Infinitely Diluted Aqueous Solutions

TABLE 1: Data Used for the Calculation of the Correlation Volumes of Alcohols at Infinite Dilution in Alcohol/Water Solutions

system	temp (K)	γ_1^∞	V_1^∞ (cm³/mol)	$-K_1(P,T)$
methanol−water	298.15	1.8	38.2	1.5
ethanol−water	298.15	4.0	55.1	4.5
1-propanol−water	303.15	16.3	70.7[a]	11.5
2-propanol−water	298.15	8.3	71.8	6.5
1-butanol−water	323.15	78.7	86.37	31
2-butanol−water	323.15	35.5	87.72	23
i-butanol−water	323.15	58.1	87.63	28
t-butanol−water	323.15	19.2	89.2	18
t-pentanol−water	328.15	78.1	100.9[b]	27

[a] $T = 298.15$ K. [b] Calculated from excess volume data[28] at $T = 303.15$ K.

TABLE 2: Data Used for the Calculation of the Correlation Volumes of Water at Infinite Dilution in Alcohol/Water Solutions

system	temp (K)	γ_2^∞	V_2^∞ (cm³/mol)	$-K_2(P,T)$
methanol−water	298.15	1.5	14.48	0.5
ethanol−water	298.15	2.7	13.81	1.3
1-propanol−water	303.15	5.6	15.09[a]	3.2
2-propanol−water	298.15	3.6	14.51	1.6
1-butanol−water	323.15	5.2	16.90[b]	2.3
2-butanol−water	323.15	4.6	17.53[b]	2.4
i-butanol−water	323.15	5.3	16.99[b]	2.6
t-butanol−water	323.15	4.9	16.05[b]	1.7
t-pentanol−water	328.15	3.43	14.7[b]	2.0

[a] $T = 298.15$ K. [b] $T = 318.15$ K.

TABLE 3: Data Used for the Calculation of the Correlation Volumes of Hydrocarbon at Infinite Dilution in Hydrocarbon/Water Solutions

system	temp (K)	$\ln \gamma_1^\infty$	V_1^∞ (cm³/mol)	$-K_1(P,T)$
propane−water	298.15	8.35[a]	70.7	80.7
n-butane−water	298.15	9.99	76.6	112.6
isobutane−water	298.15	9.86	81.3	107.4
n-pentane−water	298.15	11.6	92.3	150.4
n-hexane−water	298.15	13.1	110.0	190.4
n-heptane−water	298.15	14.5	129.4	231.1
n-octane−water	298.15	16.1	145.2	288.1
n-decane−water	298.15	18.9	176.8	395.4
n-dodecane−water	298.15	21.7	209.5	523.5
cyclopentane−water	298.15	10.1	84.5	100.5
cyclohexane−water	298.15	11.3	98.8	120.2
cycloheptane−water	298.15	12.1	105.5	131.1
benzene−water	298.15	7.8	82.5	42.5
toluene−water	298.15	9.2	97.7	61.9

[a] This value was taken from ref 19.

TABLE 4: Correlation Volumes and Intermolecular Interaction Energy Parameters in Alcohol/Water Systems at Infinite Dilution[a]

system	temp (K)	$-\Gamma_{12}$ (J/mol)	$V_{cor}^{1,0}$ (cm³/mol)	$-\Gamma_{21}$ (J/mol)	$V_{cor}^{2,0}$ (cm³/mol)
methanol + water	298.15	16 750	270	17 000	80
ethanol + water	298.15	17 560	490	18 500	110
1-propanol + water	303.15	18 620	860	19 900	220
2-propanol + water	298.15	18 020	570	19 360	120
1-butanol + water	323.15	19 010	1900	20 240	180
2-butanol + water	323.15	18 070	1600	19 600	210
i-butanol + water	323.15	18 660	1800	19 940	210
t-butanol + water	323.15	17 600	1500	19 010	140
t-pentanol + water	328.15	17 960	1700	18 860	140

[a] Component 1 is alcohol, component 2 is water and $V_{cor}^{i,0} = \lim_{x_i \to 0} V_{cor}^i$.

TABLE 5: Correlation Volumes and Intermolecular Interaction Energy Parameters in Hydrocarbon/Water Systems at Infinite Dilution[a]

system	temp (K)	$-\Gamma_{12}$ (J/mol)	$V_{cor}^{1,0}$ (cm³/mol)
propane−water	298.15	14 160	2900
n-butane−water	298.15	14 970	3760
isobutane−water	298.15	14 660	3710
n-pentane−water	298.15	15 710	5020
n-hexane−water	298.15	16 470	6350
n-heptane−water	298.15	17 240	8000
n-octane−water	298.15	18 090	9990
n-decane−water	298.15	19 930	13630
n-dodecane−water	298.15	21 950	18100
cyclopentane−water	298.15	16 120	2900
cyclohexane−water	298.15	16 810	3500
cycloheptane−water	298.15	17 750	3700
benzene−water	298.15	16 800	1440
toluene−water	298.15	17 510	2870

[a] Component 1 is hydrocarbon, component 2 is water and $V_{cor}^{i,0} = \lim_{x_i \to 0} V_{cor}^i$.

TABLE 6: Comparison between the Radius of the Correlation Volume at Infinite Dilution of Alcohol in Alcohol/Water Systems and Size of Clusters at Small Concentration of Alcohol in These Systems Obtained by Different Physical Methods[a]

		radius of cluster obtained by different physical methods			
system	radius[b] of correlation volume (Å)	radius (Å)	molar fraction, x_1	method	ref
1-propanol−water	7.0	7.3		LS	33
		~4[c]	0.05	SAXS	34
1-butanol−water	9.1	7.5	0.018	SAXS	35
2-butanol−water	8.6	8.2	0.041	SAXS	35
i-butanol−water	8.9	11.8	0.019	SAXS	35
t-butanol−water	8.4	~5[c]	0.05	SAXS	36
		8.3	0.04	LS	37

[a] LC = light scattering; SAXS = small-angle X-ray scattering. [b] The correlation volume is approximated as a sphere. [c] Data evaluated from the figures of the corresponding papers.

molecules do, the structure of water. Table 4 also reveals that the correlation volume is much larger when the alcohol is the solute. The larger value is due to the higher disturbance produced by the alcohol molecule in the structure of water solvent than that produced by a water molecule in the structure of alcohol solvent.

The correlation volume was not accounted in the Butler's scheme. Butler's scheme for dissolution in water[1] accounted only for the formation of a cavity, introduction of the solute molecule in that cavity, and its interactions with the nearest-neighbor water molecules. He assumed, however, that the water molecules are distributed around a solute molecule as randomly as in its absence. One more step should be added, namely, the formation of a "hydrophobic layer" of volume V_{cor}^1 around the cavity, in which the water molecules are reorganized and are no longer randomly distributed (Figure 2). While this layer is similar to that suggested by Frank and Evans in their "iceberg" model,[2] the present approach provides also the size of the region of water affected by the presence of the solute.

The estimation of the thickness of the water layer affected by the presence of a solute molecule was made for two geometries: (1) the cavity containing the solute and the correlation volume have the shape of a sphere, (2) both have the shape of a cylinder. The results of the calculations are listed in Table 7, which shows that the water layer is formed of several molecular shells (between 4 and 8, Table 7). The correlation volumes for cyclic hydrocarbon are much lower than for aliphatic hydrocarbons, but, among the cyclic hydrocarbons, the

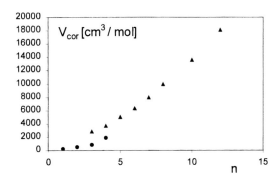

Figure 1. Dependence of correlation volume at infinite dilution on the number of hydrocarbon atoms n. Hydrocarbon in water (▲) and alcohol in water (●).

TABLE 7: Thickness of the Hydrophobic Layer (l) Covering an Aliphatic Hydrocarbon Molecule in Water at Infinite Dilution[a]

n	l_1 (Å)	l_2 (Å)	N
3	7.4	7.9	~4
4	8.3	8.7	≥4
5	9.3	9.6	~5
6	10.1	10.4	≥5
7	11.0	11.2	~6
8	12.0	12.0	≥6
10	13.4	13.4	~7
12	14.9	14.7	≥7

[a] n is the number of carbon atoms, l_1 is the thickness of the hydrophobic layer when the cavity and correlation volume are approximated as spheres and l_2 when they are approximated as cylinders, and N is the estimated number of water shells. This estimation was made on the basis of X-ray scattering data of cold water.[31]

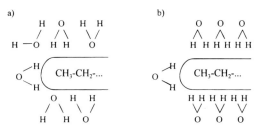

Figure 2. Schematic of a hydrocarbon molecule covered by water molecules at infinite dilution: (a) random distribution of water molecules around a hydrocarbon molecule; (b) water molecules with an ordered structure.

aromatic hydrocarbons have especially low correlation volumes (Table 5). These results should be attributed to the favorable interactions of the π electrons of the aromatic hydrocarbon with the surrounding water molecules. For this reason, the structure of water is less perturbed by the aromatic hydrocarbons than by the aliphatic ones.

Tables 4 and 5 also list the values of the energy interaction parameters Γ_{ij} for the alcohol/water and hydrocarbon/water systems. For the alcohol/water systems, the parameters were calculated for both dilute solutions of alcohol in water and dilute solutions of water in alcohol. For hydrocarbon/water systems, the calculations were carried out only for dilute solutions of hydrocarbon in water, because no experimental information could be found for the solutions of water in hydrocarbons. Figure 3 presents a plot of Γ_{12} versus the number of carbon atoms in molecules for normal alcohols and hydrocarbons.

Figure 3 and Tables 4 and 5 show that Γ_{12} increases when the number of carbon atoms increases and that Γ_{12} is greater for alcohols than for the corresponding hydrocarbons. This result

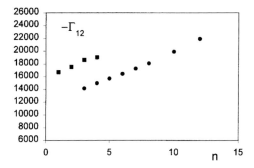

Figure 3. Energetic parameter Γ_{12} (J/mol) as a function of the number of carbon atoms in alcohol and hydrocarbon molecules (n): (■) alcohol/water systems; and (●) hydrocarbon/water systems.

is as expected, because the alcohol molecules have hydroxyl groups, which interact strongly with the water molecules. It is interesting to compare the value Γ_{ij} from both sides (when $x_1 \rightarrow 0$ and $x_2 \rightarrow 0$) in the alcohol/water systems. The data listed in Table 4 show that Γ_{12} and Γ_{21} are different. This indicates that the interactions between alcohol and water are not the same in the two limiting cases and hence that the usual assumption that $\Gamma_{12} = \Gamma_{21}$ is not a good approximation. Of course, the present calculations involve a number of approximations and it is not yet clear how accurate is the above conclusion.

References and Notes

(1) Butler, J. A. V. *Trans. Faraday Soc.* **1937**, *33*, 229.
(2) Frank, H. S.; Evans, M. W. *J. Chem. Phys.* **1945**, *13*, 507.
(3) Nemethy, G.; Scheraga, H. A. *J. Chem. Phys.* **1962**, *36*, 3401.
(4) Kirkwood, J. G.; Buff, F. P. *J. Chem. Phys.* **1951**, *19*, 774.
(5) Shulgin, I.; Ruckenstein, E. *J. Phys. Chem. B* **1999**, *103*, 872.
(6) Debenedetti, P. G.; Kumar, S. K. *AIChE J.* **1986**, *32*, 1253.
(7) Lee, K.-H.; Sandler, S. I.; Patel, N. C. *Fluid Phase Equilib.* **1986**, *25*, 31.
(8) Aranovich, G. L.; Donohue, M. D. *J. Chem. Phys.* **1996**, *105*, 7059.
(9) Wu, D. W.; Cui, Y.; Donohue, M. D. *Ind. Eng. Chem. Res.* **1998**, *37*, 2936.
(10) Wilson, G. M. *J. Am. Chem. Soc.* **1964**, *86*, 127.
(11) Flory, P. J. *J. Chem. Phys.* **1941**, *9*, 660.
(12) Huggins, M. L. *J. Chem. Phys.* **1941**, *9*, 440.
(13) Prausnitz, J. M.; Lichtenhaler, R. N.; Gomes de Azevedo, E. *Molecular Thermodynamics of Fluid-Phase Equilibria*, 2nd ed.; Prentice-Hall: Englewood Cliffs, NJ, 1986.
(14) Renon, H.; Prausnitz, J. M. *AIChE J.* **1968**, *14*, 135.
(15) Panayiotou, C.; Vera, J. H. *Fluid Phase Equilib.* **1980**, *5*, 55.
(16) Panayiotou, C.; Vera, J. H. *Can. J. Chem. Eng.* **1981**, *59*, 501.
(17) Guggenheim, E. A. *Mixtures*; Oxford University Press: Oxford, UK, 1952.
(18) Reid, R. C.; Prausnitz, J. M.; Poling, B. E. *The Properties of Gases and Liquids*, 4th ed.; McGraw-Hill: New York, 1987.
(19) Kojima, K.; Zhang, S.; Hiaki, T. *Fluid Phase Equilib.* **1997**, *131*, 145.
(20) Fischer, K.; Gmehling, J. *J. Chem. Eng. Data* **1994**, *39*, 309.
(21) Fischer, K.; Shulgin, I.; Rarey, J.; Gmehling, J. *Fluid Phase Equilib.* **1996**, *120*, 143.
(22) Christensen, S. P. *Fluid Phase Equilib.* **1998**, *150−151*, 763.
(23) Franks, F.; Desnoyers, J. E. *Water Sci. Rev.* **1985**, *1*, 171.
(24) Friedman, M. E.; Scheraga, H. A. *J. Phys. Chem.* **1965**, *69*, 3795.
(25) Kim, E. S.; Marsh, K. N. *J. Chem. Eng. Data* **1988**, *33*, 288.
(26) Sakurai, M. and Nakagawa, T. *J. Chem. Thermodyn.* **1984**, *16*, 171.
(27) Sakurai, M.; Nakamura, K.; Takenaka, N. *Bull. Chem. Soc. Jpn.* **1994**, *67*, 352.
(28) Nakanishi, K.; Kato N.; Maruyama, M. *J. Phys. Chem.* **1967**, *71*, 814.
(29) Gmehling, J. et al. *Vapor-Liquid Equilibrium Data Collection*; DECHEMA Chemistry Data Series, Vol. I, 19 Parts; DECHEMA: Frankfurt, 1977−1996.
(30) Liu, H.; Ruckenstein, E. *J. Phys. Chem. B* **1998**, *102*, 1005.
(31) Narten, A. H.; Levy, H. A. In *Water—A Comprehensive Treatise*; Plenum Press: New York, 1972; Vol. 1, pp 311−332.
(32) Vahvaselka, K. S.; Serimaa, R.; Torkkeli, M. *J. Appl. Crystallogr.* **1995**, *28*, 189.

(33) Grossmann, G. H.; Ebert, K. H. *Ber. Bunsen-Ges. Phys. Chem.* **1981**, *85*, 1026.
(34) Hayashi, H.; Udagawa, Y. *Bull. Chem. Soc. Jpn.* **1992**, *65*, 155.
(35) Shulgin, I.; Serimaa, R.; Torkkeli, M. Unpublished data, Helsinki, 1991.
(36) Nishikawa, K.; Hayashi, H.; Iijima, T. *J. Phys. Chem.* **1989**, *93*, 6559.
(37) Bender, T. M.; Pekora, R. *J. Phys. Chem.* **1986**, *90*, 1700.
(38) Abraham, M. H. *J. Chem. Soc., Faraday Trans. 1* **1984**, *80*, 153.
(39) Abraham, M. H.; Matteoli, E. *J. Chem. Soc., Faraday Trans. 1* **1988**, *84*, 1985.
(40) Tanford, C. *The Hydrophobic Effect: Formation of Micelles and Biological Membranes*, 2nd ed.; Wiley-Interscience: New York, 1980.
(41) Shinoda, K. *Principles of Solution and Solubility*; M. Decker: New York, 1977.
(42) Shinoda, K. *J. Phys. Chem.* **1977**, *81*, 1300.
(43) Ruckenstein, E. *Colloid Surf.* **1992**, *65*, 95.
(44) Ruckenstein, E. *J. Dispersion Sci. Technol.* **1998**, *19*, 329.

Aggregation in Binary Solutions Containing Hexafluorobenzene

E. Ruckenstein* and I. Shulgin[†]

Department of Chemical Engineering, State University of New York at Buffalo, Amherst, New York 14260

Received: June 3, 1999; In Final Form: September 24, 1999

Some information about the aggregation in aromatic fluorocarbon—aromatic hydrocarbon systems was obtained on the basis of the Kirkwood—Buff theory of solution. The Kirkwood—Buff integrals and the excess (or deficit) number of molecules aggregated around a central one were calculated for the following systems: hexafluorobenzene—benzene, hexafluorobenzene—toluene, hexafluorobenzene—cyclohexane, benzene—toluene, and benzene—cyclohexane. It was found that the composition dependence of the excess (or deficit) number of molecules aggregated around a central one for the systems hexafluorobenzene—benzene and hexafluorobenzene—toluene has a nontypical character since it changes sign for a mole fraction of about 0.4 for the former and 0.2 for the latter. It was found that such compositions correspond to extrema in the activity coefficients and inflection points in the excess Gibbs energy. The excess (or deficit) number of molecules aggregated around a central one in the systems investigated allowed us to conclude that in the mixtures hexafluorobenzene—benzene and hexafluorobenzene—toluene there are some like (enriched in the same component as the central one) aggregates for mole fractions of hexafluorobenzene smaller than 0.4 and 0.2, respectively, and some unlike aggregates for larger values of mole fraction of hexafluorobenzene. The change of aggregation from like to unlike as well as its moderate temperature dependence suggest that the interactions involved are not too strong. In addition, because the radius of the correlation volume (the volume affected by the presence of a molecule) is relatively large ($\sim$10 Å), the interactions are relatively long range. In the other mixtures investigated, only like aggregates were formed and for the mixture hexafluorobenzene—cyclohexane relatively large like aggregation of hexafluorobenzene was found.

Introduction

The unusual thermodynamic properties of binary mixtures containing an aliphatic or an alicyclic fluorocarbon have been known since the 1950s.[1,2] A review written by R. L. Scott[1] in 1958 "The Anomalous Behavior of Fluorocarbon Solutions" suggested various directions for further studies. Intensive studies of such systems started in the 1960s[3–10] and comprehensive reviews for the 1960s and 1970s were written by Swinton.[11–12] It was found that binary mixtures of aromatic fluorocarbon—aromatic hydrocarbon behave very differently when compared to similar mixtures containing an aliphatic or an alicyclic fluorocarbon.[12] Unusual properties were found particularly for the hexafluorobenzene (C_6F_6)—benzene (C_6H_6) system. Indeed, this system exhibits unusual dependencies of the viscosity and the excess thermal pressure coefficient on composition, and a rare type of vapor—liquid equilibrum with two azeotropic points in a wide range of pressures and temperatures.[3–4,8–10] Only a few (less than 10) double-azeotropic mixtures among more than 17 500 mixtures investigated are known.[13] The fact that these substances form an equimolar molecular compound in the solid state[6,14–15] suggested to explain their behavior by assuming that such a complex is formed in solution as well. Donor—acceptor interactions involving π electrons, the hexafluorobenzene molecule being the acceptor and the benzene molecule the donor,[14] or electrostatic interactions between their quadrupole moments, which are equal in magnitude and of opposite sign[16,17] have been considered responsible for the formation of the complex.

For the thermodynamic treatment of the problem, it was supposed that the excess free energy consists of two parts, a nonspecific part due to the "physical" fluorocarbon—hydrocarbon interactions and a specific or "chemical" part due to the complex formation between the two components.[5,11,18–19] The equilibrium constant for complex formation was calculated from the experimental data regarding the excess Gibbs energy.[5]

Several experimental studies aimed at finding experimental evidence for intermolecular complexes in aromatic fluorocarbon/aromatic hydrocarbon systems[20–27] have been carried out. The analysis of the nuclear spin—lattice relaxation in such systems suggested the existence of clusters of fluorocarbon molecules.[28] The neutron and X-ray diffraction studies were interpreted by assuming that quadrupole—quadrupole interactions constitute the main factor in these systems and that a 1:1 complex between the two species might form.[20] However, a number of spectroscopic studies regarding the formation of unlike intermolecular complexes in the system hexafluorobenzene—benzene have proven ambiguous.[21–26] Wang et al.[26] concluded that the results of spectroscopic studies "cannot be explained on the basis of a simplified model of association and reorientation." Similarly, the recently published molecular dynamic studies of the system C_6F_6—C_6H_6 found that only heterodimers cannot explain the experimental data and that long-range interactions of the order of 7–11 Å are present because of quasistacked heterodimers.[27]

In this paper, the system C_6F_6—C_6H_6 and several other systems such as C_6F_6—$C_6H_5CH_3$, C_6F_6—c-C_6H_{12}, C_6H_6—C_6H_5-CH_3, and C_6H_6—c-C_6H_{12} will be analyzed and compared on the basis of the Kirkwood—Buff (KB) theory of solutions.[29] The KB theory of solutions allows information about the excess (or deficit) number of molecules, of the same or different kind,

---

* To whom the correspondence should be addressed. E-mail: fealiru@acsu.buffalo.edu. Fax: (716) 645-3822.
[†] Current e-mail address: ishulgin@eng.buffalo.edu.

around a given molecule to be extracted from macroscopic thermodynamic properties, such as phase equilibria, excess volume, and isothermal compressibilities. The excess (or deficit) number of molecules is a result of the differences in intermolecular interactions between molecules of the same and different kind and can be calculated from the so-called Kirkwood–Buff integrals.[30] These quantities will allow some information about the local structure of the solution to be obtained.[31–34] The calculation could be carried out because reliable thermodynamic properties for aromatic fluorocarbon/aromatic hydrocarbon mixtures are available.[11–12]

Theory and Formulas

1. The Kirkwood–Buff Integrals. The Kirkwood–Buff integrals (KBI) are defined as[29]

$$G_{ij} = \int_0^\infty (g_{ij} - 1) 4\pi r^2 \, dr \quad i,j = 1, 2 \quad (1)$$

where g_{ij} is the radial distribution function between species i and j, r is the distance between the centers of molecules i and j, and G_{ij} is expressed as volume per molecule. Ben-Naim was the first to calculate, by using the Kirkwood–Buff equations, the KBIs from measured thermodynamic properties.[30] For binary mixtures, the Kirkwood–Buff equations acquire the form[35]

$$G_{12} = G_{21} = RTk_T - \frac{V_1 V_2}{VD} \quad (2)$$

$$G_{ii} = G_{12} + \frac{1}{x_i}\left(\frac{V_j}{D} - V\right) \quad i \neq j \quad (3)$$

where

$$D = \left(\frac{\partial \ln \gamma_i}{\partial x_i}\right)_{P,T} x_i + 1 \quad (4)$$

In the above equations, the G_{ij} are expressed in volume per mole, k_T is the isothermal compressibility, V_i is the partial molar volume of component i, x_i is the mole fraction of component i, V is the molar volume of the mixture, T is the absolute temperature, R is the universal gas constant, and γ_i is the activity coefficient of component i.

2. The Excess (or Deficit) Number of Molecules around a Central One. The key quantities in the Kirkwood–Buff approach are the $c_i G_{ij}$, the excess (or deficit) number of molecules i around a central molecule j, where c_i is the molar concentration of species i in the mixture.[30] However, as noted by Matteoli and Lepori, $c_i G_{ij}$ have nonzero values for ideal systems,[36–37] even though they should be considered nonaggregated. For this reason, Matteoli and Lepori suggested using the ideal system as a reference system and calculating the excess (or deficit) number of molecules i aggregated around a central molecule j as $c_i(G_{ij} - G_{ij}^{id})$,[36–37] with the KBIs for the ideal systems (G_{ij}^{id}) given by the expressions[36–37]

$$G_{12}^{id} = RTk_T^{id} - \frac{V_1^0 V_2^0}{V^{id}} = RTk_{T,2}^0 - V_1^0 -$$
$$\varphi_1(V_2^0 - V_1^0 + RTk_{T,2}^0 - RTk_{T,1}^0) \quad (5)$$

$$G_{11}^{id} = G_{12}^{id} + V_2^0 - V_1^0 \quad (6)$$

$$G_{22}^{id} = G_{12}^{id} - (V_2^0 - V_1^0) \quad (7)$$

where k_T^{id} and V^{id} are the isothermal compressibility and the molar volume of an ideal solution, respectively, φ_i is the volume fraction of component i in solution, defined on the basis of the molar volumes of the pure components, $k_{T,i}^0$ is the isothermal compressibility of the pure component i, and V_i^0 is the molar volume of the pure component i.

However, because "the volume occupied by the excess j molecules aggregated around an i molecule must be equal to the volume left free by the i molecules around the same i molecule",[36] the following equality should be satisfied:

$$V_j c_j (G_{ji} - G_{ji}^{id}) = -V_i c_i (G_{ii} - G_{ii}^{id}) \quad (8)$$

Nevertheless, eq 8 is not satisfied identically by the above expressions. Indeed, using eqs 2, 3, 5, 6, and 7, eq 8 becomes

$$RTV(k_T - k_T^{id}) = V_i V + V_i(V_j^0 - V^{id}) - \frac{V_i^0 V_j^0 V}{V^{id}} \quad (9)$$

Only if G_{ij}^{id} is replaced by G_{ij}^V (eqs 10 and 11), which is obtained from the former by substituting k_T^{id}, V_i^0, and V^{id} with k_T, V_i, and V, respectively, can eq 8 be satisfied identically.[33]

$$G_{12}^V = G_{21}^V = RTk_T - \frac{V_1 V_2}{V} \quad (10)$$

$$G_{ii}^V = G_{12}^V + V_j - V_i \quad i \neq j \quad (11)$$

Consequently, the excess (or deficit) number of molecules Δn_{ij} with respect to a reference state are given by the expressions[33]

$$\Delta n_{12} = c_1(G_{12} - G_{12}^V) = c_1 \Delta G_{12} = c_1 \Delta G_{21} = -\frac{c_1 V_1 V_2}{V}\left(\frac{1-D}{D}\right) \quad (12)$$

$$\Delta n_{21} = c_2(G_{12} - G_{12}^V) = c_2 \Delta G_{12} = c_2 \Delta G_{21} = -\frac{c_2 V_1 V_2}{V}\left(\frac{1-D}{D}\right) \quad (13)$$

$$\Delta n_{ii} = c_i \Delta G_{ii} = c_i(G_{ii} - G_{ii}^V) = \frac{c_i x_j V_j^2}{x_i V}\left(\frac{1-D}{D}\right) \quad i \neq j \quad (14)$$

In other words, a more appropriate reference state, one which is compatible with the volume conservation condition, is characterized by G_{ij}^V, and the differences $G_{ij} - G_{ij}^V$ constitute better measures of aggregation than G_{ij}.

A comparison between G_{ij}^V and G_{ij}^{id} for the methanol–water system showed that in that case they were close to each other.[34] However, as was already pointed out,[36–37] the use of a reference state is particularly important for the systems with small deviations from the Raoult law. In such cases, the reference state based on G_{ij}^V provides results very different from those based on G_{ij}^{id}.

3. The Correlation Volume and Local Composition. Let us consider a molecule of species i and its surrounding correlation volume (V_{cor}^i) in which the composition and/or the structure differ from the bulk one. The total number of aggregated molecules i and j in this volume is given by the expressions

$$n_{ii} = \Delta n_{ii} + c_i V_{cor}^i \quad (15)$$

and

$$n_{ji} = \Delta n_{ji} + c_j V_{cor}^i \quad (16)$$

According to the local composition (LC) concept, the composition of solution in the vicinity of any molecule differs from the overall (bulk) composition. For binary mixtures composed of components 1 and 2 with mole fractions x_1 and x_2, respectively, four LCs can be considered: the local mole fractions of components 1 and 2 around a central molecule 1 (x_{11} and x_{21}) and the local mole fractions of components 1 and 2 around a central molecule 2 (x_{12} and x_{22}). In terms of the total number of molecules i and j in the correlation volume, the local mole fractions are given by

$$x_{ii} = \frac{n_{ii}}{n_{ii} + n_{ji}} = \frac{\Delta n_{ii} + c_i V_{cor}^i}{\Delta n_{ii} + \Delta n_{ji} + (c_i + c_j)V_{cor}^i} \quad (17)$$

and

$$x_{ji} = \frac{n_{ji}}{n_{ii} + n_{ji}} = \frac{\Delta n_{ji} + c_j V_{cor}^i}{\Delta n_{ii} + \Delta n_{ji} + (c_i + c_j)V_{cor}^i} \quad (18)$$

Using eqs 12–14, eqs 17 and 18 become

$$x_{ii} = \frac{x_j V_j^2(1-D) + x_i V_{cor}^i VD}{x_j V_j^2(1-D) - x_j V_i V_j(1-D) + V_{cor}^i VD} \quad (19)$$

and

$$x_{ji} = \frac{-x_j V_i V_j(1-D) + x_j V_{cor}^i VD}{x_j V_j^2(1-D) - x_j V_i V_j(1-D) + V_{cor}^i VD} \quad (20)$$

Equations 19 and 20 coupled with expressions for the local compositions and for the activity coefficients at infinite dilution allowed the evaluation of the correlation volume and the unlike interaction energy parameter at infinite dilution.[38]

Treatment Procedure and Data Sources

The values of G_{ij}, ΔG_{ij}, and Δn_{ij} were calculated for the following five systems: hexafluorobenzene (HFB)–benzene (B), HFB–toluene (TL), HFB–cyclohexane (CH), B–TL, and B–CH. For the system HFB–B, the calculations were carried out at five temperatures, while for the other systems, the calculations were carried out at a single temperature. To perform the calculations, information is needed about the dependence on composition of the following quantities: D (defined by eq 4), molar volume V, partial molar volumes V_1 and V_2, and the isothermal compressibility k_T. Because the dependence of k_T on composition is not known for the systems investigated and because of the small contribution of $RT k_T$ to the KBIs,[35] the dependence of k_T on composition will be approximated by

$$k_T = k_{T,1}^0 \varphi_1 + k_{T,2}^0 \varphi_2 \quad (21)$$

The isothermal compressibilities of benzene, toluene, and cyclohexane are available in ref 39. The isothermal compressibility of hexafluorobenzene is not available in the literature, but values for the isentropic compressibility (k_S^0) are available.[8] Because the contribution of the compressibility to the values of the KBIs is small,[35] we assumed that the difference between the isothermal and isentropic compressibilities ($k_T^0 - k_S^0$) are the same for benzene and hexafluorobenzene. The sources of the data for the calculation of D, molar volume V, partial molar volumes V_1 and V_2 are listed in Table 1. The values of the molar volume V and of the partial molar volumes V_1 and V_2 were

TABLE 1: Sources of Data Used for Calculating the Kirkwood–Buff Integrals[a]

system	D data	ref	V, V_2, V_2 data	ref
HFB–B	P-x-y T = 303.15 K	5	V^E T = 313.15 K	7
	P-x-y T = 313.15 K	5		
	P-x-y T = 323.15 K	5		
	P-x-y T = 333.15 K	5		
	P-x-y T = 343.15 K	5		
HFB–TL	P-x-y T = 303.15 K	5	V^E T = 313.15 K	7
HFB–CH	P-x-y T = 303.15 K	4	V^E T = 313.15 K	7
B–TL	P-x T = 303.15 K	40	V^E T = 298.15 K	41
B–CH	P-x-y T = 298.15 K	42	V^E T = 298.15 K	43

[a] P is the pressure and x and y are equilibrium mole fractions in liquid and vapor; V^E is the excess volume.

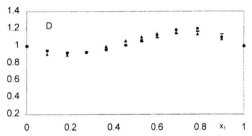

Figure 1. Values of D for the system HFB–B at 313.15 K calculated using various expressions for the activity coefficient: (–) the NRTL equation; (▲) the Van Ness–Abbott equation; (●) the equation proposed by the authors in ref 52.

obtained from the molar volumes of the pure components[4,5,39] and the excess volume data V^E (the sources of these data are listed in Table 1), which have been fitted using the Redlich–Kister equation.[44]

The analysis of the possible errors in the calculation of KBIs clearly indicated that the main error is introduced through the parameter D[35,45] and is caused by the uncertainty in the derivative of the activity coefficient. In the present paper, isothermal vapor–liquid equilibrium (VLE) data were used to calculate D. The data were taken from Gmehling's VLE compilation[46] and were selected if they satisfied the integral and the point[46] thermodynamic consistency tests. The selected VLE data were treated by the Barker method,[47,48] using only liquid mole fraction–pressure ($x-P$) data. The vapor phase nonideality was taken into account, and the total pressure was calculated using the equation

$$P = \frac{x_1 \gamma_1 P_1^0}{F_1} + \frac{x_2 \gamma_2 P_2^0}{F_2} \quad (22)$$

where

$$F_1 = \exp\frac{(B_{11} - V_1^0)(P - P_1^0) + P y_2^2 d_{12}}{RT} \quad (23)$$

$$F_2 = \exp\frac{(B_{22} - V_2^0)(P - P_2^0) + P y_1^2 d_{12}}{RT} \quad (24)$$

P_i^0 is the vapor pressure of the pure component i, B_{ii} is the second virial coefficient of component i, $d_{12} = 2B_{12} - B_{11} - B_{22}$ and B_{12} is the crossed second virial coefficient of the binary mixture. The virial coefficients of the pure components were taken from ref 5 and the vapor pressures of the pure components P_i^0 from refs 4, 5, 40, and 42. The crossed second virial coefficients of the binary mixtures were taken from refs

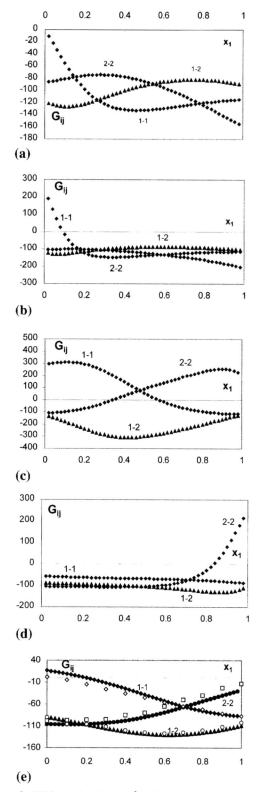

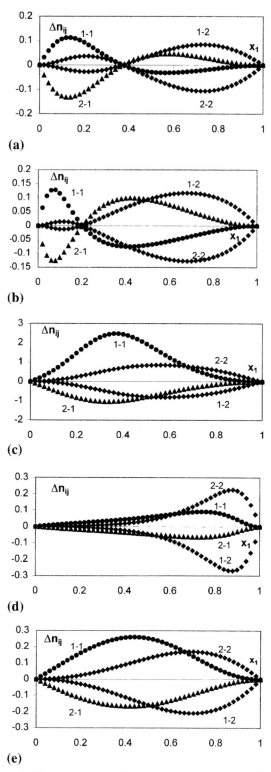

Figure 2. KB integrals (G_{ij} in [cm³/mol]) in the systems investigated: (a) the system HFB−B at 303.15 K; (b) the system HFB−TL at 313.15 K; (c) the system HFB−CH at 303.15 K; (d) the system B−TL at 303.15 K; (e) the system B−CH at 298.15 K. Figure 2e also contains the KBIs from ref 37 at 303.15 K: G_{11} (◇); G_{22} (□); G_{12} (○).

49−50 or were evaluated using the expression[5] $B_{12} = (B_{11} + B_{22})/2$. The NRTL,[51] the Van Ness−Abbott,[48] and the authors equation[52] were used to express the activity coefficients in eq 22. The expressions for the activity coefficients provided by

Figure 3. The excess (or deficit) molecules near a central one (Δn_{ij} in [mol/mol]): (a) the system HFB−B at 303.15 K; (b) the system HFB−TL at 313.15 K; (c) the system HFB−CH at 303.15 K; (d) the system B−TL at 303.15 K; (e) the system B−CH at 298.15 K.

these three methods were differentiated analytically, and the obtained derivatives were used to calculate D. There is good agreement between the values of D obtained with the three expressions. The agreement is illustrated in Figure 1 for the system HFB−B at 303.15 K.

Results and Discussion

The results of the calculations are presented in Figures 2 and 3. For the system benzene−hexafluorobenzene, the calculations were carried out at 303.15, 313.15, 323.15, 333.15, and 343.15 K. The results for 303.15 K are given in Figures 2a and 3a, and the temperature dependence is examined later in the paper. Except the system benzene−cyclohexane, for which the KBIs were calculated previously at 303.15 K,[37] all the other KBIs were calculated in this paper for the first time. There is good agreement between our calculations at 298.15 K and those calculated in the literature[37] at 303.15 K for the benzene−cyclohexane mixture (Figure 2e).

All the KBIs of the system HFB−B are negative at all compositions (G_{ij} in Figure 2a) and all the KBIs of the system HFB−TL are negative (G_{ij} in Figure 2b) for $x_1 \geq 0.1$. They are also negative for ideal mixtures in the entire composition range. Because the ideal mixtures should be considered nonaggregated, the above systems are good illustrations for the need of a reference state for the calculation of the excess (or deficit) number of molecules around a given molecule due to aggregation. The values of ΔG_{ij} for HFB−B and HFB−TL, calculated with eqs 2−3 and eqs 10−11, show that for these mixtures all the ΔG_{ij} are only slightly different from zero, hence from ideal behavior, except ΔG_{11} for $x_1 < 0.2$ and ΔG_{22} for $x_2 < 0.2 - 0.3$.

The Δn_{ij} values for the systems HFB−B and HFB−TL (parts a and b of Figure 3) change sign at $x_1^* \approx 0.4$ and at $x_1^* \approx 0.2$, respectively. At compositions x_1^*, all $\Delta n_{ij} = 0$ and all local compositions are equal to the bulk ones. To obtain more information about this point, let us note that from eqs 12−14 one finds that $\Delta n_{ij} = 0$ when $D = 1$ and hence (see eq 4) that

$$\left(\frac{\partial \ln \gamma_1}{\partial x_1}\right)_{P,T,x_1=x_1^*} = \left(\frac{\partial \ln \gamma_2}{\partial x_1}\right)_{P,T,x_1=x_1^*} = 0 \quad (25)$$

On the other hand, one can write for the molar excess Gibbs energy of a binary system the following relation:

$$\frac{\Delta G^E}{RT} = x_1 \ln \gamma_1 + x_2 \ln \gamma_2 \quad (26)$$

which differentiated gives

$$\frac{\partial \left(\frac{\Delta G^E}{RT}\right)}{\partial x_1} = \ln \gamma_1 - \ln \gamma_2 + x_1 \frac{\partial \ln \gamma_1}{\partial x_1} + x_2 \frac{\partial \ln \gamma_2}{\partial x_1} \quad (27)$$

Taking into account the Gibbs−Duhem equation for isothermal conditions

$$x_1 \frac{\partial \ln \gamma_1}{\partial x_1} = x_2 \frac{\partial \ln \gamma_2}{\partial x_2} \quad (28)$$

eq 27 leads to

$$\left[\frac{\partial^2 \left(\frac{\Delta G^E}{RT}\right)}{\partial x_1^2}\right]_T = \left(\frac{\partial \ln \gamma_1}{\partial x_1}\right)_T - \left(\frac{\partial \ln \gamma_2}{\partial x_1}\right)_T \quad (29)$$

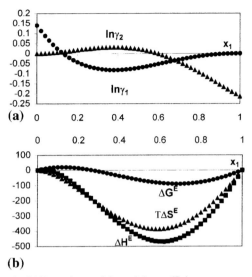

Figure 4. (a) Dependence of the activity coefficients on composition at 313.15 K for the $C_6F_6-C_6H_6$ system obtained from experimental vapor−liquid equilibrium data[5] correlated with the Van Ness−Abbott equation.[48] (b) The dependence of excess functions (J/mol) on composition at 313.15 K. ΔG^E was calculated from vapor−liquid equilibria data,[5] ΔH^E was taken from ref 53, and $T\Delta S^E$ was calculated as the difference between ΔG^E and ΔH^E.

From eqs 25 and 29 it is clear that

$$\left[\frac{\partial^2 \left(\frac{\Delta G^E}{RT}\right)}{\partial x_1^2}\right]_{T,x_1=x_1^*} = 0 \quad (30)$$

This means that for $\Delta n_{ij} = 0$, the activity coefficients reach extrema and the excess Gibbs energy exhibits an inflection point. To check this conclusion the logarithm of the activity coefficients and the excess Gibbs energy for the system HFB−B are plotted against concentration in parts a and b of Figure 4. Comparing Figure 3a and parts a and b of Figure 4, one can see that, in the points where $\Delta n_{ij} = 0$, the activity coefficients reach extrema and the excess Gibbs energy exhibits an inflection point.

The values of ΔG_{12} for the mixtures HFB−B and HFB−TL being close to zero, it is clear that in those cases the unlike interactions are comparable to the like ones. Some conclusions can be also reached from the analysis of Δn_{ij}. Indeed, one can see from parts a and b of Figure 3 that Δn_{11} and Δn_{22} are positive and small when $x_1 < x_1^*$ and they become negative and small when $x_1 > x_1^*$. The situation is inverted for Δn_{12} and Δn_{21}. Δn_{11}, Δn_{22}, Δn_{12}, and Δn_{21} have the common feature that their absolute values do not exceed 0.15−0.2 and that they change sign for $x_1 = x_1^*$. Some like (enriched in the same component as the central one) aggregates are formed when $x_1 \leq x_1^*$ and some unlike aggregates are formed when $x_1 \geq x_1^*$. This aggregation involves, however, not too strong interactions. If strong interactions would have been involved, then the Δn_{ij} should not have changed sign. It is of interest to compare HFB−B or HFB−TL with the system HFB−CH. In the latter case, Figures 2c and 3c show that like aggregations of HFB and CH, but not the unlike ones, occur over the entire composition range. The results for the system, B−CH (Figures 2e and 3e) are similar to those for HFB−CH with one exception, the like aggregation of HFB is much stronger than that of B. This means that the interactions between the HFB molecules are stronger than those between the B molecules. The mixtures

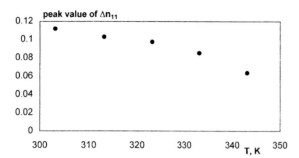

Figure 5. Temperature dependence of the peak value of Δn_{11} [mol/mol] for the system hexafluorobenzene−benzene.

containing CH behave so differently because CH is not aromatic and therefore $\pi-\pi$ interactions cannot occur with either HFB or B. It is clear that, in the latter two cases, the like interactions are stronger than the unlike one.

The correlation volume and the energy of interaction between two unlike molecules in the systems HFB−B were calculated as for the binary aqueous solutions of alcohols and hydrocarbons,[38] using eqs 19−20 and expressions for the local compositions and activity coefficients at infinite dilution. It should be however emphasized that the calculation procedure[38] is not very accurate when the activity coefficients at infinite dilution are close to unity. For the system HFB (1)−B (2) at 40 °C, $\gamma_1^\infty = 1.2$ and $\gamma_2^\infty = 0.8$; these values were calculated from VLE data[5] using the Van Ness−Abbott equation.[48] The calculations also showed that the energies of interaction between unlike molecules are almost the same for both concentration limits ($x_1 \rightarrow 0$ and $x_2 \rightarrow 0$). The correlation volume has the following values: $V_{cor}^1 \approx 14\,000$ [cm^3/mol] and $V_{cor}^2 \approx 3000$ [cm^3/mol]. Assuming a spherical shape, these values correspond to radii of 10−15 Å, which are in agreement with the neutron diffraction experiments,[27] which indicated a structuring over about 7−11 Å. The KBIs were also calculated for the system B−TL (Figure 2d). This figure shows that this system exhibits small deviations from ideality which are somewhat larger for benzene mole fractions greater than 0.6. The comparison of this system with the HFB−TL one shows that the replacement of benzene with hexafluorobenzene leads to a completely different composition dependence of the KBIs and Δn_{ij}.

The values of Δn_{ij} for the HFB−B system were calculated at 303.15, 313.15, 323.15, 333.15, and 343.15 K. The calculations indicated that all the Δn_{ij} decrease in absolute value with increasing temperature. Figure 5 presents the peak values of Δn_{11} against temperature and shows that they decrease moderately with increasing temperature, indicating again that the interactions involved are not too strong.

Conclusion

The Kirkwood−Buff theory of solutions was applied to the systems aromatic fluorocarbon/aromatic hydrocarbon. The Kirkwood−Buff integrals and the excess (or deficit) number of molecules around a central one were calculated for five systems (hexafluorobenzene−benzene, hexafluorobenzene−toluene, hexafluorobenzene−cyclohexane, benzene−toluene, benzene−cyclohexane).

The values of KBIs and the excess (or deficit) number of molecules Δn_{ij} near a central one with respect to a reference state showed that the local structure in solutions of hexafluorobenzene with benzene or toluene slightly differs from the bulk one and that this is caused, mainly, by moderate interactions.

It was found that a particular composition exists in the systems of hexafluorobenzene−benzene and hexafluorobenzene−toluene for which all Δn_{ij} are zero and that at this composition the activity coefficients exhibit extrema and the excess Gibbs energy an inflection point.

References and Notes

(1) Scott, R. L. *J. Phys. Chem.* **1958**, *62*, 136.
(2) Hildebrand, J. H.; Scott, R. L. *Regular Solution*; Prentice Hall: New York, 1962.
(3) Gaw, W. J.; Swinton, F. L. *Nature* **1966**, *212*, 283.
(4) Gaw, W. J.; Swinton, F. L. *Trans. Faraday Soc.* **1968**, *64*, 637.
(5) Gaw, W. J.; Swinton, F. L. *Trans. Faraday Soc.* **1968**, *64*, 2023.
(6) Duncan, W. A.; Swinton, F. L. *Trans. Faraday Soc.* **1966**, *62*, 1082.
(7) Duncan, W. A.; Sheridan, J. P.; Swinton, F. L. *Trans. Faraday Soc.* **1966**, *62*, 1090.
(8) Masood, A. K. M.; North, A. M.; Pethrick, R. A.; Towland, M.; Swinton, F. L. *J. Chem. Thermodyn.* **1977**, *9*, 133.
(9) Dymond, J. H.; Robertson, J. *Int. J. Thermophys.* **1985**, *6*, 21.
(10) Kogan, I. V.; Morachevsky, A. G. *Zh. Prikl. Khim.* **1972**, *45*, 1885.
(11) Swinton, F. L. In *Molecular Complexes*; Foster, R., Ed.; Paul Elek: London, 1974; Vol. 2.
(12) Swinton, F. L. In *Chemical Thermodynamics: A Special Periodic Report*; 1978; Vol. 2, Chapter 5.
(13) Gmehling, J.; Menke, J.; Krafczyk, J.; Fischer, K. *Azeotropic Data*; VCH: Verlagsgesellschaft, Weinheim, 1994.
(14) Patrick, C. R.; Prosser, G. S. *Nature* **1960**, *187*, 1021.
(15) Goates, J. R.; Ott, J. B.; Reeder, J. *J. Chem. Thermodyn.* **1973**, *5*, 135.
(16) Battaglia, M. R.; Buckingham, A. D.; Williams, J. H. *Chem. Phys. Lett.* **1983**, *78*, 421.
(17) Vrbancich, J.; Ritchie, G. L. D. *Chem. Phys. Lett.* **1983**, *94*, 63.
(18) Fenby, D. V.; Scott, R. L. J. *J. Phys. Chem.* **1966**, *70*, 602.
(19) Fenby, D. V.; Scott, R. L. J. *J. Phys. Chem.* **1967**, *71*, 4103.
(20) Bartsch, E.; Bertagnolli, H.; Chieux, P. *Ber. Bunsen-Ges. Phys. Chem.* **1986**, *90*, 34.
(21) Baur, M. E.; Horsma, D. A.; Knobler, C. M.; Perez, P. *J. Phys. Chem.* **1969**, *73*, 641.
(22) Bauer, D. R.; Brauman, J. I.; Pecora, R. *J. Chem. Phys.* **1975**, *63*, 53.
(23) Barrett, R. M.; Gill, E. B.; Steele, D. *J. Chem. Soc., Faraday Trans. 2* **1975**, *71*, 532.
(24) Tanabe, K.; Hiraishi, J. *J. Raman Spectrosc.* **1982**, *12*, 274.
(25) Suhm, M. A.; Mueller, K. J.; Weingaertner, H. *Z. Phys. Chem. Neue Folge* **1987**, *155*, 101.
(26) Wang, K. S.; Yuan, P.; Schwartz, M. *Spectrochim. Acta* **1993**, *49A*, 1035.
(27) Cabaco, M. I.; Danten, Y.; Besnard, M.; Guissani, Y.; Guillot, B. *J. Phys. Chem. B* **1998**, *102*, 10712.
(28) Watkins, C. L.; Breey, W. S. *J. Phys. Chem.* **1970**, *74*, 235.
(29) Kirkwood, J. G.; Buff, F. P. *J. Chem. Phys.* **1951**, *19*, 774.
(30) Ben-Naim, A. *J. Chem. Phys.* **1977**, *67*, 4884.
(31) Ben-Naim, A. *Pure Appl. Chem.* **1990**, *62*, 25.
(32) Ben-Naim, A. *Cell Biophys.* **1988**, *12*, 3694.
(33) Shulgin, I.; Ruckenstein, E. *J. Phys. Chem. B* **1999**, *103*, 872.
(34) Shulgin, I.; Ruckenstein, E. *J. Phys. Chem. B* **1999**, *103*, 2496.
(35) Matteoli, E.; Lepori, L. *J. Chem. Phys.* **1984**, *80*, 2856.
(36) Matteoli, E.; Lepori, L. *J. Chem. Soc., Faraday Trans.* **1995**, *91*, 431.
(37) Matteoli, E. *J. Phys. Chem. B* **1997**, *101*, 9800.
(38) Shulgin, I.; Ruckenstein, E. *J. Phys. Chem. B* **1999**, *103*, 4900.
(39) Lide, D. R., Ed. *Handbook of Chemistry and Physics*, 77th ed.; CRC Press: Boca Raton, 1996−1997.
(40) Schmidt, G. C. *Z. Phys. Chem. B* **1926**, *121*, 221.
(41) Benson, G. C. *Int. Data. Ser., Selec. Data Mixtures, Ser. A* **1982**, *2*, 126.
(42) Tasic, A.; Djordjevic, B.; Grozdanic, D. *Chem. Eng. Sci.* **1978**, *33*, 189.
(43) Tanaka, R. *Int. Data Ser., Selec. Data Mixtures, Ser. A* **1988**, *1*, 59.
(44) Redlich, O.; Kister, A. T. *Ind. Eng. Chem.* **1948**, *40*, 345.
(45) Zaitsev, A. L.; Petrenko, V. E.; Kessler, Y. M. *J. Solution Chem.* **1989**, *18*, 115.
(46) Gmehling, J. et al. *Vapor−Liquid Equilibrium Data Collection*; DECHEMA Chemistry Data Series, Vol. I, 19 Parts; DECHEMA: Frankfurt, 1977−1996.
(47) Barker, J. A. *Aust. J. Chem.* **1953**, *6*, 207.
(48) Abbott, M. M.; Van Ness, H. C. *AIChE J.* **1975**, *21*, 62.
(49) Dantzler, E. M.; Knobler, C. M. *J. Phys. Chem.* **1969**, *73*, 1602.
(50) Pasco, N. F.; Handa, Y. P.; Scott, R. L. *Int. Data Ser., Selec. Data Mixtures, Ser. A* **1986**, *2*, 135.
(51) Renon, H.; Prausnitz, J. M. *AIChE J.* **1968**, *14*, 135.
(52) Ruckenstein, E.; Shulgin, I. *Ind. Eng. Chem. Res.* **1999**, *38*, 4092.
(53) Andrews, A.; Morcom, K. W.; Duncan, W. A.; Swinton, F. L.; Pollock, J. M. *J. Chem. Thermodyn.* **1970**, *2*, 95.

Hydrophobic self-assembling in dilute aqueous solutions of alcohols and hydrocarbons

E. Ruckenstein*, I. Shulgin

Chemical Engineering Department, University of Buffalo, Buffalo, NY 14260-4200, USA

Abstract

Microheterogeneities, such as local compositions which differ from the average ones, are typical features of aqueous solutions of organic substances. A method for computing the sizes of microheterogeneities at infinite dilution (correlation volumes) is proposed, which is based on a combination between the Kirkwood–Buff theory of solution, an expression for the local concentration, ab initio quantum chemical calculations of the interaction energies and available thermodynamic data. The correlation volumes are calculated for aqueous solutions of methanol, ethanol, propanol and isopropanol. The results obtained for propanol and isopropanol could be compared with the sizes determined via small-angle X-ray scattering and good agreement between the calculated and experimental values found. In addition, the mutual affinity of solute molecules (their self-assembling), which is one of the important characteristics of hydrophobic interactions in aqueous solutions, was evaluated for a number of infinitely dilute aqueous solutions of alcohols and hydrocarbons. The calculations provided an additional argument that the hydrophobic self-assembling of hydrocarbons is mainly a result of the preference of the water molecules to interact among themselves via hydrogen bonding than a result of the interactions between hydrocarbon molecules. © 2001 Elsevier Science Ltd. All rights reserved.

Keywords: Hydrophobic interactions; Self-assembling; Correlation volume

1. Introduction

The aqueous solutions have received attention for more than a century because they are of industrial, environmental and scientific importance. One of the features of the aqueous solutions of organic substances is their microheterogeneity, reflected in the fact that the local concentration differs from the bulk concentration. The microheterogeneities in solutions can be characterized by the following nanometer-level parameters: (1) the correlation volume, i.e. the volume in which the concentration differs from the average concentration, (2) the excess (or deficit) number of molecules in the correlation volume compared to the number of molecules when they are randomly distributed, and (3) the inter molecular interactions between the molecules in the above volume.

The microheterogeneities in aqueous solutions have been investigated by various physical methods. For example, for the alcohol/water systems, microheterogeneities could be determined in solutions of the higher alcohols (PrOH, BuOH and higher) by small-angle X-ray scattering (SAXS) (Bale, Sherpler, & Sorgen, 1968; Koga, 1984; Nishikawa, Kodera, & Iijima, 1987; Nishikawa, Hayashi, & Iijima, 1989; Hayashi, Nishikawa, & Iijima, 1990; Shulgin, Serimaa, & Torkkeli, 1991; Hayashi & Udagawa, 1992), small-angle neutron scattering (SANS) (D'Arrigo & Teixeira, 1990), light scattering (LS) (Grossmann & Ebert, 1981; Euliss & Sorensen, 1984; Ito, Kato, & Fujiyma, 1981; Bender & Pekora, 1986). The microheterogeneities were, however, too small to be detected in the aqueous solutions of the lower alcohols, methanol and ethanol (Donkersloot, 1979; Nishikawa & Iijima, 1993). There is experimental and theoretical evidence that such microheterogeneities are also present in aqueous solutions of hydrocarbons and other organic substances (Tanford, 1980).

---

*Corresponding author. Tel.: +1-716-645-2911; fax: +1-716-645-3822.

E-mail addresses: feaeliru@acsu.buffalo.edu (E. Ruckenstein), ishulgin@eng.buffalo.edu (I. Shulgin).

The microheterogeneity was frequently used to explain the properties of aqueous solutions. Frank and Evans (1945) suggested that around a molecule of a nonpolar solute, there is a layer of more structured water molecules ("iceberg"). This concept was helpful in explaining, via the "melting" of icebergs, the unusually large specific heats of the solutions in water of simple hydrocarbons (Gill, Dec, Olofsson, & Wadsö, 1985; Kauzmann, 1987).

However, a theoretical method to determine the sizes of such microheterogeneities and their properties, such as the local composition, was proposed only recently (Shulgin & Ruckenstein, 1999a), by combining, for infinite dilution, the Kirkwood–Buff theory of solution with expressions for the activity coefficient and local composition. The expression for the activity coefficient provided the interaction energies between molecules. In this paper, a variant of that method is suggested, based on a synthesis between the Kirkwood–Buff theory of solution, an expression for the local composition, and ab initio quantum mechanical calculation of the interaction energies. The effect of various aliphatic alcohols on the structure of water at infinite dilution will be thus investigated. In addition, the previous approach (Shulgin & Ruckenstein, 1999a) will be used to analyze the self-assembling of solute molecules in dilute aqueous solutions. It will be shown that the results thus obtained provide information about hydrophobic bonding.

2. Theory and formulas

2.1. The correlation volume

We consider a molecule of species i and its surrounding correlation volume V_{cor}^i in which the composition and/or the structure differ from the bulk one. The total number of molecules i and j in this volume is given by the expressions

$$n_{ii} = \Delta n_{ii} + c_i V_{\text{cor}}^i \tag{1}$$

and

$$n_{ji} = \Delta n_{ji} + c_j V_{\text{cor}}^i, \tag{2}$$

where Δn_{ii} and Δn_{ji} are the excess (or deficit) number of molecules i and j around a central molecule i. Using the Kirkwood–Buff theory (KB) (Kirkwood & Buff, 1951) of solutions and a reference state for the excess quantities, Δn_{ii} and Δn_{ji} can be expressed as follows (Shulgin & Ruckenstein, 1999b):

$$\Delta n_{12} = -\frac{c_1 V_1 V_2}{V}\left(\frac{1-D}{D}\right), \tag{3}$$

$$\Delta n_{21} = -\frac{c_2 V_1 V_2}{V}\left(\frac{1-D}{D}\right) \tag{4}$$

and

$$\Delta n_{ii} = \frac{c_i x_j V_j^2}{x_i V}\left(\frac{1-D}{D}\right) \quad i \neq j, \tag{5}$$

where c_i is the average molar concentration of species i in the mixture, x_i is the average mole fraction of component i in solution, V is the molar volume of the solution, V_i and V_j are the partial molar volumes of components i and j, respectively, $D - 1 = x_i(\frac{\partial \ln \gamma_i}{\partial x_i})_{P,T}$, γ_i is the activity coefficient of component i, P is the pressure and T is the absolute temperature. Since the local mole fractions are given by

$$x_{ii} = \frac{n_{ii}}{n_{ii} + n_{ji}} = \frac{\Delta n_{ii} + c_i V_{\text{cor}}^i}{\Delta n_{ii} + \Delta n_{ji} + (c_i + c_j)V_{\text{cor}}^i} \tag{6}$$

and

$$x_{ji} = \frac{n_{ji}}{n_{ii} + n_{ji}} = \frac{\Delta n_{ji} + c_j V_{\text{cor}}^i}{\Delta n_{ii} + \Delta n_{ji} + (c_i + c_j)V_{\text{cor}}^i}, \tag{7}$$

one obtains

$$x_{ii} = \frac{x_j V_j^2(1-D) + x_i V_{\text{cor}}^i V D}{x_j V_j^2(1-D) - x_j V_i V_j(1-D) + V_{\text{cor}}^i V D} \tag{8}$$

and

$$x_{ji} = \frac{-x_j V_i V_j(1-D) + x_j V_{\text{cor}}^i V D}{x_j V_j^2(1-D) - x_j V_i V_j(1-D) + V_{\text{cor}}^i V D}, \tag{9}$$

where x_{ji} is the local mole fraction of component j in the vicinity of a central molecule i.

Due to the dilute region one can consider that (Debenedetti & Kumar, 1986) $(\partial \ln \gamma_i/\partial x_i)_{P,T}$ is independent of x_i; consequently

$$D = K_i(P,T)x_i + 1, \tag{10}$$

where

$$K_i(P,T) = \left(\frac{\partial \ln \gamma_i}{\partial x_i}\right)_{P,T,x_i \to 0}.$$

At infinite dilution, Eq. (8) leads to

$$\lim_{x_i \to 0} \frac{x_{ii}}{x_i} = 1 - \frac{V_j^0 K_i(P,T)}{V_{\text{cor}}^{i,0}} \tag{11}$$

where V_j^0 is the molar volume of pure component j and $V_{\text{cor}}^{i,0} = \lim_{x_i \to 0} V_{\text{cor}}^i$. One may note that $\lim_{x_i \to 0} x_{ii}/x_i > 1$ when $K_i(P,T) < 0$ (or $(\partial \ln \gamma_i/\partial x_i)_{P,T,x_i \to 0} < 0$). To calculate the correlation volume $V_{\text{cor}}^{i,0}$ (the size of the region affected by a solute molecule) using Eq. (11), an additional expression for $\lim_{x_i \to 0} x_{ii}/x_i$ is needed. Such an expression, based on the local composition concept will be provided in the next section.

2.2. The local composition concept

According to the local composition (LC) concept, the composition in the vicinity of any molecule differs from the overall composition. If a binary mixture is composed of components 1 and 2 with mole fractions x_1 and x_2, respectively, four LCs should be considered: the local mole fractions of components 1 and 2 near a central molecule 1 (x_{11} and x_{21}) and the local mole fractions of components 1 and 2 near a central molecule 2 (x_{12} and x_{22}). Many attempts have been made to express LC in terms of bulk compositions and intermolecular interaction parameters (Wilson, 1964; Renon & Prausnitz, 1968; Panayiotou & Vera, 1980, 1981; Lee, Sandler, & Patel, 1986; Aranovich & Donohue, 1996; Wu, Cui, & Donohue, 1998; Ruckenstein & Shulgin, 1999). In the calculations that follow, the Aranovich and Donohue (1996) expressions will be employed, because they have a theoretical basis. These expressions are

$$x_{ii} = \frac{x_i \exp(x_j \Delta/RT)}{x_j + x_i \exp(x_j \Delta/RT)} \quad (12)$$

and

$$x_{ji} = \frac{x_j}{x_j + x_i \exp(x_j \Delta/RT)}, \quad (13)$$

where $\Delta = 2e_{12} - e_{11} - e_{22}$ and e_{ij} is the interaction energy parameter between molecules i and j.

The following expression for $\lim_{x_i \to 0} x_{ii}/x_i$ is obtained from Eq. (12):

$$\lim_{x_i \to 0} \frac{x_{ii}}{x_i} = \exp\left[\frac{2e_{12} - e_{11} - e_{22}}{RT}\right]. \quad (14)$$

Let us analyze the expressions written for $\lim_{x_i \to 0} x_{ii}/x_i$ (Eqs. (11) and (14)). Eq. (11) contains the variables: $\lim_{x_i \to 0} x_{ii}/x_i$, V_j^0, K_i and $V_{\text{cor}}^{i,0}$. The quantities V_j^0 and K_i can be obtained from the density of the pure component j and phase equilibrium data, respectively (Debenedetti & Kumar, 1986; Liu & Ruckenstein, 1999; Ruckenstein & Shulgin, 2000); the correlation volume, which characterizes the solution nanostructure, is an unknown quantity, which can be calculated by combining Eq. (14) with (11) and providing values for the intermolecular interaction energies. These values can be obtained either indirectly, using an expression for the activity coefficient at infinite dilution as in our previous paper (Shulgin & Ruckenstein, 1999a), or directly, by calculating them via quantum mechanics.

3. Evaluations of the correlation volume in aqueous solutions of alcohols and hydrocarbons

By coupling the equations obtained in the previous section with expressions for the activity coefficient at infinite dilution (Shulgin & Ruckenstein, 1999a) one could calculate the correlation volumes of alcohol and hydrocarbon molecules in (infinitely) dilute aqueous solutions. The results obtained (Shulgin & Ruckenstein, 1999a) were in good agreement with experimental data.

A more direct method consists in combining Eq. (11) with (14), and with the intermolecular energies calculated via the quantum mechanical ab initio method (Levine, 1991; Szabo & Ostlund, 1996). The quantum mechanical method was recently employed to calculate the interaction energies between molecules for water/alcohol mixtures by Sum and Sandler (1999) and their results will be employed in what follows.

The combination of Eqs. (11) and (14) yields

$$\exp\left[\frac{2e_{12} - e_{11} - e_{22}}{RT}\right] = 1 - \frac{V_2^0 K_1(P,T)}{V_{\text{cor}}^{1,0}}. \quad (15)$$

Eq. (15) will be used to calculate the correlation volume. The values of V_2^0 and K_1 were taken from our previous paper (Shulgin and Ruckenstein, 1999a), and the values of the intermolecular interaction energies from the paper of Sum and Sandler (1999). The results are listed in Table 1. Calculations were carried out only for

Table 1
Size of microheterogeneities in dilute aqueous solutions of alcohols

System	Calculated[a] correlation volume		Experimental values of size of microheterogeneities		
	r(Å) present paper	r(Å) Shulgin and Ruckenstein (1999a)	Method	R(Å)	References
MeOH(1)/H$_2$O(2)	−4.4	4.7	SAXS	No microheterogeneity identified	Donkersloot (1979)
EtOH(1)/H$_2$O(2)	3.9	5.8	SAXS	No microheterogeneity identified	Nishikawa and Iijima (1993)
1-PrOH(1)/H$_2$O(2)	4.2	7.0	SAXS	4 at $x_1 = 0.05$	Hayashi and Udagawa (1992)
2-PrOH(1)/H$_2$O(2)	3.7	6.1	SAXS	2.6 at $x_1 = 0.1$	Hayashi and Udagawa (1992)

[a]The correlation volume is approximated by a sphere.

Table 2
The values of $\lim_{x_1 \to 0} x_{11}/x_1$ in infinitely dilute aqueous solutions of alcohols and hydrocarbons

System	Temperature (K)	$\lim_{x_1 \to 0} \frac{x_{11}}{x_1}$
Methanol/water	298.15	1.10
Ethanol/water	298.15	1.17
1-propanol/water	303.15	1.24
2-propanol/water	298.15	1.21
1-butanol/water	323.15	1.29
2-butanol/water	323.15	1.26
i-butanol/water	323.15	1.28
t-butanol/water	323.15	1.22
t-pentanol/water	328.15	1.29
Propane/water	298.15	1.50
n-butane/water	298.15	1.54
Isobutane/water	298.15	1.52
n-pentane/water	298.15	1.54
n-hexane/water	298.15	1.54
n-heptane/water	298.15	1.52
n-octane/water	298.15	1.52
n-decane/water	298.15	1.52
n-dodecane/water	298.15	1.52
Cyclopentane/water	298.15	1.63
Cyclohexane/water	298.15	1.62
Cycloheptane/water	298.15	1.64
Benzene/water	298.15	1.53
Toluene/water	298.15	1.39

the systems MeOH/H$_2$O, EtOH/H$_2$O, 1-PrOH/H$_2$O and 2-PrOH/H$_2$O, for which the intermolecular interaction energies were available (Sum & Sandler, 1999). It is difficult to explain the negative value obtained for the correlation volume around methanol at infinite dilution. One can, however, show that the values of the interaction energies used to calculate that correlation volume also lead to an unreasonable value for $\lim_{x_i \to 0} x_{ii}/x_i$, which is less than unity. This means that the local composition of molecules i around a central molecule i is less than the bulk composition. However, it is well known (Kozak, Knight, & Kauzmann, 1968) that self-aggregation of water and alcohol molecules are typical for the aqueous solutions of alcohols. It is worth mentioning that for the other alcohol/water systems considered in this paper the above limit is larger than unity.

4. Hydrophobic interactions and self-assembling at infinite dilution

The method suggested in the present paper provides the correlation volume. The thickness of the layer of water thus calculated, which is affected by a solute molecule, indicates how deeply a single solute molecule has perturbed the structure of the vicinal water molecules.

Another feature of the hydrophobic interactions is the self-assembling of hydrophobic molecules in water. At infinite dilution, the value of $\lim_{x_i \to 0} x_{ii}/x_i$ can serve as a measure of self-aggregation of molecule i at infinite dilution. $\lim_{x_i \to 0} x_{ii}/x_i$ can be computed using Eq. (11) or (14). Consequently, the method suggested in the current paper allows one to estimate not only the thickness of the layer of water that is affected by a solute molecule, but also the mutual affinity of solute molecules at infinite dilution. The compounds selected, hydrocarbons and alcohols, allowed one to investigate the hydrophobic effect for pure hydrophobic molecules (hydrocarbons) and for less hydrophobic substances such as the alcohols.

To evaluate $\lim_{x_i \to 0} x_{ii}/x_i$, either the interaction energies between molecules (Eq. (14)) or the values of V_j^0, K_i and $V_{cor}^{i,0}$ (Eq. (11)) are required. Because, the interaction energies were calculated by quantum mechanics only for a few systems, Eq. (11) will be used to calculate $\lim_{x_i \to 0} x_{ii}/x_i$. All the required data (V_j^0, K_i and $V_{cor}^{i,0}$) were taken from our previous paper (Shulgin & Ruckenstein, 1999a) and the results are listed in Table 2.

5. Discussion

The present paper is focused on the nanostructure of dilute binary aqueous solutions of organic substances, or, in other words, on how the addition of a small amount of an organic compound such as an alcohol or a hydrocarbon affects the nanostructure of liquid water. The results obtained regarding the correlation volumes around 1-PrOH and 2-PrOH molecules in dilute aqueous solutions are in good agreement with the experimental data obtained by small-angle X-ray diffraction. The correlation volumes around 1-PrOH and 2-PrOH allowed one to evaluate the number of ambient water layers affected by a single molecule of 1-PrOH or 2-PrOH. Using X-ray scattering data for cold water (Narten & Levy, 1972), one can find that 2–3 vicinal layers of water are affected by a single 1-PrOH or 2-PrOH molecule at infinite dilution. The thickness of the layer is even larger for infinitely dilute solution of hydrocarbons (Shulgin & Ruckenstein, 1999a). In his classical scheme, Butler (1937) suggested that "the process of bringing a solute molecule into a solvent may be supposed to consist of two steps: (1) making a cavity in the solvent large enough to hold the solute molecule; (2) introducing the solute molecule into the cavity". In the calculation of the contribution of the second step to the enthalpy of hydration, Butler considered only the interactions between the solute molecule and the nearest neighboring water molecules. He also assumed that the water molecules are distributed around a solute molecule as randomly as in its absence. It should be noted that Nemethy and Scheraga (1962) also considered that only one layer of water is affected by the solute molecule. Our results revealed, however, that, as suggested by Frank and Evans (1945), several layers of water molecules are reorganized around a solute molecule as "icebergs". In contrast to Frank and Evans, the present approach allows one to evaluate the number of layers of water which are affected by the presence of the solute.

One of the most important features of hydrophobic interactions is the self-assembling of hydrophobic molecules in aqueous solutions. Kauzmann was the first to focus the attention on self-assembling in his paper about the folding of protein molecules (Kauzmann, 1959).

We propose here to consider the quantity $\lim_{x_i \to 0} x_{ii}/x_i$ as a measure of self-assembling of solute (alcohol or hydrocarbon) molecules in infinitely dilute aqueous solutions. The results of our calculation of $\lim_{x_i \to 0} x_{ii}/x_i$ (Table 2) showed that for the alcohol molecules the above ratio increased somewhat with increasing size of the hydrophobic radical. These results are in agreement with literature observations (Kozak et al., 1968). However, the values of $\lim_{x_i \to 0} x_{ii}/x_i$ for aliphatic hydrocarbons are almost independent of the length of the molecule. For instance, the value of the limit is 1.50 for propane and 1.52 for dodecane. This provides an additional argument that the process of self-assembling of hydrocarbon molecules in infinitely dilute aqueous solutions depends mainly on the water and much less on the interactions between the solute molecules. This observation is in agreement with the idea that the hydrophobic interaction is a result of the preference of the water molecules for their hydrogen bonding than the preferences of the hydrocarbon molecules to interact among themselves (Tanford, 1980; Ruckenstein, 1998 and references therein).

References

Aranovich, G. L., & Donohue, M. D. (1996). A new model for lattice systems. *Journal of Chemical Physics, 105*, 7059–7063.

Bale, H. D., Sherpler, R. E., & Sorgen, D. K. (1968). Small-angle X-ray scattering from tert-butyl alcohol-water mixtures. *Physics and Chemistry of Liquids, 1*, 181–190.

Bender, T. M., & Pekora, R. (1986). A dynamic light scattering study of the tert-butyl alcohol-water system. *Journal of Physical Chemistry, 90*, 1700–1706.

Butler, J. A. V. (1937). The energy and entropy of hydration of organic compounds. *Transactions of the Faraday Society, 33*, 229–236.

D'Arrigo, G., & Teixeira, J. (1990). Small-angle neutron-scattering study of D2O-alcohol solutions. *Journal of the Chemical Society Faraday Transactions, 86*, 1503–1509.

Debenedetti, P. G., & Kumar, S. K. (1986). Infinite dilution fugacity coefficients and the general behavior of dilute binary systems. *A.I.Ch.E. Journal, 32*, 1253–1262.

Donkersloot, M. C. A. (1979). Concentration dependence of the zero-angle X-ray scattering from liquid mixtures of water and methanol. *Chemical Physics Letters, 60*, 435–438.

Euliss, G. W., & Sorensen, C. M. (1984). Dynamic light scattering studies of concentration fluctuations in aqueous tert-butyl alcohol solutions. *Journal of Chemical Physics, 80*, 4767–4773.

Frank, H. S., & Evans, M. W. (1945). Free volume and entropy in condensed systems. *Journal of Chemical Physics, 13*, 507–532.

Gill, S. J., Dec, S. F., Olofsson, G., & Wadsö, I. (1985). Anomalous heat capacity of hydrophobic solvation. *Journal of Physical Chemistry, 89*, 3758–3761.

Grossmann, G. H., & Ebert, K. H. (1981). Formation of clusters in 1-propanol/water-mixtures. *Ber. Bunsenges. Physical Chemistry, 85*, 1026–1029.

Hayashi, H., Nishikawa, K., & Iijima, T. (1990). Small-angle X-ray scattering study of fluctuations in 1-propanol-water and 2-propanol-water systems. *Journal of Physical Chemistry, 94*, 8334–8338.

Hayashi, H., & Udagawa, Y. (1992). Mixing state of 1-propanol aqueous solutions studied by small-angle X-ray scattering—a new parameter reflecting the shape of SAXS curve. *Bulletin of the Chemical Society of Japan, 65*, 155–159.

Ito, N., Kato, T., & Fujiyma, T. (1981). Determination of local structure and moving unit formed in binary solution of t-butyl alcohol and water. *Bulletin of the Chemical Society of Japan, 54*, 2573–2578.

Kirkwood, J. G., & Buff, F. P. (1951). The statistical mechanical theory of solution. I. *Journal of Chemical Physics, 19*, 774–777.

Kauzmann, W. (1959). Some factors in the interpretation of protein denaturation. *Advances in Protein Chemistry, 14*, 1–63.

Kauzmann, W. (1987). Thermodynamics of unfolding. *Nature, 325*, 763–764.

Koga, Y. (1984). A SAXS study of concentration fluctuations in tert-butanol–water system. *Chemical Physics Letters, 111*, 176–180.

Kozak, J. J., Knight, W. S., & Kauzmann, W. (1968). Solute–solute interactions in aqueous solutions. *Journal of Chemical Physics, 48*, 675–690.

Lee, K.-H., Sandler, S. I., & Patel, N. C. (1986). The generalized Van-der-Waals partition function .3. Local composition models for a mixture of equal size square-well molecules. *Fluid Phase Equilibria, 25*, 31–49.

Levine, I. N. (1991). *Quantum Chemistry*, 4th ed. NJ: Prentice-Hall: Englewood Cliffs.

Liu, H. Q., & Ruckenstein, E. (1999). Aggregation of hydrocarbons in dilute aqueous solutions. *Journal of Physical Chemistry, B, 102*, 1005–1012.

Narten, A. H., & Levy, H. A. (1972). Liquid water: Scattering of X-rays. In F. Franks (Ed.), *Water-A comprehensive treatise* (Vol.1) (pp. 311–332). New York: Plenum Press.

Nemethy, G., & Scheraga, H. A. (1962). Structure of water and hydrophobic bonding in proteins. II. Model for the thermodynamic properties of aqueous solutions of hydrocarbons. *Journal of Chemical Physics, 36*, 3401–3417.

Nishikawa, K., Hayashi, H., & Iijima, T. (1989). Temperature dependence of the concentration fluctuation, the Kirkwood–Buff parameters, and the correlation length of tert-butyl alcohol and water mixtures studied by small-angle X-ray scattering. *Journal of Physical Chemistry, 93*, 6559–6565.

Nishikawa, K., & Iijima, T. (1993). Small-angle X-ray scattering study of fluctuations in ethanol and water mixtures. *Journal of Physical Chemistry, 97*, 10,824–10,828.

Nishikawa, K., Kodera, Y., & Iijima, T. (1987). Fluctuations in the particle number and concentration and the Kirkwood–Buff parameters of tert-butyl alcohol and water mixtures studied by small-angle X-ray scattering. *Journal of Physical Chemistry, 91*, 3694–3699.

Panayiotou, C., & Vera, J. H. (1980). The quasi-chemical approach for non-randomness in liquid mixtures. Expressions for local composition with an application to polymer solution. *Fluid Phase Equilibria, 5*, 55–80.

Panayiotou, C., & Vera, J. H. (1981). Local composition and local surface area fractions: a theoretical discussion. *Canadian Journal of Chemical Engineering, 59*, 501–505.

Renon, H., & Prausnitz, J. M. (1968). Local composition in thermodynamic excess functions for liquid mixtures. *A.I.Ch.E. Journal, 14*, 135–144.

Ruckenstein, E. (1998). On the phenomenological thermodynamics of hydrophobic bonding. *Journal of Dispersion Science Technolology, 19*, 329–338.

Ruckenstein, E., & Shulgin, I. (1999). Modified local composition and Flory-Huggins equations for nonelectrolyte solutions. *Industrial Engineering and Chemical Research, 38*, 4092–4099.

Ruckenstein, E., & Shulgin, I. (2000). On density microheterogeneities in dilute supercritical solutions. *Journal of Physical Chemistry, B, 104*, 2540–2545.

Shulgin, I., Serimaa, R., & Torkkeli, M. (1991). Small-angle X-ray scattering from aqueous 1-propanol solutions. *Report Series in Physics*, HU-P-256: Helsinki.

Shulgin, I., & Ruckenstein, E. (1999a). Range and energy of interaction at infinite dilution in aqueous solutions of alcohols and hydrocarbons. *Journal of Physical Chemistry B, 103*, 4900–4905.

Shulgin, I., & Ruckenstein, E. (1999b). Kirkwood–Buff integrals in aqueous alcohol systems: Aggregation, correlation volume, and local composition. *Journal of Physical Chemistry B, 103*, 872–877.

Sum, A. K., & Sandler, S. I. (1999). A novel approach to phase equilibria predictions using *ab initio* methods. *Industrial Engineering and Chemical Research, 38*, 2849–2855.

Szabo, A., & Ostlund, N. S. (1996). *Modern quantum chemistry. Introduction to advanced electronic structure theory*. New York: Dover Publication.

Tanford, C. (1980). *The hydrophobic effect: Formation of micelles and biological membranes*. New York: Wiley.

Wilson, G. M. (1964). Vapor–liquid equilibrium. XI: A new expression for the excess free energy of mixing. *Journal of the American Chemical Society, 86*, 127–130.

Wu, D. W., Cui, Y., & Donohue, M. D. (1998). Local composition models for lattice mixtures. *Industrial Engineering and Chemical Research, 37*, 2936–2946.

Fluid Phase Equilibria 180 (2001) 281–297

www.elsevier.nl/locate/fluid

Effect of a third component on the interactions in a binary mixture determined from the fluctuation theory of solutions

E. Ruckenstein*, I. Shulgin

Department of Chemical Engineering, State University of New York at Buffalo, Amherst, NY 14260, USA

Received 5 June 2000; accepted 17 January 2001

Abstract

The Kirkwood–Buff (KB) theory of solution is applied to a ternary mixture by deriving explicit expressions for the various Kirkwood–Buff integrals (KBIs) and the corresponding excesses of the number of molecules around central ones. However, the ideal solution should be considered non-aggregated, and the above expressions for the excesses provide non-zero values for such a case. For this reason, in order to obtain information about clustering one must subtract from the traditional excesses those which correspond to a reference state, thus ensuring that for an ideal mixture the excesses are zero. The expressions derived for the latter excesses have been applied to the investigation of the N,N-dimethylformamide–methanol–water mixture, to conclude that: (i) in the vicinity of the water molecules there are excesses of water and N,N-dimethylformamide molecules and a deficit of methanol molecules; (ii) in the vicinity of the methanol molecules there are excesses of methanol and N,N-dimethylformamide molecules and a deficit of water molecules; (iii) in the vicinity of the N,N-dimethylformamide molecules there are excesses of methanol and water molecules and a deficit of N,N-dimethylformamide molecules; (iiii) the excesses of N,N-dimethylformamide around water and methanol molecules and those around N,N-dimethylformamide are weakly dependent on the concentration of the third component in a large range of concentrations of the latter, and these results are compatible with the existence of N,N-dimethylformamide–water and N,N-dimethylformamide–alcohol complexes. © 2001 Elsevier Science B.V. All rights reserved.

Keywords: Fluctuation theory; Kirkwood–Buff integrals; N,N-dimethylformamide; Methanol; Water; Binary mixtures; Ternary mixtures

1. Introduction

The Kirkwood–Buff (KB) theory of solution [1] (often called fluctuation theory) was originally developed in 1951. This theory connects the macroscopic properties of solutions, such as the isothermal compressibility, the concentration derivatives of the chemical potentials and the partial molar volumes

---

* Corresponding author. Tel.: +1-716-645-2911/ext. 2214; fax: +1-716-645-3822.
E-mail addresses: feaeliru@acsu.buffalo.edu (E. Ruckenstein), ishulgin@eng.buffalo.edu (I. Shulgin).

0378-3812/01/$20.00 © 2001 Elsevier Science B.V. All rights reserved.
PII: S0378-3812(01)00365-X

to their microscopic characteristics in the form of spatial integrals involving the radial distribution functions. This theory provides a unique opportunity to extract some microscopic characteristics of mixtures from measurable thermodynamic quantities. However, in spite of its attractiveness, the KB theory was not often used [2–4] in the first three decades after its appearance, for two main reasons: (1) the lack of precise data (in particular regarding the composition dependence of the chemical potentials); and (2) the difficulty to interpret the results obtained. Only after Ben–Naim indicated how to calculate numerically the Kirkwood–Buff integrals (KBIs) for binary systems [5], this theory was used more frequently. Since then, the KBIs have been calculated for numerous binary systems [6–12] and the results were used to examine the solution behavior with regard to: (1) local composition; (2) various models for phase equilibrium; (3) preferential solvation and others [13–15]. Because the KB theory of solution is an exact statistical mechanical formalism, it can also be used for near critical and supercritical binary mixtures [14,16–19], whereas the conventional theories fail under these conditions [20]. As noted by Matteoli and Lepori [10,11] and later by the authors [12], information about aggregation in such mixtures should be obtained by subtracting from the KBIs the KBIs of a reference state. Matteoli and Lepori selected the ideal solution as the reference state, because an ideal solution should be considered non-aggregated (hence its excesses should be zero) and its KBIs differ from zero. The authors have modified this reference state [12] to take into account the volume conservation condition (this issue is discussed later in the paper).

The KB theory of solution was also used to obtain information about the intermolecular interactions in ternary mixtures [10,21–26]. However, only Matteoli and Lepori [10,11,26] used a reference state in their calculations, while Zielkiewiecz [24,25] did not. The KB theory was also applied to a number of ternary mixtures regarding alloys [27], electrolyte solutions [28] and supercritical mixtures [29,30].

Equations for the KBIs in ternary mixtures are available in matrix form [2]. Explicit equations are obtained here which will allow us to analyze interesting features of ternary mixtures, such as the effect of a third component on the phase behavior of a binary mixture and the effect of a cosolvent (entrainer) on supercritical binary mixtures. Only the former problem is examined in the present paper. The calculations will be carried out for an interesting ternary mixture, namely N,N-dimethylformamide–methanol–water, in order to extract information about the intermolecular interactions. In the next section explicit equations for the KB integrals will be derived and applied to the above ternary mixture. Finally, the results obtained will be used to shed some light on the local structure and the intermolecular interactions in the above mixture.

2. Theory and formulas

2.1. The Kirkwood–Buff integrals (KBI)

The KBIs are given by the expressions [1]

$$G_{\alpha\beta} = \int_0^\infty (g_{\alpha\beta} - 1) 4\pi r^2 \, dr \tag{1}$$

where $g_{\alpha\beta}$ is the radial distribution function between species α and β, and r the distance between the centers of molecules α and β. Equations for KBIs of binary mixtures are available in literature, including those for an ideal mixture (see Appendix A).

The expressions for the KBIs of ternary systems derived from the general KB equations (in volume per mole) have the following form:

$$G_{ii} = RTk_T - \frac{V}{x_i} + \frac{RT(x_1^2 \bar{\mu}_{11} l_{ii} + x_2^2 \bar{\mu}_{22} m_{ii} + x_3^2 \bar{\mu}_{33} n_{ii})}{VQ} \qquad (2)$$

and

$$G_{ij} (i \neq j) = RTk_T + \frac{RT(x_1^2 \bar{\mu}_{11} l_{ij} + x_2^2 \bar{\mu}_{22} m_{ij} + x_3^2 \bar{\mu}_{33} n_{ij})}{VQ} \qquad (3)$$

where R denotes the universal gas constant, T represents the absolute temperature, k_T denotes the isothermal compressibility, x_i is the mole fraction of component i and V is the molar volume. The expressions of $\bar{\mu}_{ii}$, l_{ij}, m_{ij}, n_{ij} and Q are given in Appendix B. Eqs. (2) and (3) will be used to calculate the KBIs of ternary mixtures.

3. Excess number of molecules near a central one

3.1. Binary mixture

The conventional excess number of j molecules around a central molecule i is given by [5]

$$\Delta n'_{ji} = c_j G_{ji} \qquad (4)$$

where c_j is the molar concentration of component j.

Matteoli and Lepori [10,11] noted that the $c_j G_{ji}$ are non-zero for ideal mixtures, which should be considered non-aggregated, and suggested that the aggregation is better reflected in the excesses (the ML excesses) defined as

$$\Delta n''_{ji} = c_j (G_{ji} - G_{ji}^{id}) \qquad (5)$$

which provide zero values for the excesses in ideal mixtures. In Eq. (5), G_{ji}^{id} are the KBs integrals of an ideal binary system. In other words, the ideal binary mixture was considered as a reference state. However, $\Delta n''_{ii}$ and $\Delta n''_{ji}$ are not independent quantities. The relation between the two can be obtained by first observing that the thermodynamic definition of partial molar volumes implies a uniform distribution of the molecules. Indeed, $V = \sum V_i N_i$, where V_i and N_i are the partial molar volume and the number of molecules of species i, respectively. In reality, around a molecule of a given species there is an excess of one species and a deficit of the other species. To maintain the same volume V for the real system it is, therefore, necessary for the volume occupied by excess i molecules (brought from the bulk) around an i molecule to be equal to the volume left free by j molecules (displaced to the bulk) around the same i molecule. This leads to the following relation [10]

$$V_j \Delta n''_{ji} = -V_i \Delta n''_{ii} \qquad (6)$$

A derivation of Eq. (6) involving the KB theory is presented in Appendix C. It should be, however, noted that if the number of molecules clustered around a central one is calculated as $c_j(G_{ji} - G_{ji}^{id})$, Eq. (6)

cannot be satisfied identically [12,31]. Indeed, the insertion of expressions 5, with G_{ij} and G_{ij}^{id} listed in Appendix A, in Eq. (6) leads to

$$RTV(k_T - k_T^{id}) = V_i V + V_i(V_j^0 - V^{id}) - \frac{V_i^0 V_j^0 V}{V^{id}} \tag{7}$$

where the superscript 'id' indicates an ideal mixture and the superscript '0' refers to the pure components. It is clear that Eq. (5) do not satisfy Eq. (6). However, Eq. (7) indicates that the volume conservation condition can be satisfied if G_{ij}^{id} is replaced by G_{ij}^V, which is obtained from the former by replacing k_T^{id}, V_i^0 and V^{id} by k_T, V_i and V, respectively., Hence, the excesses (the SR excesses) compatible with Eq. (6) are given by

$$\Delta n_{ji} = c_j(G_{ji} - G_{ji}^V) \tag{8}$$

The new reference state reduces to G_{ij}^{id} in the limiting case of an ideal mixture, but also satisfies the volume conservation condition. The following differences exist between the ML and SR excesses: the ML excesses have non-zero values if either the partial molar volumes differ from the ideal ones or $D = 1 + x_i(\partial \ln \gamma_i/\partial x_i)_{P,T} \neq 1$, where P represents the pressure and γ_i is the activity coefficient of component i; the SR excesses have non-zero values only if $D \neq 1$. The present reference state is a hypothetical one similar to the ideal state, in which the molar volume, the partial molar volumes and the isothermal compressibility are the real ones.

3.2. Ternary mixtures

The KBIs for an ideal ternary system can be obtained from Eqs. (2) and (3), by taking into account that in this case: (1) $\gamma_i = 1$ ($i = 1, 2, 3$); (2) $V = V^{id} = \sum_{j=1}^{3} x_j V_j^0$; (3) $k_T = k_T^{id} = \sum_{j=1}^{3} \varphi_j k_{T,j}^0$; (4) $V_j = V_j^0$ ($j = 1, 2, 3$); (5) $k_{T,j} = k_{T,j}^0$ ($j = 1, 2, 3$), where V_j^0 is the molar volume of the pure component j and φ_j is the volume fraction of component j. The following simple expressions are thus obtained (in volume per mole):

$$G_{11}^{id} = RTk_T^{id} - \frac{V^{id}}{x_1} + \frac{x_2(V_2^0)^2 + x_3(V_3^0)^2 - x_2 x_3(V_2^0 - V_3^0)^2}{x_1 V^{id}} \tag{9}$$

$$G_{22}^{id} = RTk_T^{id} - \frac{V^{id}}{x_2} + \frac{x_1(V_1^0)^2 + x_3(V_3^0)^2 - x_1 x_3(V_1^0 - V_3^0)^2}{x_2 V^{id}} \tag{10}$$

$$G_{33}^{id} = RTk_T^{id} - \frac{V^{id}}{x_3} + \frac{x_1(V_1^0)^2 + x_2(V_2^0)^2 - x_1 x_2(V_1^0 - V_2^0)^2}{x_3 V^{id}} \tag{11}$$

$$G_{12}^{id} = RTk_T^{id} + \frac{x_3 V_3^0(V_3^0 - V_1^0 - V_2^0) - V_1^0 V_2^0(1 - x_3)}{V^{id}} \tag{12}$$

$$G_{13}^{id} = RTk_T^{id} + \frac{x_2 V_2^0(V_2^0 - V_1^0 - V_3^0) - V_1^0 V_3^0(1 - x_2)}{V^{id}} \tag{13}$$

and

$$G_{23}^{id} = RTk_T^{id} + \frac{x_1 V_1^0(V_1^0 - V_2^0 - V_3^0) - V_2^0 V_3^0(1 - x_1)}{V^{id}} \tag{14}$$

As for binary mixtures, $G_{ji} - G_{ji}^{id}$ does not satisfy the volume conservation condition. However, if G_{ji}^{id} is replaced by G_{ji}^{V} the conservation condition becomes satisfied.

For a ternary mixtures the volume conservation condition (6) should be recast as follows (for any central molecule $i = 1, 2, 3$)

$$\sum_{j=1}^{3} V_j \Delta n_{ji} = 0 \tag{15}$$

where $\Delta n_{ji} = c_j (G_{ji} - G_{ji}^{V})$ and the expressions for G_{ji}^{V} can be obtained from Eqs. (8)–(13) by substituting k_T^{id}, V^{id} and V_i^0 by k_T, V and V_i, respectively.

To illustrate the above method for calculating the excess number of molecules near a central one, the ternary system N,N-dimethylformamide–methanol–water and the corresponding binary mixtures will be considered. This mixture was previously examined in the framework of the KB theory of solutions [24,25]. However, in [24,25] the calculations have been carried out without an appropriate reference state. The use of a reference state is important for this particular mixture, because it deviates slightly from ideality [10] and, consequently, $|G_{ji}^{id}|$ and $|G_{ji}^{V}|$ are not negligible compared to $|G_{ji}|$.

4. Source of data and calculation procedure

All calculations were carried out at $T = 313.15$ K. The vapor–liquid equilibrium (VLE) data for the ternary mixture and the corresponding binaries were taken from [32]. The excess volume data for the ternary mixture N,N-dimethylformamide–methanol–water and binary mixtures N,N-dimethylformamide–methanol and methanol–water were taken from [33], and the excess volume data for the binary mixture N,N-dimethylformamide–water from [34]. There are no isothermal compressibility data for the ternary mixture, but the contribution of compressibility to the binary KBIs is almost negligible far from the critical point [6]. For this reason, the compressibilities in binary and ternary mixtures were taken to be equal to the ideal compressibilities, and were calculated from the isothermal compressibilities of the pure components as follows:

$$k_T = k_T^{id} = \sum_j \varphi_j k_{T,j}^0 \tag{16}$$

where $k_{T,j}^0$ is the isothermal compressibility of the pure component j. The isothermal compressibilities of pure water and methanol were taken from [35], and the isothermal compressibility of N,N-dimethylformamide from [36]. The VLE in binary mixtures was treated using the Barker method [37]. The vapor phase non-ideality was taken into account and the total pressure was calculated using the equation

$$P = \frac{x_1 \gamma_1 P_1^0}{F_1} + \frac{x_2 \gamma_2 P_2^0}{F_2} \tag{17}$$

where

$$F_1 = \exp\left(\frac{(B_{11} - V_1^0)(P - P_1^0) + P y_2^2 d_{12}}{RT}\right) \tag{18}$$

$$F_2 = \exp\left(\frac{(B_{22} - V_2^0)(P - P_2^0) + Py_1^2 d_{12}}{RT}\right) \tag{19}$$

In the above equations y_i denotes the vapor mole fraction of component i, P_i^0 is the vapor pressure of the pure component i, B_{ii} is the second virial coefficient of component i, $d_{12} = 2B_{12} - B_{11} - B_{22}$ and B_{12} is the crossed second virial coefficient of the binary mixture. The vapor pressures, the virial coefficients of the pure components and the crossed second virial coefficients of the binary mixtures were taken from [32]. The Wilson [38], NRTL [39] and the Van Ness–Abbott [40] equations were used for the activity coefficients in Eq. (17). The expressions for the activity coefficients provided by these three methods were differentiated analytically and the obtained derivatives were used to calculate $D = 1 + x_i(\partial \ln \gamma_i/\partial x_i)_{P,T}$. There is good agreement between the values of D obtained with the three expressions for the systems N,N-dimethylformamide–methanol and methanol–water. For the system N,N-dimethylformamide–water, the D values calculated with the Van Ness–Abbott equation [40] were found in good agreement with those obtained with the NRTL equation, but the agreement with the Wilson expression was less satisfactory.

The derivatives $(\partial \mu_1/\partial x_1)_{x_2}$, $(\partial \mu_2/\partial x_2)_{x_1}$ and $(\partial \mu_2/\partial x_1)_{x_2}$ for the ternary mixture were obtained from the VLE data [32]. The ternary VLE data were again treated by the Barker method [37]. The total pressure was calculated using the equation

$$P = \frac{x_1 \gamma_1 P_1^0}{F_1} + \frac{x_2 \gamma_2 P_2^0}{F_2} + \frac{x_3 \gamma_3 P_3^0}{F_3} \tag{20}$$

and the vapor phase non-ideality factors F_i were obtained from the expressions

$$F_1 = \exp\left(\frac{(B_{11} - V_1^0)(P - P_1^0) + P(y_2^2 d_{12} + y_2 y_3(d_{12} + d_{13} - d_{23}) + y_3^2 d_{13})}{RT}\right) \tag{21}$$

$$F_2 = \exp\left(\frac{(B_{22} - V_2^0)(P - P_2^0) + P(y_1^2 d_{12} + y_1 y_3(d_{12} + d_{23} - d_{13}) + y_3^2 d_{23})}{RT}\right) \tag{22}$$

and

$$F_3 = \exp\left(\frac{(B_{33} - V_3^0)(P - P_3^0) + P(y_1^2 d_{13} + y_1 y_2(d_{13} + d_{23} - d_{12}) + y_2^2 d_{23})}{RT}\right) \tag{23}$$

where $d_{ij} = d_{ji} = 2B_{ij} - B_{ii} - B_{jj}$.

The excess Gibbs energy of the ternary mixture was expressed through the Wilson [38], NRTL [39] and Zielkiewicz [32] expressions. Because of the agreement between the latter two expressions, detailed results are presented only for the more simple NRTL expression. The parameters in the NRTL equation were found by fitting x–P (the composition of liquid phase–pressure) experimental data [32]. The derivatives $(\partial \mu_1/\partial x_1)_{x_2}$, $(\partial \mu_2/\partial x_2)_{x_1}$ and $(\partial \mu_2/\partial x_1)_{x_2}$ in the ternary mixture were found by the analytical differentiation of the NRTL equation. The excess molar volume (V^E) in the binary mixtures (i–j) was expressed via the Redlich–Kister equation

$$V^E = x_i x_j \sum_{m=1} K_m^{ij}(x_i - x_j)^{m-1} \tag{24}$$

where the adjustable parameters K were taken from [33,34]. The excess molar volume (V^E) in the ternary mixtures was expressed as [34]

$$V^E = \sum_i \sum_{i>j} V_{ij}^E + V_{ter}^E \tag{25}$$

where the values of V_{ij}^E were calculated for every binary mixture using Eq. (24) (with x_i and x_j the mole fractions of components i and j in the ternary mixture) and

$$V_{ter}^E = x_1 x_2 x_3 (e_1 + e_2(2x_1 - 1) + e_3(2x_1 - 1)^2 + \cdots) \tag{26}$$

e_i being parameters provided by [33]. The derivatives of the excess molar volumes in binary and ternary mixtures can be calculated using Eqs. (24)–(26) and thus the partial molar volumes can be obtained.

4.1. Analysis of intermolecular interactions in binary and ternary mixtures

The KBIs allow to calculate the Δn_{ij} around any central molecule j. The composition dependence of Δn_{ij} in binary and ternary mixtures provides information about the intermolecular interactions. It is useful to carry out such an analysis by comparing binary mixture (ij) with quasi-binary ternary mixture (ij) for a constant mole fraction of the third component. The presentation will be restricted to two quasi-binary ternary mixtures starting from the binary mixtures of methanol–water and N,N-dimethylformamide–methanol.

Firstly, the binary mixture methanol–water and the corresponding quasi-binary ternary mixtures will be analyzed. There are several papers in which the KBIs for the mixture methanol–water were calculated [6,7,12,41,42]. The comparison made in Fig. 1 shows good agreement between the present and previous calculations.

The letters F, M and W in subscript indicate N,N-dimethylformamide, methanol and water, respectively.

Figs. 2–4 represent Δn_{ij} around methanol, water and N,N-dimethylformamide as a central molecule in the binary and quasi-binary methanol–water mixtures. As already noted [6,12,43], there is affinity between like molecules in aqueous solutions of alcohols. Two types of clusters (water-enriched around the water molecules and alcohol-enriched around the alcohol molecules) could be detected [12]. While for higher alcohols clusters could be identified by small-angle X-ray and light scattering, the clusters in the aqueous

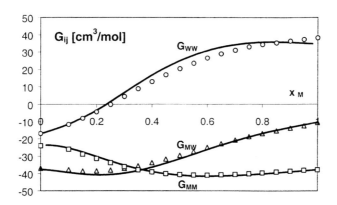

Fig. 1. The KBIs in the methanol–water mixture. The present results given by the solid line ($T = 313.15$ K) are compared with the G_{ij} obtained in [12] ($T = 298.15$ K).

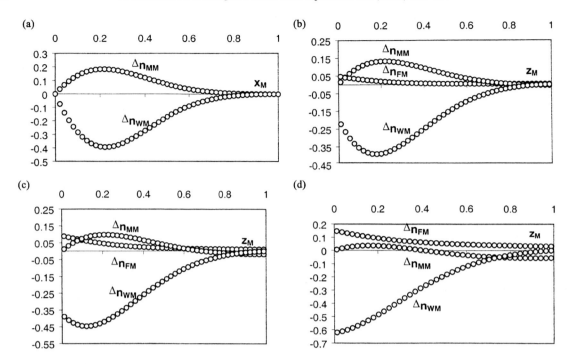

Fig. 2. The composition dependence of Δn_{ij} (mol/mol) around a methanol molecule for the (a) binary system methanol–water; and for the quasi-binary system N,N-dimethylformamide–methanol–water (b) $x_F = 0.04$; (c) $x_F = 0.08$; (d) $x_F = 0.16$. $z_M = (x_M/(x_M + x_W))$.

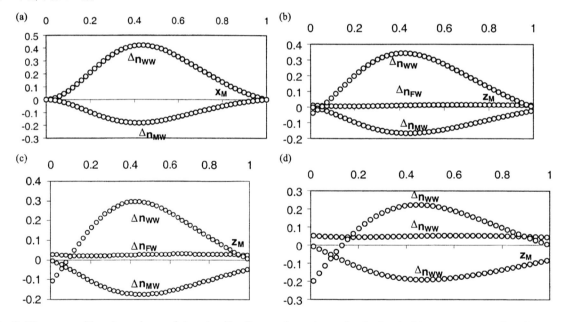

Fig. 3. The composition dependence of Δn_{ij} (mol/mol) around a water molecule for the binary system methanol–water (a); and for the quasi-binary system N,N-dimethylformamide–methanol–water (b) $x_F = 0.04$; (c) $x_F = 0.08$; (d) $x_F = 0.16$. $z_M = (x_M/(x_M + x_W))$.

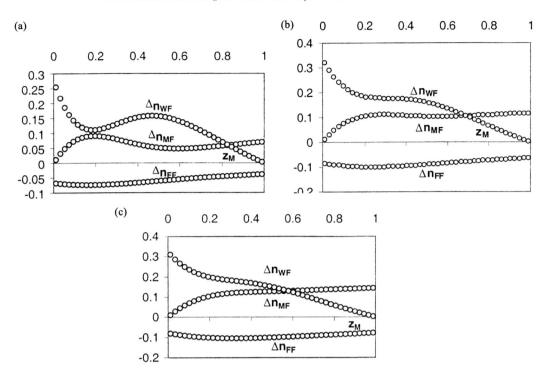

Fig. 4. The composition dependence of Δn_{ij} (mol/mol) around a N,N-dimethylformamide molecule for the quasi-binary system N,N-dimethylformamide–methanol–water; (a) $x_F = 0.04$; (b) $x_F = 0.08$; (c) $x_F = 0.16$. $z_M = (x_M/(x_M + x_W))$.

solutions of methanol and ethanol are too small to be detected by the above techniques [44–56]. Figs. 2a and 3a show that the Δn_{ii} are positive and small (the peak values are less than 1) in the latter cases. The effect of the addition of N,N-dimethylformamide to a binary methanol–water mixture can be seen from Figs. 2b–d and 3b–d, which show that the excess number of methanol molecules around a central methanol molecule (Δn_{MM}) decreases when N,N-dimethylformamide is added (Fig. 2b–d) and becomes even zero at $x_F \approx 0.25$. The excess number of water molecules around a central water molecule (Δn_{WW}) also decreases, but water-enriched clusters still exist at $x_F \approx 0.25$. Hence, the addition of N,N-dimethylformamide to a binary methanol–water mixture leads to a decay of the enrichment in methanol around a methanol molecule and in water around a water molecule. Fig. 5 presents the peak values of Δn_{MM} ($x_F = \text{const}$) and Δn_{WW} ($x_F = \text{const}$) as a function of x_F. The N,N-dimethylformamide molecules preferentially solvate the unlike species (water or methanol) than themselves. This is obvious from Fig. 4, which gives Δn_{ij} in the vicinity of a central N,N-dimethylformamide molecule. Fig. 4a–c shows that Δn_{FF} is always negative in the quasi-binary methanol–water system. On the contrary, Δn_{MF} and Δn_{WF} are always positive, clearly demonstrating that the N,N-dimethylformamide molecules prefer to solvate methanol and water molecules than themselves. This is compatible with the findings by various physical and chemical methods [26,57–64] that N,N-dimethylformamide forms complexes with water and methanol.

Figs. 6–8 represent Δn_{ij} around methanol, water and N,N-dimethylformamide as central molecules in the binary mixture N,N-dimethylformamide–methanol and the corresponding quasi-binary ternary mixtures. One can see from Figs. 6a and 7a, that the interactions between unlike species are dominant

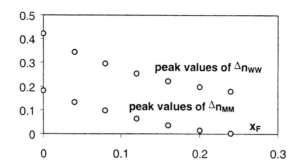

Fig. 5. The peak values of Δn_{ii} (i = M, W) for the quasi-binary methanol–water mixture as a function of the N,N-dimethylformamide mole fraction.

in the binary mixture of N,N-dimethylformamide–methanol. This behavior is qualitatively different from the aqueous solutions of alcohols, where the interactions between like species prevailed [6,12,43]. It is interesting to note that the addition of water up to a mole fraction of 0.25 affects little the peak values of Δn_{MF} and Δn_{FM} around both N,N-dimethylformamide and methanol as central molecules (Fig. 9). These results are compatible with the literature findings [62–64] that a complex is formed between N,N-dimethylformamide and methanol.

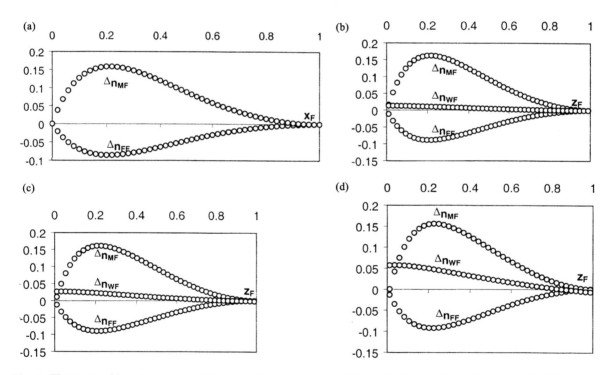

Fig. 6. The composition dependence of Δn_{ij} (mol/mol) around a N,N-dimethylformamide molecule for the binary system N,N-dimethylformamide–methanol (a) and for the quasi-binary system N,N-dimethylformamide–methanol–water; (b) $x_W = 0.04$; (c) $x_W = 0.08$; (d) $x_W = 0.16$. $z_F = (x_F/(x_F + x_M))$.

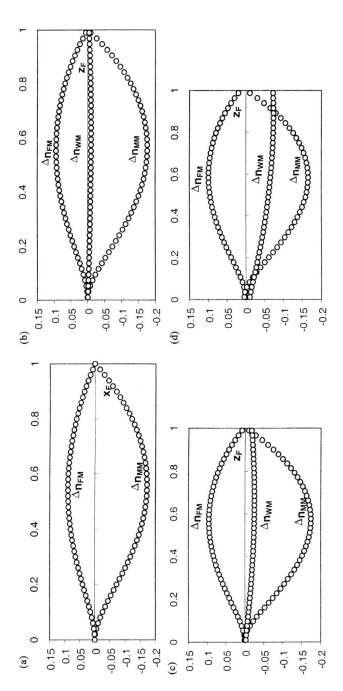

Fig. 7. The composition dependence of Δn_{ij} (mol/mol) around a methanol molecule for the binary system N,N-dimethylformamide–methanol (a) and for the quasi-binary system N,N-dimethylformamide–methanol–water; (b) $x_W = 0.04$; (c) $x_W = 0.08$; (d) $x_W = 0.16$. $z_F = (x_F/(x_F + x_M))$.

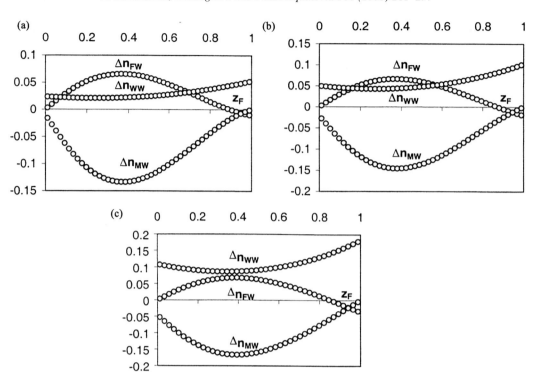

Fig. 8. The composition dependence of Δn_{ij} (mol/mol) around a water for the quasi-binary system N,N-dimethylformamide–methanol–water; (a) $x_W = 0.04$; (b) $x_W = 0.08$; (c) $x_W = 0.16$. $z_F = (x_F/(x_F + x_M))$.

Fig. 6b–d shows that there is an excess of water molecules in the cluster around the N,N-dimethylformamide molecule and (Fig. 7b–d) that there is a deficiency of water molecules in the cluster around a central methanol molecule. This observation is in agreement with the behavior of these molecules in the quasi-binary ternary mixtures corresponding to the binary methanol–water mixture.

Fig. 8a–c represents the Δn_{ij} in the vicinity of a water molecule as a central one for the quasi-binary ternary mixtures corresponding to the binary N,N-dimethylformamide–methanol mixture. One can see

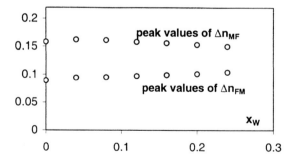

Fig. 9. The peak values of Δn_{ij} for the quasi-binary N,N-dimethylformamide–methanol mixture as a function of the water mole fraction.

from Fig. 8a–c that Δn_{FW} and Δn_{WW} are positive and Δn_{MW} is negative over the almost entire composition range considered. Again the peak value of Δn_{FW} is weakly dependent on the amount of water added up to a mole fraction of 0.25 of the latter and this result is compatible with a N,N-dimethylformamide–water complex. Figs. 2–8 show that there is preferential solvation of N,N-dimethylformamide molecules by water and methanol for all concentrations investigated. The same conclusion was reached in [24], but only for low concentrations of N,N-dimethylformamide (mole fraction of N,N-dimethylformamide less than 0.15).

5. Conclusion

Explicit expressions for the excess of the number of various species around central ones were derived on the basis of the KB theory of solutions for ternary mixtures. Because for ideal mixtures the clustering should be zero, they were obtained by subtracting from the conventional excesses calculated using the KBIs, those for a reference state. The latter were obtained from those valid for ideal mixtures, corrected to account for the volume conservation condition. The expressions thus obtained for the excess provide information about clustering. The KBIs and the excess of the number of molecules around central ones were calculated for the ternary mixture of N,N-dimethylformamide–methanol–water and the corresponding binary mixtures. The obtained results were used to discuss the local structure in the above mixtures.

Appendix A

The KBIs for binary systems are given by the expressions [6]:

$$G_{12} = RTk_T - \frac{V_1 V_2}{VD} \tag{A.1}$$

$$G_{11} = G_{12} + \frac{1}{x_1}\left(\frac{V_2}{D} - V\right) \tag{A.2}$$

and

$$G_{22} = G_{12} + \frac{1}{x_2}\left(\frac{V_1}{D} - V\right) \tag{A.3}$$

where R denotes the universal gas constant, V is the molar volume, V_i represents the partial molar volume, x_i is the molar fraction of component i, $D = 1 + x_i(\partial \ln \gamma_i/\partial x_i)_{P,T}$ and γ_i is the activity coefficient of component i.

The KBIs for ideal binary systems are given by the expressions [11]:

$$G_{12}^{id} = RTk_T^{id} - \frac{V_1^0 V_2^0}{V^{id}} \tag{A.4}$$

$$G_{11}^{id} = G_{12}^{id} + V_2^0 - V_1^0 \tag{A.5}$$

and

$$G_{22}^{id} = G_{12}^{id} - (V_2^0 - V_1^0) \tag{A.6}$$

where k_T^{id} and V^{id} are the isothermal compressibility and the molar volume of an ideal mixture and V_i^0 is the molar volume of the pure component i.

The expressions for G_{ij}^V are [31]

$$G_{12}^V = G_{21}^V = RTk_T - \frac{V_1 V_2}{V} \tag{A.7}$$

$$G_{ii}^V = G_{12}^V + V_j - V_i, \quad i \neq j \tag{A.8}$$

Appendix B

The expressions for l_{ij}, m_{ij}, n_{ij} and Q in Eqs. (2) and (3) are

$$Q = x_1^4 \bar{\mu}_{11}^2 + x_2^4 \bar{\mu}_{22}^2 + x_3^4 \bar{\mu}_{33}^2 - 2x_1^2 x_2^2 \bar{\mu}_{11} \bar{\mu}_{22} - 2x_1^2 x_3^2 \bar{\mu}_{11} \bar{\mu}_{33} - 2x_2^2 x_3^2 \bar{\mu}_{22} \bar{\mu}_{33} \tag{A.9}$$

$$l_{11} = 4x_2 x_3 V_2 V_3, \quad m_{11} = -4x_2 x_3 V_2 V_3 - 4x_3^2 V_3^2, \quad n_{11} = -4x_2 x_3 V_2 V_3 - 4x_2^2 V_2^2 \tag{A.10}$$

$$l_{22} = -4x_1 x_3 V_1 V_3 - 4x_3^2 V_3^2, \quad m_{22} = 4x_1 x_3 V_1 V_3, \quad n_{22} = -4x_1 x_3 V_1 V_3 - 4x_1^2 V_1^2 \tag{A.11}$$

$$l_{33} = -4x_1 x_2 V_1 V_2 - 4x_2^2 V_2^2, \quad m_{33} = -4x_1 x_2 V_1 V_2 - 4x_1^2 V_1^2, \quad n_{33} = 4x_1 x_2 V_1 V_2 \tag{A.12}$$

$$l_{12} = -2x_1 x_3 V_1 V_3 + 2x_2 x_3 V_2 V_3 - 2x_3^2 V_3^2, \quad m_{12} = 2x_1 x_3 V_1 V_3 - 2x_2 x_3 V_2 V_3 - 2x_3^2 V_3^2,$$
$$n_{12} = 4x_1 x_2 V_1 V_2 + 2x_1 x_3 V_1 V_3 + 2x_2 x_3 V_2 V_3 + 2x_3^2 V_3^2 \tag{A.13}$$

$$l_{13} = -2x_1 x_2 V_1 V_2 + 2x_2 x_3 V_2 V_3 - 2x_2^2 V_2^2, \quad n_{13} = 2x_1 x_2 V_1 V_2 - 2x_2 x_3 V_2 V_3 - 2x_2^2 V_2^2,$$
$$m_{13} = 4x_1 x_3 V_1 V_3 + 2x_1 x_2 V_1 V_2 + 2x_2 x_3 V_2 V_3 + 2x_2^2 V_2^2 \tag{A.14}$$

and

$$l_{23} = 4x_2 x_3 V_2 V_3 + 2x_1 x_2 V_1 V_2 + 2x_1 x_3 V_1 V_3 + 2x_1^2 V_1^2,$$
$$m_{23} = -2x_1 x_2 V_1 V_2 + 2x_1 x_3 V_1 V_3 - 2x_1^2 V_1^2,$$
$$n_{23} = 2x_1 x_2 V_1 V_2 - 2x_1 x_3 V_1 V_3 - 2x_1^2 V_1^2 \tag{A.15}$$

In the above expressions R is the universal gas constant, x_i denotes the molar fraction of component i, V represents the molar volume and V_i is the partial molar volume of component i.

The expressions for $\bar{\mu}_{ij}$ ($\bar{\mu}_{ij} = (N_1 + N_2 + N_3)\mu_{ij}$) where $\mu_{ij} = \left(\frac{\partial \mu_i}{\partial x_j}\right)_{x_k \neq x_j}$ and μ_i is the chemical potential component i are given in [22].

The derivatives of the chemical potentials can be expressed in terms of the activity coefficients (γ_i) as follows [22]:

$$\left(\frac{\partial \mu_1}{\partial x_1}\right)_{x_2} = RT \left[\frac{1}{x_1} + \left(\frac{\partial \ln \gamma_1}{\partial x_1}\right)_{x_2}\right] \tag{A.16}$$

$$\left(\frac{\partial \mu_2}{\partial x_2}\right)_{x_1} = RT \left[\frac{1}{x_2} + \left(\frac{\partial \ln \gamma_2}{\partial x_2}\right)_{x_1}\right] \tag{A.17}$$

$$\left(\frac{\partial \mu_1}{\partial x_2}\right)_{x_1} = RT\left(\frac{\partial \ln \gamma_1}{\partial x_2}\right)_{x_1} \tag{A.18}$$

and

$$\left(\frac{\partial \mu_2}{\partial x_1}\right)_{x_2} = RT\left(\frac{\partial \ln \gamma_2}{\partial x_1}\right)_{x_2} \tag{A.19}$$

Appendix C

The volume conservation condition can be derived from the KB equation for the partial molar volume in a binary mixture [1]

$$V_1 = \frac{1 + (G_{22} - G_{12})c_2}{c_1 + c_2 + c_1 c_2 (G_{11} + G_{22} - 2G_{12})} \tag{A.20}$$

Applied to our reference state G_{ij}^V, Eq. (A.20) leads to

$$V_1 = \frac{1 + (G_{22}^V - G_{12}^V)c_2}{c_1 + c_2 + c_1 c_2 (G_{11}^V + G_{22}^V - 2G_{12}^V)} \tag{A.21}$$

One can verify that Eq. (A.21) is satisfied identically by inserting the expressions ((A.7) and (A.8)) for G_{ij}^V.

Eqs. (A.20) and (A.21) yield

$$c_1 V_1 (\Delta G_{11} + \Delta G_{22} - 2\Delta G_{12}) = (\Delta G_{22} - \Delta G_{12}) \tag{A.22}$$

where $\Delta G_{ij} = G_{ij} - G_{ij}^V$. From Eq. (A.22) and relation $c_1 V_1 + c_2 V_2 = 1$, one obtains

$$\Delta V_1 \equiv V_1 \Delta n_{11} + V_2 \Delta n_{21} = V_2 \Delta n_{22} + V_1 \Delta n_{12} \equiv -\Delta V_2 \tag{A.23}$$

Since the volume V should remain constant, one must have

$$N_1 \Delta V_1 + N_2 \Delta V_2 = \Delta V_1 (N_1 - N_2) = 0 \tag{A.24}$$

Eq. (A.24) leads to $\Delta V_1 = \Delta V_2 = 0$, the volume conservation equation.

One can also demonstrate that the volume conservation condition for a ternary mixture (Eq. (15)) is satisfied if Δn_{ij} is expressed as $\Delta n_{ij} = c_i (G_{ij} - G_{ij}^V)$, with G_{ij} given by Eqs. (2) and (3) and G_{ij}^V obtained from Eqs. (9)–(14) by substituting k_T^{id}, V_i^0 and V^{id} by k_T, V_i and V, respectively.

References

[1] J.G. Kirkwood, F.P. Buff, J. Chem. Phys. 19 (1951) 774.
[2] J.P. O' Connell, Mol. Phys. 20 (1971) 27.
[3] A. Ben-Naim, Water and Aqueous Solutions, Plenum Press, New York, 1974.
[4] A. Ben-Naim, J. Chem. Phys. 63 (1975) 2064.
[5] A. Ben-Naim, J. Chem. Phys. 67 (1977) 4884.
[6] E. Matteoli, L. Lepori, J. Chem. Phys. 80 (1984) 2856.

[7] M.C.A. Donkersloot, J. Solution Chem. 8 (1979) 293.
[8] K.J. Patil, J. Solution Chem. 10 (1981) 315.
[9] A.L. Zaitsev, Y.M. Kessler, V.E. Petrenko, Zh. Fiz. Khim. 59 (1985) 2728.
[10] E. Matteoli, L. Lepori, J. Chem. Soc., Faraday Trans. 91 (1995) 431.
[11] E.J. Matteoli, Phys. Chem. B 101 (1997) 9800.
[12] I. Shulgin, E.J. Ruckenstein, Phys. Chem. B 103 (1999) 2496.
[13] A. Ben-Naim, Pure Appl. Chem. 62 (1990) 25.
[14] D.M. Pfund, L.L. Lee, H.D. Cochran, Fluid Phase Equilibria 39 (1988) 161.
[15] R.G. Rubio, M.G. Prolongo, M. Diaz Pena, J.A.R. Renuncio, J. Phys. Chem. 91 (1987) 1177.
[16] P.G. Debenedetti, Chem. Eng. Sci. 42 (1987) 2203.
[17] H.D. Cochran, L.L. Lee, in: K.P. Johnston, J.M.L. Penninger (Eds.), Supercritical Fluid Science and Technology, ACS Symposium Series No. 406, American Chemical Society, Washington, DC, 1989, Chapter 3, p. 27.
[18] D.B. McGuigan, P.A. Monson, Fluid Phase Equilibria 57 (1990) 227.
[19] H.D. Cochran, L.L. Lee, D.M. Pfund, in: E. Matteoli, G.A. Mansoori (Eds.), Fluctuation Theory of Mixtures, Taylor & Francis, New York, 1990, p. 69.
[20] J.F. Brennecke, C.A. Eckert, AIChE J. 35 (1989) 1409.
[21] E. Matteoli, L. Lepori, J. Mol. Liq. 47 (1990) 89.
[22] E. Matteoli, L. Lepori, in: E. Matteoli, G.A. Mansoori (Eds.), Fluctuation Theory of Mixtures, Taylor & Francis, New York, 1990, p. 259.
[23] J. Zielkiewiecz, J. Phys. Chem. 99 (1995) 3357.
[24] J. Zielkiewiecz, J. Phys. Chem. 99 (1995) 4787.
[25] J. Zielkiewiecz, J. Chem. Soc., Faraday Trans. 94 (1998) 1713.
[26] E. Matteoli, L. Lepori, J. Mol. Liq. 77 (1999) 101.
[27] P.J. Tumidajski, Can. J. Chem. 69 (1991) 458.
[28] K.E. Newman, in: E. Matteoli, G.A. Mansoori (Eds.), Fluctuation Theory of Mixtures, Taylor & Francis, New York, 1990, p. 373.
[29] D.A. Jonah, H.D. Cochran, Fluid Phase Equilibria 92 (1994) 107.
[30] A. Chialvo, J. Phys. Chem. 97 (1993) 2740.
[31] I. Shulgin, E. Ruckenstein, J. Phys. Chem. B 103 (1999) 872.
[32] J. Zielkiewicz, P. Oracz, Fluid Phase Equilibria 59 (1990) 279.
[33] J. Zielkiewicz, J. Chem. Thermodyn. 27 (1995) 415.
[34] J. Zielkiewicz, J. Chem. Thermodyn. 26 (1994) 1317.
[35] D.R. Lide (Ed.), Handbook of Chemistry and Physics, 77th Edition, CRC Press, Boca Raton, FL, 1996–1997.
[36] S. Miyanaga, K. Tamura, S. Murakami, J. Chem. Thermodyn. 24 (1994) 1077.
[37] J.A. Barker, Aust. J. Chem. 6 (1953) 207.
[38] G.M. Wilson, J. Am. Chem. Soc. 86 (1964) 127.
[39] H. Renon, J.M. Prausnitz, AIChE J. 14 (1968) 135.
[40] M.M. Abbott, H.C. Van Ness, AIChE J. 21 (1975) 62.
[41] E. Matteoli, Z. Phys. Chem. 185 (1994) (Part 2) 177.
[42] Y. Marcus, Phys. Chem. Chem. Phys. 1 (1999) 2975.
[43] J.J. Kozak, W.S. Knight, W. Kauzmann, J. Chem. Phys. 48 (1968) 675.
[44] H.D. Bale, R.E. Sherpler, D.K. Sorgen, Phys. Chem. Liq. 1 (1968) 181.
[45] Y. Koga, Chem. Phys. Lett. 111 (1984) 176.
[46] K. Nishikawa, Y. Kodera, T. Iijima, J. Phys. Chem. 91 (1987) 3694.
[47] K. Nishikawa, H. Hayashi, T. Iijima, J. Phys. Chem. 93 (1989) 6559.
[48] H. Hayashi, K. Nishikawa, T. Iijima, J. Phys. Chem. 94 (1990) 8334.
[49] H. Hayashi, Y. Udagawa, Bull. Chem. Soc. Jpn. 65 (1992) 155.
[50] I. Shulgin, R. Serimaa, M. Torkkeli, Report Series in Physics, HU-P-256, Helsinki, 1991.
[51] G. D'Arrigo, J. Teixeira, J. Chem. Soc. Faraday Trans. 86 (1990) 1503.
[52] G.H. Grossmann, K.H. Ebert, Ber. Bunsenges. Phys. Chem. 85 (1981) 1026.
[53] G.W. Euliss, C.M. Sorensen, J. Chem. Phys. 80 (1984) 4767.
[54] T.M. Bender, R. Pekora, J. Phys. Chem. 90 (1986) 1700.

[55] M.C.A. Donkersloot, Chem. Phys. Lett. 60 (1979) 435.
[56] K. Nishikawa, T. Iijima, J. Phys. Chem. 97 (1993) 10824.
[57] E.D. Schmid, E. Brodbek, Can. J. Chem. 63 (1985) 1365.
[58] O.E.S. Godhino, E. Greenhow, Anal. Chem. 57 (1985) 1725.
[59] B.E. Geller, Zh. Fiz. Khim. 35 (1961) 1105.
[60] J. Bougard, R. Jadot, J. Chem. Thermodyn. 7 (1975) 1185.
[61] A.M. Zaichikov, G.A. Krestov, Zh. Fiz. Khim. 69 (1995) 389.
[62] C.M. Kinart, J.W. Kinart, Pol. J. Chem. 68 (1994) 349.
[63] G. Eaton, M.C.R. Symons, J. Chem. Soc. Faraday Trans. 1 84 (1988) 3459.
[64] L. Pikkarainen, Thermochim. Acta 178 (1991) 311.

J. Phys. Chem. B **2006**, *110*, 12707−12713

The Kirkwood−Buff Theory of Solutions and the Local Composition of Liquid Mixtures

Ivan L. Shulgin[†] and Eli Ruckenstein*

Department of Chemical & Biological Engineering, State University of New York at Buffalo, Amherst, New York 14260

Received: January 30, 2006; In Final Form: April 6, 2006

The present paper is devoted to the local composition of liquid mixtures calculated in the framework of the Kirkwood−Buff theory of solutions. A new method is suggested to calculate the excess (or deficit) number of various molecules around a selected (central) molecule in binary and multicomponent liquid mixtures in terms of measurable macroscopic thermodynamic quantities, such as the derivatives of the chemical potentials with respect to concentrations, the isothermal compressibility, and the partial molar volumes. This method accounts for an inaccessible volume due to the presence of a central molecule and is applied to binary and ternary mixtures. For the ideal binary mixture it is shown that because of the difference in the volumes of the pure components there is an excess (or deficit) number of different molecules around a central molecule. The excess (or deficit) becomes zero when the components of the ideal binary mixture have the same volume. The new method is also applied to methanol + water and 2-propanol + water mixtures. In the case of the 2-propanol + water mixture, the new method, in contrast to the other ones, indicates that clusters dominated by 2-propanol disappear at high alcohol mole fractions, in agreement with experimental observations. Finally, it is shown that the application of the new procedure to the ternary mixture water/protein/cosolvent at infinite dilution of the protein led to almost the same results as the methods involving a reference state.

1. Introduction

The Kirkwood−Buff (KB) theory of solution[1] (often called fluctuation theory) employs the grand canonical ensemble to relate macroscopic properties, such as the derivatives of the chemical potentials with respect to concentrations, the isothermal compressibility, and the partial molar volumes, to microscopic properties in the form of spatial integrals involving the radial distribution function. This theory allows one to obtain information regarding some microscopic characteristics of multicomponent mixtures from measurable macroscopic thermodynamic quantities. However, despite its attractiveness, the KB theory was rarely used[2−4] in the first three decades after its publication for two main reasons: (1) the lack of precise data (in particular regarding the composition dependence of the chemical potentials) and (2) the difficulty to interpret the results obtained. Only after Ben-Naim indicated how to calculate numerically the Kirkwood−Buff integrals (KBIs) for binary systems[5] was this theory used more frequently.

So far the KBIs have been calculated for numerous binary systems,[6−17] and the results were used to examine the solution behavior with regard to (1) local composition, (2) various models for phase equilibrium, (3) preferential solvation, and others.[6−22] One should also mention the use of the KB theory for supercritical fluids and mixtures containing supercritical components[23−27] and for biochemical issues such as the behavior of a protein in aqueous mixed solvents.[28−38]

The present paper is focused on the application of the KB theory to the local composition. The key quantity related to the local composition in the KB theory is the excess (or deficit) number of molecules around a central molecule. The conventional method of calculating the excess (or deficit) number of molecules around a central molecule was developed by Ben-Naim.[3,5,18−19,39−40] Let us consider a binary mixture 1−2. The excess (or deficit) number of molecules i ($i = 1, 2$) around a central molecule j ($j = 1, 2$) was defined by Ben-Naim as[3,5,18−19,39−40]

$$\Delta n_{ij}^{BN} = c_i G_{ij} \quad (1)$$

where c_i is the molar concentration of species i in the mixture and G_{ij} is the Kirkwood−Buff integral defined as (analytical expressions for the KBIs in a binary mixture are given in the Appendix)

$$G_{ij} = \int_0^\infty (g_{ij} - 1) 4\pi r^2 \, dr \quad i, j = 1, 2 \quad (2)$$

where g_{ij} is the radial distribution function between species i and j and r is the distance between the centers of molecules i and j.

However, the following objections can be brought to the use of eq 1 for calculating the excess (or deficit) number of molecules around a central molecule. A first objection (1) is that there are many systems for which all the KBIs (G_{11}, G_{12}, and G_{22}) are negative at least in certain ranges of composition.[22,41] As a result, in such cases all Δn_{ij}^{BN} calculated with eq 1 (Δn_{11}^{BN} and Δn_{21}^{BN} around the central molecule 1 and Δn_{12}^{BN} and Δn_{22}^{BN} around the central molecule 2) would be negative, and this is not plausible because then the density around any molecule in liquid will become lower than that in the bulk. A second objection examined in detail in the next section is that (2) eq 1 does not provide the true excess.

Two other methods to calculate the excess (or deficit) number of molecules around a central molecule have been suggested

* To whom correspondence should be addressed. E-mail: feaeliru@acsu.buffalo.edu. Phone: (716) 645-2911/ext. 2214. Fax: (716) 645-3822.
[†] E-mail: ishulgin@eng.buffalo.edu.

which are based on a reference state. Matteoli and Lepori[10,11] observed that the excesses (or deficits) calculated with eq 1 provide nonzero values for ideal binary mixtures and expressed the opinion that they (the excesses (or deficits)) should be in that case equal to zero. For this reason they proposed that the excess (or deficit) number of molecules around a central molecule should be calculated with respect to a reference state (the ideal binary mixture)

$$\Delta n_{ij}^{ML} = c_i(G_{ij} - G_{ij}^{id}) \quad (3)$$

where G_{ij}^{id} is the Kirkwood−Buff integral for an ideal binary mixture (expressions for G_{ij}^{id} are given in the Appendix). In addition, Matteoli and Lepori[10,11] introduced what could be called the volume conservation condition, which for a binary mixture can be formulated as follows: "the volume occupied by the excess of i molecules around a j molecule must be equal to the volume left free by the j molecules around the same j molecule". One can show that the excesses and deficits calculated using the ideal mixture as reference state do not satisfy the above volume conservation condition. For this reason, a new reference state was suggested by Shulgin and Ruckenstein[12] in which all the activity coefficients were taken as equal to unity and no constraints on the partial molar volumes of the components were imposed. This reference state satisfies the volume conservation condition and provides as that of Matteoli and Lepori[10,11] zero excesses for ideal mixtures for both binary and ternary mixtures.[12,42] The excess (or deficit) number of molecules around a central molecule can be obtained using the expression

$$\Delta n_{ij}^{SR} = c_i(G_{ij} - G_{ij}^{SR}) \quad (4)$$

where G_{ij}^{SR} is the KBI calculated for the reference state suggested by Shulgin and Ruckenstein[12] (expressions for G_{ij}^{SR} can also be found in the Appendix).

However, there is a basic issue regarding the suggested reference states (Matteoli−Lepori and Shulgin−Ruckenstein): Is the excess (or deficit) number of molecules around a central molecule equal to zero in ideal mixtures? The considerations of the above authors imply that the distribution of components in an ideal mixture is random and therefore all excess (or deficit) number of molecules around a central molecule should be zero. However, because the volumes of the components are different, there is no absolute randomness. A new treatment is suggested below, which accounts for a volume which is not accessible to the molecules surrounding a central molecule and which reveals that for ideal mixtures the excesses (or deficits) are not zero. The inaccessible volume thus introduced could be considered a kind of reference state which, however, does not correspond to an ideal mixture.

2. A New Procedure to Calculate the Excess (Or Deficit) Number of Molecules around a Central Molecule

The average number of molecules i (n_{ij}) in a sphere of radius R around central molecules j can be calculated using the expression[40,43]

$$n_{ij} = c_i \int_0^R g_{ij} 4\pi r^2 \, dr \qquad i,j = 1, 2 \quad (5)$$

which can be recast in the form

$$n_{ij} = c_i \int_0^R (g_{ij} - 1) 4\pi r^2 \, dr + c_i \int_0^R 4\pi r^2 \, dr \quad (6)$$

As soon as R becomes large enough for $g_{ij} \approx 1$, eq 6 can be rewritten as[18−19,40]

$$n_{ij} = c_i \int_0^\infty (g_{ij} - 1) 4\pi r^2 \, dr + c_i \int_0^R 4\pi r^2 \, dr = \\ c_i G_{ij} + \frac{c_i 4\pi R^3}{3} \quad (7)$$

The difference between n_{ij} and $c_i 4\pi R^3/3$ was considered as the average excess (or deficit) number of molecules i ($i = 1, 2$) around a central molecule j ($j = 1, 2$), and eq 1 was thus obtained.[3,5,18−19,39−40]

However, the term $c_i 4\pi R^3/3$ includes molecules i assumed to be located in a volume inaccessible to them because of the presence of the central molecule j. Therefore, when the average excess (or deficit) number of molecules i ($i = 1, 2$) around a central molecule j is calculated, those i molecules should be subtracted from $c_i 4\pi R^3/3$, and the second integral in the right-hand side of eq 6 should be subdivided into two parts

$$n_{ij} = c_i \int_0^R (g_{ij} - 1) 4\pi r^2 \, dr + c_i \int_{R_j}^R 4\pi r^2 \, dr + c_i \int_0^{R_j} 4\pi r^2 \, dr \quad (8)$$

where R_j is the radius of a volume around the center of molecule j which is inaccessible to molecules i due to the presence of the central molecule j. The first integral in the right-hand side is the Kirkwood−Buff integral when R is sufficiently large, the second integral provides the number of molecules of species i in a bulk liquid between the radii R_j and R, and the third integral provides the number of molecules of species i in a bulk liquid from zero to the radius R_j. The true excess is given by the difference between n_{ij} and the number of molecules of species i in a bulk liquid between the radii R_j and R. Hence

$$\Delta n_{ij} = n_{ij} - c_i \int_{R_j}^R 4\pi r^2 \, dr \quad (9)$$

The third integral $\int_0^{R_j} 4\pi r^2 \, dr$ in eq 8 represents a bulk volume V^j which is not accessible to the molecules of species i.

Consequently, the expression for calculating the average excess (or deficit) number of molecules i ($i = 1, 2$) around a central molecule j ($j = 1, 2$) has the form

$$\Delta n_{ij} = c_i G_{ij} + c_i V^j \quad (10)$$

To estimate the volume V^j, eq 10 is applied to a pure component j ($i = j$, $c_j = c_j^0$). In this case, because of complete randomness one expects Δn_{jj} to be zero. Therefore combining eq 10 and the equation for G_{jj} of a pure substance (eq A-11 in the Appendix) one obtains the following expression for the volume V^j

$$(V^j)_{x_j=1} = V_j^0 - RTk_{T,j}^0 \quad (11)$$

where T is the temperature, x_j is the molar fraction of component j, V_j^0 is the molar volume of the pure component j, $k_{T,j}^0$ is the isothermal compressibility of the pure component j, and R is the universal gas constant. Hence, far from the critical point, the excluded volume V^j is equal to the molar volume of the pure component j because $V_j^0 \gg RTk_{T,j}^0$.[6]

On the basis of the above result, the following expression is suggested for the volume V^j in the entire composition range

$$V^j = V_j - RTk_T \quad (12)$$

where V_j is the partial molar volume of component j and k_T is the isothermal compressibility of the mixture. Of course, other forms for V^j compatible with eq 11 can be suggested.

Thus the average excess (or deficit) number of molecules i ($i = 1, 2$) around a central molecule j ($j = 1, 2$) can be calculated using the expression

$$\Delta n_{ij} = c_i G_{ij} + c_i V^j = c_i G_{ij} + c_i (V_j - RTk_T) \quad (13)$$

Similar relations can be written for the excess (or deficit) number of molecules i ($i = 1, 2, ...$) around a central molecule j ($j = 1, 2, ...$) for ternary and multicomponent mixtures. The new simple expression (13) will be now applied to various systems.

3. The Excess (Or Deficit) Number of Molecules around a Central Molecule Calculated Using the New Expression (13)

3.1. Ideal Binary Mixture.
Using eq 13 and relations (A-4 and A-5) from the Appendix, one can write the following expressions for the excesses Δn_{ij} values of ideal binary mixtures around a central molecule 1

$$\Delta n_{11}^{id} = \frac{x_1 x_2 V_2^0 (V_2^0 - V_1^0)}{(x_1 V_1^0 + x_2 V_2^0)^2} \quad (14)$$

and

$$\Delta n_{21}^{id} = \frac{x_1 x_2 V_1^0 (V_1^0 - V_2^0)}{(x_1 V_1^0 + x_2 V_2^0)^2} \quad (15)$$

These equations show that the Δn_{ij} values for an ideal binary mixture become zero only when the molar volumes of the pure components are the same, otherwise the excesses and deficits have nonzero values and can be calculated with eqs 14 and 15.

One can also demonstrate that the volume conservation conditions ($V_i \Delta n_{ij} + V_j \Delta n_{jj} = 0$; $i, j = 1, 2$, and $i \neq j$) are satisfied by an ideal mixture when the excesses (or deficits) are calculated with eqs 14 and 15. We calculated the Δn_{ij} around a central molecule 1 with eqs 14 and 15 for $V_1^0 = 30$ cm³/mol and $V_2^0 = 60$ cm³/mol. A comparison between the Δn_{ij} calculated with eq 1 and eqs 14 and 15 is presented in Figure 1. Figure 1 shows that both Δn_{11} and Δn_{21} calculated with the conventional eq 1 are negative, whereas those calculated with eqs 14 and 15 provide $\Delta n_{11} > 0$ and $\Delta n_{21} < 0$ in the entire composition range. One should note that the species with lower molar volume (component 1) is in excess. (A similar result was obtained when molecule 2 was the central molecule).

3.2. Binary System Methanol (1)–Water (2).
The KBIs for this system ($T = 313$ K)[42] are presented in Figure 2a which shows that, in the composition range ($0 < x_{MeOH} \leq 0.25$), all three KBIs (G_{11}, G_{12}, and G_{22}) are negative. As a result, in this composition range (Figure 2, parts b and c) all four Δn_{ij}^{BN} (excesses or deficits calculated with eq 1) (Δn_{11}^{BN} and Δn_{21}^{BN} around a central methanol molecule and Δn_{12}^{BN} and Δn_{22}^{BN} around a central water molecule) are negative. This means that the densities around any central molecule (methanol or water) are less than those in the bulk. However, the calculations based on eq 13 (Figure 2, parts d and e) provide more reasonable values regarding the excess (or deficit) number of molecules around a central molecule:

(1) Around a central methanol molecule, $\Delta n_{11} > 0$ and $\Delta n_{21} < 0$ for $0 < x_{MeOH} \leq 0.35$, and $\Delta n_{11} < 0$ and $\Delta n_{21} > 0$ for $x_{MeOH} > 0.35$. Therefore, in the composition range $0 < x_{MeOH}$

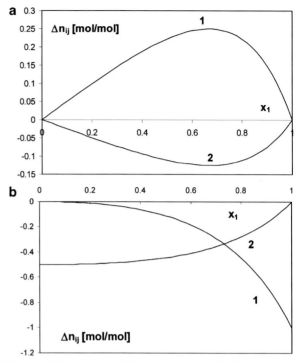

Figure 1. The excess (or deficit) number of molecules i ($i = 1, 2$) around a central molecule 1 for a binary ideal mixture with $V_1^0 = 30$ cm³/mol and $V_2^0 = 60$ cm³/mol. Line 1 is Δn_{11}, and line 2 is Δn_{21}. Shown are (a) Δn_{ij} values of an ideal binary mixture calculated with the new eq 13, and (b) Δn_{ij} values of an ideal binary mixture calculated with eq 1 (the KBIs were provided by eqs A-4 and A-5 in which k_T^{id} was taken as zero).

≤ 0.35, methanol is in excess in the vicinity of a central methanol molecule. However, at sufficiently high methanol mole fractions ($x_{MeOH} > 0.35$) the water molecules become in excess in the vicinity of a central methanol molecule. This result differs somewhat from the previous findings based on a reference state,[12,14] where methanol was found to be in excess in the vicinity of a central methanol molecule over the entire composition range.

(2) Around a central water molecule, $\Delta n_{22} > 0$ and $\Delta n_{12} < 0$ for $0 \leq x_{MeOH} \leq 1$. This means that the water molecules are in excess around a central water molecule over the entire composition range.

3.3. Binary System 2-Propanol (1)–Water (2).
The KBIs for this system ($T = 298.15$ K)[12] are presented in Figure 3a. In the composition range ($0.33-0.35 < x_{i-PrOH} \leq 1$), Δn_{11}^{BN} and Δn_{21}^{BN} around a central 2-propanol molecule are negative (Figure 3b). This means that in the above composition range the density around a 2-propanol molecule is less than in the bulk. The calculations based on eq 13 (Figure 3, parts d and e) provide more reasonable values regarding the excess (or deficit) number of molecules around a central molecule: around the 2-propanol molecule $\Delta n_{11} > 0$ and $\Delta n_{21} < 0$ for $0 < x_{i-PrOH} \leq 0.55-0.57$, and $\Delta n_{11} < 0$ and $\Delta n_{21} > 0$ for $0.55-0.57 < x_{i-PrOH} \leq 1$. Therefore, in the composition range $0 < x_{i-PrOH} \leq 0.55-0.57$ the 2-propanol is in excess in the vicinity of a central 2-propanol molecule. However, at higher 2-propanol mole fractions ($0.55-0.57 < x_{i-PrOH} \leq 1$) the water molecules are in slight excess in the vicinity of a 2-propanol molecule. These results differ somewhat from the previous findings based on a reference state,[12,14] where 2-propanol was found to be in excess in the vicinity of a central 2-propanol molecule over the entire

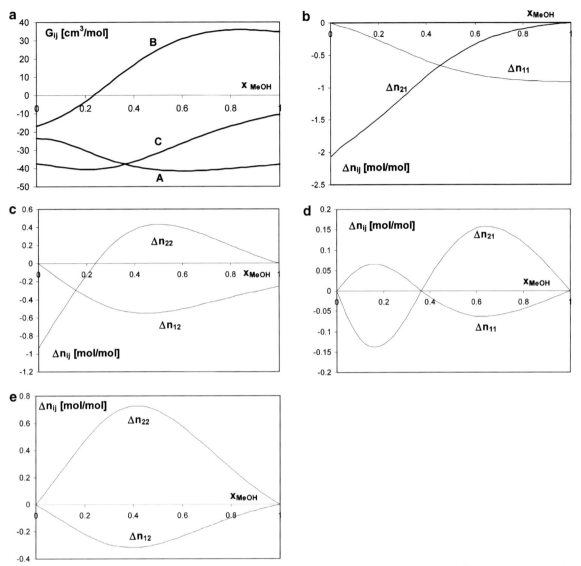

Figure 2. The KBIs and excesses (or deficits) for methanol (1)/water (2) mixtures ($T = 313.15$ K). (a) The KBIs[42]: G_{11} (A), G_{22} (B), and G_{12}(C); (b) excesses (or deficits) around a central methanol molecule calculated with eq 1; (c) excesses (or deficits) around a central water molecule calculated with eq 1; (2) excesses (or deficits) around a central methanol molecule calculated with eq 13; (e) excesses (or deficits) around a central water molecule calculated with eq 13.

composition range. There is some experimental evidence that clusters dominated by 2-propanol molecules disappear at high 2-propanol concentrations. Hayashi, et al.[44] using small-angle X-ray scattering found clusters dominated by 2-propanol molecules, with sizes from 1.5 to 6.5 Å, in the range of alcohol mole fractions from 0.1 to 0.3 that disappeared at higher mole fractions. Around a central water molecule, $\Delta n_{22} > 0$ and $\Delta n_{12} < 0$ for $0 \leq x_{i-\text{PrOH}} \leq 1$. This means that the water molecules are in excess around a central water molecule over the entire composition range.

3.4. Comparison between Equation 13 and Equations 3 and 4 for Binary Mixtures. The results obtained on the basis of eq 13 are comparable numerically to those obtained with eqs 3 and 4 when the molar volumes of the pure components or the partial molar volumes become comparable. Indeed, in this case ($V_1^0 \approx V_2^0$ or $V_1 \approx V_2$) and hence one can write, using eqs (A-4 to A-7) from the Appendix, that

$$\Delta n_{ij}^{\text{ML}} = c_i(G_{ij} - G_{ij}^{\text{id}}) \approx c_i G_{ij} + c_i(V_j^0 - RTk_T^{\text{id}}) \approx \Delta n_{ij} \quad (16)$$

and

$$\Delta n_{ij}^{\text{SR}} = c_i(G_{ij} - G_{ij}^{\text{SR}}) \approx c_i G_{ij} + c_i(V_j - RTk_T) = \Delta n_{ij} \quad (17)$$

To compare the present method with the Matteoli–Lepori and Shulgin–Ruckenstein methods based on reference states, a hypothetical binary mixture (We are indebted to one of the referees for the suggestion to use such an hypothetical binary mixture for analyzing the new method) with the volumes of the pure components $V_1^0 = 30$, $V_2^0 = 70$ and the volumes at infinite dilution $V_1^\infty = 26$ and $V_2^\infty = 66$ (all volumes in [cm^3/mol]) will be considered. In addition, the contribution of the isothermal compressibility will be neglected, and the partial molar volumes will be assumed to be linear functions of the mole fractions. Let us first observe that eqs 3, 4, and 13 can be formally written in the following unified form

$$\Delta n_{ij} = c_i(G_{ij} - \Gamma) \quad (18)$$

where $\Gamma = -V_j + RTk_T$ for the new method, $\Gamma = G_{ij}^{\text{id}}$ for

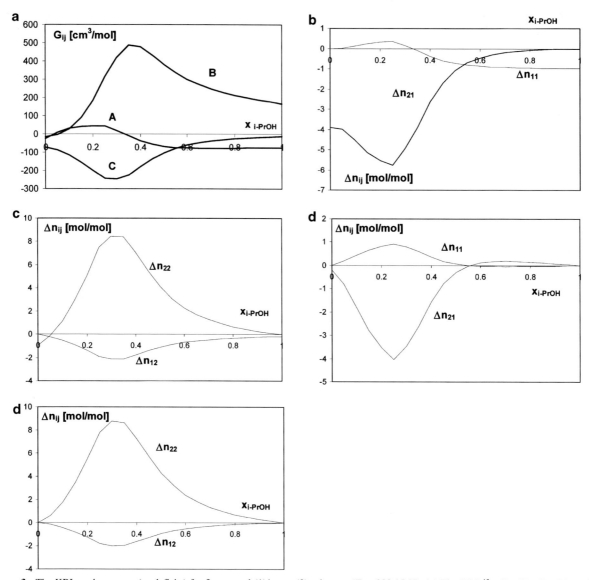

Figure 3. The KBIs and excesses (or deficits) for 2-propanol (1)/water (2) mixtures ($T = 298.15$ K). (a) The KBIs[12]: G_{11} (A), G_{22} (B), and G_{12} (C); (b) excesses (or deficits) around a central 2-propanol molecule calculated with eq 1; (c) excesses (or deficits) around a central water molecule calculated with eq 1; (3) excesses (or deficits) around a central 2-propanol molecule calculated with eq 13; (e) excesses (or deficits) around a central water molecule calculated with eq 13.

the Matteoli−Lepori reference state, and $\Gamma = G_{ij}^{SR}$ for the Shulgin−Ruckenstein reference state. The calculated values of V_1, V_2, $-G_{ij}^{id}$, and $-G_{ij}^{SR}$ are plotted in Figure 4. Several observations can be made regarding Figure 4 and eqs 3, 4, and 13: (1) All three methods provide almost the same values of Γ when the molar volumes of the pure components become comparable. (2) In the new method, $\Gamma = -V_j + RTk_T$ is the same for both Δn_{ij} ($i \neq j$) and Δn_{jj} excesses, whereas for the methods involving reference states, $\Gamma = G_{ij}^{ref}$ for Δn_{ij} ($i \neq j$) and $\Gamma = G_{jj}^{ref}$ for Δn_{jj} (where ref = id or SR). As well known and as shown in Figure 4, G_{ij}^{ref} ($i \neq j$) $\neq G_{jj}^{ref}$ for both reference states. (3) The values Δn_{jj} for pure components are zero in all three methods. (4) The differences between G_{ij}^{id} and G_{ij}^{SR} are small but not negligible, and this proves the observation[12,15] that far from critical conditions, where the differences are large, the two reference states methods provide similar results. In addition, Figure 4 indicates that the excesses calculated using the two reference states are close to each other. (5) The volume conservation conditions[10,11] ($V_i \Delta n_{ij} + V_j \Delta n_{jj} =$

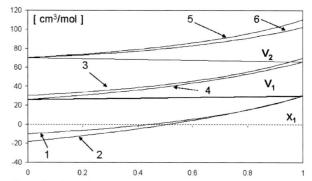

Figure 4. The partial molar volumes V_1 and V_2 for the artificial mixture (see text), and their comparison with the KBIs for Matteoli−Lepori and Shulgin−Ruckenstein reference states: $-G_{11}^{id}$ (1), $-G_{11}^{SR}$ (2), $-G_{12}^{id}$ (3), $-G_{12}^{SR}$ (4), $-G_{22}^{id}$ (5), and $-G_{22}^{SR}$ (6).

0; $i, j = 1, 2$ and $i \neq j$) are satisfied by the present method and by the method involving the reference state introduced by Shulgin and Ruckenstein.[12] For the present method, this can be demonstrated using eq 13 for Δn_{jj} and Δn_{ij} together with expressions (A-1 to A-3) for the corresponding KBIs from the Appendix.

3.5. Ternary Mixture: a Biomolecule in a Mixed Solvent. In the last several years the KB theory was applied to the systems water (1)/protein (2)/cosolvent (or salt) (3) mixtures.[28–38] Experimental information regarding the partial molar volumes and the preferential binding parameters were used to calculate G_{12} and G_{23} at infinite dilution of the protein. Further the calculated KBIs were used to obtain the excesses (or deficits) of the constituents of the mixed solvent near a protein molecule (with respect to the bulk concentrations of a protein-free mixed solvent).[34–35,38] The use of the conventional eq 1 provided unreasonable results. For example,[34] for an infinitely dilute ribonuclease A (2) in water (1) + glycerol (3) mixtures, one obtains that for a volume fraction of glycerol of 30%: $c_1 G_{12} = -341.8$ mol of water which has a volume of -6.18 L, and $c_3 G_{23} = -48.1$ mol of water which has a volume of -3.52 L. Therefore, the use of the conventional eq 1 provided large deficits of water and glycerol in the vicinity of ribonuclease A. In contrast, many experiments confirmed that in this mixture water is in excess and glycerol is in deficit in the vicinity of a protein molecule.[45–47] The use of the reference state G_{ij}^{SR} (eq 4) led to the following excesses (or deficits): $\Delta n_{12} \cong 33.3$ mol of water which has a volume of 0.602 L, and $\Delta n_{23} \cong -8.2$ mol of glycerol which has a volume of 0.602 L.

It is worth noting that the application of the new method based on eq 13 to the mixture water (1)/protein (2)/cosolvent (or salt) (3) provides the same results as those based on a reference state (eqs 3 and 4). This occurs, because the protein molecule is much larger than the water molecule and typical cosolvents molecules, $V_2^\infty \gg V_i$, $i = 1, 3$. For this reason one can write the following expressions for G_{12} and G_{23} at infinite dilution of the protein[34]

$$G_{12}^{SR} = RT k_T - V_3 c_3 (V_1 - V_3) - V_2^\infty \approx RT k_T - V_2^\infty \quad (19)$$

and

$$G_{23}^{SR} = RT k_T - V_1 c_1 (V_1 - V_3) - V_2^\infty \approx RT k_T - V_2^\infty \quad (20)$$

Hence, one can write for the excesses (or deficits) of water and cosolvent around a protein molecule

$$\Delta n_{12}^{SR} = c_1 (G_{12} - G_{12}^{SR}) \approx c_1 G_{12} + c_1 (RT k_T - V_2^\infty) = \Delta n_{12} \quad (21)$$

and

$$\Delta n_{23}^{SR} = c_3 (G_{23} - G_{23}^{SR}) \approx c_3 G_{23} + c_3 (RT k_T - V_2^\infty) = \Delta n_{23} \quad (22)$$

Therefore, for the system water (1)/protein (2)/cosolvent (or salt) (3), the new method for calculating the excesses (or deficits) around a biomolecule at its infinite dilution (eq 13) leads to the same results as those based on a reference state (eqs 3 and 4). The method based on eq 1 leads to erroneous results because it does not reflect the true excesses (deficits) of water and cosolvent around an infinitely dilute biomolecule.

4. Discussion and Conclusion

One of the most important applications of the KB theory consists of its use to extract some microscopic characteristics of liquid mixtures from measurable macroscopic thermodynamic quantities. The excess (or deficit) number of molecules of various species around a selected (central) molecule is a key quantity in the analysis of the microscopic characteristics of liquid mixtures. Therefore, the correct estimation of the above excesses (or deficits) is important, and the present paper provides a procedure for calculating the excess (or deficit) number of molecules of various species around a selected (central) molecule in binary and multicomponent mixtures.

The conventional method based on eq 1 provides unreasonable results, such as nonzero excesses (or deficits) for single components, all negative excesses for an ideal binary mixture A–B when all three KBIs are negative, and all negative excesses in some concentration ranges for some real binary mixtures.

It is shown in this paper that the number of bulk molecules in the conventional method based on eq 1 was overestimated because the inaccessible volume due to the presence of the central molecule was not taken into account.

The new method eliminates the above inconsistencies: It provides a zero excess for pure components, and excesses (or deficits) which satisfy the volume conservation condition (for both ideal and real mixtures). The derived eq 13 allows one to calculate the excess (or deficit) for an ideal binary mixture (Figure 1) and shows that they become zero only when the molar volumes of the components are equal.

It is clear (see Figures 1–3) that the excesses (or deficits) calculated with the new eq 13 are always very different from those obtained with eq 1. However, eq 13 and eqs 3 and 4 provide comparable results for binary mixtures when the molar volumes of the components are approximately the same. The results obtained using eq 1 and those obtained from eq 13 for methanol/water and 2-propanol/water mixtures are very different. In contrast to the methods based on a reference state, the new method predicts that the alcohols are preferentially hydrated at high alcohol mole fractions. For the 2-propanol/water mixtures there are experimental observations which support this prediction.

In the application of the KB theory to the system water (1)/ protein (2)/cosolvent (or salt) (3), eq 13 provided results, comparable with experiment. Because in such cases the molar volume of the protein is much larger than those of water and cosolvent (salt), eqs 3 and 4 provided almost the same results. In contrast, eq 1 failed to provide plausible values for the excesses (or deficits) of water and cosolvent (salt) in the vicinity of a protein surface.

Acknowledgment. The authors are indebted to Dr. E. Matteoli (CNR, Istituto Processi Chimico-Fisici, Pisa, Italy) for constructive criticism and helpful comments.

Appendix

The purpose of this Appendix is to provide expressions for calculating the KBIs for binary mixtures from measurable thermodynamic quantities such as the derivatives of the chemical potentials with respect to concentrations, the isothermal compressibility, and the partial molar volumes.

The main formulas for the KBIs are[6, 10–11]

$$G_{12} = G_{21} = RT k_T - \frac{V_1 V_2}{VD} \quad (A-1)$$

$$G_{ii} = G_{12} + \frac{1}{x_i}\left(\frac{V_j}{D} - V\right) \quad i \neq j \quad (A-2)$$

where

$$D = \left(\frac{\partial \ln \gamma_i}{\partial x_i}\right)_{P,T} x_i + 1 \quad (A\text{-}3)$$

P is the pressure, T is the temperature in K, k_T is the isothermal compressibility, V_i is the partial molar volume of component i, x_i is the molar fraction of component i, V is the molar volume of the mixture, γ_i is the activity coefficient of component i and R is the universal gas constant.

The KBIs for an ideal mixture G_{ij}^{id} are provided by the expressions:[10-11]

$$G_{12}^{id} = RTk_T^{id} - \frac{V_1^0 V_2^0}{V^{id}} \quad (A\text{-}4)$$

and

$$G_{ii}^{id} = G_{12}^{id} + V_j^0 - V_i^0 \quad i \neq j \quad (A\text{-}5)$$

where V_i^0, k_T^{id}, and V^{id} are the molar volume of the pure component i, the isothermal compressibility, and the molar volume of an ideal solution, respectively.

The KBIs for the Shulgin–Ruckenstein reference state are provided by the expressions[12]

$$G_{12}^{SR} = RTk_T - \frac{V_1 V_2}{V} \quad (A\text{-}6)$$

and

$$G_{ii} = G_{12}^{SR} + V_j - V_i \quad i \neq j \quad (A\text{-}7)$$

At infinite dilution, or for pure components, the following limiting expressions can be obtained for the KBIs[6,10-11]

$$\lim_{x_i \to 0} G_{12} = RTk_{T,j}^0 - V_i^\infty \quad i \neq j \quad (A\text{-}8)$$

$$\lim_{x_i \to 1} G_{12} = RTk_{T,i}^0 - V_j^\infty \quad i \neq j \quad (A\text{-}9)$$

$$\lim_{x_i \to 0} G_{ii} = RTk_{T,j}^0 + V_j^0 - 2V_i^\infty - V_j^0 \left(\frac{\partial \ln \gamma_i}{\partial x_i}\right)_{P,T,x_i=0} \quad i \neq j \quad (A\text{-}10)$$

and

$$\lim_{x_i \to 1} G_{ii} = RTk_{T,i}^0 - V_i^0 \quad (A\text{-}11)$$

where $k_{T,i}^0$ is the isothermal compressibility of the pure component i and V_i^∞ is the partial molar volume of component i at infinite dilution.

References and Notes

(1) Kirkwood, J. G.; Buff, F. P. *J. Chem. Phys.* **1951**, *19*, 774−777.
(2) O'Connell, J. P. *Molecular Physics* **1971**, *20*, 27−33.
(3) Ben-Naim, A. *Water and Aqueous Solutions*; Plenum Press: New York, 1974.
(4) Ben-Naim, A. *J. Chem. Phys.* **1975**, *63*, 2064−2073.
(5) Ben-Naim, A. *J. Chem. Phys.* **1977**, *67*, 4884−4890.
(6) Matteoli, E.; Lepori, L. *J. Chem. Phys.* **1984**, *80*, 2856−2863.
(7) Donkersloot, M. C. A. *J. Solution Chem.* **1979**, *8*, 293−307.
(8) Patil, K. J. *J. Solution Chem.* **1981**, *10*, 315−320.
(9) Zaitsev, A. L.; Kessler, Y. M.; Petrenko, V. E. *Zh. Fiz. Khim.* **1985**, *59*, 2728−2732.
(10) Matteoli, E.; Lepori, L. *J. Chem. Soc., Faraday Trans.* **1995**, *91*, 431−436.
(11) Matteoli, E. *J. Phys. Chem. B* **1997**, *101*, 9800−9810.
(12) Shulgin, I.; Ruckenstein, E. *J. Phys. Chem. B* **1999**, *103*, 2496−2503.
(13) Marcus, Y. *Phys. Chem. Chem. Phys.* **2000**, *2*, 4891−4896.
(14) Marcus, Y. *Monatsh. Chem.* **2001**, *132*, 1387−1411.
(15) Chitra, R.; Smith, P. E. *J. Phys. Chem. B* **2002**, *106*, 1491−1500.
(16) Vergara, A.; Paduano, L.; Sartorio, R. *Phys. Chem. Chem. Phys.* **2002**, *4*, 4716−4723.
(17) Marcus, Y. *Phys. Chem. Chem. Phys.* **2002**, *4*, 4462−4471.
(18) Ben-Naim, A. *Cell Biophys.* **1988**, *12*, 255−269.
(19) Ben-Naim, A. *J. Chem. Phys.* **1989**, *93*, 3809−3813.
(20) Ben-Naim, A. *Pure & Appl. Chem.* **1990**, *62*, 25−34.
(21) Pfund, D. M.; Lee, L. L.; Cochran, H. D. *Fluid Phase Equilib.* **1988**, *39*, 161−192.
(22) Rubio, R. G.; Prolongo, M. G.; Diaz Pena, M.; Renuncio, J. A. R. *J. Phys. Chem.* **1987**, *91*, 1177−1184.
(23) Debenedetti, P. G. *Chem. Eng. Sci.* **1987**, *42*, 2203−2212.
(24) Chialvo, A. *J. Phys. Chem.* **1993**, *97*, 2740−2744.
(25) Jonah, D. A.; Cochran, H. D. *Fluid Phase Equilib.* **1994**, *92*, 107−137.
(26) Ruckenstein, E.; Shulgin, I. *Fluid Phase Equilib.* **2001**, *180*, 345−359.
(27) Ruckenstein, E.; Shulgin, I. *Fluid Phase Equilib.* **2002**, *200*, 53−67.
(28) Shimizu, S. *Proc. Natl. Acad. Sci. U.S.A.* **2004**, *101*, 1195−1199.
(29) Shimizu, S. *J. Chem. Phys.* **2004**, *120*, 4989−4990.
(30) Shimizu, S.; Smith, D. J. *Chem. Phys.* **2004**, *121*, 1148−1154.
(31) Shimizu, S.; Boon, C. L. *J. Chem. Phys.* **2004**, *121*, 9147−9155.
(32) Smith, P. E. *J. Phys. Chem. B* **2004**, *108*, 16271−16278.
(33) Smith, P. E. *J. Phys. Chem. B* **2004**, *108*, 18716−18724.
(34) Shulgin, I.; Ruckenstein, E. *J. Chem. Phys.* **2005**, *123*, 054909.
(35) Shulgin, I.; Ruckenstein, E. *Biophys. Chem.* **2005**, *118*, 128−134.
(36) Schurr, J. M.; Rangel, D. P.; Aragon, S. R. *Biophys. J.* **2005**, *89*, 2258−2276.
(37) Rosgen, J.; Pettitt, B. M.; Bolen, D. W. *Biophys. J.* **2005**, *89*, 2988−2997.
(38) Shulgin, I. L.; Ruckenstein, E. *Biophys. J.* **2006**, *90*, 704−707.
(39) Ben-Naim, A. *Solvation Thermodynamics*; Plenum Press: New York, 1987.
(40) Ben-Naim, A. *Statistical Thermodynamics for Chemists and Biochemists*; Plenum Press: New York, 1992.
(41) Marcus, Y. *J. Chem. Soc., Faraday Trans.* **1995**, *91*, 427−430.
(42) Ruckenstein, E.; Shulgin, I. *Fluid Phase Equilib.* **2001**, *180*, 281−297.
(43) Hill, T. L. *Statistical Mechanics: Principles and Selected Applications*; McGraw-Hill: New York, 1956.
(44) Hayashi, H.; Nishikawa, K.; Iijima, T. *J. Phys. Chem.* **1990**, *94*, 8334−8338.
(45) Gekko, K.; Timasheff, S. N. *Biochemistry* **1981**, *20*, 4667−4676.
(46) Timasheff, S. N. *Adv. Protein Chem.* **1998**, *51*, 355−432.
(47) Courtenay, E. S.; Capp, M. W.; Anderson, C. F.; Record, J. M. T. *Biochemistry* **2000**, *39*, 4455−4471.

Excess around a central molecule with application to binary mixtures

Ivan L. Shulgin and Eli Ruckenstein*

Received 23rd August 2007, Accepted 26th November 2007
First published as an Advance Article on the web 14th December 2007
DOI: 10.1039/b713026k

It was shown by us (*J. Phys. Chem. B*, 2006, **110**, 12707) that the excess (deficit) of any species i around a central molecule j in a binary mixture is not provided by $c_i G_{ij}$ (where c_i is the molar concentration of species i in the mixture and G_{ij} are the Kirkwood–Buff integrals) as usually considered and that an additional term, involving a volume V^j which is inaccessible to molecules of species i because of the presence of the central molecule j, must be included. In this paper, the new expression is applied to various binary mixtures and used to establish a simple criterion for preferential solvation in a binary system. First, it is applied to binary Lennard-Jones fluids. The conventional expression for the excess (deficit) in binary mixtures, $c_i G_{ij}$, provides always deficits around any central molecule in such fluids. In contrast, the new expression provides excess for one species and deficit for the other one. In addition, two kinds of binary mixtures involving weak (argon/krypton) and strong (alcohols/water) intermolecular interactions were considered. Again, the conventional expression for the excess (deficit) in a binary mixture, $c_i G_{ij}$, provides always deficits for any central molecule in the argon/krypton mixture, whereas the new expression provides excess for argon (a somewhat smaller molecule) and deficit for krypton. Three alcohol/water binary mixtures (1-propanol/water, *tert*-butanol/water and methanol/water) with strong intermolecular interactions were considered and compared with the available experimental information regarding the molecular clustering in solutions. We found (for 1-propanol/water and *tert*-butanol/water) a large excess of alcohols around a central alcohol molecule and a large excess of water around a central water molecule. For both mixtures the maximum of the calculated excess with respect to the concentration corresponds to the maximum in the cluster size found experimentally, and the range of alcohol concentrations in which the calculated excess becomes very small corresponds to the composition range in which no clusters could be identified experimentally.

1. Introduction

The composition in the vicinity of a molecule is called local composition (LC). Because of differences in intermolecular interactions between various species and their sizes, the LC differs from the overall (bulk) composition. Usually the thickness of LC region in ordinary liquid mixtures does not exceed, far from critical conditions, a few molecular diameters. The knowledge of the liquid mixture organization at atomic level together with the interaction between various species is critical in the understanding of the properties of the mixtures and can be helpful in correlating and even predicting thermodynamic properties. The local composition concept is now widely used in modeling phase equilibrium,[1,2] in the theory of supercritical mixtures,[3,4] in the solution chemistry of large molecules such as polymers,[5,6] proteins,[7,8] *etc*. In the theory of aqueous solutions one routinely examines the preferential hydration (when the local composition of water is larger than the bulk composition of water).

It is difficult to determine experimentally the LC and the thickness of the corresponding layer. When the volume where the LC is appreciably different from the bulk composition is large enough it can be determined experimentally by small-angle X-ray scattering (SAXS),[9–15] small-angle neutron scattering (SANS),[16] light scattering (LS)[17–20] and other methods.[21–23] In such cases there are large microheterogeneities in the solution. However, until now it was not possible to experimentally "measure" the thickness of the LC region for most "normal" liquid mixtures.

Therefore, it is important to have a theoretical tool which allows one to examine (or even predict) the thickness of the LC region and the value of the LC on the basis of more easily available experimental information regarding liquid mixtures. A powerful and most promising method for this purpose is the fluctuation theory of Kirkwood and Buff (KB).[24] The KB theory of solutions allows one to extract information about the excess (or deficit) number of molecules, of the same or different kind, around a given molecule, from macroscopic thermodynamic properties, such as the composition dependence of the activity coefficients, molar volume, partial molar volumes and isothermal compressibilities. This theory was developed for both binary and multicomponent solutions and is applicable to any conditions including the critical and supercritical mixtures.

Department of Chemical & Biological Engineering, State University of New York at Buffalo, Amherst, NY 14260, USA. E-mail: feaeliru@acsu.buffalo.edu, ishulgin@eng.buffalo.edu; Fax: +1 (716) 645-3822; Tel: +1 (716) 645-2911/ext. 2214

The excess (or deficit) number of molecules is a result of the differences in the intermolecular interactions between molecules of the same and different kinds and can be calculated using the so-called Kirkwood–Buff integrals (KBIs):[24]

$$G_{ij} = \int_0^\infty (g_{ij} - 1) 4\pi r^2 \mathrm{d}r \quad (1)$$

where g_{ij} is the radial distribution function between species i and j, and r is the distance between the centers of molecules i and j.

The first method for calculating the excess (or deficit) number of molecules around a central molecule was suggested by Ben-Naim.[25–30] Let us consider a binary mixture 1–2. The excess (or deficit) number of molecules i ($i = 1, 2$) around a central molecule j ($j = 1, 2$) was calculated by Ben-Naim as:

$$\Delta n_{ij}^{BN} = c_i G_{ij} \quad (2)$$

where c_i is the molar concentration of species i in the mixture and G_{ij} are the Kirkwood–Buff integrals (analytical expressions for the KBIs of a binary mixture are provided in Appendix 1).

Expression (2) was used by Ben-Naim in numerous publications[25–30] (even in his most recent one[31]) and by many authors. However, as we recently demonstrated[32] this expression is incomplete and usually provides physically meaningless results for the excesses (or deficits) number of molecules i ($i = 1, 2$) around a central molecule j ($j = 1, 2$).

To explain why we consider eqn (2) incomplete, let us examine its derivation by Ben-Naim.[25,26] The average number of molecules i (n_{ij}) in a sphere of radius R (we consider only spherical volumes; however, similar consideration can be employed for nonspherical volumes and leads to the same conclusions) around a central molecules j can be calculated using the expression[30,33]

$$n_{ij} = c_i \int_0^R g_{ij} 4\pi r^2 \mathrm{d}r \quad i,j = 1,2 \quad (3)$$

which can be written in the form

$$n_{ij} = c_i \int_0^R (g_{ij} - 1) 4\pi r^2 \mathrm{d}r + c_i \int_0^R 4\pi r^2 \mathrm{d}r \quad (4)$$

As soon as R becomes large enough for g_{ij} to become unity, eqn (4) can be rewritten as[27,28,30]

$$n_{ij} = c_i \int_0^\infty (g_{ij} - 1) 4\pi r^2 \mathrm{d}r + c_i \int_0^R 4\pi r^2 \mathrm{d}r$$
$$= c_i G_{ij} + \frac{c_i 4\pi R^3}{3} \quad (5)$$

$\frac{c_i 4\pi R^3}{3}$ was identified[25–30] as the total number of molecules of species i around the center of a central molecule j for a concentration equal to the bulk concentration c_i. Therefore, the difference between n_{ij} and $\frac{c_i 4\pi R^3}{3}$ was considered as the average excess (or deficit) number of molecules i ($i = 1, 2$) around a central molecule j ($j = 1, 2$) and eqn (2) was thus obtained.[25–30]

However, the term $\frac{c_i 4\pi R^3}{3}$ overestimates the total number of molecules i at bulk concentration surrounding the central molecule j because it includes molecules i assumed to be located in a volume inaccessible to them because of the presence of the central molecule j. Therefore, when the average excess (or deficit) number of molecules i ($i = 1, 2$) around a central molecule j is calculated, those i molecules should be subtracted from $\frac{c_i 4\pi R^3}{3}$, and the second integral in the right hand side of eqn (4) should be subdivided into two parts

$$n_{ij} = c_i \int_0^R (g_{ij} - 1) 4\pi r^2 \mathrm{d}r + c_i \int_{R_j}^R 4\pi r^2 \mathrm{d}r + c_i \int_0^{R_j} 4\pi r^2 \mathrm{d}r \quad (6)$$

where R_j is the radius of the volume around the center of molecule j which is inaccessible to molecules i due to the presence of the central molecule j. For R sufficiently large for g_{ij} to become unity, the first integral in the right hand side becomes the Kirkwood–Buff integral; the second term provides the number of molecules of species i in a bulk liquid located between the radii R_j and R and the third term provides the number of molecules of species i in a bulk liquid located between zero and the radius R_j. The true excess is provided by the difference between n_{ij} and the number of molecules of species i in a bulk liquid between the radii R_j and R. Hence

$$\Delta n_{ij} = n_{ij} - c_i \int_{R_j}^R 4\pi r^2 \mathrm{d}r \quad (7)$$

The third integral $\int_0^{R_j} 4\pi r^2 \mathrm{d}r$ in eqn (6) represents a bulk volume V^j which is not accessible to the molecules of species i.

Consequently, the expression which provides the average excess (or deficit) number of molecules i ($i = 1, 2$) around a central molecule j ($j = 1, 2$) has the form

$$\Delta n_{ij} = c_i G_{ij} + c_i V^j \quad (8)$$

Comparing eqn (8) with eqn (2) one can observe that the former contains an additional term $c_i V^j$ which, as it will be shown later in the paper, is not in most cases negligible. It is now understandable why eqn (2) provides frequently unphysical results.[32,34–36] For example,[32,36] there are numerous systems for which all the KBIs (G_{11}, G_{12} and G_{22}) are negative at least in certain ranges of composition, and in such cases all Δn_{ij}^{BN} calculated with eqn (2) (Δn_{11}^{BN} and Δn_{21}^{BN} around the central molecule 1 and Δn_{12}^{BN} and Δn_{22}^{BN} around the central molecule 2) are negative, and this is not plausible because then the density around any molecule in the liquid will become lower than that in the bulk.

Another example is the ideal mixture A–B with equal molar volumes of components A and B.[32] Eqn (2) leads in this case to negative excesses around molecules A and B (Fig. 1), even though they are expected to be zero because of the complete randomness of the equal size molecules.

Unfortunately eqn (2), which has been used for decades, provides erroneous results, particularly when G_{ij} and V^j are comparable in absolute values and have different signs or $|V^j|$

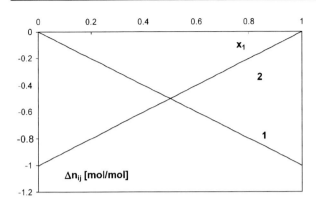

Fig. 1 Application of eqn (2) to an ideal binary mixture A–B with equal volumes of the components. Excess (or deficit) number of molecules i (i = A, B) around a central molecule A. 1 is Δn_{AA}^{BN} and 2 is Δn_{BA}^{BN} (KBIs were provided by eqns (A1-4) and (A1-5) in Appendix 1, where k_T^{id} was taken equal to zero). x_1 is the mole fraction of component A.

is large compared to $|G_{ij}|$. In such cases (which are typical, for example, of polymer or protein solutions in low molecular weights solvents), eqn (2) provides large deficits of solvent(s) around a polymer or protein molecule,[32] contradicting both common sense and the experimental information.

2. A simple criterion for preferential solvation in a binary system

The concept of preferential solvation[28–30] is closely connected with the excesses (deficits) around a central molecule. Eqn (8) can provide simple criteria for preferential solvation in a binary system.

Let us consider the binary mixture α–β. One can define local compositions in the vicinities of central molecules α (in mole fraction scale $x_{\alpha\alpha}$ and $x_{\beta\alpha}$) and β ($x_{\alpha\beta}$ and $x_{\beta\beta}$). The local compositions can be compared with the bulk compositions (x_α and x_β) and on this basis one can conclude if there is preferential solvation.

Component α is preferentially solvated by itself if

$$G_{\alpha\alpha} + V^\alpha > 0 \quad (9A)$$

and component α is preferentially solvated by component β if

$$G_{\alpha\beta} + V^\alpha > 0 \quad (9B)$$

Component β is preferentially solvated by itself if

$$G_{\beta\beta} + V^\beta > 0 \quad (10A)$$

and component β is preferentially solvated by component α if

$$G_{\alpha\beta} + V^\beta > 0 \quad (10B)$$

Inequalities (9) and (10) provide simple criteria for preferential solvation in a binary system. They reveal that, contrary to the prevailing opinion, the KBIs ($G_{\alpha\beta}$) alone can not provide information about preferential solvation.

3. Evaluation of volume V^j

To calculate Δn_{ij} with expression (8), the volume V^j inaccessible to molecules of species i because of the presence of the central molecule j must be known. Minimal and maximal estimates have been suggested for the volume of a single molecule; the van der Waals volume V_j^{vdW} as the minimal estimate and the volume of the pure component V_j^0 as the maximal estimate.[37,38]

In our previous paper,[32] eqn (8) was first applied to a pure component j ($i = j$, $c_j = c_j^0$). In this case, because of complete randomness, one expects Δn_{ij} to be zero. By combining eqn (8) with the equation for the G_{ij} of a pure substance (eqn (A1-9) in Appendix 1) we obtained the following expression for the volume V^j

$$(V^j)_{x_j=1} = V_j^0 - R_0 T k_{T,j}^0 \quad (11)$$

where T is the temperature, x_j is the molar fraction of component j, $k_{T,j}^0$ is the isothermal compressibility of the pure component j and R_0 is the universal gas constant. Hence, far from the critical point, where $V_{T,j}^0 \gg R_0 T k_{T,j}^0$, the inaccessible volume V^j is equal to the molar volume of the pure component j. This result was extrapolated to mixtures by considering that

$$V^j = V_j - R_0 T k_T \quad (12)$$

where V_j is the partial molar volume of component j and k_T is the isothermal compressibility of the mixture.

By combining eqns (8) and (12), one obtains for the average excess (or deficit) number of molecules i (i = 1, 2) around a central molecule j (j = 1, 2) the expression

$$\Delta n_{ij} = c_i G_{ij} + c_i V^j = c_i(G_{ij} + V_j - R_0 T k_T) \quad (13)$$

Similar relations can be written for the excess (or deficit) number of molecules i (i = 1, 2, ...) around a central molecule j (j = 1, 2, ...) when the volume of a pure component V_j^0 or the van der Waals volume V_j^{vdW} is used instead of V_j in eqn (13). In these cases one can write the following relations

$$\Delta n_{ij} = c_i(G_{ij} + V_j^0 - R_0 T k_T) \quad (14)$$

and

$$\Delta n_{ij} = c_i(G_{ij} + V_j^{vdW} - R_0 T k_T) \quad (15)$$

Only in eqn (13), V^j depends, through both V_j and k_T, on the mixture composition. In eqns (14) and (15) only k_T is composition dependent.

Let us apply eqns (13), (14) and (15) to a real system and compare the results. Fig. 2 provides such a comparison for the binary system isopropanol (1)–water (2) (The Kirkwood–Buff integrals were taken from literature[36] and the van der Waals volumes were calculated as suggested in ref. 39–41). Fig. 2 shows that the excesses (deficits) calculated using all three equations (eqns (13), (14) and (15)) provide quite comparable results for both central isopropanol and water molecules. The differences between the excesses (deficits) calculated with eqns (13) and (14) are small.

However, it is worth mentioning that:

(1) While for pure components, eqns (13) and (14) provide zero excesses (deficits), eqn (15) provides nonzero excesses (deficits),

(2) At infinite dilution of component j, the excesses (deficits) Δn_{ij} ($i \neq j$) are zero when eqn (13) is used and non-zero when eqns (14) and (15) are used.

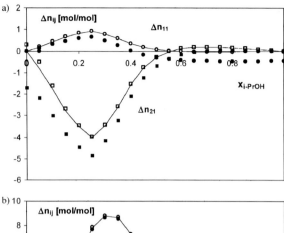

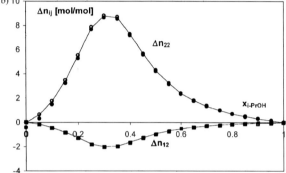

Fig. 2 Excesses (or deficits) for isopropanol (1)/water (2) mixtures ($T = 298.15$ K). (a) Excesses (or deficits) around a central isopropanol molecule (eqn (13) – solid line, eqn (14) – open circles and open squares and eqn (15) – dark circles and dark squares), (b) excesses (or deficits) around a central water molecule (eqn (13) – solid line, eqn (14) – open circles and open squares and eqn (15) – dark circles and dark squares which coincide with the open squares).

(3) The differences between the excesses (deficits) calculated with eqns (13) and (14) are small. Eqns (13) and (14) should be preferred to eqn (15) because they provide zero excesses (deficits) for pure components.

Additional physical arguments can be brought in favor of eqn (13). Namely, any central molecule in the mixture α–β can be preferentially solvated by either molecule α or β. Therefore if one species is in excess the other one must be in deficit. In terms of eqns (9) and (10), one can write that if, for instance, $G_{\alpha\alpha} + V^\alpha > 0$ (or $G_{\beta\beta} + V^\beta > 0$) holds, then the inequality $G_{\alpha\beta} + V^\alpha \leq 0$ (or $G_{\alpha\beta} + V^\beta \leq 0$) should also be fulfilled. By using the identity[32]

$$V_i c_i (G_{ij} + V_j - R_0 T k_T) + V_j c_j (G_{jj} + V_j - R_0 T k_T) = 0;$$

$$i, j = 1, 2 \text{ and } i \neq j$$

one can prove that the excesses (deficits) calculated with eqn (13) satisfy the above conditions. However, the excesses (deficits) calculated with eqns (14) or (15) do not satisfy them in the entire range of concentrations.

In what follows eqn (13) will be employed. However, all three eqns (13)–(15) provide excesses (deficits) which significantly differ from those obtained with eqn (2), the differences being larger when the volume V^j is large.[32] We already used eqn (13) to calculate the excesses (deficits) for several binary mixtures.[32]

In the present paper, the new approach will be used to calculate the excesses (deficits) in binary Lennard-Jones (LJ) fluids and for several low-molecular-weight binary solutions. In one of these mixtures, there are weak interactions between the components (argon (1)/krypton (2)). The components of this mixture are apolar and have comparable molar volumes (33 and 34 cm^3 mol^{-1}, respectively).[35] Further, three mixtures (1-propanol (1)/water (2), *tert*-butanol (1)/water (2) and methanol (1)/water (2)), which possess strong intermolecular interactions, including H-bonding, will be considered. The mixtures 1-propanol/water and *tert*-butanol have been selected because experimental information regarding their microheterogeneities is available for comparison. The results for methanol/water could also be compared with some experimental data.

4. Various binary mixtures

4.1 Lennard-Jones fluids

The LJ fluids[42] are fluids for which the interactions between the molecules can be described by LJ potentials

$$U_{ij}(r) = 4\varepsilon_{ij}\left[\left(\frac{\sigma_{ij}}{r}\right)^{12} - \left(\frac{\sigma_{ij}}{r}\right)^{6}\right] \qquad (16)$$

where ε_{ij} and σ_{ij} are the energy and size parameters of the molecular species involved.[42]

The KBIs for the LJ fluids can be calculated with eqn (1) using for the radial distribution function g_{ij} the Percus–Yevick equation.[43] The KBIs obtained in this manner by Kojima, Kato and Nomura[43] have been employed.

4.1.1 The same energy parameters but different size parameters.
First, the excesses (deficits) were calculated for LJ fluids with different size parameters ($\sigma_{11} = 1$ Å and $\sigma_{22} = 1.35$ Å), but with the same energy parameters ε_{ij} ($\varepsilon_{11} = \varepsilon_{22} = kT/1.2$, where k is the Boltzmann constant). The results of the calculations are presented in Fig. 3 and details regarding the calculations are presented in Appendix 2. The results of Fig. 3 demonstrate that the LJ molecules with smaller size (component 1) are in excess and those with a larger size (component 2) are in deficit in the vicinity of both central molecules 1 and 2. Therefore, for the same interaction energy parameters, their sizes play a major role in their distribution near central molecules, the smaller molecules being preferred to the larger ones. Eqn (2) provides physically unreasonable values for the excesses (deficits), predicting that all Δn_{ij}^{BN} are negative at all compositions.

4.1.2 Different energy parameters but the same size parameters.
Second, the excesses (deficits) were calculated for molecules with equal size parameters ($\sigma_{11} = \sigma_{22} = 1$ Å) but different energy parameters ε_{ij} ($\varepsilon_{11} = kT/1.2$, $\varepsilon_{22} = kT/1.8$ and $\varepsilon_{12} = (\varepsilon_{11} \cdot \varepsilon_{22})^{1/2}$). The results are presented in Fig. 4. In this case, the molecule with the larger energy parameter (component 1) is in excess and the molecule with the smaller energy parameter (component 2) in deficit around both central 1 and 2 molecules.

By comparing Fig. 3 and 4 one can notice that the absolute values of the excesses (deficits) in Fig. 3 are much larger than

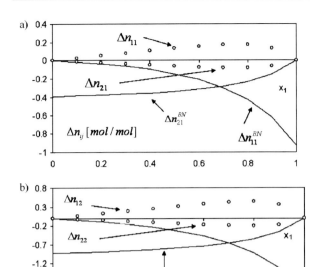

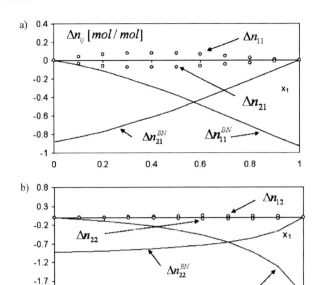

Fig. 3 Excesses (or deficits) around central molecules 1 (Fig. 3a) and 2 (Fig. 3b) for LJ fluids with molecules of different sizes ($\sigma_{11} = 1$ Å and $\sigma_{22} = 1.35$ Å) but with the same energy parameter ε_{ij} ($\varepsilon_{11} = \varepsilon_{22} = kT/1.2$). ○ – Excesses (or deficits) calculated with eqn (13), solid lines – excesses (or deficits) calculated using eqn (2).

Fig. 4 Excesses (or deficits) around central molecules 1 (Fig. 4a) and 2 (Fig. 4b) for the LJ fluid with molecules of equal sizes ($\sigma_{11} = \sigma_{22} = 1$ Å) but with different energy parameters ε_{ij} ($\varepsilon_{11} = kT/1.2$ and $\varepsilon_{22} = kT/1.8$). ○ – Excesses (or deficits) calculated with eqn (13), solid lines – excesses (or deficits) calculated with eqn (2).

in Fig. 4. This occurs because for given parameters the effect of the size of the molecule on the excesses (deficits) is greater than that of the energy parameter.

Again, eqn (2) provides unreasonable values for the excesses (deficits), all Δn_{ij}^{BN} being negative at all compositions.

One should note that there is an important difference between the considered LJ fluid and liquid mixtures under ambient conditions. For liquid mixtures under ambient conditions, the compressibility term in eqns (14) and (15) can usually be neglected because it is small compared to the other terms. However, for LJ fluids, this term becomes comparable to the other terms on the right hand side of eqns (14) and (15) and its omission leads to errors in calculating the excesses (or deficits).

4.2 Argon (1)/krypton (2) mixture

The KBIs of the argon (1)/krypton (2) mixture at 115.77 K have been calculated by Matteoli[35] on the basis of data[44] on the excess Gibbs energy and density. We repeated these calculations using the same literature data[44] and the obtained KBIs have been used to calculate the excesses (deficits) for the argon (1)/krypton (2) mixture at 115.77 K using eqn (13). The results of these calculations are compared with those obtained with eqn (2) in Fig. 5. One can see that argon (a somewhat smaller molecule than krypton) is in excess (the excess is small) over the entire composition range around both central molecules. It is noteworthy that the same conclusion was reached for ideal mixtures[32] and for LJ fluids (see section 4.1). The values provided by Δn_{ij}^{BN} are always negative for all i–j pairs in the entire composition range, results which are implausible.

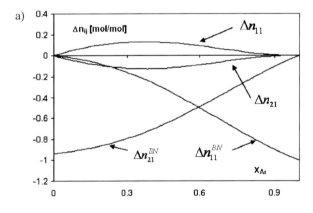

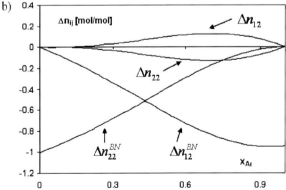

Fig. 5 Excesses (or deficits) for argon (1)/krypton (2) mixtures ($T = 115.77$ K). (a) Excesses (or deficits) around a central argon molecule calculated with eqns (2) and (13), (b) excesses (or deficits) around a central krypton molecule calculated with eqns (2) and (13).

4.3 1-Propanol (1)/water (2) mixture

The KBIs for the 1-propanol (1)/water (2) mixture are available in the literature,[36,45–48] and there is agreement between the KBIs obtained in various calculations.[36,45–48] In the present paper, the excesses (deficits) for the 1-propanol (1)/water (2) mixture have been calculated with eqn (13) by using the KBIs already calculated by us.[36] The partial molar volumes for the 1-propanol (1)/water (2) mixture have been calculated from density data[49] and the isothermal compressibilities have been evaluated using the expression:

$$k_T = k_{T,1}^0 \varphi_1 + k_{T,2}^0 \varphi_2 \qquad (17)$$

where $\varphi_i = x_i V_i^0 / (x_1 V_1^0 + x_2 V_2^0)$ is the volume fraction of component i in solution and $k_{T,i}^0$ is the isothermal compressibility of the pure component i. The isothermal compressibilities of the pure compounds were taken from ref. 50. The calculated excesses (deficits) for the 1-propanol (1)/water (2) mixture are plotted in Fig. 6.

One can see from Fig. 6 that 1-propanol is in excess around a central 1-propanol molecule whereas water is in excess around a central water molecule. Hence both components of this mixture have self-aggregation tendencies and two types of clusters are present in this mixture: one of them is enriched in alcohol and the other one in water. All Δn_{ij} have extremes around $x_1 \approx 0.25$. At large alcohol concentrations ($x_1 > 0.4$–0.5), the excesses (deficits) become small, indicating the absence of large microheterogeneities.

There are plenty of experimental results provided by SAXS,[13–15] SANS[16] and LS[17] regarding the microheterogeneities (or clustering) present in the 1-propanol (1)/water (2) mixture. They reveal that: (i) there are large microheterogeneities in the dilute aqueous solutions of 1-propanol with a maximum in size at about $x_1 \approx 0.15$–0.25 (where x_1 is the mole fraction of 1-propanol), (ii) there are no microheterogeneities (or they are too small to be detected) at high 1-propanol concentrations ($x_1 > 0.4$–0.6). The Landau–Placzek ratio (which is a measure of the microscopic aggregation) determined from Rayleigh–Brillouin scattering[51] provides a very similar picture of clustering in this mixture. One can see that our results (Fig. 6) are in agreement with the experimental data regarding the microheterogeneities in the 1-propanol (1)/water (2) mixture.

The calculations of the excesses (deficits) for the 1-propanol/water mixture carried out with eqn (2) provide results comparable to those obtained with eqn (13). This occurs because the KBIs for the 1-propanol/water mixture are very large in a wide range of compositions and the contribution of V^j to the excesses (deficits) for this mixture is small. Such cases are, however, rare.

4.4 *tert*-Butanol (1)/water (2) mixture

The KBIs for aqueous *tert*-butanol (*t*-butanol) mixtures are available in literature[36,45–48] and there is agreement between the KBIs calculated by various authors.[36,45,47,48] In the present paper, the excesses (deficits) for the *t*-butanol/water mixture have been calculated with eqn (13) using the KBIs already calculated by us.[36] The partial molar volumes for this mixture were calculated from density data.[52] The isothermal compressibilities were evaluated with eqn (17), and the isothermal compressibilities of the pure compounds were taken from ref. 50 and 53. The calculated excesses (deficits) are plotted in Fig. 7, which shows that the excesses (deficits) for this mixture are comparable to those for the 1-propanol (1)/water (2) mixture. Indeed, *t*-butanol is in excess around a central *t*-butanol molecule whereas water is in excess around a central water molecule. Hence both components exhibit self-aggregation tendencies and two types of clusters are present in the mixture: the first is enriched in alcohols and the second in water. All Δn_{ij} have extrema around $x_1 \approx 0.15$. At large alcohol concentrations $x_1 > 0.4$–0.6) the excesses (deficits) become small indicating the absence of large microheterogeneities.

The microheterogeneities in the *t*-butanol/water mixture were investigated experimentally with SAXS[10,12] and LS.[18,20] Microheterogeneities were found in the following ranges of *t*-butanol mole fractions: (i) by SAXS[10,12] for $x_1 < 0.35$–0.4, with a maximum size for $x_1 \approx 0.15$, (ii) by LS[18,20] for $x_1 \leq 0.3$. The Landau–Placzek ratio [51] revealed that microheterogeneities are present in this mixture at $x_1 < 0.4$–0.5 with a maximum at $x_1 \approx 0.15$.

One can see that our predictions of microheterogeneities in the *t*-butanol/water mixture (Fig. 7) are in agreement with the experimental data available for this mixture. The theory

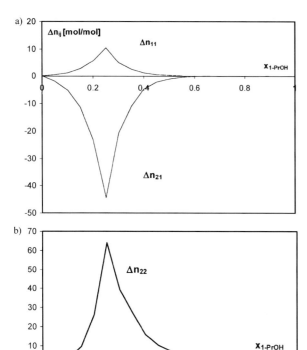

Fig. 6 Excesses (or deficits) for 1-propanol (1)/water (2) mixtures ($T = 303.15$ K). (a) Excesses (or deficits) around a central 1-propanol molecule calculated with eqn (13), (b) excesses (or deficits) around a central water molecule calculated with eqn (13).

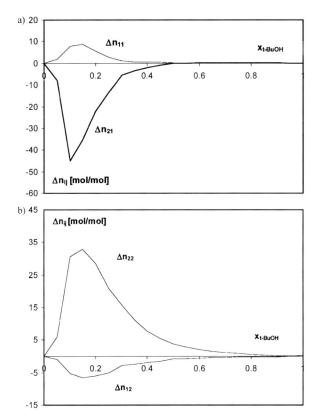

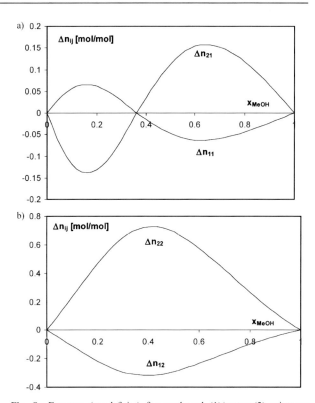

Fig. 7 Excesses (or deficits) for t-butanol (1)/water (2) mixtures ($T = 323.15$ K). (a) Excesses (or deficits) around a central t-butanol molecule calculated with eqn (13), (b) excesses (or deficits) around a central water molecule calculated using eqn (13).

Fig. 8 Excesses (or deficits) for methanol (1)/water (2) mixtures ($T = 323.15$ K).[32] (a) Excesses (or deficits) around a central methanol molecule calculated with eqn (13), (b) excesses (or deficits) around a central water molecule calculated with eqn (13).

indicates that clusters enriched in alcohols and enriched in water are present in the t-butanol/water mixture.

4.5 Methanol (1)/water (2) mixture

The excesses (deficits) in the methanol (1)/water (2) mixture were calculated in our previous paper.[32] The results (Fig. 8) are not typical of alcohol/water mixtures. One can see from Fig. 8a that methanol is in small excess around a central methanol molecule for $x_1 \leq 0.37$ and water is in small excess around a central methanol molecule for $x_1 > 0.37$. However, Fig. 8b shows that water is in excess around a central water molecule at all concentrations. According to previous calculations of the excesses (deficits) in the alcohol/water mixtures,[36] the alcohol is in excess around a central alcohol molecule at all compositions and the magnitude of the excess increases from methanol to higher alcohols. It is interesting to note that recent experimental results[54–56] (combined mass spectra and X-ray diffraction) indicated that there is a structural change in the methanol/water mixture at $x_1 \approx 0.3$. This change can be interpreted as follows: in the composition range $x_1 \leq 0.3$, the structure of the methanol/water mixture consists of a network of water molecules but at higher methanol concentrations ($x_1 > 0.3$) the water network is gradually replaced by methanol chains and finally by a network of methanol molecules.

5. Discussion and conclusion

A new expression (8) for the excess (or deficit) around any central molecule in a liquid mixture, which was derived by us in a recent paper,[32] is employed to examine various binary mixtures. Eqn (8) involves a KBI as well as a volume V^j due to the presence of the central molecule, which is inaccessible to the considered component of the mixture. In our previous paper, it was suggested to equate V^j with $(V_j - R_0 T k_T)$, where V_j is the partial molar volume of component j. In this paper, additional options are suggested for V^j, namely the molar volume of the pure components j (V_j^0) (eqn (14)), or the van der Waals volumes V_j^{vdW} (eqn (15)). The excesses (or deficits) have been calculated for the isopropanol–water mixture using all three eqns (13)–(15). Fig. 2 shows that eqns (13)–(15) lead to comparable results for the excesses (or deficits) for the isopropanol–water mixture.

Further, eqn (13) was used to calculate the excesses (deficits) for various binary mixtures. First, the new approach has been applied to binary LJ fluids. The excesses (deficits) for two binary LJ fluids were examined: (1) in the first, the same energy parameters but different size parameters have been employed, (2) in the second, different energy parameters but the same size parameters have been used (see Fig. 3 and 4). In both cases, the KBIs (G_{11}, G_{12} and G_{22}) were negative at all compositions[43] and provided negative values for all Δn_{ij}^{BN} calculated with eqn (2) (Δn_{11}^{BN} and Δn_{21}^{BN} around the central molecule 1 and Δn_{12}^{BN} and Δn_{22}^{BN} around the central molecule 2). In contrast, eqn (13) provided more physically meaningful results, namely: in the first case (Fig. 3), an excess for molecules of smaller size and a deficit for molecules of larger size in the vicinity of both central molecules 1 and 2. Similar results have been obtained previously for ideal binary mixtures as well.[32] In the second case (Fig. 4), the molecules with larger energy parameter (component 1) were in excess and those with smaller energy parameter (component 2) in deficit around both central 1 and 2 molecules. For binary LJ fluids, the excesses (deficits) are relatively small compared to those in mixtures with strong intermolecular interactions (see below).

Second, a mixture with very weak interactions between components (argon/krypton) was considered. The components of these mixture are apolar and possess comparable molar volumes (33 and 34 cm^3 mol^{-1}, respectively[35]). One can see from Fig. 5 that argon (a somewhat smaller molecule than krypton) is present in excess (which is small) over the entire composition range around both argon and krypton molecules. The effect of the intermolecular interactions on the local compositions in this mixture is smaller than that of the difference (even small) in the sizes of the components. Again, all KBIs (G_{11}, G_{12} and G_{22}) are negative for all compositions and the values of all Δn_{ij}^{BN} are always negative for all i–j pairs in the entire composition range.

Finally, the excesses (or deficits) were examined for three mixtures (1-propanol/water, *tert*-butanol/water and methanol/water) with strong intermolecular interaction including H-bonding. Fig. 6 and 7 show that there is high self-aggregation for both alcohol and water in the 1-propanol/water and *tert*-butanol/water mixtures. These self-aggregations occur for mole fractions of alcohol less than 0.4–0.6, with a maximum at 0.2–0.25 for 1-propanol/water and 0.15–0.2 for *tert*-butanol/water. These results are in agreement with the experimental findings regarding the microheterogeneities present in these mixtures, identified by SAXS, SANS and LS techniques. In addition, the excesses (or deficits) in methanol/water mixture were compared with experimental results regarding some structural features of this mixture.

Appendix 1

The purpose of this Appendix is to provide expressions for the KBIs of binary mixtures in terms of measurable thermodynamic quantities such as the derivatives of the chemical potentials with respect to concentrations, the isothermal compressibility and the partial molar volumes.

The main formulas for the KBIs are[34,35,45]

$$G_{12} = G_{21} = R_0 T k_T - \frac{V_1 V_2}{VD} \quad (A1\text{-}1)$$

$$G_{ii} = G_{12} + \frac{1}{x_i}\left(\frac{V_j}{D} - V\right) i \neq j \quad (A1\text{-}2)$$

where

$$D = \left(\frac{\partial \ln \gamma_i}{\partial x_i}\right)_{P,T} x_i + 1 \quad (A1\text{-}3)$$

P is the pressure, T is the temperature in K, k_T is the isothermal compressibility, V_i is the partial molar volume of component i, x_i is the molar fraction of component i, V is the molar volume of the mixture, γ_i is the activity coefficient of component i and R_0 is the universal gas constant.

The KBIs for an ideal mixture G_{ij}^{id} are provided by the expressions:[34,35]

$$G_{12}^{id} = R_0 T k_T^{id} - \frac{V_1^0 V_2^0}{V^{id}} \quad (A1\text{-}4)$$

and

$$G_{ii}^{id} = G_{12}^{id} + V_j^0 - V_i^0, \quad i \neq j \quad (A1\text{-}5)$$

where V_i^0, k_T^{id} and V^{id} are the molar volume of the pure component i, the isothermal compressibility and the molar volume of an ideal solution, respectively.

At infinite dilution, or for pure components, the following limiting expressions can be obtained for the KBIs[34,35,45]

$$\lim_{x_i \to 0} G_{12} = R_0 T k_{T,j}^0 - V_i^\infty \quad i \neq j \quad (A1\text{-}6)$$

$$\lim_{x_i \to 1} G_{12} = R_0 T k_{T,i}^0 - V_j^\infty \quad i \neq j \quad (A1\text{-}7)$$

$$\lim_{x_i \to 0} G_{ii} = R_0 T k_{T,j}^0 + V_j^0 - 2V_i^\infty - V_j^0 \left(\frac{\partial \ln \gamma_i}{\partial x_i}\right)_{P,T,x_i=0} \quad i \neq j \quad (A1\text{-}8)$$

and

$$\lim_{x_i \to 1} G_{ii} = R_0 T k_{T,i}^0 - V_i^0 \quad (A1\text{-}9)$$

where $k_{T,i}^0$ is the isothermal compressibility of the pure component i and V_i^∞ is the partial molar volume of component i at infinite dilution.

Appendix 2

The purpose of this Appendix is to provide details for calculating the excesses (deficits) for binary Lennard-Jones fluid.

The excesses (deficits) were calculated with eqn (13). The KBIs for binary Lennard-Jones fluids were found in literature.[43] The partial molar volumes and the isothermal compressibility were calculated using the Kirkwood–Buff expressions:[24]

$$v_1 = \frac{1 + (G_{22} - G_{12})c_2}{c_1 + c_2 + c_1 c_2 (G_{11} + G_{22} - 2G_{12})} \quad (A2\text{-}1)$$

$$v_2 = \frac{1 + (G_{11} - G_{12})c_1}{c_1 + c_2 + c_1 c_2 (G_{11} + G_{22} - 2G_{12})} \quad (A2\text{-}2)$$

$$kTk_T = \frac{1 + c_1 G_{11} + c_2 G_{22} + c_1 c_2 (G_{11} G_{22} - G_{12}^2)}{c_1 + c_2 + c_1 c_2 (G_{11} + G_{22} - 2G_{12})} \quad (A2\text{-}3)$$

The excesses (deficits) for binary Lennard-Jones fluids were calculated for two different cases:

(1) The same energy parameters but different size parameters:

We used the KBIs[43] for $\varepsilon_{11} = \varepsilon_{22} = kT/1.2$ and $\sigma_{11} = 1$ Å and $\sigma_{22} = 1.35$ Å.

(2) The same size parameters but different energy parameters:

We used the KBIs[43] for $\sigma_{11} = \sigma_{22} = 1$ Å and $\varepsilon_{11} = kT/1.2$ and $\varepsilon_{22} = kT/1.8$.

σ_{12} and ε_{12} were assumed[43] to be provided by the Lorenz-Berthelot rules

$$\sigma_{12} = (\sigma_{11} + \sigma_{22})/2 \qquad \text{(A2-5)}$$

and

$$\varepsilon_{12} = (\varepsilon_{11} \cdot \varepsilon_{22})^{1/2} \qquad \text{(A2-6)}$$

The KBIs[43] were calculated at constant packing fraction

$$\eta = \pi(c_1 + c_2)(x_1\sigma_{11}^3 + x_2\sigma_{22}^3)/6 \qquad \text{(A2-7)}$$

References

1. G. M. Wilson, *J. Am. Chem. Soc.*, 1964, **86**, 127.
2. J. M. Prausnitz, R. N. Lichtenthaler and E. Gomes de Azevedo, *Molecular thermodynamics of fluid—Phase equilibria*, Prentice-Hall, Englewood Cliffs, NJ, 2nd edn, 1986.
3. S. Kim and K. P. Johnston, *Ind. Eng. Chem. Res.*, 1987, **26**, 1206.
4. D. J. Phillips and J. F. Brennecke, *Ind. Eng. Chem. Res.*, 1993, **32**, 943.
5. C. Panayiotou and J. H. Vera, *Fluid Phase Equilib.*, 1980, **5**, 55.
6. R. G. Rubio and J. A. R. Renuncio, *Macromolecules*, 1980, **13**, 1508.
7. S. N. Timasheff, *Biochemistry*, 2002, **41**, 13473.
8. D. J. Felitsky and M. T. Record, *Biochemistry*, 2004, **43**, 9276.
9. H. D. Bale, R. E. Sherpler and D. K. Sorgen, *Phys. Chem. Liquids*, 1968, **1**, 181.
10. Y. Koga, *Chem. Phys. Lett.*, 1984, **111**, 176.
11. K. Nishikawa, Y. Kodera and T. Iijima, *J. Phys. Chem.*, 1987, **91**, 3694.
12. K. Nishikawa, H. Hayashi and T. Iijima, *J. Phys. Chem.*, 1989, **93**, 6559.
13. H. Hayashi, K. Nishikawa and T. Iijima, *J. Phys. Chem.*, 1990, **94**, 8334.
14. H. Hayashi and Y. Udagawa, *Bull. Chem. Soc. Jpn.*, 1992, **65**, 155.
15. I. Shulgin, R. Serimaa and M. Torkkeli, *Rep. Ser. Phys.*, HU-P-256, Helsinki, 1991.
16. G. D'Arrigo and J. Teixeira, *J. Chem. Soc., Faraday Trans.*, 1990, **86**, 1503.
17. G. H. Grossmann and K. H. Ebert, *Ber. Bunsen-Ges. Phys. Chem.*, 1981, **85**, 1026.
18. G. W. Euliss and C. M. Sorensen, *J. Chem. Phys.*, 1984, **80**, 4767.
19. N. Ito, T. Kato and T. Fujiyma, *Bull. Chem. Soc. Jpn.*, 1981, **54**, 2573.
20. T. M. Bender and R. Pekora, *J. Phys. Chem.*, 1986, **90**, 1700.
21. R. Zana and M. J. Eljebari, *J. Phys. Chem.*, 1993, **97**, 11134.
22. S. Mashimoto, T. Umehara and H. Redin, *J. Chem. Phys.*, 1991, **95**, 6257.
23. T. Marosi, I. Dekany and G. Lagaly, *Colloid Polym. Sci.*, 1994, **272**, 1136.
24. J. G. Kirkwood and F. P. Buff, *J. Chem. Phys.*, 1951, **19**, 774.
25. A. Ben-Naim, *Water and aqueous solutions*, Plenum Press, New York, 1974.
26. A. Ben-Naim, *J. Chem. Phys.*, 1977, **67**, 4884.
27. A. Ben-Naim, *Cell Biophys.*, 1988, **12**, 255.
28. A. Ben-Naim, *J. Phys. Chem.*, 1989, **93**, 3809.
29. A. Ben-Naim, *Solvation thermodynamics*, Plenum Press, New York, 1987.
30. A. Ben-Naim, *Statistical Thermodynamics for Chemists and Biochemists*, Plenum Press, New York, 1992.
31. A. Ben-Naim, *J. Phys. Chem. B*, 2007, **111**, 2896.
32. I. L. Shulgin and E. Ruckenstein, *J. Phys. Chem. B*, 2006, **110**, 12707.
33. T. L. Hill, *Statistical mechanics: principles and selected applications*, McGraw-Hill, New York, 1956.
34. E. Matteoli and L. Lepori, *J. Chem. Soc., Faraday Trans.*, 1995, **91**, 431.
35. E. Matteoli, *J. Phys. Chem. B*, 1997, **101**, 9800.
36. I. Shulgin and E. Ruckenstein, *J. Phys. Chem. B*, 1999, **103**, 2496.
37. A. Y. Meyer, *Chem. Soc. Rev.*, 1986, **15**, 449.
38. Y. Marcus, *J. Phys. Org. Chem.*, 2003, **16**, 398.
39. A. Bondi, *J. Phys. Chem.*, 1964, **68**, 441.
40. A. Bondi, *Physical Properties of Molecular Crystals, Liquids, and Glasses*, Wiley, New York, 1968.
41. P. G. Kusalik and I. M. Svishchev, *Science*, 1994, **265**, 1219.
42. L. L. Lee, *Molecular Thermodynamics of Nonideal Fluids*, Butterworth Publishers, Boston, MA, 1988.
43. K. Kojima, T. Kato and H. Nomura, *J. Solution Chem.*, 1984, **13**, 151.
44. R. H. Davies, A. G. Duncan, G. Saville and L. A. K. Staveley, *Trans. Faraday Soc.*, 1967, **63**, 855.
45. E. Matteoli and L. Lepori, *J. Chem. Phys.*, 1984, **80**, 2856.
46. Y. Cheng, M. Page and C. Jolicoeur, *J. Phys. Chem.*, 1993, **93**, 7359.
47. Y. Marcus, *Monatsh. Chem.*, 2001, **132**, 1387.
48. A. Perera, F. Sokolic, L. Almasy and Y. Koga, *J. Chem. Phys.*, 2005, **124**, Art. No. 124515.
49. C. Dethlefsen, P. G. Sorensen and A. Hvidt, *J. Solution Chem.*, 1984, **13**, 191.
50. *CRC Handbook of Chemistry and Physics*, ed. D. R. Lide, CRC Press, New York, 78th edn, 1997.
51. Y. Amo and Y. Tominago, *Chem. Phys. Lett.*, 2000, **332**, 521.
52. E. S. Kim and K. N. Marsh, *J. Chem. Eng. Data*, 1988, **33**, 288.
53. T. Moriyoshi and H. Inubushi, *J. Chem. Thermodyn.*, 1977, **9**, 587.
54. T. Takamuku, T. Yamaguchi, M. Asato, M. Matsumoto and N. Nishi, *Z. Naturforsch.*, 2000, **55**, 513.
55. K. Yoshida and T. Yamaguchi, *Z. Naturforsch.*, 2001, **56a**, 529.
56. T. Takamuku, K. Saisho, S. Aoki and T. Yamaguchi, *Z. Naturforsch.*, 2002, **57**, 982.

Reply to "Comment on 'The Kirkwood–Buff Theory of Solutions and the Local Composition of Liquid Mixtures'"

Ivan L. Shulgin[†] and Eli Ruckenstein*

Department of Chemical & Biological Engineering, State University of New York at Buffalo, Amherst, New York 14260

Received: October 8, 2007; In Final Form: December 29, 2007

The Kirkwood–Buff (KB) theory of solutions[1] relates the local properties of solutions, expressed through the KB integrals, to macroscopic thermodynamic quantities. An important application of this theory is to the excess (or deficit) number of molecules of any type around a central molecule. The calculation of this excess (or deficit) is the matter of our disagreement with the preceding Ben-Naim comment.[2]

First, let us note that the Ben-Naim assertion that we "sought a 'correction' to the interpretation of the Kirkwood-Buff integrals" is not accurate. The latter theory does not require any correction. We do not disagree with Kirkwood and Buff, but we disagree with Ben-Naim, namely with the expression he has employed for calculation of the excess (or deficit).

According to the numerous publications of Ben-Naim (as cited in our paper[3]), in a binary mixture of components α and β, the average excess (or deficit) ($\Delta N_{\alpha\beta}$) of the number of α molecules [for the sake of simplicity only a binary mixture is considered] around a central β molecule is provided by the expression

$$\Delta N_{\alpha\beta} = c_\alpha G_{\alpha\beta} \qquad (1)$$

where c_α is the bulk molar concentration of component α and $G_{\alpha\beta}$ is the KB integral.

It should be emphasized that $\Delta N_{\alpha\beta}$ was considered by Ben-Naim to represent the difference between the number of molecules of a given species around a central molecule and the number of molecules of the same species at the bulk concentration in the same volume surrounding the central molecule. Indeed, in his 1977 paper,[4] he writes "Clearly, $c_\alpha G_{\alpha\beta}$ reflects the total average excess (or deficiency) of α molecules in the entire surrounding of a β molecule".

We recently[3] demonstrated that expression (1) for the excess (or deficit) as defined above is not correct, and we provided a correct expression that involves the KB integral. Ben-Naim in his comment[2] disagrees. As a reply to his comment, we present again, in a very simple manner, our arguments.

Let us consider a binary mixture $\alpha-\beta$ and a central molecule β. The excess (or deficit) of α molecules in a sphere of radius R around a central molecule β (see Figure 1) is obviously provided by the difference between $n_{\alpha\beta}$, the average number of α molecules around a central molecule β in the volume between R_β and R, and $n_{\alpha\beta}^{bulk}$, the number of α molecules at the bulk concentration in the same volume for which $n_{\alpha\beta}$ was calculated. In Figure 1, R_β is the radius of a volume that is not accessible to α molecules because of the presence of the central molecule

---

* To whom correspondence should be addressed. E-mail: feaeliru@acsu.buffalo.edu. Fax: (716) 645-3822. Phone: (716) 645-2911/ext. 2214.
[†] E-mail: ishulgin@eng.buffalo.edu.

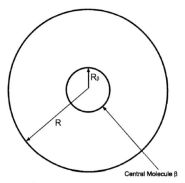

Figure 1. Illustration to the calculation of excess (or deficit) molecules α around a central molecule β.

β (for the sake of simplicity the shapes of the molecules are considered spherical but our considerations are valid for molecules of any shape).

Let us now express the above considerations in more details. The number of molecules $n_{\alpha\beta}$ is given by[5–6]

$$n_{\alpha\beta} = c_\alpha \int_0^R g_{\alpha\beta} 4\pi r^2 \, dr \qquad (2)$$

where $g_{\alpha\beta}$ is the radial distribution function between species α and β.

$n_{\alpha\beta}^{bulk}$ is obviously provided by

$$n_{\alpha\beta}^{bulk} = c_\alpha \int_{R_\beta}^R 4\pi r^2 \, dr \qquad (3)$$

Because[5–6] $g_{\alpha\beta} = 0$ for $r \leq R_\beta$, $n_{\alpha\beta}$ and $n_{\alpha\beta}^{bulk}$ are calculated for the same volume between the radii R_β and R. As it will be shown later, eq 1 of Ben-Naim can be obtained if one takes $R_\beta = 0$ in eq 3. Thus, Ben-Naim's expression (1) provides the difference between the number of α molecules around a central β molecule and the number of α molecules at the concentration of the bulk in a sphere of radius R that includes the volume of radius R_β. In other words, in his approach, the volumes involved in the two terms are not the same. It is clear that the physically meaningful and relevant excess is that which involves the same volume surrounding the central molecule. As already noted, Ben-Naim intended to calculate the excess as defined after eq 1, but obtained instead an "excess" corresponding to $R_\beta = 0$ in eq 3.

The basic equations of the two approaches can be derived starting from eqs 2 and 3.

The Ben-Naim's "excess" of α molecules in a sphere of radius R around a central molecule β is obtained by taking $R_\beta = 0$ in eq 3

$$\Delta N_{\alpha\beta}^{(R)} = c_\alpha \int_0^R g_{\alpha\beta} 4\pi r^2 \, dr - c_\alpha \int_0^R 4\pi r^2 \, dr \qquad (4)$$

which can be rewritten as

$$\Delta N_{\alpha\beta}^{(R)} = c_\alpha \int_0^R (g_{\alpha\beta} - 1) 4\pi r^2 \, dr \qquad (5)$$

Because for sufficiently large R, $g_{\alpha\beta} = 1$, eq 5 leads to his eq 1

$$\Delta N_{\alpha\beta} = c_\alpha G_{\alpha\beta} \qquad (1)$$

where $G_{\alpha\beta} = \int_0^\infty (g_{\alpha\beta} - 1) 4\pi r^2 \, dr$ is the KB integral.

Comments

In our approach, the excess of α molecules in a sphere of radius R around a central molecule β is

$$\Delta n_{\alpha\beta}^{(R)} = n_{\alpha\beta} - n_{\alpha\beta}^{\text{bulk}} =$$
$$c_\alpha \int_0^R g_{\alpha\beta} 4\pi r^2 \, dr - c_\alpha \int_{R_\beta}^R 4\pi r^2 \, dr =$$
$$c_\alpha \int_0^R (g_{\alpha\beta} - 1) 4\pi r^2 \, dr + c_\alpha \int_0^{R_\beta} 4\pi r^2 \, dr \quad (6)$$

For sufficiently large values of R, $g_{\alpha\beta} = 1$, and eq 6 becomes

$$\Delta n_{\alpha\beta} = c_\alpha G_{\alpha\beta} + c_\alpha 4\pi R_\beta^3/3 = c_\alpha G_{\alpha\beta} + c_\alpha V^\beta \quad (7)$$

where V^β is the volume inaccessible to α molecules because of the presence of the central molecule β. Comparing eqs 1 and 7, it is clear that in the Ben-Naim treatment the volume V^β is ignored. The difference between the two quantities $\Delta n_{\alpha\beta}$ and $\Delta N_{\alpha\beta}$ is particularly important for large central molecules such as the biomolecules.

Let us consider a very instructive example:[3] a protein molecule, ribonuclease A (denoted 2), at infinite dilution in a binary mixture of water (denoted 1) and glycerol (denoted 3). Is the ribonuclease A molecule hydrated by water or solvated by glycerol? We calculated[3] the excesses using both our approach and Ben-Naim's eq 1. The excess calculated with our expression (for a glycerol volume fraction of 30%) is positive for water and negative for glycerol ($\Delta n_{12} \cong 33.3$ mol of water per mole of protein and $\Delta n_{32} \cong -8.2$ mol of glycerol/mol of protein). With the Ben-Naim expression both are negative ($\Delta N_{12} \cong -341.8$ mol of water/mol of protein and $\Delta N_{32} \cong -48.1$ mol of glycerol/mol of protein). Therefore, the use of Ben-Naim's eq 1 provides large deficits for both water and glycerol in the vicinity of ribonuclease A. In contrast, many experiments[7-9] confirmed that in this mixture water is in excess and glycerol is in deficit in the vicinity of the protein molecule. It is clear why the Ben-Naim excesses are negative: eq 1 ignores the term $c_\alpha V^\beta$ which is comparable in magnitude in the considered example with $c_\alpha G_{\alpha\beta}$ ($\beta = 2$ and $\alpha = 1,3$); when this term is ignored, the number of molecules subtracted ($c_\alpha \int_0^R 4\pi r^2 \, dr$) from the number of molecules $n_{\alpha\beta}$ around a central molecule β becomes too large (see eq 4).

There are additional erroneous comments in the Ben-Naim paper that are less relevant. Let us mention one of them. He is not accurate in attributing to us "the claim that a negative KBI is not plausible". We have not made such an assertion. On the contrary, we provided[3] several examples of real mixtures for which all the KBIs are negative. We use such examples because they show that Ben-Naim's eq 1 can provide negative values for all of the excesses around any central molecule, and we said that this is not plausible.

Unfortunately, Ben-Naim's eq 1 has been used in many papers published over the last three decades. All of them considered erroneously that it provides an excess with respect to the bulk concentration in the surrounding of a central molecule.

It should be emphasized that Matteoli and Lepori[10] and Matteoli[11] were the first to criticize the Ben-Naim excess by introducing a reference state. For a detailed comparison between their approach and ours, one can see ref 3.

To close our reply, we emphasize the contradictions between the first and the last parts of Ben-Naim's comments, as well as the contradictions between most of his published papers concerning this topic (some cited in our paper[3]) and the present comment. In his previously published papers, he defined the excess as noted after eq 1, but derived an incorrect result: eq 1; in the present paper he interpreted the quantity $c_\alpha G_{\alpha\beta}$ in a correct manner, but failed to mention that $c_\alpha G_{\alpha\beta}$ is not the excess defined after eq 1.

References and Notes

(1) Kirkwood, J. G.; Buff, F. P. *J. Chem. Phys.* **1951**, *19*, 774.
(2) Ben-Naim, A. *J. Phys. Chem. B* **2008**, *112*, 5874.
(3) Shulgin, I. L.; Ruckenstein, E. *J. Phys. Chem. B* **2006**, *110*, 12707.
(4) Ben-Naim, A. *J. Chem. Phys.* **1977**, *67*, 4884.
(5) Hill, T. L. *Statistical Mechanics: Principles and Selected Applications*; McGraw-Hill: New York, 1956.
(6) Ben-Naim, A. *Statistical Thermodynamics for Chemists and Biochemists*; Plenum Press: New York, 1992.
(7) Gekko, K.; Timasheff, S. N. *Biochemistry* **1981**, *20*, 4667.
(8) Timasheff, S. N. *Adv. Protein Chem.* **1998**, *51*, 355.
(9) Courtenay, E. S.; Capp, M. W.; Anderson, C. F.; Record, J. M. T. *Biochemistry* **2000**, *39*, 4455.
(10) Matteoli, E.; Lepori, L. *J. Chem. Soc. Faraday Trans.* **1995**, *91*, 431.
(11) Matteoli, E. *J. Phys. Chem. B* **1997**, *101*, 9800.

GENERAL RESEARCH

An Improved Local Composition Expression and Its Implication for Phase Equilibrium Models

Ivan L. Shulgin and Eli Ruckenstein*

Department of Chemical & Biological Engineering, State University of New York at Buffalo, Amherst, New York 14260

A modified local composition (LC) expression is suggested, which accounts for the recent finding that the LC in an ideal binary mixture should be equal to the bulk composition only when the molar volumes of the two pure components are equal. However, the expressions available in the literature for the LCs in binary mixtures do not satisfy this requirement. Some LCs are examined including the popular LC-based NRTL model, to show how the above inconsistency can be eliminated. Further, the emphasis is on the modified NRTL model. The newly derived activity coefficient expressions have three adjustable parameters as the NRTL equations do, but contain, in addition, the ratio of the molar volumes of the pure components, a quantity that is usually available. The correlation capability of the modified activity coefficients was compared to the traditional NRTL equations for 42 vapor−liquid equilibrium data sets from two different kinds of binary mixtures: (i) highly nonideal alcohol/water mixtures (33 sets), and (ii) mixtures formed of weakly interacting components, such as benzene, hexafluorobenzene, toluene, and cyclohexane (9 sets). The new equations provided better performances in correlating the vapor pressure than the NRTL for 36 data sets, less well for 4 data sets, and equal performances for 2 data sets. Similar modifications can be applied to any phase equilibrium model based on the LC concept.

1. Introduction

Generally speaking, the structure of a solution at the nanometer scale and the intermolecular interactions between various constituents constitute the key to the understanding and even predicting both its microscopic and its macroscopic properties.

A typical feature of the liquid solutions is their microheterogeneity, that is, the presence of microparts of solution where the local concentration differs from the bulk concentration. These microheterogeneities can be characterized by the following nanometer-level quantities: (i) the correlation volume, that is, the volume in which the concentration differs from the average concentration, (ii) the excess (or deficit) number of molecules in the correlation volume with respect to the number of molecules corresponding to the bulk concentration, and (iii) the intermolecular interactions between molecules in the above volume.

The concentrations of the components in the vicinity of any molecule are usually called local compositions (LCs). According to the LC concept, the composition in the vicinity of any molecule differs from the overall composition. If a binary mixture is composed of components 1 and 2 with overall mole fractions x_1 and x_2, respectively, four LCs can be defined: local mole fractions of components 1 and 2 near a central molecule 1 (x_{11} and x_{21}) and local mole fractions of components 1 and 2 near a central molecule 2 (x_{12} and x_{22}). Many attempts have been made to express LC in terms of the bulk compositions and some intermolecular interaction parameters.[1−9] Wilson was the first[1] to suggest expressions for the local mole fractions and to derive on their basis expressions for the activity coefficients of binary mixtures. Since then, many expressions for LC were suggested, and the LC concept proved to be a very effective method in the prediction and correlation of phase equilibria in binary and multicomponent mixtures.

In this Article, a modified LC expression is suggested. This modification is a result of the observation that the traditional expressions for the LCs[1−8] are inconsistent with the expressions for the excesses around molecules in ideal binary mixtures (see the next section). The new LCs will be used to obtain expressions for the activity coefficients of binary mixtures using the NRTL equations[2] for illustration. The traditional and corrected NRTL equations will be used to correlate the vapor−liquid equilibria (VLE) for alcohol + water binary mixtures and binary mixtures containing benzene, hexafluorobenzene, toluene, and cyclohexane. It is shown that the modified LCs provide a moderate improvement of the NRTL results.

2. The Local Composition in an Ideal Mixture

On the basis of the Kirkwood−Buff theory of solution,[10] one can show that the excess (or deficit) number of molecules i ($i = 1, 2$) around a central molecule j ($j = 1, 2$), Δn_{ij}, in a binary liquid mixture can be obtained by using the expression:[11]

$$\Delta n_{ij} = c_i G_{ij} + c_i(V_j - RTk_T) \quad (1)$$

where c_i is the bulk molar concentration of species i in the mixture, G_{ij} is the Kirkwood−Buff integral,[10] T is the temperature, V_j is the partial molar volume of component j, k_T is the isothermal compressibility of the mixture, and R is the universal gas constant. For an ideal mixture (according to thermodynamics,[9] an ideal mixture is a mixture for which the activity coefficients of the components are equal to unity at any concentration, temperature, and pressure), the above expression leads for the excesses around a central molecule 1 (similar expressions can be obtained for a central molecule 2) to[11]

---

* To whom correspondence should be addressed. Tel.: (716) 645-2911ext. 2214. Fax: (716) 645-3822. E-mail: feaeliru@acsu.buffalo.edu.

$$\Delta n_{11}^{id} = \frac{x_1 x_2 V_2^0 (V_2^0 - V_1^0)}{(x_1 V_1^0 + x_2 V_2^0)^2} \quad (2)$$

and

$$\Delta n_{21}^{id} = \frac{x_1 x_2 V_1^0 (V_1^0 - V_2^0)}{(x_1 V_1^0 + x_2 V_2^0)^2} \quad (3)$$

where V_i^0 is the molar volume of the pure component i, and x_i is the bulk molar fraction of component i. These equations show that the excesses Δn_{ij} for an ideal binary mixture become zero only when the molar volumes of the pure components are the same; otherwise the excesses and deficits have nonzero values. Such a behavior provides a criterium for the LC; the LCs of an ideal binary mixture should be equal to the bulk compositions only when the molar volumes of the pure components are the same. However, the existing expressions[1-8] for LC fail to satisfy this criterium. Let us consider as an example the NRTL model.[2] According to this model, the local mole fractions around a central molecule i ($i = 1, 2$) can be expressed as:[2]

$$x_{ii} = \frac{x_i}{x_i + x_j \exp\left(-\alpha_{12}\frac{\lambda_{ij} - \lambda_{ii}}{RT}\right)} \quad (4)$$

and

$$x_{ji} = \frac{x_j \exp\left(-\alpha_{12}\frac{\lambda_{ij} - \lambda_{ii}}{RT}\right)}{x_i + x_j \exp\left(-\alpha_{12}\frac{\lambda_{ij} - \lambda_{ii}}{RT}\right)} \quad (5)$$

where λ_{ij} are interaction parameters between molecules i and j, and α_{12} is an additional dimensionless parameter.

According to the NRTL model,[2] the activity coefficients can be expressed as:

$$\ln \gamma_1 = x_2^2 \left(\tau_{21} \frac{\exp(-2\alpha_{12}\tau_{21})}{[x_1 + x_2(-\alpha_{12}\tau_{21})]^2} + \tau_{12} \frac{\exp(-\alpha_{12}\tau_{12})}{[x_2 + x_1(-\alpha_{12}\tau_{12})]^2} \right) \quad (6)$$

and

$$\ln \gamma_2 = x_1^2 \left(\tau_{12} \frac{\exp(-2\alpha_{12}\tau_{12})}{[x_2 + x_1(-\alpha_{12}\tau_{12})]^2} + \tau_{21} \frac{\exp(-\alpha_{12}\tau_{21})}{[x_1 + x_2(-\alpha_{12}\tau_{21})]^2} \right) \quad (7)$$

where $\tau_{12} = (\lambda_{12} - \lambda_{22})/RT$ and $\tau_{21} = (\lambda_{21} - \lambda_{11})/RT$.

It is obvious from eqs 4 and 5 that $x_{ji} = x_j$ only when $\lambda_{ij} = \lambda_{ii}$ ($i = 1, 2$ and $j = 1, 2$), and this leads (see eqs 6 and 7) to $\ln \gamma_1 = 0$ and $\ln \gamma_2 = 0$, which imply that the mixtures with $x_{ji} = x_j$ are ideal. One can easily show that by assuming $\ln \gamma_1 = 0$ and $\ln \gamma_2 = 0$, one obtains from eqs 6 and 7 that $x_{ji} = x_j$ ($i = 1, 2$ and $j = 1, 2$). Consequently, from the point of view of the NRTL equations, the ideality of a binary mixture and equality of local and bulk compositions are equivalent results. However, there is a contradiction between eqs 2 and 3 on one side and eqs 4 and 5 on the other side, because for ideal mixtures and $x_{ji} = x_j$, the former require V_1^0 to be equal to V_2^0.

3. A Modification of the NRTL Equation

The present modification of the NRTL equations for the activity coefficients is based on a correction of LC, which eliminates the above inconsistency. This correction consists of replacing the bulk molar fractions x_i in eqs 4 and 5 by the volume fractions $\varphi_i = V_i^0 x_i/(V_1^0 x_1 + V_2^0 x_2)$ ($i = 1, 2$). One thus obtains

$$x_{ii} = \frac{V_i^0 x_i}{V_i^0 x_i + V_j^0 x_j \exp\left(-\alpha_{12}\frac{\lambda_{ij} - \lambda_{ii}}{RT}\right)} \quad (8)$$

and

$$x_{ji} = \frac{V_j^0 x_j \exp\left(-\alpha_{12}\frac{\lambda_{ij} - \lambda_{ii}}{RT}\right)}{V_i^0 x_i + V_j^0 x_j \exp\left(-\alpha_{12}\frac{\lambda_{ij} - \lambda_{ii}}{RT}\right)} \quad (9)$$

One can easily verify that eqs 8 and 9 are no longer in contradiction with eqs 2 and 3. Indeed, $x_{ji} = x_j$ for $\lambda_{ij} = \lambda_{ii}$ and $V_1^0 = V_2^0$.

The new activity coefficient expressions are obtained using the method of Renon and Prausnitz[2] and eqs 8 and 9 for the LC. The resulting expressions are

$$\ln \gamma_1 = x_2^2 \left(\tau_{21} \frac{W^2 \exp(-2\alpha_{12}\tau_{21})}{[x_1 + Wx_2(-\alpha_{12}\tau_{21})]^2} + \tau_{12} \frac{W \exp(-\alpha_{12}\tau_{12})}{[Wx_2 + x_1(-\alpha_{12}\tau_{12})]^2} \right) \quad (10)$$

and

$$\ln \gamma_2 = x_1^2 \left(\tau_{12} \frac{\exp(-2\alpha_{12}\tau_{12})}{[Wx_2 + x_1(-\alpha_{12}\tau_{12})]^2} + \tau_{21} \frac{W \exp(-\alpha_{12}\tau_{21})}{[x_1 + Wx_2(-\alpha_{12}\tau_{21})]^2} \right) \quad (11)$$

where $W = V_2^0/V_1^0$.

One can see that, as NRTL, the new eqs 10 and 11 are three-parameter equations, but contain, in addition, a new quantity $W = V_2^0/V_1^0$, which is easily available.

4. Comments on Other Expressions for the Local Concentrations

A number of additional expressions have been derived for the local concentration.[3-8] They have been presented in our previous work.[8] They exhibit the same inconsistency as the NRTL expression does with regard to the ideal mixtures, and the contradiction can be eliminated in a similar way. For illustration, let us consider the Panagiotou and Vera[3,4] expressions derived on the basis of the quasi chemical approach:

$$x_{ii} = 1 - \frac{2x_j}{1 + [1 - 4x_i x_j(1 - G_{12})]^{1/2}}$$

$$x_{ji} = \frac{2x_j}{1 + [1 - 4x_i x_j(1 - G_{12})]^{1/2}}$$

where $G_{12} = \exp[(\varepsilon_{11} + \varepsilon_{22} - 2\varepsilon_{12})/RT]$, and ε_{ij} is the interaction energy between the pair of molecules i and j. The above expressions are not consistent with eqs 2 and 3 because they lead for ideal mixtures ($G_{12} = 1$) to $x_{ii} = x_i$ and $x_{ji} = x_j$. To make them consistent, one should replace x_i and x_j by the corresponding volume fractions $V_i^0 x_i/(V_i^0 x_i + V_j^0 x_j)$ and $V_j^0 x_j/(V_i^0 x_i + V_j^0 x_j)$, respectively. Indeed, then $x_{ii} = x_i$ when $G_{12} = 1$ and $V_i^0 = V_j^0$. Another possible solution is to replace x_j in the numerator by the corresponding volume fraction and keep $x_i x_j$ at the denominator unchanged. Of course, the consistency criterium cannot provide a choice between the two. Intuition suggests that the first choice is more plausible.

Table 1. Information about Vapor−Liquid Data in Alcohol/Water Mixtures Used in Calculations

system	total number of points	temperature range, K	references
MeOH/H$_2$O	79	273.15−333.15	12−17
EtOH/H$_2$O	337	298.15−381.35	18−26
1-PrOH/H$_2$O	104	303.15−363.15	14, 24, 27−29
t-BuOH/H$_2$O	91	298.15−323.15	30−33

Table 2. Information about Vapor−Liquid Data in Binary Systems Containing Benzene, Hexafluorobenzene, Toluene, and Cyclohexane

system	total number of points	temperature range, K	references
HFB/B	59	303.15−343.15	35
HFB/TL	12	303.15	35
HFB/CH	12	303.15	36
B/TL	11	303.15	37
B/CH	13	298.15	38

In the next section, the emphasis will be on eqs 10 and 11, which are employed to correlate the VLEs of various binary mixtures, and the results will be compared to those obtained with the traditional NRTL equations.

5. A Vapor−Liquid Equilibrium Correlation with Equations 10 and 11

5.1. Systems Employed. Two kinds of binary mixtures were selected to test eqs 10 and 11.

The first kind are the alcohol/water mixtures. These mixtures are highly nonideal with strong intermolecular interactions, including H-bonds between the components. Thirty-three binary alcohol−water VLE sets for methanol (MeOH), ethanol (EtOH), 1-propanol (1-PrOH), and t-butanol (t-BuOH) were employed. The information about the alcohol/water mixtures used in calculations is listed in Table 1.

The second kind of binary systems contain benzene (B), hexafluorobenzene (HFB), toluene (TL), and cyclohexane (CH): HFB/B (for five different temperatures), HFB/TL, HFB/CH, B/TL, and B/CH. In contrast to the first group, these systems have small deviations from ideality,[34] but some of them have more complex thermodynamic behaviors such as double azeotropy.[35] The information about these mixtures is listed in Table 2.

5.2. Calculation Procedure. The selected VLE data were treated by the Barker method,[39,40] using only liquid mole fraction−pressure (x−P) data. The vapor phase nonideality was taken into account, and the total pressure was calculated using the equation

$$P = \frac{x_1 \gamma_1 P_1^0}{F_1} + \frac{x_2 \gamma_2 P_2^0}{F_2} \quad (12)$$

where

$$F_1 = \exp \frac{(B_{11} - V_1^0)(P - P_1^0) + P y_2^2 d_{12}}{RT} \quad (13)$$

$$F_2 = \exp \frac{(B_{22} - V_2^0)(P - P_2^0) + P y_1^2 d_{12}}{RT} \quad (14)$$

where y_i is the mole fraction of component i in the vapor phase, P_i^0 is the vapor pressure of the pure component i, B_{ii} is the second virial coefficient of component i, $d_{12} = 2B_{12} - B_{11} - B_{22}$, and B_{12} is the crossed second virial coefficient of the binary mixture, which was evaluated using for B_{12} the expression $B_{12} = (B_{11} + B_{22})/2$. The virial coefficients of the pure components and the vapor pressures of the pure components P_i^0 were obtained from references listed in our previous work.[8,34]

The VLE data listed in Tables 1 and 2 were treated by the NRTL, and the new equations for the activity coefficients and the results were compared.

5.3. Results of Calculations. The root-mean-square deviations (rmsd) of the calculated pressures with respect to the experimental ones are listed in Tables 3 and 4. The comparison of the rmsd for alcohol/water mixtures indicates that the new equations provide the best performances for 29 VLE and NRTL for 5 sets. A similar conclusion was reached for the systems containing benzene, hexafluorobenzene, toluene, and cyclohexane: the new equations provide the best performances for 7 sets, and the new equations and the NRTL provide almost the same results for 2 sets.

The vapor compositions were also calculated and compared to the available experimental data. The rmsd values of the calculated vapor compositions with respect to the experimental ones are listed in Tables 5 and 6. For the alcohol/water mixtures, the new equations perform better for 15 of the 17 sets, and NRTL performs better for 2 sets. For the systems containing benzene, hexafluorobenzene, toluene, and cyclohexane, the new equations provide the best performance for 6 VLE sets, and the new equations and the NRTL provide almost the same results for 2 sets.

Table 3. Comparison between the Calculated Pressure Obtained via the NRTL and the New Equations for Aqueous Systems of Alcohols and Experiment

system	total number of points	rmsd, mm Hg	
		NRTL	new equations
MeOH/H$_2$O	79	1.23	1.16
EtOH/H$_2$O	337	0.73	0.72
1-PrOH/H$_2$O	104	2.03	1.00
t-BuOH/H$_2$O	91	1.22	1.21

Table 4. Comparison between the Calculated Pressure Obtained via NRTL and the New Equations for Binary Systems Containing Benzene, Hexafluorobenzene, Toluene, and Cyclohexane

system	total number of points	rmsd, mm Hg	
		NRTL	new equations
HFB/B	59	0.79	0.66
HFB/TL	12	0.30	0.28
HFB/CH	12	0.27	0.24
B/TL	11	0.52	0.52
B/CH	13	0.03	0.03

Table 5. Comparison between the Calculated Vapor Composition Obtained via NRTL and the New Equations for Aqueous Systems of Alcohols

system	total number of points	rmsd, molar percent	
		NRTL	new equations
MeOH/H$_2$O	36	0.62	0.58
EtOH/H$_2$O	75	0.75	0.73
1-PrOH/H$_2$O	77	1.57	1.29
t-BuOH/H$_2$O	35	2.08	2.08

Table 6. Comparison between the Calculated Vapor Composition Obtained via NRTL and the New Equation for Binary Systems Containing Benzene, Hexafluorobenzene, Toluene, and Cyclohexane

system	total number of points	rmsd, molar percent	
		NRTL	new equations
HFB/B	59	0.35	0.29
HFB/TL	12	0.63	0.62
HFB/CH	12	0.15	0.15
B/CH	13	0.16	0.16

6. Conclusion

The contradiction between the expressions obtained by the authors for the excesses and the LC expressions is resolved by suggesting a simple modification of the latter expressions. The modification is based on the observation that the LCs for an ideal binary mixture should be equal to the bulk composition only when the components of the mixture have equal molar volumes. The new expressions for the LCs are used to improve the NRTL expressions for the activity coefficients.

The new equations for the activity coefficients of binary mixtures were used to correlate the VLE for 42 binary mixtures. The new equations provide a better performance than does NRTL for most mixtures. However, the difference between the two methods is relatively small.

Literature Cited

(1) Wilson, G. M. Vapor−liquid equilibrium. XI: A new expression for the excess free energy of mixing. *J. Am. Chem. Soc.* **1964**, *86*, 127−130.

(2) Renon, H.; Prausnitz, J. M. Local composition in thermodynamic excess functions for liquid mixtures. *AIChE J.* **1968**, *14*, 135−144.

(3) Panayiotou, C.; Vera, J. H. The quasi-chemical approach for non-randomness in liquid mixtures. Expressions for local composition with an application to polymer solution. *Fluid Phase Equilib.* **1980**, *5*, 55−80.

(4) Panayiotou, C.; Vera, J. H. Local composition and local surface area fractions: a theoretical discussion. *Can. J. Chem. Eng.* **1981**, *59*, 501−505.

(5) Lee, K.-H.; Sandler, S. I.; Patel, N. C. The generalized van-der-Waals partition function 0.3. Local composition models for a mixture of equal size square-well molecules. *Fluid Phase Equilib.* **1986**, *25*, 31−49.

(6) Aranovich, G. L.; Donohue, M. D. A new model for lattice systems. *J. Chem. Phys.* **1996**, *105*, 7059−7063.

(7) Wu, D. W.; Cui, Y.; Donohue, M. D. Local composition models for lattice mixtures. *Ind. Eng. Chem. Res.* **1998**, *37*, 2936−2946.

(8) Ruckenstein, E.; Shulgin, I. Modified local composition and Flory-Huggins equations for nonelectrolyte solutions. *Ind. Eng. Chem. Res.* **1999**, *38*, 4092−4099.

(9) Prausnitz, J. M.; Lichtenthaler, R. N.; Gomes de Azevedo, E. *Molecular Thermodynamics of Fluid−Phase Equilibria*, 2nd ed.; Prentice-Hall: Englewood Cliffs, NJ, 1986.

(10) Kirkwood, J. G.; Buff, F. P. The statistical mechanical theory of solution. I. *J. Chem. Phys.* **1951**, *19*, 774−782.

(11) Shulgin, I. L.; Ruckenstein, E. The Kirkwood−Buff theory of solutions and the local composition of liquid mixtures. *J. Phys. Chem. B* **2006**, *110*, 12707−12713.

(12) Yarym-Agaev, N. L.; Kotsarenko, A. A. Designing the phase diagram of 2- component system under the absence of experimental data on equilibrium data composition. *Zh. Fiz. Khim.* **1990**, *64*, 2349−2353.

(13) Kooner, Z. S.; Phutela, R. C.; Fenby, D. V. Determination of the equilibrium constants in water-methanol deuterium exchange reaction from vapor pressure measurements. *Aust. J. Chem.* **1980**, *33*, 9−13.

(14) Ratcliff, G. A.; Chao, K. L. Prediction of thermodynamic properties of polar mixtures by a group solution model. *Can. J. Chem. Eng.* **1969**, *47*, 148−153.

(15) Zielkiewicz, J.; Oracz, P. Vapor−liquid equilibrium in the ternary system *N,N*-dimethylformamide−methanol−water at 313.15 K. *Fluid Phase Equilib.* **1990**, *59*, 279−290.

(16) Zharov, V. T.; Pervukhin, O. K. Structure of liquid−vapor equilibrium diagrams in systems with chemical interaction. II. Methanol−formic acid−methyl formate−water system. *Zh. Fiz. Khim.* **1972**, *46*, 1970−1973.

(17) Broul, M.; Hlavaty, K.; Linek, J. Liquid−vapor equilibriums in systems of electrolytic components. V. The system $CH_3OH-H_2O-LiCl$ at 60 °C. *Collect. Czech. Chem. Commun.* **1969**, *34*, 3428−3435.

(18) Phutela, R. C.; Kooner, Z. S.; Fenby, D. V. Vapor pressure study of deuterium exchange reaction in water-ethanol system: equilibrium constant determination. *Aust. J. Chem.* **1979**, *32*, 2353−2359.

(19) Dobson, H. J. E. The partial pressures of aqueous ethyl alcohol. *J. Chem. Soc.* **1925**, 2866−2873.

(20) Hall, D. J.; Mash, C. J.; Pemberton, R. C. Vapor liquid equilibriums for the systems water−methanol, water−ethanol, and water−methanol−ethanol at 298.15 K determined by a rapid transpiration method. *NPL Report Chem.* **1979**, *95*, 36.

(21) Rarey, J. R.; Gmehling, J. Computer-operated differential static apparatus for the measurement of vapor−liquid equilibrium data. *Fluid Phase Equilib.* **1993**, *83*, 279−287.

(22) Pemberton, R. C.; Mash, C. J. Thermodynamic properties of aqueous nonelectrolytes mixtures. II. Vapor pressure and excess Gibbs energy for water + ethanol at 303.15 to 363.15 K determined by an accurate static method. *J. Chem. Thermodyn.* **1978**, *10*, 867−888.

(23) Mertl, I. Liquid−vapor equilibrium. II. Phase equilibriums in the ternary system ethyl acetate−ethanol−water. *Collect. Czech. Chem. Commun.* **1972**, *37*, 366−374.

(24) Zielkiewicz, J.; Konitz, P. Vapor−liquid equilibrium in ternary system *N,N*- dimethylformamide−water−ethanol at 313.15 K. *Fluid Phase Equilib.* **1991**, *63*, 129−139.

(25) Chaudhry, M. M.; Van Ness, H. C.; Abbott, M. M. Excess thermodynamic functions for ternary systems. 6. Total-pressure data and G^E for acetone−ethanol−water at 50 °C. *J. Chem. Eng. Data* **1980**, *25*, 254−257.

(26) Pemberton, R. C. Thermodynamic properties of aqueous nonelectrolytes mixtures. Vapor pressures for the system water + ethanol 303.15−363.15 K determined by an accurate static method. *Conf. Int. Thermodyn. Chim. Montpellier* **1975**, *6*, 137−144.

(27) Udovenko, V. V.; Mazanko, T. F. Liquid−vapor equilibrium in propyl alcohol−water and propyl alcohol−benzene systems. *Izv. Vyssh. Ucheb. Zaved. Khim. Khim. Tekhnol.* **1972**, *15*, 1654−1658.

(28) Vrevsky, M. S. Composition and vapor tension of solutions. *Zh. Russ. Fiz. Khim. Obshch.* **1910**, *42*, 1−35.

(29) Schreiber, E.; Schuettau, E.; Rant, D.; Schuberth, H. Extent to which a metalchloride can influence the behavior of isothermal phase equilibrium in *n*-propanol−water and *n*-butanol−water systems. *Z. Phys. Chem. (Leipzig)* **1971**, *247*, 23−40.

(30) Brown, A. C.; Ives, D. J. G. *tert*-BuOH−water system. *J. Chem. Soc. London* **1962**, 1608−1619.

(31) Edwards, D.; Marucco, J.; Ratouis, M.; Dode, M. Thermodynamic activities of aliphatic alcohols in water and in Ringer solution. II. 2-Methyl-2-propanol. *J. Chim. Phys. Phys.-Chim. Biol.* **1966**, *63*, 239−241.

(32) Koga, Y.; Siu, W. W. Y.; Wong, T. Y. H. Excess partial molar free energies and entropies in aqueous tert-butyl alcohol solutions at 25 °C. *J. Phys. Chem.* **1990**, *94*, 7700−7706.

(33) Fischer, K.; Gmehling, J. $P-x$ and γ^∞ data for the different binary butanol−water systems at 50 °C. *J. Chem. Eng. Data* **1994**, *39*, 309−315.

(34) Ruckenstein, E.; Shulgin, I. Aggregation in binary solutions containing hexafluorobenzene. *J. Phys. Chem. B* **1999**, *103*, 10266−10271.

(35) Gaw, W. J.; Swinton, F. L. Thermodynamic properties of binary systems containing hexafluorobenzene. Part 4. Excess Gibbs free energies of the three systems hexafluorobenzene + benzene, touene, and *p*-xylene. *Trans. Faraday Soc.* **1968**, *64*, 2023−2034.

(36) Gaw, W. J.; Swinton, F. L. Thermodynamic properties of binary systems containing hexafluorobenzene. Part 3. Excess Gibbs free energy of the system hexafluorobenzene + cyclohexane. *Trans. Faraday Soc.* **1968**, *64*, 637−647.

(37) Schmidt, G. C. Binary mixtures. *Z. Phys. Chem. B* **1926**, *121*, 221−253.

(38) Tasic, A.; Djordjevic, B.; Grozdanic, D. Vapor−liquid-equilibria of systems acetone−benzene, benzene−cyclohexane and acetone−cyclohexane 25 °C. *Chem. Eng. Sci.* **1978**, *33*, 189−197.

(39) Barker, J. A. Determination of activity coefficients from total pressure measurements. *Aust. J. Chem.* **1953**, *6*, 207−210.

(40) Abbott, M. M.; Van Ness, H. C. Vapor−liquid-equilibrium. 3. Data reduction with precise expressions for GE. *AIChE J.* **1975**, *21*, 62−71.

Received for review June 6, 2008
Revised manuscript received August 4, 2008
Accepted August 5, 2008

IE800897J

Chapter 2

Supercritical mixtures

2.1 On density microheterogeneities in dilute supercritical solutions.

2.2 Why density augmentation occurs in dilute supercritical solutions.

2.3 Fluctuations in dilute binary supercritical mixtures.

2.4 Entrainer effect in supercritical mixtures.

2.5 The solubility of solids in mixtures composed of a supercritical fluid and an entrainer.

2.6 A simple equation for the solubility of a solid in a supercritical fluid cosolvent with a gas or another supercritical fluid.

2.7 Cubic equation of state and local composition mixing rules: correlations and predictions. Application to the solubility of solids in supercritical solvents.

Introduction to Chapter 2

Chapter 2 deals with supercritical fluids and their mixtures. This Chapter consists of two parts: The first is concerned with the structure of the supercritical fluids and their mixtures (2.1–2.3); The second is concerned with the modeling of the physico-chemical properties of supercritical fluids and their mixtures (2.4–2.7).

Supercritical fluids and their mixtures have many applications in pharmaceutical, food, and other industries. In addition, they constitute a challenge for many experimentalists and theoreticians, because they exhibit unusual physico-chemical properties that cannot be explained using the methods usually employed for mixtures far from the critical point. One of the most striking examples of such a behavior is the concentration dependence of the partial molar volume of the solute in mixtures consisting of a supercritical fluid + solute. Indeed, the partial molar volume of the solute at infinite dilution is usually negative and large in magnitude, which is not typical for nonelectrolyte solutions far from the critical point in which the partial molar volume is usually positive and has a value comparable to the molar volume of the pure solute (see 2.3)). This behavior of the partial molar volume of the solute and the corresponding density augmentation observed experimentally was previously believed to result from the clustering of the solvent about individual solute molecules. In contrast, we found that the augmentation is not caused by the solute, but rather it is due to the preexisting near-critical fluctuations in the pure solvent and the preference of the solute for the high density regions of the solvent (2.1–2.2). Our explanation is in agreement with experimental data regarding both pure supercritical solvents and dilute mixtures of supercritical solvent + solute.

The high solubility of solid substances in supercritical fluids compared to those in ideal gases (enhancement factors of 10^4–10^8 are common) allows their use as solvents in pharmaceutical, biomedical and food industries. Sections 2.4–2.7 are devoted to predictions of the entrainer effect, and of solubility in supercritical fluids with and without entrainer. Reliable predictive methods for solid solubilities in mixtures of a supercritical solvent + cosolvent were developed (2.4–2.6). These apply not only to the usual cosolvents such as organic liquids (2.4–2.5), but also to cases in which the cosolvent is a gas or another supercritical fluid (2.6). Our methods provided good agreement with experimental data in all of these cases (2.4–2.6).

On Density Microheterogeneities in Dilute Supercritical Solutions

E. Ruckenstein* and I. Shulgin[†]

Department of Chemical Engineering, State University of New York at Buffalo, Amherst, New York 14260

Received: October 27, 1999; In Final Form: January 3, 2000

The dilute supercritical mixtures were examined in the framework of the Kirkwood–Buff theory of solutions. Various expressions were employed for the excess number of aggregated molecules of solvent around individual solute molecules to conclude that at infinite dilution the above mentioned excess is zero. This suggested that the density enhancement observed when small amounts of a solute were added to a solvent near the critical point of the latter may not be caused by the aggregation of the solvent molecules around individual solute molecules as usually considered. Further, comparing experimental results, it was shown that the density enhancement caused by the near critical fluctuations in a pure solvent are almost the same, in a wide range of pressures, as those in dilute supercritical mixtures near the critical point of the solvent.

Introduction

During the last three decades the use of supercritical fluids (SCF) has dramatically increased because of their applications to extraction, chromatography, and as media for chemical reactions.[1–6] The most interesting applications of SCFs occur in the following ranges of pressure and temperature:[3,7] $1 < P/P_c < 2$ and $1 < T/T_c < 1.1$, where P and T are the pressure and temperature and P_c and T_c are their critical values. Under these conditions, SCFs (such as CO_2, C_2H_4, CHF_3, etc.) are typically less dense than the liquids by factors of 1.5–3.[3] Since these intervals of pressure and temperature are close to the critical point, the compressibility of SCFs is high (at the critical point it becomes infinite).[3] Experiment also shows that for a SCF near the critical point of the solvent, the partial molar volume of the solute at infinite dilution becomes usually negative and large in magnitude;[8] this behavior is not, however, typical for a nonelectrolyte solution far from the critical region. Because of the high compressibility of fluids near the critical point, their density and dissolving power can be tuned through small changes in pressure.[3] The high dissolving power of the SCF leads to unusually high solubilities of solids in the SCFs. Compared to that of a dilute gas, the solubility enhancement can be as large as 10^{12} (ref 7). A large body of research was aimed at understanding the unusual properties of SCFs at the molecular level, and comprehensive reviews were recently published.[3–6] One of the key problems regarding intermolecular interactions in a dilute solution involving a SCF as solvent (such a solution will be called SCR mixture) is the so-called local density enhancement induced by the addition of a small amount of solute into the solvent near the critical point of the latter.[3–6] There have been many attempts to explain this effect.[3–6] The large negative infinite dilution partial molar volume of the solute suggested that the aggregation of solvent molecules around individual molecules of solute is responsible for the enhancement.[7,9] This explanation was challenged with the argument that because of the high compressibility near the critical point of the solvent, the addition of a small amount of solute causes a large change in volume, which is responsible for the negative partial molar volume of the solute at high dilution.[10]

In this paper, some recent experimental results regarding the density fluctuations in pure SCF[11–15] are used to show that the local density enhancement in dilute SCR mixtures is mainly due to the near critical fluctuations in the solvent and an explanation is suggested for the negative partial molar volume of the solute. This conclusion was also strengthened by a discussion, presented in the following section, based on the Kirkwood–Buff (KB) theory of solution.[16] First, the problem will be examined in the framework of the Kirkwood–Buff theory of solution. Second, using experimental results about the near critical fluctuations in pure SCF, it will be shown that the density enhancement in dilute SCR mixtures is mainly caused by the near critical density fluctuations in pure SCF.

Theory and Formulas

1. The Kirkwood–Buff Integrals. The Kirkwood–Buff theory of solution relates the so-called Kirkwood–Buff integral (defined below) to macroscopic quantities, such as the compressibility, partial molar volumes, and the composition derivative of the activity coefficient.

The Kirkwood–Buff integrals (KBIs) are given by the expressions

$$G_{ij} = \int_0^\infty (g_{ij} - 1) 4\pi r^2 \, dr \quad i,j = 1,2 \quad (1)$$

where g_{ij} is the radial distribution function between species i and j and r is the distance between the centers of molecules i and j. The KBIs in binary mixtures can be calculated using the expressions:[17]

$$G_{12} = G_{21} = RTk_T - \frac{V_1 V_2}{VD} \quad (2)$$

and

$$G_{ii} = G_{12} + \frac{1}{x_i}\left(\frac{V_j}{D} - V\right) \quad i \neq j \quad (3)$$

where

---

* Author to whom correspondence should be addressed. Fax: (716) 645-3822. E-mail: feariu@acsu.buffalo.edu.
† Current E-mail address: ishulgin@eng.buffalo.edu.

$$D = \left(\frac{\partial \ln \gamma_i}{\partial x_i}\right)_{P,T} x_i + 1 \quad (4)$$

In eqs 2−4, T is the absolute temperature, P is the pressure, k_T is the isothermal compressibility, V_i is the partial molar volume of component i, x_i is the molar fraction of component i, V is the molar volume of the mixture, R is the universal gas constant, and γ_i is the activity coefficient of component i. In the following considerations, expressions for the KBIs at the extreme concentrations ($x_1 \to 0$ and $x_2 \to 0$), as well as expressions for KBIs for ideal systems, will be needed. All these expressions are provided in Appendix 1.

2. Excess Number of Molecules Near a Central One. Ben-Naim[20] suggested to calculate the excess (the BN excess) number of j molecules around a central molecule i as

$$\Delta n_{ji} = c_j G_{ji} \quad (5)$$

and Debenedetti[21] considered this excess as a measure of aggregation. Matteoli and Lepori[18,22] noted that the $c_j G_{ji}$ are nonzero for ideal mixtures, which should be considered nonaggregated and suggested that the effects due to aggregation are better reflected in excesses (the ML excesses) defined as

$$\Delta n'_{ji} = c_j (G_{ji} - G^{id}_{ji}) \quad (6)$$

where G^{id}_{ji} are the KBs integrals for ideal systems (see Appendix 1 for the expressions of G^{id}_{ij}). In other words, the ideal mixture was considered as a reference system.

However, $\Delta n'_{ii}$ and $\Delta n'_{ji}$ are not independent quantities, because the volume occupied by the excess i molecules aggregated around an i molecule should be equal to the volume left free by the j molecules around the same i molecule.[22] This leads to the following relation:

$$V_j \Delta n'_{ji} = -V_i \Delta n'_{ii} \quad (7)$$

It was, however, noted that if the number of molecules clustered around a central one is calculated as $c_j(G_{ji} - G^{id}_{ji})$, eq 7 can not be satisfied identically.[19,23] Equation 7 can be satisfied only if G^{id}_{ij} is replaced by another reference state G^V_{ij}, which for ideal mixtures reduces to G^{id}_{ij} (see Appendix 1 for the expressions of G^V_{ij}). Hence, the excess (the SR excess) which satisfies eq 7 is given by

$$\Delta n''_{ji} = c_j (G_{ji} - G^V_{ji}) \quad (8)$$

In the next section, the density enhancement provided by the BN, ML, and SR excesses will be examined in more detail at low dilution.

3. Density Enhancement in SCR Mixtures through the KB Theory of Solutions. Usually, the local densities in SCR mixtures were determined in very dilute solutions (molar fractions of solute between 10^{-4} and 10^{-8}), in order to avoid the experimental and computational complications caused by solute−solute interactions.[24,25] The experimental data are provided in refs 24 and 25 as either the density augmentation $\Delta \rho^{(2)}$ around a solute molecule[25] (the solvent is denoted as component 1 and the solute as component 2), or the local density $\rho^{(2)}$ around a solute molecule.[25]

The density augmentation is provided by the expression

$$\Delta \rho^{(2)} = \frac{\Delta n_{12} + \Delta n_{22}}{V^{(2)}_{cor}} \quad (9)$$

for the BN excesses and similar expressions in which Δn_{ji} are replaced by $\Delta n'_{ji}$ and $\Delta n''_{ji}$ for the ML and SR excesses, respectively. In eq 9, $V^{(2)}_{cor}$ is the correlation volume, i.e., the volume around a solute molecule where the density differs from that in the bulk. Usually, the density augmentation $\Delta \rho^{(2)}$ or the local density $\rho^{(2)}$ is compared with $\rho^{(2)}_{bulk}$, the bulk density around a solute molecule (which can be taken as the density of the pure SCF).

Now expressions for $\Delta \rho^{(2)}$ will be derived when the molar fraction of the solute is very small ($x_2 \approx 10^{-4}-10^{-8}$) for the three expressions written above for the aggregated excesses. For small x_2 one can suppose that $V_1 = V^0_1$, $V_2 = V^\infty_2$, $k_T = k^0_{T,1}$, where the superscripts (0) and (∞) refer to the pure component and infinite dilution, respectively, and $D = (\partial \ln \gamma_2/\partial x_2)_{P,T} x_2 + 1 \equiv K_2 x_2 + 1$, where K_2 is independent of composition.[26]

For the BN excess one thus obtains

$$\Delta n_{12} = c_1 G_{12} = c_1 (RT k^0_{T,1} - V^\infty_2) \quad (10a)$$

$$\Delta n_{22} = c_2 G_{22} = c_2 (RT k^0_{T,1} + V^0_1 - 2V^\infty_2 - K_2 V^0_1) \quad (10b)$$

which for infinite solute dilution become

$$\lim_{x_2 \to 0} \Delta n_{12} = \lim_{x_2 \to 0} c_1 G_{12} = c^0_1 (RT k^0_{T,1} - V^\infty_2) \quad (10c)$$

and

$$\lim_{x_2 \to 0} \Delta n_{22} = \lim_{x_2 \to 0} c_2 G_{22} = 0 \quad (10d)$$

The ML excesses are given by

$$\Delta n'_{12} = c_1 (G_{12} - G^{id}_{12}) = c_1 (V^0_2 - V^\infty_2) \quad (11a)$$

$$\Delta n'_{22} = c_2 (G_{22} - G^{id}_{22}) = c_2 (2V^0_2 - 2V^\infty_2 - K_2 V^0_1) \quad (11b)$$

At infinite dilution they become

$$\lim_{x_2 \to 0} \Delta n'_{12} = \lim_{x_2 \to 0} c_1 (G_{12} - G^{id}_{12}) = c^0_1 (V^0_2 - V^\infty_2) \quad (11c)$$

$$\lim_{x_2 \to 0} \Delta n'_{22} = \lim_{x_2 \to 0} c_2 (G_{22} - G^{id}_{22}) = 0 \quad (11d)$$

The SR excesses have the form

$$\Delta n''_{12} = c_1 (G_{12} - G^V_{12}) = c_1 V^\infty_2 K_2 x_2 \quad (12a)$$

$$\Delta n''_{22} = c_2 (G_{22} - G^V_{22}) = -c_2 V^0_1 K_2 \quad (12b)$$

and for infinite solute dilution

$$\lim_{x_2 \to 0} \Delta n''_{12} = \lim_{x_2 \to 0} c_1 (G_{12} - G^V_{12}) = 0 \quad (12c)$$

$$\lim_{x_2 \to 0} \Delta n''_{22} = \lim_{x_2 \to 0} c_2 (G_{22} - G^V_{22}) = 0 \quad (12d)$$

The local density enhancement for the BN excesses (eqs 10a and 10b) is given by

$$\Delta \rho^{(2)} = \frac{c_1 (RT k^0_{T,1} - V^\infty_2) + c_2 (RT k^0_{T,1} + V^0_1 - 2V^\infty_2 - K_2 V^0_1)}{V^{(2)}_{cor}} \quad (13)$$

for the ML excesses (eqs 11a−11b) by

$$\Delta\rho^{(2)} = \frac{c_1(V_2^0 - V_2^\infty) + c_2(2V_2^0 - 2V_2^\infty - K_2V_1^0)}{V_{cor}^{(2)}} \quad (14)$$

and for the SR excesses (eqs 12a and 12b) by

$$\Delta\rho^{(2)} = \frac{c_1 V_2^\infty K_2 x_2 - c_2 V_1^0 K_2}{V_{cor}^{(2)}} \quad (15)$$

The above three equations (eqs 13−15) will be further used to evaluate the correlation volume.

Calculations

1. Source of Data. There are only a few local density data, and partial molar volumes of solutes at infinite dilution are scarce as well. Only two systems could be identified for which data for the calculation of the correlation volume are available: CO_2 + naphthalene and CO_2 + pyrene. The augmented local density data in these systems were taken from ref 24 and the partial molar volume of the solute at infinite dilution in CO_2 + naphthalene system from ref 8. Because the partial molar volume of the solute at infinite dilution for the CO_2 + pyrene system was not available, it was taken equal to that for the CO_2 + phenanthrene.[27] The density and compressibility of the pure SCR CO_2 were taken from refs 1 and 24, respectively, and the solubilities of naphthalene and pyrene in SCR CO_2 from refs 28 and 29, respectively.

2. Calculation of K_2. K_2 is defined as

$$K_2 = \left(\frac{\partial \ln \gamma_2}{\partial x_2}\right)_{P,T,x_2 \to 0} \quad (16)$$

and was calculated through the fugacity coefficient of solute $\hat{\phi}_2$, which in the dilute region is given by[26]

$$\ln \hat{\phi}_2 = \ln \hat{\phi}_2^\infty + K_2 x_2 \quad (17)$$

where $\hat{\phi}_2^\infty$ is the infinite dilution fugacity coefficient. Solubility data were first used to calculate the fugacity coefficient at saturation, using the expression[30]

$$x_{2,S} = \frac{P_2^0}{P\hat{\phi}_2} \exp\left(\frac{(P - P_2^0)V_2^0}{RT}\right) \quad (18)$$

where $x_{2,S}$ is the molar fraction at saturation of the solute, and P_2^0 and V_2^0 are the saturation vapor pressure and molar volume of the solid solute, respectively. The values of P_2^0 and V_2^0 for naphthalene and pyrene were taken from ref 31. The equation for the fugacity coefficient based on the Soave−Redlich−Kwong equation of state was then employed to calculate the binary interaction parameter k_{12}, which appears in one of the mixing rules (see Appendix 2).

Further, the fugacity coefficient was calculated as a function of x_2, and the value of K_2 was obtained from the slope of the curve $\ln \hat{\phi}_2$ against x_2 (for additional details see Appendix 2). The calculated values of K_2 for the CO_2 + naphthalene and CO_2 + pyrene systems are plotted in Figure 1. Similar calculations were carried out using the Peng−Robinson (PR) EOS.[33] Good agreement was found between the values of K_2 obtained from the two equations of state.

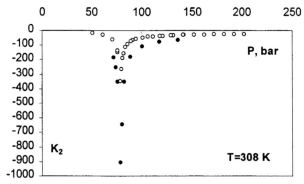

Figure 1. K_2 values calculated with the SRK EOS; (○) CO_2 + naphthalene, (●) CO_2 + pyrene.

TABLE 1: The Correlation Volume and Excess Number of Molecules around a Solute Molecule in Mixtures of Naphthalene and CO_2 Based on BN Excesses

P, bar	$V_{cor}^{(2)}$, cm³	$r_{cor}^{(2)}$, Å	$n_1 = c_1 V_{cor}^{(2)}$, number of bulk molecules	Δn_{12}, molecules
System CO_2 + Naphthalene, $x_2 = 3.5 \times 10^{-5}$, T = 308 K				
141.1	760	6.7	13.9	0.4
129.6	1930	9.1	34.5	1.0
106.8	2020	9.3	34.5	3.6
94.7	6910	14.0	111.3	6.9
89.2	10590	16.1	157.8	9.5
83.9	18000	19.3	250.7	45.0
81.9	34090	23.8	397.8	75.0
79.9	33860	23.8	329.0	115.1
78.7	19510	19.8	169.0	70.2
75.8	8790	15.2	58.6	24.6
System CO_2 + Pyrene, $x_2 = 3.0 \times 10^{-8}$, T = 308 K				
136.3	17	19	0.3	0.7
118.1	700	6.5	12.3	2.2
100.7	1950	9.2	32.8	6.4
89.0	5390	12.9	81.4	14.0
82.9	14180	17.8	181.4	69.5
80.4	16320	18.6	148.4	104.5
78.7	10020	15.8	77.4	61.1
75.6	4800	12.4	30.0	23.1
73.9	4040	11.7	23.0	18.2
71.6	4430	12.1	23.7	14.2

The correlation volume ($V_{cor}^{(2)}$) and correlation radius ($r_{cor}^{(2)} = (3V_{cor}^{(2)}/4\pi)^{1/3}$) were calculated using eqs 13−15. The results are summarized in Tables 1−3.

The calculations based on eqs 13 and 14 indicated that in the dilute region (molar fraction of solute between 10^{-4} and 10^{-8}), the second term in the numerator ($c_2(RTk_{T,1}^0 + V_1^0 - 2V_2^\infty - K_2V_1^0)$ in eq 13 and $c_2(2V_2^0 - 2V_2^\infty - K_2V_1^0)$ in eq 14) is negligible and that the relations valid at infinite dilution (eqs 10c and 11c) can be used. Equation 13, whose numerator at infinite dilution becomes $c_1^0(RTk_{T,1}^0 - V_2^\infty)$, was used to calculate the solvent cluster size around a solute molecule at infinite dilution.[21] The cluster size for a solution of naphthalene in SCF CO_2 at 308.39 K was thus found to vary between 20 and 100 molecules of CO_2 per molecule of solute in the pressure range 75−90 bars. The terms $RTk_{T,1}^0$ and V_2^∞ provided comparable contributions to the cluster size, because not too far from the critical point of the solvent, the compressibility $k_{T,1}^0$ has large positive values and the partial molar volume at infinite dilution of the solute is negative and large in absolute value. The above values are in agreement with the solvent cluster sizes around a solute molecule evaluated from the experimental partial molar volume of the solute at infinite dilution.[9] However, eq 10c and the equation for $\lim_{x_2 \to 0} \Delta n_{11}$ ($\lim_{x_2 \to 0} \Delta n_{11} = \lim_{x_2 \to 0} c_1 G_{11} = c_1^0(RTk_{T,1}^0 - V_1^0)$) should be valid not only for SCR mixtures, but also for

Density Microheterogeneities in Dilute SCF

TABLE 2: The Correlation Volume and Excess Number of Molecules around a Solute Molecule in Mixtures of Naphthalene and CO_2 based on ML Excesses

P, bar	$V_{cor}^{(2)}$, cm^3	$r_{cor}^{(2)}$, Å	$n_1 = c_1 V_{cor}^{(2)}$, number of bulk molecules	$\Delta n'_{12}$, molecules
\multicolumn{5}{c}{System CO_2 + Naphthalene, $x_2 = 3.5 \times 10^{-5}$, T = 308 K}
141.1	2700	10.2	49.2	1.6
129.6	3940	11.6	70.5	2.0
106.8	2140	9.5	36.6	3.8
94.7	5800	13.2	93.4	5.8
89.2	7620	14.4	113.5	6.9
83.9	14540	17.9	202.6	36.4
81.9	24450	21.3	285.3	53.8
79.9	21750	20.5	211.3	73.9
78.7	13500	17.5	116.9	48.6
75.8	5500	13.0	36.7	15.4
\multicolumn{5}{c}{System CO_2 + Pyrene, $x_2 = 3.0 \times 10^{-8}$, T = 308 K}
136.3	45	2.6	0.8	1.8
118.1	910	7.1	16.2	2.8
100.7	1830	9.0	30.8	6.0
89.0	4120	11.8	62.3	10.7
82.9	11000	16.4	140.8	53.9
80.4	10810	16.2	98.2	69.2
78.7	6980	14.0	53.9	42.6
75.6	3120	10.7	19.5	15.0
73.9	2660	10.2	15.1	12.0
71.6	2850	10.4	15.2	9.2

TABLE 3: The Correlation Volume and Excess Number of Molecules around a Solute Molecule in Mixtures of Naphthalene and CO_2 based on SR excesses

P, bar	$V_{cor}^{(2)}$, cm^3	$r_{cor}^{(2)}$, Å	$n_1 = c_1 V_{cor}^{(2)}$, number of bulk molecules	$\Delta n''_{12}$, molecules
\multicolumn{5}{c}{System CO_2 + Naphthalene, $x_2 = 3.5 \times 10^{-5}$, T = 308 K}
141.1	1.0	0.7	0.02	0
129.6	2.2	1.0	0.04	0
106.8	2.4	1.0	0.04	0
94.7	9.9	1.6	0.2	0
89.2	17.4	1.9	0.3	0.02
83.9	56.5	2.9	0.8	0.2
81.9	131.6	3.7	1.5	0.3
79.9	199.3	4.3	1.9	0.7
78.7	162.6	4.0	1.4	0.6
75.8	26.8	2.2	0.2	0.1
\multicolumn{5}{c}{System CO_2 + Pyrene, $x_2 = 3.0 \times 10^{-8}$, T = 308 K}
136.3	4×10^{-5}	0.02	7×10^{-7}	13×10^{-8}
118.1	0.001	0.08	2×10^{-5}	10^{-5}
100.7	0.005	0.12	9×10^{-5}	10^{-5}
89.0	0.02	0.20	0.0003	0.0001
82.9	0.1	0.4	0.001	0.001
80.4	0.2	0.4	0.002	0.001
78.7	0.2	0.4	0.001	0.002
75.6	0.03	0.2	0.0002	0.0002
73.9	0.02	0.2	0.0001	0.0001
71.6	0.02	0.2	9×10^{-5}	10^{-5}

any binary mixture such as alcohol/water far from the critical point. In the latter cases $RTk_{T,1}^0$ is negligible compared to V_2^∞ (which is positive) and V_1^0. This means that at infinite dilution ($x_2 \rightarrow 0$) and far from the critical point, $\lim_{x_2 \rightarrow 0} c_1 G_{12}$ and $\lim_{x_2 \rightarrow 0} c_1 G_{11}$ for all "normal" binary mixtures are negative and hence that there are deficits of molecules 1 in the aggregates around both 1 and 2 molecules compared to the bulk numbers, regardless of the nature of the interactions between molecules.

The results listed in Tables 1 and 2 also indicate that the sizes of the solvent aggregates predicted from the BN and ML excesses are in agreement with those evaluated from the large negative partial molar volume of the solute at infinite dilution.[9] However, as already noted the BN excesses provide unreasonable results for the aggregate sizes in mixtures far from the

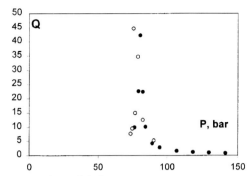

Figure 2. $Q = \langle(\Delta N_1)^2\rangle/\langle N_1\rangle$ in SCF CO_2 at 308 K: (○) experimental data[13] ($T = 307.1-307.2$ K), (●) calculated with eq 19.

critical point at infinite dilution. While the ML excesses provide reasonable values, they do not satisfy the volume conservation condition. The SR excesses do satisfy the volume conservation condition and lead at infinite dilution to zero excess solvent molecules aggregated around individual solute molecules. The latter result stimulated us to look for a different interpretation of the density enhancement. The comparison between the experimental results regarding the near critical density enhancement in the pure SCF and the density enhancement caused by the addition of extremely small amounts of solute appears to show that the former constitute the main effect. Details are provided in the next section.

Density Enhancement as a Result of the Near Critical Fluctuations in the Pure SCF

A number of experimental results are available regarding the quantity $Q = \langle(\Delta N_1)^2\rangle/\langle N_1\rangle$ (which involves the average of the square of the fluctuation ΔN_1 of the number of molecules N_1 and the average of the number of molecules N_1) and the correlation length in pure SCF.[11-15] This information was obtained via the small-angle X-ray scattering (SAXS) and refs 13-14 provide details about the pressure and temperature dependencies of $Q = \langle(\Delta N_1)^2\rangle/\langle N_1\rangle$ and correlation length in pure SCF CO_2. It is worth noting that $\langle(\Delta N_1)^2\rangle/\langle N_1\rangle$ can be also calculated using the expression[34]

$$\frac{\langle(\Delta N_1)^2\rangle}{\langle N_1\rangle} = c_1^0 RT k_{T,1}^0 \quad (19)$$

and Figure 2 shows that there is agreement between the calculated and experimental values of $Q = \langle(\Delta N_1)^2\rangle/\langle N_1\rangle$.

Let us compare the density enhancements generated by the near critical density fluctuations in a pure solvent to those observed when small amounts of solute were added to the solvent. The density enhancements in pure CO_2 and pure CHF_3 were calculated from the experimental data provided by refs 13-15 regarding the correlation lengths and the fluctuations of the number of molecules given by $Q = \langle(\Delta N_1)^2\rangle/\langle N_1\rangle$. The calculations are compared with the experimental results obtained for CO_2 + naphthalene and CO_2 + pyrene[24] in Figure 3 and for CHF_3 + pyrene[24] in Figure 4. Figures 3 and 4 clearly reveal that the density enhancements in these mixtures are almost identical to those in the pure solvent. We are therefore tempted to conclude that the density enhancement in dilute SCR mixtures is mainly caused by the near critical density fluctuations in the pure SCF. An explanation for the negative partial molar volume of the solute at infinite dilution, which differs from that which involves the aggregation of the solvent about the solute,[9] was provided by Economou and Donohue.[10] Because of the high

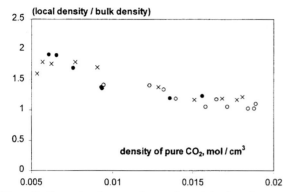

Figure 3. Comparison between the experimental density enhancement in pure SCF CO_2 and density enhancement in the SCR mixtures CO_2 + naphthalene and CO_2 + pyrene: (●) calculated from experimental data[13,14] ($T = 307.1–307.2$ K) for pure CO_2, (○) experimental data[24] for the system CO_2 + naphthalene ($T = 308$ K and $x_2 = 3.5 \times 10^{-5}$), (×) experimental data[24] for the system CO_2 + pyrene ($T = 308$ K and $x_2 = 3.0 \times 10^{-8}$).

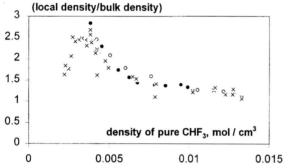

Figure 4. Comparison between the density enhancement in pure SCF CHF_3 and density enhancement in SCR mixtures CHF_3 + pyrene: (●) calculated from experimental data[15] ($T = 310.8–311.5$ K) for pure CHF_3, (○) experimental data[24] for the system CHF_3 + pyrene ($T = 303$ K and $x_2 = 5 \times 10^{-6}$), (×) experimental data[24] for the system CHF_3 + pyrene ($T = 303$ K and $x_2 = 3 \times 10^{-7}$). For this solvent (CHF_3) there are no data for densities lower than ≈ 0.004. However, the experimental data for pure SCF Ar indicate[35] a shape of the curve very similar to the experimental curves for the mixture CHF_3 + pyrene over the entire range of densities.

compressibility near the critical point of the solvent the addition of a small amount of solute causes a large decrease in the volume of the system.

Conclusion

One first demonstrates that in a dilute supercritical mixture the aggregation number of the solvent around individual solute molecules tends to zero at infinite solute dilution. These theoretical considerations suggested that it is necessary to revise the conventional explanation, which considers that the clustering of solvent molecules around individual solute molecules is responsible for the density augmentation observed when small amounts of solute were added to a solvent near the critical point of the latter. By comparing the density fluctuations determined by small-angle X-ray scattering for pure supercritical solvents and the density augmentation determined experimentally for dilute supercritical mixtures we noted that they almost coincide. In conclusion, it is likely that the near critical fluctuations are mainly responsible for the local density augmentation in dilute supercritical mixtures.

Appendixes

Appendix 1. (1) The KBIs for ideal systems are given by the expressions[18]

$$G_{12}^{id} = RTk_T^{id} - \frac{V_1^0 V_2^0}{V^{id}} \quad \text{(A1-1)}$$

$$G_{11}^{id} = G_{12}^{id} + V_2^0 - V_1^0 \quad \text{(A1-2)}$$

$$G_{22}^{id} = G_{12}^{id} - (V_2^0 - V_1^0) \quad \text{(A1-3)}$$

where k_T^{id} and V^{id} are the isothermal compressibility and the molar volume of an ideal mixture and V_i^0 is the molar volume of the pure component i.

(2) The KBIs at infinite dilution have the forms[17]

$$\lim_{x_i \to 0} G_{12} = RTk_{T,j}^0 - V_i^\infty \quad \text{(A1-4)}$$

$$\lim_{x_i \to 1} G_{12} = RTk_{T,i}^0 - V_j^\infty \quad \text{(A1-5)}$$

$$\lim_{x_i \to 0} G_{ii} = RTk_{T,j}^0 + V_j^0 - 2V_i^\infty - V_j^0 \left(\frac{\partial \ln \gamma_i}{\partial x_i}\right)_{P,T,x_i=0} \quad \text{(A1-6)}$$

$$\lim_{x_i \to 1} G_{ii} = RTk_{T,i}^0 - V_i^0 \quad \text{(A1-7)}$$

where V_i^∞ is the partial molar volume of component i at infinite dilution, $k_{T,i}^0$ is the isothermal compressibility of the pure component i.

(3) The expressions for G_{ij}^V are[19]

$$G_{12}^V = G_{21}^V = RTk_T - \frac{V_1 V_2}{V} \quad \text{(A1-8)}$$

$$G_{ii}^V = G_{12}^V + V_j - V_i \quad i \neq j \quad \text{(A1-9)}$$

Appendix 2. The Soave−Redlich−Kwong (SRK) EOS[32]

$$P = \frac{RT}{V - b} - \frac{a(T)}{V(V + b)} \quad \text{(A2-1)}$$

was selected to calculate the fugacity coefficient. In eq A2-1, V is the molar volume of the mixture and a and b are the EOS mixture parameters. The latter quantities are provided by the usual mixing rules:

$$a = x_1^2 a_1 + x_2^2 a_2 + 2x_1 x_2 (a_1 a_2)^{0.5}(1 - k_{12}) \quad \text{(A2-2)}$$

and

$$b = x_1 b_1 + x_2 b_2 \quad \text{(A2-3)}$$

where a_i, b_i are the EOS parameters for the pure component i, and k_{12} is the binary interaction parameter, which was determined by fitting the experimental solubility data.

For the fugacity coefficient of the solute $\hat{\phi}_2$ the following expressions was used:

$$RT \ln \hat{\phi}_2 = \frac{b_2}{b}(Z-1) - \ln(Z-B) +$$
$$\frac{A}{B}\left(\frac{2(x_2 a_2 + x_1(a_1 a_2)^{0.5}(1-k_{12}))}{a} - \frac{b_2}{b}\right) \times \ln\left(\frac{Z}{Z+B}\right) \quad \text{(A2-4)}$$

where $A = aP/(RT)^2$, $B = bP/RT$, and Z is the compressibility factor.

By fitting the experimental solubility data with the help of eqs 18 and A2-4, the binary interaction parameters k_{12} were determined (at 308 K, $k_{12} = 0.09803$ for CO_2 + naphthalene and $k_{12} = 0.07619$ for CO_2 + pyrene). These values of k_{12} were used to calculate K_2 at different pressures using eqs 17 and A2-4.

References and Notes

(1) Taylor, L. T. *Supercritical Fluid Extraction*; John Wiley & Sons: New York, 1996.
(2) McHugh, M.; Krukonis, V. *Supercritical Fluid Extraction*; Butterworth-Heinemann: Boston, 1994.
(3) Eckert, C. A.; Knutson, B. L.; Debenedetti, P. G. *Nature* **1996**, *383*, 313.
(4) Tucker, S. C.; Maddox, M. W. *J. Phys. Chem. B* **1998**, *102*, 2437.
(5) Tucker, S. C. *Chem. Rev.* **1999**, *99*, 391.
(6) Kajimoto, O. *Chem. Rev.* **1999**, *99*, 355.
(7) Brennecke, J. F.; Eckert, C. A. *AIChE J.* **1989**, *35*, 1409.
(8) Eckert, C. A.; Ziger, D. H.; Johnston, K. P.; Kim, S. *J. Phys. Chem.* **1986**, *90*, 2738.
(9) Eckert, C. A.; Ziger, D. H.; Johnston, K. P.; Ellison, T. K. *Fluid Phase Equil.* **1983**, *14*, 167.
(10) Economou, I. G.; Donohue, M. D. *AIChE J.* **1990**, *36*, 1920.
(11) Pfund, D. M.; Zemanian, T. S.; Linehan, J. C.; Fulton, J. L.; Yonker, C. R. *J. Phys. Chem.* **1994**, *98*, 11846.
(12) Nishikawa, K.; Tanaka, I. *Chem. Phys. Lett.* **1995**, *244*, 149.
(13) Nishikawa, K.; Tanaka, I.; Amemiya, Y. *J. Phys. Chem.* **1996**, *100*, 418.
(14) Nishikawa, K.; Morita, T. *J. Supercrit. Fluids* **1998**, *13*, 143.
(15) Nishikawa, K.; Morita, T. *J. Phys. Chem.* **1997**, *101*, 1413.
(16) Kirkwood, J. G.; Buff, F. P. *J. Chem. Phys.* **1951**, *19*, 774.
(17) Matteoli, E.; Lepori, L. *J. Chem. Phys.* **1984**, *80*, 2856.
(18) Matteoli, E. *J. Phys. Chem. B* **1997**, *101*, 9800.
(19) Shulgin, I.; Ruckenstein, E. *J. Phys. Chem. B* **1999**, *103*, 872.
(20) Ben-Naim, A. *J. Chem. Phys.* **1977**, *67*, 4884.
(21) Debenedetti, P. G. *Chem. Eng. Sci.* **1987**, *42*, 2203.
(22) Matteoli, E.; Lepori, L. *J. Chem. Soc. Faraday Trans.* **1995**, *91*, 431.
(23) Shulgin, I.; Ruckenstein, E. *J. Phys. Chem. B* **1999**, *103*, 2496.
(24) Brennecke, J. F.; Tomasko, D. L.; Peshkin, J.; Eckert, C. A. *Ind. Eng. Chem. Res.* **1990**, *29*, 1682.
(25) Knutson, B. L.; Tomasko, D. L.; Eckert, C. A.; Debenedetti, P. G.; Chialvo, A. A. In *Supercritical Fluid Technology. Theoretical and Applied Approaches in Analytical Chemistry*; Bright, F. V., McNally, M. E. P., Eds.; Am. Chem. Soc.: Washington, DC, 1992; Chapter 5.
(26) Debenedetti, P. G.; Kumar, S. K. *AIChE J.* **1986**, *32*, 1253.
(27) Shim, J.-J.; Johnston, K. P. *J. Phys. Chem.* **1991**, *95*, 353.
(28) Tsekhanskaya, Yu. V.; Iomtev, M. B.; Mushkina, E. V. *Russ. J. Phys. Chem.* **1964**, *38*, 1173.
(29) Bartle, K. D.; Clifford, A. A.; Jafar, S. A. *J. Chem. Eng. Data* **1990**, *35*, 355.
(30) Prausnitz, J. M.; Lichtenthaler, R. N.; Gomes de Azevedo, E. *Molecular Thermodynamics of Fluid - Phase Equilibria*, 2nd ed.; Prentice - Hall: Englewood Cliffs, NJ, 1986.
(31) Garnier, S.; Neau, E.; Alessi, P.; Cortesi, A.; Kikic, I. *Fluid Phase Equilibria* **1999**, *158–160*, 491.
(32) Soave, G. *Chem. Eng. Sci.* **1972**, *27*, 1197.
(33) Peng, D. Y.; Robinson, D. B. *Ind. Eng. Chem. Fundam.* **1976**, *15*,
(34) Landau, L. D.; Lifshitz, E. M. *Statistical Physics*, Pt. 1; Pergamon: Oxford, 1980.
(35) Carlier, C.; Randolph, T. W. *AIChE J.* **1993**, *39*, 876.

17 November 2000

Chemical Physics Letters 330 (2000) 551–557

CHEMICAL PHYSICS LETTERS

www.elsevier.nl/locate/cplett

Why density augmentation occurs in dilute supercritical solutions

Eli Ruckenstein *, Ivan L. Shulgin

Department of Chemical Engineering, State University of New York at Buffalo, Amherst, NY 14260, USA

Received 28 July 2000; in final form 12 September 2000

Abstract

The conventional explanation of the density augmentation in a supercritical solvent, observed spectroscopically when a small amount of a solute was added, involved the clustering of the solvent about individual solute molecules. Here it is suggested that the augmentation is not caused by the solute, but rather it is due to the preexisting near critical fluctuations in the pure solvent and the preference of the solute for the high density regions of the supercritical solvent. It is also shown that the local composition of the solute molecules about a solute molecules is enhanced compared to its bulk composition. © 2000 Elsevier Science B.V. All rights reserved.

1. Introduction

Dilute fluid mixtures near the critical point of the solvent exhibit a peculiar behavior because of their high sensitivity to changes in temperature and pressure. Among the supercritical fluids (SCF), CO_2 is widely used in many areas of technology because, in addition to the high sensitivity noted above, it has a critical temperature just above the room temperature, it is nontoxic and nonflammable and can therefore be used as an environmentally benign solvent in separation processes and as a medium for chemical reactions [1]. Numerous investigations have been carried out to study the unusual thermodynamic properties of dilute mixtures containing SCFs as solvents (SCR mixtures). One of them is the negative and large (compared to the molar volume of the pure solute) in magnitude partial molar volume of the solute at infinite dilution [2–4]. In contrast, the partial molar volumes at infinite dilution for typical nonelectrolyte mixtures far from the critical point have positive values comparable to the molar volumes of the respective pure components. Another unusual characteristic is the density enhancement in the solvent, observed spectroscopically when a small amount of solute is added [3]. To explain these observations, a clustering of the solvent molecules about individual solute molecules was suggested [3,5–7]. Other explanations involving the high compressibility near the critical point [8,9] and the long-range contribution to the partial molar properties of the solute were also suggested [10]. We recently noted that the augmented density in the solvent, observed spectroscopically when a small amount of a solute was added, almost coincides with that due to the near critical fluctuations in the pure solvent [11]. We suggest here that the enhanced density around a solute molecule at high dilution is caused by the

---

* Corresponding author. Fax: +1-716-645-3822.
 E-mail addresses: faeliru@acsu.buffalo.edu (E. Ruckenstein), ishulgin@eng.buffalo.edu (I.L. Shulgin).

0009-2614/00/$ - see front matter © 2000 Elsevier Science B.V. All rights reserved.
PII: S0009-2614(00)01131-3

higher solubility of the solute in the preexisting enhanced density regions of the near critical solvent. In other words, the enhanced density is not caused by the condensation of the solvent molecules upon the solute molecules, but by the preexisting density fluctuations in the solvent and the preference of the solute molecules for the high density regions of the latter. In addition, it is noted that the local concentration of the solute molecules about a solute molecule is higher than the bulk concentration. The preference of the solute for the high density regions of the solvent, the solute clustering and the high compressibility near the critical point are likely responsible for the peculiar partial molar properties of the solute at infinite dilution.

2. Fluctuations in SCR mixtures

Let us consider an open region of volume v (grand canonical ensemble) of a binary mixture in which there are N_1 molecules of type 1 and N_2 molecules of type 2. The average number of molecules of type i ($i = 1, 2$) is denoted by $\overline{N}_i$ and the local deviation from the average by ΔN_i. Thus the mean total number of molecules is $\overline{N} = \overline{N}_1 + \overline{N}_2$ and the local deviation is $\Delta N = \Delta N_1 + \Delta N_2$. If x_i is the bulk mole fraction of component i, then the local deviation from the bulk mole fraction is $\Delta x_i = (x_j \Delta N_i - x_i \Delta N_j)/\overline{N}$, where $i, j = 1, 2$ and $i \neq j$. Three kinds of fluctuations are relevant: the mean-square fluctuation in concentration $\overline{N}\langle(\Delta x_1)^2\rangle$, the mean-square fluctuation in particle number (often-called density fluctuation) $\langle(\Delta N)^2\rangle/\overline{N}$ and their correlation $\langle(\Delta x_1)(\Delta N)\rangle$ [12]. The following expressions can be written for these quantities [12]:

$$\overline{N}\langle(\Delta x_1)^2\rangle = \frac{\overline{N} k_B T}{(\partial^2 G/\partial x_1^2)_{T,P,\overline{N}}}, \quad (1)$$

$$\langle \Delta x_1 \Delta N \rangle = -\delta \overline{N} \langle(\Delta x_1)^2\rangle, \quad (2)$$

$$\langle(\Delta N)^2\rangle/\overline{N} = \frac{\overline{N}}{v} k_B T k_T + \delta^2 \overline{N} \langle(\Delta x_1)^2\rangle, \quad (3)$$

where T is the absolute temperature, P the pressure, G the molar Gibbs energy, k_T the isothermal compressibility, k_B the Boltzmann constant, $\delta = (\overline{N}/v)(v_1 - v_2)$ is the dilatation factor and v_1 and v_2 are the partial molar volumes per molecule of the two species of the mixture. For one component system ($x_1 = 1$) the expression for the density fluctuation becomes [12,13]

$$\langle(\Delta N_1)^2\rangle/\overline{N}_1 = \frac{\overline{N}_1}{v} k_B T k_T^0, \quad (4)$$

where k_T^0 is the isothermal compressibility of the pure component. Because k_T^0 becomes large near the critical point (at the critical point k_T^0 becomes infinite), the density fluctuations are large near the critical point of the solvent in both pure SCF and dilute SCR mixtures. It should be noted that $\langle \Delta N_1^2 \rangle/\overline{N}_1$ represents the average of the local density fluctuation. Indeed [14],

$$\langle(\Delta N_1)^2\rangle/\overline{N}_1 = \rho_1^0 \int_0^\infty (g_{11}^0 - 1) 4\pi r^2 \, dr + 1, \quad (5)$$

where the first term in the right-hand side represents the excess number of molecules in a pure liquid around any given central molecule and the second term accounts for the central molecule. Both together provide the average density fluctuation. In Eq. (5), g_{11}^0 is the radial distribution function of the pure liquid and ρ_1^0 is the density of the pure liquid.

3. Experimental determination of the local density microheterogeneities in pure SCF

Light scattering, small-angle X-ray scattering (SAXS) and small-angle neutron scattering (SANS) can be employed to determine the microheterogeneities caused by the near critical fluctuations. SAXS experiments were carried out with pure CO_2 and pure CHF_3 [15–17]. They provided the mean-square fluctuation in the particle number and also the correlation length, which characterizes the size of the volume in which those fluctuations occur. From the above experimental information the density enhancement in pure SCF could be calculated.

4. Local density augmentation in SCR mixtures

Numerous experimental papers have been devoted to the determination of the augmented density in the solvent, when a small amount of a solute was added. However, only two of them, based on fluorescence spectroscopy measurements, provided numerical values for these densities [18,19]. Augmented densities in SCR mixtures were determined in very dilute solutions (mole fractions of solute between 10^{-4} and 10^{-8}), in order to avoid the experimental and computational complications caused by solute–solute interactions [18,19]. In the present Letter, the augmented densities in dilute SCR mixtures [18,19] (CO_2 + naphthalene, CO_2 + pyrene and CHF_3 + pyrene) are compared to those in pure SCFs (CO_2 and CHF_3) caused by near critical fluctuations. The latter ones were calculated for SCF CO_2 and CHF_3 using the mean-square fluctuations in the particle number and the correlation lengths determined experimentally [16,17]. The density augmentation was calculated as $[\langle(\Delta N)^2\rangle/\overline{N}]/(\frac{4}{3}\pi r^3)$, where the radius of the microheterogeneities is related to the correlation length (l) via the relation [20] $l = 1.1 r$. The correlation lengths for pure CO_2 ($T = 307.1$–307.2 K, $74.1 < P < 90.4$ bar) are in the range 9–19 Å. The comparison is made in Figs. 1 and 2 and reveals that the pressure dependence

of augmented densities in dilute SCR mixtures and those in pure SCF are almost the same in a wide range of pressures. This means that the augmented densities in dilute SCR mixtures and those in pure SCF solvents have the same origin and are caused by the preexisting near critical solvent density fluctuations, and not by the clustering of solvent molecules around individual solute molecules. It is also worth noting that the experimental augmented densities in SCR mixtures CO_2 + naphthalene and CO_2 + pyrene (Fig. 1) are almost the same and hence do not depend on the nature of the solute. Besides, Fig. 2 indicates that the pressure dependence of the augmented density is almost independent of the solute mole fraction. Because the average augmented densities in the pure solvent and in the dilute solute system are almost the same, it is reasonable to conclude that the solute molecules prefer the higher density regions of the supercritical solvent, hence that the solute dissolves preferentially and samples the high-density regions of the solvent. It is of interest to notice that indeed the solubility of a solute in a SCF increases with the density of the latter [22,23]. Kumar and Johnston [24] proposed the following

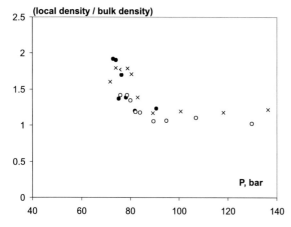

Fig. 1. Comparison between the experimental augmented densities in pure SCF CO_2 and in the SCR mixtures CO_2 + naphthalene and CO_2 + pyrene: (•) calculated from experimental data [16] ($T = 307.1$–307.2 K) for pure CO_2; (○) experimental data [18] for the system CO_2 + naphthalene ($T = 308$ K and $x_2 = 3.5 \times 10^{-5}$); (×) experimental data [18] for the system CO_2 + pyrene ($T = 308$ K and $x_2 = 3.0 \times 10^{-8}$).

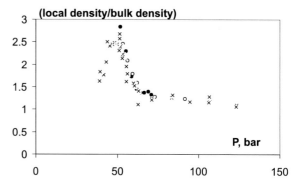

Fig. 2. Comparison between the augmented densities in pure SCF CHF_3 and in SCR mixtures CHF_3 + pyrene: (•) calculated from experimental data [17] ($T = 310.8$–311.5 K) for pure CHF_3; (○) experimental data [18] for the system CHF_3 + pyrene ($T = 303$ K and $x_2 = 5 \times 10^{-6}$); (×) experimental data [18] for the system CHF_3 + pyrene ($T = 303$ K and $x_2 = 3 \times 10^{-7}$). For pure CHF_3 there are no data for pressures lower than ≈ 50 bar. However the experimental data for pure SCF Ar indicate [21] a shape of the curve very similar to the experimental curves for the mixtures CHF_3 + pyrene over the entire range of pressures.

semiempirical expressions for the solubility x_2 of the solute in SCF

$$\ln x_2 = A + B\rho_r, \quad (6)$$

$$\ln x_2 = D + E \ln \rho_r, \quad (7)$$

where ρ_r is the reduced density of the solvent and A, B, D and E are constants.

In order to estimate the distribution of the solute between the average and the enhanced density regions, Eq. (7) will be employed for a CO_2 + naphthalene mixture. In this case $E = 2.74$ at 308 K [24] and Fig. 1 provides the ratio between the enhanced and bulk densities. For example, at $P = 75$ bar this ratio is about 1.9 and Eq. (7) provides the ratio of 6 between the solubilities of naphthalene in the dense and bulk regions. If instead of the bulk density, the density of the depleted region of the solvent would have been considered (assuming the same difference with changed sign between the depleted and bulk densities as between the enhanced and bulk ones), then the ratio of solubilities would be very large (about 3200). This means that most solute molecules are located in the enhanced densities regions of the solvent. This also contributes to the augmentation of the solute local composition in those regions of the SCF, issue which will be discussed in Section 5.

It is important to emphasize that a fluorescent solute molecule samples a large number of microheterogeneities. Indeed, because the excited-state fluorescence lifetime τ is relatively long (for pyrene [25] it has a value of the order of 1×10^{-7}–4×10^{-7} s) and the diffusion coefficient D [26] is on the order of 10^{-3} cm^2/s, the solute samples a region characterized by the length $\Delta \approx (D\tau)^{1/2} \approx 10^{-5}$ cm $= 10^3$ Å. The correlation length being of the order of 10 Å, it is clear that the solute samples a large number of microheterogeneities. Because of its tendency to locate in the high density regions, it is reasonable to consider that the solute fluorescence provides the average of the higher density regions.

5. Solute–solute interactions

As pointed out by Brennecke and Eckert [27] solute–solute interactions take place even when the solute mole fraction is as small as 10^{-5} or 10^{-6}. They concluded from spectroscopic studies that about 50 % of the solute molecules in the system CO_2 + pyrene were present as excimers because of the solute–solute interactions. While the interpretation of their data is in doubt [28], molecular dynamics simulations [29] indicated that the solute molecules prefer to be located in the neighborhood of other solute molecules, hence that the local solute concentration around a solute molecule is greater than the bulk concentration. There is some experimental evidence that such clustering occurs. Indeed, Randolph et al. [30] using high-pressure EPR spectroscopy found that in the system CO_2 + cholesterol, the cholesterol molecules did aggregate. In order to obtain more information about the solute–solute interactions in dilute SCR mixtures, the solute excess around a solute molecule will be calculated on the basis of the Kirkwood–Buff (KB) theory of solution [31]. The excess will be further employed to obtain the local mole fraction of component j in the vicinity of component i (binary system will be considered), using the expressions

$$x_{ji} = \frac{n_{ji}}{n_{ii} + n_{ji}} = \frac{\Delta n_{ji} + c_j V_{cor,i}}{\Delta n_{ii} + \Delta n_{ji} + (c_i + c_j)V_{cor,i}} \quad (8)$$

and

$$x_{ii} = \frac{n_{ii}}{n_{ii} + n_{ji}} = \frac{\Delta n_{ii} + c_i V_{cor,i}}{\Delta n_{ii} + \Delta n_{ji} + (c_i + c_j)V_{cor,i}}, \quad (9)$$

where n_{ji} is the number of moles of component j in the correlation volume $V_{cor,i}$ (in which the local composition differs from the bulk one), c_j the overall molar composition of component j, Δn_{ji} is the excess (or deficit) of number of molecules j in the vicinity of molecule i. The excesses Δn_{ji} can be expressed in terms of the KB integrals (G_{ij}) [32–34]. Several expressions for Δn_{ji} based on the KB theory of solution have been proposed [32–34].

The conventional excess number of j molecules around a central molecule i is given by [32]

$$\Delta n'_{ji} = c_j G_{ji} = c_j \int_0^\infty (g_{ij} - 1) 4\pi r^2 \, dr, \quad (10)$$

where g_{ij} is the radial distribution function between species i and j and r is the distance between the centers of molecules i and j. The expressions for KB integrals (G_{ji}) of a binary mixture are given in Appendix A.

Matteoli [33] noted that the $c_j G_{ji}$ are nonzero for ideal mixtures, which should be considered nonaggregated and suggested that the aggregation is better reflected in excesses (the ML excesses) defined as

$$\Delta n''_{ji} = c_j(G_{ji} - G^{id}_{ji}), \qquad (11)$$

where G^{id}_{ji} are the KBs integrals for ideal binary systems (see Appendix A for the expressions of G^{id}_{ij} for binary mixtures). In other words, the ideal binary system constitutes a background reference state. However, $\Delta n''_{ii}$ and $\Delta n''_{ji}$ are not independent quantities, because the volume occupied by $\Delta n''_{ii}$ of i molecules around an i molecule should be equal to the volume left free by $\Delta n''_{ji}$ of j molecules around the same i molecule [33]. This leads to the relation

$$V_j \Delta n''_{ji} = -V_i \Delta n''_{ii}, \qquad (12)$$

where V_i is the partial molar volume of component i.

It was however noted that if the excess number of molecules is calculated as $c_j(G_{ji} - G^{id}_{ji})$, Eq. (12) can not be satisfied identically [34]. Eq. (12) can be satisfied only if G^{id}_{ij} is replaced by another reference state G^{V}_{ij}, which for ideal binary mixtures reduces to G^{id}_{ij} (See Appendix A for the expressions of G^{V}_{ij} for binary mixtures.) One can verify that the expressions for G^{V}_{ij} can be obtained from those for G^{id}_{ij} by replacing the volumes of the pure components by the partial molar volumes and the ideal compressibility by the real one. Hence, the excess (the SR excess) which satisfies Eq. (12) is given by

$$\Delta n_{ji} = c_j(G_{ji} - G^{V}_{ji}). \qquad (13)$$

Expressions for the local compositions can be found from any of the Eqs. (10) and (11) or Eq. (13). In the present paper, the local mole fraction of the solute around a central solute molecule in dilute SCR solutions is of interest. The following expressions are obtained using the appropriate excess and the equations from Appendix A.

The conventional excess (Eq. (10)) provides the following expression for the local composition x'_{22}

$$\lim_{x_2 \to 0} \frac{x'_{22}}{x_2} = \frac{RTk^0_{T,1} + V^0_1 - 2V^\infty_2 - V^0_1 K_2 + V_{cor,2}}{RTk^0_{T,1} - V^\infty_2 + V_{cor,2}}. \qquad (14)$$

The ML excess (Eq. (11)) leads to the following expression for the local composition x''_{22}

$$\lim_{x_2 \to 0} \frac{x''_{22}}{x_2} = \frac{2V^0_2 - 2V^\infty_2 - V^0_1 K_2 + V_{cor,2}}{V^0_2 - V^\infty_2 + V_{cor,2}} \qquad (15)$$

while the SR excess leads to the following expression for the local composition x_{22}

$$\lim_{x_2 \to 0} \frac{x_{22}}{x_2} = 1 - \frac{V^0_1 K_2}{V_{cor,2}}. \qquad (16)$$

In Eqs. (14)–(16), V^0_i is the molar volume of the pure component i, V^∞_i is the partial molar volume of component i at infinite dilution,

$$K_2 = \lim_{x_2 \to 0} \left(\frac{\partial \ln \gamma_2}{\partial x_2} \right)_{P,T},$$

where γ_2 is the activity coefficient of the solute. Because the latter expression (Eq. (16)) incorporates the fact that the ideal mixture is nonaggregated and involves the volume conservation condition (Eq. (12)), it provides a better measure of aggregation than Eqs. (14) and (15).

The calculation of

$$\lim_{x_2 \to 0} \frac{X_{22}}{x_2} \quad (\text{with } X_{22} = x'_{22}, x''_{22} \text{ or } x_{22})$$

was carried out for the SCR mixture CO_2 + pyrene using Eqs. (14)–(16). All the necessary values ($K_2, V^0_i, V^\infty_i, k^0_{T,i}$) were taken from our previous paper [11]. The correlation volume ($V_{cor,2}$) in Eqs. (12)–(14) was taken to be equal to a sphere with a radius of 7.4 Å, as suggested in Ref. [19]. The pressure dependence of $\lim_{x_2 \to 0} X_{22}/x_2$ is presented in Fig. 3, which shows that $\lim_{x_2 \to 0} x_{22}/x_2$ increases when the pressure approaches the critical value, but that the $\lim_{x_2 \to 0} x'_{22}/x_2$ and $\lim_{x_2 \to 0} x''_{22}/x_2$ become negative (and this is unphysical) near the critical point. Because the SR excess is a better measure of aggregation than the other two excesses, it is likely for the local concentration of the solute around a solute molecule to be larger than the bulk concentration, particularly near the critical point.

The increased local concentration of the solute, the tendency of the solute to be located in the high

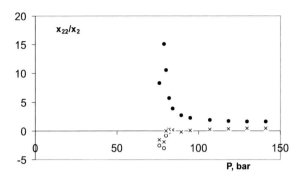

Fig. 3. The pressure dependence of $\lim_{x_2 \to 0} X_{22}/x_2$ ($X_{22} = x'_{22}, x''_{22}$ or x_{22}) in $CO_2(1)$ + pyrene (2) mixture, calculated using the three excesses (Eqs. (14)–(16)): (×) calculated with Eq. (14); (○) calculated with Eq. (15); (●) calculated with Eq. (16).

density regions of the near critical solvent and the rapid increase in compressibility near the critical point are most likely responsible for the unusual thermodynamic properties of dilute supercritical mixtures. Details in this direction will be examined in a forthcoming paper.

6. Conclusion

Evidence was brought that the density enhancement in a supercritical solvent, determined spectroscopically when small amounts of solute molecules are added, almost coincides with the preexisting density enhancement in the pure solvent. Consequently, the density enhancement is not caused by the presence of the solute, but is due to the preexisting near critical fluctuations in the solvent. The higher solubility of the solute in the high density regions of the solvent allows one to determine via fluorescence spectroscopy the average enhancement in the high density regions of the solvent. In addition, there is an enhanced local concentration of solute molecules about a solute molecule.

Appendix A

(1) The KBIs for binary systems are given by the expressions [35]

$$G_{12} = RTk_T - \frac{V_1 V_2}{VD}, \quad (A.1)$$

$$G_{11} = G_{12} + \frac{1}{x_1}\left(\frac{V_2}{D} - V\right) \quad (A.2)$$

and

$$G_{22} = G_{12} + \frac{1}{x_2}\left(\frac{V_1}{D} - V\right), \quad (A.3)$$

where R is the universal gas constant, k_T the isothermal compressibility, V the molar volume, V_i the partial molar volume, x_i the molar fraction of component i,

$$D = 1 + x_i\left(\frac{\partial \ln \gamma_i}{\partial x_i}\right)_{P,T}$$

and γ_i is the activity coefficient of component i.

(2) The KBIs for ideal binary systems are given by the expressions [33]

$$G_{12}^{id} = RTk_T^{id} - \frac{V_1^0 V_2^0}{V^{id}}, \quad (A.4)$$

$$G_{11}^{id} = G_{12}^{id} + V_2^0 - V_1^0 \quad (A.5)$$

and

$$G_{22}^{id} = G_{12}^{id} - (V_2^0 - V_1^0), \quad (A.6)$$

where k_T^{id} and V^{id} are the isothermal compressibility and the molar volume of an ideal mixture and V_i^0 is the molar volume of the pure component i.

(3) The expressions for G_{ij}^V are [34]

$$G_{12}^V = G_{21}^V = RTk_T - \frac{V_1 V_2}{V}, \quad (A.7)$$

$$G_{ii}^V = G_{12}^V + V_j - V_i, \quad i \neq j. \quad (A.8)$$

References

[1] M. McHugh, V. Krukonis, Supercritical Fluid Extraction, Butterworth-Heinemann, Boston, 1994.
[2] N.E. Khazanova, E.E. Sominskaya, Russ. J. Phys. Chem. 42 (1968) 676.
[3] C.A. Eckert, D.H. Ziger, K.P. Johnston, T.K. Ellison, Fluid Phase Equilibria 14 (1983) 167.
[4] C.A. Eckert, D.H. Ziger, K.P. Johnston, S. Kim, J. Phys. Chem. 90 (1986) 2738.

[5] P.G. Debenedetti, Chem. Eng. Sci. 42 (1987) 2203.
[6] C.A. Eckert, B.L. Knutson, P.G. Debenedetti, Nature 383 (1996) 313.
[7] O. Kajimoto, Chem. Rev. 99 (1999) 355.
[8] I.G. Economou, M.D. Donohue, AIChE J. 36 (1990) 1920.
[9] R. Fernandez-Prini, Pure Appl. Chem. 67 (1995) 519.
[10] A.A. Chialvo, P.T. Cummings, AIChE J. 40 (1994) 1558.
[11] E. Ruckenstein, I. Shulgin, J. Phys. Chem. B 104 (2000) 2540.
[12] A.B. Bhatia, D.E. Thornton, Phys. Rev. B 2 (1970) 3004.
[13] L.D. Landau, E.M. Lifshitz, Statistical Physics, Pt.1, Pergamon Press, Oxford, 1980.
[14] D.A. McQuarrie, Statistical Mechanics, Harper & Row, New York, 1975.
[15] D.M. Pfund, T.S. Zemanian, J.C. Linehan, J.L. Fulton, C.R. Yonker, J. Phys. Chem. 98 (1994) 11846.
[16] K. Nishikawa, I. Tanaka, Y. Amemiya, J. Phys. Chem. 100 (1996) 418.
[17] K. Nishikawa, T. Morita, J. Phys. Chem. 101 (1997) 1413.
[18] J.F. Brennecke, D.L. Tomasko, J. Peshkin, C.A. Eckert, Ind. Eng. Chem. Res. 29 (1990) 1682.
[19] B.L. Knutson, D.L. Tomasko, C.A. Eckert, P.G. Debenedetti, A. Chialvo, in: F.V. Bright, M.E.P. McNally (Eds.), Supercritical Fluid Technology. Theoretical and Applied Approaches in Analytical Chemistry, ACS Symp. Ser. No. 488, American Chemical Society, Washington, DC, 1992, p. 60 (Chapter 5).
[20] K. Nishikawa, Y. Kodera, T. Iijima, J. Phys. Chem. 91 (1987) 3694.
[21] C. Carlier, T.W. Randolph, AIChE J. 39 (1993) 876.
[22] Yu.V. Tsekhanskaya, M.B. Iomtev, E.V. Mushkina, Russ. J. Phys. Chem. 38 (1964) 1173.
[23] C. Schultze, M.D. Donohue, Fluid Phase Equilibria 116 (1996) 465.
[24] S.K. Kumar, K.P. Johnston, J. Supercrit. Fluids 1 (1988) 15.
[25] J.R. Lakowicz, Principles of Fluorescence Spectroscopy, Kluwer Academic Publishers/Plenum Press, New York, 1999.
[26] M.D. Palmieri, J. Chem. Edu. 10 (1988) A254.
[27] J.F. Brennecke, C.A. Eckert, in: K.P. Johnston, J.M.L. Penninger (Eds.), Supercritical Fluid Science and Technology, ACS Symp. Ser. No. 406, American Chemical Society, Washington, DC, 1988, p. 14 (Chapter 2).
[28] J. Zagrobelny, F.V. Bright, in: F.V. Bright, M.E.P. McNally (Eds.), Supercritical Fluid Technology. Theoretical and Applied Approaches in Analytical Chemistry, ACS Symp. Ser. No. 488, American Chemical Society, Washington, DC, 1992, p. 73.
[29] P.G. Debenedetti, A.A. Chialvo, J. Chem. Phys. 97 (1992) 504.
[30] T.W. Randolph, D.S. Clark, H.W. Blanch, J.M. Prausnitz, Proc. Natl. Acad. Sci. USA 85 (1988) 2979.
[31] J.G. Kirkwood, F.P. Buff, J. Chem. Phys. 19 (1951) 774.
[32] A. Ben-Naim, J. Chem. Phys. 67 (1977) 4884.
[33] E. Matteoli, J. Phys. Chem. B 101 (1997) 9800.
[34] I. Shulgin, E. Ruckenstein, J. Phys. Chem. B 103 (1999) 2496.
[35] E. Matteoli, L. Lepori, J. Chem. Phys. 80 (1984) 2856.

Journal of Molecular Liquids 95 (2002) 205–226
www.elsevier.com/locate/molliq

Fluctuations in dilute binary supercritical mixtures

I. Shulgin [1] and E. Ruckenstein [*]

Department of Chemical Engineering, State University of New York at Buffalo, Amherst, NY 14260, USA

Received 11 July 2001; accepted 25 October 2001

The density fluctuation and the fluctuations in the number of particles of a binary mixture are expressed in terms of measurable thermodynamic quantities and the obtained equations are used in the dilute range to compare the behaviors of the CO_2 / naphthalene and CO_2 / ethane mixtures. When a small amount of naphthalene is added to the supercritical CO_2, the density fluctuation and those in the number of particles exhibit sharp increases, with increasing naphthalene concentration, compared to the corresponding ideal mixture. In contrast, the fluctuations in the mixture CO_2 / ethane almost coincide to those of the corresponding ideal mixture. The opposite behaviors regarding the partial molar volume in dilute solutions of ethane in CO_2 and CO_2 in ethane, which have positive partial molar volumes for the solute, and naphthalene in CO_2, which has a negative partial molar volume for the solute, are explained on the basis of general statistical thermodynamic equations. The degrees of nonideality are responsible for the different behaviors of the above mixtures.
© 2002 Elsevier Science B.V. All rights reserved.

1. INTRODUCTION

The supercritical fluids (SCF) have a wide range of applications, which are described in several books [1,2] and reviews [3-5]. The solutions involving a SCF as solvent (they will

---

[*] Correspondence author. E-mail: feaeliru@acsu.buffalo.edu. Fax: (716) 645-3822.

[1] E-mail address: ishulgin@eng.buffalo.edu

0167-7322/02/$ - see front matter © 2002 Elsevier Science B.V. All rights reserved.
PII S0167-7322(01) 00348-8

be called SC solutions) exhibit very unusual thermodynamic properties [6-8]. For example, the partial molar volume of naphthalene at infinite dilution in a mixture of naphthalene and supercritical carbon dioxide is -2300 cm^3/mol at P=76 bar and T=308.38 K [9], while the partial molar volume of ethane at infinite dilution in a mixture of ethane and supercritical carbon dioxide is $+350$ cm^3/mol at P=73.97 bar and T=305.65 K [7]. The partial molar volumes at infinite dilution for typical nonelectrolyte mixtures far from the critical point have positive values, comparable to the molar volumes of the corresponding pure components. It is worth noting that the partial molar volume of naphthalene at infinite dilution diverges negatively and that of ethane diverges positively at the critical point of the solvent. According to the classification of Debenedetti and Mohamed [10] the above two mixtures (CO_2/ naphthalene and CO_2/ ethane) belong to different types of supercritical mixtures; CO_2/ naphthalene is an "attractive" mixture and CO_2/ ethane is a "weakly attractive" mixture.

Petsche and Debenedetti [11] using various equations of state, which are generally used far from the critical point, related the behavior at infinite dilution of these systems to the values of the parameters involved. The goal of this paper is to analyze the problem by examining all types of fluctuations for dilute (not infinitely dilute) solutions without employing a model, but using only general statistical thermodynamic equations. On this basis the behavior of dilute supercritical solutions will be associated to their degree of nonideality. First, expressions for various fluctuations in dilute binary mixtures will be written, which will be combined with experimental information regarding the partial molar volumes, isothermal compressibilities and densities to obtain the concentration dependence of the fluctuations. Further, general statistical thermodynamic equations will be employed to explain the behavior of the partial molar volumes of the solutes in the above mixtures at high dilution.

The importance of fluctuations in the understanding the properties of SC mixtures in the dilute region was recently emphasized by the authors [12]. Arguments were brought that the density augmentation in a supercritical solvent, observed spectroscopically when an extremely small amount of a solute is added, is not caused by the solute, but rather it is due to the preexisting near critical fluctuations in the pure solvent and the preference of the solute for the high density regions of the supercritical solvent [12].

2. THEORY AND FORMULAS

2.1. Expressions for fluctuations

The fluctuation in the number of particles (or density fluctuation) in one-component and multicomponent mixtures can be determined experimentally by light scattering, small-angle X-ray scattering (SAXS) and small-angle neutron scattering (SANS). These experimental methods also allowed one to determine the correlation length, which constitutes a measure of the size of the microheterogeneities. SAXS and SANS methods were used to obtain information about the density fluctuation and correlation length in near critical CO_2 and CHF_3 [13-18]. There are no experimental data about the density fluctuation in dilute SC mixtures. However, this fluctuation can be calculated on the basis of available thermodynamic information.

Let us consider an open region of volume v (grand canonical ensemble) of a binary mixture in which there are N_1 molecules of type 1 and N_2 molecules of type 2. The average number of molecules of type i (i=1, 2) is denoted by $\langle N_i \rangle$ and the deviation from the average by ΔN_i. Thus the mean of the total number of molecules is $\langle N \rangle = \langle N_1 \rangle + \langle N_2 \rangle$ and the deviation is $\Delta N = \Delta N_1 + \Delta N_2$. The fluctuation in concentration for a binary mixture is given by [19]

$$N \langle (\Delta x_i)^2 \rangle = \frac{RT}{(\partial^2 G / \partial x_i^2)_{T,P}} \qquad (1)$$

where T is the absolute temperature, P is the pressure, x_i is the bulk mole fraction of component i, G is the molar Gibbs energy and R is the universal gas constant.

General expressions for the fluctuations in the number of particles in the binary mixtures can be obtained on the basis of the following relations [20]

$$G_{ij} = v\left[\frac{\langle(\Delta N_i)(\Delta N_j)\rangle}{\langle\langle N_i\rangle\rangle\langle\langle N_j\rangle\rangle} - \frac{\delta_{ij}}{\langle N_i\rangle}\right] \tag{2}$$

where G_{ij} is the Kirkwood-Buff integral [20] ($G_{ij} = \int_0^\infty (g_{ij}-1)4\pi r^2 dr$), δ_{ij} is the Kroneker symbol, g_{ij} is the radial distribution function between species i and j, and r is the distance between the centers of species i and j. Since the Kirkwood-Buff integrals can be expressed in terms of thermodynamic parameters [21], one can write the following relations for the fluctuations in a binary mixture

$$\left[\frac{\langle(\Delta N_1)^2\rangle}{\langle N_1\rangle}\right] = \left(RTk_T - \frac{V_1V_2}{VD} + \frac{V_2}{x_1 D}\right)\frac{x_1}{V} \tag{3}$$

$$\left[\frac{\langle(\Delta N_2)^2\rangle}{\langle N_2\rangle}\right] = \left(RTk_T - \frac{V_1V_2}{VD} + \frac{V_1}{x_2 D}\right)\frac{x_2}{V} \tag{4}$$

$$\left[\frac{\langle\Delta N_1\Delta N_2\rangle}{\langle N_1\rangle}\right] = \left(RTk_T - \frac{V_1V_2}{VD}\right)\frac{x_2}{V} \tag{5}$$

and

$$\left[\frac{\langle\Delta N_1\Delta N_2\rangle}{\langle N_2\rangle}\right] = \left(RTk_T - \frac{V_1V_2}{VD}\right)\frac{x_1}{V} \tag{6}$$

where

$$D = \left(\frac{\partial \ln \gamma_i}{\partial x_i}\right)_{P,T} x_i + 1 \tag{7}$$

In equations (3-7), k_T is the isothermal compressibility of a binary mixture, V_i is the partial molar volume of component i, V is the molar volume of the mixture and γ_i is the activity coefficient of component i. The expressions at infinite dilution of the fluctuations in the number of particles can be easily derived from eqs 3-6 and are listed in the Appendix.

For an ideal mixture, one can obtain from eq 1 the following expression for the concentration fluctuation

$$N\langle(\Delta x_i)^2\rangle_{id} = x_1 x_2 \tag{8}$$

The fluctuations in the number of particles in ideal mixtures can be also obtained from eqs 3-6 and are listed in the Appendix.

While eqs. 1, 3-6 have a general character, it is difficult to apply them to SC mixtures because the values of k_T, D, V_i and V are usually unknown. However, for very dilute (but not infinitely dilute) SC mixtures information about those parameters can be obtained. For the dilute region of component 2, one can write

$$k_T = \varphi_1 k_{T,1}^0 + \varphi_2 k_{T,2}^0 \tag{9}$$

$$V = x_1 V_1^0 + x_2 V_2^\infty \tag{10}$$

and

$$D = 1 + x_2 k_{22} \tag{11}$$

where $k_{T,i}^0$ is the isothermal compressibility of the pure component i, V_i^0 is the molar volume of the pure component i, V_2^∞ is the partial molar volume of component 2 at infinite dilution, $\varphi_i = V_i^0 x_i / (V_1^0 x_1 + V_2^0 x_2)$ is the volume fraction of component i and

$$k_{22} = \lim_{x_2 \to 0} \left(\frac{\partial \ln \gamma_2}{\partial x_2}\right)_{P,T}$$

is a composition independent parameter [22]. Using eqs. 9-11, eqs. 1, 3-6 become

$$N\langle(\Delta x_1)^2\rangle = \frac{x_1 x_2}{1+x_2 k_{22}} \quad (12)$$

$$\left[\frac{\langle(\Delta N_1)^2\rangle}{\langle N_1\rangle}\right] = \frac{x_1 RT(\varphi_1 k_{T,1}^0 + \varphi_2 k_{T,2}^0)}{x_1 V_1^0 + x_2 V_2^\infty} + \frac{x_2 (V_2^\infty)^2}{(x_1 V_1^0 + x_2 V_2^\infty)^2 (1+x_2 k_{22})} \quad (13)$$

$$\left[\frac{\langle(\Delta N_2)^2\rangle}{\langle N_2\rangle}\right] = \frac{x_2 RT(\varphi_1 k_{T,1}^0 + \varphi_2 k_{T,2}^0)}{x_1 V_1^0 + x_2 V_2^\infty} + \frac{x_1 (V_1^0)^2}{(x_1 V_1^0 + x_2 V_2^\infty)^2 (1+x_2 k_{22})} \quad (14)$$

$$\left[\frac{\langle\Delta N_1 \Delta N_2\rangle}{\langle N_1\rangle}\right] = \frac{x_2 RT(\varphi_1 k_{T,1}^0 + \varphi_2 k_{T,2}^0)}{x_1 V_1^0 + x_2 V_2^\infty} - \frac{x_2 V_1^0 V_2^\infty}{(x_1 V_1^0 + x_2 V_2^\infty)^2 (1+x_2 k_{22})} \quad (15)$$

and

$$\left[\frac{\langle\Delta N_1 \Delta N_2\rangle}{\langle N_2\rangle}\right] = \frac{x_1 RT(\varphi_1 k_{T,1}^0 + \varphi_2 k_{T,2}^0)}{x_1 V_1^0 + x_2 V_2^\infty} - \frac{x_1 V_1^0 V_2^\infty}{(x_1 V_1^0 + x_2 V_2^\infty)^2 (1+x_2 k_{22})} \quad (16)$$

One can see from eqs 12-16 that the concentration fluctuation and the fluctuations in the number of particles can be calculated in the dilute region if the values of $k_{T,1}^0$, $k_{T,2}^0$, V_1^0, V_2^∞ and k_{22} are available. For the systems CO_2/ naphthalene and CO_2/ ethane such values are either available [7,9] or can be calculated as indicated in the next section.

3. CALCULATION OF FLUCTUATIONS OF NEAR-CRITICAL MIXTURES IN THE DILUTE REGION

The fluctuations were calculated for the following dilute SC mixtures: 1) the dilute mixture of CO_2 in supercritical ethane at P=50.73 bar and t=32.5 ^0C, 2) the dilute mixture of ethane in supercritical CO_2 at P=73.97 bar and t=32.5 ^0C, 3) the dilute mixture of naphthalene in supercritical CO_2 at P=75.99 bar and t=35 ^0C.

The partial molar volumes of a solute at infinite dilution were taken from Refs 7 and 9 (for the mixture CO_2/ethane they were estimated by graphical extrapolation). The molar volumes of the pure SCF (CO_2 or ethane) were calculated using multi-parameter equations of state, which are very accurate in the critical region [23, 24]. The isothermal compressibilities of CO_2 and ethane at the above pressures and temperatures were calculated using the same multi-parameter equations of state [23, 24]. Because the isothermal compressibilities of solids are negligible compared to those of SCF, the isothermal compressibility of naphthalene in eqs. 13-16 was taken equal to zero. The parameters k_{22} for the mixture CO_2/naphthalene were calculated with the Soave-Redlich-Kwong equation of state as in our previous paper [25], using experimental data [26] regarding the solubility of naphthalene in supercritical CO_2. The parameters k_{22} for the mixture CO_2/ethane were calculated on the basis of the linear dependence of the logarithm of the fugacity coefficient on the mole fraction in the dilute region [22]. The fugacity coefficient was calculated with the Soave-Redlich-Kwong equation of state as in a previous paper [25]. The binary interaction parameter (K_{12}) of the van der Waals mixing rules was calculated separately for the dilute mixture of ethane in CO_2 and for the dilute mixture of CO_2 in ethane, using P-V-T-x data from literature [27]. First, the dilute mixture of ethane in CO_2 was examined. The experimental P-V data [27] for a mole fraction of ethane of 0.0225 at t=32.5 ^0C were used to calculate the binary interaction parameter (K_{12}). The linear composition dependence of the logarithm of the fugacity coefficient in the dilute region provided the values of k_{22}. The same procedure was used for the dilute mixture of CO_2 in ethane. The experimental P-V data for a mole fraction of CO_2 of 0.0175 were used to calculate the binary interaction parameter (K_{12}). All the parameters ($k_{T,1}^0$, $k_{T,2}^0$, V_1^0, V_2^∞ and k_{22}) used for calculating the fluctuations are listed in Table 1. The data from Table 1 show that the systems CO_2/ethane and CO_2/naphthalene differ not only through the values of the partial molar volumes of solutes at infinite dilution, but also through the values of the parameter k_{22}. Indeed, the parameter k_{22} for the mixture CO_2/naphthalene is negative and large in magnitude, while its value is small, around zero, for the mixture

CO_2/ethane. Because for an ideal mixture the parameter k_{22} is zero, the SC mixture CO_2/ethane behaves quite similar to an ideal mixture (see the next section for a detailed discussion).

Table 1

The numerical values of V_1^0, V_2^∞, $k_{T,1}^0$, $k_{T,2}^0$ and k_{22} used in the calculations of the fluctuations.

Mixture	V_1^0, cm^3/mol, V_2^0, cm^3/mol	V_2^∞, cm^3/mol	$k_{T,1}^0$, bar^{-1}	$k_{T,2}^0$, bar^{-1}	k_{22}
CO_2/ethane [A]	108.05 355.87	700	0.031	0.029	-1.59
CO_2/ethane [B]	145.35 87.6	350	0.085	0.0035	-2.72
CO_2/naphthalene [C]	150.83 110	-2300	0.061	0	-145

[A]- dilute mixture of CO_2 in supercritical ethane at P=50.73 bar and t=32.5 ^0C, component 1 is ethane and component 2 is CO_2; [B]- dilute mixture of ethane in supercritical CO_2 at P=73.97 bar and t=32.5 ^0C, component 1 is CO_2 and component 2 is ethane; [C]- P=75.99 bar and t=35 ^0C, component 1 is CO_2 and component 2 is naphthalene.

4. THE FLUCTUATIONS IN THE DILUTE REGION OF NEAR-CRITICAL MIXTURES: RESULTS AND DISCUSSION

The calculated fluctuations $N\langle(\Delta x_i)^2\rangle$, $\dfrac{\langle(\Delta N_i)^2\rangle}{\langle N_i\rangle}$ and $\dfrac{\langle\Delta N_1 \Delta N_2\rangle}{\langle N_i\rangle}$ in the mixtures investigated and in ideal mixtures are plotted in Figs.1-4. Fig. 1 presents the concentration fluctuation in dilute mixtures of naphthalene and ethane in SCF CO_2 near the critical point of the solvent. It shows that the addition of a small amount of ethane to SCF CO_2 leads to a concentration fluctuation, which almost coincides with that of an ideal mixture. In contrast, the addition of a small amount of naphthalene to SCF CO_2 leads to a sharp increase in the concentration fluctuation in the dilute mixture compared to the ideal one.

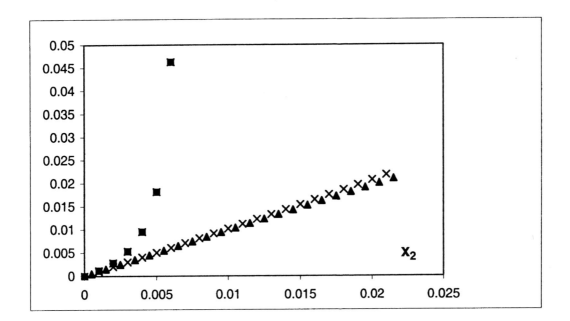

Fig. 1. Concentration fluctuation $N\langle(\Delta x_1)^2\rangle$ in the following mixtures:

(■)-dilute mixture of naphthalene (2) in CO_2 (1) at P=75.99 bar and t=35 ^0C,

(x)- dilute mixture of ethane (2) in CO_2 (1) at P=73.97 bar and t=32.5 ^0C, (▲)-ideal mixture.

$N\langle(\Delta x_i)^2\rangle$ can serve as a measure of the local composition due to fluctuations. Therefore, one can expect large changes in the local compositions due to the fluctuations in the CO_2/ naphthalene dilute mixture; in contrast, the local compositions due to the fluctuations in the CO_2/ ethane dilute mixture are expected to be almost the same as those in the corresponding ideal mixtures. It is well known (from spectroscopic measurements [28] and from integral equation calculations [29]) that the local compositions in the dilute mixture CO_2 / naphthalene greatly differ from the bulk ones. However, no experimental data about the local composition in the dilute CO_2 / ethane mixture are available. Our calculations indicate that it should be near the bulk concentration.

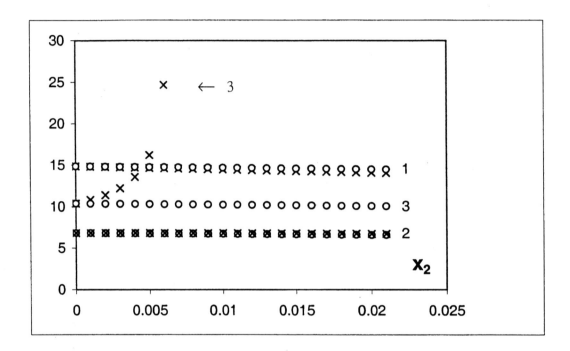

Fig. 2. Fluctuation in the number of particles $\dfrac{\langle (\Delta N_1)^2 \rangle}{\langle N_1 \rangle}$ in the following mixtures:

1-dilute mixture of ethane (2) in supercritical CO_2 (1) at P=73.97 bar and t=32.5 ^0C,

2-dilute mixture of CO_2 (2) in supercritical ethane (1) at P=50.73 bar and t=32.5 ^0C, 3-dilute mixture of naphthalene (2) in supercritical CO_2 (1) at P=75.99 bar and t=35 ^0C.

(o)-ideal mixture, (x)-real mixture.

The fluctuations in the number of particles $\dfrac{\langle (\Delta N_i)^2 \rangle}{\langle N_i \rangle}$ exhibit behaviors similar to those of the concentration fluctuations. In the mixture CO_2 / ethane, these fluctuations for both the dilute mixture of ethane in supercritical CO_2 and the dilute mixture of CO_2 in supercritical ethane almost coincide with the fluctuations in the number of particles $\frac{\langle(\Delta N_i)^2\rangle}{\langle N_i \rangle}$ in the corresponding ideal mixtures.

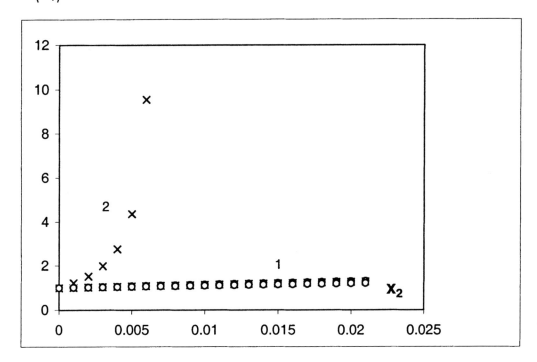

Fig. 3. Fluctuation in the number of particles $\frac{\langle(\Delta N_2)^2\rangle}{\langle N_2 \rangle}$ in the following mixtures:

1- dilute mixture of ethane (2) in supercritical CO_2 (1) at P=73.97 bar and t=32.5 ^0C, 2- dilute mixture of naphthalene (2) in supercritical CO_2 (1) at P=75.99 bar and t=35 ^0C. (o)-ideal mixture, (x)-real mixture.

However, the fluctuation in the number of particles $\frac{\langle(\Delta N_i)^2\rangle}{\langle N_i \rangle}$ in the dilute mixture CO_2/ naphthalene sharply increases when a small amount of naphthalene is added to the pure SCF. The fluctuations in the number of particles of CO_2 ($\frac{\langle(\Delta N_1)^2\rangle}{\langle N_1 \rangle}$) and naphthalene ($\frac{\langle(\Delta N_2)^2\rangle}{\langle N_2 \rangle}$) also increase sharply with an increase in the mole fraction of naphthalene.

216

a)

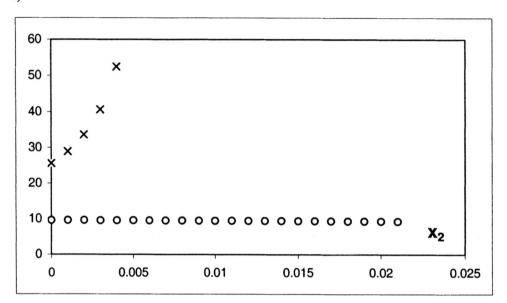

b)

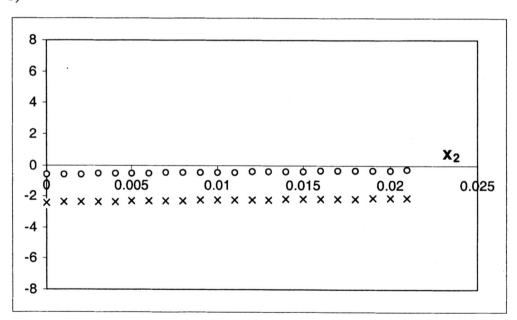

Fig. 4.

c)

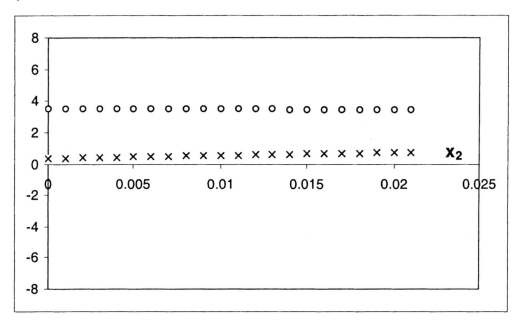

Fig. 4 (continued). Fluctuations $\dfrac{\langle (\Delta N_1)(\Delta N_2) \rangle}{\langle N_2 \rangle}$.

4a) dilute mixture of naphthalene (2) in supercritical CO_2 (1) at P=75.99 bar and t=35 ^0C ;

4b) dilute mixture of ethane (2) in supercritical CO_2 (1) at P=73.97 bar and t=32.5 ^0C ;

4c) dilute mixture of CO_2 (2) in supercritical ethane (1) at P=50.73 bar and t=32.5 ^0C.

(o)-ideal mixture, (x)-real mixture.

The fluctuations $\dfrac{\langle \Delta N_1 \Delta N_2 \rangle}{\langle N_2 \rangle}$ are plotted in Fig. 4.

For comparison purposes, the fluctuations $\dfrac{\langle (\Delta N_1)^2 \rangle}{\langle N_1 \rangle}$, $\dfrac{\langle (\Delta N_2)^2 \rangle}{\langle N_2 \rangle}$ and $\dfrac{\langle \Delta N_1 \Delta N_2 \rangle}{\langle N_2 \rangle}$ were also calculated far from the critical conditions for the methanol / water mixture at T=313.15 K and P=1 atm and presented in Fig. 5.

The calculations were carried out using eq. 2 and the values of the corresponding Kirkwood-Buff integrals obtained in Ref. 30. One can see from Fig. 5 that the fluctuations $\frac{\langle(\Delta N_i)^2\rangle}{\langle N_i\rangle}$ are small when x_i is small and that the correlation $\frac{\langle\Delta N_1\Delta N_2\rangle}{\langle N_2\rangle}$ is negative for all compositions. This situation is typical for liquid mixtures far from the critical conditions, because in such cases [21] $RTk_T \ll \frac{V_1V_2}{VD}$ (see eq. 6). However, under critical conditions $\frac{\langle\Delta N_1\Delta N_2\rangle}{\langle N_2\rangle}$ can become positive when the system moves toward the critical conditions (Figs. 4 and 5). This is important for the interpretation of the behavior of the partial molar volumes.

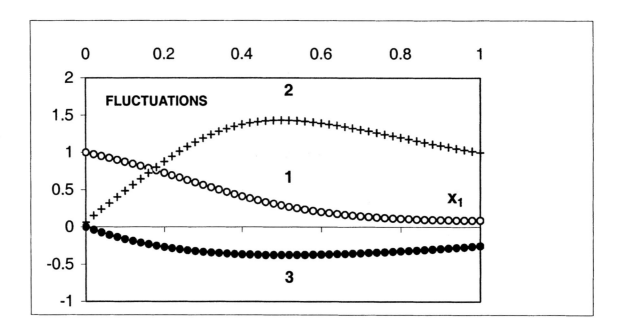

Fig. 5. The fluctuations: (1) is $\frac{\langle(\Delta N_1)^2\rangle}{\langle N_1\rangle}$; (2) is $\frac{\langle(\Delta N_2)^2\rangle}{\langle N_2\rangle}$; (3) is $\frac{\langle(\Delta N_1)(\Delta N_2)\rangle}{\langle N_2\rangle}$ in methanol (1) / water (2) mixtures at T=313.15 K and P=1 atm.

5. THE BEHAVIOR OF THE PARTIAL MOLAR VOLUME

The partial molar volume of the solute (component 2) can be expressed in terms of the Kirkwood-Buff integrals as follows [20]:

$$v_2 = \frac{1+(G_{11}-G_{12})c_1}{c_1+c_2+c_1c_2(G_{11}+G_{22}-2G_{12})} \qquad (17)$$

where v_2 is the partial molar volume (per particle) of the solute and c_i is the molecular concentration of component i.

Using eqs 2-6, eq. 17 becomes

$$v_2 = \frac{\left(\dfrac{\langle(\Delta N_1)^2\rangle}{\langle N_1\rangle} - \dfrac{\langle\Delta N_1 \Delta N_2\rangle}{\langle N_2\rangle}\right)}{c_1+c_2+c_1c_2(G_{11}+G_{22}-2G_{12})} \qquad (18)$$

which, because [20]

$$\left(\frac{\partial \mu_2}{\partial x_2}\right)_{T,P} = \frac{kT}{x_2(1+x_2c_1(G_{11}+G_{22}-2G_{12}))} , \qquad (19)$$

can be rewritten as

$$v_2 = \frac{\dfrac{\langle(\Delta N_1)^2\rangle}{\langle N_1\rangle} - \dfrac{\langle\Delta N_1 \Delta N_2\rangle}{\langle N_2\rangle}}{\dfrac{(c_1+c_2)}{x_2\left(\dfrac{\partial \mu_2}{\partial x_2}\right)_{P,T}}} \qquad (20)$$

where μ_2 is the chemical potential of the solute.

However, according to one of the stability criteria [31] $\left(\frac{\partial \mu_2}{\partial x_2}\right)_{P,T} > 0$; consequently, the sign of the partial molar volume of a solute is determined by the sign of $\left(\frac{\langle(\Delta N_1)^2\rangle}{\langle N_1 \rangle} - \frac{\langle \Delta N_1 \Delta N_2 \rangle}{\langle N_2 \rangle}\right)$. Fig. 5 shows that far from critical conditions $\frac{\langle(\Delta N_1)^2\rangle}{\langle N_1 \rangle} > 0$ and $\frac{\langle \Delta N_1 \Delta N_2 \rangle}{\langle N_2 \rangle} < 0$ and hence, as expected, $v_2 > 0$. The situation can be different not far from the critical conditions, because the sign of the numerator in the right hand of eq. 20 can become negative (Fig. 6).

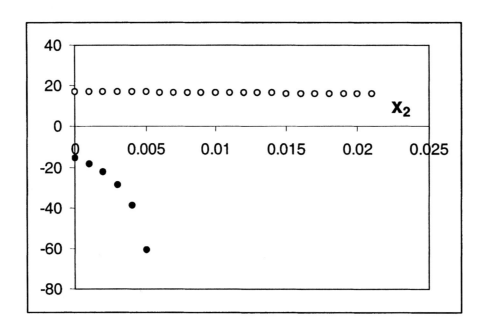

Fig.6. Difference $\left(\frac{\langle(\Delta N_1)^2\rangle}{\langle N_1 \rangle} - \frac{\langle \Delta N_1 \Delta N_2 \rangle}{\langle N_2 \rangle}\right)$ in the following mixtures:

(•)-dilute mixture of naphthalene (2) in CO_2 (1) at P=75.99 bar and t=35 ^0C.

(o)- dilute mixture of ethane (2) in CO_2 (1) at P=73.97 bar and t=32.5 ^0C.

Examination of literature data [32] has shown that most SC mixtures exhibit the same behavior as CO_2 / naphthalene regarding the partial molar volume of the solute; when the mixture moves towards the critical conditions, the partial molar volume of the solute decreases and under certain conditions becomes zero and further becomes negative and at infinite dilution diverges negatively at the critical point of the SCF. A few mixtures, such as CO_2 / ethane, exhibit a different behavior. When the mixture approaches the critical conditions, the partial molar volume of the solute remains positive and at infinite dilution diverges positively at the critical point of the SCF.

These behaviors can be understood qualitatively if one takes into account that the CO_2 / ethane mixture behaves almost as an ideal mixture and the CO_2 / naphthalene mixture as a highly nonideal one. If the interactions between different species are comparable to those between the species of the same kind, then the fluctuations ΔN_1 and ΔN_2 in the number of particles of species 1 and 2 will be moderately correlated and $\dfrac{\langle \Delta N_1 \Delta N_2 \rangle}{\langle N_2 \rangle}$ can be relatively small in absolute value. Hence, the difference $\left(\dfrac{\langle (\Delta N_1)^2 \rangle}{\langle N_1 \rangle} - \dfrac{\langle \Delta N_1 \Delta N_2 \rangle}{\langle N_2 \rangle}\right)$ can be positive, and in the limit of infinite dilution at the critical point of the solvent v_2 can diverge positively. This is the case of CO_2 / ethane mixtures. If the interactions between the two species are strong, then the fluctuations (ΔN_1) will have the same sign as (ΔN_2) and the average $\langle \Delta N_1 \Delta N_2 \rangle$ will be positive. In this case the difference $\left(\dfrac{\langle (\Delta N_1)^2 \rangle}{\langle N_1 \rangle} - \dfrac{\langle \Delta N_1 \Delta N_2 \rangle}{\langle N_2 \rangle}\right)$ can become negative, and at infinite dilution at the critical point of the solvent v_2 can diverge negatively. This is the case of the CO_2 / naphthalene mixture. Finally, if the interactions between the two species are much smaller than those between the species of the same kind, the correlation $\dfrac{\langle \Delta N_1 \Delta N_2 \rangle}{\langle N_2 \rangle}$ will be small, the difference

222

$(\dfrac{\langle(\Delta N_1)^2\rangle}{\langle N_1\rangle}-\dfrac{\langle\Delta N_1\Delta N_2\rangle}{\langle N_2\rangle})$ can be positive, and at infinite dilution v_2 can diverge positively. We do not have an example for this third case.

Consequently, in the first and third cases the systems behave in the same way, while in the second case the system behaves differently. It is appropriate to note that in the first case k_{22} is small in absolute value, in the third case it is positive and large, while in the second case it is large in absolute value but negative. Obviously, the quantity k_{22} is a measure of the nonideality of the system. In the first case, the system behaves almost as an ideal one; in the second case it behaves like a nonideal system with a strong positive deviation from ideality and in the third case like a nonideal system with a strong negative deviation from ideality. In the framework of the Kirkwood-Buff theory of solution the degree of nonideality can be characterized through the parameter Δ [33, 34]

$$\Delta = G_{11} + G_{22} - 2G_{12} \qquad (21)$$

which can be related to k_{22}.

For an ideal solution $\Delta = 0$ for all compositions; for positive deviations from ideality $\Delta > 0$ and for negative deviations $\Delta < 0$. A connection between k_{22} and Δ can be established through eq. 19, which leads to the following relation

$$\left(\dfrac{\partial \ln \gamma_2}{\partial x_2}\right)_{T,P} = -\dfrac{c_1 \Delta}{1 + x_2 c_1 \Delta}, \qquad (22)$$

At infinite dilution, eq. 22 becomes

$$k_{22} = -c_1^0 \lim_{x_2 \to 0} \Delta \qquad (23)$$

where $c_1^0 = \lim\limits_{x_2 \to 0} c_1$.

6. CONCLUSION

Two different types of dilute supercritical mixtures were compared from the viewpoint of fluctuations. The first type (a typical example being the CO_2 / naphthalene mixture) represents most of the SC mixtures and the analysis showed that the addition of a small amount of solute to the SCF led to sharp increases of the fluctuations in the composition and in the number of the particles compared to those in the corresponding ideal mixture. The reverse picture was found for the second type of supercritical mixtures, which are represented by the CO_2 / ethane mixture. The fluctuations in these mixtures were almost identical to those in the corresponding ideal mixtures.

It was also shown that the sign of the partial molar volume of a solute is determined by the sign of the difference $(\frac{\langle (\Delta N_1)^2 \rangle}{\langle N_1 \rangle} - \frac{\langle \Delta N_1 \Delta N_2 \rangle}{\langle N_2 \rangle})$. Far from critical conditions, this difference is positive because $\frac{\langle (\Delta N_1)^2 \rangle}{\langle N_1 \rangle} > 0$ and $\frac{\langle \Delta N_1 \Delta N_2 \rangle}{\langle N_2 \rangle} < 0$. The SC mixtures of the first type (such as the CO_2 / ethane mixture) behave almost as the corresponding ideal mixtures. In these cases, the correlation $\frac{\langle \Delta N_1 \Delta N_2 \rangle}{\langle N_2 \rangle}$ oscillates around zero remaining small in absolute value compared to the fluctuation in the number of particles $\frac{\langle (\Delta N_1)^2 \rangle}{\langle N_1 \rangle}$, and the difference $(\frac{\langle (\Delta N_1)^2 \rangle}{\langle N_1 \rangle} - \frac{\langle \Delta N_1 \Delta N_2 \rangle}{\langle N_2 \rangle})$ and the partial molar volume of the solute remain positive as for the corresponding ideal mixtures. For supercritical conditions of the second type (such as the CO_2 / naphthalene mixture), the fluctuation $\frac{\langle \Delta N_1 \Delta N_2 \rangle}{\langle N_2 \rangle}$ is positive and the difference $(\frac{\langle (\Delta N_1)^2 \rangle}{\langle N_1 \rangle} - \frac{\langle \Delta N_1 \Delta N_2 \rangle}{\langle N_2 \rangle})$ and the partial molar volume of the solute are negative.

APPENDIX

Expressions for the infinitely dilute systems can be derived from eqs 3-6 to obtain

$$\lim_{x_2 \to 0}\left[\frac{\langle(\Delta N_1)^2\rangle}{\langle N_1\rangle}\right] = \frac{RTk_{T,1}^0}{V_1^0} \qquad (A\text{-}1)$$

$$\lim_{x_2 \to 0}\left[\frac{\langle(\Delta N_2)^2\rangle}{\langle N_2\rangle}\right] = 1 \qquad (A\text{-}2)$$

$$\lim_{x_2 \to 0}\left[\frac{\langle\Delta N_1 \Delta N_2\rangle}{\langle N_1\rangle}\right] = 0 \qquad (A\text{-}3)$$

$$\lim_{x_2 \to 0}\left[\frac{\langle\Delta N_1 \Delta N_2\rangle}{\langle N_2\rangle}\right] = \frac{RTk_{T,1}^0}{V_1^0} - \frac{V_2^\infty}{V_1^0} \qquad (A\text{-}4)$$

where V_1^0 is the molar volume of the pure component 1, V_2^∞ is the partial molar volume of component 2 at infinite dilution. These limiting expressions were also obtained in Ref. 35. The fluctuations in the number of particles in an ideal mixture can be also obtained from eqs 3-6. Taking into account that in an ideal mixture $V_i = V_i^0$ and $D = 1$, one can find after simple algebra that

$$\left[\frac{\langle(\Delta N_1)^2\rangle}{\langle N_1\rangle}\right]_{id} = \frac{x_1 RTk_T^{id}}{V^{id}} + x_2\left(\frac{V_2^0}{V^{id}}\right)^2 \qquad (A\text{-}5)$$

$$\left[\frac{\langle(\Delta N_2)^2\rangle}{\langle N_2\rangle}\right]_{id} = \frac{x_2 RTk_T^{id}}{V^{id}} + x_1\left(\frac{V_1^0}{V^{id}}\right)^2 \qquad (A\text{-}6)$$

$$\left[\frac{\langle \Delta N_1 \Delta N_2 \rangle}{\langle N_1 \rangle}\right]_{id} = \frac{x_2 RT k_T^{id}}{V^{id}} - \frac{x_2 V_1^0 V_2^0}{(V^{id})^2} \quad \text{(A-7)}$$

or

$$\left[\frac{\langle \Delta N_1 \Delta N_2 \rangle}{\langle N_2 \rangle}\right]_{id} = \frac{x_1 RT k_T^{id}}{V^{id}} - \frac{x_1 V_1^0 V_2^0}{(V^{id})^2} \quad \text{(A-8)}$$

where $V^{id} = \sum_{j=1}^{2} x_j V_j^0$ and $k_T^{id} = \sum_{j=1}^{2} \varphi_j k_{T,j}^0$ are the molar volume and the isothermal compressibility of an ideal mixture.

REFERENCES

1. M. McHugh, V. Krukonis "Supercritical Fluid Extraction", Butterworth-Heinemann, Boston, 1994.
2. "Chemical engineering at supercritical fluid conditions", edited by M. E. Paulaitis, J. M. L. Penninger, R. D. Gray, P. Davidson; Ann Arbor Science, 1983.
3. J. F. Brennecke, C. A. Eckert, AIChE Journal, 35 (1989) 1409.
4. C. A. Eckert, B. L. Knutson, P. G. Debenedetti, Nature, 383 (1996) 313.
5. J. F. Brennecke, Chemistry & Industry, 21 (1996) 831.
6. I. R. Krichevskii, Zh. Fiz. Khim., 41 (1967) 2458.
7. N. E. Khazanova, E. E. Sominskaya, Zh. Fiz. Khim., 42 (1968) 1289.
8. A. M. Rozen, Zh. Fiz. Khim., 50 (1976) 1381.
9. C. A. Eckert, D. H. Ziger, K. P. Johnston, S. Kim, J. Phys. Chem., 90 (1986) 2738.
10. P. G. Debenedetti, R. S. Mohamed, J. Chem. Phys., 90 (1989) 4528.
11. I. B. Petsche, P. G. Debenedetti, J. Phys. Chem., 95 (1991) 386.
12. E. Ruckenstein, I. L. Shulgin, Chem. Phys. Letters, 330 (2000) 551.
13. J. D., Londono, V. M. Shah, G. D. Wignall, H. D. Cochran, P. R. Bienkowski, J.

Chem. Phys., 99 (1993) 466.

14. D. M. Pfund, T. S. Zemanian, J. C. Linehan, J. L. Fulton, C. R. Yonker, J. Phys. Chem., 98 (1994) 11846.
15. K. Nishikawa, I. Tanaka, Chem. Phys. Letters, 244 (1995) 149.
16. K. Nishikawa, I. Tanaka, Y. Amemiya, J. Phys. Chem., 100 (1996) 418.
17. K. Nishikawa, T. Morita, J. Supercrit. Fluids, 13 (1998) 143.
18. K. Nishikawa, T. Morita, J. Phys. Chem., 101 (1997) 1413.
19. L. D. Landau, E. M. Lifshitz, "Statistical Physics", Pt.1; Pergamon, Oxford, 1980.
20. J. G. Kirkwood, F. P. Buff, J. Chem. Phys., 19 (1951) 774.
21. E. Matteoli, L. Lepori, J. Chem. Phys., 80 (1984) 2856.
22. P. G. Debenedetti, S. K. Kumar, AIChE Journal, 32 (1986) 1253.
23. F. H. Huang, F. T. H. Chung, L. L. Lee, M. H. Li, K. E. Starling, J. Chem. Eng. Japan, 18 (1985) 490.
24. B. A. Younglove, J. F. Ely, J. Phys. Chem. Ref. Data, 16 (1987) 577.
25. E. Ruckenstein, I. Shulgin, J. Phys. Chem. B, 104 (2000) 2540.
26. Yu. V.; Tsekhanskaya, M. B. Iomtev, E. V. Mushkina, Russ. J. Phys. Chem., 38 (1964) 1173.
27. E. E. Sominskaya, N. E. Khazanova, Zh. Fiz. Khim., 44 (1970) 1657.
28. J. F. Brennecke, D. L. Tomasko, J. Peshkin, C. A. Eckert, Ind. Eng. Chem. Res., 29 (1990) 1682.
29. J. W. Tom, P. G. Debenedetti, Ind. Eng. Chem. Res., 32 (1993) 2118.
30. E. Ruckenstein, I. Shulgin, Fluid Phase Equilibria, 180 (2001) 345.
31. H. C. Van Ness, M. M. Abbott "Classical thermodynamics of non-electrolyte solutions with application to phase equilibria" McGraw-Hill Book Company, New York, 1982.
32. H. Q. Liu, E. Macedo, Ind. Eng. Chem. Res., 34 (1995) 2029.
33. A. Ben-Naim "Water and aqueous solutions; introduction to a molecular theory", New York, Plenum Press, 1974.
34. A. Ben-Naim, J. Chem. Phys., 67 (1977) 4884.
35. P. G. Debenedetti, Chem. Eng. Sci., 42 (1987) 2203.

Entrainer effect in supercritical mixtures

E. Ruckenstein*, I. Shulgin

Department of Chemical Engineering, State University of New York at Buffalo, Amherst, NY 14260, USA

Received 14 September 2000; received in revised form 16 January 2001; accepted 25 January 2001

Abstract

The objective of this paper is to propose a predictive method for the estimation of the change in the solubility of a solid in a supercritical solvent when another solute (entrainer) or a cosolvent is added to the system. To achieve this goal, the solubility equations were coupled with the Kirkwood–Buff (KB) theory of dilute ternary solutions. In this manner, the solubility of a solid in a supercritical fluid (SCF) in the presence of an entrainer or a cosolvent could be expressed in terms of only binary data. The obtained predictive method was applied to six ternary SCF–solute–cosolute and two SCF–solute–cosolvent systems. In the former case, the agreement with experiment was very good, whereas in the latter, the agreement was only satisfactory, because the data were not for the very dilute systems for which the present approach is valid. © 2001 Elsevier Science B.V. All rights reserved.

Keywords: Cosolvent; Entrainer; Supercritical fluid; Fluctuation theory; Solubility prediction

1. Introduction

The addition of a small amount (usually less than 5 mol%) of a volatile cosolvent to a supercritical fluid (SCF) can lead to a very large enhancement in the solubility of a solute (up to several hundred percent) [1]. Similarly, the use of multiple solutes (usually binary) can enhance their solubilities [1]. These phenomena, often called the cosolvent and entrainer effects, have attracted attention both among experimentalists [2–7] and theoreticians [8–12]. Two types of ternary supercritical mixtures are of interest.

1. Mixtures containing a cosolvent: SCF (component 1)–solute (usually solid) (2)–cosolvent (usually a subcritical liquid) (3).
2. Mixtures containing two solutes: SCF (1)–solute (2)–cosolute (3).

For the solubility of a solid in an SCF (binary supercritical mixture SCF (1) + solid solute(2)), one can write the well-known relation [13]

$$x_2^{\text{bin}} = \frac{P_2^0}{P\phi_2^{\text{bin}}} \exp \frac{(P - P_2^0)V_2^0}{RT} \tag{1}$$

---

* Corresponding author. Tel.: +1-716-645-2911; fax: +1-716-645-3822.
E-mail addresses: feaeliru@acsu.buffalo.edu (E. Ruckenstein), ishulgin@eng.buffalo.edu (I. Shulgin).

where P is the pressure, T the temperature (K), R the universal gas constant, ϕ_2^{bin} the fugacity coefficient of the solute in the binary mixture, x_2^{bin} the mole fraction at saturation of the solute, P_2^0 and V_2^0 the saturation vapor pressure and molar volume of the solid solute, respectively. A similar expression can be written for the solubility of a solute in a ternary supercritical mixture

$$x_2^{ter} = \frac{P_2^0}{P\phi_2^{ter}} \exp \frac{(P - P_2^0)V_2^0}{RT} \qquad (2)$$

where ϕ_2^{ter} is the fugacity coefficient of the solute and x_2^{ter} the mole fraction at saturation of the solute. When there are two solutes, the equation for the solubility of the cosolute (component 3) has the form

$$x_3^{ter} = \frac{P_3^0}{P\phi_3^{ter}} \exp \frac{(P - P_3^0)V_3^0}{RT} \qquad (3)$$

where P_3^0 and V_3^0 are the saturation vapor pressure and molar volume of the cosolute, respectively, ϕ_3^{ter} the fugacity coefficient of the cosolute and x_3^{ter} the mole fraction at saturation of the cosolute.

Eqs. (1)–(3) show that the solubilities of solids in an SCF depend among others on their fugacity coefficients: ϕ_2^{bin}, ϕ_2^{ter}, ϕ_3^{ter} and the calculations indicated that these coefficients were responsible for the large solubilities of solids in supercritical solvents. These solubilities are much larger than those in ideal gases, and enhancement factors of 10^4–10^8 are not uncommon [1]; they are, however, still relatively small and usually do not exceed several mole percent. Consequently, these supercritical solutions can be considered dilute and the expressions for the fugacity coefficients in binary and ternary supercritical mixtures simplified accordingly.

For a binary dilute mixture, Debenedetti and Kumar [14] suggested the following expression for the fugacity coefficient

$$\ln \phi_2^{bin} = \ln \phi_2^{bin,\infty} - k_{22}x_2 \qquad (4)$$

where

$$k_{22} = -\left(\frac{\partial \ln \gamma_2}{\partial x_2}\right)_{P,T,x_2 \to 0} = -\left(\frac{\partial \ln \phi_2^{bin}}{\partial x_2}\right)_{P,T,x_2 \to 0} \qquad (5)$$

In the above equations, x_2 and γ_2 are the mole fraction and the activity coefficient of component 2, and $\phi_2^{bin,\infty}$ the fugacity coefficient at infinite dilution.

The Debenedetti–Kumar expression was extended to multicomponent mixtures in the form [10,12]

$$\ln \phi_i = \ln \phi_i^\infty + x_2 \left(\frac{\partial \ln \phi_i}{\partial x_2}\right)_{P,T,x_3} + x_3 \left(\frac{\partial \ln \phi_i}{\partial x_3}\right)_{P,T,x_2} \qquad (6)$$

where $i = 2, 3$ and components 2 and 3 are at high dilution.

For the ternary mixture SCF (1)–solute (2)–cosolute (3), one can write for the fugacity coefficients of the solute (2) and cosolute (3) the expressions

$$\ln \phi_2^{ter} = \ln \phi_2^{ter,\infty} + x_2 \left(\frac{\partial \ln \phi_2^{ter}}{\partial x_2}\right)_{P,T,x_3} + x_3 \left(\frac{\partial \ln \phi_2^{ter}}{\partial x_3}\right)_{P,T,x_2} \equiv \ln \phi_2^{ter,\infty} - x_2 K_{22} - x_3 K_{23} \qquad (7)$$

and

$$\ln \phi_3^{\text{ter}} = \ln \phi_3^{\text{ter},\infty} + x_2 \left(\frac{\partial \ln \phi_3^{\text{ter}}}{\partial x_2}\right)_{P,T,x_3} + x_3 \left(\frac{\partial \ln \phi_3^{\text{ter}}}{\partial x_3}\right)_{P,T,x_2} \equiv \ln \phi_3^{\text{ter},\infty} - x_2 K_{32} - x_3 K_{33} \quad (8)$$

Consequently, for a dilute supercritical mixture SCF (1)–solute (2)–cosolute (3), one can write the following system of equations for the solubilities of the solute and cosolute

$$x_2^{\text{ter}} = \frac{P_2^0}{P\phi_2^{\text{ter},\infty} \exp(-x_2^{\text{ter}} K_{22} - x_3^{\text{ter}} K_{23})} \exp \frac{(P - P_2^0)V_2^0}{RT} \quad (9)$$

and

$$x_3^{\text{ter}} = \frac{P_3^0}{P\phi_3^{\text{ter},\infty} \exp(-x_2^{\text{ter}} K_{23} - x_3^{\text{ter}} K_{33})} \exp \frac{(P - P_3^0)V_3^0}{RT} \quad (10)$$

where it was taken into account that $K_{23} = K_{32}$.

From Eqs. (9) and (10) one can see, that the calculation of the solubilities of solids in a SCF in the presence of an entrainer (cosolute or cosolvent) requires information about the properties of the pure components, the fugacity coefficients at infinite dilution and the values of $K_{\alpha\beta}$.

Chimowitz and coworkers [9,10] emphasized the synergism caused by the cross exponential terms $\exp(-x_i K_{23})$ on the entrainer effect, and Jonah and Cochran [12] related the coefficients $K_{\alpha\beta}$ to the limiting values of the Kirkwood–Buff (KB) integrals and used the conformal solution theory to discuss the entrainer effect.

In the present paper it is shown, on the basis of the KB theory of solution [15], that there are conditions under which the main parameters in Eqs. (9) and (10), namely, K_{22}, K_{33} and K_{23}, can be expressed in terms of the parameters for the two binary mixtures formed with the SCF solvent. Finally, the solubilities in ternary mixtures are predicted using solubility data for the above binary mixtures.

2. Theory and formulas

2.1. The Kirkwood–Buff theory of solution for ternary mixtures

The KB theory of solution [15] (often called fluctuation theory of solution) employed the grand canonical ensemble to relate macroscopic properties, such as the derivatives of the chemical potentials with respect to concentrations, the isothermal compressibility and the partial molar volumes to microscopic properties in the form of spatial integrals involving the radial distribution function.

Kirkwood and Buff [15] obtained expressions for those quantities in compact matrix forms. For binary mixtures, Kirkwood and Buff provided the results listed in Appendix A. Starting from the matrix form and employing the algebraic software Mathematica [16], analytical expressions for the partial molar volumes, the isothermal compressibility and the derivatives of the chemical potentials for ternary mixtures were obtained by us. They are listed in Appendix B together with the expressions at infinite dilution for the partial molar volumes and isothermal compressibility.

An important quantity in what follows (see Appendix B, Eq. (B.5)) is

$$\Delta_{\alpha\beta} = G_{\alpha\alpha} + G_{\beta\beta} - 2G_{\alpha\beta}, \quad \alpha \neq \beta \quad (11)$$

where $G_{\alpha\beta}$ is the KB integral given by [15]

$$G_{\alpha\beta} = \int_0^\infty (g_{\alpha\beta} - 1) 4\pi r^2 \, dr \tag{12}$$

$g_{\alpha\beta}$ is the radial distribution function between species α and β and r the distance between the centers of molecules α and β.

As suggested by Ben-Naim [17], $\Delta_{\alpha\beta}$ is a measure of the nonideality of the binary mixture $\alpha - \beta$, because for an ideal system $\Delta_{\alpha\beta} = 0$. For a ternary mixture 1–2–3, Δ_{123} defined in Appendix B (Eq. (B.6)) also constitutes a measure of nonideality. Indeed, inserting $G_{\alpha\beta}^{\text{id}}$ for an ideal mixture (they are listed in Appendix C) into the expression of Δ_{123}, one obtains that for an ideal ternary mixture $\Delta_{123} = 0$.

The chemical potential of component α in a multicomponent mixture can be expressed as

$$\mu_\alpha = \mu_\alpha^0(P, T) + kT \ln x_\alpha \gamma_\alpha \tag{13}$$

where k is the Boltzmann constant, μ_α and $\mu_\alpha^0(P, T)$ the chemical potential and the standard chemical potential per molecule of species α, respectively, and γ_α the activity coefficient of component α. Consequently

$$\left(\frac{\partial \mu_\alpha}{\partial x_\beta}\right)_{P,T,x_3} = \frac{kT}{x_\alpha}\left(\frac{\partial x_\alpha}{\partial x_\beta}\right)_{P,T,x_3} + kT\left(\frac{\partial \ln \gamma_\alpha}{\partial x_\beta}\right)_{P,T,x_3} \tag{14}$$

where $\beta \neq 3$.

Combining Eq. (14) with Eqs. (B.19)–(B.24) of Appendix B allowed us to obtain expressions at infinite dilution for the derivatives of the activity coefficients with respect to the mole fractions in ternary mixtures. They are

$$K_{11} = -\lim_{\substack{x_2 \to 0 \\ x_3 \to 0}} \left(\frac{\partial \ln \gamma_1}{\partial x_1}\right)_{P,T,x_3} = 0 \tag{15}$$

$$K_{12} = -\lim_{\substack{x_2 \to 0 \\ x_3 \to 0}} \left(\frac{\partial \ln \gamma_1}{\partial x_2}\right)_{P,T,x_3} = 0 \tag{16}$$

$$K_{22} = -\lim_{\substack{x_2 \to 0 \\ x_3 \to 0}} \left(\frac{\partial \ln \gamma_2}{\partial x_2}\right)_{P,T,x_3} = c_1^0(G_{11}^0 + G_{22}^\infty - 2G_{12}^\infty) \tag{17}$$

$$K_{21} = -\lim_{\substack{x_2 \to 0 \\ x_3 \to 0}} \left(\frac{\partial \ln \gamma_2}{\partial x_1}\right)_{P,T,x_3} = -c_1^0(G_{11}^0 + G_{22}^\infty - 2G_{12}^\infty) \tag{18}$$

$$K_{31} = -\lim_{\substack{x_2 \to 0 \\ x_3 \to 0}} \left(\frac{\partial \ln \gamma_3}{\partial x_1}\right)_{P,T,x_3} = -c_1^0(G_{11}^0 + G_{23}^\infty - G_{12}^\infty - G_{13}^\infty) \tag{19}$$

and

$$K_{32} = -\lim_{\substack{x_2 \to 0 \\ x_3 \to 0}} \left(\frac{\partial \ln \gamma_3}{\partial x_2}\right)_{P,T,x_3} = c_1^0(G_{11}^0 + G_{23}^\infty - G_{12}^\infty - G_{13}^\infty) \tag{20}$$

where

$$G_{11}^0 = \lim_{\substack{x_2 \to 0 \\ x_3 \to 0}} G_{11}, \qquad G_{\alpha\beta}^{\infty} = \lim_{\substack{x_2 \to 0 \\ x_3 \to 0}} G_{\alpha\beta}, \quad (\alpha \neq \beta \neq 1), \qquad K_{\alpha\beta} = K_{\beta\alpha}, \qquad c_{\alpha} = \frac{N_{\alpha}}{v}$$

N_{α} is the number of molecules of species α in volume v, and c_{α}^0 the value of c_{α} for the pure fluid α.

The expressions of $(\partial \ln \gamma_{\alpha}/\partial x_{\beta})_{P,T,x_3}$ and of their infinitely dilute limits are important for the thermodynamics of dilute ternary solutions [20,21], especially for dilute supercritical ternary solutions [9–12]. The limiting expressions (17) and (20) were already derived in a different way by Jonah and Cochran [12] and Chailvo [11]. In the next section of the paper, the above expressions will be applied to ternary supercritical solutions.

2.2. The prediction of entrainer and cosolvent effects

In this section, a method for predicting the entrainer (cosolvent) effect on the basis of binary solubility data will be suggested. The equations for the solubilities in ternary mixtures (x_2^{ter} and x_3^{ter}) can be combined with those in the corresponding binaries (x_2^{bin} and x_3^{bin}) (Eqs. (1)–(3)) to obtain

$$\frac{x_2^{\text{ter}}}{x_2^{\text{bin}}} = \frac{\phi_2^{\text{bin}}}{\phi_2^{\text{ter}}} \tag{21}$$

$$\frac{x_3^{\text{ter}}}{x_3^{\text{bin}}} = \frac{\phi_3^{\text{bin}}}{\phi_3^{\text{ter}}} \tag{22}$$

The entrainer effect can be calculated if the fugacity coefficients of the solutes are known. Eqs. (21) and (22) can be rewritten in a more convenient form, by observing that k_{22} for a binary mixture, defined by Eq. (5), is given by (see Eq. (A.4) in Appendix A)

$$k_{22} = c_1^0 \lim_{x_2^{\text{bin}} \to 0} (G_{11} + G_{22} - 2G_{12}) \tag{23}$$

Because

$$\lim_{x_2^{\text{bin}} \to 0} G_{\alpha\beta}^{\text{bin}} = \lim_{\substack{x_2^{\text{ter}} \to 0 \\ x_3^{\text{ter}} \to 0}} G_{\alpha\beta}^{\text{ter}}$$

comparing Eqs. (17) and (23) one can conclude that $k_{22} = K_{22}$ and $k_{33} = K_{33}$. Finally, because

$$\lim_{x_2^{\text{bin}} \to 0} \phi_2^{\text{bin}} = \lim_{\substack{x_2^{\text{ter}} \to 0 \\ x_3^{\text{ter}} \to 0}} \phi_2^{\text{ter}} \quad \text{and} \quad \lim_{x_2^{\text{bin}} \to 0} \phi_3^{\text{bin}} = \lim_{\substack{x_2^{\text{ter}} \to 0 \\ x_3^{\text{ter}} \to 0}} \phi_3^{\text{ter}}$$

Eqs. (4) and (7) allow us to rewrite Eqs. (21) and (22) as

$$x_2^{\text{ter}} = x_2^{\text{bin}} \exp\left[k_{22}(x_2^{\text{ter}} - x_2^{\text{bin}}) + K_{23}x_3^{\text{ter}}\right] \tag{24}$$

and

$$x_3^{\text{ter}} = x_3^{\text{bin}} \exp\left[k_{33}(x_3^{\text{ter}} - x_3^{\text{bin}}) + K_{23}x_2^{\text{ter}}\right] \tag{25}$$

Table 1
The pressure dependence of $k_{22} = -(\partial \ln \gamma_2/\partial x_2)_{P,T,x_2\to 0}$ for a binary system SCF CO_2 (1) + solid solute (2) at $T = 308.15$ K[a]

P (bar)	For (k_{22}) various SCF CO_2 (1) + solid solute (2) mixtures				
	PH	NA	2,3-DMN	2,6-DMN	BA
120	63.6	34.6	45.3	46.8	64.9
140	51.6	29.5	37.9	39.9	53.0
160	45.0	26.7	34.1	36.3	46.3
180	41.0	25.0	31.8	34.2	41.8
200	38.2	23.8	30.3	32.9	38.5
220	36.1	23.0	29.3	32.0	36.1
240	34.5	22.3	28.6	31.3	34.1
260	33.2	21.9	28.0	30.9	32.5
280	32.3	21.5	27.7	30.6	31.2
300	31.4	21.2	27.4	30.4	30.1
320	30.8	20.9	27.2	30.2	29.1

[a] PH: phenanthrene; NA: naphthalene; 2,3-DMN: 2,3-dimethylnaphthalene; 2,6-DMN: 2,6-dimethylnaphthalene; BA: benzoic acid.

This system of equations can be solved for x_2^{ter} and x_3^{ter}, if the solubilities of the solute and the cosolute in the corresponding binary mixtures (x_2^{bin} and x_3^{bin}) and the values of k_{22}, k_{33} and K_{23} are known. It should be noted that only Eq. (24) is necessary for the analysis of the mixture SCF (1)–solute (2)–cosolvent (3), when the ternary mixture containing the cosolvent is not in equilibrium with the pure cosolvent.

The parameters k_{22} and k_{33} can be easily found from binary solubility data (for the binary mixtures used in this paper, the parameters $k_{\alpha\alpha}$ are listed in Table 1). The main difficulty consists in obtaining information about the parameter K_{23}; its evaluation, suggested below, is based on the KB theory of solutions.

Eq. (20) can be rearranged in the form

$$K_{23} = K_{32} = -\lim_{\substack{x_2\to 0\\x_3\to 0}}\left(\frac{\partial \ln \gamma_3}{\partial x_2}\right)_{P,T,x_3} = c_1^0(G_{11}^0 + G_{23}^\infty - G_{12}^\infty - G_{13}^\infty) \equiv c_1^0 \lim_{\substack{x_2\to 0\\x_3\to 0}} \frac{\Delta_{12} + \Delta_{13} - \Delta_{23}}{2} \quad (26)$$

where (see Eq. (11)) $\Delta_{\alpha\beta} = G_{\alpha\alpha} + G_{\beta\beta} - 2G_{\alpha\beta}$. As already noted, for a binary $\alpha - \beta$ mixture, $\Delta_{\alpha\beta}$ is a measure of nonideality. Eq. (23) and the limiting expressions written after it indicate that in the limit $x_2 \to 0$ and $x_3 \to 0$, $c_1^0 \Delta_{12}$ and $c_1^0 \Delta_{13}$ become equal to k_{22} and k_{33} (defined by Eq. (5)), respectively, which can be calculated from binary data. While for $x_2 \to 0$ and $x_3 \to 0$, Δ_{23} is not equal to Δ_{23} of the corresponding binary, it is plausible to consider that if the latter Δ_{23} is small, the former will also be small. If the binary mixtures of the solvent and each of the solutes are much more nonideal than the mixture of solutes and Δ_{12} and Δ_{13} have the same sign (they are positive and large for binary mixtures involving an SCF), then

$$\Delta_{12} + \Delta_{13} - \Delta_{23} \approx \Delta_{12} + \Delta_{13} \quad (27)$$

and

$$K_{23} = c_1^0 \lim_{\substack{x_2\to 0\\x_3\to 0}} \frac{\Delta_{12} + \Delta_{13} - \Delta_{23}}{2} \approx c_1^0 \lim_{\substack{x_2\to 0\\x_3\to 0}} \frac{\Delta_{12} + \Delta_{13}}{2} = \frac{k_{22} + k_{33}}{2} \quad (28)$$

Estimations indicated that the above approximation is reasonable. Indeed, let us consider the ternary supercritical mixture CO_2 (1) + phenanthrene (PH) (2) + naphthalene (NA) (3). The parameters k_{22} for the binary mixtures CO_2 (1) + PH (2) and CO_2 (1) + NA (2) in the interval of pressures 120–320 bar (Table 1) have values in the range 20–60 and their sum varies in the range 50–100. The parameters for the binary mixture PH (1) + NA (2), estimated from solid–liquid equilibrium data [22] (see Appendix D), are $(\partial \ln \gamma_1/\partial x_1)_{T,x_1\to 0} = -0.076$ and $(\partial \ln \gamma_2/\partial x_2)_{T,x_2\to 0} = -0.011$, hence much smaller than the above values.

3. Calculations

First, the entrainer effect in dilute ternary mixture with two solutes will be considered. The following ternary mixtures for which there are solubility data (for both ternary and binary constituents [2,3,25–27]) were selected: CO_2 +PH + NA, CO_2 +PH + benzoic acid (BA), CO_2 +PH + 2,3-dimethylnaphthalene (2,3-DMN), CO_2 +PH + 2,6-dimethylnaphthalene (2,6-DMN), CO_2 +NA + BA, and CO_2 +2,3-DMN +

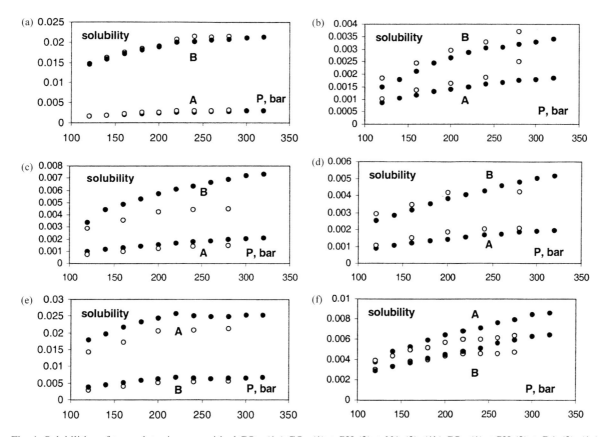

Fig. 1. Solubilities of two solutes in supercritical CO_2: (1a) CO_2 (1) + PH (2) + NA (3); (1b) CO_2 (1) + PH (2) + BA (3); (1c) CO_2 (1) + PH (2) + 2,3-DMN (3); (1d) CO_2 (1) + PH (2) + 2,6-DMN (3); (1e) CO_2 (1) + NA (2) + BA (3) and (1f) CO_2 (1) + 2,3-DMN (2) + 2,6-DMN (3). The solubilities are given in mole fractions: (○), experimental values [2]; (●), calculated solubilities; A, solute (component 2); B, cosolute (component 3).

2,6-DMN. The binary solubilities allowed us to calculate k_{22} and k_{33} using a procedure based on an equation of state (EOS) [14,28]. The Soave–Redlich–Kwong (SRK) [29] EOS was selected, and the $k_{\alpha\alpha}$ were calculated as in our previous paper [28]. The results of the calculations are listed in Table 1.

The solubilities for the above-listed ternary systems were calculated using Eqs. (24) and (25) with K_{23} given by Eq. (28). The predicted solubilities and those found experimentally [2] are compared in Fig. 1, which demonstrates a good agreement. Even for the system CO_2 (1) + PH + 2,3-DMN, for which there are the largest differences between predictions and experiment, the agreement is good.

The accurate prediction of the solubilities of the two solutes requires: (1) the solubilities in both binary and ternary mixtures to be small, because Eq. (4) for binary mixtures and Eqs. (6)–(8) for ternary mixtures are valid only for very dilute mixtures and (2) the inequality $|\Delta_{12} + \Delta_{13}| \gg |\Delta_{23}|$ to be fulfilled.

Only Eq. (24) is necessary to estimate the effect of a cosolvent, which is not in equilibrium with the mixture, on the solubility of a solute. In the latter case

$$x_2^{\text{ter}} = x_2^{\text{bin}} \exp\left[k_{22}(x_2^{\text{ter}} - x_2^{\text{bin}}) + \frac{k_{22} + k_{33}}{2} x_3^{\text{ter}}\right] \qquad (29)$$

where x_2^{bin} is the solubility of the solute in a cosolvent free binary mixture SCF (1)–solute (2) and x_3^{ter} the known mole fraction of the cosolvent in the ternary mixture [5–7]. The values of k_{22} for the solutes are listed in Table 1 and those of k_{33} were calculated for the binary mixture SCF–cosolvent at infinite dilution of the cosolvent. Two ternary mixtures: CO_2 +PH + methanol (M) and CO_2 +PH + acetone (A) were selected, because experimental solubilities were available [7]. Since the solubilities of methanol and acetone in supercritical CO_2 are not available, the values of k_{33} were estimated from the critical loci of the binary mixtures CO_2 +methanol and CO_2 +acetone [30] (Appendix E). The experimental and predicted solubilities are compared in Fig. 2 for (CO_2 +PH + M) and in Fig. 3 for (CO_2 +PH + A). Figs. 2 and 3 show that Eq. (24) combined with Eq. (28) can satisfactorily predict the cosolvent effect for low cosolvent

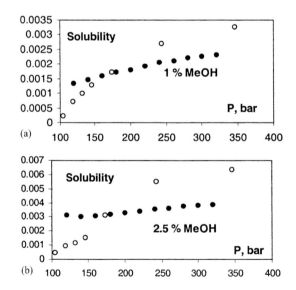

Fig. 2. Solubility (mole fraction) of PH in supercritical CO_2 at $T = 308.15$ K, when a cosolvent (methanol) was added: (○), experimental solubility [7]; (●), calculated using Eq. (24); (2a) 1 mol% methanol added; (2b) 2.5 mol% methanol added.

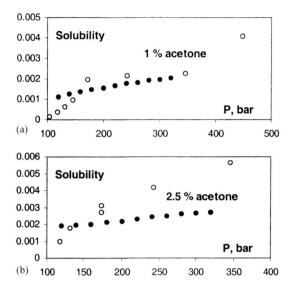

Fig. 3. Solubility (mole fraction) of PH in supercritical CO_2 at $T = 308.15$ K, when a cosolvent (acetone) was added: (○), experimental solubility [7]; (●), calculated using Eq. (24); (3a) 1 mol% acetone added; (3b) 2.5 mol% acetone added.

concentrations. For a cosolvent (methanol or acetone) concentration of 1 mol%, the predictions are better than for a concentration of 2.5 mol%, because only for sufficiently low mole fractions of the solute and cosolvent, the infinite dilute approximation is applicable. It should be also mentioned that the present approach is valid only if $|\Delta_{12} + \Delta_{13}| \gg |\Delta_{23}|$.

4. Conclusion

Predictive equations for the entrainer (cosolvent) effect are derived by combining solubility equations with the KB theory of solution of dilute ternary mixtures.

Explicit expressions for the isothermal compressibility, the partial molar volumes and the derivatives of the chemical potentials with respect to concentrations are obtained in terms of the KB integrals. These equations are employed to derive expressions at infinite dilution, which are relevant in the calculation of the solubilities in ternary mixtures in terms of those in the binary constituents.

Two types of systems: (i) SCF (1)–solute (2)–cosolute (3) and (ii) SCF (1)–solute (2)–cosolvent (3), were considered. The established equations are valid under two conditions: (i) the solution of the two solutes (or solute and cosolvent) must be dilute; (ii) the absolute value of the sum of the nonidealities (represented by the quantities $\Delta_{\alpha\beta}$) of the two binary mixture between the supercritical solvent and each of the two solutes must be much larger than the absolute value of the nonideality of the mixture of the two solutes.

For the mixtures of the type, SCF (1)–solute (2)–cosolute (3), good agreement between predictions and experiment was found. For the mixtures of the type, SCF (1)–solute (2)–cosolvent (3), the agreement was better (as expected) at low cosolvent concentrations.

Appendix A

For binary mixtures, Kirkwood and Buff [15] obtained the following expressions for the partial molar volumes, the isothermal compressibility and the derivatives of the chemical potentials with respect to concentrations.

$$v_1 = \frac{1 + (G_{22} - G_{12})c_2}{c_1 + c_2 + c_1 c_2 (G_{11} + G_{22} - 2G_{12})} \tag{A.1}$$

$$v_2 = \frac{1 + (G_{11} - G_{12})c_1}{c_1 + c_2 + c_1 c_2 (G_{11} + G_{22} - 2G_{12})} \tag{A.2}$$

$$kTk_T = \frac{1 + c_1 G_{11} + c_2 G_{22} + c_1 c_2 (G_{11} G_{22} - G_{12}^2)}{c_1 + c_2 + c_1 c_2 (G_{11} + G_{22} - 2G_{12})} \tag{A.3}$$

$$\left(\frac{\partial \mu_2}{\partial x_2}\right)_{T,P} = \frac{kT}{x_2(1 + x_2 c_1 (G_{11} + G_{22} - 2G_{12}))} \tag{A.4}$$

$$\left(\frac{\partial \mu_1}{\partial c_1}\right)_{T,P} = \frac{kT}{c_1(1 + c_1(G_{11} - G_{12}))} \tag{A.5}$$

$$\left(\frac{\partial \mu_2}{\partial c_2}\right)_{T,P} = \frac{kT}{c_2(1 + c_2(G_{22} - G_{12}))} \tag{A.6}$$

where v_α is the partial molar volume per molecule of species α, k_T the isothermal compressibility, and c_α the bulk molecular concentration of component α.

Appendix B

For a ternary mixture, the following expressions for the partial molar volumes and the isothermal compressibility were obtained by us.

$$v_1 = \frac{\begin{array}{c}1 + c_2(G_{22} - G_{12}) + c_3(G_{33} - G_{13}) + c_2 c_3(-G_{13}G_{22} + G_{13}G_{23} - G_{23}^2 \\ + G_{22}G_{33} + G_{12}G_{23} - G_{12}G_{33})\end{array}}{c_1 + c_2 + c_3 + c_1 c_2 \Delta_{12} + c_1 c_3 \Delta_{13} + c_2 c_3 \Delta_{23} + c_1 c_2 c_3 \Delta_{123}} \tag{B.1}$$

$$v_2 = \frac{\begin{array}{c}1 + c_1(G_{11} - G_{12}) + c_3(G_{33} - G_{23}) + c_1 c_3(-G_{12}G_{33} + G_{12}G_{13} - G_{13}^2 \\ + G_{13}G_{23} + G_{11}G_{33} - G_{11}G_{23})\end{array}}{c_1 + c_2 + c_3 + c_1 c_2 \Delta_{12} + c_1 c_3 \Delta_{13} + c_2 c_3 \Delta_{23} + c_1 c_2 c_3 \Delta_{123}} \tag{B.2}$$

$$v_3 = \frac{\begin{array}{c}1 + c_1(G_{11} - G_{13}) + c_2(G_{22} - G_{23}) + c_1 c_2(-G_{13}G_{22} + G_{12}G_{13} - G_{12}^2 \\ + G_{12}G_{23} + G_{11}G_{22} - G_{11}G_{23})\end{array}}{c_1 + c_2 + c_3 + c_1 c_2 \Delta_{12} + c_1 c_3 \Delta_{13} + c_2 c_3 \Delta_{23} + c_1 c_2 c_3 \Delta_{123}} \tag{B.3}$$

$$k_T = \frac{1 + c_1 G_{11} + c_2 G_{22} + c_3 G_{33} + c_1 c_2 F_{12} + c_1 c_3 F_{13} + c_2 c_3 F_{23} + c_1 c_2 c_3 F_{123}}{kT(c_1 + c_2 + c_3 + c_1 c_2 \Delta_{12} + c_1 c_3 \Delta_{13} + c_2 c_3 \Delta_{23} + c_1 c_2 c_3 \Delta_{123})} \tag{B.4}$$

where

$$\Delta_{\alpha\beta} = G_{\alpha\alpha} + G_{\beta\beta} - 2G_{\alpha\beta}, \quad \alpha \neq \beta \tag{B.5}$$

$$\Delta_{123} = G_{11}G_{22} + G_{11}G_{33} + G_{22}G_{33} + 2G_{12}G_{13} + 2G_{12}G_{23} + 2G_{13}G_{23} - G_{12}^2 - G_{13}^2$$
$$- G_{23}^2 - 2G_{11}G_{23} - 2G_{22}G_{13} - 2G_{33}G_{12} \tag{B.6}$$

$$F_{\alpha\beta} = G_{\alpha\alpha}G_{\beta\beta} - G_{\alpha\beta}^2, \quad \alpha \neq \beta \tag{B.7}$$

$$F_{123} = G_{11}G_{22}G_{33} + 2G_{12}G_{13}G_{23} - G_{13}^2 G_{22} - G_{12}^2 G_{33} - G_{23}^2 G_{11} \tag{B.8}$$

At infinite dilution, Eqs. (B.2)–(B.4) become

$$\lim_{\substack{x_2 \to 0 \\ x_3 \to 0}} v_2 = \frac{1}{c_1^0} + G_{11}^0 - G_{12}^\infty \tag{B.9}$$

$$\lim_{\substack{x_2 \to 0 \\ x_3 \to 0}} v_3 = \frac{1}{c_1^0} + G_{11}^0 - G_{13}^\infty \tag{B.10}$$

$$kTk_{T,1}^0 = kT \lim_{\substack{x_2 \to 0 \\ x_3 \to 0}} k_T = \frac{1}{c_1^0} + G_{11}^0 \tag{B.11}$$

where

$$G_{11}^0 = \lim_{\substack{x_2 \to 0 \\ x_3 \to 0}} G_{11} \quad \text{and} \quad G_{1\alpha}^\infty = \lim_{\substack{x_2 \to 0 \\ x_3 \to 0}} G_{1\alpha} \quad (\alpha = 2, 3)$$

Eq. (B.11) is the compressibility equation for a fluid [19].

By applying the general KB equations to a ternary mixture and taking into account Eqs. (B.1)–(B.4), the following expressions for the derivatives of the chemical potential with respect to concentrations under isothermal–isobaric conditions were obtained

$$\left(\frac{\partial \mu_1}{\partial N_1}\right)_{T,P,N_{\gamma \neq 1}} = \frac{kT(c_2 + c_3 + c_2 c_3 \Delta_{23})}{vc_1(c_1 + c_2 + c_3 + c_1 c_2 \Delta_{12} + c_1 c_3 \Delta_{13} + c_2 c_3 \Delta_{23} + c_1 c_2 c_3 \Delta_{123})} \tag{B.12}$$

$$\left(\frac{\partial \mu_1}{\partial N_2}\right)_{T,P,N_{\gamma \neq 2}} = \frac{-kT(1 + c_3(G_{12} + G_{33} - G_{13} - G_{23}))}{v(c_1 + c_2 + c_3 + c_1 c_2 \Delta_{12} + c_1 c_3 \Delta_{13} + c_2 c_3 \Delta_{23} + c_1 c_2 c_3 \Delta_{123})} \tag{B.13}$$

$$\left(\frac{\partial \mu_1}{\partial N_3}\right)_{T,P,N_{\gamma \neq 3}} = \frac{-kT(1 + c_2(G_{13} + G_{22} - G_{12} - G_{23}))}{v(c_1 + c_2 + c_3 + c_1 c_2 \Delta_{12} + c_1 c_3 \Delta_{13} + c_2 c_3 \Delta_{23} + c_1 c_2 c_3 \Delta_{123})} \tag{B.14}$$

$$\left(\frac{\partial \mu_2}{\partial N_2}\right)_{T,P,N_{\gamma \neq 2}} = \frac{kT(c_1 + c_3 + c_1 c_3 \Delta_{13})}{vc_2(c_1 + c_2 + c_3 + c_1 c_2 \Delta_{12} + c_1 c_3 \Delta_{13} + c_2 c_3 \Delta_{23} + c_1 c_2 c_3 \Delta_{123})} \tag{B.15}$$

$$\left(\frac{\partial \mu_2}{\partial N_3}\right)_{T,P,N_{\gamma \neq 3}} = \frac{-kT(1 + c_1(G_{23} + G_{11} - G_{12} - G_{13}))}{v(c_1 + c_2 + c_3 + c_1 c_2 \Delta_{12} + c_1 c_3 \Delta_{13} + c_2 c_3 \Delta_{23} + c_1 c_2 c_3 \Delta_{123})} \tag{B.16}$$

$$\left(\frac{\partial \mu_3}{\partial N_3}\right)_{T,P,N_{\gamma \neq 3}} = \frac{kT(c_1 + c_2 + c_1 c_2 \Delta_{12})}{v c_3 (c_1 + c_2 + c_3 + c_1 c_2 \Delta_{12} + c_1 c_3 \Delta_{13} + c_2 c_3 \Delta_{23} + c_1 c_2 c_3 \Delta_{123})} \tag{B.17}$$

where N_β is the number of molecules of species β in the volume v.

Expressions for $(\partial \mu_\alpha / \partial x_\beta)_{T,P,x_\gamma}$ at $x_3 = $ constant can be obtained by combining Eqs. (B.12)–(B.17) with the expression [12]

$$\left(\frac{\partial \mu_\alpha}{\partial x_\beta}\right)_{T,P,x_3} = N \left(\frac{\partial \mu_\alpha}{\partial N_\beta}\right)_{T,P,N_{\gamma \neq \beta}} - N \left(\frac{\partial \mu_\alpha}{\partial N_\theta}\right)_{T,P,N_{\gamma \neq \theta}} \tag{B.18}$$

where $\alpha = 1, 2, 3$, $\beta \neq 3$, $\theta \neq \beta, 3$ and $N = N_1 + N_2 + N_3$.

This yields

$$\left(\frac{\partial \mu_1}{\partial x_1}\right)_{T,P,x_3} = \frac{kT(c_1 + c_2 + c_3 + c_1 c_3 (G_{12} + G_{33} - G_{13} - G_{23}) + c_2 c_3 \Delta_{23})}{x_1 (c_1 + c_2 + c_3 + c_1 c_2 \Delta_{12} + c_1 c_3 \Delta_{13} + c_2 c_3 \Delta_{23} + c_1 c_2 c_3 \Delta_{123})} \tag{B.19}$$

$$\left(\frac{\partial \mu_1}{\partial x_2}\right)_{T,P,x_3} = -\frac{kT(c_1 + c_2 + c_3 + c_1 c_3 (G_{12} + G_{33} - G_{13} - G_{23}) + c_2 c_3 \Delta_{23})}{x_1 (c_1 + c_2 + c_3 + c_1 c_2 \Delta_{12} + c_1 c_3 \Delta_{13} + c_2 c_3 \Delta_{23} + c_1 c_2 c_3 \Delta_{123})} \tag{B.20}$$

$$\left(\frac{\partial \mu_2}{\partial x_2}\right)_{T,P,x_3} = \frac{kT(c_1 + c_2 + c_3 + c_2 c_3 (G_{12} + G_{33} - G_{13} - G_{23}) + c_1 c_3 \Delta_{13})}{x_2 (c_1 + c_2 + c_3 + c_1 c_2 \Delta_{12} + c_1 c_3 \Delta_{13} + c_2 c_3 \Delta_{23} + c_1 c_2 c_3 \Delta_{123})} \tag{B.21}$$

$$\left(\frac{\partial \mu_2}{\partial x_1}\right)_{T,P,x_3} = -\frac{kT(c_1 + c_2 + c_3 + c_2 c_3 (G_{12} + G_{33} - G_{13} - G_{23}) + c_1 c_3 \Delta_{13})}{x_2 (c_1 + c_2 + c_3 + c_1 c_2 \Delta_{12} + c_1 c_3 \Delta_{13} + c_2 c_3 \Delta_{23} + c_1 c_2 c_3 \Delta_{123})} \tag{B.22}$$

$$\left(\frac{\partial \mu_3}{\partial x_1}\right)_{T,P,x_3} = \frac{kT(c_1 c_3 (G_{11} + G_{23} - G_{12} - G_{13}) + c_2 c_3 (G_{12} + G_{23} - G_{13} - G_{22}))}{x_3 (c_1 + c_2 + c_3 + c_1 c_2 \Delta_{12} + c_1 c_3 \Delta_{13} + c_2 c_3 \Delta_{23} + c_1 c_2 c_3 \Delta_{123})} \tag{B.23}$$

and

$$\left(\frac{\partial \mu_3}{\partial x_2}\right)_{T,P,x_3} = -\frac{kT(c_1 c_3 (G_{11} + G_{23} - G_{12} - G_{13}) + c_2 c_3 (G_{12} + G_{23} - G_{13} - G_{22}))}{x_3 (c_1 + c_2 + c_3 + c_1 c_2 \Delta_{12} + c_1 c_3 \Delta_{13} + c_2 c_3 \Delta_{23} + c_1 c_2 c_3 \Delta_{123})} \tag{B.24}$$

It should be noted that the derivatives $(\partial \mu_\alpha / \partial x_\beta)_{T,P,x_{\gamma \neq \beta}}$ depend on which of the x_γ is kept constant.

Appendix C

The KB integrals for an ideal ternary system have been obtained in our previous paper [18]. They are (vol/mol)

$$G_{11}^{id} = RT k_T^{id} - \frac{V^{id}}{x_1} + \frac{x_2 (V_2^0)^2 + x_3 (V_3^0)^2 - x_2 x_3 (V_2^0 - V_3^0)^2}{x_1 V^{id}} \tag{C.1}$$

$$G_{22}^{id} = RT k_T^{id} - \frac{V^{id}}{x_2} + \frac{x_1 (V_1^0)^2 + x_3 (V_3^0)^2 - x_1 x_3 (V_1^0 - V_3^0)^2}{x_2 V^{id}} \tag{C.2}$$

$$G_{33}^{id} = RT k_T^{id} - \frac{V^{id}}{x_3} + \frac{x_1 (V_1^0)^2 + x_2 (V_2^0)^2 - x_1 x_2 (V_1^0 - V_2^0)^2}{x_3 V^{id}} \tag{C.3}$$

$$G_{12}^{id} = RTk_T^{id} + \frac{x_3 V_3^0 (V_3^0 - V_1^0 - V_2^0) - V_1^0 V_2^0 (1 - x_3)}{V^{id}} \tag{C.4}$$

$$G_{13}^{id} = RTk_T^{id} + \frac{x_2 V_2^0 (V_2^0 - V_1^0 - V_3^0) - V_1^0 V_3^0 (1 - x_2)}{V^{id}} \tag{C.5}$$

and

$$G_{23}^{id} = RTk_T^{id} + \frac{x_1 V_1^0 (V_1^0 - V_2^0 - V_3^0) - V_2^0 V_3^0 (1 - x_1)}{V^{id}} \tag{C.6}$$

where x_i is the molar fraction of component i in the ternary mixture, V_j^0 and $k_{T,j}^0$ the molar volume and the isothermal compressibility of the pure component j, respectively, $V^{id} = \sum_{j=1}^{3} x_j V_j^0$ and $k_T^{id} = \sum_{j=1}^{3} \varphi_j k_{T,j}^0$ with φ_j the volume fraction of component j.

Appendix D

To obtain the values of $k_{\alpha\alpha}$ for the binary mixture PH (1) + NA (2), the solid–liquid equilibrium data for this system [22] were used. The mole fraction of component α in the liquid phase was expressed through the modified Schröder equation [23].

$$x_\alpha \gamma_\alpha = \exp\left[\frac{\Delta H_\alpha^0}{R}\left(\frac{1}{T_\alpha^0} - \frac{1}{T}\right)\right] \tag{D.1}$$

where R is the universal gas constant, x_α and γ_α the molar fraction and activity coefficient of component α in the liquid phase, respectively, T_α^0 and ΔH_α^0 the melting temperature and the molar enthalpy of fusion of component α and T the melting temperature of the mixture. The activity coefficients were expressed through the Wilson equation [24].

$$\ln \gamma_1 = -\ln(x_1 + x_2 \Lambda_{12}) + x_2 \left(\frac{\Lambda_{12}}{x_1 + x_2 \Lambda_{12}} - \frac{\Lambda_{21}}{x_2 + x_1 \Lambda_{21}}\right) \tag{D.2}$$

and

$$\ln \gamma_2 = -\ln(x_2 + x_1 \Lambda_{21}) - x_1 \left(\frac{\Lambda_{12}}{x_1 + x_2 \Lambda_{12}} - \frac{\Lambda_{21}}{x_2 + x_1 \Lambda_{21}}\right) \tag{D.3}$$

where Λ_{12} and Λ_{21} are the temperature-dependent Wilson parameters.

$$\Lambda_{12} = \frac{V_2^0}{V_1^0} \exp\left(-\frac{\lambda_{12} - \lambda_{11}}{RT}\right) \tag{D.4}$$

$$\Lambda_{21} = \frac{V_1^0}{V_2^0} \exp\left(-\frac{\lambda_{21} - \lambda_{22}}{RT}\right) \tag{D.5}$$

where $(\lambda_{12} - \lambda_{11})$ and $(\lambda_{21} - \lambda_{22})$ are temperature independent parameters and V_α^0 the molar volume of the pure component α. The parameters $(\lambda_{12} - \lambda_{11})$ and $(\lambda_{21} - \lambda_{22})$ for the binary mixture PH + NA were found by fitting Eq. (D.1) to the $x_\alpha - T$ experimental data [22]. They are $\lambda_{12} - \lambda_{11} = 71.695$ (J/mol) and $\lambda_{21} - \lambda_{22} = 98.897$ (J/mol).

Eqs. (D.2) and (D.3) yield

$$\left(\frac{\partial \ln \gamma_\alpha}{\partial x_\alpha}\right)_{T, x_\alpha \to 0} = -\frac{2}{\Lambda_{\alpha\beta}} + 1 + \Lambda_{\beta\alpha}^2 \tag{D.6}$$

where $\alpha, \beta = 1, 2$ and $\alpha \neq \beta$. The insertion of the parameters Λ_{12} and Λ_{21} calculated at the temperature $T = 308.15$ K into Eq. (D.6) gives for the binary mixture PH (1) + NA (2), $(\partial \ln \gamma_1/\partial x_1)_{T, x_1 \to 0} = -0.076$ and $(\partial \ln \gamma_2/\partial x_2)_{T, x_2 \to 0} = -0.011$.

Appendix E

The values of k_{22} for the systems SCF CO_2 + methanol and SCF CO_2 + acetone were calculated using the SRK EOS [29]

$$P = \frac{RT}{V - b} - \frac{a(T)}{V(V + b)} \tag{E.1}$$

where V is the molar volume of the mixture and a and b are provided by the usual mixing rules

$$a = x_1^2 a_1 + x_2^2 a_2 + 2x_1 x_2 (a_1 a_2)^{0.5}(1 - q_{12}) \tag{E.2}$$

and

$$b = x_1 b_1 + x_2 b_2 \tag{E.3}$$

where a_α, b_α are the SRK EOS parameters of the pure components, and q_{12} the binary interaction parameter.

The method of calculation of k_{22} is presented in detail in [14] and [28] and requires the EOS parameters of the pure components, and the binary interaction parameter (q_{12}).

The SRK EOS parameters of the pure components can be calculated in terms of their critical pressure and temperature [29]. The binary interaction parameter q_{12} can be found from phase equilibria data for the binary mixture. Because, such data are not available, the critical loci data for the systems CO_2 (1) + methanol (2) and CO_2 (1) + acetone (2) [30] were used to calculate q_{12} (Reference [30]), provided the binary critical data in the form: $x_2 - P_{cr} - T_{cr}$, where x_2 is the molar fraction of component 2 in the critical mixture, P_{cr} the critical pressure and T_{cr} the critical temperature of the mixture. The mixture parameter a (a') in the SRK EOS was calculated for every $x_2 - P_{cr} - T_{cr}$ point using the expression [29]

$$a' = 0.42747 \frac{R^2 T_{cr}^2}{P_{cr}} \tag{E.4}$$

The parameter a was also calculated using Eq. (E.2) for every x_2 as a function of the binary interaction parameter q_{12}. The binary interaction parameter q_{12} was calculated by minimizing the sum $\sum_{\text{all } x_2} (a' - a)^2$.

References

[1] C.A. Eckert, B.L. Knutson, P.G. Debenedetti, Nature 383 (1996) 313–318.
[2] R.T. Kurnik, R.C. Reid, Fluid Phase Equilib. 8 (1982) 93–105.

[3] J.M. Dobbs, J.M. Wong, K.P. Johnston, J. Chem. Eng. Data 31 (1986) 303–308.
[4] W.J. Schmitt, R.C. Reid, Fluid Phase Equilib. 32 (1986) 77–99.
[5] J.M. Dobbs, J.M. Wong, R.J. Lahiere, K.P. Johnston, Ind. Eng. Chem. Res. 26 (1987) 56–65.
[6] R.M. Lemert, K.P. Johnston, Ind. Eng. Chem. Res. 30 (1991) 1222–1231.
[7] J.G. Van Alsten, C.A. Eckert, J. Chem. Eng. Data 38 (1993) 605–610.
[8] J.M. Walsh, G.D. Ikonomou, M.D. Donohue, Fluid Phase Equilib. 33 (1987) 295–314.
[9] F. Munoz, E.H. Chimowitz, J. Chem. Phys. 99 (1993) 5450–5461.
[10] F. Munoz, T.W. Li, E.H. Chimowitz, AIChE J. 41 (1995) 389–401.
[11] A. Chialvo, J. Phys. Chem. 97 (1993) 2740–2744.
[12] D.A. Jonah, H.D. Cochran, Fluid Phase Equilib. 92 (1994) 107–137.
[13] J.M. Prausnitz, R.N. Lichtenthaler, E. Gomes de Azevedo, Molecular Thermodynamics of Fluid-Phase Equilibria, 2nd Edition, Prentice-Hall, Englewood Cliffs, NJ, 1986.
[14] P.G. Debenedetti, S.K. Kumar, AIChE J. 32 (1986) 1253–1262.
[15] J.G. Kirkwood, F.P. Buff, J. Chem. Phys. 19 (1951) 774–777.
[16] S. Wolfram, Mathematica: a system for doing mathematics by computer, in: Advanced Book Program, Addison-Wesley, Redwood City, CA, 1991.
[17] A. Ben-Naim, J. Chem. Phys. 67 (1977) 4884–4890.
[18] E. Ruckenstein, I. Shulgin, Fluid Phase Equilib., 180 (1–2), pp. 271–287.
[19] D.A. McQuarrie, Statistical Mechanics, Harper & Row, New York, 1975.
[20] J.P. O' Connell, Mol. Phys. 20 (1971) 27–33.
[21] J.P. O' Connell, AIChE J. 17 (1971) 658–663.
[22] Y.F. Klochko-Zhovnir, Zh. Prikl. Khim. 22 (1949) 848–852.
[23] J.R. Elliot, C.T. Lira, Introductory Chemical Engineering Thermodynamics, Prentice-Hall, Upper Saddle River, NJ, 1999.
[24] G.M. Wilson, J. Am. Chem. Soc. 86 (1964) 127–130.
[25] Yu.V. Tsekhanskaya, M.B. Iomtev, E.V. Mushkina, Russ. J. Phys. Chem. 38 (1964) 1173–1176.
[26] R.T. Kurnik, S.J. Holla, R.C. Reid, J. Chem. Eng. Data 26 (1981) 47–51.
[27] W.J. Schmitt, R.C. Reid, J. Chem. Eng. Data 31 (1986) 204–212.
[28] E. Ruckenstein, I. Shulgin, J. Phys. Chem. B 104 (2000) 2540–2545.
[29] G. Soave, Chem. Eng. Sci. 27 (1972) 1197–1203.
[30] G.S. Gurdial, N.R. Foster, S.L.J. Yun, K.D. Tilly, in: E. Kiran, J.F. Brennecke (Eds.), Supercritical Fluid Engineering Science Fundamentals and Application, American Chemical Society, Washington, DC, 1993, pp. 34–45.

Fluid Phase Equilibria 200 (2002) 53–67

www.elsevier.com/locate/fluid

The solubility of solids in mixtures composed of a supercritical fluid and an entrainer

E. Ruckenstein*, I. Shulgin

Department of Chemical Engineering, State University of New York at Buffalo, Clifford C. Furnas Hall, Box 604200, Amherst, NY 14260-4200, USA

Received 17 August 2001; accepted 15 December 2001

Abstract

The goal of this paper is to develop a predictive method for the solubility of a solid in a supercritical fluid containing an entrainer at any concentration. The main difficulty consists in the derivation of an expression for the fugacity coefficient of the solid solute in the binary solvent. A method based on the Kirkwood–Buff formalism was employed and expressions for the derivatives of the fugacity coefficient of the solute in a ternary mixture with respect to the mole fractions were obtained. On the basis of these expressions an algebraic equation was derived, which allowed one to predict the solubility of a solid solute in terms of its solubilities in the supercritical fluid and in the entrainer. The equation was compared with the experimental results available in the literature regarding the solubility in a mixture of supercritical fluids and good agreement was obtained. © 2002 Elsevier Science B.V. All rights reserved.

Keywords: Fluctuation theory; Entrainer; Solubility; Supercritical fluid

1. Introduction

The addition of an entrainer to a supercritical (SC) solvent can lead to a very large enhancement in the solubility of a solute (up to several hundred percent) [1–3]. This phenomenon, often called the entrainer effect, has relevance in the SC fluid technology. The addition of a small amount of entrainer can increase the solubility of a solid much more than a pressure increase of several hundred bars [4]. Because the entrainer effect depends upon the nature of the solute it can be used to enhance the selectivity of a SC fluid for certain compounds.

The entrainer can be a liquid, a gas, a solid or a supercritical fluid [2]. Does the enhancement have a physical origin, being caused by the physical interactions among the molecules, or a chemical one, being a result of the formation of a complex between the solute and entrainer? In a number of cases treated in a previous paper [5], the enhancement caused in the solubility of a solid by the addition to the SC solvent of a small amount of a cosolvent was explained in terms of physical interactions. In what follows it will be

---

* Corresponding author. Tel.: +1-716-645-2911; fax: +1-716-645-3822.
E-mail addresses: feaeliru@acsu.buffalo.edu (E. Ruckenstein), ishulgin@eng.buffalo.edu (I. Shulgin).

0378-3812/02/$ – see front matter © 2002 Elsevier Science B.V. All rights reserved.
PII: S0378-3812(02)00012-2

shown that the enhancement produced by the addition of a cosolvent when its concentration is large can be also a result of physical interactions. This does not mean that there are no cases in which the chemical interactions play a role. An example is the mixture CO_2 + hydroquinone + tri-n-butylphosphate [6]. In this mixture, the intermolecular interactions between solute and entrainer generate a charge transfer complex between hydroquinone and tri-n-butylphosphate [6]. Another example of chemical (specific) interactions in ternary supercritical mixtures involves hydrogen bonding. For instance, the H-bonding equilibrium between perfluoro-$tert$-butyl alcohol and dimethyl ether in supercritical SF_6 was investigated [7] and an equilibrium constant for the resulting H-bonded complexes was evaluated.

Recently, a method [5] for the prediction of the solubility of a solute in a SC fluid in the presence of an entrainer has been proposed. The method, based on the Kirkwood–Buff (KB) formalism, was however developed for cases in which the entrainer was in dilute amounts. The present paper is focused on the solubility of a solid in a non-dilute mixture of a SC fluid and an entrainer. The theoretical treatment, which is more complex than for the dilute case, is also based on the KB formalism. In this paper the following aspects will be addressed: (1) general equations for the solubility in binary and ternary mixtures will be written for the cases involving a small amount of solute; (2) the KB formalism will be used to obtain expressions for the derivatives of the fugacity coefficients in a ternary mixture with respect to mole fractions; (3) these expressions will be employed to derive an equation for the solubility of a solute in a SC fluid containing an entrainer at any concentration; (4) a predictive method for this solubility will be proposed in terms of the solubilities of the solute in the SC fluid and in the entrainer; (5) the derived equation will be compared with experimental results from literature regarding the solubility of a solute in a mixture of two SC fluids.

2. Theory

2.1. General equations for the solubility of a solid in SC fluids

The solubility of a solid in a SC fluid (binary mixture SC fluid (1) + solid solute (2)) can be calculated using the equation [8]

$$x_2^{bin} = \frac{P_2^0}{P\phi_2^{bin}} \exp\left(\frac{(P - P_2^0)V_2^0}{RT}\right) \quad (1)$$

where R is the universal gas constant, P the pressure, T the temperature in K, x_2^{bin} and ϕ_2^{bin} the mole fraction and fugacity coefficient of a solute in the gaseous phase at equilibrium, and P_2^0 and V_2^0 are the saturation vapor pressure and molar volume of the pure solute. A similar expression can be written for the solubility of a solute in a ternary supercritical mixture (SC fluid (1) + solid solute (2) + entrainer (3))

$$x_2^{ter} = \frac{P_2^0}{P\phi_2^{ter}} \exp\left(\frac{(P - P_2^0)V_2^0}{RT}\right) \quad (2)$$

where ϕ_2^{ter} and x_2^{ter} are the fugacity coefficient and the mole fraction of the solute in the gaseous phase at equilibrium.

For a binary dilute mixture, Debenedetti and Kumar [9] used the following series expansion for the fugacity coefficient of the solute:

$$\ln \phi_2^{bin} = \ln \phi_2^{bin,\infty} - k_{22}x_2 \quad (3)$$

where $\phi_2^{\text{bin},\infty}$ is the fugacity coefficient at infinite dilution and x_2 is the mole fraction of the solute, and

$$k_{22} = -\left(\frac{\partial \ln \gamma_2^{\text{bin}}}{\partial x_2}\right)_{P,T,x_2\to 0} = -\left(\frac{\partial \ln \phi_2^{\text{bin}}}{\partial x_2}\right)_{P,T,x_2\to 0} \qquad (4)$$

where γ_2^{bin} is the activity coefficient of the solute.

The above expression was extended to ternary mixtures, containing a solute and an entrainer, both at high dilution, in the form [10–12]:

$$\ln \phi_2^{\text{ter}} = \ln \phi_2^{\text{ter},\infty} + x_2 \left(\frac{\partial \ln \phi_2^{\text{ter}}}{\partial x_2}\right)_{P,T,x_3,0} + x_3 \left(\frac{\partial \ln \phi_2^{\text{ter}}}{\partial x_3}\right)_{P,T,x_2,0} \equiv \ln \phi_2^{\text{ter},\infty} - x_2 K_{22} - x_3 K_{23} \qquad (5)$$

and

$$\ln \phi_3^{\text{ter}} = \ln \phi_3^{\text{ter},\infty} + x_2 \left(\frac{\partial \ln \phi_3^{\text{ter}}}{\partial x_2}\right)_{P,T,x_3,0} + x_3 \left(\frac{\partial \ln \phi_3^{\text{ter}}}{\partial x_3}\right)_{P,T,x_2,0} \equiv \ln \phi_3^{\text{ter},\infty} - x_2 K_{32} - x_3 K_{33} \qquad (6)$$

where x_i is the molar fraction of component i in the ternary mixture, subscript 0 indicates that the derivative should be calculated at infinite dilution of components 2 and 3 and the coefficients K_{ij} in the ternary mixture are defined as

$$K_{ij} = -\lim_{x_2\to 0, x_3\to 0} \left(\frac{\partial \ln \gamma_i}{\partial x_j}\right)_{P,T,x_\alpha \neq 1 \text{ and } j} \qquad (7)$$

It should be noted that the k_{ii} are for binary mixtures and the K_{ij} for the ternary ones. Consequently, for a dilute mixture SC fluid (1)–solute (2)–entrainer (3), one can write the following equation for the solubility of a solute (x_2^{ter}):

$$x_2^{\text{ter}} = \frac{P_2^0}{P\phi_2^{\text{ter},\infty} \exp(-x_2^{\text{ter}} K_{22} - x_3^{\text{ter}} K_{23})} \exp\left(\frac{(P-P_2^0)V_2^0}{RT}\right) \qquad (8)$$

On the basis of the KB theory of solution [13], it was shown [5], that K_{22} and K_{23} in Eq. (8) can be expressed in terms of the parameters for the two binary mixtures formed by the solute and entrainer with the SC solvent. Consequently, the solubility of a solute in a binary mixture could be predicted in terms of binary data.

For the cases in which only the solute concentration is small, the derivation of an expression for the fugacity coefficient ϕ_2^{ter} (see Eq. (2)) is still critical for the prediction of the solubility x_2^{ter}. Let us consider those compositions of the ternary mixture which are located on the line between the points ($x_1^{\text{ter}} = 0$, $x_2^{\text{ter}} = 1$, $x_3^{\text{ter}} = 0$) and ($x_1^{\text{ter}} = x_1^0$, $x_2^{\text{ter}} = 0$, $x_3^{\text{ter}} = x_3^0$) in the Gibbs triangle. This line connects the pure component 2 and the binary mixtures 1–3 with a mole fraction of component 1 equal to x_1^0. Physically speaking, this line represents the locus of compositions of the ternary mixtures, formed by adding a solute to a binary mixture of a SC fluid and an entrainer.

On the above line, the following relation holds:

$$\left(\frac{x_1^{\text{ter}}}{x_3^{\text{ter}}}\right) = \left(\frac{x_1^0}{x_3^0}\right) = \alpha \qquad (9)$$

Because $x_1^{\text{ter}} + x_2^{\text{ter}} + x_3^{\text{ter}} = 1$, one can write that

$$x_1^{\text{ter}} = \alpha \frac{1 - x_2^{\text{ter}}}{1 + \alpha} \qquad (10)$$

and

$$x_3^{\text{ter}} = \frac{1 - x_2^{\text{ter}}}{1 + \alpha} \tag{11}$$

For the fugacity coefficient of a solute, whose solubility x_2^{ter} is small, one can write, at constant temperature and pressure, near the concentration $x_1^{\text{ter}} = x_1^0$, $x_2^{\text{ter}} = 0$, $x_3^{\text{ter}} = x_3^0$, the following expression:

$$\ln \phi_2^{\text{ter}} = \ln \phi_2^{\text{ter}}(x_1^0, 0, x_3^0) + x_2^{\text{ter}} \left(\frac{\partial \ln \phi_2^{\text{ter}}}{\partial x_2^{\text{ter}}} \right)_{P,T,\alpha(x_1^0,0,x_3^0)} \tag{12}$$

where the subscript $(x_1^0, 0, x_3^0)$ indicates that the derivative should be calculated at infinite dilution of the solute.

If, at a given pressure and temperature, the mole fractions of components 1 and 3 are taken as independent variables, one can rewrite Eq. (12) under the form

$$\ln \phi_2^{\text{ter}} = \ln \phi_2^{\text{ter}}(x_1^0, 0, x_3^0) + x_2^{\text{ter}} \left[\left(\frac{\partial \ln \phi_2^{\text{ter}}}{\partial x_1^{\text{ter}}} \right)_{P,T,x_3^{\text{ter}}(x_1^0,0,x_3^0)} \left(\frac{\partial x_1^{\text{ter}}}{\partial x_2^{\text{ter}}} \right)_\alpha \right.$$
$$\left. + \left(\frac{\partial \ln \phi_2^{\text{ter}}}{\partial x_3^{\text{ter}}} \right)_{P,T,x_1^{\text{ter}}(x_1^0,0,x_3^0)} \left(\frac{\partial x_3^{\text{ter}}}{\partial x_2^{\text{ter}}} \right)_\alpha \right] \tag{13}$$

which, taking into account Eqs. (10) and (11), becomes

$$\ln \phi_2^{\text{ter}} = \ln \phi_2^{\text{ter}}(x_1^0, 0, x_3^0) - \frac{x_2^{\text{ter}}}{1+\alpha} \left[\alpha \left(\frac{\partial \ln \phi_2^{\text{ter}}}{\partial x_1^{\text{ter}}} \right)_{P,T,x_3^{\text{ter}}(x_1^0,0,x_3^0)} + \left(\frac{\partial \ln \phi_2^{\text{ter}}}{\partial x_3^{\text{ter}}} \right)_{P,T,x_1^{\text{ter}}(x_1^0,0,x_3^0)} \right] \tag{14}$$

or equivalently

$$\ln \phi_2^{\text{ter}} = \ln \phi_2^{\text{ter}}(x_1^0, 0, x_3^0) - x_2^{\text{ter}} \left[x_1^0 \left(\frac{\partial \ln \phi_2^{\text{ter}}}{\partial x_1^{\text{ter}}} \right)_{P,T,x_3^{\text{ter}}(x_1^0,0,x_3^0)} + x_3^0 \left(\frac{\partial \ln \phi_2^{\text{ter}}}{\partial x_3^{\text{ter}}} \right)_{P,T,x_1^{\text{ter}}(x_1^0,0,x_3^0)} \right] \tag{15}$$

The above equations can become useful if expressions for the two partial derivatives can be obtained.

2.2. Expressions for the derivatives $(\partial \ln \phi_2^{\text{ter}}/\partial x_1^{\text{ter}})_{P,T,x_3^{\text{ter}}}$ and $(\partial \ln \phi_2^{\text{ter}}/\partial x_3^{\text{ter}})_{P,T,x_1^{\text{ter}}}$ through the Kirkwood–Buff formalism

It was shown previously [5,14] that the KB theory of solution can be used to relate the thermodynamic properties of ternary mixtures, such as the partial molar volumes, the isothermal compressibility and the derivatives of the chemical potentials to the KB integrals. In particular for the derivatives of the activity coefficients (γ_2^{ter}) one can write the following rigorous relations [5]:

$$\left(\frac{\partial \ln \gamma_2^{\text{ter}}}{\partial x_1^{\text{ter}}} \right)_{T,P,x_3^{\text{ter}}} = \frac{-c_2 c_3 (G_{12} + G_{33} - G_{13} - G_{23}) + c_1 c_2 \Delta_{12} + c_2 c_3 \Delta_{23} + c_1 c_2 c_3 \Delta_{123}}{x_2^{\text{ter}}(c_1 + c_2 + c_3 + c_1 c_2 \Delta_{12} + c_1 c_3 \Delta_{13} + c_2 c_3 \Delta_{23} + c_1 c_2 c_3 \Delta_{123})} \tag{16}$$

and

$$\left(\frac{\partial \ln \gamma_2^{\text{ter}}}{\partial x_3^{\text{ter}}}\right)_{T,P,x_1^{\text{ter}}} = \frac{-c_1 c_2 (G_{11} + G_{23} - G_{12} - G_{13}) + c_1 c_2 \Delta_{12} + c_2 c_3 \Delta_{23} + c_1 c_2 c_3 \Delta_{123}}{x_2^{\text{ter}}(c_1 + c_2 + c_3 + c_1 c_2 \Delta_{12} + c_1 c_3 \Delta_{13} + c_2 c_3 \Delta_{23} + c_1 c_2 c_3 \Delta_{123})} \quad (17)$$

where c_k is the bulk molecular concentration of component k and $G_{\alpha\beta}$ is the KB integral given by

$$G_{\alpha\beta} = \int_0^\infty (g_{\alpha\beta} - 1) 4\pi r^2 \, dr \quad (18)$$

In the above expressions, $g_{\alpha\beta}$ is the radial distribution function between species α and β, r the distance between the centers of molecules α and β, and $\Delta_{\alpha\beta}$ and Δ_{123} are defined as follows:

$$\Delta_{\alpha\beta} = G_{\alpha\alpha} + G_{\beta\beta} - 2G_{\alpha\beta}, \quad \alpha \neq \beta \quad (19)$$

and

$$\Delta_{123} = G_{11}G_{22} + G_{11}G_{33} + G_{22}G_{33} + 2G_{12}G_{13} + 2G_{12}G_{23} + 2G_{13}G_{23} - G_{12}^2 - G_{13}^2 - G_{23}^2 \\ - 2G_{11}G_{23} - 2G_{22}G_{13} - 2G_{33}G_{12} \quad (20)$$

One can show that the factors in brackets in the numerators of Eqs. (16) and (17) and Δ_{123} can be expressed in terms of $\Delta_{\alpha\beta}$ as follows:

$$G_{12} + G_{33} - G_{13} - G_{23} = \tfrac{1}{2}(\Delta_{13} + \Delta_{23} - \Delta_{12}) \quad (21)$$

$$G_{11} + G_{23} - G_{12} - G_{13} = \tfrac{1}{2}(\Delta_{12} + \Delta_{13} - \Delta_{23}) \quad (22)$$

and

$$\Delta_{123} = -\tfrac{1}{4}((\Delta_{12})^2 + (\Delta_{13})^2 + (\Delta_{23})^2 - 2\Delta_{12}\Delta_{13} - 2\Delta_{12}\Delta_{23} - 2\Delta_{13}\Delta_{23}) \quad (23)$$

The insertion of Eqs. (21)–(23) into Eqs. (16) and (17) provides rigorous expressions for the derivatives $(\partial \ln \phi_2^{\text{ter}}/\partial x_1^{\text{ter}})_{P,T,x_3^{\text{ter}}}$ and $(\partial \ln \phi_2^{\text{ter}}/\partial x_3^{\text{ter}})_{P,T,x_1^{\text{ter}}}$ in terms of $\Delta_{\alpha\beta}$ and concentrations.

It is worth noting that Δ_{ij} is a measure of nonideality [15] of the binary mixture α–β, because for an ideal mixture $\Delta_{\alpha\beta} = 0$. For a ternary mixture 1–2–3, Δ_{123} also constitutes a measure of nonideality. Indeed, inserting $G_{\alpha\beta}^{\text{id}}$ for an ideal mixture [14] into the expression of Δ_{123}, one obtains that for an ideal ternary mixture $\Delta_{123} = 0$. One should also mention that the nonideality parameter Δ_{12} in a binary mixture is connected to the parameter k_{22} (see Eq. (4)). Indeed, at infinite dilution one can write the following expression:

$$k_{22} = -\left(\frac{\partial \ln \gamma_2}{\partial x_2}\right)_{P,T,x_2 \to 0} = -\left(\frac{\partial \ln \phi_2^{\text{bin}}}{\partial x_2}\right)_{P,T,x_2 \to 0} = c_1^{0,\text{bin}} \lim_{x_2^{\text{bin}} \to 0} (G_{11} + G_{22} - 2G_{12}) \\ = c_1^{0,\text{bin}} \lim_{x_2^{\text{bin}} \to 0} \Delta_{12} \quad (24)$$

where $c_1^{0,\text{bin}} = \lim_{x_2^{\text{bin}} \to 0} c_1^0$ and c_1^0 is the bulk molecular concentration of component 1 in the binary mixture 1–2.

2.3. The solubility of a solid in a binary mixture of SC fluids

In this case, the derivatives of the activity coefficient with respect to mole fractions when $x_2^{ter} \to 0$ are needed, and Eqs. (16) and (17) lead to

$$\lim_{x_2 \to 0} \left(\frac{\partial \ln \gamma_2^{ter}}{\partial x_1^{ter}} \right)_{T,P,x_3^{ter}} = \frac{(c_1^0 + c_3^0)\{(c_1^0 + 0.5c_3^0)\Delta_{12} + 0.5c_3^0\Delta_{23} - 0.5c_3^0\Delta_{13}\}_{x_2^{ter}=0}}{c_1^0 + c_3^0 + c_1^0 c_3^0 \Delta_{13}}$$
$$- \frac{c_1^0 c_3^0 (c_1^0 + c_3^0)\{(\Delta_{12})^2 + (\Delta_{13})^2 + (\Delta_{23})^2 - 2\Delta_{12}\Delta_{13} - 2\Delta_{12}\Delta_{23} - 2\Delta_{13}\Delta_{23}\}_{x_2^{ter}=0}}{4(c_1^0 + c_3^0 + c_1^0 c_3^0 \Delta_{13})} \quad (25)$$

and

$$\lim_{x_2 \to 0} \left(\frac{\partial \ln \gamma_2^{ter}}{\partial x_3^{ter}} \right)_{T,P,x_1^{ter}} = \frac{(c_1^0 + c_3^0)\{0.5c_1^0\Delta_{12} + (0.5c_1^0 + c_3^0)\Delta_{23} - 0.5c_1^0\Delta_{13}\}_{x_2^{ter}=0}}{c_1^0 + c_3^0 + c_1^0 c_3^0 \Delta_{13}}$$
$$- \frac{c_1^0 c_3^0 (c_1^0 + c_3^0)\{(\Delta_{12})^2 + (\Delta_{13})^2 + (\Delta_{23})^2 - 2\Delta_{12}\Delta_{13} - 2\Delta_{12}\Delta_{23} - 2\Delta_{13}\Delta_{23}\}_{x_2^{ter}=0}}{4(c_1^0 + c_3^0 + c_1^0 c_3^0 \Delta_{13})} \quad (26)$$

The value of the parameter Δ_{13} in a gas mixture can be calculated from PVT data using any traditional EOS. For the mixtures that obey the Lewis–Randall rule [16] (the fugacity of a species in a gaseous mixture is the product of its mole fraction and the fugacity of the pure gaseous component at the same temperature and pressure), the fugacity coefficients of the components of the mixture are independent of composition. In such cases, the KB equation [13] for the binary mixtures 1–3:

$$\left(\frac{\partial \ln \phi_3^{bin}}{\partial x_3^0} \right)_{P,T} = -\frac{c_1^0 \Delta_{13}}{1 + c_1^0 x_3^0 \Delta_{13}} \quad (27)$$

leads to $\Delta_{13} = 0$. For numerous gaseous mixtures the Lewis–Randall rule is valid at both low and high pressures [16], and as shown in Appendix A, it can be applied to the mixtures SC CO_2 and SC ethane of interest in the present paper. When, however, an entrainer is a liquid or a solid, the parameter Δ_{13} is different from zero and should be calculated from binary data.

For the Lewis–Randall mixtures, one can therefore write that

$$\lim_{x_2 \to 0} \left(\frac{\partial \ln \gamma_2^{ter}}{\partial x_1^{ter}} \right)_{T,P,x_3^{ter}} = \left[(c_1^0 + 0.5c_3^0)\Delta_{12} + 0.5c_3^0\Delta_{23} - c_1^0 c_3^0 \frac{(\Delta_{12} - \Delta_{23})^2}{4} \right]_{x_2^{ter}=0} \quad (28)$$

and

$$\lim_{x_2 \to 0} \left(\frac{\partial \ln \gamma_2^{ter}}{\partial x_3^{ter}} \right)_{T,P,x_1^{ter}} = \left[0.5c_1^0 \Delta_{12} + (0.5c_1^0 + c_3^0)\Delta_{23} - c_1^0 c_3^0 \frac{(\Delta_{12} - \Delta_{23})^2}{4} \right]_{x_2^{ter}=0} \quad (29)$$

and Eq. (15) becomes

$$\phi_2^{ter} = \phi_2^{ter}(x_1^0, 0, x_3^0) \exp\left\{ -x_2^{ter} \left[c_1^0 \Delta_{12} + c_3^0 \Delta_{23} - c_1^0 c_3^0 \frac{(\Delta_{12} - \Delta_{23})^2}{4} \right] \right\}_{x_2^{ter}=0} \quad (30)$$

For the fugacity coefficients in binary mixtures of a solid (2) at high dilution in a SC fluid (1) and in an entrainer (3), one can write the following equations [9] (see Eq. (3)):

$$\ln \phi_2^{\text{bin},1} = \ln \phi_2^{\text{bin},1,\infty} - \overline{k_{22}} x_2^{\text{bin},1} \tag{31}$$

and

$$\ln \phi_2^{\text{bin},3} = \ln \phi_2^{\text{bin},3,\infty} - \overline{k_{33}} x_2^{\text{bin},3} \tag{32}$$

where

$$\overline{k_{22}} = -\left(\frac{\partial \ln \gamma_2^{\text{bin},1}}{\partial x_2^{\text{bin},1}}\right)_{P,T,x_2^{\text{bin},1} \to 0} = -\left(\frac{\partial \ln \phi_2^{\text{bin},1}}{\partial x_2^{\text{bin},1}}\right)_{P,T,x_2^{\text{bin},1} \to 0} \tag{33}$$

and

$$\overline{k_{33}} = -\left(\frac{\partial \ln \gamma_2^{\text{bin},3}}{\partial x_2^{\text{bin},3}}\right)_{P,T,x_2^{\text{bin},3} \to 0} = -\left(\frac{\partial \ln \phi_2^{\text{bin},3}}{\partial x_2^{\text{bin},3}}\right)_{P,T,x_2^{\text{bin},3} \to 0} \tag{34}$$

However, from Eqs. (1) and (2), one can easily obtain that

$$\frac{x_2^{\text{ter}}}{x_2^{\text{bin},1}} = \frac{\phi_2^{\text{bin},1}}{\phi_2^{\text{ter}}} \tag{35}$$

$$\frac{x_2^{\text{ter}}}{x_2^{\text{bin},3}} = \frac{\phi_2^{\text{bin},3}}{\phi_2^{\text{ter}}} \tag{36}$$

For the fugacity coefficient of a solute at infinite dilution in a mixed gaseous solvent the following expression:

$$\phi_2^{\text{ter}}(x_1^0, 0, x_3^0) = x_1^0 \phi_2^{\text{bin},1,\infty} + x_3^0 \phi_2^{\text{bin},3,\infty} \tag{37}$$

will be employed.

Combining Eqs. (31), (32) and (35)–(37), yields

$$\frac{1}{x_2^{\text{ter}}} \exp\left\{ x_2^{\text{ter}} \left[c_1^0 \Delta_{12} + c_3^0 \Delta_{23} - c_1^0 c_3^0 \frac{(\Delta_{12} - \Delta_{23})^2}{4} \right]_{x_2^{\text{ter}}=0} \right\}$$
$$= \frac{x_1^0}{x_2^{\text{bin},1}} \exp(\overline{k_{22}} x_2^{\text{bin},1}) + \frac{x_3^0}{x_2^{\text{bin},3}} \exp(\overline{k_{33}} x_2^{\text{bin},3}) \tag{38}$$

Eq. (38) can be further simplified making use of the following approximations $\lim_{x_2 \to 0} \Delta_{12} = \lim_{x_2 \to 0, x_3 \to 0} \Delta_{12}$ and $\lim_{x_2 \to 0} \Delta_{23} = \lim_{x_2 \to 0, x_3 \to 0} \Delta_{23}$. Consequently, Eq. (24) leads to

$$\lim_{x_2 \to 0} \Delta_{12} = V_1^0 \overline{k_{22}} \tag{39}$$

and

$$\lim_{x_2 \to 0} \Delta_{23} = V_3^0 \overline{k_{33}} \tag{40}$$

where V_i^0 is the molar volume of the pure component i (SC fluid or SC entrainer). The molar volume of the 1–3 gas mixtures will be calculated using the expression:

$$V = x_1^0 V_1^0 + x_3^0 V_3^0 \qquad (41)$$

With these simplifications, Eq. (38) becomes

$$\frac{1}{x_2^{\text{ter}}} \exp\left\{x_2^{\text{ter}}\left[\varphi_1 \overline{k_{22}} + \varphi_3 \overline{k_{33}} - \frac{(x_3^0 \varphi_1 \overline{k_{22}} - x_1^0 \varphi_3 \overline{k_{33}})^2}{4 x_1^0 x_3^0}\right]\right\}$$
$$= \frac{x_1^0}{x_2^{\text{bin},1}} \exp(\overline{k_{22}} x_2^{\text{bin},1}) + \frac{x_3^0}{x_2^{\text{bin},3}} \exp(\overline{k_{33}} x_2^{\text{bin},3}) \qquad (42)$$

where $\varphi_i = x_i^0 V_i^0 / (x_1^0 V_1^0 + x_3^0 V_3^0)$ is the volume fraction of component i ($i = 1, 3$) in the mixture of the SC fluid and SC entrainer.

Consequently, the solubility of a solid in the mixture of a SC fluid and a SC entrainer can be calculated from Eq. (42) if the properties of the pure fluids and the binary solubilities $x_2^{\text{bin},1}$ and $x_2^{\text{bin},3}$ are known ($\overline{k_{22}}$ and $\overline{k_{33}}$ can be calculated from the binary solubilities).

The multi-parameter EOS [17,18] allows one to accurately calculate the densities (V_i^0) at any pressure and temperature. The parameters $\overline{k_{22}}$ and $\overline{k_{33}}$ can be calculated [9,19] using any traditional EOS, such as the Soave–Redlich–Kwong [20] or the Peng–Robinson [21] EOS.

3. Calculations

The equation derived in the preceding section of the paper will be now compared with the experimental solubilities of solids in binary mixtures of SC fluids.

3.1. Source of data

Three ternary mixtures: (a) CO_2 (1) + naphthalene (2) + ethane (3); (b) CO_2 (1) + benzoic acid (2) + ethane (3) and (c) CO_2 (1) + phenanthrene (2) + ethane (3) will be considered. The information [22–26] available regarding the solubilities of the above solids in ternary and binary mixtures (references, pressure range and temperature) is summarized in Table 1.

3.2. The calculation of V_i^0, $\overline{k_{22}}$ and $\overline{k_{33}}$

The densities of CO_2 and ethane were calculated using multi-parameter EOS [17,18], which are accurate near the critical point. The parameters $\overline{k_{22}}$ and $\overline{k_{33}}$ were obtained from solubility data in binary mixtures (solid/SC fluid and solid/SC entrainer). The Soave–Redlich–Kwong [20] EOS was employed in combination with the classical van der Waals mixing rules as in our previous paper [5].

3.3. Results

The comparison between predicted and experimental solubilities is presented in Fig. 1 for naphthalene, in Fig. 2 for benzoic acid and in Fig. 3 for phenanthrene. One can see that there is excellent agreement

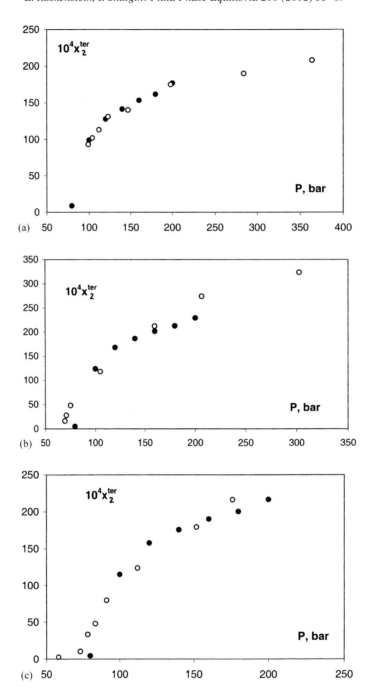

Fig. 1. Comparison between experimental (○) and predicted (●) solubilities of naphthalene in the mixtures of SC CO_2 and SC ethane: (a) $T = 308.15$ K, $x_1^0 = 0.938$, experimental data from [2]; (b) $T = 308$ K, $x_1^0 = 0.399$, experimental data from [22]; (c) $T = 308$ K, $x_1^0 = 0.496$, experimental data from [22]; (d) $T = 307.9$ K, $x_1^0 = 0.412$, experimental data from [23]; (e) $T = 318.15$ K, $x_1^0 = 0.938$, experimental data from [2]; (f) $T = 318$ K, $x_1^0 = 0.254$, experimental data from [22]; (g) $T = 318$ K, $x_1^0 = 0.474$, experimental data from [22].

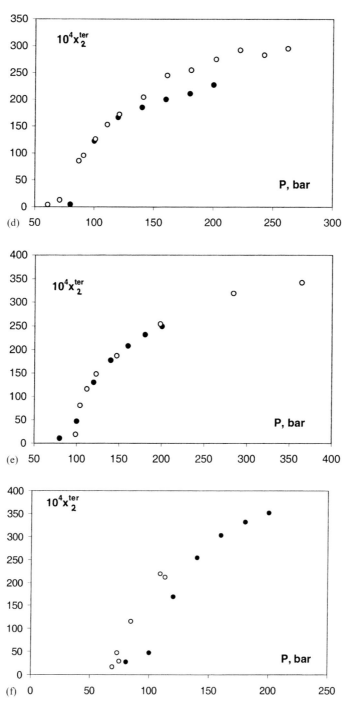

Fig. 1. (*Continued*).

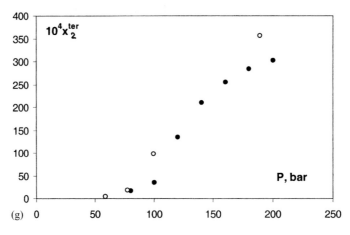

(g)

Fig. 1. (*Continued*).

Table 1
Information about the ternary and binary SC mixtures employed in the calculations

Mixture	T (K)	Pressure range (bar)	x_1^0	Reference
CO_2 (1) + naphthalene (2) + ethane (3)	308.15	99–364	0.938	[2]
	318.15	99–364	0.938	[2]
	308	69.9–302.6	0.399	[22]
	308	58.7–176.1	0.496	[22]
	318	69.0–117.5	0.254	[22]
	318	58.7–188.8	0.474	[22]
	307.9	61.7–265.5	0.412	[23]
CO_2 + naphthalene	308.15	60.8–364		[24]
	318.15	62.8–314.1		[24]
Ethane + naphthalene	308.15	50.9–362		[25]
	318.15	51–364		[25]
CO_2 (1) + benzoic acid (2) + ethane (3)	328.15	116–364	0.938	[2]
	343.15	116–364	0.938	[2]
CO_2 + benzoic acid	328.15	101–363.3		[25]
	343.15	101–364.1		[25]
Ethane + benzoic acid	328.15	54.1–361.8		[25]
	343.15	66–363.5		[25]
CO_2 (1) + phenanthrene (2) + ethane (3)	313	110–350	0.95	[26]
	313	110–350	0.9	[26]
CO_2 + phenanthrene	313	110–350		[26]
Ethane + phenanthrene	313	110–350		[26]

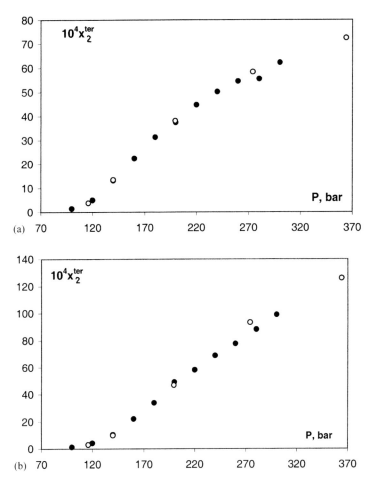

Fig. 2. Comparison between experimental (○) and predicted (●) solubilities of benzoic acid in the mixtures of SC CO_2 and SC ethane: (a) $T = 328.15$ K, $x_1^0 = 0.938$, experimental data from [2]; (b) $T = 343.15$ K, $x_1^0 = 0.938$, experimental data from [2].

between the predicted and experimental values. It is worth noting that the experimental solubilities for naphthalene were obtained by a number of authors under different conditions (temperature, pressure range and composition of the binary SC mixture).

4. Discussion and conclusion

The predictive method suggested in this paper allows one to calculate the solubility of a solid in a binary mixture of SC fluids. The solubilities of three solids were predicted using only experimental data regarding the solubilities in the constituent binary mixtures (solid/SC fluid and solid/SC entrainer). Very good agreement was found between the experimental and predicted solubilities. For the solubilities of naphthalene and benzoic acid, the prediction of Eq. (42) provided even better agreement than the correlation [2] of experimental data based on the Peng–Robinson EOS with parameters determined from ternary data.

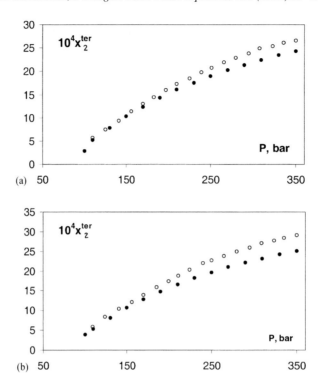

Fig. 3. Comparison between experimental (○) and predicted (●) solubilities of phenanthrene in the mixtures of SC CO_2 and SC ethane: (a) $T = 313$ K, $x_1^0 = 0.95$, experimental data from [26]; (b) $T = 313$ K, $x_1^0 = 0.9$, experimental data from [26].

It should be noted that, while the expressions for the derivatives of the fugacity (activity) coefficients in ternary mixtures with respect to mole fractions (Eqs. (25) and (26)) employed to derive Eq. (42) are rigorous, Eq. (42) for the solubility also implies that $\lim_{x_2 \to 0} \Delta_{13} = 0$. The latter approximation is probably not entirely accurate at very high pressures when the intermolecular interactions between the SC fluids become comparable to those in liquids. Indeed, the predicted solubilities of phenanthrene in the mixture of SC CO_2 and SC ethane (Fig. 3) deviate somewhat from the experimental data [26] at high pressures. This deviation might have been caused also by the experimental solubilities of phenanthrene in SC ethane [26] employed in our calculations, which are quite different from other experimental data [27].

As a rule [1–3], the addition of an entrainer or cosolvent to a SC fluid enhances the solubility of a solid compared to its solubility in a pure SC fluid. However, this is no longer true when the entrainer (cosolvent) is a SC fluid or a gas. Indeed, the addition of the SC ethane to the SC CO_2 enhanced the solubility of naphthalene or phenanthrene compared to that in the pure carbon dioxide; however, the addition of the SC CO_2 to the SC ethane decreased the solubility of naphthalene or phenanthrene compared to that in the pure ethane [2,22,23,26]. These results can be explained in the framework of the present approach. For this purpose, Eq. (42) will be first simplified by assuming that (1) all the solubilities (x_2^{ter}, $x_2^{bin,1}$ and $x_2^{bin,3}$) are small and hence the exponential expressions in Eq. (42) can be expanded in series up to the first term in x; (2) the molar volumes of the pure components 1 and 3 are equal. Then, Eq. (42) acquires the simple form

$$-x_1^0 x_3^0 \frac{(\overline{k_{22}} - \overline{k_{33}})^2}{4} + \frac{1}{x_2^{\text{ter}}} = \frac{x_1^0}{x_2^{\text{bin},1}} + \frac{x_3^0}{x_2^{\text{bin},3}} \qquad (43)$$

The latter equation shows that for an ideal mixture (when all $\overline{k_{ii}}$ are equal to zero), the solubility in a binary solvent ($x_2^{\text{ter,id}}$) has an intermediate value between the solubilities in the binary constituents $x_2^{\text{bin},1}$ and $x_2^{\text{bin},3}$. When $\overline{k_{22}}$ and $\overline{k_{33}}$ have non-zero values, then $x_2^{\text{ter}} \leq x_2^{\text{ter,id}}$ and in some cases x_2^{ter} can be smaller than both binary solubilities ($x_2^{\text{bin},1}$ and $x_2^{\text{bin},3}$). Therefore, one can conclude that when an "inert" gas (i.e. a gas in which the solubility of the solid is much smaller than in a SC fluid) is added to a SC fluid it should decrease the solubility of the solid compared to that in the pure SC fluid. Such a decrease in solubility was found experimentally for caffeine in the SC CO_2 entrained by nitrogen [28,29].

Appendix A

The composition dependence of the parameter Δ_{13} was calculated for the mixture SC CO_2 and SC ethane at $T = 350$ K and $P = 10$ MPa. For this purpose, precise PVT data [30] were treated using the Soave–Redlich–Kwong [20] EOS and the classical van der Waals mixing rules. The binary interaction parameter q_{12} was calculated by minimizing the sum $\sum_{i=1}^{N}(V^{i,\text{exp}} - V^{i,\text{calc}})^2$, where $V^{i,\text{exp}}$ and $V^{i,\text{calc}}$ are the experimental and calculated molar volumes of the mixture and N is the number of experimental points. It was found that for the selected pressure and temperature $q_{12} = 0.043$. The calculated fugacity coefficients have a weak linear composition dependence from which one could obtain $(\partial \ln \phi_2^{\text{bin}}/\partial x_2)_{P=10\,\text{MPa},T=350\,\text{K}} = 0.09$. The composition dependence of the parameter Δ_{13} can be calculated using Eq. (27). For $x_1^0 = 0.5$, $\Delta_{13} \approx -30$ (cm^3/mol), while for the mixtures [5] CO_2 (1) + naphthalene (2) and CO_2 (1) + phenanthrene (2) at $T = 308$ K and $P = 10$ MPa, the nonideality parameters for $x_2 \to 0$, calculated with Eqs. (39) and (40), are $\Delta_{12} = 9 \times 10^3$ (cm^3/mol) and $\Delta_{12} = 4 \times 10^3$ (cm^3/mol), respectively. Fugacity data [31] also indicated that the fugacity coefficients of CO_2 and ethane are independent of composition at $T = 308$ K in a wide range of pressures. Consequently, the Lewis–Randall rule constitutes a good approximation for the mixture SC CO_2 and SC ethane.

References

[1] G. Brunner, Fluid Phase Equilib. 10 (1983) 289–298.
[2] W.J. Schmitt, R.C. Reid, Fluid Phase Equilib. 32 (1986) 77–99.
[3] C.A. Eckert, B.L. Knutson, P.G. Debenedetti, Nature 383 (1996) 313–318.
[4] S. Kim, K.P. Johnston, Am. Inst. Chem. Eng. J. 33 (1987) 1603–1611.
[5] E. Ruckenstein, I. Shulgin, Fluid Phase Equilib. 180 (2001) 345–359.
[6] R.M. Lemert, K.P. Johnston, Ind. Eng. Chem. Res. 30 (1991) 1222–1231.
[7] S.G. Kazarian, R.B. Gupta, M.J. Clarke, K.P. Johnston, M. Poliakoff, J. Am. Chem. Soc. 115 (1993) 11099–11109.
[8] J.M. Prausnitz, R.N. Lichtenthaler, E. Gomes de Azevedo, Molecular Thermodynamics of Fluid-Phase Equilibria, 2nd Edition, Prentice-Hall, Englewood Cliffs, NJ, 1986.
[9] P.G. Debenedetti, S.K. Kumar, Am. Inst. Chem. Eng. J. 32 (1986) 1253–1262.
[10] A. Chialvo, J. Phys. Chem. 97 (1993) 2740–2744.
[11] D.A. Jonah, H.D. Cochran, Fluid Phase Equilib. 92 (1994) 107–137.
[12] F. Munoz, T.W. Li, E.H. Chimowitz, Am. Inst. Chem. Eng. J. 41 (1995) 389–401.

[13] J.G. Kirkwood, F.P. Buff, J. Chem. Phys. 19 (1951) 774–782.
[14] E. Ruckenstein, I. Shulgin, Fluid Phase Equilib. 180 (2001) 281–297.
[15] A. Ben-Naim, J. Chem. Phys. 67 (1977) 4884–4890.
[16] S.I. Sandler, Chemical and Engineering Thermodynamics, 3rd Edition, Wiley, New York, 1999.
[17] F.H. Huang, F.T.H. Chung, L.L. Lee, M.H. Li, K.E. Starling, J. Chem. Eng. Jpn. 18 (1985) 490–496.
[18] B.A. Younglove, J.F. Ely, J. Phys. Chem. Ref. Data 16 (1987) 577–798.
[19] E. Ruckenstein, I. Shulgin, J. Phys. Chem. B 104 (2000) 2540–2545.
[20] G. Soave, Chem. Eng. Sci. 27 (1972) 1197–1203.
[21] D.Y. Peng, D.B. Robinson, Ind. Eng. Chem. Fundam. 15 (1976) 59–64.
[22] W.E. Hollar, P. Ehrlich, J. Chem. Eng. Data 35 (1990) 271–275.
[23] G.R. Smith, C.J. Wormald, Fluid Phase Equilib. 57 (1990) 205–222.
[24] Yu.V. Tsekhanskaya, M.B. Iomtev, E.V. Mushkina, Russ. J. Phys. Chem. 38 (1964) 1173–1176.
[25] W.J. Schmitt, R.C. Reid, J. Chem. Eng. Data 31 (1986) 204–212.
[26] G. Anitescu, L.L. Tavlarides, J. Supercrit. Fluids 11 (1997) 37–51.
[27] K.P. Johnston, C.A. Eckert, D.H. Ziger, Ind. Eng. Chem. Fundam. 21 (1982) 191–197.
[28] H.J. Gahrs, Ber. Bunsen Phys. Chem. 88 (1984) 894–897.
[29] G. Brunner, in: J.A. Marinsky, Y. Marcus (Eds.), Ion Exchange and Solvent Extraction, Vol. 10, Marcel Dekker, New York, 1988, pp. 105–140.
[30] W.W.R. Lau, C.A. Hwang, J.C. Holste, K.P. Hall, B.E. Gammon, K.N. Marsh, J. Chem. Eng. Data 42 (1997) 900–902.
[31] N.E. Khazanova, E.E. Sominskaya, M.B. Rozovskii, Zh. Fiz. Khim. 52 (1978) 912–914.

A Simple Equation for the Solubility of a Solid in a Supercritical Fluid Cosolvent with a Gas or Another Supercritical Fluid

E. Ruckenstein* and I. Shulgin[†]

Department of Chemical Engineering, State University of New York at Buffalo, Amherst, New York 14260

A simple equation for calculating the solubilities of solid substances in gaseous mixtures of a supercritical fluid with another supercritical one or an inert gas was derived. This equation involves only the solubilities of the solid in the individual constituents of the gaseous mixture and the molar volumes of the latter. The equation was tested for the solubilities of naphthalene, benzoic acid, caffeine, cholesterol, and soybean oil in gaseous mixtures containing at least one supercritical fluid. In all cases good agreement with experimental data was found.

1. Introduction

The addition of a miscible compound to a supercritical (SC) solvent can lead to a dramatic change (in most cases to an enhancement) in the solubility of a solute.[1–3] This phenomenon, known as the entrainer effect, is relevant in the SC fluid technology. According to a common definition, an entrainer is a compound that has a higher volatility than the solute.[1] As a consequence, the entrainer can be a liquid, a solid, a gas, or a SC fluid.[2] While the cases in which the entrainer was a solid or a liquid were frequently investigated both experimentally and theoretically, those in which the entrainer was a gas or a SC fluid have received less attention. The former cases usually provided large enhancements in the solubility of a solute (up to several hundred percent).[1–3] In contrast, as first noted by Gährs,[4] a gas entrainer can decrease the solubility of a solute compared to that in a pure SC solvent; the addition of even a small amount of an inert gas can dramatically reduce the solute solubility (for instance, 5 vol % of N_2 decreases the solubility of caffeine in SC CO_2 by 50%).[4] The results of Gährs were later qualitatively discussed by Brunner.[5]

A similar effect was observed for the solubility of a solid in a gas mixture composed of two SC fluids.[2,6–9] In the latter case, experiments[2,6–9] have revealed that the solubility has a value intermediate between those recorded in the individual SC fluids. Thus, the addition of SC ethane to SC carbon dioxide enhanced the solute solubility compared to that found in pure CO_2 but decreased it relative to that in pure C_2H_6. King and co-workers[10,11] also investigated the effect of the addition of helium on the solubilities of cholesterol and soybean oil in SC CO_2 and found that the addition reduces them dramatically. To date, a theory that can predict the effect of a gaseous entrainer on the solute solubility has not yet been developed. The aim of the present research was to derive an equation able to predict the solubility of a solid in a SC fluid + entrainer mixture, when the entrainer is another SC fluid or an inert gas. For this purpose, the Kirkwood–Buff formalism[12] for ternary mixtures[13] was used. In previous papers,[13,14] the Kirkwood–Buff formalism for ternary mixtures was utilized to describe the entrainer effect; however, those methods are not applicable to gaseous entrainers, and the aim of the present paper is to propose a method for such cases.

2. Theory

2.1. Thermodynamic Relations for the Solubility of Solids in SC Fluids and Their Mixtures. Let us denote a SC fluid as component 1, a solid as component 2, and an entrainer as component 3. For the solubility of a solid in individual SC fluids and their mixture, one can write[15] the relations

$$z_2^{b,1} = \frac{P_2^0}{P\phi_2^{b,1}} \exp\left[\frac{(P - P_2^0)V_2^0}{RT}\right] \quad (1)$$

$$z_2^{b,3} = \frac{P_2^0}{P\phi_2^{b,3}} \exp\left[\frac{(P - P_2^0)V_2^0}{RT}\right] \quad (2)$$

and

$$z_2^t = \frac{P_2^0}{P\phi_2^t} \exp\left[\frac{(P - P_2^0)V_2^0}{RT}\right] \quad (3)$$

where R is the universal gas constant, P is the pressure, T is the temperature in K, P_2^0 and V_2^0 are the saturation vapor pressure and molar volume of the pure solute, $z_2^{b,i}$ and $\phi_2^{b,i}$ ($i = 1, 3$) are the mole fractions and fugacity coefficients of a solute in the gaseous phase of the binary mixtures SC fluid (i) + solute (2) at equilibrium, and ϕ_2^t and z_2^t are the fugacity coefficient and the mole fraction of the solute in the gaseous phase of a ternary mixture [SC fluid (1) + solid solute (2) + SC entrainer (3)] at equilibrium. The main difficulties in predicting the solubility of a solid in the pure SC fluids and their mixtures are the calculations of the fugacity coefficients of the solid in the binary and ternary mixtures. This can be done by using the Kirkwood–Buff formalism for ternary mixtures,[13] which allows one to express the fugacity coefficient of a solid in a ternary mixture in terms of the fugacity coefficients of the solid in the binary mixtures solid + individual SC fluids.

2.2. Expression for the Fugacity Coefficient of a Solid in a Ternary Mixture at Infinite Dilution by the Kirkwood–Buff Formalism. The following

---

* Corresponding author. E-mail: feaeliru@acsu.buffalo.edu. Fax: (716) 645-3822. Phone: (716) 645-2911 ext. 2214.
† E-mail address: ishulgin@eng.buffalo.edu.

expression for the derivative of the fugacity coefficient of a solid in a ternary mixture at infinite dilution can be written[13]

$$\lim_{x_2^t \to 0} \left(\frac{\partial \ln \phi_2^t}{\partial x_3^t} \right)_{T,P,x_2^t} = $$
$$- \frac{(c_1^0 + c_3^0)[(c_1^0 + c_3^0)(\Delta_{12} - \Delta_{23})_{x_2^t=0} + (c_1^0 - c_3^0)(\Delta_{13})_{x_2^t=0}]}{2(c_1^0 + c_3^0 + c_1^0 c_3^0 (\Delta_{13})_{x_2^t=0})} \quad (4)$$

where x_i^t is the mole fraction of component i in the ternary mixture, c_k^0 ($k = 1, 3$) is the bulk molecular concentration of component k in the binary gaseous mixture containing the constituents 1 and 3, denoted further by binary mixture 1–3, and $G_{\alpha\beta}$ is the Kirkwood–Buff integral given by

$$G_{\alpha\beta} = \int_0^\infty (g_{\alpha\beta} - 1) 4\pi r^2 \, dr \quad (5)$$

In the above expressions, $g_{\alpha\beta}$ is the radial distribution function between species α and β, r is the distance between the centers of molecules α and β, and $\Delta_{\alpha\beta}$ are defined as follows:

$$\Delta_{\alpha\beta} = G_{\alpha\alpha} + G_{\beta\beta} - 2G_{\alpha\beta}, \quad \alpha \neq \beta \quad (6)$$

It should be noted that for an ideal mixture $\Delta_{\alpha\beta} = 0$. Consequently, $\Delta_{\alpha\beta}$ is a measure of the nonideality[16] of the binary mixture $\alpha - \beta$.

Furthermore, for a binary 1–3 mixture, one can write the relation[12]

$$\left(\frac{\partial \ln \phi_1^{b,1-3}}{\partial x_3^{b,1-3}} \right)_{P,T} = \frac{c_3^0 \Delta_{13}}{1 + c_1^0 x_3^{b,1-3} \Delta_{13}} \quad (7)$$

where $x_i^{b,1-3}$ and $\phi_i^{b,1-3}$ are the mole fraction and the fugacity coefficient of component i ($i = 1, 3$), respectively, in the binary gaseous mixture 1–3 (the superscript b,1–3 indicates the binary 1–3 mixture).

Introducing Δ_{13} from eq 7 in eq 4 and integrating the obtained expression yields

$$\ln \phi_2^{t,\infty} = - \int (c_1^0 + c_3^0) \frac{(\Delta_{12} - \Delta_{23})_{x_2^t=0}}{2} \times$$
$$\left[1 + x_3^{b,1-3} \left(\frac{\partial \ln \phi_3^{b,1-3}}{\partial x_3^{b,1-3}} \right)_{P,T} \right] dx_3^{b,1-3} +$$
$$\frac{1}{2} \int \frac{x_1^{b,1-3} - x_3^{b,1-3}}{x_1^{b,1-3}} \left(\frac{\partial \ln \phi_3^{b,1-3}}{\partial x_3^{b,1-3}} \right)_{P,T} dx_3^{b,1-3} + A \quad (8)$$

where $\phi_2^{t,\infty} = \lim_{x_2^t \to 0} \phi_2^t$ is the fugacity coefficient of a solute at infinite dilution in the ternary mixture 1–2–3 and A is an integration constant.

For an ideal binary gaseous mixture 1–3, $\phi_3^{b,1-3} = 1$ and

$$V = x_1^{b,1-3} V_1^0 + x_3^{b,1-3} V_3^0 \quad (9)$$

where V is the molar volume of the binary gaseous mixture 1–3 and V_1^0 and V_3^0 are the molar volumes of the individual fluids 1 and 3. Under such conditions, eq 8 becomes

$$\ln \phi_2^{t,\infty} = - \int \frac{(\Delta_{12} - \Delta_{23})_{x_2^t=0}}{2(x_1^{b,1-3} V_1^0 + x_3^{b,1-3} V_3^0)} dx_3^{b,1-3} + A \quad (10)$$

The main approximation of this paper is the assumption that $(\Delta_{12})_{x_2^t=0} = (G_{11} + G_{22} - 2G_{12})_{x_2^t=0}$ and $(\Delta_{23})_{x_2^t=0} = (G_{22} + G_{33} - 2G_{23})_{x_2^t=0}$ are independent of the composition of the gaseous 1–3 mixture. Consequently, eq 10 becomes

$$\ln \phi_2^{t,\infty} = A(P,T) - \frac{B(P,T) \ln(x_1^{b,1-3} V_1^0 + x_3^{b,1-3} V_3^0)}{V_3^0 - V_1^0} \quad (11)$$

where $B(P,T) = (\Delta_{12} - \Delta_{23})_{x_2^t=0}/2$.

Expressions for the constants $A(P,T)$ and $B(P,T)$ can be obtained using the following limiting relationships for the fugacity coefficient of a solute at infinite dilution in the binary gaseous mixture 1–3:

$$(\ln \phi_2^{t,\infty})_{x_1^{b,1-3}=0} = \ln \phi_2^{b,3,\infty} \quad (12)$$

and

$$(\ln \phi_2^{t,\infty})_{x_3^{b,1-3}=0} = \ln \phi_2^{b,1,\infty} \quad (13)$$

Combining eqs 11–13 yields the final result

$$\ln \phi_2^{t,\infty} = $$
$$\frac{(\ln V - \ln V_3^0) \ln \phi_2^{b,1,\infty} + (\ln V_1^0 - \ln V) \ln \phi_2^{b,3,\infty}}{\ln V_1^0 - \ln V_3^0} \quad (14)$$

Equation 14 allows one to calculate the fugacity coefficient of a solute at infinite dilution in the binary mixture of two SC fluids, in terms of the fugacity coefficients of the solute at infinite dilution for each of the SC fluids. This expression will be used in the next section to derive an expression for the solubility of a solid in a gaseous mixture of two SC fluids.

2.3. Expression for the Solubility of a Solid in a Gaseous Mixture Formed of Two SC Fluids. Assuming that the solute solubilities are very small and that the fugacity coefficients have the same values as those at infinite dilution, eqs 1–3 can be recast as

$$\ln \phi_2^{t,\infty} = \ln E - \ln z_2^t \quad (15)$$

$$\ln \phi_2^{b,1,\infty} = \ln E - \ln z_2^{b,1} \quad (16)$$

$$\ln \phi_2^{b,3,\infty} = \ln E - \ln z_2^{b,3} \quad (17)$$

where $E = (P_2^0/P) \exp[(P - P_2^0) V_2^0/RT]$ is a composition-independent quantity.

By combining eqs 14–17, one obtains the following expression for the solubility of a solid in a gaseous mixture formed of two SC fluids:

$$\ln z_2^t = \frac{(\ln V - \ln V_3^0) \ln z_2^{b,1} + (\ln V_1^0 - \ln V) \ln z_2^{b,3}}{\ln V_1^0 - \ln V_3^0} \quad (18)$$

This simple equation does not require information about the ternary mixture 1−2−3 or about its binary constituents; it requires only the solubilities of the solid in the individual fluids and the molar volumes of the latter.

3. Calculations

3.1. Systems Used in the Calculations. As noted previously, two different types of systems are considered:

(a) The entrainer is a SC fluid. For comparison with experiment, we selected the solubilities of naphthalene,[6,7] benzoic acid,[2] and cholesterol[8] in the binary mixture of SC CO_2 and SC C_2H_6.

(b) The entrainer is an inert gas. For comparison with experiment, we selected the solubility[4] of caffeine in the binary mixture of SC CO_2 and N_2 and those of cholesterol[10,11] and soybean oil[10,11] in the binary mixture of SC CO_2 and He.

3.2. Calculation Procedure. The molar volumes (V_i^0) of the pure SC fluids at a given pressure and temperature were calculated using multiparameter equations of state,[17,18] which are accurate even in the critical region. The molar volumes (V_i^0) of the inert gases (N_2 and He) were calculated as for an ideal gas ($V_i^0 = RT/P$). The solubilities of the solids in the individual SC fluids were obtained from refs 4, 8, 10, 11, and 19−21. The solubilities of the solids in the inert gases were estimated[22] as those in an ideal gas, i.e., using eqs 1 or 2 with the fugacity coefficients $\phi_2^{b,i} = 1$. The saturated vapor pressures of caffeine, cholesterol, and soybean oil, as well as their molar volumes, were available in the literature.[23−26]

4. Results

4.1. Entrainer as a SC Fluid. The predicted and experimental solubilities of naphthalene, benzoic acid, and cholesterol in the mixture of SC CO_2 and SC C_2H_6 are presented in Figures 1−3, respectively. For the solubility of cholesterol, the calculations were carried out at $T = 328.1$ K (because experimental data[8] were available only for this temperature). However, no data on the solubility of cholesterol in pure SC CO_2 at this temperature were found, and we used the solubility at $T = 333.15$ K.[21]

4.2. Entrainer as an Inert Gas. The predicted and experimental solubilities of caffeine in SC CO_2 entrained with N_2 are compared in Figure 4. Explicit numerical values for the experimental solubilities[4] of caffeine in SC CO_2 and in SC CO_2 entrained with N_2 were not available; they could be, however, evaluated from Figures 1 and 3 of ref 4. The predicted and experimental solubilities of cholesterol and soybean oil in SC CO_2 entrained with He are compared in Figures 5 and 6, respectively.

5. Discussion

Figures 1−6 show that eq 18 provides accurate predictions for the solubilities of solid substances in a SC fluid entrained with another SC fluid or with an inert gas, such as nitrogen or helium. Because the selected systems involve different entrainers and wide ranges of pressure and temperature and because the solubilities were determined by different authors (with inevitable experimental errors), the predictions (Figures 1−6) can be considered as excellent. While there is some

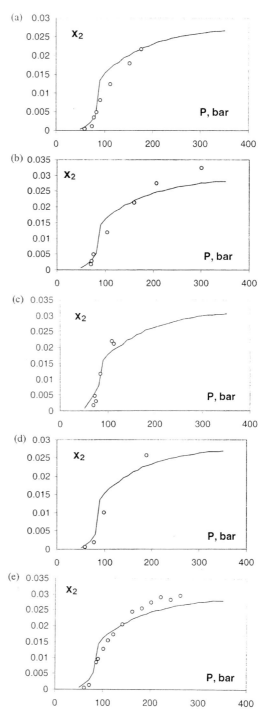

Figure 1. Comparison between experimental (○) and predicted (solid lines) solubilities of naphthalene in SC CO_2 and SC ethane mixtures: (a) $T = 308$ K, the mole fraction of SC CO_2 in the solute-free gaseous mixture $x_1^{b,1-3} = 0.496$; (b) $T = 308$ K, $x_1^{b,1-3} = 0.399$; (c) $T = 318$ K, $x_1^{b,1-3} = 0.254$; (d) $T = 318$ K, $x_1^{b,1-3} = 0.474$; (e) $T = 307.9$ K, $x_1^{b,1-3} = 0.412$. The experimental data are taken from ref 6 for parts a−d and from ref 7 for part e.

disagreement regarding the experimental and calculated solubilities of cholesterol in SC CO_2 entrained with He (Figure 5), it should be pointed out that there are large deviations among the literature data for this system. Indeed, we used in the calculations for the solubility of cholesterol in neat SC CO_2 the values[11] $z_2^t = 6.2 \times 10^{-5}$ at $T = 313.15$ K and $P = 241.3$ bar and z_2^t

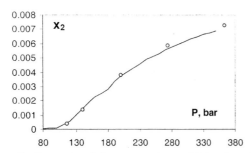

Figure 2. Comparison between experimental (○) and predicted (solid lines) solubilities of benzoic acid in a SC CO_2 and SC ethane mixture: $T = 328.15$ K, the mole fraction of SC CO_2 in the solute-free gaseous mixture $x_1^{b,1-3} = 0.938$. The experimental data are taken from ref 2.

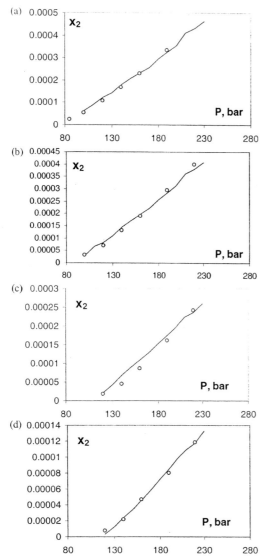

Figure 3. Comparison between experimental (○) and predicted (solid lines) solubilities of cholesterol in SC CO_2 and SC ethane mixtures: (a) $T = 328.1$ K, the mole fraction of SC CO_2 in the solute-free gaseous mixture $x_1^{b,1-3} = 0.035$; (b) $T = 328.1$ K, $x_1^{b,1-3} = 0.14$; (c) $T = 328.1$ K, $x_1^{b,1-3} = 0.5$; (d) $T = 328.1$ K, $x_1^{b,1-3} = 0.965$. The experimental data are taken from ref 8.

$= 8.4 \times 10^{-5}$ at $T = 323.15$ K and $P = 275.8$ bar, whereas in the literature, one can also find that[21] $z_2^t = 9.37 \times 10^{-5}$ at $T = 313.15$ K and $P = 250.0$ bar and z_2^t

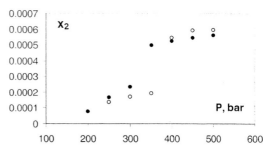

Figure 4. Comparison between experimental (○) and predicted (●) solubilities of caffeine (x_2 is the mole fraction of caffeine, $T = 353.15$ K) in SC CO_2 entrained with 5 vol % N_2.

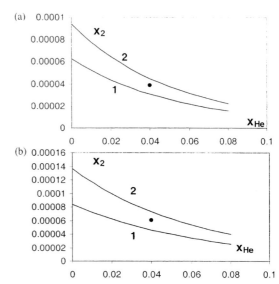

Figure 5. Comparison between experimental[11] (●) and predicted (solid lines) solubilities of cholesterol in the SC CO_2 entrained with He: (a) $T = 313.15$ K, $P = 241.3$ bar; (b) $T = 323.15$ K, $P = 275.8$ bar. Curve 1: solubility based on binary data from ref 11. Curve 2: solubility based on binary data from ref 21.

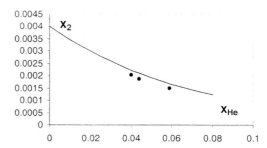

Figure 6. Comparison between experimental[11] (●) and predicted (solid line) solubilities of soybean oil in the SC CO_2 entrained with He. x_2 is the mole fraction of soybean oil, and x_{He} is the mole fraction of helium in the solute-free gaseous mixture. $T = 343.15$ K, and $P = 655.0$ bar.

$= 12.4 \times 10^{-5}$ at $T = 323.15$ K and $P = 250.0$ bar, as well as[24] $z_2^t = 19.2 \times 10^{-5}$ at $T = 313.15$ K and $P = 252.7$ bar, and $z_2^t = 32.4 \times 10^{-5}$ at $T = 333.15$ K and $P = 253.6$ bar. The results in Figure 5 clearly show that the accuracy of the binary data employed is important in predicting the solubility in the mixture.

One of the assumptions made in the derivation of eq 18 is that the mixture of the two gases can be considered ideal. This assumption is reasonable because it was shown[14] that $|\Delta_{\alpha\beta}|$ for a mixture composed of SC CO_2 and SC C_2H_6 is much smaller than those for the mixtures of a solid and SC CO_2 or SC C_2H_6; this inequality becomes even stronger when the gaseous mixture is composed of a SC fluid and an inert gas.

Equation 18 indicates the factors which affect the solubility of a solid in a SC fluid in the presence of a gaseous entrainer. The Brunner rule[5] that an entrainer with a lower (higher) critical temperature than that of a SC fluid decreases (increases) the solubility of a solid can predict only qualitatively when an entrainer will increase or decrease the solubility. The solubility of a solid in a SC fluid could be often correlated with the density of the fluid.[27] However, as mentioned by King and co-workers,[10,11] the change of the density of a gaseous mixture cannot provide alone an explanation for the change in the solubility when a SC fluid is entrained with another SC fluid or an inert gas. Equation 18 clearly shows that the solubility of a solid in a SC fluid in the presence of a gaseous entrainer depends not only on the densities of the constituents of the gaseous mixture but also on the solubilities of the solid in the SC fluid and in the gaseous entrainer.

6. Conclusion

This paper deals with the entrainer effect on the solubility of a solid in a mixture composed of a SC fluid and an entrainer, when the latter is a SC fluid or an inert gas. The main thermodynamic difficulty in the treatment of the entrainer effect is the calculation of the fugacity coefficient of a solute in a binary gaseous mixture. In this paper, the Kirkwood–Buff formalism for ternary mixtures was used to derive an expression for this fugacity coefficient in terms of the fugacity coefficients of the solid in each of the constituents of the gaseous binary mixtures. One of the main assumptions is that the nonideality of the gaseous mixture (a SC fluid + another SC fluid or an inert gas) can be neglected compared to the nonidealities between the solid and each of the components of the gaseous mixture. Based on this reasonable simplification, a simple equation for the solubility of a solid in a SC fluid in the presence of a gaseous entrainer was derived. According to this equation (eq 18), the solubility in the presence of a gaseous entrainer can be calculated in terms of the solubilities of the solid in each of the individual constituents of the gaseous mixture and the molar volumes of the constituents of the gaseous mixture. The comparison with experimental data demonstrated that the derived equation provides excellent agreement.

Literature Cited

(1) Brunner, G. Selectivity of supercritical compounds and entrainers with respect to model substances. *Fluid Phase Equilib.* **1983**, *10*, 289–298.

(2) Schmitt, W. J.; Reid, R. C. The use of entrainers in modifying the solubility of phenanthrene and benzoic acid in supercritical carbon-dioxide and ethane. *Fluid Phase Equilib.* **1986**, *32*, 77–99.

(3) Eckert, C. A.; Knutson, B. L.; Debenedetti, P. G. Supercritical fluids as solvents for chemical and materials processing. *Nature* **1996**, *383*, 313–318.

(4) Gährs, H. J. Applications of atmospheric gases in high-pressure extraction. *Ber. Bunsen-Ges.* **1984**, *88*, 894–897.

(5) Brunner, G. In *Ion Exchange and Solvent Extraction*; Marinsky, J. A., Marcus, Y., Eds.; Marcel Dekker: New York and Basel, 1988; Vol. 10, pp 105–140.

(6) Hollar, W. E.; Ehrlich, P. Solubility of naphthalene in mixtures of carbon dioxide and ethane. *J. Chem. Eng. Data* **1990**, *35*, 271–275.

(7) Smith, G. R.; Wormald, C. J. Solubilities of naphthalene in (CO_2 + C_2H_6) and (CO_2 + C_3H_8) up to 333 K and 17.7 MPa. *Fluid Phase Equilib.* **1990**, *57*, 205–222.

(8) Singh, H.; Yun, S. L. J.; Macnaughton, S. J.; Tomasko, D. L.; Foster, N. R. Solubility of cholesterol in supercritical ethane and binary gas-mixtures containing ethane. *Ind. Eng. Chem. Res.* **1993**, *32*, 2841–2848.

(9) Anitescu, G.; Tavlarides, L. L. Solubilities of solids in supercritical fluids. 2. Polycyclic aromatic hydrocarbons (PAHs) plus CO_2/cosolvent. *J. Supercrit. Fluids* **1997**, *11*, 37–51.

(10) King, J. W.; Johnson, J. H.; Eller, F. J. Effect of supercritical carbon dioxide pressurized with helium on solute solubility during supercritical fluid extraction. *Anal. Chem.* **1995**, *67*, 2288–2291.

(11) Zhang, Z.; King, J. W. The effect of dissolved helium on the density and solvation power of supercritical carbon dioxide. *J. Chromatogr. Sci.* **1997**, *35*, 483–488.

(12) Kirkwood, J. G.; Buff, F. P. Statistical mechanical theory of solutions. I. *J. Chem. Phys.* **1951**, *19*, 774–782.

(13) Ruckenstein, E.; Shulgin, I. Entrainer effect in supercritical mixtures. *Fluid Phase Equilib.* **2001**, *180*, 345–359.

(14) Ruckenstein, E.; Shulgin, I. The solubility of solids in mixtures containing a supercritical fluid and an entrainer. *Fluid Phase Equilib.* **2002**, *200*, 53–67.

(15) Prausnitz, J. M.; Lichtenthaler, R. N.; Gomes de Azevedo, E. *Molecular Thermodynamics of Fluid–Phase Equilibria*, 2nd ed.; Prentice-Hall: Englewood Cliffs, NJ, 1986.

(16) Ben-Naim, A. Inversion of the Kirkwood–Buff theory of solutions: application to the water–ethanol system. *J. Chem. Phys.* **1977**, *67*, 4884–4890.

(17) Huang, F. H.; Chung, F. T. H.; Lee, L. L.; Li, M. H.; Starling, K. E. An accurate equation of state for carbon dioxide. *J. Chem. Eng. Jpn.* **1985**, *18*, 490–496.

(18) Younglove, B. A.; Ely, J. F. Thermophysical properties of fluids. II. Methane, ethane, propane, isobutane, and normal butane. *J. Phys. Chem. Ref. Data* **1987**, *16*, 577–798.

(19) Tsekhanskaya, Yu. V.; Iomtev, M. B.; Mushkina, E. V. Solubility of naphthalene in ethylene and carbon dioxide under pressure. *Russ. J. Phys. Chem.* **1964**, *38*, 1173–1176.

(20) Schmitt, W. J.; Reid, R. C. Solubility of monofunctional organic solids in chemically diverse supercritical fluids. *J. Chem. Eng. Data* **1986**, *31*, 204–212.

(21) Yun, S. L. J.; Liong, K. K.; Gurdial, G. S.; Foster, N. R. Solubility of cholesterol in supercritical ethane carbon dioxide. *Ind. Eng. Chem. Res.* **1991**, *30*, 2476–2482.

(22) McHugh, M.; Krukonis, V. *Supercritical Fluid Extraction*; Butterworth-Heinemann: Boston, 1994.

(23) Li, S.; Varadarajan, G. S.; Hartland, S. Solubilities of theobromine and caffeine in supercritical carbon dioxide: correlation with density-based models. *Fluid Phase Equilib.* **1991**, *68*, 263–280.

(24) Wong, J. M.; Johnston, K. P. Solubilization of biomolecules in carbon dioxide based supercritical fluids. *Biotechnol. Prog.* **1986**, *2*, 29–39.

(25) Perry, E. S.; Weber, W. H.; Daubert, B. F. Vapor pressures of phlegmatic liquids. I. Simple and mixed triglycerides. *J. Am. Chem. Soc.* **1949**, *71*, 3720–3726.

(26) Acosta, G. M.; Smith, R. L.; Arai, K. High-pressure PVT behavior of natural fats and oils, trilaurin, triolein, and *n*-tridecane from 303 K to 353 K from atmospheric pressure to 150 MPa. *J. Chem. Eng. Data* **1996**, *41*, 961–969.

(27) Kumar, S. K.; Johnston, K. P. Modelling the solubility of solids in supercritical fluids with density as the independent variable. *J. Supercrit. Fluids* **1988**, *1*, 15–22.

Received for review November 1, 2002
Revised manuscript received January 7, 2003
Accepted January 9, 2003

IE020871C

Cubic Equation of State and Local Composition Mixing Rules: Correlations and Predictions. Application to the Solubility of Solids in Supercritical Solvents

E. Ruckenstein* and I. Shulgin[†]

Department of Chemical Engineering, State University of New York at Buffalo, Buffalo, New York 14260

A family of new mixing rules for the cubic equation of state through a synthesis between the classical van der Waals mixing rule and the local composition concept is proposed. The binary interaction parameter in the van der Waals mixing rule, which is devoid of physical meaning, is thus replaced with a more physically meaningful parameter. The new mixing rules were used to correlate the dependence on pressure of the solubilities of a number of solid substances, including penicillins, in supercritical fluids. Because the new mixing rules contain physically meaningful parameters, dependent on the energies of the binary intermolecular interactions, the calculations of those energies can allow one to predict the solubilities of solids in supercritical fluids. Such predictions were made for the solubility of solid CCl_4 in the supercritical CF_4. The required binary intermolecular energies were computed with the help of quantum mechanical ab initio calculations, and a good agreement between the predicted and experimental solubilities was obtained.

Introduction

In a number of areas of modeling phase equilibria, the cubic equation of state (EOS) provided equal or even better results than the traditional approach based on the activity coefficient concept. In fact, for certain types of phase equilibria, the EOS is the only method that provided acceptable results. The solubility of solids in a supercritical fluid (SCF) constitutes such a case. For the solubility of a solid in a SCF [SCF (1) + solid solute (2)], one can write the well-known relation[1]

$$y_2 = \frac{P_2^0}{P\phi_2} \exp\left[\frac{(P - P_2^0)V_2^0}{RT}\right] \quad (1)$$

where P is the pressure, T is the temperature in K, R is the universal gas constant, ϕ_2 is the fugacity coefficient of the solute in the binary mixture, y_2 is the mole fraction at saturation of the solute, and P_2^0 and V_2^0 are the saturation vapor pressure and molar volume of the solid solute, respectively.

Equation 1 shows that the solubility of a solid in SCF depends among others on the fugacity coefficient ϕ_2, and as is well-known, this coefficient is responsible for the unusually large values of the solubility. These solubilities are much larger than those in ideal gases, and enhancement factors of 10^4-10^8 are not uncommon.[2] They are, however, still relatively small and usually do not exceed several mole percent.

The fugacity coefficient can be calculated using a suitable EOS. The Soave−Redlich−Kwong[3] EOS (SRK EOS) will be employed in this paper. Starting from the SRK EOS

$$P = \frac{RT}{V-b} - \frac{a(T)}{V(V+b)} \quad (2)$$

the following expression for the fugacity coefficient was obtained:[4]

$$RT \ln \phi_i = -RT \ln \frac{V-b}{V} + \left[\frac{RT}{V-b} - \frac{a}{b(V+b)} + \frac{a}{b^2}\ln\frac{V+b}{V}\right]\frac{\partial(nb)}{\partial n_i} - \frac{1}{nb}\left(\ln\frac{V+b}{V}\right)\frac{\partial(n^2 a)}{\partial n_i} - RT \ln z \quad (3)$$

In eqs 2 and 3, V is the molar volume, z is the compressibility factor, n is the total number of moles in the system, and n_i is the number of moles of component i. Equations 2 and 3 show that the fugacity coefficient ϕ_2 at a given pressure and temperature can be calculated if the parameters a and b and their derivatives with respect to the number of moles of solute are known. While near the critical point the fluctuations are important and an EOS involving them should be used,[1] we neglect for the time being their effect.

The mixture parameters a and b can be expressed in terms of those for the pure components a_{ii} and b_{ii}, using a variety of mixing rules starting from those of van der Waals[4] to the modern ones.[5,6] For the solubility of a solid in a SCF, the van der Waals mixing rules are most often used. They have the form

$$a = \sum_i \sum_j y_i y_j a_{ij} \quad (4)$$

and

$$b = \sum_i \sum_j y_i y_j b_{ij} \quad (5)$$

where y_i is the mole fraction of component i.

---

* Correspondence author. E-mail: feaeliru@acsu.buffalo.edu. Fax: (716) 645-3822.
[†] E-mail address: ishulgin@eng.buffalo.edu.

When the parameters are applied to a binary mixture and considering that $a_{12} = \sqrt{a_{11}a_{22}}(1 - k_{12})$ and $b_{12} = (b_{11} + b_{22})/2$, they become

$$a = y_1^2 a_{11} + y_2^2 a_{22} + 2y_1 y_2 (a_{11}a_{22})^{0.5}(1 - k_{12}) \quad (6)$$

and

$$b = y_1 b_{11} + y_2 b_{22} \quad (7)$$

where k_{12} is the interaction parameter. Using these mixing rules, one easily obtains

$$RT \ln \phi_1 = -RT \ln \frac{V - b}{V} + \left[\frac{RT}{V - b} - \frac{a}{b(V + b)} + \frac{a}{b^2} \ln \frac{V + b}{V} \right] b_{11} - \frac{1}{b} \left(\ln \frac{V + b}{V} \right) [2y_1 a_{11} + 2y_2(a_{11}a_{22})^{0.5}(1 - k_{12})] - RT \ln z \quad (8)$$

Among the drawbacks of the van der Waals mixing rules, an important one concerns the interaction parameter k_{12}. This quantity being purely empirical cannot be predicted on a physical basis. In addition, being temperature-dependent, it must be calculated for each isotherm. The aim of this paper is to suggest mixing rules based on the local composition (LC) concept[1] that no longer involve empirical parameters such as k_{12} but parameters with a more clear physical meaning. The application of such rules to supercritical mixtures is most natural, because the near critical density fluctuations generate large local density and composition changes.[7]

Theory and Formulas

1. LC Concept. According to the LC concept, the composition in the vicinity of any molecule differs from the overall composition. If a binary mixture is composed of components 1 and 2 with mole fractions y_1 and y_2, respectively, four LCs can be defined: the local mole fractions of components 1 and 2 near a central molecule 1 (y_{11} and y_{21}) and the local mole fractions of components 1 and 2 near a central molecule 2 (y_{12} and y_{22}). Numerous attempts have been made to express the LCs in terms of the bulk compositions and intermolecular interaction parameters.[8-15] The idea of LC acquired acceptance starting with Wilson's paper on phase equilibria,[8] where the following expressions for the LCs were suggested:

$$y_{ii} = \frac{y_i}{y_i + y_j \exp\left(-\frac{\lambda_{ij} - \lambda_{ii}}{RT}\right)} \quad (9)$$

and

$$y_{ji} = \frac{y_j \exp\left(-\frac{\lambda_{ij} - \lambda_{ii}}{RT}\right)}{y_i + y_j \exp\left(-\frac{\lambda_{ij} - \lambda_{ii}}{RT}\right)} \quad (10)$$

with λ_{ij} being the interaction parameter between molecules i and j.[1,8]

Vera and Panayiotou used the quasi-chemical approach and suggested the following expressions for the LCs:[10,11]

$$y_{ii} = 1 - \frac{2y_j}{1 + [1 - 4y_1 y_2(1 - G_{12})]^{1/2}} \quad (11)$$

and

$$y_{ji} = \frac{2y_j}{1 + [1 - 4y_1 y_2(1 - G_{12})]^{1/2}} \quad (12)$$

where $G_{12} = \exp[(\epsilon_{11} + \epsilon_{22} - 2\epsilon_{12})/RT]$ and ϵ_{ij} is the interaction energy parameter for the pair i and j.

On the basis of a lattice model, the following expressions for LC were derived:[12-14]

$$y_{ii} = \frac{y_i \exp\left(y_j \frac{\Delta}{RT}\right)}{y_j + y_i \exp\left(y_j \frac{\Delta}{RT}\right)} \quad (13)$$

and

$$y_{ji} = \frac{y_j}{y_j + y_i \exp\left(y_j \frac{\Delta}{RT}\right)} \quad (14)$$

where $\Delta = 2e_{12} - e_{11} - e_{22}$ and e_{ij} is an interaction energy parameter between molecules i and j.

Recently,[15] a modification of the Wilson expressions for LCs was proposed by assuming that the interaction energy parameter between molecules of different types depends on the composition, and the following expressions were obtained:

$$y_{ii} = \frac{y_i}{y_i + y_j \exp\left[-\frac{y_i(\lambda_{ji}^0 - \lambda_{ij}^0) + \lambda_{ij}^0 - \lambda_{ii}^0}{RT}\right]} \quad (15)$$

and

$$y_{ji} = \frac{y_j \exp\left[-\frac{y_i(\lambda_{ji}^0 - \lambda_{ij}^0) + \lambda_{ij}^0 - \lambda_{ii}^0}{RT}\right]}{y_i + y_j \exp\left[-\frac{y_i(\lambda_{ji}^0 - \lambda_{ij}^0) + \lambda_{ij}^0 - \lambda_{ii}^0}{RT}\right]} \quad (16)$$

where λ_{ij}^0 is the interaction energy parameter between molecules i and j ($i \neq j$) when $y_i \to 0$ and λ_{ji}^0 is the interaction energy parameter between molecules j and i ($i \neq j$) when $y_j \to 0$.

2. LC Mixing Rules. There were attempts[16-20] to express the mixture parameters a and b in terms of the LC. However, most of the suggested mixing rules belong to the so-called density-dependent mixing rules (with the mixture parameter a being a function of the mixture density) and require information about the above density. Our considerations will be restricted to density-independent mixing rules.

The van der Waals parameters a and b are measures of the attractive energy in intermolecular interactions and size, respectively. It is, therefore, reasonable to express for a mixture these parameters in terms of LCs.

In the present paper we suggest a family of LC mixing rules in which some or all of the bulk compositions in expressions (4) and (5) are replaced by LCs. Numerous expressions are possible, and calculations have been performed with many of them. We provide in the tables only results obtained with the following mixing rules, which appear to be more meaningful from a physical point of view:

For the parameter a

$$a = y_1^2 a_1 + y_2^2 a_2 + 2y_{11}y_{22}(a_1 a_2)^{0.5} \quad (17a)$$

$$a = y_{11}^2 a_1 + y_{22}^2 a_2 + 2y_{11}y_{22}(a_1 a_2)^{0.5} \quad (17b)$$

and

$$a = y_{11}^2 a_1 + y_{22}^2 a_2 + 2y_{12}y_{21}(a_1 a_2)^{0.5} \quad (17c)$$

and for the parameter b

$$b = y_1 b_1 + y_2 b_2 \quad (18a)$$

$$b = y_{11} b_1 + y_{22} b_2 \quad (18b)$$

and

$$b = y_{12} b_1 + y_{21} b_2 \quad (18c)$$

where $a_i = a_{ii}$ and $b_i = b_{ii}$. Many of the other combinations provided comparable results.

If the LCs are expressed through eqs 11–14, then the expressions for the mixture parameters a and b will contain only one unknown parameter, G_{12} or Δ, instead of the interaction parameter k_{12}. It should be noted that, in contrast to the interaction parameter k_{12}, G_{12} and Δ have a clear physical meaning connected with the intermolecular interaction energies. Furthermore, it was recently[21,22] shown that the latter parameters can be calculated independently through an ab initio quantum mechanical calculation.

Wilson's expressions for the LCs (eqs 9 and 10) can also be used to generate mixing rules: the combination of eqs 9 and 10 with eqs 17a and 18b leads to two-parameter mixing rules, with parameters $\lambda_{12} - \lambda_{11}$ and $\lambda_{21} - \lambda_{22}$. Three-parameter mixing rules with parameters $\lambda_{12}^0 - \lambda_{11}$, $\lambda_{21}^0 - \lambda_{22}$, and $\lambda_{21}^0 - \lambda_{12}^0$ can be obtained by combining eqs 15 and 16 for the LCs and eqs 17a and 18b. This flexibility allows one to use two- or three-parameter mixing rules for systems for which the one-parameter mixing rules fail.

It should be also noted that eqs 18b and 18c provide a temperature dependence for parameter b. In contrast, the conventional expressions (eq 18b), used in numerous mixing rules,[6] provide no temperature dependence of parameter b, even though the direct calculation of a and b from experimental data indicated that b is slightly temperature-dependent.[23]

This family of LC mixing rules will now be applied to binary supercritical mixtures, but of course they can be applied to any kind of phase equilibria.

Correlation of Solubility Data

1. Testing New Mixing Rules for the Supercritical Mixture CO_2 + Naphthalene. The mixture CO_2 + naphthalene was selected to test the new mixing rules because the solubility of naphthalene in supercritical CO_2 was determined in numerous papers and reliable data at several temperatures are available. The calculations were carried out at three different temperatures (308, 318, and 328 K). The critical temperatures and pressures of naphthalene and CO_2 were taken from refs 24 and 25 and the values of P_2^0 and V_2^0 for naphthalene

Table 1. Comparison between the LC Mixing Rules and the van der Waals Mixing Rules for the Correlation of Solubilities[a] of Naphthalene in Supercritical CO_2 at Three Different Temperatures

a	b	LC	adjustable parameter[b]	AAD,[c] %
		$T = 308$ K		
eq 6	eq 7		0.09803	11.36
eq 17b	eq 18a	eqs 13 and 14	−243.32	10.80
eq 17b	eq 18c	eqs 13 and 14	−220.99	10.84
eq 17b	eq 18a	eqs 11 and 12	−243.63	11.00
eq 17b	eq 18c	eqs 11 and 12	−220.88	10.84
eq 17c	eq 18a	eqs 11 and 12	−236.90	11.58
		$T = 318$ K		
eq 6	eq 7		0.10047	19.39
eq 17b	eq 18a	eqs 13 and 14	−249.43	15.91
eq 17b	eq 18c	eqs 13 and 14	−224.65	13.91
eq 17b	eq 18a	eqs 11 and 12	−243.63	15.87
eq 17b	eq 18c	eqs 11 and 12	−224.51	13.90
eq 17c	eq 18a	eqs 11 and 12	−239.67	12.27
		$T = 328$ K		
eq 6	eq 7		0.09524	14.56
eq 17b	eq 18a	eqs 13 and 14	−222.87	8.03
eq 17b	eq 18c	eqs 13 and 14	−198.33	8.29
eq 17b	eq 18a	eqs 11 and 12	−222.40	7.99
eq 17b	eq 18c	eqs 11 and 12	−198.15	8.30
eq 17c	eq 18a	eqs 11 and 12	−209.60	11.28

[a] Experimental solubilities of naphthalene in supercritical CO_2 were taken from: Tsekhanskaya, Yu. V.; Iomtev, M. B.; Mushkina, E. V. *Russ. J. Phys. Chem.* **1964**, *38*, 1173. [b] Adjustable parameter k_{12} is dimensionless, and adjustable parameters $\epsilon_{11} + \epsilon_{22} - 2\epsilon_{12}$ in eqs 11 and 12 and Δ in eqs 13 and 14 are given in J/mol. [c] AAD(%) = $100[\sum_{i=1}^{n} \text{abs}(y_2^{\text{exp}} - y_2^{\text{calc}})/(ny_2^{\text{exp}})]$, where n is the number of experimental points, y_2^{exp} is the experimental solubility, and y_2^{calc} is the calculated solubility.

from ref 25. Table 1 compares the new mixing rules with one adjustable parameter and the van der Waals mixing rule in a wide range of pressures and at three different temperatures.

One can see from Table 1 that SRK EOS with the one-parameter new mixing rules describes the solubility of naphthalene in supercritical CO_2 somewhat better (at 308 K) or better (at 318 and 328 K) than the van der Waals mixing rules. It should be noted that the energy parameter exhibits a weak temperature dependence.

2. Correlation of the Solubility of Solids in Various SCF. The new LC mixing rules were also tested for the solubility of a large number of solid solutes in various SCFs. The critical temperatures and pressures of solids and SCFs were taken from refs 24 and 25. The molar volumes of the solids and their saturated vapor pressures were taken from ref 25. The saturated vapor pressure of perylene was found in ref 26. The results are compared with the van der Waals mixing rules in Table 2, which shows that they are comparable. The parameters of SRK EOS (a and b) can be expressed by combining one of eqs 17a−c with one of eqs 18a−c. Only a few combinations have been included in Table 2; the other ones have also been tested and provided comparable results.

3. Correlation of the Solubilities of Penicillins V and G in SCF CO_2. As already mentioned, the LC mixing rules can contain one, two, or three adjustable parameters. This flexibility has proven to be useful in representing the solubilities of antibiotic penicillins in SCF CO_2. These solubilities could not be satisfactorily correlated by the cubic EOS with the conventional mixing rules,[27,28] and several empirical expressions containing up to seven parameters were employed to correlate them.[28] The LC mixing rules were used by us

Table 2. Comparison between the LC Mixing Rules and the van der Waals Mixing Rule for the Solubilities of Solids in SCFs at Various Temperatures[a]

		AAD (%) with different mixing rules		
system	T, K	M1	M2	M3
NP + C_2H_4[b]	285.15	23.86	23.56	23.56
NP + C_2H_4[b]	298.15	25.25	24.84	24.89
NP + C_2H_4[b]	308.15	23.92	23.51	23.58
NP + C_2H_4[b]	318.15	29.90	29.37	29.48
2,6-D + CO_2[c]	308.2	21.60	21.12	21.02
2,6-D + CO_2[c]	328.2	33.62	7.65	8.00
2,7-D + CO_2[c]	308.2	6.84	6.95	7.51
2,7-D + CO_2[c]	328.2	12.03	9.47	10.04
PR + CO_2[d]	308.2	10.26	10.24	10.23
PR + CO_2[d]	318.2	8.23	8.22	8.23
PR + CO_2[d]	323.2	6.55	6.48	6.49
PR + CO_2[d]	338.2	7.98	7.92	7.94
AT + CO_2[e]	313	14.97	15.0	15.0
AT + CO_2[e]	323	13.34	13.39	13.40
AT + CO_2[e]	333	5.56	5.58	5.58
AT + C_2H_6[e]	313	10.13	10.15	10.15
AT + C_2H_6[e]	323	7.47	7.50	7.49
AT + C_2H_6[e]	333	8.79	8.82	8.82
PH + CO_2[e]	313	4.34	5.14	5.14
PH + CO_2[e]	323	4.08	4.88	4.89
PH + CO_2[e]	333	8.56	7.88	7.88
PH + C_2H_6[e]	313	5.25	5.20	5.20
PH + C_2H_6[e]	323	2.09	2.12	2.12
PH + C_2H_6[e]	333	3.23	3.11	3.11
PL + CO_2[e]	323	10.23	10.22	10.22
PL + CO_2[e]	333	15.74	15.74	15.74
PL + C_2H_6[e]	333	19.20	19.20	19.20

[a] In Table 2 the following abbreviations were used: NP, naphthalene; 2,6-D, 2,6-dimethylnaphthalene; 2,7-D, 2,7-dimethylnaphthalene; PR, pyrene; AT, anthracene; PH, phenanthrene; PL, perylene; M1, van der Waals mixing rules (eqs 6 and 7); M2, LC mixing rule (eqs 17b and 18a with LCs given by eqs 13 and 14); M3, LC mixing rules (eqs 17b and 18a with LCs given by eqs 11 and 12). [b] Tsekhanskaya, Yu. V.; Iomtev, M. B.; Mushkina, E. V. *Russ. J. Phys. Chem.* **1964**, *38*, 1173. [c] Iwai, Y.; Mori, Y.; Hosotani, H.; et al. *J. Chem. Eng. Data* **1993**, *38*, 509. [d] Bartle, K. D.; Clifford, A. A.; Jafar, S. A. *J. Chem. Eng. Data* **1990**, *35*, 355. [e] Anitescu, G.; Tavlarides, L. L. *J. Supercrit. Fluids* **1997**, *10*, 175.

to correlate those data. The critical temperatures and pressures and the acentric factor ω of penicillins V and G estimated in refs 27 and 28 were used. The saturated pressure of penicillin V was taken from ref 27, and that of penicillin G was estimated using the Lee and Kesler correlation.[29] The results of the calculations are listed in Table 3, which shows that SRK EOS with one adjustable parameter LC mixing rules provided values comparable to those obtained with the van der Waals mixing rule. However, the two- and three-parameter LC mixing rules provided improvements in the correlations of experimental solubilities. While the improvement was achieved by adding additional parameters, it should be emphasized that the six-parameter empirical equation[28] provided for the solubility of penicillin G in SCF CO_2 the same accuracy as the new mixing rule with only three adjustable parameters. In addition, those three parameters have clear physical meaning, and there is the possibility for their prediction on the basis of quantum mechanical ab initio calculations, as shown for a more simple case below.

4. LCs Obtained during the Calculations Listed in Table 4 for the CO_2/Naphthalene Mixture.

One may note that y_1 is somewhat smaller than y_{12}, hence, that there is some enrichment of the solvent around a solute molecule. Experimental[30] and integral equation studies[31] have shown that the local density of the solvent around a solute is higher than its bulk density. The calculation of the densities from the mole fractions involves the correlation volume (where the local density differs from the bulk one) which is not available. Consequently, either the enhancement is mainly a density effect or the results reflect a limitation of the model employed.

Prediction of the Solubility of Solid Substances in SCF

To our knowledge, no successful prediction of the solubilities of solid substances in SCFs has been made. Because of the physical meaning of the parameters contained in the LC mixing rules, such an attempt becomes possible. Indeed, the new parameters depend on the intermolecular energies which can be calculated independently.

Quantum mechanical ab initio calculations were performed recently to calculate the interaction energies between pairs of molecules in binary systems of water and alcohols or other organic compounds.[21,22] These energies were used to calculate the Wilson and UNIQUAC parameters and then to successfully predict the activity coefficients. The interaction energies were calculated as follows:[21,22] (a) A cluster composed of eight molecules (four of each kind) was considered to represent a dense fluid. (b) The cluster geometry was identified by an optimization procedure involving the PM3 semiempirical method, followed by the Hartree−Fock method with a 6-31 G** basis set. (c) Interacting molecular pairs (like and unlike pairs) were selected from the above optimized cluster. (d) The interaction energy of each molecular pair was computed using the Hartree−Fock method for the separation distances and orientation obtained in the previous steps.

A similar approach with the following modifications was used in the present paper: (1) The more rigorous Møller−Plesset (MP) perturbation theory[32,33] was selected instead of the Hartree−Fock method. (2) Clusters of two molecules were employed for the geometry

Table 3. Solubilitity of Penicillins V and G in Supercritical CO_2, Described by SRK EOS with Various Mixing Rules[a]

		one-parameter mixing rules, AAD (%)		two-parameter mixing rules, AAD (%)	three-parameter mixing rules, AAD (%)
system	temp, K	M1	M2	M3	M4
penicillin V + CO_2[b]	314.85	39.09	39.52	33.50	33.16
penicillin V + CO_2[b]	324.85	41.42	41.51	41.46	40.78
penicillin V + CO_2[b]	334.85	51.96	51.96	51.81	46.51
penicillin G + CO_2[c]	313.15	29.08	29.06	29.06	24.32
penicillin G + CO_2[c]	323.15	28.37	28.31	28.30	21.50
penicillin G + CO_2[c]	333.15	41.87	41.82	41.81	27.40

[a] M1 = van der Waals mixing rule (eqs 6 and 7), M2 = LC mixing rule (eqs 17b and 18a with LCs given by eqs 13 and 14), M3 = LC mixing rule (eqs 17b and 18a with LCs given by eqs 9 and 10), and M4 = LC mixing rule (eqs 17a and 18c with LCs given by eqs 15 and 16). [b] Experimental solubility data were taken from ref 27. [c] Experimental solubility data were taken from ref 28.

Table 4. Comparison between Local and Bulk Mole Fractions in CO_2 (1) + Naphthalene (2) Mixture at $T = 308.15$ K Found from Solubility Data[a]

pressure, bar	y_1	y_{12}
60.8	0.999	0.999
79.5	0.997	0.997
80.6	0.996	0.996
92.2	0.992	0.993
106.4	0.990	0.990
152.0	0.985	0.986
192.5	0.983	0.985
243.2	0.982	0.984
293.8	0.982	0.984
334.4	0.982	0.984

[a] The calculations for Table 4 were carried out using the LC mixing rules 17b and 18a with the LCs given by eqs 13 and 14. The calculations for the mixtures CO_2 (1) + pyrene (2) at $T = 308.15$ K and C_2H_6 (1) + phenanthrene (2) at $T = 313.15$ K indicated a similar behavior.

Table 5. Interaction Energies by Ab Initio Calculations Using the MP2 Method with a 6-31G Basis Set[a]

system	E_{11} (kJ/mol)	E_{22} (kJ/mol)	E_{12} (kJ/mol)	Δu_{12} (J/mol)	Δu_{21} (J/mol)
CF_4 (1) + CCl_4 (2)	−1.968	−5.528	−2.944	−976	2584

[a] E_{ij} is the energy of intermolecular interactions between molecules i and j, and $\Delta u_{ij} = \lambda_{ij} - \lambda_{ii}$ are the Wilson parameters in eqs 9 and 10.

optimization and calculation of the intermolecular energies. Of course, it would have been preferable to consider larger clusters. However, the computational cost of such calculations for clusters containing seven to eight molecules by the MP method would have been prohibitively high. For the sake of illustration, the relatively simple system of nonpolar and symmetrical components [CF_4 (1) + CCl_4 (2)] was selected. Each pair (CF_4 + CCl_4, CF_4 + CF_4, and CCl_4 + CCl_4) was treated using the MP2 method (with the 6-31G basis sets) available in the standard Gaussian software.[34] The results of these calculations are summarized in Table 5 and used to predict the solubility of solid CCl_4 in the SCF CF_4. Experimental data regarding the solubility of solid CCl_4 in the SCF CF_4 are available in the literature.[35] The required data for the pure-component properties were taken from refs 24 and 36. The solubilities of solid CCl_4 in the SCF CF_4 were predicted for three different temperatures. Parameter a was calculated using eq 17a and parameter b using eq 18a. The LCs were expressed through the Wilson equations (eqs 9 and 10). A comparison between the predicted and experimental solubilities is presented in Figures 1–3, which show that the suggested method provides excellent predictions regarding the pressure dependence of the solubility of solid CCl_4 in the SCF CF_4 at 244 and 249 K but only satisfactory agreement at 234 K. However, the authors of ref 35 pointed out that the experiments at 234 K "proved to be unexpectedly difficult" and may have been affected by the presence of a third liquid phase.

Conclusion

A family of mixing rules for the cubic EOS was suggested in which the empirical binary interaction parameter k_{12} in the van der Waals mixing rule was replaced by a physically more meaningful parameter. In the new mixing rules, some mole fractions in the expressions of parameters a and b in the van der Waals mixing rules were replaced with various expressions for

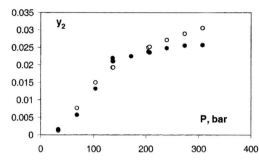

Figure 1. Comparison between predicted (○) and experimental[35] (●) solubilities of solid CCl_4 in the SCF CF_4 at 249 K. Parameter a is given by eq 17a and parameter b by eq 18a. The LCs are given by the Wilson equations (eqs 9 and 10).

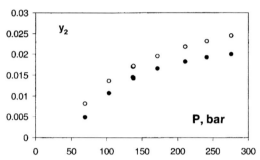

Figure 2. Comparison between predicted (○) and experimental[35] (●) solubilities of solid CCl_4 in the SCF CF_4 at 244 K. Parameter a is given by eq 17a and parameter b by eq 18a. The LCs are given by the Wilson equations (eqs 9 and 10).

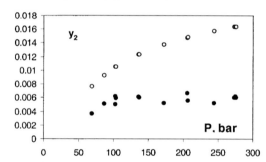

Figure 3. Comparison between predicted (○) and experimental[35] (●) solubilities of solid CCl_4 in the SCF CF_4 at 234 K. Parameter a is given by eq 17a and parameter b by eq 18a. The LCs are given by the Wilson equations (eqs 9 and 10).

the local mole fractions. The family of the new mixing rules can contain one, two, or even three adjustable parameters. The mixing rules were applied to the correlation of the solubilities of a number of solids in SCFs. One of the advantages of the new mixing rules is their flexibility regarding the number of adjustable parameters. In particular, it was shown that the new mixing rules with two or three adjustable parameters provided better correlations of the experimental data for the solubilities of the antibiotic penicillins in SCF CO_2 than the conventional mixing rules or the empirical expressions containing many more parameters.

Another attractive feature of the new mixing rules is that they allow one to predict the solubilities of solids in SCFs using only data for the pure components and the intermolecular interactions. In this paper, the solubilities of solid CCl_4 in the SCF CF_4 were predicted for three different temperatures. The energies of the intermolecular interactions (CF_4 + CCl_4, CF_4 + CF_4, and

$CCl_4 + CCl_4$) were computed using quantum mechanical ab initio calculation. A good agreement was obtained.

Acknowledgment

The authors are indebted to the Center for Computational Research (CCR) of the University at Buffalo for the use of its facilities and to Dr. J. L. Tilson (CCR) for his help concerning the Gaussian software. We are also indebted to Mr. Amadeu Sum and Prof. S. I. Sandler (University of Delaware) for useful discussions regarding the ab initio calculations.

Literature Cited

(1) Prausnitz, J. M.; Lichtenthaler, R. N.; Gomes de Azevedo, E. *Molecular Thermodynamics of Fluid-Phase Equilibria*, 2nd ed.; Prentice-Hall: Englewood Cliffs, NJ, 1986.

(2) Eckert, C. A.; Knutson, B. L.; Debenedetti, P. G. Supercritical fluids as solvents for chemical and materials processing. *Nature* **1996**, *383*, 313−318.

(3) Soave, G. Equilibrium constants from a modified Redlich−Kwong equation of state. *Chem. Eng. Sci.* **1972**, *27*, 1197−1203.

(4) Elliot, J. R.; Lira, C. T. *Introductory chemical engineering thermodynamics*; Prentice-Hall PTR: Upper Saddle River, NJ, 1999.

(5) Fischer, K.; Gmehling, J. Further development, status and results of the PSRK method for the prediction of vapor−liquid equilibria and gas solubilities. *Fluid Phase Equilib.* **1996**, *121*, 185−206.

(6) Orbey, H.; Sandler, S. I. *Modeling Vapor−Liquid Equilibria. Cubic Equation of State and Their Mixing Rules*; Cambridge University Press: Cambridge, U.K., 1998.

(7) Ruckenstein, E.; Shulgin, I. On density microheterogeneities in dilute supercritical solutions. *J. Phys. Chem. B* **2000**, *104*, 2540−2545.

(8) Wilson, G. M. Vapor−liquid equilibrium. XI: A new expression for the excess free energy of mixing. *J. Am. Chem. Soc.* **1964**, *86*, 127−130.

(9) Renon, H.; Prausnitz, J. M. Local composition in thermodynamic excess functions for liquid mixtures. *AIChE J.* **1968**, *14*, 135−144.

(10) Panayiotou, C.; Vera, J. H. The quasi-chemical approach for non-randomness in liquid mixtures. Expressions for local composition with an application to polymer solution. *Fluid Phase Equilib.* **1980**, *5*, 55−80.

(11) Panayiotou, C.; Vera, J. H. Local composition and local surface area fractions: a theoretical discussion. *Can. J. Chem. Eng.* **1981**, *59*, 501−505.

(12) Lee, K.-H.; Sandler, S. I.; Patel, N. C. The generalized van der Waals partition function. 3. Local composition models for a mixture of equal size square-well molecules. *Fluid Phase Equilib.* **1986**, *25*, 31−49.

(13) Aranovich, G. L.; Donohue, M. D. A new model for lattice systems. *J. Chem. Phys.* **1996**, *105*, 7059−7063.

(14) Wu, D. W.; Cui, Y.; Donohue, M. D. Local composition models for lattice mixtures. *Ind. Eng. Chem. Res.* **1998**, *37*, 2936−2946.

(15) Ruckenstein, E.; Shulgin, I. Modified local composition and Flory−Huggins equations for nonelectrolyte solutions. *Ind. Eng. Chem. Res.* **1999**, *38*, 4092−4099.

(16) Mollerup, J. A note on excess Gibbs energy models, equations of state and the local composition concept. *Fluid Phase Equilib.* **1981**, *7*, 121−138.

(17) Whiting, W. B.; Prausnitz, J. M. Equations of state for strongly nonideal fluid mixtures: application of local compositions toward density-dependent mixing rules. *Fluid Phase Equilib.* **1982**, *9*, 119−147.

(18) Lee, R. J.; Chao, K. C. Cubic chain-of-rotators equations of state with toward density-dependent local composition mixing rules. *Fluid Phase Equilib.* **1986**, *29*, 475−484.

(19) Lee, R. J.; Chao, K. C. Local composition of square-well molecules of diverse energies and sizes. *Fluid Phase Equilib.* **1987**, *37*, 325−336.

(20) Kim, S.; Lee, Y. G.; Park, Y. O. New local composition model and mixing rule for the mixtures asymmetric both in size and energy. *Fluid Phase Equilib.* **1997**, *140*, 1−16.

(21) Sum, A. K.; Sandler, S. I. A novel approach to phase equilibria predictions using ab initio methods. *Ind. Eng. Chem. Res.* **1999**, *38*, 2849−2855.

(22) Sum, A. K.; Sandler, S. I. Use of ab initio methods to make phase equilibria predictions using activity coefficient models. *Fluid Phase Equilib.* **1999**, *160*, 375−380.

(23) Fischer, K.; Shulgin, I.; Gmehling, J. Direct calculation of SRK parameters from experimental PvT data in the saturation state. *Fluid Phase Equilib.* **1995**, *103*, 1−10.

(24) Reid, R. C.; Prausnitz, J. M.; Poling, B. E. *The Properties of Gases and Liquids*, 4th ed.; McGraw-Hill: New York, 1987.

(25) Garnier, S.; Neau, E.; Alessi, P.; Cortesi, A.; Kikic, I. Modelling solubility of solids in supercritical fluids using fusion properties. *Fluid Phase Equilib.* **1999**, *158−160*, 491−500.

(26) Oja, V.; Suuberg, E. M. Vapor pressures and enthalpies of sublimation of polycyclic aromatic hydrocarbons and their derivatives. *J. Chem. Eng. Data* **1998**, *43*, 486-492.

(27) Ko, M.; Shah, V.; Bienkowski, P. R.; Cochran, H. D. Solubility of the antibiotic Penicillin V in supercritical CO_2. *J. Supercrit. Fluids* **1991**, *4*, 32−39.

(28) Gordillo, M. D.; Blanco, M. A.; Molero, A.; de la Ossa, E. M. Solubility of the antibiotic Penicillin G in supercritical carbon dioxide. *J. Supercrit. Fluids* **1999**, *15*, 183−190.

(29) Lee, B. I.; Kesler, M. G. A generalized thermodynamic correlation based on three-parameter corresponding states. *AIChE J.* **1975**, *21*, 510−527.

(30) Brennecke, J. F.; Tomasko, D. L.; Peshkin, J.; Eckert, C. A. Fluorescence spectroscopy studies of dilute supercritical solutions. *Ind. Eng. Chem. Res.* **1990**, *29*, 1682−1690.

(31) Tom, J. W.; Debenedetti, P. G. Integral equation study of microstructure and solvation in model attractive. *Ind. Eng. Chem. Res.* **1993**, *32*, 2118−2128.

(32) Levine, I. N. *Quantum Chemistry*, 4th ed.; Prentice-Hall: Englewood Cliffs, NJ, 1991.

(33) Szabo, A.; Ostlund, N. S. *Modern Quantum Chemistry. Introduction to Advanced Electronic Structure Theory*; Dover Publication: New York, 1996.

(34) Frisch, M. J.; Trucks, G. W.; Schlegel, H. B.; et al. *Gaussian 94*, Revision E.2; Gaussian, Inc.: Pittsburgh, PA, 1995.

(35) Barber, V. A.; Cochran, H. D.; Bienkowski, P. R. Solubility of solid CCl_4 in supercritical CF_4. *J. Chem. Eng. Data* **1991**, *36*, 99−102.

(36) *International critical tables of numerical data, physics, chemistry and technology*; compiled by C. J. West, with the collaboration of C. Hull; National Research Council (U.S.): New York, 1933; index, Vol. III.

Received for review November 8, 2000
Revised manuscript received March 27, 2001
Accepted April 9, 2001

IE000955Q

Chapter 3

Solubility of gases in mixed solvents

3.1 Henry's constant in mixed solvents from binary data.
3.2 Salting-out or -in by fluctuation theory.
3.3 The solubility of binary mixed gases by the fluctuation theory.
3.4 Prediction of the solubility of gases in binary polymer + solvent mixtures.
3.5 Ideal multicomponent liquid solution as a mixed solvent.
3.6 Solubility and local structure around a dilute solute molecule in an aqueous solvent: From gases to biomolecules.

Introduction to Chapter 3

The solubilities of gases in binary, ternary or more complex multicomponent solvents are good examples in which the Kirkwood–Buff theory of solutions provides excellent results that cannot be obtained using the methods of traditional thermodynamics. Thermodynamics cannot provide explicit pressure, temperature, and composition dependence of the thermodynamic functions, such as the activity coefficients of the components. Therefore, various assumptions regarding the activity coefficients must be made. In contrast, the Kirkwood–Buff theory of solution allows one to establish, in some cases, relations between multicomponent and binary mixtures (see 3.5). Although these relations are not simple, they could be applied to ternary (3.1–3.4) and quaternary (3.5) mixtures to derive relations for the activity coefficients.

Our results regarding the gas solubility in binary mixed solvents are presented in papers (3.1–3.2, 3.4). The new expression for the composition dependence of Henry's constant in mixed solvents, which requires only the solubilities in the pure solvents (3.1), and the new analysis of the gas solubility in aqueous salt solutions, which provides new criteria for salting-in or salting-out, should be noted (3.2). In addition, a method for predicting Henry's constant in multicomponent (ternary and higher) mixed solvents was developed and compared with experiment (3.5). Our method also allows one to predict the solubility of a binary (or multicomponent) gas mixture in individual solvents (3.4).

Structural and energetic characteristics of infinitely dilute solutions of gases in water and aqueous mixed solvent are also examined in (3.6). The Kirkwood–Buff theory of solution was used to extract from solubility data information about the local composition of the solvent around a gas molecule (3.6). It is worth mentioning that such information could be obtained until now only experimentally by small-angle X-ray scattering, small-angle neutron scattering, light scattering, and related methods.

Henry's Constant in Mixed Solvents from Binary Data

I. Shulgin[†] and E. Ruckenstein[*]

Department of Chemical Engineering, State University of New York at Buffalo, Buffalo, New York 14260

The Kirkwood–Buff formalism was used to derive an expression for the composition dependence of the Henry's constant in a binary solvent. A binary mixed solvent can be considered as composed of two solvents, or one solvent and a solute, such as a salt, polymer, or protein. The following simple expression for the Henry's constant in a binary solvent (H_{2t}) was obtained when the binary solvent was assumed ideal: $\ln H_{2t} = [\ln H_{2,1}(\ln V - \ln V_3^0) + \ln H_{2,3}(\ln V_1^0 - \ln V)]/(\ln V_1^0 - \ln V_3^0)$. In this expression, $H_{2,1}$ and $H_{2,3}$ are the Henry's constants for the pure single solvents 1 and 3, respectively; V is the molar volume of the ideal binary solvent 1–3; and V_1^0 and V_3^0 are the molar volumes of the pure individual solvents 1 and 3. The comparison with experimental data for aqueous binary solvents demonstrated that the derived expression provides the best predictions among the known equations. Even though the aqueous solvents are nonideal, their degree of nonideality is much smaller than those of the solute gas in each of the constituents. For this reason, the ideality assumption for the binary solvent constitutes a most reasonable approximation even for nonideal mixtures.

1. Introduction

The solubility of a gas in a mixture of solvents is a problem of interest in many industrial applications. One example is the removal of acidic compounds from industrial and natural gases.[1] The solubility of a gas in a binary mixture containing water has particular importance because it is connected with the solubility of gases in blood, seawater, rainwater, and many other aqueous solutions of biological and environmental significance.[2] Therefore, it is important to be able to predict the gas solubility in a mixture in terms of the solvent composition and the solubilities in the individual constituents of the solvent or in one pure component and a selected composition of the mixed solvent.

The oldest and simplest relationship between the Henry's constant in a binary solvent and those in the individual solvents [throughout this paper, only binary mixtures of solvents will be considered, and the following subscripts for the components will be used: 1, first solvent; 2, solute (gas); 3, second solvent] is that proposed by Krichevsky[3]

$$\ln H_{2,t} = x_1^{b,1-3} \ln H_{2,1} + x_3^{b,1-3} \ln H_{2,3} \quad (1)$$

where $H_{2,t}$, $H_{2,1}$, and $H_{2,3}$ are the Henry's constants in the binary solvent 1–3 and in the individual solvents 1 and 3, respectively, and $x_1^{b,1-3}$ and $x_3^{b,1-3}$ are the mole fractions of components 1 and 3, respectively, in the binary solvent 1–3.

The rigorous thermodynamic equation[4] for the excess Henry's constant in a binary solvent has the form

$$\ln H_{2,t}^E = \ln \gamma_{2,t}^\infty - x_1^{b,1-3} \ln \gamma_{2,1}^\infty - x_3^{b,1-3} \ln \gamma_{2,3}^\infty \quad (2)$$

where $\gamma_{2,t}^\infty$, $\gamma_{2,1}^\infty$, $\gamma_{2,3}^\infty$ ($\gamma_{2,t}^\infty = \lim_{x_2 \to 0} \gamma_{2,t}$, $\gamma_{2,1}^\infty = \lim_{x_2 \to 0} \gamma_2^{b,1-2}$, and $\gamma_{2,3}^\infty = \lim_{x_2 \to 0} \gamma_2^{b,2-3}$) are the activity coefficients of the solute in the ternary ($\gamma_{2,t}$) and binary ($\gamma_{2,1}$ and $\gamma_{2,3}$) mixtures at infinite dilution. It shows that Krichevsky's relationship (eq 1) is valid when

$$\ln \gamma_{2,t}^\infty = x_1^{b,1-3} \ln \gamma_{2,1}^\infty + x_3^{b,1-3} \ln \gamma_{2,3}^\infty \quad (3)$$

Although, in principle, eq 3 can be satisfied by some nonideal systems, it is surely valid when the ternary mixture and the binary mixtures (1–2 and 2–3) are ideal.

Equation 2 was used as the starting point for the prediction of the Henry's constant in a binary solvent mixture in terms of binary data, by expressing the activity coefficients at infinite dilution ($\gamma_{2,t}^\infty$, $\gamma_{2,1}^\infty$, and $\gamma_{2,3}^\infty$) through the Wilson[5,6] or the van Laar equations.[7,8] The Kirkwood–Buff theory of solution[9] was used by O'Connell to develop a semiempirical expression for the prediction of the Henry's constant in binary solvents from binary data.[4] He found that his expression provided better results than that of Krichevsky and that obtained[5] by combining the Wilson equation with eq 2. Using a corresponding state method, Campanella et al.[10] calculated the Henry's constant for mixed solvents and obtained good agreement when the excess volume of the nonideal solvent was taken into account.

The present authors employed the Kirkwood–Buff theory of solution to obtain expressions for the derivatives of the activity coefficients in a ternary mixture with respect to the mole fractions and applied them to ternary mixtures when the composition(s) of one (or two) component(s) was (were) small.[11,12] That approach will be used here to derive new expressions that can predict the Henry's constant in a binary solvent mixture in terms of binary data.

2. Theory

2.1. Expressions for the Derivatives of the Activity Coefficients in a Ternary Mixture with Respect to Mole Fractions through the Kirkwood–Buff Theory of Solution. For the present purpose, the following two derivatives, obtained in a previous paper,[11] are useful

---

[*] Correspondence author. E-mail: feaeliru@acsu.buffalo.edu. Fax: (716) 645-3822. Phone: (716) 645-2911, ext. 2214.
[†] E-mail: ishulgin@eng.buffalo.edu.

$$\left(\frac{\partial \ln \gamma_{2,t}}{\partial x_3^t}\right)_{T,P,x_2^t} =$$
$$-\frac{(c_1+c_2+c_3)(c_1[G_{11}+G_{23}-G_{12}-G_{13}]+c_3[-G_{12}-G_{33}+G_{13}+G_{23}])}{c_1+c_2+c_3+c_1c_2\Delta_{12}+c_1c_3\Delta_{13}+c_2c_3\Delta_{23}+c_1c_2c_3\Delta_{123}} \quad (4)$$

and

$$\left(\frac{\partial \ln \gamma_{1,t}}{\partial x_2^t}\right)_{T,P,x_3^t} =$$
$$-\frac{(c_1+c_2+c_3)(c_3[G_{11}+G_{23}-G_{12}-G_{13}]+c_2\Delta_{12}+c_2c_3\Delta_{123})}{c_1+c_2+c_3+c_1c_2\Delta_{12}+c_1c_3\Delta_{13}+c_2c_3\Delta_{23}+c_1c_2c_3\Delta_{123}} \quad (5)$$

where x_2^t is the mole fraction of component 2 in the ternary mixture, c_k is the bulk molecular concentration of component k in the ternary mixture 1–2–3, and $G_{\alpha\beta}$ is the Kirkwood–Buff integral given by

$$G_{\alpha\beta} = \int_0^\infty (g_{\alpha\beta}-1)4\pi r^2\, dr \quad (6)$$

In the above expression, $g_{\alpha\beta}$ is the radial distribution function between species α and β, r is the distance between the centers of molecules α and β, and $\Delta_{\alpha\beta}$ and Δ_{123} are defined as

$$\Delta_{\alpha\beta} = G_{\alpha\alpha} + G_{\beta\beta} - 2G_{\alpha\beta}, \quad \alpha \neq \beta \quad (7)$$

and

$$\Delta_{123} = G_{11}G_{22} + G_{11}G_{33} + G_{22}G_{33} + 2G_{12}G_{13} +$$
$$2G_{12}G_{23} + 2G_{13}G_{23} - G_{12}^2 - G_{13}^2 - G_{23}^2 -$$
$$2G_{11}G_{23} - 2G_{22}G_{13} - 2G_{33}G_{12} \quad (8)$$

One can verify that the factors in the square brackets in the numerators of eqs 4 and 5 and Δ_{123} can be expressed in terms of $\Delta_{\alpha\beta}$ as follows

$$G_{12} + G_{33} - G_{13} - G_{23} = \frac{\Delta_{13}+\Delta_{23}-\Delta_{12}}{2} \quad (9)$$

$$G_{11} + G_{23} - G_{12} - G_{13} = \frac{\Delta_{12}+\Delta_{13}-\Delta_{23}}{2} \quad (10)$$

and

$$\Delta_{123} =$$
$$-\frac{(\Delta_{12})^2+(\Delta_{13})^2+(\Delta_{23})^2-2\Delta_{12}\Delta_{13}-2\Delta_{12}\Delta_{23}-2\Delta_{13}\Delta_{23}}{4} \quad (11)$$

The insertion of eqs 9–11 into eqs 4 and 5 provides rigorous expressions for the derivatives $(\partial \ln \gamma_{2,t}/\partial x_3^t)_{T,P,x_2^t}$ and $(\partial \ln \gamma_{1,t}/\partial x_2^t)_{T,P,x_3^t}$ in terms of $\Delta_{\alpha\beta}$ and concentrations.

It should be noted that Δ_{ij} is a measure of the nonideality[13] of the binary mixture α–β because, for an ideal mixture, $\Delta_{\alpha\beta} = 0$. For the ternary mixture 1–2–3, Δ_{123} also constitutes a measure of nonideality. Indeed, inserting $G_{\alpha\beta}^{id}$ for an ideal mixture[14] into the expression Δ_{123}, one obtains that for, an ideal ternary mixture, $\Delta_{123} = 0$.

At infinite dilution of component 2, eqs 4 and 5 become

$$\lim_{x_2^t \to 0}\left(\frac{\partial \ln \gamma_{2,t}}{\partial x_3^t}\right)_{T,P,x_2^t} =$$
$$-\frac{(c_1^0+c_3^0)[(c_1^0+c_3^0)(\Delta_{12}-\Delta_{23})_{x_2^t=0}+(c_1^0-c_3^0)(\Delta_{13})_{x_2^t=0}]}{2[c_1^0+c_3^0+c_1^0c_3^0(\Delta_{13})_{x_2^t=0}]} \quad (12)$$

and

$$\lim_{x_2^t \to 0}\left(\frac{\partial \ln \gamma_{1,t}}{\partial x_2^t}\right)_{T,P,x_3^t} = \frac{c_3^0(c_1^0+c_3^0)(\Delta_{12}+\Delta_{13}-\Delta_{23})_{x_2^t=0}}{2[c_1^0+c_3^0+c_1^0c_3^0(\Delta_{13})_{x_2^t=0}]} \quad (13)$$

In eqs 12 and 13, c_1^0 and c_3^0 represent the bulk molecular concentrations of components 1 and 3, respectively, in the gas-free binary solvent 1–3. In addition to eqs 12 and 13, the following expression[9] for the derivative of the activity coefficient in a binary mixture with respect to the mole fractions will be used in the next section to derive the basic equation for the Henry's constant for mixed solvents

$$\left(\frac{\partial \ln \gamma_1^{b,1-3}}{\partial x_3^{b,1-3}}\right)_{P,T} = \frac{c_3^0 \Delta_{13}}{1+c_1^0 x_3^{b,1-3}\Delta_{13}} \quad (14)$$

where $x_3^{b,1-3}$ and $\gamma_1^{b,1-3}$ are the mole fraction of component 3 and the activity coefficient of component 1, respectively, in the gas-free binary solvent 1–3.

2.2. Composition Dependence of the Henry's Constant for a Binary Solvent. To obtain the composition dependence of the Henry's constant in a binary solvent, one should consider either the derivative $(\partial \ln H_{2t}/\partial x_3^t)_{P,T,x_2^t=0}$ or the derivative $(\partial \ln H_{2t}/\partial x_1^t)_{P,T,x_2^t=0}$. To obtain the above derivatives, one can start from the following expression for the Henry's constant in a binary solvent[4]

$$\ln H_{2,t} = \lim_{x_2^t \to 0} \ln \gamma_{2,t} + \ln f_2^0(P,T) \quad (15)$$

where $f_i^0(P,T)$ is the fugacity of component i.[15] The combination of eqs 12 and 15 leads to the result

$$\left(\frac{\partial \ln H_{2t}}{\partial x_3^t}\right)_{P,T,x_2^t=0} =$$
$$-\frac{(c_1^0+c_3^0)[(c_1^0+c_3^0)(\Delta_{12}-\Delta_{23})_{x_2^t=0}+(c_1^0-c_3^0)(\Delta_{13})_{x_2^t=0}]}{2[c_1^0+c_3^0+c_1^0c_3^0(\Delta_{13})_{x_2^t=0}]} \quad (16)$$

Integration of eq 16 provides the following relation for the composition dependence of the Henry's constant in a binary solvent mixture at constant temperature and pressure

$$\ln H_{2t} = -\int \frac{(c_1^0+c_3^0)(\Delta_{12}-\Delta_{23})_{x_2^t=0}}{2[1+c_1^0 x_3^{b,1-3}(\Delta_{13})_{x_2^t=0}]} dx_3^{b,1-3} -$$
$$\int \frac{(c_1^0-c_3^0)(\Delta_{13})_{x_2^t=0}}{2[1+c_1^0 x_3^{b,1-3}(\Delta_{13})_{x_2^t=0}]} dx_3^{b,1-3} + A \quad (17)$$

where $A(P,T)$ is a composition-independent constant of integration.

The last equation can be also derived starting from the Gibbs−Duhem equation for a ternary mixture (see Appendix 1).

By eliminating $(\Delta_{13})_{x'_2=0}$ with the help of eq 14, eq 17 becomes

$$\ln H_{2t} = -\int (c_1^0 + c_3^0)\frac{(\Delta_{12} - \Delta_{23})_{x'_2=0}}{2} \times \left[1 + x_3^{b,1-3}\left(\frac{\partial \ln \gamma_3^{b,1-3}}{\partial x_3^{b,1-3}}\right)_{P,T}\right] dx_3^{b,1-3} + \frac{1}{2}\int \frac{(x_1^{b,1-3} - x_3^{b,1-3})}{x_1^{b,1-3}}\left(\frac{\partial \ln \gamma_3^{b,1-3}}{\partial x_3^{b,1-3}}\right)_{P,T} dx_3^{b,1-3} + A \quad (18)$$

The first term on the right-hand side of eq 18 involves the ternary mixture through the limiting value $(\Delta_{12} - \Delta_{23})_{x'_2=0}$, whereas the second involves the gas-free binary solvent. Equation 18 can be transformed using the Gibbs−Duhem equation for a binary system in the second integral on the right-hand side of eq 18. The Gibbs−Duhem equation for a binary system at constant temperature and pressure has the form

$$x_1^{b,1-3} \, d \ln \gamma_1^{b,1-3} = -x_3^{b,1-3} \, d \ln \gamma_3^{b,1-3}$$

Therefore, one can write

$$\frac{(x_1^{b,1-3} - x_3^{b,1-3})}{x_1^{b,1-3}}\left(\frac{\partial \ln \gamma_3^{b,1-3}}{\partial x_3^{b,1-3}}\right)_{P,T} = \left(\frac{\partial \ln \gamma_1^{b,1-3}}{\partial x_3^{b,1-3}}\right)_{P,T} + \left(\frac{\partial \ln \gamma_3^{b,1-3}}{\partial x_3^{b,1-3}}\right)_{P,T}$$

which, introduced into eq 18, leads to

$$\ln H_{2t} = -\int (c_1^0 + c_3^0)\frac{(\Delta_{12} - \Delta_{23})_{x'_2=0}}{2} \times \left[1 + x_3^{b,1-3}\left(\frac{\partial \ln \gamma_3^{b,1-3}}{\partial x_3^{b,1-3}}\right)_{P,T}\right] dx_3^{b,1-3} + \frac{1}{2}(\ln \gamma_1^{b,1-3} + \ln \gamma_3^{b,1-3}) + A \quad (19)$$

It should be noted that this equation can be applied to the solubility of a gas in various kinds of binary mixtures: a mixture of two solvents, or a mixture of a solvent and a solute (salt, polymer, or protein).

As already mentioned, the Krichevsky equation (eq 1) is valid when the binary mixtures 1−2 and 2−3 (gas solute/pure solvents) and the ternary mixture 1−2−3 are ideal. However, these conditions are often far from reality. Let us consider, for example, the solubility of a hydrocarbon in a water−alcohol solvent (for instance, water−methanol, water−ethanol, etc.). The activity coefficient[16] of propane in water at infinite dilution is $\sim 4 \times 10^3$, whereas the activity coefficients of alcohols and water in aqueous solutions of simple alcohols seldom exceed 10. It is therefore clear that the main contribution to the nonideality of the ternary gas−binary solvent mixture comes from the nonidealities of the gas solute in the individual solvents, which are neglected in the Krichevsky equation.

For this reason, in a first step, it will be assumed that only the binary solvent (1−3) behaves as an ideal mixture. One can therefore write that $\gamma_3^{b,1-3} = 1$ and

$$V = x_1^{b,1-3} V_1^0 + x_3^{b,1-3} V_3^0 \quad (20)$$

where V is the molar volume of the binary mixture 1−3 and V_1^0 and V_3^0 are the molar volumes of the individual solvents 1 and 3, respectively. Under these conditions, eq 19 becomes

$$\ln H_{2t} = -\int \frac{(\Delta_{12} - \Delta_{23})_{x'_2=0}}{2(x_1^{b,1-3} V_1^0 + x_3^{b,1-3} V_3^0)} dx_3^{b,1-3} + A \quad (21)$$

The main single approximation of this paper is the assumption that $(\Delta_{12})_{x'_2=0} = (G_{11} + G_{22} - 2G_{12})_{x'_2=0}$ and $(\Delta_{23})_{x'_2=0} = (G_{22} + G_{33} - 2G_{23})_{x'_2=0}$ are independent of the composition of the solvent mixture. Consequently, eq 21 becomes

$$\ln H_{2t} = A(P,T) - \frac{B(P,T) \ln(x_1^{b,1-3} V_1^0 + x_3^{b,1-3} V_3^0)}{V_3^0 - V_1^0} \quad (22)$$

where $B(P,T) = (\Delta_{12} - \Delta_{23})_{x'_2=0}/2$.

The constants $A(P,T)$ and $B(P,T)$ can be obtained using the following extreme expressions

$$(\ln H_{2t})_{x_1^{b,1-3}=0} = \ln H_{2,3} \quad (23)$$

and

$$(\ln H_{2t})_{x_3^{b,1-3}=0} = \ln H_{2,1} \quad (24)$$

Combining eqs 22−24 yields the final result

$$\ln H_{2t} = \frac{\ln H_{2,1}(\ln V - \ln V_3^0) + \ln H_{2,3}(\ln V_1^0 - \ln V)}{\ln V_1^0 - \ln V_3^0} \quad (25)$$

Equation 25 provides the Henry's constant for a binary solvent in terms of those for the individual solvents and the molar volumes of the pure solvents. This simple equation was obtained using less restrictive approximations than those involved in the Krichevsky equation by assuming that only the binary solvent 1−3 is an ideal mixture.

This assumption is reasonable because, as already noted, the nonideality of the binary solvent is much lower than the nonidealities of the solute gas and each of the constituents of the solvent.

Equation 19 can, however, be integrated using any of the analytical expressions available for the activity coefficient $\ln \gamma_3^{b,1-3}$, such as the van Laar, Margules, Wilson, NRTL, etc. To take into account the nonideality of the molar volume, one can use the expression

$$V = x_1^{b,1-3} V_1^0 + x_3^{b,1-3} V_3^0 + V^E \quad (26)$$

where V^E is the excess molar volume.

When the integration in eq 19 cannot be performed analytically, one can first perform the integration numerically between $0 < x_3^{b,1-3} < 1$ to obtain the expression

$$\ln H_{2,3} - \ln H_{2,1} =$$
$$-B\int_0^1 \frac{1}{V}\left[1 + x_3^{b,1-3}\left(\frac{\partial \ln \gamma_3^{b,1-3}}{\partial x_3^{b,1-3}}\right)_{P,T}\right]dx_3^{b,1-3} +$$
$$\frac{1}{2}(\ln \gamma_{1,3}^\infty - \ln \gamma_{3,1}^\infty) \quad (27)$$

where $\gamma_{1,3}^\infty = \lim_{x_1\to 0} \gamma_1^{b,1-3}$ and $\gamma_{3,1}^\infty = \lim_{x_3\to 0} \gamma_3^{b,1-3}$.

Equation 27 allows for the determination of the constant B. Further, eq 19 can be integrated between $0 < x_3^{b,1-3} < x$ to obtain the Henry's constant for the mole fraction x in the binary solvent

$$\ln H_{2,t} =$$
$$\ln H_{2,1} - B\int_0^x \frac{1}{V}\left[1 + x_3^{b,1-3}\left(\frac{\partial \ln \gamma_3^{b,1-3}}{\partial x_3^{b,1-3}}\right)_{P,T}\right]dx_3^{b,1-3} +$$
$$\frac{1}{2}[\ln \gamma_1^{b,1-3}(x) + \ln \gamma_3^{b,1-3}(x) - \ln \gamma_{3,1}^\infty] \quad (28)$$

This procedure allows one to account for the nonideality (activity coefficients and molar volume) of the binary solvent.

The analytical expressions obtained using the two-suffix Margules equations[15] for the activity coefficients and eq 20 for the molar volume are given in Appendix 2.

The extension of eq 25 to a multicomponent solvent is not straightforward and requires additional investigation.

3. Calculations and Comparison with Experimental Data

For comparison, we selected the solubilities of gases in aqueous binary solvents because, as noted in ref 6, the prediction of the Henry's constant for such mixtures is the most difficult and the available methods are not reliable. The results of the calculations are presented and compared in Table 1 and Figure 1 with the Krichevsky equation and an empirical correlation for aqueous mixtures that provided the best results[6] among the existing expressions.

All of the necessary experimental data [V_i^0, $H_{2,1}$, $H_{2,3}$, and E (Margules parameter)] were taken from the original publications (indicated as footnotes to Table 1) or calculated using the data from Gmehling's vapor–liquid equilibrium data compilation.[21] Figure 1 and Table 1 show that the present eq 25 is in much better agreement with experiment than Krichevsky's eq 1 and equations A2-3–5 from Appendix 2, which involve the Margules expression for the activity coefficient. The new eq 25 provides predictions that are comparable to those of an empirical correlation for aqueous mixtures of solvents,[6] which involves three adjustable parameters.

However, none of the expressions available, including eq 25, can represent the extremum in the mixed Henry's constant found in some experiments at low alcohol concentrations.[17] Perhaps only very accurate representations of the activity coefficients and excess molar volume of the mixed solvent in the dilute region can explain this anomaly.

Table 1 also shows that equations A2-3–5, based on the two-suffix Margules equation, provide results that are comparable to those of Krichevsky's eq 1 but much less accurate than those of the new eq 25. Numerical calculations based on eqs 27–28 showed that the use of the Wilson equation for the activity coefficients in binary solvent mixtures improved the results obtained via eqs A2-3–5 only slightly. It seems that eqs 27–28, which contain the derivatives of the activity coefficient with respect to the mole fraction, require a much more accurate representation of the vapor–liquid equilibrium than that provided by the two-suffix Margules or Wilson equations. As noted in the literature,[4] the above equa-

Table 1. Comparison between Experimental and Calculated Henry's Constants for a Binary Solvent Mixture

system	T (K)	Krichevsky's eq 1	empirical correlation[b]	predictions of this work eq 25	predictions of this work eq A2-3
argon (2)– acetone (1)– water (3)[c]	288.15	17.0	–	15.3	37.9
	298.15	28.3	–	7.1	39.4
	308.15	29.6	–	8.4	39.3
helium (2)– methanol (1)– water (3)[d]	298.15	5.7	–	10.8	7.3
helium (2)– ethanol (1)– water (3)[d]	298.15	5.6	–	12.1	7.8
helium (2)– 1-propanol (1)– water (3)[d]	298.15	18.2	–	5.6	24.1
helium (2)– 2-propanol (1)– water (3)[d]	298.15	13.5	–	7.9	8.9
oxygen (2)– methanol (1)– water (3)[e]	273.15	13.9	7.6	22.4	19.8
	293.15	7.0	6.2	13.3	9.8
	313.15	10.4	10.8	8.9	7.4
oxygen (2)– ethanol (1)– water (3)[e]	273.15	12.4	12.2	21.9	20.9
	293.15	16.7	6.2	10.3	16.4
	313.15	31.5	6.0	8.1	20.5
oxygen (2)– 1-propanol (1)– water (3)[e]	273.15	20.4	7.9	13.8	32.7
	293.15	45.5	8.3	12.5	35.7
	313.15	53.5	13.0	13.9	36.1
oxygen (2)– 2-propanol (1)– water (3)[e]	273.15	26.6	14.3	19.2	31.0
	293.15	41.4	12.6	16.2	27.5
	313.15	67.7	9.2	12.6	30.6
nitrogen (2)– methanol (1)– water (3)[e]	293.15	7.8	9.9	17.5	15.2
	313.15	7.7	10.0	9.7	8.5
nitrogen (2)– ethanol (1)– water (3)[e]	293.15	17.3	8.8	7.8	15.0
	313.15	22.7	14.2	6.5	19.5
nitrogen (2)– 1-propanol (1)– water (3)[e]	293.15	43.9	9.5	9.1	29.9
	313.15	73.5	14.5	20.3	40.0
nitrogen (2)– 2-propanol (1)– water (3)[e]	293.15	44.6	14.5	8.9	28.1
	313.15	61.2	14.0	11.8	28.3
carbon dioxide (2)– 1-propanol (1)– water (3)[e]	283.15	14.1	–	12.1	26.4
	293.15	18.3	–	6.1	25.2
	303.15	23.4	–	3.7	24.1
	313.15	33.7	–	6.9	28.8
carbon dioxide (2)– 2-propanol (1)– water (3)[e]	283.15	14.0	–	16.1	21.6
	293.15	21.7	–	10.2	22.5
	303.15	23.4	–	8.5	23.4
	313.15	30.5	–	6.4	22.5
methane (2)– methanol (1)– water (3)[f]	293.15	5.3	25.5	15.0	11.7
methane (2)– ethanol (1)– water (3)[f]	293.15	33.7	30.3	16.3	24.5
average (%)		25.8	12.1	11.6	23.5

[a] Defined as $(100/m)\sum_i |H_{2,t(\text{exp})}^{(i)} - H_{2,t(\text{calc})}^{(i)}|/H_{2,t(\text{exp})}^{(i)}$, where m is the number of experimental points. [b] Taken directly from ref 6. [c] Experimental data from ref 18. [d] Experimental data from ref 19. [e] Experimental data from ref 17. [f] Experimental data from ref 20.

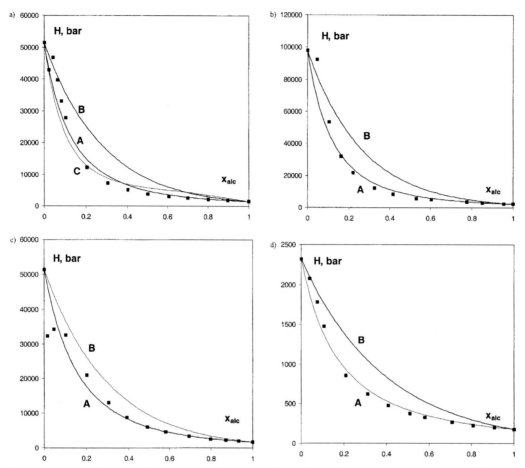

Figure 1. Henry's constants of gases in binary solvent mixtures at 760 mmHg partial pressure (■, experimental;[17] A, calculated with the new eq 25; B, calculated with Krichevsky eq 1; in Figure 1a, C represents eq A2-3): (a) oxygen (2) in 1-propanol (1)−water (3) at 40 °C, (b) nitrogen (2) in 2-propanol (1)−water (3) at 40 °C, (c) oxygen (2) in ethanol (1)−water (3) at 40 °C, and (d) carbon dioxide (2) in 1-propanol (1)−water (3) at 40 °C. x_{alc} is the mole fraction of the alcohol in the gas-free mixture of solvents.

tions applied to complex mixtures, such as aqueous mixtures, can lead to results in the wrong direction for the Henry's constant.

Conclusion

In this paper, the Kirkwood−Buff formalism was used to relate the Henry's constant for a binary solvent mixture to the binary data and the composition of the solvent. A general equation describing the above dependence was obtained, which can be solved (analytically or numerically) if the composition dependence of the molar volume and the activity coefficients in the gas-free mixed solvent are known. A simple expression was obtained when the mixture of solvents was considered to be ideal. In this case, the Henry's constant for a binary solvent mixture could be expressed in terms of the Henry's constants for the individual solvents and the molar volumes of the individual solvents. The agreement with experiment for aqueous solvents is better than that provided by any other expression available, including an empirical one involving three adjustable parameters. Even though the aqueous solvents considered are nonideal, their degrees of nonideality are much lower than those of the solute gas in each of the constituent solvents. For this reason, the assumption that the binary solvent behaves as an ideal mixture constitutes a reasonable approximation.

Appendix 1

On the basis of the Gibbs−Duhem equations for a ternary mixture, Krichevsky and Sorina[22] derived the following equation

$$\left(\frac{\partial \ln H_{2t}}{\partial x_3^t}\right)_{P,T,x_2^t \to 0} = \left(\frac{\partial \ln \gamma_1^{b,1-3}}{\partial x_3^{b,1-3}}\right)_{P,T} - \frac{1}{x_3^{b,1-3}} \lim_{x_2 \to 0}\left(\frac{\partial \ln \gamma_1^t}{\partial x_2}\right)_{P,T,x_3^t} \quad (A1\text{-}1)$$

Using eq 14 for the first term of the right-hand side of eq A1-1 and eq 13 for the second term, eq A1-1 becomes

$$\left(\frac{\partial \ln H_{2t}}{\partial x_3^t}\right)_{P,T,x_2^t \to 0} = \frac{c_3^0 \Delta_{13}}{1 + c_1^0 x_3^0 \Delta_{13}} - \frac{1}{x_3^{b,1-3}} \frac{c_3^0(c_1^0 + c_3^0)\left(\frac{\Delta_{12} + \Delta_{13} - \Delta_{23}}{2}\right)_{x_2^t=0}}{c_1^0 + c_3^0 + c_1^0 c_3^0 \Delta_{13}} \quad (A1\text{-}2)$$

which after some algebraic transformation acquires the form

$$\left(\frac{\partial \ln H_{2t}}{\partial x_3^t}\right)_{P,T,x_2^t \to 0} =$$
$$\frac{-(c_1^0 + c_3^0)(\Delta_{12} - \Delta_{23})_{x_2^t=0} - (c_1^0 - c_3^0)\Delta_{13}}{2(1 + c_1^0 x_3^{b,1-3} \Delta_{13})} \quad \text{(A1-3)}$$

which coincides with eq 16 in the text.

Appendix 2

The activity coefficients of the binary mixture 1−3 can be expressed through the two-suffix Margules equations[15] as

$$\ln \gamma_1^{b,1-3} = E(x_3^{b,1-3})^2 \quad \text{(A2-1)}$$

$$\ln \gamma_3^{b,1-3} = E(x_1^{b,1-3})^2 \quad \text{(A2-2)}$$

where E is a temperature-dependent constant.

Insertion of eq A2-2 into eq 19 provides the following result for the Henry's constant in the mixed solvent (the molar volume of the mixed solvent being expressed through eq 20)

$$\ln H_{2,t} = A - B\left[\frac{\ln V}{V_3^0 - V_1^0} + \frac{2V_1^0 V_3^0 E \ln V}{(V_3^0 - V_1^0)^3} + \frac{E x_3^{b,1-3}}{(V_3^0 - V_1^0)^2}(x_3^{b,1-3} V_3^0 - x_1^{b,1-3} V_3^0 - 2V_3^0)\right] - 2E[x_3^{b,1-3} - (x_1^{b,1-3})^2] \quad \text{(A2-3)}$$

The composition-independent constants A and B can be obtained by combining eq A2-3 with eqs 23 and 24. One thus obtains the following expressions for A and B

$$A = \frac{\ln V_3^0 \ln H_{2,1} - \ln V_1^0 \ln H_{2,3}}{\ln V_3^0 - \ln V_1^0} \quad \text{(A2-4)}$$

and

$$B = \frac{(\ln H_{2,1} - \ln H_{2,3})(V_3^0 - V_1^0)^3}{(\ln V_3^0 - \ln V_1^0)[(V_3^0 - V_1^0)^2 + 2V_1^0 V_3^0 E]} \quad \text{(A2-5)}$$

Using eqs A2-3−5, one can calculate the Henry's constant for any composition of the solvent.

Note Added after ASAP Posting

This article was released ASAP on 2/21/02 with errors in eqs A1-1 and A1-2 and in footnote a of Table 1. The correct version was posted on 2/25/02.

Literature Cited

(1) Li, Y. G.; Mather, A. E. Correlation and prediction of the solubility of N$_2$O in mixed solvents. *Fluid Phase Equilib.* **1994**, *96*, 119−142.

(2) Wilhelm, E.; Battino, R.; Wilcock, R. J. Low-pressure solubility of gases in liquid water. *Chem. Rev.* **1977**, *77*, 219−262.

(3) Krichevsky, I. R. Thermodynamics of an infinitely dilute solution in mixed solvents. I. The Henry's coefficient in a mixed solvent behaving as an ideal solvent. *Zh. Fiz. Khim.* **1937**, *9*, 41−47.

(4) O'Connell, J. P. Molecular thermodynamics of gases in mixed solvents. *AIChE J.* **1971**, *17*, 658−663.

(5) Prausnitz, J. M.; Eckert, C. A.; Orye, R. V.; O'Connell, J. P. *Computer Calculations for Multicomponent Vapor−Liquid Equilibria*; Prentice Hall: Englewood Cliffs, NJ, 1967.

(6) Kung, J. K.; Nazario, F. N.; Joffe, J.; Tassios, D. Prediction of Henry's constants in mixed solvents from binary data. *Ind. Eng. Chem. Res.* **1984**, *23*, 170−175.

(7) Prausnitz, J. M.; Chueh, P. L. *Computer Calculations for High-Pressure Vapor−Liquid Equilibria*; Prentice Hall: Englewood Cliffs, NJ, 1968.

(8) Boublik, T.; Hala, E. Solubility of gases in mixed solvents. *Collect. Czech. Chem. Commun.* **1966**, *31*, 1628−1635.

(9) Kirkwood, J. G.; Buff, F. P. Statistical mechanical theory of solutions. I. *J. Chem. Phys.* **1951**, *19*, 774−782.

(10) Campanella, E. A.; Mathias, P. M.; O'Connell, J. P. Equilibrium properties of liquids containing supercritical substances. *AIChE J.* **1987**, *33*, 2057−2066.

(11) Ruckenstein, E.; Shulgin, I. Entrainer effect in supercritical mixtures. *Fluid Phase Equilib.* **2001**, *180*, 345−359.

(12) Ruckenstein, E.; Shulgin, I. The solubility of solids in mixtures containing a supercritical fluid and an entrainer. *Fluid Phase Equilib.* **2002**, *200*, 53−67.

(13) Ben-Naim, A. Inversion of the Kirkwood−Buff theory of solutions: Application to the water−ethanol system. *J. Chem. Phys.* **1977**, *67*, 4884−4890.

(14) Ruckenstein, E.; Shulgin, I. Effect of a third component on the interactions in a binary mixture determined from the fluctuation theory of solutions. *Fluid Phase Equilib.* **2001**, *180*, 281−297.

(15) Prausnitz, J. M.; Lichtenthaler, R. N.; Gomes de Azevedo, E. *Molecular Thermodynamics of Fluid-Phase Equilibria*, 2nd ed.; Prentice Hall: Englewood Cliffs, NJ, 1986.

(16) Kojima, K.; Zhang, S. J.; Hiaki, T. Measuring methods of infinite dilution activity coefficients and a database for systems including water. *Fluid Phase Equilib.* **1997**, *131*, 145−179.

(17) Tokunaga, J. Solubilities of oxygen, nitrogen, and carbon dioxide in aqueous alcohol solutions. *J. Chem. Eng. Data* **1975**, *20*, 41−46.

(18) Yamamoto, H.; Tokunaga, J.; Koike, K. Solubility of argon in acetone plus water mixed solvent at 288.15, 298.15 and 308.15 K. *Can. J. Chem. Eng.* **1994**, *72*, 541−545.

(19) Yamamoto, H.; Ichikawa, K.; Tokunaga, J. Solubility of helium in methanol plus water, ethanol plus water, 1-propanol plus water, and 2-propanol plus water solutions at 25 °C. *J. Chem. Eng. Data* **1994**, *39*, 155−157.

(20) Tokunaga, J.; Kawai, M. Solubilities of methane in methanol−water and ethanol−water solutions. *J. Chem. Eng. Jpn.* **1975**, *8*, 326−327.

(21) Gmehling, J. et al. *Vapor−Liquid Equilibrium Data Collection*; DECHEMA Chemistry Data Series; DECHEMA: Frankfurt, Germany, 1977−1996; Vol. I.

(22) Krichevsky, I. R.; Sorina, G. A. What thermodynamics can state about Henry's coefficient in a mixed solvent. *Dokl. Akad. Nauk* **1992**, *325*, 325−328.

Received for review November 12, 2001
Revised manuscript received January 11, 2002
Accepted January 15, 2002

IE010911X

Salting-Out or -In by Fluctuation Theory

E. Ruckenstein* and I. Shulgin[†]

Department of Chemical Engineering, State University of New York at Buffalo, Amherst, New York 14260

In this paper, the Kirkwood–Buff formalism was used to examine the effect of the addition of a salt on the gas solubility. A general expression for the derivative of the Henry constant with respect to the salt concentration was thus derived. The obtained equation was used to correlate the experimental solubilities as a function of the salt molality. The correlation involves one parameter, which has to be determined from the experimental data. In addition, it requires information about the molar volume of the salt solution and the mean activity coefficient of the salt. It has been shown that the experimental solubilities can be well correlated when an accurate expression for the mean activity coefficient of the salt is used. It was also shown that the well-known Sechenov equation constitutes a particular case of the obtained expression. The general expression allowed one to find a criterion for the prediction of the kind of salting (salting-in or salting-out) for dilute salt solutions. According to this criterion, the kind of salting depends mainly on the molar volume of the salt at infinite dilution. This explains the literature observations that the salts with large molar volumes at infinite dilution usually increase the gas solubility compared to that in pure water.

1. Introduction

The gas solubilities in aqueous solutions of salts or organic substances constitute most useful information in many areas of chemical and biochemical engineering, hydrometallurgy, geochemistry, etc. There are many processes, such as the biological and organic reactions, the corrosion and oxidation of materials, the aerobic fermentation, the petroleum and natural gas exploitation, the petroleum refining, the coal gasification, the gas antisolvent crystallization, the formation of gas hydrates, etc., in which the salting effect is relevant.[1] The solubility of naturally occurring and atmospheric gases in sodium chloride solutions, and its dependence on pressure, temperature, and the salt concentration, is of interest not only to the physical chemist but also to the geochemist because the aqueous salt solutions containing gases can simulate the brine in the earth's crust.

The information about the solubility of atmospheric gases in water and seawater is also relevant to the understanding of the ecological balance between the freshwater and seawater systems.[1]

The addition of an electrolyte to water decreases in many cases the gas solubility (salting-out). Similarly, the addition of an organic compound to water can decrease or increase the solubility of gases compared to that in pure water.

The present paper is devoted to the examination of the effect of the addition of an inorganic substance, mainly a salt, to water on the gas solubility. Usually the effect of the salt addition on the solubility has been attributed to the greater attraction between the ions and the water molecules than between the nonpolar or slightly polar gas molecules and water.[2] Therefore, the interactions between the ions and the water molecules should decrease the number of "free" water molecules available to dissolve the gas.[3] This explanation is, however, oversimplified because it ignores the effect of the ions on the water structure.[3] Indeed, some electrolytes with large ions, which disorganize the structure of water, increase the solubility (salting-in);[4–7] in contrast, electrolytes with small ions, which organize the structure of water, decrease the solubility (salting-out). An interesting observation regarding the effect of an ion on the water structure was made in ref 8, where it was found that the "salting-out" and "salting-in" salts affect differently the partial molar volume of the gas in an electrolyte solution, compared to that in pure water. While the "salting-out" salts decrease this volume, the "salting-in" salts increase it. However, so far no criterion for the prediction of whether a salt generates "salting-out" or "salting-in" was suggested.

The oldest but, nevertheless, the most popular equation for the representation of the gas solubility in the presence of a salt is the Sechenov equation[9]

$$\ln(H_{2,t}/H_{21}) = k_S m \quad (1)$$

where $H_{2,t}$ and H_{21} are the Henry constants in the salt solution and pure water, respectively, k_S is the Sechenov coefficient, and m is the molality. The Sechenov coefficient depends on the nature of the salt and solute and the temperature. The following subscripts will be used throughout this paper: component 1 is the solvent, 2 is the gas, and 3 is the salt.

Of course, the Sechenov equation (1) can represent both the salting-out and the salting-in; for salting-out the Sechenov coefficient is positive and for salting-in negative. Because of its simplicity, the Sechenov equation has become a very popular tool for the correlation of the gas solubility in salt solutions. It has attracted the attention of theoreticians,[5,7,10–11] and several modifications have been suggested.[3,12,13] However, it has failed to correlate the gas solubility at high salt concentrations and also was not satisfactory in correlating the solubility of carbon dioxide in a number of salt solutions.[14] In addition, recent investigations[15,16] indicated that the salting effect is more complex. For

---

* Corresponding author. E-mail: feaeliru@acsu.buffalo.edu. Fax: (716) 645-3822. Phone: (716) 645-2911/ext. 2214.
[†] E-mail: ishulgin@eng.buffalo.edu.

example, the kind of salting effect can be inverted with a change in the gas partial pressure, and this inversion cannot be described by the Sechenov equation. Last but not least, because of its empirical character, the Sechenov equation cannot predict whether a salt will increase the solubility (salting-in) or will decrease it (salting-out).

The aim of the present paper is to develop a theoretical approach for the description of the gas solubility in a solvent containing a salt. To achieve this goal, the Kirkwood–Buff formalism[17] for ternary mixtures will be used. Recently, such a formalism has been used to predict the gas solubility in mixed solvents[18] (mixture of two nonelectrolytes) in terms of the solubilities in the individual solvents. A similar approach will be employed here.

The paper is organized as follows: first, the Kirkwood–Buff formalism will be used to derive general expressions for the derivatives of the activity coefficients in ternary mixtures with respect to the mole fractions. Then, the obtained expressions will be applied to the gas solubility in dilute and concentrated salt solutions. Numerical calculation will be carried out for several mixtures, particularly for those for which the Sechenov equation failed to provide an accurate correlation. Finally, a criterion will be proposed for the a priori prediction of the kind of salting (salting-in or salting-out).

2. The Henry Constant in a Salt Solution

The Henry constant in a binary solvent is given by the following expression:[19]

$$\ln H_{2,t} = \lim_{x_2^t \to 0} \ln \gamma_{2,t} + \ln f_2^0(P,T) \quad (2)$$

where x_i^t and $\gamma_{i,t}$ are the mole fraction and the activity coefficient of component i in the ternary mixture, respectively, $f_i^0(P,T)$ is the fugacity of the pure component i,[3] P is the pressure, and T is the temperature in K.

For the derivative of the Henry constant in a binary solvent with respect to the mole fraction of the electrolyte, one can write

$$\left(\frac{\partial \ln H_{2,t}}{\partial x_3^t}\right)_{P,T,x_2^t=0} = \lim_{x_2^t \to 0}\left(\frac{\partial \ln \gamma_{2,t}}{\partial x_3^t}\right)_{T,P,x_2^t} \quad (3)$$

In a previous paper,[20] the Kirkwood–Buff formalism was applied to ternary mixtures and explicit expressions for the partial molar volumes, isothermal compressibility, and the derivatives of the activity coefficients with respect to the mole fractions derived. In particular, the following expression for the derivative $(\partial \ln \gamma_{2,t}/\partial x_3^t)_{T,P,x_2^t}$ was obtained:

$$\left(\frac{\partial \ln \gamma_{2,t}}{\partial x_3^t}\right)_{T,P,x_2^t} =$$
$$-\{(c_1 + c_2 + c_3)(c_1[G_{11} + G_{23} - G_{12} - G_{13}] + c_3[-G_{12} - G_{33} + G_{13} + G_{23}])\}/(c_1 + c_2 + c_3 + c_1c_2\Delta_{12} + c_1c_3\Delta_{13} + c_2c_3\Delta_{23} + c_1c_2c_3\Delta_{123}) \quad (4)$$

where c_k is the bulk molecular concentration of component k in the ternary mixture 1–2–3 and $G_{\alpha\beta}$ is the Kirkwood–Buff integral given by

$$G_{\alpha\beta} = \int_0^\infty (g_{\alpha\beta} - 1)4\pi r^2 \, dr \quad (5)$$

In the above expression, $g_{\alpha\beta}$ is the radial distribution function between species α and β, r is the distance between the centers of molecules α and β, and $\Delta_{\alpha\beta}$ and Δ_{123} are the following combinations of the Kirkwood–Buff integrals

$$\Delta_{\alpha\beta} = G_{\alpha\alpha} + G_{\beta\beta} - 2G_{\alpha\beta}, \quad \alpha \neq \beta \quad (6)$$

and

$$\Delta_{123} = G_{11}G_{22} + G_{11}G_{33} + G_{22}G_{33} + 2G_{12}G_{13} + 2G_{12}G_{23} + 2G_{13}G_{23} - G_{12}^2 - G_{13}^2 - G_{23}^2 - 2G_{11}G_{23} - 2G_{22}G_{13} - 2G_{33}G_{12} \quad (7)$$

One can show[18] that the factors in the square brackets in eq 4 and Δ_{123} can be expressed in terms of $\Delta_{\alpha\beta}$ as follows:

$$G_{12} + G_{33} - G_{13} - G_{23} = \frac{\Delta_{13} + \Delta_{23} - \Delta_{12}}{2} \quad (8)$$

$$G_{11} + G_{23} - G_{12} - G_{13} = \frac{\Delta_{12} + \Delta_{13} - \Delta_{23}}{2} \quad (9)$$

and

$$\Delta_{123} = -(\Delta_{12}^2 + \Delta_{13}^2 + \Delta_{23}^2 - 2\Delta_{12}\Delta_{13} - 2\Delta_{12}\Delta_{23} - 2\Delta_{13}\Delta_{23})/4 \quad (10)$$

The substitution of eqs 8–10 into eq 4, and considering infinite dilution of the solute (component 2), yields the following rigorous expressions for the derivative $(\partial \ln \gamma_{2,t}/\partial x_3^t)_{T,P,x_2^t}$:

$$\lim_{x_2^t \to 0}\left(\frac{\partial \ln \gamma_{2,t}}{\partial x_3^t}\right)_{T,P,x_2^t} = -[(c_1^0 + c_3^0)((c_1^0 + c_3^0)(\Delta_{12} - \Delta_{23})_{x_2^t=0} + (c_1^0 - c_3^0)(\Delta_{13})_{x_2^t=0}]/2(c_1^0 + c_3^0 + c_1^0 c_3^0 (\Delta_{13})_{x_2^t=0}) \quad (11)$$

In eq 11 c_1^0 and c_3^0 represent the bulk molecular concentrations of components 1 and 3 in the gas-free binary solvent 1–3. In addition to eq 11, the following expression[17] for the derivative of the activity coefficient in a binary mixture 1–3 with respect to the mole fraction of the electrolyte can be written:

$$\left(\frac{\partial \ln \gamma_1^{b,1-3}}{\partial x_3^{b,1-3}}\right)_{P,T} = \frac{c_3^0 \Delta_{13}}{1 + c_1^0 x_3^{b,1-3} \Delta_{13}} \quad (12)$$

where $x_3^{b,1-3}$ and $\gamma_1^{b,1-3}$ are the mole fraction of component 3 and the activity coefficient of component 1 in the gas-free binary solvent 1–3, respectively.

The combination of eqs 3, 11, and 12 provides the following expression for the derivative of the Henry constant in a binary solvent:[18]

$$\left(\frac{\partial \ln H_{2,t}}{\partial x_3^t}\right)_{P,T,x_2^t=0} =$$
$$-(c_1^0 + c_3^0)\frac{(\Delta_{12} - \Delta_{23})_{x_2^t=0}}{2}\left[1 + x_3^{b,1-3}\left(\frac{\partial \ln \gamma_3^{b,1-3}}{\partial x_3^{b,1-3}}\right)_{P,T}\right] +$$
$$\frac{1}{2}\frac{x_1^{b,1-3} - x_3^{b,1-3}}{x_1^{b,1-3}}\left(\frac{\partial \ln \gamma_3^{b,1-3}}{\partial x_3^{b,1-3}}\right)_{P,T} \quad (13)$$

The first term on the right-hand side of eq 13 involves the ternary mixture through the limiting value $(\Delta_{12} - \Delta_{23})_{x_2^t=0}$, while the second involves the gas-free binary solvent.

It is important to emphasize that $\Delta_{\alpha\beta}$ constitutes a measure of the nonideality[21] of the binary mixtures $\alpha-\beta$ because for an ideal mixture $\Delta_{\alpha\beta} = 0$.

Two cases will be examined in what follows: (a) the case of dilute salt solutions and (b) the case of concentrated ones:

(a) Dilute Salt Solutions. In the dilute range, it will be assumed that $\Delta_{13} = 0$ (in other words, the mixed solvent 1-3 behaves like an ideal mixture). From eq 12 one can find that in this case

$$\left(\frac{\partial \ln \gamma_i^{b,1-3}}{\partial x_i^{b,1-3}}\right)_{P,T} = 0 \quad i = 1, 3 \quad (14)$$

In a previous paper[18] regarding the gas solubility in mixtures of two nonelectrolytes, the ideality approximation for the binary solvent was employed to obtain an expression for the gas solubility. The ideality of the mixed solvents constituted a good approximation because usually the nonideality of the mixture of two nonelectrolytes is much lower than those between each of them and the gas. A similar assumption can be made for dilute aqueous salt solutions. Indeed, the data regarding the activity coefficient of water (γ_w) in dilute aqueous solutions of sodium chloride[22] indicate that $|(\partial \ln \gamma_w/\partial x_w)_{P,T}| < 0.01$ for a molality of sodium chloride smaller than 0.8. Considering, in addition, that $(\Delta_{12} - \Delta_{23})_{x_2^t=0}$ is independent of composition, eq 13 becomes

$$\ln H_{2,t} - \ln H_{21} = -B\int_0^{x_3}\frac{dx_3^{b,1-3}}{V} \quad (15)$$

where $B = (\Delta_{12} - \Delta_{23})_{x_2^t=0}/2$ and V is the molar volume of the binary gas-free mixture. However, for very dilute electrolyte solutions, one can write[23]

$$V = x_3^{b,1-3} V_3^{\infty} + (1 - x_3^{b,1-3})V_1^0 \quad (16)$$

where V_1^0 and V_3^{∞} are the molar volume of the pure solvent and the partial molar volume of the electrolyte at infinite dilution, respectively.

Using eq 16, eq 15 becomes

$$\ln\frac{H_{2,t}}{H_{21}} = -\frac{B(P,T)\ln[1 + (V_3^{\infty}/V_1^0 - 1)x_3]}{V_3^{\infty} - V_1^0} \quad (17)$$

Because, for very small values of y, $\ln(1 + y) \approx y$, eq 17 can be rewritten in the form of the Sechenov equation:

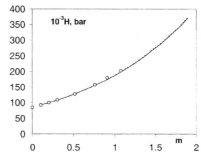

Figure 1. The Henry constant of nitrogen in an aqueous solution of sodium sulfate at 25 °C: (○) experimental data;[12] (dashed line) the Henry constant calculated with eq 15; (solid line) the Henry constant calculated with the Sechenov equation.

$$\ln\frac{H_{2,t}}{H_{21}} \approx -\frac{B(P,T)}{V_1^0}x_3 \approx -\frac{\rho_w B(P,T)}{1000}m = C(P,T)\,m \quad (18)$$

where ρ_w is the water density and $C(P,T)$ is a composition-independent constant

$$C(P,T) = -\frac{\rho_w B(P,T)}{1000} = -\frac{\rho_w(\Delta_{12} - \Delta_{23})_{x_2^t=0}}{2000} \quad (19)$$

Equation 15 can be used to correlate the gas solubility in the presence of a salt if the composition dependence of the molar volume of the binary electrolyte–water mixture is known. Such data are available in the literature for numerous aqueous salt solutions.[24] Like that of Sechenov, eq 15 is a one-parameter equation whose parameter B has to be determined from the solubility data. The two equations provide almost the same results (see Figure 1).

(b) Concentrated Solutions of Electrolytes. Because the mean activity coefficient of a salt ($K_{\nu^+}A_{\nu^-}$, where ν^+ and ν^- are respectively the number of cations and anions in the salt molecule) is usually expressed in terms of the salt molality, eq 13 will be converted to the molality scale. First, the Gibbs–Duhem equation for the binary mixture water (1)–electrolyte (3) allows one to rewrite eq 13 as follows:

$$\left(\frac{\partial \ln H_{2,t}}{\partial x_3^t}\right)_{P,T,x_2^t=0} =$$
$$-(c_1^0 + c_3^0)\frac{(\Delta_{12} - \Delta_{23})_{x_2^t=0}}{2}\left[1 + x_1^{b,1-3}\left(\frac{\partial \ln \gamma_1^{b,1-3}}{\partial x_1^{b,1-3}}\right)_{P,T}\right] +$$
$$\frac{1}{2}\frac{x_1^{b,1-3} - x_3^{b,1-3}}{x_3^{b,1-3}}\left(\frac{\partial \ln \gamma_1^{b,1-3}}{\partial x_1^{b,1-3}}\right)_{P,T} \quad (20)$$

Second, for the water activity coefficient in the binary mixture water (1)–electrolyte (3), one can use the relation[2]

$$\nu m\, d\ln(m\gamma_{\pm}) = -\frac{1000}{M_1}d\ln x_1^{b,1-3} - \frac{1000}{M_1}d\ln \gamma_1^{b,1-3} \quad (21)$$

where $\gamma_{\pm}$ is the mean activity coefficient of the salt, $\nu = \nu^+ + \nu^-$, M_1 is the molecular weight of water, and m is the molality of the salt. By combining eqs 20 and 21 with the obvious relations

3 Solubility of gases in mixed solvents

Table 1. Information about the Mixtures Used in the Calculations

gas and salt solution		temp, K	composition range, m	ref
gas	salt solution			
O_2	$Na_2SO_4 + H_2O$	298.15	0–1.517	a
O_2	$Na_2SO_4 + H_2O$	308.15	0–1.656	a
N_2	$Na_2SO_4 + H_2O$	298.15	0–1.070	a
CO_2	$NaCl + H_2O$	298.15	0–5.096	b
CO_2	$Na_2SO_4 + H_2O$	298.15	0–2.205	b
O_2	$NaCl + H_2O$	298.15	0–5.4	c

[a] Yasunishi, A. *J. Chem. Eng. Jpn.* **1977**, *10*, 89. [b] Yasunishi, A.; Yoshida, F. *J. Chem. Eng. Data* **1979**, *24*, 11. [c] Mishina, T. A.; Avdeeva, O. I.; Bozhovskaya, T. K. *Mater. Vses. Nauchno-Issled. Geol. Inst.* **1961**, *46*, 93 (as given in *Solubility Data Series*; Pergamon: New York, 1981; Vol. 7).

$$x_1^{b,1-3} = \frac{1000}{1000 + M_1 m} \quad (22)$$

$$dx_1^{b,1-3} = -\frac{1000 M_1}{(1000 + M_1 m)^2} dm \quad (23)$$

and integrating, one obtains an equation for the Henry constant at a given molality m (again B was considered independent of composition):

$$\ln\left(\frac{H_{2,t}}{H_{21}}\right) = -B \int_0^m \frac{M_1 \nu \left[1 + m\left(\frac{\partial \ln \gamma_\pm}{\partial m}\right)_{P,T}\right]}{1000(1 + 0.001 M_1 m) V} dm + \frac{1}{2}\int_0^m \frac{1 - 0.001 M_1 m}{1 + 0.001 M_1 m}\left(\nu(1 + 0.001 M_1 m)\left[1 + m\left(\frac{\partial \ln \gamma_\pm}{\partial m}\right)_{P,T}\right] - 1\right) dm \quad (24)$$

This equation contains a parameter B that must be calculated from the experimental data. In addition, information about the molar volume of the mixture and the mean activity coefficient of the salt on the molality in the binary mixture water (1)–salt (3) is necessary.

3. One-Parameter Gas Solubility in Aqueous Salt Mixtures: Comparison with Experiment

Several aqueous salt mixtures were selected to verify eq 24. They are listed in Table 1. Some mixtures (oxygen or carbon dioxide with water + sodium sulfate or sodium chloride) have been selected because the Sechenov equation did not provide an accurate correlation of the solubility in these mixtures at high molalities. Other mixtures, listed in Table 1, have also been considered.

Accurate density data for many aqueous salt mixtures are available in the literature.[24] The following expression was used for the analytical representation of the molar volume of aqueous salt solutions:[25]

$$V = x_3 \phi + x_1 V_1 = \frac{1000 V_1^0}{1000 + M_1 m} + \beta \frac{M_1 m^{1.5}}{1000 + M_1 m} + \frac{M_1 m \phi^0}{1000 + M_1 m} \quad (25)$$

where ϕ is the apparent molar volume of the electrolyte, $\phi^0 = V_3^{b,\infty}$, V_1 is the partial molar volume of water in the binary electrolyte solution, M_1 and V_1^0 are the molecular weight and molar volume of the pure water,

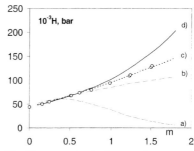

Figure 2. The Henry constant of oxygen in aqueous solutions of sodium sulfate at 25 °C: (○) experimental data; (a) the Henry constant calculated with eq 24 using for the mean activity coefficient of dissolved salt the Debye–Hückel equation; (b) the Henry constant calculated with eq 24 using for the mean activity coefficient of dissolved salt the extended Debye–Hückel equation; (c) the Henry constant calculated with eq 24 using for the mean activity coefficient of dissolved salt the Bromley equation; (d) the Henry constant calculated with eq 15.

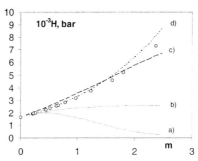

Figure 3. The Henry constant of carbon dioxide in aqueous solutions of sodium sulfate at 25 °C: (○) experimental data; (a) the Henry constant calculated with eq 24 using for the mean activity coefficient of dissolved salt the Debye–Hückel equation; (b) the Henry constant calculated with eq 24 using for the mean activity coefficient of dissolved salt the extended Debye–Hückel equation; (c) the Henry constant calculated with eq 24 using for the mean activity coefficient of dissolved salt the Bromley equation; (d) the Henry constant calculated with eq 15.

Table 2. Constant β and ϕ^0 in Equation 25 for the Mixtures Investigated

mixture	temp, K	composition range, m	ϕ^0, cm^3/mol	β, (cm^3 kg$^{0.5}$)/mol$^{1.5}$
$NaCl + H_2O$	298.15	0–6.1	16.04	2.09
$Na_2SO_4 + H_2O$	298.15	0–2	10.43	12.66

and β is a constant; β and ϕ^0 were evaluated from experimental density data,[24] and their values are listed in Table 2.

For the mean activity coefficient of the salt, several expressions have been used, such as the Debye–Hückel equation,[2] the extended Debye–Hückel equation,[2] and the Bromley equation.[2] The Bromley equation was selected because of its simplicity and its accuracy; of course, other accurate equations[2] are also available. The values of the parameter B for all cases examined are listed in Table 3.

A comparison between the experimental solubilities and those calculated with eq 24 is presented in Figures 2–5. They show that the values calculated using eq 24 are highly dependent on the activity coefficient employed. Indeed, the correlation based on the Debye–Hückel equation is very poor, particularly at high molalities (see curves a in Figures 2–5). The extended Debye–Hückel equation (see curves b in Figures 2–5) provides a better agreement but is not yet satisfactory at high molalities. Equation 15 [based on the ideal 1

Table 3. Values of B Obtained from Experimental Solubility Data (for the Sources of Experimental Solubility Data, See Table 1)

	parameter B, cm³/mol			
mixture	eq 15	eq 24 with the Debye–Hückel equation	eq 24 with the extended Debye–Hückel equation	eq 24 with the Bromley equation
oxygen in $Na_2SO_4 + H_2O$ ($T = 298.15$ K)	−838	−136	−221	−149
oxygen in $Na_2SO_4 + H_2O$ ($T = 308.15$ K)	−822	−337	−209	−141
nitrogen in $Na_2SO_4 + H_2O$	−822	−337	−209	−141
carbon dioxide in $Na_2SO_4 + H_2O$	−708	−98	−39	−151
oxygen in $NaCl + H_2O$	−3.8	150	184	175
carbon dioxide in $NaCl + H_2O$	−6.42	−7.53	36.9	166.4

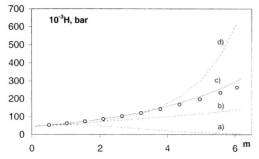

Figure 4. The Henry constant of oxygen in aqueous solutions of sodium chloride at 25 °C: (O) experimental data; (a) the Henry constant calculated with eq 24 using for the mean activity coefficient of dissolved salt the Debye–Hückel equation; (b) the Henry constant calculated with eq 24 using for the mean activity coefficient of dissolved salt the extended Debye–Hückel equation; (c) the Henry constant calculated with eq 24 using for the mean activity coefficient of dissolved salt the Bromley equation; (d) the Henry constant calculated with eq 15.

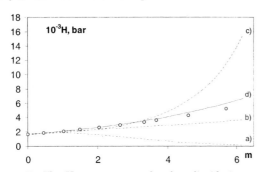

Figure 5. The Henry constant of carbon dioxide in aqueous solutions of sodium chloride at 25 °C: (O) experimental data; (a) the Henry constant calculated with eq 24 using for the mean activity coefficient of dissolved salt the Debye–Hückel equation; (b) the Henry constant calculated with eq 24 using for the mean activity coefficient of dissolved salt the extended Debye–Hückel equation; (c) the Henry constant calculated with eq 24 using for the mean activity coefficient of dissolved salt the Bromley equation; (d) the Henry constant calculated with eq 15.

(water)–3 (electrolyte) mixture approximation] provides a better agreement for not too high molalities [the Sechenov equation and eq 15 provide in all of the cases very similar results (see Figure 1)]. Only when the more accurate Bromley equation has been employed for the mean activity coefficient of the salt was the agreement between experiment and calculation very good. Additional systems, also listed in Table 1, have been examined, and very good agreement was obtained when the Bromley equation was used for the activity coefficient. We have also employed other accurate equations for the mean activity coefficient of the salt, such as the Pitzer equation,[2,3] and the agreement was as good as that provided by the Bromley equation. Figures 2–5 examine only some of the systems investigated. Information about the other systems is contained in Table 3. For NaCl, the results are not too accurate at high molalities. However, it is well-known[3] that most models fail to represent accurately the mean activity coefficient for the NaCl + H_2O mixtures at high molalities. One can, therefore, conclude that the gas solubility in aqueous salt solutions can be well described by eq 24 when accurate expressions for the mean activity coefficient of the salt in the binary water + salt mixtures are used.

4. Salting-In or Salting-Out?

It is worth mentioning that no criterion is presently available to answer the question of whether the addition of a salt will increase or decrease the solubility of a gas in a solvent.

Let us consider eq 15, which is valid for sufficiently small salt concentrations. Of course, the sign of B accounts for the type of salting, because the integral always has positive values: consequently, B should be negative for salting-out and positive for salting-in.

Using eq 6, one obtains the following relation for B:

$$B = \frac{(G_{11} - G_{33} - 2G_{12} + 2G_{23})_{x_2^t=0}}{2} \quad (26)$$

The partial molar volume of the solute (component 2) at infinite dilution in the ternary mixture 1–2–3 is given by[20]

$$\lim_{x_2^t \to 0} V_2^t = V_2^{t,\infty} =$$
$$\lim_{x_2^t \to 0} \{[1 + c_1(G_{11} - G_{12}) + c_3(G_{33} - G_{23}) + c_1c_3(-G_{12}G_{33} + G_{12}G_{13} - G_{13}^2 + G_{13}G_{23} + G_{11}G_{33} - G_{11}G_{23})]/(c_1 + c_3 + c_1c_3\Delta_{13})\} \quad (27)$$

Assuming that the mixture 1–3 behaves like an ideal one and using eq 6, one can write

$$\Delta_{13} = 0 \quad \text{and} \quad 2G_{13} = G_{11} + G_{33} \quad (28)$$

The combination of eqs 26–28 yields the following expression for B in the dilute region ($c_3 \ll c_1$ and $c_1 \approx c_1^0$) (for details, see the appendix):

$$B = -\frac{V_2^{t,\infty} - V_2^{b,\infty} - c_3 V_3^{b,\infty}(V_3^{b,\infty} - V_1^0)}{c_3 V_3^{b,\infty}} \quad (29)$$

where $V_2^{b,\infty}$ is the partial molar volume of the solute gas (2) in water at infinite dilution, $V_3^{b,\infty}$ is the partial molar volume of the electrolyte (3) in water at infinite dilution, and V_1^0 is the molar volume of pure water.

Table 4. Aqueous Mixtures with Salting-In[a]

gas	salt	$V_3^{b,\infty}$,[b] cm^3/mol	gas	salt	$V_3^{b,\infty}$,[b] cm^3/mol
CH$_4$[c]	(Me)$_4$NBr	~114	O$_2$, He, Kr	(Et)$_4$NBr	~174
CH$_4$[c]	(Et)$_4$NBr	~174	He	(Bu)$_4$NBr	~301
CH$_4$[c]	(Pr)$_4$NBr	~240	Ar	(Et)$_4$NI	~185
CH$_4$[c]	(Bu)$_4$NBr	~301	Ar	(Pr)$_4$NI	~250

[a] As given in ref 7. [b] Data regarding $V_3^{b,\infty}$ ($T = 298.15$ K) obtained from: Millero, F. J. In *Water and Aqueous Solutions: Structure, Thermodynamics and Transport Processes*; Horne, R. A., Ed.; Wiley: London, 1972; Chapter 13. [c] Salting-in was also observed for C$_2$H$_6$, C$_3$H$_8$, and C$_4$H$_{10}$.

For small values of c_3, one can, however, write

$$V_2^{t,\infty} = V_2^{b,\infty} + \left(\frac{\partial V_2^{t,\infty}}{\partial c_3}\right)_{P,T} c_3 \quad (30)$$

and consequently eq 29 becomes

$$B = \frac{-\left(\frac{\partial V_2^{t,\infty}}{\partial c_3}\right)_{P,T} + V_3^{b,\infty}(V_3^{b,\infty} - V_1^0)}{V_3^{b,\infty}} \quad (31)$$

The available information[8] indicates that $\alpha = (\partial V_2^{t,\infty}/\partial c_3)_{P,T}$ is usually small;[8] e.g., for Ar in CaCl$_2$, KCl, KI, (Me)$_4$NBr, and (Bu)$_4$NBr, α is approximately equal to $-1, -0.5, -0.3, +0.4,$ and $+1.5$ (cm^3/mol)^2, respectively. The values of α for methane, oxygen, and hydrogen are also small.[8] However, when the difference $V_3^{b,\infty} - V_1^0$ is small, the value of α can affect the sign of B.

When α is small and can be neglected, one obtains a very simple criterion for salting-out and salting-in. Namely, salting-out will occur when

$$V_3^{b,\infty} < V_1^0 \quad (32)$$

and salting-in when

$$V_3^{b,\infty} > V_1^0 \quad (33)$$

The above criterion indicates that salting-in occurs for salts with large values of $V_3^{b,\infty}$ and salting-out for salts with relatively small or negative values of $V_3^{b,\infty}$. The literature data are in agreement with this conclusion (see Table 4). Salting-in was also observed[16] (not for all conditions) for aqueous solutions of CH$_3$COONH$_4$, CH$_3$COONa ($V_3^{b,\infty} \cong 40$ cm^3/mol),[26] and CH$_3$COOH ($V_3^{b,\infty} \cong 52$ cm^3/mol).[27] To our knowledge, salting-in was not observed for salts with relatively small (or negative) $V_3^{b,\infty}$, for example, for aqueous solutions of NaOH ($V_3^{b,\infty} \cong -5$ cm^3/mol)[26] or KOH ($V_3^{b,\infty} \cong 3-4$ cm^3/mol).[26]

The above criterion can also be extended to nonelectrolytes. Indeed, the addition of an alcohol (for alcohols[28] $V_3^{b,\infty} > V_1^0$: for instance, for methanol $V_3^{b,\infty} = 38.2$ cm^3/mol, for ethanol $V_3^{b,\infty} = 55.1$ cm^3/mol, and for 1-propanol $V_3^{b,\infty} = 70.7$ cm^3/mol) increases the solubilities of oxygen, nitrogen, and carbon dioxide in water.[29]

However, it should be noted that the above criteria (32) and (33) pertain only to very low salt concentrations and involve the approximation of ideal behavior for dilute solutions of salt in water.

5. Discussion and Conclusion

The Kirkwood−Buff formalism was used to derive a general expression for the derivative of the activity coefficient $\gamma_{2,t}$ of the gas in a ternary mixture with respect to the mole fraction x_3^t of the salt. The derived expression was used to obtain the composition dependence of the Henry constant for a gas dissolved in a mixed solvent. It should be pointed out that the mixed solvent can be composed of two nonelectrolytes or a solvent and a solute, such as a salt. In this paper the emphasis is on the latter case.

A general expression for the Henry constant in a salt solution was also obtained, which contains as a particular case the Sechenov equation and which, like the Sechenov equation, is a one-parameter equation. This equation requires information about the molar volume and the mean activity coefficient of the salt in the binary water + salt mixture. The Sechenov equation and the new ones (eqs 15 and 24) have been compared with experimental data. The results obtained with the Sechenov equation underestimate, in agreement with literature observations,[13] the gas solubility at high salt concentrations. In contrast, the expressions based on the Debye−Hückel equation or the extended Debye−Hückel equation for the mean activity coefficient of the salt overestimate the gas solubility. When the derived eq 24 has been combined with an accurate equation for the mean activity coefficient of the dissolved salt, such as the Bromley equation, an accurate correlation for the oxygen solubility in an aqueous solution of sodium sulfate could be obtained. However, even the Bromley equation is not accurate enough to represent the mean activity coefficient of the salt for the NaCl + H$_2$O mixture at high molalities, and this explains the less good prediction obtained for the gas solubilities in NaCl solutions at high molalities.

The main advantage of the new equations in comparison with that of Sechenov and its modifications is their clear physical meaning, which allowed one to derive a criterion for predicting the kind of salting. The obtained criterion predicted salting-in for "large" ions and salting-out for relatively small ions, in agreement with the available experimental information.

Appendix: Expression for Coefficient B in Equation 15

Equation 27 for the partial molar volume of component 2 at infinite dilution in a binary mixture water (1) + electrolyte (3)[20] for $\Delta_{13} = 0$, $c_3 \ll c_1$, and $c_1 \approx c_1^0$ can be recast in the form

$$V_2^{t,\infty} = [1 + c_1(G_{11} - G_{12})_{x_2^t=0} + c_3(G_{33} - G_{23})_{x_2^t=0} + c_1 c_3 B(G_{33} - G_{11})_{x_2^t=0}]/(c_1 + c_3) \quad (A1-1)$$

From eq 26, one obtains

$$(G_{23})_{x_2^t=0} = \frac{B}{2} - \frac{(G_{11} - G_{33} - 2G_{12})_{x_2^t=0}}{2} \quad (A1-2)$$

which, introduced in eq A1-1, provides the following expression for B:

$$B = -[2c_1^0 V_2^{t,\infty} - 2 + 2c_1^0(G_{12} - G_{11})_{x_2^t=0} + c_3(2G_{12} - G_{33} - G_{11})_{x_2^t=0}]/ [c_3(1 - c_1^0 G_{33} + c_1^0 G_{11})_{x_2^t=0}] \quad (A1-3)$$

For small values of c_3, $(G_{12})_{x_2^t=0}$ can be replaced by[30]

$$\lim_{\substack{x_2^t \to 0}} G_{12} = \lim_{\substack{x_2^t \to 0 \\ x_3^t \to 0}} G_{12} = G_{12}^\infty = RTk_{T,1}^0 - V_2^{b,\infty} \quad \text{(A1-4)}$$

The Kirkwood–Buff integrals for the mixture water–electrolyte, when one assumes ideal behavior in the dilute region, can be expressed as follows:[30]

$$G_{13} = G_{13}^{id} = Rk_{T,1}^0 - V_1^0 - \varphi_1(V_3^{b,\infty} - V_1^0) \quad \text{(A1-5)}$$

$$G_{11} = G_{11}^{id} = G_{13}^{id} + V_3^{b,\infty} - V_1^0 \quad \text{(A1-6)}$$

and

$$G_{33} = G_{33}^{id} = G_{13}^{id} - (V_3^{b,\infty} - V_1^0) \quad \text{(A1-7)}$$

In the above expressions, R is the universal gas constant, $k_{T,1}^0$ is the isothermal compressibility of the pure solvent, $V_i^{b,\infty}$ is the partial molar volume of component i at infinite dilution in the binary mixture of component i and solvent, and φ_1 is the volume fraction of water.

It should be noted that expressions (A1-5)–(A1-7) involve $\Delta_{13} = 0$ and $(\partial \ln \gamma_3^{b,1-3}/\partial x_3^{b,1-3})_{P,T} = 0$ (see eq 14).

Combining eqs A1-4–A1-7 with eq A1-3 leads to eq 31 in the text.

Literature Cited

(1) Wilhelm, E.; Battino, R.; Wilcock, R. J. Low-pressure solubility of gases in liquid water. *Chem. Rev.* **1977**, *77*, 219–262.

(2) Zemaitis, J. F.; Clark, D. M.; Rafal, M.; Scrivner, N. C. *Handbook of Aqueous Electrolyte Thermodynamics*; AIChE: New York, 1986.

(3) Prausnitz, J. M.; Lichtenthaler, R. N.; Gomes de Azevedo, E. *Molecular Thermodynamics of Fluid-Phase Equilibria*, 2nd ed.; Prentice-Hall: Englewood Cliffs, NJ, 1986.

(4) Desnoyers, J. E.; Pelletier, G. E.; Jolicoeur, C. Salting-in by quaternary ammonium salts. *Can. J. Chem.* **1965**, *43*, 3232–3237.

(5) Masterton, W. L.; Bolocofsky, D.; Lee, T. P. Ionic radii from scaled particle theory of the salt effect. *J. Phys. Chem.* **1971**, *75*, 2809–2815.

(6) Feillolay, A.; Lucas, M. Solubility of helium and methane in aqueous tetrabutylammonium bromide solutions at 25 and 35 °C. *J. Phys. Chem.* **1972**, *76*, 3068–3072.

(7) Krishnan, C. V.; Friedman, H. L. Model calculations for Setchenow coefficients. *J. Solution Chem.* **1974**, *3*, 727–744.

(8) Tiepel, E. W.; Gubbins, K. E. Partial molal volumes of gases dissolved in electrolyte solutions. *J. Phys. Chem.* **1972**, *76*, 3044–3049.

(9) Sechenov, I. M. Über die konstitution der salzlösungen auf grund ihres verhaltens zu kohlensäure. *Z. Phys. Chem.* **1889**, *4*, 117–125.

(10) Long, F. A.; McDevit, W. F. Activity coefficients of nonelectrolyte solutes in aqueous salt solutions. *Chem. Rev.* **1952**, *51*, 119–169.

(11) Tiepel, E. W.; Gubbins, K. E. Thermodynamic properties of gases dissolved in electrolyte solutions. *Ind. Eng. Chem. Fundam.* **1973**, *12*, 18–25.

(12) Yasunishi, A. Solubilities of sparingly soluble gases in aqueous sodium sulfate and sulfite solutions. *J. Chem. Eng. Jpn.* **1977**, *10*, 89–94.

(13) Schumpe, A. The estimation of gas solubilities in saltsolutions. *Chem. Eng. Sci.* **1993**, *48*, 153–158.

(14) Yasunishi, A.; Yoshida, F. Solubility of carbon dioxide in aqueous electrolyte solutions. *J. Chem. Eng. Data* **1979**, *24*, 11–14.

(15) Sing, R.; Rumpf, B.; Maurer, G. Solubility of ammonia in aqueous solutions of single electrolytes sodium chloride, sodium nitrate, sodium acetate, and sodium hydroxide. *Ind. Eng. Chem. Res.* **1999**, *38*, 2098–2109.

(16) Xia, J. Z.; Kamps, A. P. S.; Rumpf, B.; Maurer, G. Solubility of H_2S in (H_2O + CH_3COONa) and (H_2O + CH_3COONH_4) from 313 to 393 K and at pressures up to 10 MPa. *J. Chem. Eng. Data* **2000**, *45*, 194–201.

(17) Kirkwood, J. G.; Buff, F. P. Statistical mechanical theory of solutions. I. *J. Chem. Phys.* **1951**, *19*, 774–782.

(18) Shulgin, I.; Ruckenstein, E. Henry's constant in mixed solvents from binary data. *Ind. Eng. Chem. Res.* **2002**, *41*, 1689–1694.

(19) O'Connell, J. P. Molecular thermodynamics of gases in mixed solvents. *AIChE J.* **1971**, *17*, 658–663.

(20) Ruckenstein, E.; Shulgin, I. Entrainer effect in supercritical mixtures. *Fluid Phase Equilib.* **2001**, *180*, 345–359.

(21) Ben-Naim, A. Inversion of the Kirkwood–Buff theory of solutions: application to the water–ethanol system. *J. Chem. Phys.* **1977**, *67*, 4884–4890.

(22) Perry, R. L.; Massie, J. D.; Cummings, P. T. An analytic model for aqueous-electrolyte solutions based on fluctuation solution theory. *Fluid Phase Equilib.* **1988**, *39*, 227–266.

(23) Cooney, W. R.; O'Connell, J. P. Correlation of partial molar volumes at infinite dilution of salts in water. *Chem. Eng. Commun.* **1987**, *56*, 341–349.

(24) *CRC Handbook of Chemistry and Physics*, 81st ed.; Lide, D. R., Ed.; CRC Press: Boca Raton, FL, 2000–2001.

(25) Harned, H. S.; Owen, B. B. *The physical chemistry of electrolytic solutions*, 2nd ed.; Reinhold Pub. Corp.: New York, 1950.

(26) Millero, F. J. In *Water and Aqueous Solutions: Structure, Thermodynamics and Transport Processes*; Horne, R. A., Ed.; Wiley: London, 1972; Chapter 13;

(27) Lang, W. Setchenov coefficients for oxygen in aqueous solutions of various organic compounds. *Fluid Phase Equilib.* **1996**, *114*, 123–133.

(28) Franks, F.; Desnoyers, J. E. Alcohol–water mixtures revisited. *Water Sci. Rev.* **1985**, *1*, 171–232.

(29) Tokunaga, J. Solubilities of oxygen, nitrogen, and carbon dioxide in aqueous alcohol solutions. *J. Chem. Eng. Data* **1975**, *20*, 41–46.

(30) Shulgin, I.; Ruckenstein, E. Kirkwood–Buff integrals in aqueous alcohol systems: comparison between thermodynamic calculations and X-ray scattering experiments. *J. Phys. Chem. B* **1999**, *103*, 2496–2503.

Received for review May 8, 2002
Revised manuscript received July 2, 2002
Accepted July 2, 2002

IE020348Y

The Solubility of Binary Mixed Gases by the Fluctuation Theory

I. Shulgin[†] and E. Ruckenstein[*]

Department of Chemical Engineering, State University of New York at Buffalo, Amherst, New York 14260

This paper is devoted to the solubility of mixed gases in a liquid, the goal being to predict their solubilities from binary data. Only sparingly soluble and weakly interacting gases are considered. On the basis of the Kirkwood–Buff theory of solution, two transcendental equations are derived that allow one to predict the solubility of binary mixed gases from the solubilities of pure individual gases. The suggested method was tested for the solubilities of methane–ethane, methane–n-butane, and methane–carbon dioxide gas mixtures in water at high pressures. Good agreement between experiment and predictions was found.

1. Introduction

The removal of acid gases from natural gas streams; the solubilities of hydrocarbons and natural-gas components such as CO_2 and H_2S in water under high-pressure/high-temperature conditions; and the solubilities of air and other mixed gases in water, blood, seawater, rainwater, and many other aqueous solutions are a few examples for which information about the solubility of mixed gases in a solvent is needed. This topic has attracted the attention of both experimentalists and theoreticians.[1–8] Whereas the solubilities of many individual gases in liquids have been precisely measured,[9–11] those of mixed gases have rarely been determined; even complete information about the solubility of air in water in a wide range of pressures and temperatures is not available.[9,10] So far, there is no rigorous method for predicting the solubilities of gaseous mixtures in liquids; only an empirical method for mixtures of hydrocarbons has been suggested.[2] As mentioned in the literature,[6] the usual methods for predicting vapor–liquid equilibrium, such as the Wilson, NRTL, and UNIQUAC approaches, cannot be straightforwardly extended to the solubility of mixtures of two supercritical gases. Cubic equations of state (EOS) such as the Peng–Robinson[12] and the Soave–Redlich–Kwong[13] EOS provide accurate descriptions for the solubility of single gases in liquids but can not be extended to the solubility of gaseous mixtures, especially when the solvent is polar,[13,14] because of the empirical nature of the interaction parameter in the van der Waals mixing rule. Whereas the interaction parameter can be taken zero for multicomponent gaseous mixtures containing similar compounds,[13] it cannot be predicted for unsymmetrical multicomponent mixtures, such as $CH_4 + C_2H_6$ + polar solvent, from the interaction parameters for binary (individual gas–solvent) mixtures. However, the combination of one of the above EOS with modern mixing rules and group contribution methods[14–16] seems to be promising in predicting the solubilities of gaseous mixtures in liquids. The aim of this paper is to propose a method for predicting mixed-gas solubilities from the solubilities of the constituent gases in the same solvent, without using an EOS, and to compare the obtained results with available experimental data. The fluctuation theory of Kirkwood and Buff[17] for ternary mixtures will be employed to develop the aforementioned method.

2. Theory

2.1. General Expressions for the Solubility of a Gas Mixture in a Single Solvent.

Let us consider the solubility of a mixed gas (composed of two supercritical gases: component 2 with mole fraction y_2 and component 3 with mole fraction y_3) in a single solvent (component 1). At equilibrium, the fugacities of the components in the liquid and gaseous phases should be equal. Therefore, one can write

$$f_i^{G(t)} = f_i^{L(t)} \quad (i = 2, 3) \tag{1}$$

where the superscripts G and L refer to the gaseous and liquid phases, respectively, and t indicates a ternary mixture.

The Lewis–Randall rule[18] for the fugacity of a species in a gas mixture will be adopted; hence, the fugacity of a component in a mixture is obtained by multiplying its fugacity as a pure gas with its mole fraction. In addition, for the sake of simplicity, the solubilities of both gases will be assumed small, and the concentration of the solvent in the gas phase will be neglected. Therefore, for the fugacities of the two species of the gas mixture, one can write

$$f_i^{G(t)} = f_i^0(P,T) y_i \quad (i = 2, 3) \tag{2}$$

where $f_i^0(P,T)$ is the fugacity of the pure gas i at the pressure and temperature of the system.

The fugacities of the components in the liquid phase can be expressed as[18]

$$f_i^{L(t)} = x_i^t \gamma_i^t f_i^L(T,P) \quad (i = 2, 3) \tag{3}$$

where x_i^t and γ_i^t are the mole fraction and the activity coefficient, respectively, of component i in the liquid phase and $f_i^L(T,P)$ is the fugacity of the pure component i in the (hypothetical) liquid state.

Under the same conditions, for the solubilities of the pure gases in the same solvent (neglecting the concentration of the solvent in the gaseous phase), one can write

---

[*] Corresponding author. E-mail: feaeliru@acsu.buffalo.edu. Fax: (716) 645-3822. Phone: (716) 645-2911/ext. 2214.
[†] E-mail address: ishulgin@eng.buffalo.edu.

$$f_i^0 = x_i^b \gamma_i^b f_i^L(T,P) \quad (i = 2,3) \quad (4)$$

where x_i^b and γ_i^b represent the mole fraction and the activity coefficient, respectively, of component i in the liquid phase of the binary mixture $1-i$, where i is 2 or 3.

The combination of eqs 1–4 yields the relations

$$x_2^t \gamma_2^t = y_2 x_2^b \gamma_2^b \quad (5)$$

and

$$x_3^t \gamma_3^t = y_3 x_3^b \gamma_3^b \quad (6)$$

For dilute binary and ternary mixtures (all solute mole fractions are small), one can write[19–22]

for the binary mixtures 1–2 and 1–3[19]

$$\ln \gamma_2^b = \ln \gamma_2^{b,\infty} - k_{22} x_2^b \quad (7)$$

and

$$\ln \gamma_3^b = \ln \gamma_3^{b,\infty} - k_{33} x_3^b \quad (8)$$

where $\gamma_i^{b,\infty}$ is the activity coefficient of component i ($i = 2, 3$) at infinite dilution of component i in the binary mixture $1-i$ and

$$k_{22} = -\left(\frac{\partial \ln \gamma_2^b}{\partial x_2^b}\right)_{P,T,x_2 \to 0} \quad (9)$$

and

$$k_{33} = -\left(\frac{\partial \ln \gamma_3^b}{\partial x_3^b}\right)_{P,T,x_3 \to 0} \quad (10)$$

for the ternary mixture 1–2–3 at high dilutions of components 2 and 3[20–22]

$$\ln \gamma_2^t = \ln \gamma_2^{t,\infty} + x_2^t \left(\frac{\partial \ln \gamma_2^t}{\partial x_2^t}\right)_{P,T,x_3,0} + x_3^t \left(\frac{\partial \ln \gamma_2^t}{\partial x_3^t}\right)_{P,T,x_2,0}$$

$$\equiv \ln \gamma_2^{t,\infty} - x_2^t K_{22} - x_3^t K_{23} \quad (11)$$

and

$$\ln \gamma_3^t = \ln \gamma_3^{t,\infty} + x_2^t \left(\frac{\partial \ln \gamma_3^t}{\partial x_2^t}\right)_{P,T,x_3,0} + x_3^t \left(\frac{\partial \ln \gamma_3^t}{\partial x_3^t}\right)_{P,T,x_2,0}$$

$$\equiv \ln \gamma_3^{t,\infty} - x_2^t K_{32} - x_3^t K_{33} \quad (12)$$

where $\gamma_i^{t,\infty}$ is the activity coefficient of component i ($i = 2, 3$) at infinite dilutions of components 2 and 3 in the ternary mixture 1–2–3 and the subscript 0 indicates that the derivatives should be calculated for $x_2^t \to 0$ and $x_3^t \to 0$. It should be noted that k_{ii} in eqs 7–10 refers to binary mixtures, whereas K_{ij}, defined by eqs 11 and 12, refers to ternary mixtures.

Because[22,23] $\gamma_2^{t,\infty} = \gamma_2^{b,\infty}$, $\gamma_3^{t,\infty} = \gamma_3^{b,\infty}$, and $K_{23} = K_{32}$, the combination of eqs 5 and 6 with eqs 7 and 8 and eqs 11 and 12 yields

$$x_2^t = x_2^b y_2 \exp[k_{22}(x_2^t - x_2^b) + K_{23} x_3^t] \quad (13)$$

and

$$x_3^t = x_3^b y_3 \exp[k_{33}(x_3^t - x_3^b) + K_{23} x_2^t] \quad (14)$$

Equations 13 and 14 can be used to calculate the solubilities of mixed gases if the solubilities of the pure constituent gases in the same solvent and the values of k_{22}, k_{33}, and K_{23} are known. Whereas the values of k_{22} and k_{33} can be determined from the solubilities of the individual gases,[19,24] an expression for K_{23} will be obtained below using the fluctuation theory of solution.

2.2. Expressions for the Derivative of the Activity Coefficient $(\partial \ln \gamma_2^t / \partial x_3^t)_{P,T,x_2}$ **in a Ternary Mixture through the Kirkwood–Buff Theory of Solution.** General expressions for the derivatives of the activity coefficients in a ternary mixture with respect to the mole fractions were obtained in a previous paper[23] in the form

$$\left(\frac{\partial \ln \gamma_{2,t}}{\partial x_3^t}\right)_{T,P,x_2^t} =$$
$$-(c_1 + c_2 + c_3)(c_1[G_{11} + G_{23} - G_{12} - G_{13}] + c_3[-G_{12} - G_{33} + G_{13} + G_{23}])/(c_1 + c_2 + c_3 + c_1 c_2 \Delta_{12} + c_1 c_3 \Delta_{13} + c_2 c_3 \Delta_{23} + c_1 c_2 c_3 \Delta_{123}) \quad (15)$$

where c_k is the bulk molecular concentration of component k in the ternary mixture 1–2–3 and $G_{\alpha\beta}$ is the Kirkwood–Buff integral given by

$$G_{\alpha\beta} = \int_0^\infty (g_{\alpha\beta} - 1) 4\pi r^2 \, dr \quad (16)$$

In the above expressions, $g_{\alpha\beta}$ is the radial distribution function between species α and β, r is the distance between the centers of molecules α and β, and $\Delta_{\alpha\beta}$ and Δ_{123} are defined as

$$\Delta_{\alpha\beta} = G_{\alpha\alpha} + G_{\beta\beta} - 2G_{\alpha\beta}, \quad \alpha \neq \beta \quad (17)$$

and

$$\Delta_{123} = G_{11} G_{22} + G_{11} G_{33} + G_{22} G_{33} + 2G_{12} G_{13} + 2G_{12} G_{23} + 2G_{13} G_{23} - G_{12}^2 - G_{13}^2 - G_{23}^2 - 2G_{11} G_{23} - 2G_{22} G_{13} - 2G_{33} G_{12} \quad (18)$$

One can show[25] that the factors in the square brackets in the numerator of eq 15 and Δ_{123} can be expressed in terms of $\Delta_{\alpha\beta}$ as

$$G_{12} + G_{33} - G_{13} - G_{23} = \frac{\Delta_{13} + \Delta_{23} - \Delta_{12}}{2} \quad (19)$$

$$G_{11} + G_{23} - G_{12} - G_{13} = \frac{\Delta_{12} + \Delta_{13} - \Delta_{23}}{2} \quad (20)$$

and

$$\Delta_{123} = -[(\Delta_{12})^2 + (\Delta_{13})^2 + (\Delta_{23})^2 - 2\Delta_{12} \Delta_{13} - 2\Delta_{12} \Delta_{23} - 2\Delta_{13} \Delta_{23}]/4 \quad (21)$$

The insertion of eqs 19–21 into eq 15 provides an expression for the derivatives $(\partial \ln \gamma_{2,t} / \partial x_3^t)_{T,P,x_2^t}$ in terms of $\Delta_{\alpha\beta}$ and concentrations.

It should be noted that, according to Ben-Naim,[26] $\Delta_{\alpha\beta}$ is a measure of the nonideality of the binary mixture $\alpha-\beta$ because, for an ideal mixture, $\Delta_{\alpha\beta} = 0$.

At infinite dilution of components 2 and 3, eqs 15 and 19–21 lead for K_{23} (defined by eq 11) to

$$K_{23} = -\lim_{\substack{x_2^t \to 0 \\ x_3^t \to 0}} \left(\frac{\partial \ln \gamma_{2,t}}{\partial x_3^t}\right)_{T,P,x_2^t} = c_1^0 \lim_{\substack{x_2^t \to 0 \\ x_3^t \to 0}} \left(\frac{\Delta_{12} + \Delta_{13} - \Delta_{23}}{2}\right) \quad (22)$$

where

$$c_1^0 = \lim_{\substack{x_2 \to 0 \\ x_3 \to 0}} c_1$$

When the pair 2–3 (pair of nonpolar gases) is ideal or its nonideality $|\Delta_{23}|$ is much smaller than that of the combined binary pairs 1–2 and 1–3 (solvent–gases) $|\Delta_{12} + \Delta_{13}|$, one can write

$$\Delta_{12} + \Delta_{13} - \Delta_{23} \approx \Delta_{12} + \Delta_{13} \quad (23)$$

Taking into account eq 23, eq 22 acquires the form

$$K_{23} = c_1^0 \lim_{\substack{x_2 \to 0 \\ x_3 \to 0}} \left(\frac{\Delta_{12} + \Delta_{13}}{2}\right) \quad (24)$$

Because[23] Δ_{12} is the same for a binary mixture 1–2 in the limit $x_2^b \to 0$ and for a ternary mixture in the limit $x_2^t \to 0$ and $x_3^t \to 0$

$$k_{22} = K_{22} = c_1^0 \lim_{\substack{x_2 \to 0 \\ x_3 \to 0}} \Delta_{12} \quad (25)$$

and

$$k_{33} = K_{33} = c_1^0 \lim_{\substack{x_2 \to 0 \\ x_3 \to 0}} \Delta_{13} \quad (26)$$

Consequently, eqs 13 and 14 become

$$x_2^t = x_2^b y_2 \exp\left[k_{22}(x_2^t - x_2^b) + \frac{k_{22} + k_{33}}{2} x_3^t\right] \quad (27)$$

and

$$x_3^t = x_3^b y_3 \exp\left[k_{33}(x_3^t - x_3^b) + \frac{k_{22} + k_{33}}{2} x_2^t\right] \quad (28)$$

The system of transcendental eqs 27 and 28 can be used to predict the mixed-gas solubility from the solubilities of the individual gases.

3. Calculation Procedure

Calculations were carried out for the solubilities of mixtures of hydrocarbons (methane–ethane and methane–n-butane) and for the mixture methane–carbon dioxide in water, because experimental data regarding the solubilities of binary gas mixtures and individual gases are available for these mixtures.[7,8,27]

Table 1. Dependence of the Parameter $k_{22} = -(\partial \ln \gamma_2/\partial x_2)_{P,T,x_2 \to 0}$ on Pressure at $T = 344.25$ K for the Systems Investigated

system	pressure (MPa)	k_{22}
water (1)–methane (2)	100	−14.3
	75	−14.4
	50	−14.5
	20	−14.6
water (1)–ethane (2)	100	−40.2
	75	−40.3
	50	−40.5
	20	−40.7
water (1)–n-butane (2)	100	−77.6
	75	−77.7
	50	−77.8
	20	−78.1
water (1)–carbon dioxide (2)	100	−33.7
	75	−33.8
	50	−34.0
	20	−34.3

Table 2. Comparison between Predicted and Experimental Solubilities of Methane–n-Butane Mixtures in Water at $T = 344.25$ K^a

		experimental solubilities[8]		predicted solubilities	
P (MPa)	y_2	$10^3 x_2^{t,exp}$	$10^3 x_3^{t,exp}$	$10^3 x_2^{t,calc}$	$10^3 x_3^{t,calc}$
100	0.043	0.286	0.090	0.233	0.093
100	0.230	1.148	0.075	1.232	0.071
100	0.455	2.329	0.052	2.399	0.048
75	0.043	0.233	0.070	0.213	0.098
75	0.230	1.118	0.076	1.124	0.075
75	0.455	2.118	0.048	2.192	0.051
50	0.043	0.198	0.062	0.168	0.087
50	0.230	1.003	0.081	0.889	0.068
50	0.455	1.884	0.042	1.740	0.046
20	0.043	0.127	0.084	0.096	0.091
20	0.230	0.799	0.056	0.513	0.072
20	0.455	1.441	0.037	1.008	0.049
20	0.830	1.886	0.024	1.820	0.015

^{a} $x_2^{t,exp}$ and $x_3^{t,exp}$ are experimental solubilities (mole fractions) of methane and n-butane in water and $x_2^{t,calc}$ and $x_3^{t,calc}$ are their solubilities (mole fractions) in water predicted by eqs 27 and 28.

For the prediction of the mixed-gas solubilities from the solubilities of the pure individual gases, the pressure dependence of the binary parameters k_{ii} is needed. The Peng–Robinson[12] EOS was used to determine the binary parameters k_{ii}. The binary interaction parameter q_{12} in the van der Waals mixing rule was taken from ref 28, where it was evaluated for the water-rich phases of water–hydrocarbon and water–carbon dioxide binary mixtures. The calculated binary parameters k_{ii} are listed in Table 1. One should note that, as expected for a liquid phase, the above parameters are almost independent of pressure, in contrast to their dependence on pressure in the gaseous phase near the critical point.[19,24]

4. Results and Discussion

The results of the present calculations are compared with experiment in Table 2 and Figures 1 and 2, where y_2 is the mole fraction of methane in the gas phase. One can see that there is good agreement between the two. The deviations at $P = 20$ MPa for the methane–n-butane gas mixture are possibly caused by the experimental uncertainties regarding the solubility of the pure n-butane in water.[8]

Our calculations indicate that the solubility of methane–ethane gaseous mixture in water (Figure 1) ex-

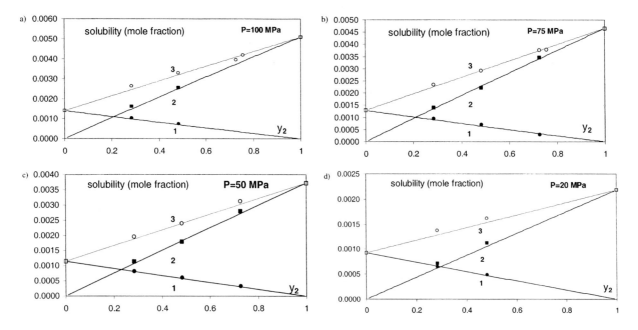

Figure 1. Solubility of the methane−ethane gas mixture in water at $T = 344.25$ K and different pressures. The solubilities calculated from eqs 27 and 28 are represented by the solid lines [(1) x_2^t, (2) x_3^t, and (3) $(x_2^t + x_3^t)$]. The experimental solubilities are taken from ref 8. (■) Mole fraction of methane, (●) mole fraction of ethane, (○) sum of the mole fractions of methane and ethane, and (□) solubilities of the pure gases.

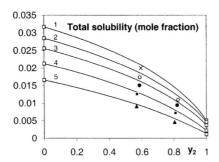

Figure 2. Solubility of the methane−carbon dioxide gas mixture in water at $T = 344.25$ K and different pressures. The total solubilities $(x_2^t + x_3^t)$ calculated from eqs 27 and 28 are represented by the solid lines [(1) $P = 100$ MPa, (2) $P = 75$ MPa, (3) $P = 50$ MPa, (4) $P = 20$ MPa and (5) $P = 10$ MPa]. The experimental solubilities are taken from ref 7. (×) 100, (○) 75, (●) 50, (■) 20, and (▲) 10 MPa and (□) solubilities of the pure gases.

hibits almost linear behavior (this means that the solubility of each constituent of the gas mixture can be determined by multiplying the solubility of the pure component by its mole fraction in the gaseous mixture). This conclusion is in full agreement with the experimental results obtained in ref 8 but in disagreement with those of ref 2, where extrema in the dependence on composition of the solubilities of hydrocarbon mixtures in water (at P, $T = $ const) were found. Our calculations also show, in agreement with experiment,[7,8] that the solubility of methane−n-butane gaseous mixtures (Table 2) exhibits a slight nonlinear behavior and that of methane−carbon dioxide mixtures (Figure 2), a nonlinear one.

For ideal binary mixtures, k_{22} and k_{33} are equal to zero, and eqs 27 and 28 reduce to $x_2^t = x_2^b y_2$ and $x_3^t = x_3^b y_3$. Of course, linear behavior can be reached when either k_{ii} and/or the solubilities x_2^b and x_3^b are small enough for

$$\left| k_{22}(x_2^t - x_2^b) + \frac{k_{22} + k_{33}}{2} x_3^t \right| \ll 1 \quad (29)$$

and

$$\left| k_{33}(x_3^t - x_3^b) + \frac{k_{22} + k_{33}}{2} x_2^t \right| \ll 1 \quad (30)$$

5. Conclusion

The purpose of this paper was to propose a predictive method for the solubilities of binary mixed gases in a liquid in terms of the individual solubilities. For this aim, the derivatives of the activity coefficients in a ternary mixture with respect to the mole fractions were derived through the fluctuation theory of solutions and used to obtain expressions for the solubility at high dilutions of both gases. The suggested method was tested at 344.25 K and in the pressure range 20−100 MPa for the solubilities of methane−ethane, methane−n-butane, and methane−carbon dioxide in water. The predicted solubilities were compared with experimental data and good agreement was found.

Literature Cited

(1) McKetta, J. J.; Katz, D. L. Methane−n-butane−water system in two- and three-phase regions. *Ind. Eng. Chem.* **1948**, *40*, 853−863.

(2) Amirijafari, B.; Campbell, J. Solubility of gaseous mixtures in water. *Soc. Pet. Eng. J.* **1972**, 21−27.

(3) Mathias, P. M.; O'Connell, J. P. Molecular thermodynamics of liquids containing supercritical compounds. *Chem. Eng. Sci.* **1981**, *36*, 1123−1132.

(4) Campanella, E. A.; Mathias, P. M.; O'Connell, J. P. Equilibrium properties of liquids containing supercritical substances. *AIChE J.* **1987**, *33*, 2057−2066.

(5) Gurikov, Yu. V. Solubility of a mixture of nonpolar gases in water. *Zh. Strukt. Khim.* **1969**, *10*, 583−588.

(6) Myers, A. K.; Myers, A. L. Prediction of mixed gas solubility at high pressure. *Fluid Phase Equilib.* **1988**, *44*, 125−144.

(7) Dhima, A.; de Hemptinne, J.-C.; Moracchini, G. Solubility of hydrocarbons and CO_2 mixtures in water under high pressure. *Fluid Phase Equilib.* **1999**, *38*, 129−150.

(8) Dhima, A.; de Hemptinne, J.-C.; Jose, J. Solubility of light hydrocarbons and their mixtures in pure water under high pressure. *Ind. Eng. Chem. Res.* **1998**, *145*, 3144−3161.

(9) Wilhelm, E.; Battino, R.; Wilcock, R. J. Low-pressure solubility of gases in liquid water. *Chem. Rev.* **1977**, *77*, 219−262.

(10) *Solubility Data Series*; Pergamon Press: Elmsford, NY, 1982; Vol. 10.

(11) *Solubility Data Series*; Pergamon Press: Elmsford, NY, 1981; Vol. 8.

(12) Peng, D.-Y.; Robinson, D. B. A new two-constant equation of state. *Ind. Eng. Chem. Fundam.* **1976**, *15*, 59−64.

(13) Soave, G. Equilibrium constants from a modified Redlich−Kwong equation of state. *Chem. Eng. Sci.* **1972**, *27*, 1197−1203.

(14) Orbey, H.; Sandler, S. I. *Modeling Vapor−Liquid Equilibria. Cubic Equations of State and Their Mixing Rules*; Cambridge University Press: New York, 1998.

(15) Holderbaum, T.; Gmehling, J. PSRK: A group contribution equation of state based on UNIFAC. *Fluid Phase Equilib.* **1991**, *70*, 251−265.

(16) Fischer, K.; Gmehling, J. Further development, status and results of the PSRK method for the prediction of vapor−liquid equilibria and gas solubilities. *Fluid Phase Equilib.* **1996**, *121*, 185−206.

(17) Kirkwood, J. G.; Buff, F. P. Statistical mechanical theory of solutions. I. *J. Chem. Phys.* **1951**, *19*, 774−782.

(18) Sandler, S. I. *Chemical and Engineering Thermodynamics*, 3rd ed.; Wiley: 1999.

(19) Debenedetti, P. G.; Kumar, S. K. Infinite dilution fugacity coefficients and the general behavior of dilute binary systems. *AIChE J.* **1986**, *32*, 1253−1262.

(20) Munoz, F.; Li, T. W.; Chimowitz, E. H. Henry's law and synergism in dilute near-critical solutions−Theory and simulation. *AIChE J.* **1995**, *41*, 389−401.

(21) Chialvo, A. J. Solute solute and solute solvent correlations in dilute near-critical ternary mixtures−Mixed-solute and entrainer effects. *J. Phys. Chem.* **1993**, *97*, 2740−2744.

(22) Jonah, D. A.; Cochran, H. D. Chemical potentials in dilute, multicomponent solutions. *Fluid Phase Equilib.* **1994**, *92*, 107−137.

(23) Ruckenstein, E.; Shulgin, I. Entrainer effect in supercritical mixtures. *Fluid Phase Equilib.* **2001**, *180*, 345−359.

(24) Ruckenstein, E.; Shulgin, I. On density microheterogeneities in dilute supercritical solutions. *J. Phys. Chem. B* **2000**, *104*, 2540−2545.

(25) Ruckenstein, E.; Shulgin, I. The solubility of solids in mixtures containing a supercritical fluid and an entrainer. *Fluid Phase Equilib.*, **2002**, *200*, 53−67.

(26) Ben-Naim, A. Inversion of the Kirkwood−Buff theory of solutions: Application to the water−ethanol system. *J. Chem. Phys.* **1977**, *67*, 4884−4890.

(27) Culberson, O. L.; McKetta, J. J. Phase equilibria in hydrocarbon−water systems. III. The solubility of methane in water at pressures to 10,000 psia. *Pet. Technol.* **1951**, *3*, 223−226.

(28) Daridon, J. L.; Lagourette, B.; Saint-Guirons, H.; Xans, P. A cubic equation of state model for phase equilibrium calculation of alkane + carbon dioxide + water using a group contribution k_{ij}. *Fluid Phase Equilib.* **1993**, *91*, 31−54.

Received for review January 8, 2002
Revised manuscript received March 18, 2002
Accepted March 18, 2002

IE020016T

Prediction of gas solubility in binary polymer + solvent mixtures

I. Shulgin, E. Ruckenstein*

Department of Chemical Engineering, State University of New York at Buffalo, Buffalo, NY 14260-4200, USA

Received 23 August 2002; received in revised form 27 September 2002; accepted 28 September 2002

Abstract

This paper is devoted to a theory of gas solubility in highly asymmetrical mixed solvents composed of a low molecular weight (such as water, alcohol, etc.) and a high molecular weight (such as polymer, protein, etc.) cosolvents. The experimental solubilities of Ar, CH_4, C_2H_6 and C_3H_8 in aqueous solutions of polypropylene glycol and polyethylene glycol were selected for comparison with the theory. The approach for predicting these solubilities is based on the Kirkwood–Buff formalism for ternary mixtures, which allowed one to derive a rigorous expression for the Henry constant in mixed solvents. Starting from this expression, the solubilities could be predicted in terms of those in each of the two constituents and the properties of the mixed solvent. This expression combined with the Flory–Huggins equation for the activity coefficient in a binary mixed solvent provided very accurate results, when the Flory–Huggins interaction parameter was used as an adjustable quantity. A simple expression in which the solubility could be predicted in terms of those in each of the two constituents and the molar volumes of the latter was also derived. While less accurate that the previous expression, it provided more than satisfactory results.
© 2002 Published by Elsevier Science Ltd.

Keywords: Gas solubility; Mixed solvents; Polymer solutions

1. Introduction

The prediction of the solubilities of gases in mixed solvents composed of water and a high molecular weight cosolvent such as a polymer, protein, detergent, biomolecule, drug, etc. is important from a practical point of view [1]. One modern example relevant to this topic is the gas antisolvent recrystallization process [2], which is widely used for refining explosives, pharmaceuticals, proteins, etc. The present paper is devoted to the development of equations able to predict the gas solubility in a mixed solvent composed of water (or any other low molecular weight solvent) and a high molecular weight constituent (cosolvent) such as polymers, proteins, etc.

The solubilities of nonpolar nonacidic gases in water are usually small compared to those in nonaqueous solvents; this behavior is usually attributed to the hydrophobic effect [3]. However, the solubility of the same gases in nonpolar solvents is much higher. The solubilities of gases in high molecular weight solvents, such as liquid (or molten) polymers differ in major ways from those in low molecular weight solvents. Indeed, Table 1 which provides a comparison (in terms of the Henry constant and Ostwald coefficient) between the solubilities of gases in different kinds of solvents, reveals that the solubilities [4,5] of nonpolar, nonacidic gases in liquid polymers, such as polypropylene glycol (PPG) and polyethylene glycol (PEG), are much higher than those in water. They are even comparable with the solubilities [6,7] of the same gases in *n*-decane. This means that the well-known principle 'like dissolves like' cannot be applied to high molecular weight solvents. Indeed, the high molecular weight compounds that possess polar groups are often very good solvents for nonpolar, nonacidic gases, and one can generally state that small size gases are fairly soluble in high molecular weight solvents. This probably can be explained by the larger free space between the polymer molecules than between the low molecular weight solvents.

In a previous paper [9] we developed an equation which could predict the gas solubility in a mixed solvent from the solubilities in the individual constituents and the properties of their mixture. This equation was applied to mixed solvents composed of small molecules. In the present paper, we will apply it to the solubility of a gas in a polymer + water mixture.

* Corresponding author. Tel.: +1-716-645-2911x2214; fax: +1-716-645-3822.
E-mail addresses: feaeliru@acsu.buffalo.edu (E. Ruckenstein), ishulgin@eng.buffalo.edu (I. Shulgin).

Table 1
The solubility of several nonpolar gases in water, decane and liquid polymers at $T = 298.15$ K and gas partial pressure 1 atm

Gas	Solubility[a], Henry constant (H, MPa)[b] (Ostwald coefficient, 10^2 L)[c]			
	Water	n-decane	PPG-400[d]	PEG-200[e]
Argon	4025 (3.4)	40.9 (31.0)	41.8 (14.9)	202.5 (6.9)
Methane	4039 (3.4)	39.3 (32.3)	21.0 (29.7)	114.9 (12.1)
Ethane	3027 (4.5)	2.8 (464)	4.1 (154)	24.0 (58.1)
Propane	3745 (3.7)	0.7 (2060)	1.3 (510)	10.6 (132)

[a] The data for the solubilities in water and polymers were taken from Refs. [4,5], and those in n-decane from Refs. [6,7].
[b] The Henry constant is defined as the limiting value of the ratio of the gas partial pressure to its mole fraction in solution as the latter tends to zero [8].
[c] The Ostwald coefficient is the ratio of the volume of gas absorbed to the volume of the absorbing liquid, both measured at the same temperature [8].
[d] Polypropylene glycol with average molecular weight of 400.
[e] Polyethylene glycol with average molecular weight of 200.

The aqueous mixtures of polymers (PEG and PPG) were selected for comparison with the theory, because accurate data [4,5] regarding the solubility of argon (Ar), methane (CH$_4$), ethane (C$_2$H$_6$) and propane (C$_3$H$_8$) in the individual constituents and the polymer + water mixtures are available. In addition, the above polymers and water are miscible in all proportions and solubility data [4,5] are available for the entire composition range. The theoretical approach regarding the solubility of gases in polymer + water mixed solvents can be extended to the correlation of their solubility in mixed solvents formed of water and pharmaceuticals, proteins, biomolecules, etc.

2. Theory

The gas solubility will be expressed in terms of the Henry constant. There are a number of expressions for the Henry constant in binary mixed solvents. The oldest and simplest relationship between the Henry constant in binary solvents and those in the individual constituents is that proposed by Krichevsky [10]:

$$\ln H_{2,t} = x_1^{b,1-3} \ln H_{2,1} + x_3^{b,1-3} \ln H_{2,3} \quad (1)$$

where $H_{2,t}$, $H_{2,1}$ and $H_{2,3}$ are the Henry constants in the binary mixed solvent 1–3 and the individual solvents 1 and 3, respectively, and $x_1^{b,1-3}$ and $x_3^{b,1-3}$ are the mole fractions of components 1 and 3 in the binary solvent 1–3 (throughout this paper the following subscripts for the components will be used: 1, high molecular weight cosolvent, 2, solute (gas), 3, low molecular weight cosolvent). Krichevsky's relationship (1) is valid when the ternary and binary mixtures (1–2 and 2–3) are ideal [9]. However, the ternary 1–2–3 and binary mixtures (1–2 and 2–3) do not always satisfy the ideality conditions. Indeed, the activity coefficients at infinite dilution for the binary mixtures gas/solvent (particularly for high molecular weight solvents) have values much larger [11,12] than unity. On the basis of the Kirkwood–Buff theory of solutions [13] for ternary mixtures [14] the authors derived [9] the following relation for the Henry constant in a binary solvent mixture

$$\ln H_{2,t} = \frac{(\ln H_{2,1})(\ln V^{ID} - \ln V_3^0) + (\ln H_{2,3})(\ln V_1^0 - \ln V^{ID})}{\ln V_1^0 - \ln V_3^0} \quad (2)$$

where V_1^0 and V_3^0 are the molar volumes of the individual solvents 1 and 3, and V^{ID} is the molar volume of the ideal binary mixture 1–3 ($V^{ID} = x_1^{b,1-3} V_1^0 + x_3^{b,1-3} V_3^0$). Eq. (2) is less restrictive than the Krichevsky Eq. (1), because it requires that only the binary mixed solvent 1–3 be an ideal mixture. Such an approximation is reasonable because the activity coefficient of water in the binary mixture PEG + water [15,16] is small and at infinite dilution [11] is about 0.5, while those between the gas and each of the constituents of the solvent are very large.

Eq. (2) does not contain any adjustable parameter and can be used to predict the gas solubility in mixed solvents in terms of the solubilities in the individual solvents (1 and 3) and their molar volumes. Eq. (2) provided a very good agreement [9] with the experimental gas solubilities in binary aqueous solutions of nonelectrolytes; a somewhat modified form correlated well the gas solubilities in aqueous salt solutions [17]. The authors also derived the following rigorous expression for the Henry constant in a binary solvent mixture [9] (Appendix A for the details of the derivation):

$$\ln H_{2,t} = -\int \frac{B}{V}\left[1 + x_3^{b,1-3}\left(\frac{\partial \ln \gamma_3^{b,1-3}}{\partial x_3^{b,1-3}}\right)_{P,T}\right]dx_3^{b,1-3}$$

$$+ \frac{1}{2}\int \frac{(x_1^{b,1-3} - x_3^{b,1-3})}{x_1^{b,1-3}}\left(\frac{\partial \ln \gamma_3^{b,1-3}}{\partial x_3^{b,1-3}}\right)_{P,T}$$

$$dx_3^{b,1-3} + A \quad (3)$$

where V is the molar volume of the binary mixed solvent 1–3, $B = (\Delta_{12} - \Delta_{23})_{x_2^t=0}/2$, x_2^t is the mole fraction of the solute (component 2) in the ternary mixture 1–2–3, $A(P,T)$ is a composition-independent constant of integration, $\gamma_3^{b,1-3}$ is the activity coefficient of component 3, the superscript b, 1–3 indicates that the activity coefficient is for the binary 1–3 mixture, and Δ_{12} and Δ_{23} are functions of the Kirkwood–Buff integrals (Appendix A). If B is considered independent of the composition of the binary mixed solvent 1–3, Eq. (3) can be rewritten in the form

$$\ln H_{2,t} = -BI_1 + \frac{I_2}{2} + A \quad (4)$$

where the integrals

$$I_1 = -\int \frac{\left[1 + x_3^{b,1-3}\left(\frac{\partial \ln \gamma_3^{b,1-3}}{\partial x_3^{b,1-3}}\right)_{P,T}\right]}{V} dx_3^{b,1-3}$$

and

$$I_2 = \int \frac{(x_1^{b,1-3} - x_3^{b,1-3})}{x_1^{b,1-3}} \left(\frac{\partial \ln \gamma_3^{b,1-3}}{\partial x_3^{b,1-3}}\right)_{P,T} dx_3^{b,1-3}$$

can be calculated if the composition dependencies of the activity coefficient and molar volume of the binary mixed solvent 1–3 are known. The combination of Eq. (4) with the following limiting expressions

$$(\ln H_{2,t})_{x_1^{b,1-3}=0} = \ln H_{2,3} \quad (5)$$

and

$$(\ln H_{2,t})_{x_3^{b,1-3}=0} = \ln H_{2,1} \quad (6)$$

allows one to obtain the constants $A(P,T)$ and $B(P,T)$. Expressions (1), (2) and (4) will be used to calculate the gas solubility in a mixed solvent composed of water and polymer.

3. The solubility of gases in binary polymer + solvent mixtures

3.1. The systems considered

The experimental solubilities [4,5] of Ar, CH_4, C_2H_6 and C_3H_8 in the aqueous solutions of PPG with the average molecular weight of 400 (PPG-400), PEG with the average molecular weight of 200 (PEG-200) and PEG with the average molecular weight of 400 (PEG-400) were selected for comparison with the theory.

3.2. The calculation procedure

The experimental Ostwald coefficients [4,5] were converted into the Henry constants. The molar volumes of the aqueous solutions of PPG-400, PEG-200 and PEG-400 were calculated from experimental densities [4,5] to conclude that they are well approximated by the expression

$$V = x_1^{b,1-3} V_1^0 + x_3^{b,1-3} V_3^0 \quad (7)$$

The Flory–Huggins equation [18,19] for the activity coefficient of water in the binary mixed solvent polymer (1) + water (3) will be employed. It has the form

$$\ln \gamma_3^{b,1-3} = \ln\left[1 - \left(1 - \frac{1}{r}\right)\varphi_1\right] + \left(1 - \frac{1}{r}\right)\varphi_1 + \chi\varphi_1^2 \quad (8)$$

where χ is the Flory–Huggins interaction parameter considered here as composition-independent, $\varphi_1 = rx_1^{b,1-3}/(rx_1^{b,1-3} + x_3^{b,1-3})$ is the volume fraction of polymer in the mixed solvent polymer + water and r is the number of segments in the polymer molecule (taken as the ratio of the molar volumes of the polymer and water, $r = V_1^0/V_3^0$). For the derivative of the activity coefficient $(\partial \ln \gamma_3^{b,1-3}/\partial x_3^{b,1-3})_{P,T}$, the Flory–Huggins equation provides the following expression

$$\left(\frac{\partial \ln \gamma_3^{b,1-3}}{\partial x_3^{b,1-3}}\right)_{P,T} = -\frac{x_1^{b,1-3}(-r^3 x_1^{b,1-3} - 3r^2 x_3^{b,1-3} + 3rx_3^{b,1-3} + 2r^2 - x_3^{b,1-3} - r + 2r^2\chi)}{(rx_1^{b,1-3} + x_3^{b,1-3})^3} \quad (9)$$

The insertion of Eqs. (7) and (9) into Eq. (4) provides an expression which can be integrated analytically (Appendix B). The Flory–Huggins interaction parameter χ can be used either as an adjustable parameter, or can be obtained from phase equilibrium data for the binary mixture polymer + water.

3.3. Results

The comparison of the experimental solubilities [4,5] of Ar, CH_4, C_2H_6 and C_3H_8 in the binary aqueous mixtures of PPG-400, PEG-200 and PEG-400 with the calculated ones is presented in Figs. 1–3 and Table 2. They show that Eq. (4) coupled with the Flory–Huggins equation, in which the interaction parameter χ is used as an adjustable parameter, is very accurate. The Krichevsky equation (1) does not provide accurate predictions. While less accurate than Eq. (4), the simple Eq. (2) provides very satisfactory results without involving any adjustable parameters. It should be noted that Eq. (4) coupled with the Flory–Huggins equation with $\chi = 0$ (athermal solutions) does not involve any adjustable parameters and provides results comparable to those of Eq. (2).

4. Discussion

A simple and reliable method for the correlation of the gas solubility in a mixed solvent composed of two cosolvents, one of high molecular weight and the other of low molecular weight, was proposed. It was shown that the well-known Krichevsky equation could not provide accurate predictions of the gas solubilities in such mixed solvents. The failure of Krichevsky's equation is not surprising since it requires the ternary 1–2–3 and the binary 1–2 and 2–3 mixtures to be ideal. Such conditions cannot be satisfied by the highly asymmetrical mixtures of a high molecular weight cosolvent (1)–gas (2)–low molecular weight cosolvent (3). Eq. (2) obtained on the basis of the Kirkwood–Buff formalism is less restrictive, because it involves the more realistic assumption [9] that the nonidealities of the gas/cosolvent mixtures are much higher

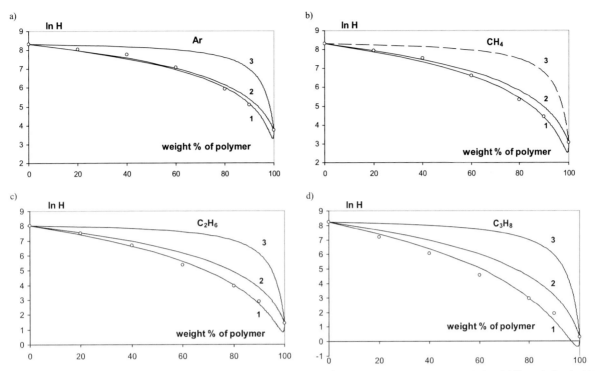

Fig. 1. The solubilities of Ar (a), CH$_4$ (b), C$_2$H$_6$ (c), and C$_3$H$_8$ (d) in PPG-400 at 25 °C. ○, experimental data [4], curve 1, the solubility calculated with Eq. (4) combined with the Flory–Huggins Eq. (8) with χ as adjustable parameter, curve 2, the solubility calculated with Eq. (2), curve 3, the solubility calculated with Krichevsky's Eq. (1).

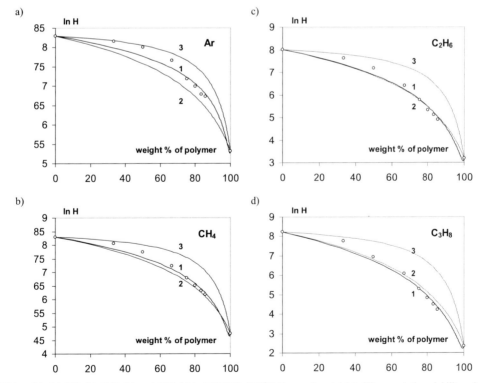

Fig. 2. The solubilities of Ar (a), CH$_4$ (b), C$_2$H$_6$ (c), and C$_3$H$_8$ (d) in PEG-200 at 25 °C. ○, experimental data [5], curve 1, the solubility calculated with Eq. (4) combined with the Flory–Huggins Eq. (8) with χ as adjustable parameter, curve 2, the solubility calculated with Eq. (2), curve 3, the solubility calculated with Krichevsky's Eq. (1).

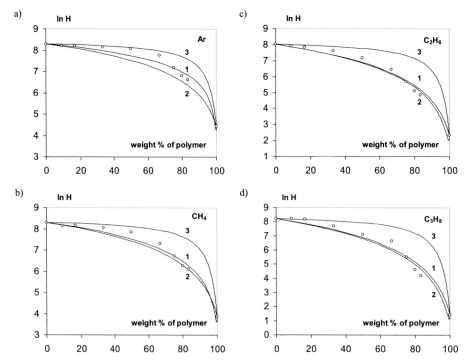

Fig. 3. The solubilities of Ar (a), CH$_4$ (b), C$_2$H$_6$ (c), and C$_3$H$_8$ (d) in PEG-400 at 25 °C. ○, experimental data [5]; curve 1, the solubility calculated with Eq. (4) combined with the Flory–Huggins Eq. (8) with χ as adjustable parameter; curve 2, the solubility calculated with Eq. (2); curve 3, the solubility calculated with Krichevsky's Eq. (1).

Table 2
Comparison between experimental and calculated Henry constants in a binary mixed solvent polymer + water at 298.15 K

Systems		Deviations (%) of experimental data [4,5] from calculations[a]			
Gas	Mixed solvent	Krichevsky's equation	Eq. (2)	Eq. (4)[b]	Eq. (4)[c]
Argone	PPG-400 + water	12.5	1.9	1.1 (−0.17)	1.2
	PEG-200 + water	3.7	3.3	1.3 (−1.77)	4.8
	PEG-400 + water	4.8	3.8	2.0 (−2.52)	5.2
Methane	PPG-400 + water	18.1	3.9	0.9 (0.55)	2.3
	PEG-200 + Water	7.1	1.8	1.2 (−0.94)	3.2
	PEG-400 + water	8.5	2.0	2.0 (−1.17)	3.3
Ethane	PPG-400 + water	36.4	11.5	1.8 (1.49)	9.3
	PEG-200 + water	13.9	2.1	2.0 (−0.29)	2.1
	PEG-400 + water	15.9	3.7	2.1 (−0.10)	2.6
Propane	PPG-400 + water	66.4	25.3	6.5 (2.20)	23.1
	PEG-200 + water	21.5	3.8	2.3 (0.30)	3.2
	PEG-400 + water	21.2	5.5	5.0 (0.18)	5.1
Average (%)		19.2	5.7	1.9	5.5

[a] Defined as

$$100 \sum_i \frac{\left| \ln(H^{(i)}_{2,t(\exp)}) - \ln(H^{(i)}_{2,t(\text{calc})}) \right|}{\ln(H^{(i)}_{2,t(\exp)})} \bigg/ m$$

where m is the number of experimental points.
[b] Eq. (4) combined with the Flory–Huggins equation with adjustable parameter χ (the value of the parameter χ is given in parenthesis).
[c] Eq. (4) combined with the Flory–Huggins equation with parameter $\chi = 0$ (athermal solution).

than those of the mixed solvents. For this reason, Eq. (2) provides more accurate predictions. This equation does not require information about the ternary mixture 1–2–3 nor about its binary constituents, requiring only the gas solubilities in the individual solvents and the molar volumes of the latter.

The most accurate correlation was obtained with the rigorous Eq. (4) coupled with the Flory–Huggins equation for the activity coefficient of water in the polymer + water binary mixture, when the Flory–Huggins interaction parameter χ was considered as an adjustable parameter independent of composition. The values of the Flory–Huggins parameter χ for the systems investigated are listed in Table 2. While, according to the theory, the interaction parameter χ should be independent of the polymer concentration and weight, in reality, in most cases χ changes considerably with both [20]. In particular, for the binary mixtures PPG + water and PEG + water, χ depends on both composition and average molecular weight of the polymer [15,16]. Therefore, the values of χ used in our calculations constitute adjustable parameters. The calculations (Table 2) show that χ depends not only on the average molecular weight of the polymer, but has different values for each of the gases considered.

The figures show that Eq. (4) coupled with the Flory–Huggins equation for the activity coefficient of water in polymer + water mixed solvents provides a minimum for the Henry constant at high (>95 wt%) polymer compositions. Because gas solubility data are not available for such high polymer compositions, one cannot determine whether this minimum is due to the empirical nature of the adjustable parameter χ or reflects an experimental feature of the gas

$$\lim_{x_2^t \to 0} \left(\frac{\partial \ln \gamma_{2,t}}{\partial x_3^t} \right)_{T,P,x_2^t} = -\frac{(c_1^0 + c_3^0)((c_1^0 + c_3^0)(\Delta_{12} - \Delta_{23})_{x_2^t=0} + (c_1^0 - c_3^0)(\Delta_{13})_{x_2^t=0})}{2(c_1^0 + c_3^0 + c_1^0 c_3^0 (\Delta_{13})_{x_2^t=0})} \quad (A2)$$

solubility in such mixed solvents at high polymer concentrations.

5. Conclusion

The Kirkwood–Buff theory of solutions for ternary mixtures was used to analyze the gas solubility in a mixed binary solvent composed of a high molecular weight and a low molecular weight cosolvent, such as the aqueous solutions of water soluble polymers. A rigorous expression for the composition derivatives of the gas activity coefficient in ternary solution was used to derive the composition dependence of the Henry constant under isobaric and isothermal conditions. The obtained expressions as well as the well-known Krichevsky equation were tested for the solubilities of Ar, CH_4, C_2H_6 and C_3H_8 in the aqueous solutions of PPG-400, PEG-200 and PEG-400. It was shown that the coupling of our Eq. (4) with the Flory–Huggins equation for the activity coefficient of the water in the binary mixed solvent provides an accurate correlation for the gas solubility with a single adjustable parameter. However, the more simple Eq. (2) has a satisfactory accuracy and is recommended because it requires only the gas solubilities in the individual solvents and the molar volumes of the latter.

Acknowledgements

We are indebted to Prof. A.D. King, Jr (Department of Chemistry, University of Georgia) for providing information regarding the densities of aqueous mixtures of polyethylene glycol and polypropylene glycol.

Appendix A

The following expression can be written for the Henry constant in a binary solvent [21]

$$\ln H_{2,t} = \lim_{x_2^t \to 0} \ln \gamma_{2,t} + \ln f_2^0(P,T) \quad (A1)$$

where $\gamma_{2,t}$ is the activity coefficient of the solute in the ternary mixture 1–2–3 and $f_i^0(P,T)$ is the fugacity of component i [20]. The Kirkwood–Buff theory of solutions [13] for ternary mixtures [14] provides the following expression for the composition derivatives of $\gamma_{2,t}$ at infinite dilution

where x_3^t is the mole fraction of component 3 in the ternary mixture, c_k^0 ($k = 1, 3$) is the bulk molecular concentration of component k in the binary mixture 1–3 and $G_{\alpha\beta}$ is the Kirkwood–Buff integral given by

$$G_{\alpha\beta} = \int_0^\infty (g_{\alpha\beta} - 1) 4\pi r^2 \, dr \quad (A3)$$

In the above expressions, $g_{\alpha\beta}$ is the radial distribution function between species α and β, r is the distance between the centers of molecules α and β, and $\Delta_{\alpha\beta}$ are defined as follows

$$\Delta_{\alpha\beta} = G_{\alpha\alpha} + G_{\beta\beta} - 2G_{\alpha\beta}, \quad \alpha \neq \beta \quad (A4)$$

It should be noted that Δ_{ij} is a measure of the nonideality [22] of the binary mixture α–β, because for an ideal mixture $\Delta_{\alpha\beta} = 0$.

The combination of Eqs. (A1) and (A2) leads to

$$\left(\frac{\partial \ln H_{2,t}}{\partial x_3^t}\right)_{P,T,x_2^t=0} = -\frac{(c_1^0 + c_3^0)((c_1^0 + c_3^0)(\Delta_{12} - \Delta_{23})_{x_2^t=0} + (c_1^0 - c_3^0)(\Delta_{13})_{x_2^t=0})}{2(c_1^0 + c_3^0 + c_1^0 c_3^0 (\Delta_{13})_{x_2^t=0})} \quad (A5)$$

In addition, for the binary 1–3 mixture one can write the following relation [13]

$$\left(\frac{\partial \ln \gamma_1^{b,1-3}}{\partial x_3^{b,1-3}}\right)_{P,T} = \frac{c_3^0 \Delta_{13}}{1 + c_1^0 x_3^{b,1-3} \Delta_{13}} \quad (A6)$$

Combination of Eqs. (A5) and (A6) leads to

$$\ln H_{2,t} = -\int (c_1^0 + c_3^0) \frac{(\Delta_{12} - \Delta_{23})_{x_2^t=0}}{2}$$

$$\times \left[1 + x_3^{b,1-3} \left(\frac{\partial \ln \gamma_3^{b,1-3}}{\partial x_3^{b,1-3}}\right)_{P,T}\right] dx_3^{b,1-3} \quad (A7)$$

$$+ \frac{1}{2} \int \frac{(x_1^{b,1-3} - x_3^{b,1-3})}{x_1^{b,1-3}} \left(\frac{\partial \ln \gamma_3^{b,1-3}}{\partial x_3^{b,1-3}}\right)_{P,T} dx_3^{b,1-3} + A$$

which is just Eq. (3) in the text.

Appendix B

The aim of this appendix is to provide analytical expressions for the integrals I_1 and I_2 of Eq. (4)

$$I_1 = -\int \frac{\left[1 + x_3^{b,1-3} \left(\frac{\partial \ln \gamma_3^{b,1-3}}{\partial x_3^{b,1-3}}\right)_{P,T}\right]}{V} dx_3^{b,1-3} \quad (B1)$$

and

$$I_2 = \int \frac{(x_1^{b,1-3} - x_3^{b,1-3})}{x_1^{b,1-3}} \left(\frac{\partial \ln \gamma_3^{b,1-3}}{\partial x_3^{b,1-3}}\right)_{P,T} dx_3^{b,1-3} \quad (B2)$$

The Flory–Huggins Eq. (8) was employed for the activity coefficient of water $\gamma_3^{b,1-3}$ in a mixed solvent polymer (1) + water (3). Because the integrated expressions require a too large space, we provide only the results obtained for $\chi = 0$ (athermal mixtures).[1] Using Eqs. (7) and (9), the integration of Eqs. (B1) and (B2) leads to the following expressions

$$I_1 = \frac{\alpha_1 + \alpha_2 \ln(ax_3^{b,1-3} + b) + \alpha_3 \ln(rx_1^{b,1-3} + x_3^{b,1-3})}{(rx_1^{b,1-3} + x_3^{b,1-3})(ra + rb - b)^2} \quad (B3)$$

and

$$I_2 = \frac{2(rx_1^{b,1-3} + x_3^{b,1-3})\ln(rx_1^{b,1-3} + x_3^{b,1-3}) + (1 + r)}{(rx_1^{b,1-3} + x_3^{b,1-3})} \quad (B4)$$

where $a = V_3^0 - V_1^0$ and $b = -V_1^0$. The coefficients α_1, α_2 and α_3 have the following forms

$$\alpha_1 = -r^2 b - r^2 a + rb \quad (B5)$$

$$\alpha_2 = r^3 a x_1^{b,1-3} + r^3 b x_1^{b,1-3} + r^2 a x_3^{b,1-3} + r^2 b x_3^{b,1-3}$$

$$- rb x_1^{b,1-3} - b x_3^{b,1-3} \quad (B6)$$

and

$$\alpha_3 = -r^3 a x_1^{b,1-3} - r^3 b x_1^{b,1-3} - r^2 a x_3^{b,1-3} - r^2 b x_3^{b,1-3}$$

$$+ rb x_1^{b,1-3} + b x_3^{b,1-3} \quad (B7)$$

References

[1] Wilhelm E, Battino R, Wilcock R.J. Chem Rev 1977;77:219.
[2] McHugh M, Krukonis V. Supercritical fluid extraction. Boston: Butterworth-Heinemann; 1994.
[3] Tanford C. The hydrophobic effect: formation of micelles and biological membranes, 2nd ed. New York: Willey-Interscience; 1980.
[4] King AD. J Colloid Interface Sci 2001;243:457.
[5] King AD. J Colloid Interface Sci 1991;144:457.
[6] Wilcock RJ, Battino R, Danforth WF, Wilhelm E. J Chem Thermodyn 1978;10:817.
[7] Jadot R. J Chim Phys 1972;69:1036.
[8] Sandler SI. Chemical and engineering thermodynamics, 3rd ed. New York: Wiley; 1999.
[9] Shulgin I, Ruckenstein E. Ind Engng Chem Res 2002;41:1689.
[10] Krichevsky IR. Zh Fiz Khim 1937;9:41.
[11] Kojima K, Zhang SJ, Hiaki T. Fluid Phase Equilibria 1997;131:145.
[12] Sandler SI. Fluid Phase Equilibria 1996;116:343.
[13] Kirkwood JG, Buff FP. J Chem Phys 1951;19:774.
[14] Ruckenstein E, Shulgin I. Fluid Phase Equilibria 2001;180:345.
[15] Malcolm GN, Rowlinson JS. Trans Faraday Soc 1957;53:921.
[16] Eliassi A, Modarress H, Mansoori GA. J Chem Engng Data 1999;44:52.
[17] Ruckenstein E, Shulgin I. Ind Engng Chem Res 2002;41:4674.
[18] Flory PJ. J Chem Phys 1941;9:660.
[19] Huggins ML. J Chem Phys 1941;9:440.
[20] Prausnitz JM, Lichtenthaler RN, Gomes de Azevedo E. Molecular thermodynamics of fluid—phase equilibria, 2nd ed. Englewood Cliffs, NJ: Prentice-Hall; 1986.
[21] O' Connell JP. AIChE J 1971;17:658.
[22] Ben-Naim A. J Chem Phys 1977;67:4884.

---

[1] The results of integration for $\chi \neq 0$ can be obtained from the authors by request.

…

Ideal Multicomponent Liquid Solution as a Mixed Solvent

E. Ruckenstein* and I. Shulgin[†]

Department of Chemical Engineering, State University of New York at Buffalo, Amherst, New York 14260

The present paper is concerned with mixtures composed of a highly nonideal solute and a multicomponent ideal solvent. A model-free methodology, based on the Kirkwood−Buff (KB) theory of solutions, was employed. The quaternary mixture was considered as an example, and the full set of expressions for the derivatives of the chemical potentials with respect to the number of particles, the partial molar volumes, and the isothermal compressibility were derived on the basis of the KB theory of solutions. Further, the expressions for the derivatives of the activity coefficients were applied to quaternary mixtures composed of a solute and an ideal ternary solvent. It was shown that the activity coefficient of a solute at infinite dilution in an ideal ternary solvent can be predicted in terms of the activity coefficients of the solute at infinite dilution in subsystems (solute + the individual three solvents, or solute + two binaries among the solvent species). The methodology could be extended to a system formed of a solute + a multicomponent ideal mixed solvent. The obtained equations were used to predict the gas solubilities and the solubilities of crystalline nonelectrolytes in multicomponent ideal mixed solvents. Good agreement between the predicted and experimental solubilities was obtained.

Introduction

The solubilities of gases, liquids, and solids in multicomponent solvents constitute important issues in science and technology. The aqueous multicomponent solutions represent a meaningful example because the overwhelming majority of solutions of biological and environmental interest are aqueous multicomponent solutions.

The experimental research on the solubilities in multicomponent solutions is tedious and time-consuming because of the large number of compositions needed to cover the concentration ranges. For example, 11 measurements are needed for different compositions in a binary solution (10 mol % steps for composition changes), 66 in ternary, 286 in quaternary, and so on.

Therefore, it is important to have a reliable and accurate method for predicting the solubility in multicomponent solutions from those in its pure or binary constituents. The main difficulty in predicting the solubility in multicomponent solutions consists of the calculation of the activity coefficient of the solute. Thermodynamics cannot provide the explicit pressure, temperature, and composition dependence of thermodynamic functions, such as the activity coefficients of the components in multicomponent mixtures. For this reason, empirical expressions such as the Wohl expansion[1] have often been used to represent thermodynamic data regarding multicomponent mixtures.

Another approach is to employ rigorous statistical thermodynamic theories. In this paper, the Kirkwood−Buff (KB) theory of solutions[2] (fluctuation theory of solutions) is employed to analyze the thermodynamics of multicomponent mixtures, with the emphasis on quaternary mixtures. This theory connects the macroscopic properties of n-component solutions, such as the isothermal compressibility, the concentration derivatives of the chemical potentials, and the partial molar volumes to the microscopic properties of solutions in the form of spatial integrals involving the radial distribution functions, namely, the KB integrals. The KB integrals are provided by the expression[2]

$$G_{\alpha\beta} = \int_0^\infty (g_{\alpha\beta} - 1) 4\pi r^2 \, dr \quad (1)$$

where $g_{\alpha\beta}$ is the radial distribution function between species α and β and r is the distance between the centers of molecules α and β.

Previously,[3] the authors have applied the KB theory of solutions to ternary mixtures. In particular, the following kinds of relations for the concentration derivatives of the activity coefficients in a ternary mixture were obtained:

$$\left(\frac{\partial \ln \gamma_{2,t}}{\partial x_3^t}\right)_{T,P,x_2^t} = -\frac{(c_1+c_2+c_3)(c_1[G_{11}+G_{23}-G_{12}-G_{13}] + c_3[-G_{12}-G_{33}+G_{13}+G_{23}])}{c_1+c_2+c_3+c_1c_2\Delta_{12}+c_1c_3\Delta_{13}+c_2c_3\Delta_{23}+c_1c_2c_3\Delta_{123}} \quad (2)$$

where P is the pressure, T is the absolute temperature, x_2^t and $\gamma_{2,t}$ are the mole fraction and activity coefficient of a solute in a ternary mixture (in this paper, component 2 designates a solute), x_3^t is the mole fraction of component 3 (one of the solvents, with the other one being component 1) in a ternary mixture, c_α is the bulk molecular concentration of component α in a ternary mixture, and $\Delta_{\alpha\beta}$ and Δ_{123} are defined by

$$\Delta_{\alpha\beta} = G_{\alpha\alpha} + G_{\beta\beta} - 2G_{\alpha\beta}, \quad \alpha \neq \beta \quad (3)$$

---

* To whom correspondence should be addressed. Tel.: (716) 645-2911/ext. 2214. Fax: (716) 645-3822. E-mail: feaeliru@acsu.buffalo.edu.
† E-mail: ishulgin@eng.buffalo.edu.

and

$$\Delta_{123} = G_{11}G_{22} + G_{11}G_{33} + G_{22}G_{33} + 2G_{12}G_{13} + 2G_{12}G_{23} + 2G_{13}G_{23} - G_{12}^2 - G_{13}^2 - G_{23}^2 - 2G_{11}G_{23} - 2G_{22}G_{13} - 2G_{33}G_{12} \quad (4)$$

The factors in the square brackets in the numerator of eq 2 and Δ_{123} can be expressed in terms of $\Delta_{\alpha\beta}$ as follows:

$$G_{12} + G_{33} - G_{13} - G_{23} = \frac{\Delta_{13} + \Delta_{23} - \Delta_{12}}{2} \quad (5)$$

$$G_{11} + G_{23} - G_{12} - G_{13} = \frac{\Delta_{12} + \Delta_{13} - \Delta_{23}}{2} \quad (6)$$

and

$$\Delta_{123} = -\frac{\Delta_{12}^2 + \Delta_{13}^2 + \Delta_{23}^2 - 2\Delta_{12}\Delta_{13} - 2\Delta_{12}\Delta_{23} - 2\Delta_{13}\Delta_{23}}{4} \quad (7)$$

The insertion of eqs 5–7 into eq 2 provides an expression for the derivative $(\partial \ln \gamma_{2,t}/\partial x_3^t)_{T,P,x_2^t}$ in terms of $\Delta_{\alpha\beta}$ and concentrations.

It should be noted that $\Delta_{\alpha\beta}$ is a measure of the nonideality[4] of the binary mixture α–β because for an ideal mixture $\Delta_{\alpha\beta} = 0$. For a ternary mixture 1–2–3, Δ_{123} also constitutes a measure of nonideality. Indeed, by inserting the KB integrals for an ideal ternary mixture[5] into the expression of Δ_{123}, one could obtain that for an ideal ternary mixture $\Delta_{123} = 0$.

Considering a solvent composed of components 1 and 3 as an ideal one ($\Delta_{13} = 0$), eq 2 at infinite dilution of a solute (component 2) leads to the relation

$$\lim_{x_2^t \to 0} \left(\frac{\partial \ln \gamma_{2,t}}{\partial x_3^t}\right)_{T,P,x_2^t} = \left(\frac{\partial \ln \gamma_2^{t,\infty}}{\partial x_3^t}\right)_{T,P,x_2^t=0} = -\frac{(c_1^0 + c_3^0)(\Delta_{12} - \Delta_{23})_{x_2^t=0}}{2} \quad (8)$$

where c_1^0 and c_3^0 are the bulk molecular concentrations of components 1 and 3 in the binary 1–3 solvent and $\gamma_2^{t,\infty}$ is the activity coefficient of a solute in a ternary mixture at infinite dilution.

When the factor $(\Delta_{12} - \Delta_{23})_{x_2^t=0}$ is considered to be constant and eq 8 is integrated, the following relation between $\gamma_2^{t,\infty}$ and the activity coefficients at infinite dilution of the solute in each of the constituents (1 and 3) of the solvent was obtained:

$$\ln \gamma_2^{t,\infty} = \frac{(\ln v - \ln V_3^0)\ln \gamma_2^{b_1,\infty} + (\ln V_1^0 - \ln v)\ln \gamma_2^{b_3,\infty}}{\ln V_1^0 - \ln V_3^0} \quad (9)$$

where V_1^0 and V_3^0 are the molar volumes of the individual solvents 1 and 3, $\gamma_2^{b_1,\infty}$ and $\gamma_2^{b_3,\infty}$ are the activity coefficients at infinite dilution of the solute in the pure solvents 1 and 3, respectively, and v is the molar volume of an ideal binary mixed solvent.

Equation 9 has proven to be useful in the representation of the gas solubility in a binary solvent,[6,7] the solubility of solids in a supercritical fluid (SCF) mixed with a gas or with another SCF,[8] and so on.

The aim of the present paper is (a) to derive relations for the activity coefficients in multicomponent mixtures in terms of the KB integrals, (b) to obtain on their basis an expression for the solubility of a solute in an ideal multicomponent solvent, (c) to use the obtained equations to predict the solubilities in real systems, and (d) to compare the predicted solubilities with the experimental ones.

Theory and Formulas

Expressions for the derivatives of the chemical potentials with respect to the number of particles, the partial molar volumes, and the isothermal compressibility were derived by Kirkwood and Buff[2] in compact matrix forms (see Appendix 1). The derivation of explicit expressions for the above quantities in multicomponent mixtures required an enormous number of algebraic transformations, which could be carried out by using a special algebraic software (Maple[9] 8 was used in the present paper). A full set of expressions for the derivatives of the chemical potentials with respect to the number of particles, the partial molar volumes, and the isothermal compressibilities in a quaternary mixture were derived. However, our main interest in this paper is related to the derivatives of the activity coefficient with respect to the mole fractions (all of the expressions for the derivatives of the chemical potentials with respect to the number of particles, the partial molar volumes, and the isothermal compressibility can be obtained from the authors at request), namely, the derivatives of the form $(\partial \ln \gamma_{2,q}/\partial x_3^q)_{T,P,x_2^q,x_4^q}$, where x_2^q, x_3^q, x_4^q, and $\gamma_{2,q}$ are the mole fractions of components 2–4 and the solute activity coefficient, respectively, in the quaternary mixture. The above derivative under isothermal–isobaric conditions could be obtained from those of the chemical potential with respect to the number of particles:

$$\left(\frac{\partial \mu_2}{\partial x_3^q}\right)_{T,P,x_2^q,x_4^q} = N\left(\frac{\partial \mu_2}{\partial N_3}\right)_{T,P,N_{\gamma \neq 3}} - N\left(\frac{\partial \mu_2}{\partial N_1}\right)_{T,P,N_{\gamma \neq 1}} \quad (10)$$

where $N = N_1 + N_2 + N_3 + N_4$, with N_i ($i = 1$–4) the number of particles of species i, $i = 2$ being the solute and 1, 3, and 4 the components of the solvent.

The final expression for $(\partial \ln \gamma_{2,q}/\partial x_3^q)_{T,P,x_2^q,x_4^q}$ has the following form:

$$\left(\frac{\partial \ln \gamma_{2,q}}{\partial x_3^q}\right)_{T,P,x_2^q,x_4^q} = -\frac{(c_1 + c_2 + c_3 + c_4)(c_1 h_1 + c_3 h_3 + c_1 c_4 h_{14} + c_3 c_4 h_{34})}{c_1 + c_2 + c_3 + c_4 + \tau_2 + \tau_3 + c_1 c_2 c_3 c_4 \Delta_{1234}} \quad (11)$$

where h_1, h_3, h_{14}, h_{34}, τ_2, τ_3, and Δ_{1234} are defined in Appendix 2.

One can demonstrate that for an ideal quaternary mixture $\Delta_{1234} = 0$ and hence that Δ_{1234} is a measure of the nonideality of the quaternary mixture 1–2–3–4. The examination of eq 11 reveals that the derivative $(\partial \ln \gamma_{2,q}/\partial x_3^q)_{T,P,x_2^q,x_4^q}$ depends on compositions and the parameters that characterize the degrees of nonidealities Δ_{1234}, Δ_{ijk}, and Δ_{ij}. Moreover, the parameters Δ_{1234}

Table 1. Information about Mixed and Individual Solvents

mixed and individual solvents	composition (mole fractions)	molar volume	activity coefficient of a solute at infinite dilution	comments
ternary 1–3–4	x_1^t, x_3^t, x_4^t	$V = x_1^t V_1^0 + x_3^t V_3^0 + x_4^t V_4^0$	$\gamma_2^{q,\infty}$	eq 19, Figure 1
binary 1–3		v	$\gamma_2^{t,\infty}$	eq 9
binary 1–4	$x_1^{b_1} = 1 - x_4^t, x_4^{b_1} = x_4^t$	$V_{b,1} = x_1^{b_1} V_1^0 + x_4^{b_1} V_4^0$	$\gamma_2^{t_1,\infty}$	eq 16, Figure 1
binary 3–4	$x_3^{b_3} = 1 - x_4^t, x_4^{b_3} = x_4^t$	$V_{b,3} = x_3^{b_3} V_3^0 + x_4^{b_3} V_4^0$	$\gamma_2^{t_3,\infty}$	eq 15, Figure 1
binary 3–4	x_3^{bi}, x_4^{bi} (eq 19)	$V_{bi} = x_3^{bi} V_3^0 + x_4^{bi} V_4^0$	$\gamma_2^{bi,\infty}$	eq 20, Figure 1
individual solvent 1		V_1^0	$\gamma_2^{b_1,\infty}$	eq 9
individual solvent 3		V_3^0	$\gamma_2^{b_3,\infty}$	eq 9
individual solvent 4		V_4^0		

and Δ_{ijk} can be expressed in terms of Δ_{ij}, as was shown before for ternary mixtures (see eq 7). Thus, the KB expression for the derivative $(\partial \ln \gamma_{2,q}/\partial x_3^q)_{T,P,x_2^q,x_4^t}$, which depends for quaternary solutions on 10 KB integrals G_{ij}, is replaced by eq 11, which contains only six parameters Δ_{ij}. Furthermore, being measures of the nonidealities, the parameters Δ_{ij} have a clear physical meaning, and this helps in the analysis of the thermodynamics of multicomponent mixtures.

At infinite dilution of the solute ($x_2^q \to 0$), eq 11 acquires the following form:

$$\left(\frac{\partial \ln \gamma_2^{q,\infty}}{\partial x_3^q}\right)_{T,P,x_2^q=0,x_4^t} = -\frac{(c_1^t + c_3^t + c_4^t)(c_1 h_1 + c_3 h_3 + c_1 c_4 h_{14} + c_3 c_4 h_{34})_{x_2^q=0}}{c_1^t + c_3^t + c_4^t + (\tau_2 + \tau_3)_{x_2^q=0}} \quad (12)$$

where $c_i^t = \lim_{x_2^q \to 0} c_i$ and $\gamma_2^{q,\infty} = \lim_{x_2^q \to 0} \gamma_{2,q}$.

By considering the ternary solvent (1–3–4) an ideal solution, hence that $\Delta_{13} = 0$, $\Delta_{14} = 0$, and $\Delta_{34} = 0$, and taking into account eqs 7 and A2-1–A2-6 in Appendix 2, eq 12 can be recast in the more simple form

$$\left(\frac{\partial \ln \gamma_2^{q,\infty}}{\partial x_3^q}\right)_{T,P,x_2^q=0,x_4^t} = -\frac{(c_1^t + c_3^t + c_4^t)(\Delta_{12} - \Delta_{23})_{x_2^q=0}}{2} \quad (13)$$

Equation 13 has the same form as expression (8) for the derivative $(\partial \ln \gamma_2^{t,\infty}/\partial x_3^t)_{T,P,x_2^t}$ in a ternary solution.

Equation 13 can be used to derive an expression for the activity coefficient $\gamma_2^{q,\infty}$ in a manner similar to that employed for the binary mixed-solvent case. Considering the factor $(\Delta_{12} - \Delta_{23})_{x_2^q=0}$ to be constant and integrating eq 13, one obtains the following relation:

$$(\ln \gamma_2^{q,\infty})_{x_4^t} = -\left(\frac{B \ln V}{V_3^0 - V_1^0}\right)_{x_4^t} + A \quad (14)$$

where $B = (\Delta_{12} - \Delta_{23})_{x_2^q=0}/2$, A is a constant of integration, and $V = x_1^t V_1^0 + x_3^t V_3^0 + x_4^t V_4^0$ is the molar volume of the ternary mixed solvent 1–3–4 (for an explanation of the notations employed, see Table 1, which provides information about the activity coefficients of a solute in mixed and individual solvents and about the compositions and molar volumes of mixed and individual solvents). The constants A and B can be determined from two limiting conditions (see Figure 1):

(1) $x_1^t = 0$. In this case, the ternary mixed solvent becomes a binary one composed of components 3 and 4,

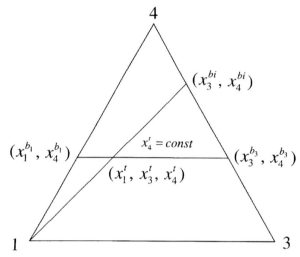

Figure 1. Various representations of a ternary mixed solvent with mole fractions x_1^t, x_3^t, and x_4^t by the combination of binaries or the combination of one pure solvent and one binary mixture.

with mole fractions $x_3^{b_3} = 1 - x_4^t$ and $x_4^{b_3} = x_4^t$ and a molar volume $V_{b,3} = x_3^{b_3} V_3^0 + x_4^{b_3} V_4^0$, and the activity coefficient of a solute in a ternary mixed solvent becomes equal to that in the binary (3–4) mixed solvent at infinite dilution $(\gamma_2^{t_3,\infty})$

$$(\gamma_2^{q,\infty})_{x_4^t,x_1^t=0} = \gamma_2^{t_3,\infty} \quad (15)$$

(2) $x_3^t = 0$. In this case, the ternary mixed solvent becomes a binary one composed of components 1 and 4, with mole fractions $x_1^{b_1} = 1 - x_4^t$ and $x_4^{b_1} = x_4^t$ and molar volume $V_{b,1} = x_1^{b_2} V_1^0 + x_4^{b_1} V_4^0$, and the activity coefficient of a solute in a ternary mixed solvent becomes equal to that in the binary (1–4) mixed solvent at infinite dilution $(\gamma_2^{t_1,\infty})$

$$(\gamma_2^{q,\infty})_{x_4^t,x_3^t=0} = \gamma_2^{t_1,\infty} \quad (16)$$

Consequently, eq 14 can be rewritten in the form

$$(\ln \gamma_2^{q,\infty})_{x_4^t} = \frac{(\ln V - \ln V_{b,3}) \ln \gamma_2^{t_1,\infty} + (\ln V_{b,1} - \ln V) \ln \gamma_2^{t_3,\infty}}{\ln V_{b,1} - \ln V_{b,3}} \quad (17)$$

Of course, the same procedure can be applied to $(\partial \ln \gamma_{2,q}/\partial x_3^q)_{T,P,x_2^q,x_1^t}$ with x_1^q = constant, instead of x_4^q = constant, and so on.

Equation 17 can also be derived using a more simple procedure. One can consider the above two mixed solvents (1–4 with mole fractions $x_1^{b_1} = 1 - x_4^t$ and $x_4^{b_1} = $

x_4^t and molar volume $V_{b,1}$, and 3–4 with mole fractions $x_3^{b_3} = 1 - x_4^t$ and $x_4^{b_3} = x_4^t$ and molar volume $V_{b,3}$ as pseudo pure components. These two "components" form an ideal pseudobinary (see Figure 1). However, in this case, one can use eq 9.

$$\ln \gamma_2^{t,\infty} = \frac{(\ln v - \ln V_3^0) \ln \gamma_2^{b_1,\infty} + (\ln V_1^0 - \ln v) \ln \gamma_2^{b_3,\infty}}{\ln V_1^0 - \ln V_3^0} \quad (9A)$$

Because in this case $\gamma_2^{t,\infty} = \gamma_2^{q,\infty}$, $\gamma_2^{b_i,\infty} = \gamma_2^{t_i,\infty}$, $v = V$, and $V_i^0 = V_{b,i}$ ($i = 1$ and 3), eq 9A leads to eq 17.

The above ideal ternary mixture with mole fractions x_1^t, x_3^t, and x_4^t and molar volume V can also be obtained as an ideal pseudobinary one by mixing the pure solvent 1 and the binary mixture with molar fractions x_3^{bi} and x_4^{bi} and molar volume V_{bi} (see Figure 1), which can be considered as a pseudo pure component. The mole fractions x_3^{bi} and x_4^{bi} can be calculated as follows. If one denotes

$$\alpha = x_3^t / x_4^t \quad (18)$$

then one can write the following relations:

$$x_3^{bi} = \frac{\alpha}{1+\alpha}$$
$$x_4^{bi} = \frac{1}{1+\alpha} \quad (19)$$

Using eq 9, the expression for the activity coefficient in an ideal ternary solvent can be written as

$$(\ln \gamma_2^{q,\infty})_{x_4^q} = \frac{(\ln V - \ln V_{bi}) \ln \gamma_2^{b_1,\infty} + (\ln V_1^0 - \ln V) \ln \gamma_2^{bi,\infty}}{\ln V_1^0 - \ln V_{bi}} \quad (20)$$

A similar procedure can be employed for the prediction of the activity coefficient of a solute in any ideal n-component mixed solvent, namely, (1) the n-component ideal mixed solvent can be represented by two ($n - 1$)-component ideal mixed solvents or by one pure solvent and a ($n - 1$)-component ideal mixed solvent, (2) the ($n - 1$)-component ideal mixed solvent can be represented by two ($n - 2$)-component ideal mixed solvents or by one pure solvent and a ($n - 2$)-component ideal mixed solvent, and so on.

One can see from eqs 17 and 20 that the activity coefficient of a solute at infinite dilution in an ideal ternary mixed solvent ($\gamma_2^{q,\infty}$) can be calculated in terms of the activity coefficients of that solute at infinite dilution in any two binaries of the solvent and their molar volumes or in terms of the activity coefficients of the solute at infinite dilution in one binary solvent and in the remaining individual solvent and their molar volumes. The activity coefficients of a solute at infinite dilution in a binary mixed solvent can be obtained experimentally or calculated. For instance, they can be calculated using eq 9 and, in this case, the activity coefficient of a solute at infinite dilution in an ideal ternary mixed solvent ($\gamma_2^{q,\infty}$) can be predicted from the activity coefficients of the solute at infinite dilution in the individual solvents (components 1, 3, and 4) and their molar volumes.

Applications

The obtained equations can be applied to numerous cases involving multicomponent solutions. In the present paper, we examine the gas solubilities and the solubilities of solid nonelectrolytes in multicomponent mixed solvents.

Gas Solubility. In this case, an expression for the composition dependence of the Henry constant will first be obtained. The Henry constants and the activity coefficients of a solute in a ternary mixed solvent (1–3–4) with mole fractions x_1^t, x_3^t, and x_4^t and in two binary mixed solvents (1–4 and 3–4) with compositions ($x_1^{b_1} = 1 - x_4^t$, $x_4^{b_1} = x_4^t$) and ($x_3^{b_3} = 1 - x_4^t$, $x_4^{b_3} = x_4^t$) (see Figure 1) are related via the expressions[10]

$$\ln H_{2,q} = \ln \gamma_2^{q,\infty} + \ln f_2^0(P,T) \quad (21)$$

$$\ln H_{2,t_1} = \ln \gamma_2^{t_1,\infty} + \ln f_2^0(P,T) \quad (22)$$

and

$$\ln H_{2,t_3} = \ln \gamma_2^{t_3,\infty} + \ln f_2^0(P,T) \quad (23)$$

where $H_{2,q}$, H_{2,t_1}, and H_{2,t_3} are the Henry constants in a ternary mixed solvent (1–3–4) with mole fractions x_1^t, x_3^t, and x_4^t, and in two binary mixed solvents (1–4) and (3–4) with compositions ($x_1^{b_1} = 1 - x_4^t$, $x_4^{b_1} = x_4^t$) and ($x_3^{b_3} = 1 - x_4^t$, $x_4^{b_3} = x_4^t$), respectively (see Figure 1), and $f_2^0(P,T)$ is the fugacity of component 2.

Inserting expressions (21)–(23) into eq 17 leads to the following expression relating the Henry constant in a ternary mixed solvent to those in two binaries of the solvent constituents and their molar volumes:

$$(\ln H_{2,q})_{x_4^q} = \frac{(\ln V - \ln V_{b,3}) \ln H_{2,t_1} + (\ln V_{b,1} - \ln V) \ln H_{2,t_3}}{\ln V_{b,1} - \ln V_{b,3}} \quad (24)$$

Similarly, the Henry constant in a ternary mixed solvent can be related to those in one binary mixed solvent, for instance, 3–4 (of course, any other pair can be taken) and in the remaining individual component (1) and their molar volumes:

$$\ln H_{2,q} = \frac{(\ln V - \ln V_{bi}) \ln H_{2,1} + (\ln V_1^0 - \ln V) \ln H_{2,bi}}{\ln V_1^0 - \ln V_{bi}} \quad (25)$$

where $H_{2,bi}$ is the Henry constant in a 3–4 mixed solvent with molar fractions x_3^{bi} and x_4^{bi} and molar volume V_{bi} (see Figure 1).

Equations 24 and 25 can be used to predict the Henry constant in a ternary mixed solvent in two different ways:

Table 2. Comparison between Predicted and Experimental Henry's Constants at $T = 298.15$ K and Atmospheric Pressure for the Solubilities of Ethane in the Ternary Mixed Solvent Acetone (1)−Methanol (3)−Water (4)

composition (gas-free) of mixed solvent (mole fraction)			Henry's constant (bar)		
x_1	x_3	expt[11]	eq 24 with Henry's constants for two binaries (calculated using eqs 26 and 27 which involve x_4^t = constant)	eq 25 with Henry's constants for the methanol/water binary (calculated using eq 28) and acetone	eq 24 with Henry's constants for two binaries (1−4 and 3−4) provided by experiment
0.3318	0.3367	348.49	370.69	370.56	368.32
0.1006	0.8022	270.16	271.15	271.03	281.13
0.1981	0.6005	309.11	313.92	313.6	327.7
0.7983	0.1028	124.72	130.18	130.43	125.84
0.6129	0.1928	177.65	189.58	189.92	181.07
0.2019	0.2009	1058.3	1034.1	1033.3	1040.8
0.0995	0.1043	3587.2	2849.9	2848.3	3520.9
deviation[a]			(a) 6.2 (b) 6.1 (c) 6.3	6.1	(a) 3.2

[a] Deviation (%) between the predicted and experimental Henry's constants is defined as $\{100\sum_{i=1}^{m}[(H_{2,q(\text{exp})}^{(i)} - H_{2,q(\text{calc})}^{(i)})/H_{2,q(\text{exp})}^{(i)}]\}/m$, where m is the number of experimental points, (a) x_4^t = constant, (b) x_3^t = constant, and (c) x_1^t = constant.

Table 3. Comparison between Predicted and Experimental Henry's Constants at $T = 298.15$ K and Atmospheric Pressure for the Solubilities of Ethylene in the Ternary Mixed Solvent Acetone (1)−Methanol (3)−Water (4)

composition (gas-free) of mixed solvent (mole fraction)			Henry's constant (bar)		
x_1	x_3	expt[11]	eq 24 with Henry's constants for two binaries (calculated using eqs 26 and 27 which involve x_4^t = constant)	eq 25 with Henry's constants for the methanol/water binary (calculated using eq 28) and acetone	eq 24 with Henry's constants for two binaries (1−4 and 3−4) provided by experiment
0.3318	0.3367	440.24	495.21	489.92	459.25
0.1006	0.8022	301.44	307.47	306.82	317.36
0.1981	0.6005	363.1	382.9	380.4	390.5
0.7983	0.1028	144.06	152.44	152.20	141.84
0.6129	0.1928	209.54	233.26	232.05	209.10
0.2019	0.2009	1520.8	1685.5	1668.6	1504.4
0.0995	0.1043	6549.3	5623.3	5595.7	6264.9
deviation[a]			(a) 8.9 (b) 8.4 (c) 9.1	8.3	(a) 3.5

[a] Deviation (%) between the predicted and experimental Henry's constants is defined as $\{100\sum_{i=1}^{m}[(H_{2,q(\text{exp})}^{(i)} - H_{2,q(\text{calc})}^{(i)})/H_{2,q(\text{exp})}^{(i)}]\}/m$, where m is the number of experimental points, (a) x_4^t = constant, (b) x_3^t = constant, and (c) x_1^t = constant.

(A) H_{2,t_1}, H_{2,t_3}, and $H_{2,\text{bi}}$ can be calculated from the Henry constants in the pure solvents (1, 3, and 4) as suggested before:[6]

$$\ln H_{2,t_1} = \frac{(\ln V_{b,1} - \ln V_4^0)\ln H_{2,1} + (\ln V_1^0 - \ln V_{b,1})\ln H_{2,4}}{\ln V_1^0 - \ln V_4^0} \quad (26)$$

$$\ln H_{2,t_3} = \frac{(\ln V_{b,3} - \ln V_4^0)\ln H_{2,3} + (\ln V_3^0 - \ln V_{b,3})\ln H_{2,4}}{\ln V_3^0 - \ln V_4^0} \quad (27)$$

and

$$\ln H_{2,\text{bi}} = \frac{(\ln V_{\text{bi}} - \ln V_4^0)\ln H_{2,3} + (\ln V_3^0 - \ln V_{\text{bi}})\ln H_{2,4}}{\ln V_3^0 - \ln V_4^0} \quad (28)$$

where $H_{2,1}$, $H_{2,3}$, and $H_{2,4}$ are the Henry constants in the individual solvents 1, 3, and 4.

(B) H_{2,t_1} and H_{2,t_3} can be determined from experimental data.

We applied both techniques to the solubilities of ethane and ethylene in the ternary mixed solvent acetone (1)−methanol (3)−water (4).[11] Information about the solubilities of ethane and ethylene in the corresponding binary mixed solvents and individual solvents was also found in the same publication.[11] The results of the solubility predictions are summarized in Tables 2 and 3.

Solubility of a Solid. For the solubilities of poorly soluble crystalline nonelectrolytes in a multicomponent mixed solvent, one can use the infinite-dilution approximation and consider that the activity coefficient of a solute in a mixed solvent is equal to the activity coefficient at infinite dilution. Therefore, one can write the following relations for the solubility of a poorly soluble crystalline nonelectrolyte in a ternary mixed solvent and in two of its binaries:[12,13]

$$f_2^S/f_2^L(T,P) = x_2^q \gamma_2^{q,\infty} \quad (29)$$

$$f_2^S/f_2^L(T,P) = x_2^{t_1} \gamma_2^{t_1,\infty} \quad (30)$$

and

$$f_2^S/f_2^L(T,P) = x_2^{t_3} \gamma_2^{t_3,\infty} \quad (31)$$

where $\gamma_2^{q,\infty}$, $\gamma_2^{t_1,\infty}$, and $\gamma_2^{t_3,\infty}$ are the activity coefficients at infinite dilution of the solute in a ternary mixed solvent (1−3−4) with mole fractions x_1^t, x_3^t, and x_4^t and in two binary mixed solvents (1−4) and (3−4) with compositions ($x_1^{b_1} = 1 - x_4^t$, $x_4^{b_1} = x_4^t$) and ($x_3^{b_3} = 1 - x_4^t$, $x_4^{b_3} = x_4^t$),

Table 4. Comparison between Predicted and Experimental Solubilities at $T = 298.15$ K and Atmospheric Pressure of Anthracene in Ternary Mixed Solvent 1-Propanol (1)−2-Propanol (3)−Cyclohexane (4)

composition (solute-free) of mixed solvent (mole fraction)			mole fraction solubilities		
x_1	x_3	expt[14]	eq 32, which involves x_4^t = constant, with the solubilities of anthracene in two binaries predicted from those in the individual solvents[15]	eq 33 with the solubilities of anthracene in one binary (3−4) (predicted from those in the individual solvents[15]) and component 1	eq 32 with solubilities of anthracene in two binaries (1−4 and 3−4) provided by experiment[16−18]
0.3804	0.357	0.000 897	0.000 697	0.000 697	0.000 884
0.165	0.722	0.000 61	0.000 52	0.000 52	0.000 604
0.299	0.2887	0.001 126	0.000 834	0.000 833	0.001 103
0.2773	0.5317	0.000 761	0.000 607	0.000 607	0.000 749
0.7297	0.117	0.000 782	0.000 674	0.000 674	0.000 779
0.7111	0.2081	0.000 661	0.000 602	0.000 601	0.000 658
0.2164	0.7038	0.000 567	0.000 503	0.000 503	0.000 562
0.1313	0.552	0.000 924	0.000 69	0.000 69	0.000 925
0.4179	0.5031	0.000 604	0.000 54	0.000 54	0.000 597
0.5153	0.4032	0.000 617	0.000 561	0.000 561	0.000 62
0.1946	0.188	0.001 395	0.001 05	0.001 05	0.001 355
0.735	0.1554	0.000 709	0.000 633	0.000 633	0.000 71
0.5465	0.2649	0.000 812	0.000 665	0.000 665	0.000 798
0.1323	0.249	0.001 389	0.001 031	0.001 031	0.001 351
0.1167	0.7275	0.000 664	0.000 545	0.000 545	0.000 662
0.2603	0.1395	0.001 389	0.001 051	0.001 05	0.001 342
0.1342	0.4639	0.001 053	0.000 777	0.000 777	0.001 06
0.4666	0.1318	0.001 107	0.000 87	0.000 87	0.001 115
0.556	0.1282	0.000 99	0.000 798	0.000 798	0.001 003
deviation[a]			(a) 18.5 (b) 18.5 (c) 18.5	18.5	(a) 1.2

[a] Deviation (%) between the predicted and experimental solubilities of anthracene is defined as $\{100\sum_{i=1}^{m}\{[(x_2^q)_{\text{exp}}^{(i)} - (x_2^q)_{\text{calc}}^{(i)}]/(x_2^q)_{\text{exp}}^{(i)}\}\}m$, where m is the number of experimental points, (a) x_4^t = constant, (b) x_3^t = constant, and (c) x_1^t = constant.

Table 5. Comparison between Predicted and Experimental Solubilities at $T = 298.15$ K and Atmospheric Pressure of Anthracene in Ternary Mixed Solvent 1-Butanol (1)−2-Butanol (3)−Cyclohexane (4)

composition (solute-free) of mixed solvent (mole fractions)			mole fraction solubilities		
x_1	x_3	expt[14]	eq 32, which involves x_4^t = constant, with the solubilities of anthracene in two binaries predicted from those in the individual solvents[15]	eq 33 with the solubilities of anthracene in one binary (3−4) (predicted from those in the individual solvents[15]) and component 1	eq 32 with solubilities of anthracene in two binaries (1−4 and 3−4) provided by experiment[16−18]
0.3531	0.3475	0.001 06	0.000 892	0.000 892	0.001 079
0.1557	0.7135	0.000 806	0.000 706	0.000 706	0.000 803
0.2727	0.2707	0.001 284	0.001 019	0.001 019	0.001 267
0.2623	0.5162	0.000 938	0.000 801	0.000 801	0.000 955
0.7231	0.1023	0.000 975	0.000 881	0.000 88	0.000 978
0.7134	0.1988	0.000 849	0.000 803	0.000 803	0.000 856
0.2038	0.7107	0.000 738	0.000 683	0.000 683	0.000 746
0.1126	0.5298	0.001 104	0.000 879	0.000 879	0.001 125
0.4087	0.5061	0.000 792	0.000 728	0.000 728	0.000 786
0.5102	0.4045	0.000 789	0.000 751	0.000 751	0.000 808
0.1728	0.1666	0.001 52	0.001 206	0.001 206	0.001 453
0.7165	0.1532	0.000 909	0.000 84	0.000 839	0.000 916
0.5219	0.2567	0.000 991	0.000 868	0.000 868	0.001 002
0.1154	0.2233	0.001 476	0.001 186	0.001 186	0.001 457
0.1074	0.7174	0.000 846	0.000 728	0.000 728	0.000 86
0.2269	0.1127	0.001 508	0.001 225	0.001 225	0.001 449
0.114	0.4305	0.001 248	0.000 97	0.000 97	0.001 253
0.4313	0.1126	0.001 307	0.001 068	0.001 068	0.001 278
0.5324	0.1103	0.001 188	0.000 999	0.000 999	0.001 179
deviation[a]			(a) 14.2 (b) 14.2 (c) 14.2	14.2	(a) 1.5

[a] Deviation (%) between the predicted and experimental solubilities of anthracene is defined as $\{100\sum_{i=1}^{m}\{[(x_2^q)_{\text{exp}}^{(i)} - (x_2^q)_{\text{calc}}^{(i)}]/(x_2^q)_{\text{exp}}^{(i)}\}\}m$, where m is the number of experimental points, (a) x_4^t = constant, (b) x_3^t = constant, and (c) x_1^t = constant.

and x_2^q and $x_2^{t_1}$, $x_2^{t_3}$ are the solubilities (mole fractions) of a poorly soluble crystalline nonelectrolyte in the ternary and binary mixed solvents, $f_2^L(T,P)$ is a hypothetical fugacity of a solid as a (subcooled) liquid at a given pressure (P) and temperature (T), and f_2^S is the fugacity of a pure solid component 2. If the solubilities of the pure and mixed solvents in the solid phase are negligible, then the left-hand sides of eqs 29−31 depend only on the properties of the solute. By inserting into eq 17 the expressions of the activity coefficients from eqs 29−31, one obtains an expression for the solubility (mole fractions) of a poorly soluble solid nonelectrolyte in a ternary mixed solvent in terms of those in two of its binary mixed solvents:

$$(\ln x_2^q)_{x_4^q} = \frac{(\ln V - \ln V_{b,3})\ln x_2^{t_1} + (\ln V_{b,1} - \ln V)\ln x_2^{t_3}}{\ln V_{b,1} - \ln V_{b,3}} \quad (32)$$

As for the gas solubility, the above equations can be used to calculate the solubility (a) from the solubilities in the pure solvents (1, 3, and 4) and (b) from experimental solubilities in two binary solvents.

The solubility of a poorly soluble crystalline nonelectrolyte in a ternary mixed solvent can also be calculated from those in one binary mixed solvent, for example, the 3–4 binary, and the remaining individual component and their molar volumes (of course, any other pair can also be selected):

$$\ln x_2^q = \frac{(\ln V - \ln V_{bi}) \ln x_2^{b1} + (\ln V_1^0 - \ln V) \ln x_2^{bi}}{\ln V_1^0 - \ln V_{bi}} \quad (33)$$

where x_2^{bi} is the solubility of the solute in the 3–4 mixed solvent with molar fractions x_3^{bi} and x_4^{bi} and x_2^{b1} is the solubility of the solute in the pure solvent 1.

We applied both techniques to two systems:[14] (a) the solubility of anthracene in 1-propanol (1)–2-propanol (3)–cyclohexane (4); (b) the solubility of anthracene in 1-butanol (1)–2-butanol (3)–cyclohexane (4). The results of the solubility predictions are summarized in Tables 4 and 5.

The results presented in Tables 2–5 show that the approach suggested in the present paper can be successfully applied to the gas solubilities and the solubilities of crystalline nonelectrolytes in multicomponent mixed solvents. The solubilities of a solute in ternary mixed solvents were predicted either from those in pure solvents or from those in two binaries of the solvent components. The predictions from the solubilities in pure solvents were satisfactory (more satisfactory in the case of gas solubilities), and the predictions were excellent when binary data were used. Even though the solvent is assumed to be an ideal mixture, such an agreement is not unexpected. Indeed, as previously emphasized,[6,15] in such highly asymmetrical systems such as a gas + liquid mixed solvent or a crystalline nonelectrolyte + liquid mixed solvent and at high dilution, the nonidealities of the pairs solute + individual solvents are much higher that those between the components of the mixed solvent.

Conclusion

In this paper, the fluctuation theory of solutions was applied to a quaternary mixture and rigorous expressions for the derivatives of the chemical potentials with respect to the number of particles, the partial molar volumes, and the isothermal compressibility were derived. Expressions for the derivatives of the activity coefficients of a solute with respect to the concentrations of the solvents in an ideal ternary mixed solvent were obtained from the derivatives of the chemical potentials with respect to the number of particles. Using the obtained expressions, an equation was derived for the activity coefficient of a solute in an ideal n-component mixed solvent in terms of those in constituent subsystems formed either of the individual solvents or of two of the binaries among the three components of the solvent and their molar volumes. Examples regarding the gas solubilities and the solubilities of crystalline nonelectrolytes in multicomponent mixed solvents were examined for illustration purposes. The comparison with the experiment revealed, in general, a good agreement and, particularly, an excellent agreement between the experimental solubilities and those predicted from accurate data for the solubilities in binary solvent mixtures.

Acknowledgment

We are indebted to Prof. W. E. Acree, Jr. (Department of Chemistry, University of North Texas, Denton, TX), for helpful discussion of some results of this paper.

Appendix 1

Kirkwood and Buff[2] expressed the concentration derivatives of the chemical potentials, the partial molar volumes, and the isothermal compressibility in compact matrix forms as follows:

$$\frac{1}{kT}\left(\frac{\partial \mu_\alpha}{\partial N_\beta}\right)_{T,V,N_{\gamma \neq \beta}} = \frac{|B|_{\alpha\beta}}{v|B|} \quad (A1\text{-}1)$$

$$v_\alpha = \frac{\sum_{\beta=1}^{n} c_\beta |B|_{\alpha\beta}}{\sum_{\beta,\gamma=1}^{n} c_\beta c_\gamma |B|_{\beta\gamma}} \quad (A1\text{-}2)$$

and

$$k_T kT = \frac{|B|}{\sum_{\alpha,\beta=1}^{n} c_\alpha c_\beta |B|_{\alpha\beta}} \quad (A1\text{-}3)$$

In eqs A1-1–A1-3, k is the Boltzmann constant, T is the absolute temperature, N_β is the number of particles of species β in the volume v, μ_α is the chemical potential per molecule of species α, v_α is the partial molar volume per molecule of species α, k_T is the isothermal compressibility, and c_α is the bulk molecular concentration of component α ($c_\alpha = N_\alpha/v$). The derivative $(\partial \mu_\alpha/\partial N_\beta)_{T,V,N_{\gamma \neq \beta}}$ is taken under isothermal–isochoric conditions and with N_γ = constant for any $\gamma \neq \beta$. $|B|_{\alpha\beta}$ represents the cofactor of $B_{\alpha\beta}$ of the determinant $|B|$. For $B_{\alpha\beta}$ the following equation can be written:

$$B_{\alpha\beta} = c_\alpha \delta_{\alpha\beta} + c_\alpha c_\beta G_{\alpha\beta} \quad (A1\text{-}4)$$

where $\delta_{\alpha\beta}$ is the Kroneker delta ($\delta_{\alpha\beta} = 1$ for $\alpha = \beta$, and $\delta_{\alpha\beta} = 0$ for $\alpha \neq \beta$) and $G_{\alpha\beta}$ is the KB integral.

Instead of the isothermal–isochoric derivative of the chemical potential $(\partial \mu_\alpha/\partial N_\beta)_{T,V,N_{\gamma \neq \beta}}$, one can introduce in eq A1-1 the isothermal–isobaric derivative $(\partial \mu_\alpha/\partial N_\beta)_{T,P,N_{\gamma \neq \beta}}$. The two derivatives are related via the expression.[2]

$$\left(\frac{\partial \mu_\alpha}{\partial N_\beta}\right)_{T,V,N_{\gamma \neq \beta}} = \left(\frac{\partial \mu_\alpha}{\partial N_\beta}\right)_{T,P,N_{\gamma \neq \beta}} + \frac{v_\alpha v_\beta}{k_T v} \quad (A1\text{-}5)$$

One can see from eqs A1-1–A1-3 that the thermodynamic quantities $(\partial \mu_\alpha/\partial N_\beta)_{T,V,N_{\gamma \neq \beta}}$ or $(\partial \mu_\alpha/\partial N_\beta)_{T,P,N_{\gamma \neq \beta}}$, v_α, and k_T for any n-component mixture can be expressed in terms of the mixture composition and the KB integrals $G_{\alpha\beta}$.

Appendix 2

The aim of Appendix 2 is to provide the expressions for h_1, h_3, h_{14}, h_{34}, τ_2, τ_3, and Δ_{1234} of eq 11. They are

$$h_1 = \frac{\Delta_{12} + \Delta_{13} - \Delta_{23}}{2} \quad \text{(A2-1)}$$

$$h_3 = \frac{\Delta_{12} - \Delta_{13} - \Delta_{23}}{2} \quad \text{(A2-2)}$$

$$h_{14} = (-\Delta_{13}\Delta_{12} + \Delta_{13}\Delta_{24} + \Delta_{14}\Delta_{13} - 2\Delta_{14}\Delta_{23} + \Delta_{14}\Delta_{12} + \Delta_{34}\Delta_{12} - \Delta_{34}\Delta_{24} + \Delta_{14}\Delta_{24} + \Delta_{34}\Delta_{14} - \Delta_{14}^2)/4 \quad \text{(A2-3)}$$

$$h_{34} = (-\Delta_{34}\Delta_{23} - \Delta_{13}\Delta_{34} + \Delta_{13}\Delta_{23} - 2\Delta_{34}\Delta_{12} - \Delta_{13}\Delta_{24} - \Delta_{14}\Delta_{23} - \Delta_{34}\Delta_{24} + \Delta_{14}\Delta_{24} - \Delta_{34}\Delta_{14} + \Delta_{34}^2)/4 \quad \text{(A2-4)}$$

$$\tau_2 = c_1c_2\Delta_{12} + c_1c_3\Delta_{13} + c_1c_4\Delta_{14} + c_2c_3\Delta_{23} + c_2c_4\Delta_{24} + c_3c_4\Delta_{34} \quad \text{(A2-5)}$$

$$\tau_3 = c_1c_2c_3\Delta_{123} + c_1c_2c_4\Delta_{124} + c_1c_3c_4\Delta_{134} + c_2c_3c_4\Delta_{234} \quad \text{(A2-6)}$$

and

$$\begin{aligned}
\Delta_{1234} = &-2G_{11}G_{22}G_{34} - 2G_{12}G_{23}G_{14} - G_{24}^2G_{11} - \\
&G_{12}^2G_{44} + 2G_{24}G_{12}G_{14} + G_{11}G_{22}G_{44} - G_{14}^2G_{22} - \\
&2G_{24}G_{12}G_{13} + 2G_{24}G_{11}G_{23} + 2G_{12}^2G_{34} - \\
&2G_{24}G_{11}G_{33} - 2G_{12}G_{13}G_{34} - 2G_{13}G_{23}G_{14} + \\
&2G_{12}G_{13}G_{44} + 2G_{11}G_{23}G_{34} + 2G_{12}G_{33}G_{14} + \\
&2G_{13}^2G_{24} - 2G_{24}G_{13}G_{14} + 2G_{24}G_{11}G_{34} - \\
&2G_{11}G_{23}G_{44} - 2G_{12}G_{14}G_{34} - G_{13}^2G_{44} - G_{14}^2G_{33} + \\
&2G_{14}^2G_{23} - G_{34}^2G_{11} + 2G_{13}G_{14}G_{34} + 2G_{23}^2G_{14} + \\
&G_{11}G_{33}G_{44} + 2G_{24}G_{12}G_{33} + 2G_{22}G_{13}G_{34} - \\
&2G_{22}G_{33}G_{14} - 2G_{24}G_{13}G_{23} + 2G_{22}G_{14}G_{34} - \\
&2G_{24}G_{12}G_{34} + 2G_{12}G_{23}G_{44} - 2G_{22}G_{13}G_{44} - \\
&2G_{24}G_{23}G_{14} - 2G_{12}G_{23}G_{34} + 2G_{13}G_{23}G_{44} + \\
&2G_{24}G_{33}G_{14} - 2G_{24}G_{13}G_{34} + 2G_{34}^2G_{12} + 2G_{24}^2G_{13} - \\
&G_{24}^2G_{33} - 2G_{23}G_{14}G_{34} - G_{34}^2G_{22} - G_{23}^2G_{44} - \\
&2G_{12}G_{33}G_{44} + G_{22}G_{33}G_{44} + 2G_{24}G_{23}G_{34} - G_{12}^2G_{33} + \\
&2G_{12}G_{13}G_{23} - G_{13}^2G_{22} + 2G_{22}G_{13}G_{14} + G_{11}G_{22}G_{33} - \\
&G_{23}^2G_{11} \quad \text{(A2-7)}
\end{aligned}$$

Literature Cited

(1) Wohl, K. J. Thermodynamic evaluation of binary and ternary liquid systems. *Trans. Am. Inst. Chem. Eng.* **1946**, *42*, 215−249.

(2) Kirkwood, J. G.; Buff, F. P. Statistical mechanical theory of solutions. I. *J. Chem. Phys.* **1951**, *19*, 774−782.

(3) Ruckenstein, E.; Shulgin, I. Entrainer effect in supercritical mixtures. *Fluid Phase Equilib.* **2001**, *180*, 345−359.

(4) Ben-Naim, A. Inversion of the Kirkwood−Buff theory of solutions: application to the water−ethanol system. *J. Chem. Phys.* **1977**, *67*, 4884−4890.

(5) Ruckenstein, E.; Shulgin, I. Effect of a third component on the interactions in a binary mixture determined from the fluctuation theory of solutions. *Fluid Phase Equilib.* **2001**, *180*, 281−297.

(6) Shulgin, I.; Ruckenstein, E. Henry's constant in mixed solvents from binary data. *Ind. Eng. Chem. Res.* **2002**, *41*, 1689−1694.

(7) Shulgin, I.; Ruckenstein, E. Prediction of the solubility of gases in binary polymer + solvent mixtures. *Polymer* **2003**, *44*, 901−907.

(8) Ruckenstein, E.; Shulgin, I. A simple equation for the solubility of a solid in a supercritical fluid cosolvent with a gas or another supercritical fluid. *Ind. Eng. Chem. Res.* **2003**, *42*, 1106−1110.

(9) Kofler, M. *Maple: An Introduction and Reference*; Addison-Wesley: Harlow, England; 1997.

(10) O'Connell, J. P. Molecular thermodynamics of gases in mixed solvents. *AIChE J.* **1971**, *17*, 658−663.

(11) Zeck, S.; Knapp, H. Solubilities of ethylene, ethane, and carbon dioxide in mixed solvents consisting of methanol, acetone, and water. *Int. J. Thermophys.* **1985**, *6*, 643−656.

(12) Acree, W. E. *Thermodynamic properties of nonelectrolyte solutions*; Academic Press: Orlando, FL, 1984.

(13) Prausnitz, J. M.; Lichtenthaler, R. N.; Gomes de Azevedo, E. *Molecular Thermodynamics of Fluid-Phase Equilibria*, 2nd ed.; Prentice-Hall: Englewood Cliffs, NJ, 1986.

(14) Deng, T. H.; Horiuchi, S.; De Fina, K. M.; Hernandez, C. E.; Acree, W. E. Solubility of anthracene in multicomponent solvent mixtures containing propanol, butanol, and alkanes. *J. Chem. Eng. Data* **1999**, *44*, 798−802.

(15) Ruckenstein, E.; Shulgin, I. Solubility of drugs in aqueous solutions. Part 1: Ideal mixed solvent approximation. *Int. J. Pharm.* **2003**, *258*, 193−201.

(16) Zvaigzne, A. I.; Teng, I. L.; Martinez, E.; Trejo, J.; Acree, W. E. Solubility of anthracene in binary alkane + 1-propanol and alkane + 1-butanol solvent mixtures. *J. Chem. Eng. Data* **1993**, *38*, 389−392.

(17) Zvaigzne, A. I.; Acree, W. E. Solubility of anthracene in binary alkane plus 2-butanol solvent mixtures. *J. Chem. Eng. Data* **1993**, *39*, 114−116.

(18) Acree, W. E.; Zvaigzne, A. I.; Tucker, S. A. Thermochemical investigations of hydrogen-bonded solutions—development of a predictive equation for the solubility of anthracene in binary hydrocarbon plus alcohol mixtures based upon mobile order theory. *Fluid Phase Equilib.* **1994**, *92*, 233−253.

Received for review April 29, 2003
Revised manuscript received June 30, 2003
Accepted July 2, 2003

IE030362Q

Fluid Phase Equilibria 260 (2007) 126–134

www.elsevier.com/locate/fluid

Solubility and local structure around a dilute solute molecule in an aqueous solvent: From gases to biomolecules

Ivan L. Shulgin*, Eli Ruckenstein*

Department of Chemical & Biological Engineering, State University of New York at Buffalo, Amherst, NY 14260, USA

Received 10 May 2006; received in revised form 26 March 2007; accepted 29 March 2007
Available online 4 April 2007

Abstract

A new approach regarding the solubility of various sparingly soluble solutes, such as proteins, drugs and gases, and the local structure around a solute molecule is presented. This approach is based on an expression for the activity coefficient derived through the Kirkwood–Buff fluctuation theory of solutions. First, an expression for the solubility of proteins in aqueous solutions in terms of the preferential binding parameter is derived and criteria for salting-out or salting-in by various cosolvents obtained. Second, the methodology developed for the solubility of proteins in water + cosolvent mixtures is extended to the solubility of sparingly soluble gases in the same kinds of solvents. The derived equation was successfully applied to the experimental data regarding the solubilities of oxygen, carbon dioxide and methane in water + sodium chloride. In addition, the excesses (or deficits) of water and sodium chloride molecules in the vicinity of a gas molecule have been calculated to conclude that the infinitely dilute solute gas molecules are preferentially hydrated.
© 2007 Elsevier B.V. All rights reserved.

Keywords: Solubility; Fluctuation theory; Solvation; Mixed solvent; Proteins; Gases

1. Introduction

It is a privilege to contribute to this volume of Fluid Phase Equilibria honoring Professor Jürgen Gmehling. One of us (ILS) had the good fortune to be a postdoctoral scientist in his Department at Oldenburg and had benefited from numerous discussions with him. One of the frequent topics of these discussions was the aqueous system, its properties and structure. The goal of our contribution to this volume of Fluid Phase Equilibria is to extend the treatment previously suggested by us for the solubility of proteins to the solubility of gases.

The overwhelming majority of liquids in nature are the aqueous mixed solvents, such as the sea and rain waters, the brine in the earth crust, the blood, the lymph, the industrial waters and so on. Therefore, the research on the solubilities of different types of solutes in the above mentioned multicomponent aqueous mixed solvents is of paramount importance in medicine, pharmaceutics, environmental science, industry, etc. One can list numerous relevant topics, such as the solubilities of gases in blood, seawater, rain water and in many other aqueous solutions of biological and environmental significance [1], as well as the solubilities of drugs, proteins and other biomolecules in biological liquids. The list of examples regarding the solubilities of molecules of biomedical and environmental significance in aqueous solvents is countless.

Generally speaking, the thermodynamic properties of these complex mixtures (solute + multicomponent aqueous solvent) depend on many factors such as the chemical natures of the solute and of the constituents of the mixed solvent, the intermolecular interactions between the components in these mixtures, the mixture composition and the pressure and temperature. In the present paper only low soluble solutes are considered. Therefore, the solutions can be considered as dilute and the intermolecular interactions between the solute molecules can be neglected. Thus, the properties of a solute-free mixed solvent and the activity coefficient of the solute at infinite dilution can describe the behavior of such dilute mixtures.

The prediction of the activity coefficients of solutes (such as gases and large molecules of biomedical and environmental significance) in saturated solutions of multicomponent mixtures constitutes the main difficulty in calculating the solute solubil-

* Corresponding authors. Tel.: +1 716 645 2911x2214; fax: +1 716 645 3822.
E-mail addresses: ishulgin@eng.buffalo.edu (I.L. Shulgin),
feaeliru@acsu.buffalo.edu (E. Ruckenstein).

0378-3812/$ – see front matter © 2007 Elsevier B.V. All rights reserved.

ity. Generally speaking, the activity coefficient of a solute in a saturated solution of a multicomponent mixture can be predicted using group-contribution methods, such as UNIFAC and ASOG [2–6]. However, the above group-contribution methods cannot provide accurate results for the activity coefficient of large molecules, such as drugs and environmental molecules, in aqueous mixed solvents [7–10].

The application of UNIFAC to the solid–liquid equilibrium of solids, such as naphthalene and anthracene, in nonaqueous mixed solvents provided quite accurate results [11]. Unfortunately, the accuracy of UNIFAC regarding the solubility of solids in aqueous solutions is low [7–9]. Large deviations from the experimental activity coefficients at infinite dilution and the experimental octanol/water partition coefficients have been reported [8,9] when the classical old version of UNIFAC interaction parameters [4] was used. To improve the prediction of the activity coefficients at infinite dilution and of the octanol/water partition coefficients of environmentally significant substances, special ad hoc sets of parameters were introduced [7–9]. The reason is that the UNIFAC parameters were determined mostly using the equilibrium properties of mixtures composed of low molecular weight molecules. Also, the UNIFAC method cannot be applied to the phase equilibrium in systems containing supercritical components [10,12], or even to gas solubilities in pure and mixed solvents. However, a group-contribution equation of state such as PSRK [12,13] could be successfully applied to predict the gas solubilities in the above solvents. Gmehling and coworkers have demonstrated that PSRK can be successfully used to predict the solubilities of various gases in water and water + salt solvents [14–17].

Another method suggested by the authors for predicting the solubility of gases and large molecules such as the proteins, drugs and other biomolecules in a mixed solvent is based on the Kirkwood–Buff theory of solutions [18]. This theory connects the macroscopic properties of solutions, such as the isothermal compressibility, the derivatives of the chemical potentials with respect to the concentration and the partial molar volumes to their microscopic characteristics in the form of spatial integrals involving the radial distribution function. This theory allowed one to extract some microscopic characteristics of mixtures from measurable thermodynamic quantities. The present authors employed the Kirkwood–Buff theory of solution to obtain expressions for the derivatives of the activity coefficients in ternary [19] and multicomponent [20] mixtures with respect to the mole fractions. These expressions for the derivatives of the activity coefficients were used to predict the solubilities of various solutes in aqueous mixed solvents, namely:

(1) the solubilities of drugs and environmentally significant molecules in binary and multicomponent aqueous mixed solvents [21–25],
(2) the solubilities of gases in binary and multicomponent aqueous mixed solvents [20,26–28],
(3) the solubilities of various proteins in aqueous solutions [29–31].

The present paper is devoted to the extension of the theory developed by the authors for the solubility of proteins to the solubility of gases, Because this theory is based on the Kirkwood–Buff fluctuation theory of solutions, the next section summarizes the expressions which are involved. This is followed by a summary of the derivation of an equation for the solubility of proteins and finally its extension to the solubility of gases.

2. The application of the Kirkwood–Buff fluctuation theory of solutions to the activity coefficients in ternary and multicomponent solutions

On the basis of the fluctuation theory, the following expression for the derivative of the activity coefficient of a solute (γ_{2t}) in a water (1)–solute (2)–cosolvent (3) mixture can be derived [19], which is valid for any kinds of solutes and cosolvents:

$$\left(\frac{\partial \ln \gamma_{2,t}}{\partial x_3^t}\right)_{T,P,x_2^t} = -\frac{(c_1+c_2+c_3)(c_1[G_{11}+G_{23}-G_{12}-G_{13}]+c_3[-G_{12}-G_{33}+G_{13}+G_{23}])}{c_1+c_2+c_3+c_1c_2\Delta_{12}+c_1c_3\Delta_{13}+c_2c_3\Delta_{23}+c_1c_2c_3\Delta_{123}} \quad (1)$$

where x_i^t is the mole fraction of component i in the ternary mixture, c_k the bulk molecular concentration of component k in the ternary mixture 1–2–3, P the pressure, T the temperature in K, and $G_{\alpha\beta}$ is the Kirkwood–Buff integral (KBI) given by

$$G_{\alpha\beta} = \int_0^\infty (g_{\alpha\beta}-1)4\pi r^2 \, dr \quad (2)$$

where $g_{\alpha\beta}$ is the radial distribution function between species α and β, r the distance between the centers of molecules α and β, and $\Delta_{\alpha\beta}$ and Δ_{123} are defined as

$$\Delta_{\alpha\beta} = G_{\alpha\alpha}+G_{\beta\beta}-2G_{\alpha\beta}, \quad \alpha\neq\beta \quad (3)$$

and

$$\Delta_{123} = G_{11}G_{22}+G_{11}G_{33}+G_{22}G_{33}+2G_{12}G_{13}+2G_{12}G_{23}$$
$$+2G_{13}G_{23}-G_{12}^2-G_{13}^2-G_{23}^2-2G_{11}G_{23}$$
$$-2G_{22}G_{13}-2G_{33}G_{12} \quad (4)$$

One can verify that the factors in the square brackets in the numerators of Eq. (1), and Δ_{123} can be expressed in terms of $\Delta_{\alpha\beta}$ as follows

$$G_{12}+G_{33}-G_{13}-G_{23} = \frac{\Delta_{13}+\Delta_{23}-\Delta_{12}}{2} \quad (5)$$

$$G_{11}+G_{23}-G_{12}-G_{13} = \frac{\Delta_{12}+\Delta_{13}-\Delta_{23}}{2} \quad (6)$$

and

$$\Delta_{123} = -\frac{(\Delta_{12})^2 + (\Delta_{13})^2 + (\Delta_{23})^2 - 2\Delta_{12}\Delta_{13} - 2\Delta_{12}\Delta_{23} - 2\Delta_{13}\Delta_{23}}{4} \quad (7)$$

The insertion of Eqs. (5)–(7) into Eq. (1) provides a rigorous expression for the derivatives $(\partial \ln \gamma_{2,t}/\partial x_3^t)_{T,P,x_2^t}$ in terms of $\Delta_{\alpha\beta}$ and concentrations.

It should be noted that $\Delta_{\alpha\beta}$ is a measure of the nonideality [32] of the binary mixture α–β, because for an ideal mixture $\Delta_{\alpha\beta} = 0$. For a ternary mixture 1–2–3, Δ_{123} also constitutes a measure of nonideality. Indeed, inserting $G_{\alpha\beta}^{id}$ for an ideal mixture [33] into the expression of Δ_{123}, one obtains that for an ideal ternary mixture $\Delta_{123} = 0$.

At infinite dilution of component 2, Eq. (1) becomes

$$\lim_{x_2^t \to 0}\left(\frac{\partial \ln \gamma_{2,t}}{\partial x_3^t}\right)_{T,P,x_2^t} = -\frac{(c_1^0 + c_3^0)((c_1^0 + c_3^0)(\Delta_{12} - \Delta_{23})_{x_2^t=0} + (c_1^0 - c_3^0)(\Delta_{13})_{x_2^t=0})}{2(c_1^0 + c_3^0 + c_1^0 c_3^0 (\Delta_{13})_{x_2^t=0})} \quad (8)$$

where c_1^0 and c_3^0 are the bulk molecular concentrations of components 1 and 3 in the solute-free binary solvent 1–3. In addition to Eq. (8), the following expression [18] for the derivative of the activity coefficient in a binary mixture with respect to the mole fraction x_1^0 can be written

$$J_{11} = \left(\frac{\partial \ln \gamma_1^0}{\partial x_1^0}\right)_{P,T} = -\frac{c_3^0 \Delta_{13}}{1 + c_1^0 x_3^0 \Delta_{13}} \quad (9)$$

where x_i^0 and γ_i^0 are the mole fraction of component i ($i = 1, 3$) and the activity coefficient of component 1 in the solute-free binary solvent 1–3.

By combining Eqs. (8) and (9), one obtains an expression for the derivative of the activity coefficient of an infinitely dilute solute with respect to the cosolvent mole fraction in terms of the characteristics of the solute-free binary solvent (J_{11}, c_1^0 and c_3^0) and the parameters Δ_{12} and Δ_{23} which characterize the interactions of an infinitely dilute solute with the components of the mixed solvent. Even though Eq. (8) constitutes a formal statistical thermodynamics relation in which all Δ_{ij} are unknown, it could be successfully used to derive analytical expressions for the solubilities of gases and large molecules of biomedical and environmental significance, including the proteins, in aqueous media [20–30]. In addition, Eq. (8) could be also used to identify whether a cosolvent is a salting-out or salting-in agent [27,29–31]. The derivative of the activity coefficient of an infinitely dilute solute in quaternary and multicomponent mixtures with respect to the cosolvent mole fraction can also be expressed in terms of KBIs [20]. The latter expressions are, of course, more complicated than Eq. (8).

3. Solubility of a protein in aqueous solutions

It is well known that the solubility of a protein in a water + cosolvent mixture depends on numerous factors such as temperature, cosolvent concentration, pH, type of buffer used, etc. [31,34–38]. Experimental measurements of the solubility of a protein in a water + cosolvent mixture are difficult and time-consuming [38]. There are in the literature a number of examples in which the same cosolvent was found to generate both salting-out and salting-in. However, reliable experimental data appear to be available only for the water (1)–lyzosyme (2)–NaCl (3) mixtures for which there is agreement between the results from various laboratories [39–43].

One should distinguish between the effects on aqueous protein solubility of two different types of cosolvents: organic cosolvent and salts. Experiment has shown that the addition of an organic cosolvent reduces the aqueous protein solubility [36,37]. Therefore, the organic cosolvents are, in general, salting-out agents. However, there are exceptions; for instance, urea increases the aqueous solubility of ribonuclease Sa [44].

Regarding the salts, the old solubility measurements [45,46] indicated that a small amount of salt increases the aqueous protein solubility, and a large one decreases the protein solubility. In contrast, the measurements regarding the protein solubility in the presence of a salt carried out in the last two decades [39–43] revealed only salting-out effects.

One of the parameters which characterizes the behavior of a protein in an aqueous mixed solvent is the preferential binding parameter [47–51]. The preferential binding parameter can be defined in various concentration scales (component 1 is water, component 2 is a protein and component 3 is a cosolvent):

(1) in molal concentrations

$$\Gamma_{23}^{(m)} \equiv \lim_{m_2 \to 0}\left(\frac{\partial m_3}{\partial m_2}\right)_{T,P,\mu_3} \quad (10)$$

where m_i is the molality of component i and μ_i is the chemical potential of component i.

(2) in molar concentrations

$$\Gamma_{23}^{(c)} \equiv \lim_{c_2 \to 0}\left(\frac{\partial c_3}{\partial c_2}\right)_{T,P,\mu_3} \quad (11)$$

where c_i is the molar concentration of component i. One should notice that $\Gamma_{23}^{(m)}$ and $\Gamma_{23}^{(c)}$ are defined at infinite dilution of the protein.

The preferential binding parameter $\Gamma_{23}^{(m)}$ was determined experimentally by sedimentation [49], dialysis equilibrium [51], etc., for numerous systems [47–51].

Both preferential binding parameters ($\Gamma_{23}^{(m)}$ and $\Gamma_{23}^{(c)}$) can be expressed in terms of the KBIs as follows [52,53]:

$$\Gamma_{23}^{(m)} = \frac{c_3^0}{c_1^0} + c_3^0(G_{23} - G_{12} + G_{11} - G_{13}) \quad (12)$$

and

$$\Gamma_{23}^{(c)} = c_3^0(G_{23} - G_{13}) \quad (13)$$

Using the expressions for G_{11}, G_{13}, G_{12} and G_{23} from Appendix A, one can obtain an expression for the derivative of the activity coefficient of an infinitely dilute solute with respect to the cosolvent mole fraction in terms of the properties of the solute-free binary solvent, and the preferential binding parameter $\Gamma_{23}^{(m)}$ [29,30]:

$$\lim_{x_2^t \to 0} \left(\frac{\partial \ln \gamma_{2,t}}{\partial x_3^t}\right)_{T,P,x_2^t}$$
$$= \frac{c_3^0(c_1^0 + c_3^0)V_1 - \Gamma_{23}^{(m)}(1 - c_3^0 V_3)(c_1^0 + c_1^0 J_{11} + c_3^0)}{c_1^0 c_3^0 V_1} \quad (14)$$

where V_1 and V_3 are the partial molar volumes of components 1 and 3 in a protein-free mixed solvent and $J_{11} = (\partial \ln \gamma_1^0 / \partial x_1^0)_{x_2^t = 0}$.

Eq. (14) was obtained from Eq. (8) by replacing the parameters Δ_{12} and Δ_{23} in terms of the preferential binding parameter $\Gamma_{23}^{(m)}$ and partial molar volumes V_1 and V_3.

Using Eq. (14), a simple criterion for salting-in or salting-out at low cosolvent concentrations can be derived. For low cosolvent concentrations ($c_3^0 \to 0$), Eq. (14) leads to [29]

$$\left(\frac{\partial \ln y_2}{\partial x_3^0}\right) = -\left(\frac{\partial \ln y_2}{\partial x_1^0}\right) = \frac{\alpha}{V_1^0} - 1 \quad (15)$$

where y_2 is the mole fraction of the protein solubility, $\alpha = \lim_{c_3^0 \to 0}(\Gamma_{23}^{(m)}/c_3^0)$ and V_1^0 is the molar volume of pure water. For low cosolvent concentrations salting-in occurs when

$$\left(\frac{\partial \ln y_2}{\partial x_3^0}\right) > 0, \quad \text{hence when } \alpha > V_1^0 \quad (16)$$

and salting-out occurs when

$$\left(\frac{\partial \ln y_2}{\partial x_3^0}\right) < 0, \quad \text{hence when } \alpha < V_1^0 \quad (17)$$

Expression (15) can be used as a consistency test of the experimental data regarding the protein solubility and the preferential binding parameter. If the experimental data regarding the solubility and the preferential binding parameter do not satisfy Eq. (15), then one of the above quantities or both of them does (do) not correspond to thermodynamic equilibrium. In some cases Eqs. (15)–(17) can help to estimate the quality of the solubility data. For example, there are controversial data regarding the solubility of lysozyme in water + polyethylene glycol (PEG) mixtures. According to numerous data [54–56] the addition of PEG decreases the protein solubility in an aqueous solution (water + PEG) compared with that in pure water; there are, however, some measurement [57] which indicate that PEG can act as a salting-in agent. Experimental measurements [58–60] of the preferential binding parameter in the system water (1) + lysozyme (2) + PEG (3) showed that in this system $\Gamma_{23}^{(m)} < 0$ and hence that inequality (17) is valid. Therefore, our criteria reveal a salting-out effect of PEG.

Eq. (14) can be also used to derive an expression for the solubility of a protein in a mixed solvent as a function of the cosolvent concentration [29,30]. Indeed, by combining Eq. (14) with the phase equilibrium condition one obtains the following equation [30]

$$\left(\frac{\partial \ln y_2}{\partial x_3^0}\right) = -\frac{1}{x_1^0} + \frac{\Gamma_{23}^{(m)}(c_1^0 + c_1^0 J_{11} + c_3^0)}{c_3^0} \quad (18)$$

Eq. (18) allows one to calculate the protein solubility in a wide range of cosolvent concentrations if information regarding (i) the composition dependence of the preferential binding parameter and (ii) the properties of the protein-free mixed solvent such as the molar volume and the activity coefficients of the components are available. In addition, one should mention that Eq. (18) was obtained for ternary mixtures (water (1)–protein (2)–cosolvent (3)). However, those mixtures contain also a buffer, the effect of which is taken into account only indirectly through the preferential binding parameter $\Gamma_{23}^{(m)}$. Another limitation of Eq. (18) is the infinite dilution approximation, which means that the protein solubility is supposed to be small enough to satisfy the infinite dilution approximation ($\gamma_2 \cong \gamma_2^\infty$, where γ_2^∞ is the activity coefficient of a protein at infinite dilution).

In the dilute region of the cosolvent, the preferential binding parameters $\Gamma_{23}^{(m)}$ and $\Gamma_{23}^{(c)}$ are linear functions of the cosolvent concentration [61–63] and Eq. (18) leads after a number of simplifications to

$$\ln \frac{y_2}{y_2^w} \approx -c_1^0 \alpha \ln a_1 \quad (19)$$

where y_2^w is the protein solubility in a cosolvent-free water plus buffer, a_1 the water activity in the protein-free mixed solvent and $\alpha = \Gamma_{23}^{(m)}/c_3^0$. Eq. (19) was successfully used to predict the protein solubility in several water + salt mixed solvents by using experimental preferential binding parameter values [29].

It is worth noting that the developed theory provides not only an equation for the protein solubility in mixed solvents, but also provides some insight into the hydration of a protein molecule in aqueous solutions and its connection with the protein solubility. In particular, it was shown [29] that the preferential hydration of a protein molecule ($\Gamma_{23}^{(m)} < 0$) is connected with the decrease of the protein solubility (salting-out) and when the water is preferentially excluded from a protein surface ($\Gamma_{23}^{(m)} > 0$), the addition of a small amount of cosolvent increases the protein solubility (salting-in).

4. Solubility of a gas in aqueous salt solutions

The Kirkwood–Buff formalism can be also used to derive the composition dependence of the Henry constant for a sparingly soluble gas dissolved in a mixed solvent containing water + electrolyte [27]. The obtained equation requires information about the molar volume and the mean activity coefficient of the electrolyte in the binary (water + electrolyte) mixture. Several expressions for the mean activity coefficient of the electrolyte were tested and it was concluded that the accuracy in predicting the Henry constant is highly dependent on the accuracy of the expression used for the mean activity coefficient of the electrolyte. In addition, a criterion for predicting the kind of salting (salting-in or salting-out) for dilute salt solutions was established [27]. According to that criterion the kind of salting depends mainly on the molar volume of the electrolyte at infinite dilution. This result explains the experimental observation that the electrolytes with large molar volumes at infinite dilution increase the gas solubility compared to that in pure water.

In the present section, the expression obtained for the protein solubility will be extended to the solubility of a sparingly soluble gas in an aqueous electrolyte solution. Eq. (14) will be used for the derivative of the activity coefficient of an infinitely dilute gas with respect to the mole fraction of the cosolvent (electrolyte). While there are no experimental results for the preferential binding parameter $\Gamma_{23}^{(m)}$ for the systems water (1) + gas (2) + electrolyte (3), Eq. (14) can be still used by assuming, however, in analogy with the protein solutions that $\Gamma_{23}^{(m)} = \varepsilon x_3^0$, where ε is a constant. Because for the derivative of the logarithm of the Henry constant H_{2t} in a binary solvent with respect to the mole fraction of the electrolyte one can write [27]

$$\left(\frac{\partial \ln H_{2t}}{\partial x_3^t}\right)_{P,T,x_2^t=0} = \lim_{x_2^t \to 0} \left(\frac{\partial \ln \gamma_{2,t}}{\partial x_3^t}\right)_{P,T} \quad (20)$$

Eq. (14) can be recast in the form

$$\left(\frac{\partial \ln H_{2t}}{\partial x_3^t}\right)_{P,T,x_2^t=0} = \left(\frac{\partial \ln H_{2t}}{\partial x_3^0}\right)_{P,T} = \frac{1}{x_1^0} - \varepsilon(1 + x_1^0 J_{11}) \quad (21)$$

which integrated leads to

$$\ln\left(\frac{H_{2t}}{H_{21}}\right)_{P,T} = -\ln x_1^0 - \varepsilon \int_0^{x_3^0} (1 + x_1^0 J_{11}) \, dx_3^0 \quad (22)$$

where H_{21} is the Henry constant in pure water.

Because $dx_3^0 = -dx_1^0$ and $a_1 = x_1^0 \gamma_1^0$, Eq. (22) becomes

$$\ln\left(\frac{H_{2t}}{H_{21}}\right)_{P,T} = -\ln x_1^0 + \varepsilon \int_0^{a_1} x_1^0 \, d\ln a_1$$

$$= -\ln x_1^0 + \varepsilon x_1^0 \ln(a_1(x_1^0)) - \varepsilon \int_0^{x_1^0} \ln a_1 \, dx_1^0 \quad (23)$$

where $a_1(x_1^0)$ is the activity of water at a mole fraction x_1^0. By taking into account that

$$\ln a_1 = \frac{-\varphi M_1 m_3 \nu}{1000} \quad (24)$$

and

$$dx_1^0 = -\frac{1000 M_1}{(1000 + M_1 m_3)^2} \, dm_3 \quad (25)$$

where φ is the osmotic coefficient, M_1 the molar weight of water, m_3 the molality of the electrolyte in the gas-free mixed solvent, and ν is the number of ions formed through the complete dissociation of the electrolyte molecule, Eq. (23) can be recast as

$$\ln\left(\frac{H_{2t}}{H_{21}}\right)_{P,T} = -\ln x_1^0 + \varepsilon x_1^0 \ln(a_1(x_1^0)) - \varepsilon \int_0^{m_3} \frac{(M_1)^2 \varphi m_3 \nu}{(1000 + M_1 m_3)^2} \, dm_3 \quad (26)$$

Eq. (26) can be used to calculate the Henry constant H_{2t} in a binary solvent water (1) + electrolyte (3) using experimental data for the osmotic coefficient and considering ε as an adjustable parameter.

Three systems were selected for examination, namely the solubilities of oxygen, carbon dioxide, and methane in water + sodium chloride. An accurate semiempirical equation [64] was used to express the composition dependence of the osmotic coefficient in water + sodium chloride. The results of the calculations are presented in Fig. 1 and Table 1. One can see that Eq. (26) provides an accurate correlation for the gas solubility in solutions of strong electrolytes. In addition, the fluctuation theory allows one to use the experimental solubility data to examine the hydration in water (1)–gas (2)–cosolvent (3) mixtures.

5. The use of experimental solubility data to analyze hydration phenomena

The excesses (or deficits) Δn_{12} and Δn_{32} were calculated using the following relations [65]:

$$\Delta n_{12} = c_1^0 G_{12} + c_1^0 (V_2^\infty - RT k_T) \quad (27)$$

Table 1
Calculation of the gas solubility with Eq. (26)

The gas and the electrolyte solution		Temperature (K)	Composition range, molality	References	Value of the parameter ε in Eq. (26)
Gas	Electrolyte solution				
O_2	NaCl + H_2O	298.15	0–6.1	a	−6.97
CO_2	NaCl + H_2O	298.15	0–5.7	b	−4.84
CH_4	NaCl + H_2O	298.15	0–2.1	c	−8.74

[a] T.A. Mishina, O.I. Avdeeva, T.K. Bozhovskaya, Materialy Vses. Nauchn. Issled. Geol. Inst. 46 (1961) 93 (as given in Solubility Data Series, vol. 7, Pergamon, 1981).
[b] F.J. Yoshida, Chem. Eng. Data 24 (1979) 11.
[c] A. Ben-Naim, M. Yaacobi, J. Phys. Chem. 78 (1974) 170.

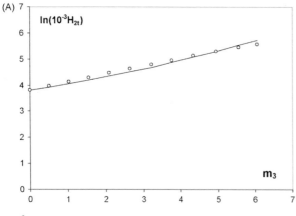
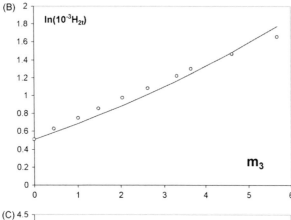
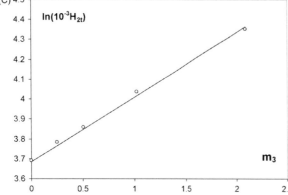

Fig. 1. The Henry constant (expressed in bars) of oxygen (A), carbon dioxide (B) and methane (C) in aqueous solutions of sodium chloride at 25 °C (○) experimental data, solid line (—) the Henry constant calculated with Eq. (26) (see Table 1 for details).

and

$$\Delta n_{32} = c_3^0 G_{23} + c_3^0 (V_2^\infty - RTk_T) \qquad (28)$$

where V_2^∞ is the partial molar volume of the solute at infinite dilution and k_T is the isothermal compressibility of the mixed solvent.

The above expressions show that to calculate Δn_{12} and Δn_{32} information about the Kirkwood–Buff integrals G_{12} and G_{23} is necessary. This information can be extracted from $\Gamma_{23}^{(m)}$ and an expression relating the partial molar volume V_2^∞ at infinite dilution of the solute to the Kirkwood–Buff integrals and to the partial molar volumes V_1 and V_3 of the components of the mixed solvent-free of solute [66]. Indeed, the preferential binding parameter $\Gamma_{23}^{(m)}$ can be expressed in terms of KBIs as follows [53]

$$\Gamma_{23}^{(m)} = \frac{c_3^0}{c_1^0} + c_3^0 (G_{23} - G_{12} + G_{11} - G_{13}) \qquad (12)$$

and the partial molar volume V_2^∞ at infinite dilution is given by [66]

$$V_2^\infty = -c_1^0 V_1 G_{12} - c_3^0 V_3 G_{23} + kTk_T \qquad (29)$$

where k is the Boltzmann constant.

Whereas G_{12} and G_{23} depend on the gas characteristics, G_{11} and G_{13} depend only on the characteristics of the gas-free mixed solvent and can be calculated from the thermodynamic properties of the gas-free mixed solvent.

The calculations have been carried out for those systems for which the solubility calculations have been performed. The dilute region of sodium chloride ($c_3^0 \leq 0.3$) was selected to ensure that the condition $\Gamma_{23}^{(m)}/x_3^0 = \varepsilon =$ constant is satisfied. The partial molar volume V_2^∞ was estimated using literature data [67–69]. According to the latter data, V_2^∞ depends weakly on c_3^0 and this dependence is linear in the dilute range [68,69]. For sodium chloride and potassium chloride, V_2^∞ decreases by at most 1 cm^3/mol when c_3^0 is changed from 0 to 2 mol/l. In our calculations, the above decrease was taken 1 cm^3/mol. On this basis the composition dependence of V_2^∞ was evaluated in the composition range $0 \leq c_3^0 \leq 0.3$. The partial molar volumes V_1 and V_3 of water and sodium chloride in the binary mixture water (1) + sodium chloride (3) were obtained from data available in the literature [70,71], and the composition dependence of the isothermal compressibility of the mixed solvent (water (1) + sodium chloride (3)) was taken from reference [71].

The Kirkwood–Buff integrals G_{11} and G_{13} in the binary mixture water (1) + sodium chloride (3) were taken from reference [71]. The values obtained for G_{12} and G_{23} from Eqs. (27) and (28) were used to calculate the excesses (or deficits) number of water and sodium chloride molecules in the vicinity of a gas molecule.

The results of the calculations are presented in Table 2, which shows that infinitely dilute oxygen, carbon dioxide and methane molecules are preferentially hydrated in dilute aqueous solutions of sodium chloride. This means that water is in excess and sodium chloride is in deficit in the vicinity of a gas molecule in contrast with the random distribution of water and sodium chloride in their binary mixtures. The values of Δn_{12} and Δn_{32} are small because only the dilute region was considered in Table 2 ($c_3^0 = 0.3$ corresponds to $x_3^0 = 0.0054$). To carry out similar calculations in the nondilute region, experimental data regarding $\Gamma_{23}^{(m)}$ and V_2^∞ are required and they are not available.

As our calculations have shown, the contribution of the compressibility to the excesses Δn_{12} and Δn_{32} is negligibly small and to V_2^∞, G_{12} and G_{23} small.

Table 2
The Kirkwood–Buff integrals G_{12} and G_{23} and the excesses (or deficits) numbers Δn_{12} and Δn_{32} in a binary mixture water (1) + gas (2) + sodium chloride (3)

c_3^0	G_{12} (cm^3/mol)	G_{23} (cm^3/mol)	Δn_{12} (mol/mol)	Δn_{32} (mol/mol)
Water (1) + oxygen (2) + sodium chloride (3)				
0.05	−33.1	−168.1	0.007	−0.007
0.1	−32.9	−168.0	0.013	−0.013
0.15	−32.8	−167.9	0.020	−0.020
0.2	−32.6	−167.8	0.027	−0.027
0.25	−32.5	−167.7	0.033	−0.034
0.3	−32.3	−167.6	0.040	−0.040
Water (1) + carbon dioxide (2) + sodium chloride (3)				
0.05	−32.7	−129.2	0.005	−0.005
0.1	−32.6	−129.1	0.009	−0.010
0.15	−32.5	−129.0	0.014	−0.015
0.2	−32.4	−129.0	0.019	−0.019
0.25	−32.3	−128.9	0.024	−0.024
0.3	−33.2	−129.8	0.029	−0.029
Water (1) + methane (2) + sodium chloride (3)				
0.05	−36.1	−203.1	0.008	−0.008
0.1	−36.0	−203.0	0.016	−0.017
0.15	−35.8	−202.9	0.024	−0.025
0.2	−35.6	−202.7	0.033	−0.033
0.25	−35.5	−202.6	0.041	−0.042
0.3	−36.4	−203.6	0.050	−0.050

Molecular dynamics simulations for the mixture water (1) + methane (2) + sodium chloride (3) revealed a similar local structure around an infinitely dilute gas molecule, namely the methane molecule is preferentially hydrated and sodium chloride is preferentially excluded from the vicinity of a methane molecule [72].

6. Discussion and conclusion

The authors of the present paper have shown previously [21–31] that the fluctuation theory of solution can provide a new approach to the solubility of gases, drugs, protein, etc., in binary and multicomponent aqueous mixed solvents.

In the present paper, the method which the authors employed previously to derive an expression for the solubility of various proteins in aqueous solutions, has been extended to the solubility of gases in mixtures of water + strong electrolytes. One parameter equation for the solubility of gases has been derived, which was used to represent the solubilities of oxygen, carbon dioxide and methane in water + sodium chloride. In additions, the developed theory could be used to examine the local composition of the solvent around a gas molecule. The results revealed that the oxygen, carbon dioxide and methane molecules are preferentially hydrated in water + sodium chloride mixtures. A similar result was obtained for the water + methane + sodium chloride by molecular dynamics simulations [72].

There is a similarity between the solubility of a protein and that of a gas in an aqueous mixed solvent. For the solution of a protein in an aqueous mixed solvent, it was shown [29] that when the protein molecule is preferentially hydrated ($\Gamma_{23}^{(m)} < 0$), the addition of a small amount of cosolvent decreases the protein solubility (salting-out) and when the water is preferentially excluded from the protein surface ($\Gamma_{23}^{(m)} > 0$), the addition of a small amount of cosolvent increases the protein solubility (salting-in). One can see from Table 2 that sodium chloride is preferentially excluded from an infinitely dilute gas molecule, hence that this gas molecule is preferentially hydrated. This occurs because Na$^+$ and Cl$^-$ prefer the environment of water and less that of the gas molecule.

List of symbols

a_1	activity of water
c_i	molar concentration of component i
c_i^0	bulk molecular concentrations of component i in the solute-free binary solvent
D	parameter in Eqs. (A.1) and (A.2) defined in Eq. (A.3)
$g_{\alpha\beta}$	radial distribution function between species α and β
$G_{\alpha\beta}$	Kirkwood–Buff integral
H_{2t}	Henry constant in a binary solvent
H_{21}	Henry constant in pure water
J_{11}	derivative of the activity coefficient of water in the water + cosolvent mixture with respect to the mole fraction of water
k	Boltzmann constant
k_T	isothermal compressibility
m_i	molality of component i
M_1	molar weight of water
Δn_{ij}	excess (or deficit) of molecule i around a central molecule j
P	pressure
r	distance between the centers of molecules α and β
R	universal gas constant
T	temperature (K)
V	molar volume of the mixture

V_i	partial molar volume of component i
V_1^0	molar volume of pure water
V_2^∞	partial molar volume of the solute at infinite dilution
x_i^t	mole fraction of component i in the ternary mixture
x_i^0	mole fraction of component i in the solute-free binary solvent
y_2	mole fraction of the protein solubility
y_2^w	protein solubility in a cosolvent-free water plus buffer

Greek letters

α	parameter in Eq. (15) defined in the text after Eq. (15)
$\Delta_{\alpha\beta}$	measure of the nonideality of the binary mixture α–β
Δ_{123}	measure of the nonideality of the ternary mixture 1–2–3
ε	constant defined before Eq. (20)
γ_i^0	activity coefficient of component i in the solute-free binary solvent
$\gamma_{2,t}$	activity coefficient of a solute in a water (1)–solute (2)–cosolvent (3) mixture
γ_2^∞	activity coefficient of a protein at infinite dilution
$\Gamma_{23}^{(c)}$	preferential binding parameter defined on the basis of molar concentration
$\Gamma_{23}^{(m)}$	preferential binding parameter defined on the basis of molal concentration
φ	osmotic coefficient
μ_i	chemical potential of component i
ν	number of ions formed through the complete dissociation of an electrolyte molecule

Appendix A

The purpose of this Appendix is to provide expressions for the KBIs for binary and ternary mixtures (G_{11}, G_{13}, G_{12} and G_{23}) in terms of measurable thermodynamic quantities such as the derivatives of the chemical potentials with respect to concentrations, the isothermal compressibility and the partial molar volumes.

The main formulas for the KBIs of binary mixtures are [73,74]

$$G_{13} = G_{31} = RTk_T - \frac{V_1 V_2}{VD} \quad (A.1)$$

and

$$G_{11} = G_{13} + \frac{1}{x_1^0}\left(\frac{V_3}{D} - V\right) \quad (A.2)$$

where

$$D = \left(\frac{\partial \ln \gamma_i^0}{\partial x_i^0}\right)_{P,T} x_i^0 + 1, \quad i = 1, 3 \quad (A.3)$$

P is the pressure, T the temperature in K, k_T the isothermal compressibility, V_i the partial molar volume of component i, x_i^0 the molar fraction of component i, V the molar volume of the mixture, γ_i^0 the activity coefficient of component i and R is the universal gas constant.

The KBIs for a ternary mixture at infinite dilution of component 2 (G_{12} and G_{23}) are provided by the expressions [52]:

$$G_{12} = kTk_T - \frac{J_{21}V_3c_3^0 + J_{11}V_2^\infty c_1^0}{c_1^0 + c_1^0 J_{11} + c_3^0}$$

$$- \frac{V_3c_3^0(c_1^0 + c_3^0)(V_1 - V_3) + V_2^\infty(c_1^0 + c_3^0)}{c_1^0 + c_1^0 J_{11} + c_3^0} \quad (A.4)$$

and

$$G_{23} = kTk_T + \frac{J_{21}V_1c_1^0 - J_{11}c_1^0 V_2^\infty}{c_1^0 + c_1^0 J_{11} + c_3^0}$$

$$+ \frac{c_1^0 V_1(c_1^0 + c_3^0)(V_1 - V_3) - V_2^\infty(c_1^0 + c_3^0)}{c_1^0 + c_1^0 J_{11} + c_3^0} \quad (A.5)$$

where

$$J_{21} = \lim_{x_2^t \to 0} \left(\frac{\partial \ln \gamma_{2,t}}{\partial x_1^t}\right)_{x_2^t, P, T}.$$

References

[1] E. Wilhelm, R. Battino, R. Wilcock, Chem. Rev. 77 (1977) 219–262.
[2] E.L. Derr, C.H. Deal, Inst. Chem. Eng. Symp. London Ser. 3 (1969) 40–53.
[3] Aa. Fredenslund, R.L. Jones, J.M. Prausnitz, AIChE J. 21 (1975) 1086–1099.
[4] Aa. Fredenslund, J. Gmehling, P. Rasmussen, Vapor–Liquid Equilibria Using UNIFAC, Elsevier, Amsterdam, 1977.
[5] K. Kojima, K. Tochigi, Prediction of Vapor–Liquid Equilibria by the ASOG Method, Elsevier, Tokyo, 1979.
[6] J.M. Prausnitz, R.M. Lichtenthaler, E.G. de Azevedo, Molecular Thermodynamics of Fluid Phase Equilibria, second ed., Prentice Hall, Englewood Cliffs, 1986.
[7] A.T. Kan, M.B. Tomson, Environ. Sci. Technol. 30 (1996) 1369–1376.
[8] G. Wienke, J. Gmehling, Toxicol. Environ. Chem. 65 (1998) 57–86.
[9] A. Li, W.J. Doucette, A.W. Andren, Chemosphere 29 (1994) 657–669.
[10] A. Fredenslund, M.J. Sorensen, in: S. Sandler (Ed.), Models for Thermodynamic and Phase Equilibria Calculations, Dekker, New York, 1994, pp. 287–361.
[11] J.G. Gmehling, T.F. Anderson, J.M. Prausnitz, Ind. Eng. Chem. Fund. 17 (1978) 269–273.
[12] T. Holderbaum, J. Gmehling, Fluid Phase Equilibr. 70 (1991) 251–265.
[13] K. Fischer, J. Gmehling, Fluid Phase Equilibr. 112 (1995) 1–22.
[14] J. Li, M. Topphoff, K. Fischer, J. Gmehling, Ind. Eng. Chem. Res. 40 (2001) 3703–3710.
[15] J. Kiepe, S. Horstmann, K. Fischer, J. Gmehling, Ind. Eng. Chem. Res. 41 (2002) 4393–4398.
[16] J. Kiepe, S. Horstmann, K. Fischer, J. Gmehling, Ind. Eng. Chem. Res. 42 (2003) 5392–5398.
[17] J. Kiepe, S. Horstmann, K. Fischer, J. Gmehling, Ind. Eng. Chem. Res. 43 (2004) 6607–6615.
[18] J.G. Kirkwood, F.P. Buff, J. Chem. Phys. 19 (1951) 774–782.
[19] E. Ruckenstein, I. Shulgin, Fluid Phase Equilibr. 180 (2001) 345–359.
[20] E. Ruckenstein, I. Shulgin, Ind. Eng. Chem. Res. 42 (2003) 4406–4413.
[21] E. Ruckenstein, I. Shulgin, Int. J. Pharmaceut. 258 (2003) 193–201.
[22] E. Ruckenstein, I. Shulgin, Int. J. Pharmaceut. 260 (2003) 283–291.
[23] E. Ruckenstein, I. Shulgin, Int. J. Pharmaceut. 267 (2003) 121–127.
[24] E. Ruckenstein, I. Shulgin, Int. J. Pharmaceut. 278 (2004) 221–229.
[25] E. Ruckenstein, I. Shulgin, Environ. Sci. Technol. 39 (2005) 1623–1631.
[26] I. Shulgin, E. Ruckenstein, Ind. Eng. Chem. Res. 41 (2002) 1689–1694.
[27] E. Ruckenstein, I. Shulgin, Ind. Eng. Chem. Res. 41 (2002) 4674–4680.
[28] E. Ruckenstein, I. Shulgin, Polymer 44 (2003) 901–907.
[29] I.L. Shulgin, E. Ruckenstein, Biophys. Chem. 118 (2005) 128–134.

[30] I.L. Shulgin, E. Ruckenstein, Biophys. Chem. 120 (2006) 188–198.
[31] E. Ruckenstein, I.L. Shulgin, Adv. Coll. Interf. Sci. 123 (2006) 97–103.
[32] A. Ben-Naim, J. Chem. Phys. 67 (1977) 4884–4890.
[33] E. Ruckenstein, I. Shulgin, Fluid Phase Equilibr. 180 (2001) 281–297.
[34] E.J. Cohn, Physiol. Rev. 5 (1925) 349–437.
[35] E.J. Cohn, J.T. Edsall, Proteins, Amino Acids and Peptides, Reinhold, New York, 1943.
[36] T. Arakawa, S.N. Timasheff, Methods Enzymol. 114 (1985) 49–74.
[37] C.H. Schein, Biotechnology (NY) 8 (1990) 308–315.
[38] M. Riès-Kautt, A. Ducruix, in: A. Ducruix, R. Giégé (Eds.), Crystallization of Nucleic Acids and Proteins: a Practical Approach, Oxford University Press, New York, 1999, pp. 287–361.
[39] M. Ataka, S. Tanaka, Biopolymers 25 (1986) 337–350.
[40] S.B. Howard, P.J. Twigg, J.K. Baird, E.J. Meehan, J. Cryst. Growth 90 (1986) 94–104.
[41] V. Mikol, R. Giege, J. Cryst. Growth 97 (1989) 324–332.
[42] M.M. Ries-Kautt, A.F. Ducruix, J. Biol. Chem. 264 (1989) 745–748.
[43] E. Cacioppo, M.L. Pusey, J. Cryst. Growth 114 (1991) 286–292.
[44] C.N. Pace, S. Trevino, E. Prabhakaran, J.M. Scholtz, Philos. Trans. R. Soc. Lond. B Biol. Sci. 359 (2004) 1225–1234.
[45] A.A. Green, J. Biol. Chem. 93 (1931) 495–516.
[46] A. Grönwall, C R Trans Labs. Carls. 24 (1942) 185–200.
[47] E.F. Casassa, H. Eisenberg, Adv. Protein Chem. 19 (1964) 287–393.
[48] J. Wyman, Adv. Protein Chem. 19 (1964) 223–286.
[49] I.D. Kuntz, W. Kauzmann, Adv. Protein Chem. 28 (1974) 239–345.
[50] S.N. Timasheff, Annu. Rev. Biophys. Biomol. Struct. 22 (1993) 67–97.
[51] S.N. Timasheff, Adv. Protein Chem. 51 (1998) 355–432.
[52] I.L. Shulgin, E. Ruckenstein, J. Chem. Phys. 123 (2005) (Art. No. 054909).
[53] I.L. Shulgin, E. Ruckenstein, Biophys. J. 90 (2006) 704–707.
[54] D.H. Atha, K.C. Ingham, J. Biol. Chem. 256 (1981) 12108–12117.
[55] I.R.M. Juckes, Biochim. Biophys. Acta 229 (1971) 535–546.
[56] H. Mahadevan, C.K. Hall, Fluid Phase Equilibr. 78 (1992) 297–321.
[57] A. Kulkarni, C. Zukoski, Langmuir 18 (2002) 3090–3099.
[58] J.C. Lee, L.L.Y. Lee, J. Biol. Chem. 256 (1981) 625–631.
[59] T. Arakawa, S.N. Timasheff, Biochemistry 24 (1985) 6756–6762.
[60] R. Bhat, S.N. Timasheff, Protein Sci. 1 (1992) 1133–1143.
[61] E.S. Courtenay, M.W. Capp, C.F. Anderson, M.T. Record, Biochemistry 39 (2000) 4455–4471.
[62] M.T. Record, W.T. Zhang, C.F. Anderson, Adv. Protein Chem. 51 (1998) 281–353.
[63] B.M. Baynes, B.L. Trout, J. Phys. Chem. B 107 (2003) 14058–14067.
[64] W.J. Hamer, Y.-C. Wu, J. Phys. Chem. Ref. Data 1 (1972) 1047–1099.
[65] I.L. Shulgin, E. Ruckenstein, J. Phys. Chem. B 110 (2006) 12707–12713.
[66] A. Ben-Naim, Statistical Thermodynamics for Chemists and Biochemists, Plenum, New York, 1992.
[67] J.C. Moore, R. Battino, T.R. Rettich, Y.P. Handa, E. Wilhelm, J. Chem. Eng. Data 27 (1982) 22–24.
[68] E.W. Tiepel, K.E. Gubbins, J. Phys. Chem. 76 (1972) 3044–3049.
[69] T.D. O'Sullivan, N.O. Smith, J. Phys. Chem. 74 (7) (1970) 1460–1466.
[70] F.J. Millero, J. Phys. Chem. 74 (1970) 356–362.
[71] R. Chitra, P.E. Smith, J. Phys. Chem. B 106 (2002) 1491–1500.
[72] N.F.A. van der Vegt, W.F. van Gunsteren, J. Phys. Chem. B 108 (2004) 1056–1064.
[73] E. Matteoli, L. Lepori, J. Chem. Phys. 80 (1984) 2856–2863.
[74] I. Shulgin, E. Ruckenstein, J. Phys. Chem. 103 (1999) 10266–10271.

Chapter 4

Solubility of pharmaceuticals and environmentally important compounds

4.1 Solubility of drugs in aqueous solutions. Part 1: Ideal mixed solvent approximation.
4.2 Solubility of drugs in aqueous solutions. Part 2: Non-ideal mixed solvent.
4.3 Solubility of drugs in aqueous solutions. Part 3: Multi-component mixed solvent.
4.4 Solubility of drugs in aqueous solutions. Part 4: Drug solubility by the dilute approximation.
4.5 Solubility of drugs in aqueous solutions. Part 5: Thermodynamic consistency test for the solubility data.
4.6 Solubility of hydrophobic organic pollutants in binary and multicomponent aqueous solvents.

Introduction to Chapter 4

Chapter 4 is devoted to the solubility of poorly soluble solids in mixed solvents (especially in aqueous solvents). Two types of solids are considered here: 1) drugs and 2) environmentally important compounds.

It is self-evident that accurate prediction and/or correlation of the solubility of drugs in mixed aqueous solutions are of great interest for pharmaceutics. One or more cosolvents can be added to water to create a desirable multicomponent solvent, which satisfies pharmaceutical requirements with respect to solubility, toxicity, stability, price, and other factors. However, the experimental determination of solubilities in multicomponent solutions is time-consuming, because of the large number of compositions which must be covered in the concentration ranges of interest, and can be very expensive because of the high prices of some modern drugs. For this reason, it is important to provide a reliable method for predicting the solubility of drugs in multicomponent mixed solvents from available experimental solubilities in subsystems such as pure solvents, binary mixed solvents, etc. In Chapter 4 we provide such methods for predicting and correlating the solubility of drugs in multicomponent mixed solvents.

First, the fluctuation theory for multicomponent solutions is coupled with the thermodynamic condition for equilibrium between a solid and a solvent, and several equations for correlating the drug solubility data are derived (4.1–4.2, 4.4). Second, experimental data regarding the solubility of drugs in binary mixed solvents are examined and compared with the equations (4.1–4.2, 4.4). The main difference between these equations and the numerous empirical expressions from the literature is that they have a theoretical basis in the fluctuation theory of solutions. In addition, they perform better than those from literature (4.1–4.2). A method for predicting drug solubility in multicomponent (binary and higher) mixed solvents is developed and compared with available experimental data in paper (4.3). This contribution is important because, to the best of our knowledge, there is no other method for predicting the solubility of drugs in multicomponent mixed solvents. This method can provide a computerized scheme for fast screening of many combinations of potential multicomponent solvents to select a desirable solvent (or solvents) that satisfies the pharmaceutical requirements with respect to the solubility.

The solubility of an organic compound (including pollutants) in water is one of the key factors that affects its environmental behavior. Because this topic is of high importance in environmental science, we apply our methods to the available experimental information about the solubility of hydrophobic organic pollutants in binary and multicomponent aqueous solvents. Our methodology provides a simple and reliable method that can be used for correlating and predicting the solubility of hydrophobic organic pollutants in binary and multicomponent aqueous solvents (4.6).

At present, the solubility of pharmaceuticals and environmentally important compounds in binary and multicomponent aqueous solvents is attracting the attention of many research laboratories, which produce a large amount of experimental data. Frequently, the results of one laboratory contradict the results from other laboratories. Therefore, a rigorous criterion for the verification of the correctness of the solubility data is important. Such a criterion, based on the Gibbs–Duhem equation for ternary mixtures, is suggested in paper (4.5).

Solubility of drugs in aqueous solutions
Part 1. Ideal mixed solvent approximation

E. Ruckenstein*, I. Shulgin

Department of Chemical Engineering, State University of New York at Buffalo, Amherst, NY 14260, USA

Received 13 December 2002; received in revised form 28 January 2003; accepted 12 March 2003

Abstract

The present paper deals with the application of the fluctuation theory of solutions to the solubility of poorly soluble drugs in aqueous mixed solvents. The fluctuation theory of ternary solutions is first used to derive an expression for the activity coefficient of a solute at infinite dilution in an ideal mixed solvent and, further, to obtain an equation for the solubility of a poorly soluble solid in an ideal mixed solvent. Finally, this equation is adapted to the solubility of poorly soluble drugs in aqueous mixed solvents by treating the molar volume of the mixed solvent as nonideal and including one adjustable parameter in its expression. The obtained expression was applied to 32 experimental data sets and the results were compared with the three parameter equations available in the literature.
© 2003 Elsevier Science B.V. All rights reserved.

Keywords: Drug solubility; Fluctuation theory; Aqueous mixed solvents

1. Introduction

It is well-known that the addition of an organic cosolvent to water can dramatically change the solubility of drugs (Yalkowsky and Roseman, 1981). This fact is important for pharmaceutics because a poor aqueous solubility can often affect the drug efficiency. For this reason, the prediction of the solubility of drugs in aqueous mixed solvents or even a reliable correlation of the available experimental data is of interest to the pharmaceutical science and industry.

The solubility of solid substances in pure and mixed solvents can be described by the usual solid–liquid equilibrium conditions (Acree, 1984; Prausnitz et al., 1986). For the solubilities of a solid substance (solute, component 2) in water (component 3), cosolvent (component 1) and their mixture (mixed solvent, 1–3), one can write the following equations:

$$\frac{f_2^S}{f_2^L(T, P)} = x_2^{b_1} \gamma_2^{b_1}(T, P, \{x\}) \qquad (1)$$

$$\frac{f_2^S}{f_2^L(T, P)} = x_2^{b_3} \gamma_2^{b_3}(T, P, \{x\}) \qquad (2)$$

$$\frac{f_2^S}{f_2^L(T, P)} = x_2^{t} \gamma_2^{t}(T, P, \{x\}) \qquad (3)$$

In Eqs. (1)–(3), $x_2^{b_1}$, $x_2^{b_3}$ and x_2^{t} are the solubilities (mole fractions) of the solid component 2 in the cosolvent, water, and their mixture, respectively, $\gamma_2^{b_1}$, $\gamma_2^{b_3}$ and γ_2^{t} are the activity coefficients of the solid in its saturated solutions in the cosolvent, water, and

* Corresponding author. Tel.: +1-716-645-2911x2214; fax: +1-716-645-3822.
E-mail addresses: feaeliru@acsu.buffalo.edu (E. Ruckenstein), ishulgin@eng.buffalo.edu (I. Shulgin).

mixed solvent, $f_2^L(T, P)$ is the hypothetical fugacity of a solid as a (subcooled) liquid at a given pressure (P) and temperature (T), f_2^S is the fugacity of a pure solid component 2, and $\{x\}$ designates that the activity coefficients of the solid depend on composition. If the solubilities of the pure and mixed solvents in the solid phase are negligible, then the left hand sides of Eqs. (1)–(3) depend only on the properties of the solute. Eqs. (1)–(3) show that the solubilities of solid substances in pure and mixed solvents can be calculated if its activity coefficients in the binary and ternary saturated solutions (1–2, 2–3, and 1–2–3) are known. The activity coefficients of a solute in a pure and mixed solvent can be calculated by group-contribution methods, such as UNIFAC or ASOG (Acree, 1984; Prausnitz et al., 1986). The application of UNIFAC to the solubility of naphthalene in nonaqueous mixed solvents provided satisfactory results when compared to experimental data (Acree, 1984). However, the accuracy of the UNIFAC was poor for the solubility of solids in aqueous solutions (Fan and Jafvert, 1997).

The activity coefficients of a solute in a mixed solvent could be also calculated by employing various well-known phase equilibria models, such as the Wilson, NRTL, Margules, etc., which using information for binary subsystems could predict the activity coefficients in ternary mixtures (Fan and Jafvert, 1997; Domanska, 1990).

Many other methods, mainly empirical and semiempirical, were suggested for the correlation and prediction of the solubility of solids in a mixed solvent. Details regarding these methods and their comparison with experiment were summarized in books and recent publications (Acree, 1984; Prausnitz et al., 1986; Barzegar-Jalali and Jouyban-Gharamaleki, 1996; Jouyban-Gharamaleki et al., 1999).

The solubility of drugs in aqueous mixed solvents often exhibits a maximum in the curve solubility versus mixed solvent composition. This "enhancement" in solubility often greatly exceeds the solubilities not only in water, which is quite natural, but also in nonaqueous cosolvents. Such a dependence could not be explained by simple equations like the log-linear model for the solubility in a mixed solvent (Yalkowsky and Roseman, 1981)

$$\ln x_2^t = \varphi_1 \ln x_2^{b_1} + \varphi_3 \ln x_2^{b_3} \tag{4}$$

where ϕ_i ($i = 1, 3$) is the volume fraction of component i in the mixed solvent 1–3. One should mention that such simple equations provided satisfactory results for systems which did not exhibit maxima.

Various other models for drug solubility in aqueous mixed solvents have been proposed and the results were compared (Barzegar-Jalali and Jouyban-Gharamaleki, 1996; Jouyban-Gharamaleki et al., 1999).

The main difficulty in predicting the solid solubility in a mixed solvent consists in calculating the activity coefficient of a solute in a ternary mixture (γ_2^t). In this paper, the Kirkwood–Buff (KB) theory of solutions (or fluctuation theory) (Kirkwood and Buff, 1951) is employed to analyze the solid (particularly drug) solubility in mixed (mainly aqueous) solvents. The analysis is based on results obtained previously regarding the composition derivatives of the activity coefficients in ternary solutions (Ruckenstein and Shulgin, 2001). These equations were successfully applied to gas solubilities in mixed solvents (Ruckenstein and Shulgin, 2002; Shulgin and Ruckenstein, 2002).

Thus, the aim of the present paper is to apply the fluctuation theory for ternary mixtures to the solubility of drugs in aqueous mixed solvents and to suggest on this basis a simple and accurate method for its correlation.

2. Theory

2.1. The Kirkwood–Buff theory of solution

The KB theory of solution (Kirkwood and Buff, 1951) connects the macroscopic properties of solutions, such as the isothermal compressibility, the concentration derivatives of the chemical potentials, and the partial molar volumes to their microscopic characteristics in the form of spatial integrals involving the radial distribution function.

The key quantities in the KB theory of solution are the so-called Kirkwood–Buff integrals (KBIs), defined as

$$G_{\alpha\beta} = \int_0^\infty (g_{\alpha\beta} - 1) 4\pi r^2 \, dr \tag{5}$$

where $g_{\alpha\beta}$ is the radial distribution function between species α and β, and r is the distance between the centers of molecules α and β. The isothermal compressibility, the concentration derivatives of the chemical potentials, and the partial molar volumes in any multicomponent mixture can be expressed in terms of the KBIs. In this paper, the attention is focused on the concentration derivatives of the chemical potentials, because they can provide useful information regarding the activity coefficient of a solute in a ternary mixture (γ_2^t).

Kirkwood and Buff (Kirkwood and Buff, 1951) obtained the following expression for the concentration derivative of the activity coefficient of component α in a binary mixture α–β:

$$\left(\frac{\partial \ln \gamma_\alpha}{\partial x_\alpha}\right)_{P,T} = \frac{c_\beta^0 (G_{\alpha\alpha} + G_{\beta\beta} - 2G_{\alpha\beta})}{1 + c_\alpha^0 x_\beta (G_{\alpha\alpha} + G_{\beta\beta} - 2G_{\alpha\beta})} \quad (6)$$

where x_i and γ_i are the mole fraction and the activity coefficient of component i in the binary mixture α–β and c_i^0 is the bulk molecular concentrations of component i. The present authors (Ruckenstein and Shulgin, 2001) established explicit expressions for the concentration derivatives of the activity coefficients in a ternary mixture. These expressions are more complicated than Eq. (6), and the only derivative which is of interest in the present paper has the form

$$\left(\frac{\partial \ln \gamma_{2,t}}{\partial x_3^t}\right)_{T,P,x_2^t}$$
$$= -\frac{(c_1 + c_2 + c_3)(c_1[G_{11} + G_{23} - G_{12} - G_{13}] + c_3[-G_{12} - G_{33} + G_{13} + G_{23}])}{c_1 + c_2 + c_3 + c_1 c_2 \Delta_{12} + c_1 c_3 \Delta_{13} + c_2 c_3 \Delta_{23} + c_1 c_2 c_3 \Delta_{123}} \quad (7)$$

where $\Delta_{\alpha\beta}$ and Δ_{123} are defined as follows:

$$\Delta_{\alpha\beta} = G_{\alpha\alpha} + G_{\beta\beta} - 2G_{\alpha\beta}, \quad \alpha \neq \beta \quad (8)$$

and

$$\Delta_{123} = G_{11}G_{22} + G_{11}G_{33} + G_{22}G_{33} + 2G_{12}G_{13}$$
$$+ 2G_{12}G_{23} + 2G_{13}G_{23} - G_{12}^2 - G_{13}^2$$
$$- G_{23}^2 - 2G_{11}G_{23} - 2G_{22}G_{13} - 2G_{33}G_{12} \quad (9)$$

The factors in the square brackets in the numerator of Eq. (7) and Δ_{123} can be expressed in terms of $\Delta_{\alpha\beta}$ as follows

$$G_{12} + G_{33} - G_{13} - G_{23} = \frac{\Delta_{13} + \Delta_{23} - \Delta_{12}}{2} \quad (10)$$

$$G_{11} + G_{23} - G_{12} - G_{13} = \frac{\Delta_{12} + \Delta_{13} - \Delta_{23}}{2} \quad (11)$$

and

$$\Delta_{123} = -\frac{(\Delta_{12})^2 + (\Delta_{13})^2 + (\Delta_{23})^2 - 2\Delta_{12}\Delta_{13} - 2\Delta_{12}\Delta_{23} - 2\Delta_{13}\Delta_{23}}{4} \quad (12)$$

The insertion of Eqs. (10)–(12) into Eq. (7) provides a rigorous expression for the derivative $(\partial \ln \gamma_{2,t}/\partial x_3^t)_{T,P,x_2^t}$ in terms of $\Delta_{\alpha\beta}$ and concentrations.

It should be noted that $\Delta_{\alpha\beta}$ is a measure of the nonideality (Ben-Naim, 1977) of the binary mixture α and β, because for an ideal mixture $\Delta_{\alpha\beta} = 0$. For a ternary mixture 1–2–3, Δ_{123} also constitutes a measure of nonideality. Indeed, inserting $G_{\alpha\beta}^{id}$ for an ideal mixture (Ruckenstein and Shulgin, 2001) into the expression of Δ_{123} one obtains that for an ideal ternary mixture $\Delta_{123} = 0$.

2.2. The activity coefficient of a solute in a mixed solvent at infinite dilution

At infinite dilution of a solute, Eq. (7) can be recast as follows:

$$\lim_{x_2^t \to 0} \left(\frac{\partial \ln \gamma_{2,t}}{\partial x_3^t}\right)_{T,P,x_2^t}$$
$$= -\frac{(c_1^0 + c_3^0)((c_1^0 + c_3^0)(\Delta_{12} - \Delta_{23})_{x_2^t=0} + (c_1^0 - c_3^0)(\Delta_{13})_{x_2^t=0})}{2(c_1^0 + c_3^0 + c_1^0 c_3^0 (\Delta_{13})_{x_2^t=0})} \quad (13)$$

where c_1^0 and c_3^0 are the bulk molecular concentrations of components 1 and 3 in the binary 1–3 solvent.

For a binary 1–3 solvent, Eq. (6) can be rewritten as follows:

$$\left(\frac{\partial \ln \gamma_3}{\partial x_3}\right)_{P,T} = \frac{c_3^0 \Delta_{13}}{1 + c_3^0 x_1 \Delta_{13}} \quad (6a)$$

Eq. (6a) allows one to obtain for Δ_{13} the following expression:

$$\Delta_{13} = \frac{(\partial \ln \gamma_3 / \partial x_3)_{P,T}}{c_3^0 - c_3^0 x_1 (\partial \ln \gamma_3 / \partial x_3)_{P,T}} \quad (6b)$$

Introducing Δ_{13} from Eq. (6b) in Eq. (13) and integrating yields

$$\ln \gamma_2^{t,\infty} = -\int (c_1^0 + c_3^0) \frac{(\Delta_{12} - \Delta_{23})_{x_2^t=0}}{2} \left[1 + x_3^{b,1-3} \left(\frac{\partial \ln \gamma_3^{b,1-3}}{\partial x_3^{b,1-3}}\right)_{P,T}\right] dx_3^{b,1-3}$$

$$+ \frac{1}{2} \int \frac{(x_1^{b,1-3} - x_3^{b,1-3})}{x_1^{b,1-3}} \left(\frac{\partial \ln \gamma_3^{b,1-3}}{\partial x_3^{b,1-3}}\right)_{P,T} dx_3^{b,1-3} + A \quad (14)$$

where $x_i^{b,1-3}$ ($i = 1, 3$) is the mole fraction of component i in the mixed solvent, $\gamma_2^{t,\infty}$ is the activity coefficient of a solute in a mixed solvent at infinite dilution and A is a constant of integration.

Eq. (14) will be used in the next section to derive an expression for the solubility of a solid in a mixed solvent.

2.3. Solubility of poorly soluble solids in an ideal mixed solvent

For poorly soluble solids one can use the infinite dilution approximation and consider that the activity coefficient of a solute in a mixed solvent is equal to the activity coefficient at infinite dilution. Thus, Eqs. (1)–(3) can be rewritten as follows:

$$\frac{f_2^S}{f_2^L(T,P)} = x_2^t \gamma_2^{t,\infty} \quad (15)$$

$$\frac{f_2^S}{f_2^L(T,P)} = x_2^{b_1} \gamma_2^{b_1,\infty} \quad (16)$$

and

$$\frac{f_2^S}{f_2^L(T,P)} = x_2^{b_3} \gamma_2^{b_3,\infty} \quad (17)$$

where $\gamma_2^{b_1,\infty}$ and $\gamma_2^{b_3,\infty}$ are the activity coefficients at infinite dilution of the solute in the pure solvents 1 and 3.

Eq. (14) is a rigorous equation applicable to any ternary mixture.

At this point, two simplifications are introduced which allow one to obtain working expressions for the solubility of poorly soluble solids in an ideal mixed solvent:

(a) $(\Delta_{12})_{x_2^t=0} = (G_{11} + G_{22} - 2G_{12})_{x_2^t=0}$ and
$(\Delta_{23})_{x_2^t=0} = (G_{22} + G_{33} - 2G_{23})_{x_2^t=0}$ are independent of the composition of the solvent mixture, and

(b) the binary solvent 1–3 is ideal and therefore $\gamma_3^{b,1-3} = 1$ and

$$V = x_1^{b,1-3} V_1^0 + x_3^{b,1-3} V_3^0 \quad (18)$$

where V is the molar volume of the binary mixture 1–3, and V_1^0 and V_3^0 are the molar volumes of the individual solvents 1 and 3.

With these two simplifications, Eq. (14) can be rewritten, when $V_1^0 \neq V_3^0$, in the form

$$\ln \gamma_2^{t,\infty} = A(P,T) - \frac{B(P,T) \ln(x_1^{b,1-3} V_1^0 + x_3^{b,1-3} V_3^0)}{V_3^0 - V_1^0} \quad (19)$$

where $B(P,T) = (\Delta_{12} - \Delta_{23})_{x_2^t=0}/2$.

The constants $A(P,T)$ and $B(P,T)$ can be obtained using the following limiting expressions:

$$(\ln \gamma_2^{t,\infty})_{x_1^{b,1-3}=0} = \ln \gamma_2^{b,2-3,\infty} \quad (20)$$

and

$$(\ln \gamma_2^{t,\infty})_{x_3^{b,1-3}=0} = \ln \gamma_2^{b,1-2,\infty} \quad (21)$$

Combining Eqs. (19)–(21) yields the following expression for the activity coefficient of a solute in an ideal mixed solvent at infinite dilution when $V_1^0 \neq V_3^0$

$$\ln \gamma_2^{t,\infty} = \frac{(\ln V - \ln V_3^0) \ln \gamma_2^{b,1-2,\infty} + (\ln V_1^0 - \ln V) \ln \gamma_2^{b,2-3,\infty}}{\ln V_1^0 - \ln V_3^0} \quad (22)$$

Inserting expressions (15)–(17) into Eq. (22) yields the following equation for the solubility of a poorly soluble solid in an ideal mixed solvent:

$$\ln x_2^t = \frac{(\ln V - \ln V_3^0) \ln x_2^{b_1} + (\ln V_1^0 - \ln V) \ln x_2^{b_3}}{\ln V_1^0 - \ln V_3^0},$$

$$V_1^0 \neq V_3^0 \quad (23)$$

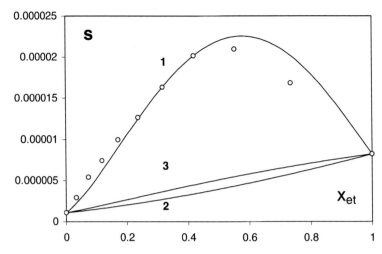

Fig. 1. Comparison between experimental (○) (Jouyban et al., 2002) and predicted (solid lines) solubilities of oxolinic acid (S is the mole fraction of oxolinic acid) in the mixed solvent water/ethanol (x_{et} is the mole fraction of ethanol) at room temperature. 1—solubility calculated using Eqs. (23) and (25), 2—solubility calculated using Eqs. (23) and (18), and 3—the solubility calculated using Eq. (4).

However, when $V_1^0 = V_3^0$, Eq. (23) leads to a nondetermination 0/0. In this case, using the same approximations as in the previous case and taking into account that $V = V_1^0 = V_3^0$, Eq. (14) leads to

$$\ln x_2^t = x_1^{b,1-3} \ln x_2^{b_1} + x_3^{b,1-3} \ln x_2^{b_3} \qquad (24)$$

Eq. (24) is similar to Eq. (4) with the difference that the volume fractions for the mixed solvent are replaced by mole fractions.

Eq. (23), which was derived using the KB theory of solutions for a ternary mixture, can predict the solubility of a poorly soluble solid in an ideal mixed solvent in terms of the solubilities of the solid in the individual constituents of the mixed solvent and their molar volumes.

However, Eq. (23) cannot describe the maximum in the curve of solubility versus mixed solvent composition which was frequently observed experimentally for the solubilities of drugs in aqueous mixed

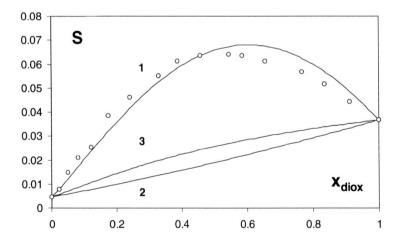

Fig. 2. Comparison between experimental (○) (Bustamante et al., 1993) and predicted (solid lines) solubilities of sulfadiazine (S is the mole fraction of sulfadiazine) in the mixed solvent water/dioxane (x_{diox} is the mole fraction of dioxane) at room temperature. 1—solubility calculated using Eqs. (23) and (25), 2—solubility calculated using Eqs. (23) and (18), and 3—the solubility calculated using Eq. (4).

solvents (Jouyban-Gharamaleki et al., 1999 and references therein). To accommodate this feature of the solubility curve, the molar volume of the mixed solvent will be replaced in Eq. (23) by

$$v = x_1^{b,1-3} V_1^0 + x_3^{b,1-3} V_3^0 + e x_1^{b,1-3} x_3^{b,1-3} \quad (25)$$

where e is an empirical parameter which is evaluated from the solubility data in a mixed solvent. One should not expect for Eq. (25) to satisfactorily represent the molar volume of the mixed solvent. The insertion of Eq. (25) into Eq. (23) leads to a one-parameter semiempirical equation for the solubility of a solid in a mixed solvent. This equation exhibits a maximum in the curve of solubility versus mixed solvent composition (Figs. 1 and 2).

3. Results and discussion

In order to verify the applicability of Eq. (23) combined with the nonideal molar volume of a mixed solvent to the solubility of a drug in an aqueous mixed solvents, 32 experimental sets were selected. Most of them were taken from the paper of

Table 1
The experimental data[a] regarding the solubilities (at room temperature) of drugs in aqueous mixed solvents used in calculations

Systems no.	Cosolvent	Solute	n^b	Reference	Value of e (cm^3/mol) in Eq. (25)
1	N,N-Dimethylformamide	Sulfadiazine	14	Martin et al. (1982)	49.3
2*	N,N-Dimethylformamide	Theophyllene	11	Gonzalez et al. (1994)	45.2
3*	N,N-Dimethylformamide	Caffeine	11	Herrador and Gonzalez (1997)	42.8
4	Dioxane	Caffeine	16	Adjei et al. (1980)	1433.9
5	Dioxane	p-Hydroxybenzoic acid	13	Wu and Martin (1983)	183.2
6	Dioxane	Paracetamol	17	Romero et al. (1996)	365.4
7	Dioxane	Phenacetin	13	Bustamante and Bustamante (1996)	249.8
8	Dioxane	Sulfadiazine	17	Bustamante et al. (1993)	325.9
9	Dioxane	Sulfadimidine	19	Bustamante et al. (1993)	220.5
10	Dioxane	Sulfamethizole	19	Reillo et al. (1995a)	678.6
11	Dioxane	Sulfamethoxazole	15	Bustamante et al. (1993)	199.0
12	Dioxane	Sulfapyridine	17	Reillo et al. (1995b)	390.5
13	Dioxane	Sulfamethoxypyridazine	19	Bustamante et al. (1993)	252.9
14	Dioxane	Sulfanilamide	16	Reillo et al. (1993)	256.3
15	Dioxane	Sulfisomidine	21	Martin et al. (1985)	536.0
16	Dioxane	Theobromine	11	Martin et al. (1981)	348.8
17	Dioxane	Theophyllene	21	Martin et al. (1980)	2317.7
18	Ethanol	Paracetamol	13	Romero et al. (1996)	108.3
19	Ethanol	Sulfamethazine	11	Bustamante et al. (1994)	152.0
20	Ethanol	Sulfanilamide	12	Bustamante et al. (1994)	113.0
21*	Ethanol	Oxolinic acid	11	Jouyban et al. (2002)	261.3
22	Ethylene glycol	Naphthalene	18	Khossravi and Connors (1992)	2.2
23	Ethylene glycol	Theophyllene	17	Khossravi and Connors (1992)	24.7
24	Methanol	Theophyllene	13	Khossravi and Connors (1992)	151.2
25	Propylene glycol	Butyl p-aminobenzoate	11	Rubino and Obeng (1991)	32.5
26	Propylene glycol	Butyl p-hydroxybenzoate	11	Rubino and Obeng (1991)	19.6
27	Propylene glycol	Ethyl p-aminobenzoate	11	Rubino and Obeng (1991)	44.5
28	Propylene glycol	Ethyl p-hydroxybenzoate	11	Rubino and Obeng (1991)	40.5
29	Propylene glycol	Methyl p-aminobenzoate	11	Rubino and Obeng (1991).	43.1
30	Propylene glycol	Methyl p-hydroxybenzoate	11	Rubino and Obeng (1991).	46.8
31	Propylene glycol	Propyl p-aminobenzoate	11	Rubino and Obeng (1991).	34.2
32	Propylene glycol	Propyl p-hydroxybenzoate	11	Rubino and Obeng (1991).	21.8

[a] Most of the references were taken from the paper of Jouyban-Gharamaleki et al. (Jouyban-Gharamaleki et al., 1999), but some additional data (*) were also included.

[b] n is the number of experimental points in each data set.

Table 2
Comparison between the drug solubilities calculated using Eqs. (23) and (25) and literature models

Number of constants	MPD (%)[a]		
	Using Eqs. (23) and (25)	MRS[b]	GSM[c]
3	14.1	15.9	15.9

[a] MPD (%) is the mean percentage deviation defined as $\frac{100\sum_{j=1}^{M}\sum_{i=1}^{N_j}|(x_i^{\exp}-x_i^{\text{calc}})/x_i^{\exp}|}{\sum_{j=1}^{M}N_j}$ where $x_i^{\exp}$ and x_i^{calc} are experimental and calculated solubilities (mole fractions), N_j is the number of experimental points in the data set j (Table 1), M is the number of experimental data sets (here 32).

[b] MRS is the mixture response surface method (Ochsner et al., 1985). The value of MPD was taken from Table 2 of the Jouyban-Gharamaleki et al. (Jouyban-Gharamaleki et al., 1999) paper.

[c] GSM is the general single model (Barzegar-Jalali and Jouyban-Gharamaleki, 1996). The value of MPD was taken from Table 2 of the Jouyban-Gharamaleki et al. (Jouyban-Gharamaleki et al., 1999) paper.

Jouyban-Gharamaleki et al. (Jouyban-Gharamaleki et al., 1999), but some additional data were also included. All selected mixtures and the results of calculations are listed in Tables 1 and 2.

There is only one adjustable parameter (e) in our equation. However, the solubilities of the solute in the individual constituents of the mixed solvent are also needed. Therefore, one can consider our equation as a three-parameter one. For this reason, our results in Table 2 are compared to the best three-parameter equations. One can see from Table 2 that Eq. (23) with the molar volume given by Eq. (25) provides slightly better results than the three-parameter equations available in literature.

Generally speaking, the correlating equations should meet the following criteria:

(a) provide an accurate enough representation of the experimental data,
(b) use a minimum number of adjustable constants,
(c) have some theoretical justification, and
(d) have predictable power.

Regarding criterion (a), 30% for the mean percentage deviation is considered an acceptable error range (Reillo et al., 1995b). Therefore, all equations listed in Table 2 satisfy criterion (a). Of course, one can achieve a much better mathematical representation of the data by using a larger number of adjustable parameters. In the paper by Jouyban-Gharamaleki et al. (Jouyban-Gharamaleki et al., 1999), equations with 4, 5, and 6 adjustable parameters were listed. However, they are devoid of any physical meaning and require numerous experimental points for the parameter estimation. The adjustable parameter (e) in our equation can be found from a single solubility measurement. Furthermore, our Eq. (23) was derived using the fluctuation theory for ternary mixtures and is rigorously valid. It is clear that the idealized model employed cannot predict some peculiar features of real systems, such as the maximum in the curve of solubility versus mixed solvent composition. However, a simple modification (Eq. (25)) enabled Eq. (23) to represent this maximum.

An inspection of the values of the parameter (e) (Table 1) shows that this parameter has always positive values for the systems investigated and depends on the natures of both the drug and cosolvent. For the solubilities of structurally related caffeine and theophyllene in aqueous N,N-dimethylformamide, the values of (e) are close to each other (45.2 and 42.8). However, the values of (e) for the structurally more different sulfonamides (sulfadiazine, sulfadimidine, sulfamethizole, sulfamethoxazole, sulfapyridine, sulfamethoxypyridazine, sulfanilamide, and sulfisomidine) in water/dioxane mixtures differ by a factor of two and even three for sulfamethizole.

The limitations of the proposed method are directly related to the simplifications made. The two most important ones are: (1) the ideality of the mixed solvents and (2) the infinite dilution approximation. Our next papers will be focused on nonideal mixed solvents and on the effect of the finite concentration of a solute.

4. Conclusion

In this paper, the fluctuation theory of solutions was applied to the solubility of drugs in aqueous mixed solvents. A rigorous expression for the composition derivative of the activity coefficient of a solute in a ternary solution (Ruckenstein and Shulgin, 2001) was used to derive an equation for the activity coefficient of a solute at infinite dilution in an ideal mixed solvent and an expression for the solubility of a poorly soluble solid in an ideal mixed solvent (Eq. (23)). This simple equation can predict the solubility in terms of those in the individual constituents of the mixed solvent and their molar volumes. However, this simple equation cannot explain the maximum observed experimentally in the curve of solubility versus mixed solvent composition. By considering that the molar volume of the mixed solvent is nonideal and that the excess volume depends on its composition, the above equation was modified by including one adjustable parameter. This modified equation can be considered a three parameter equation (parameter (e) in Eq. (25) and the two solubilities of the solid in the individual constituents). The semiempirical equation proposed was compared with other three parameter equations for the solubility of drugs in an aqueous mixed solvent.

Acknowledgements

We are indebted to Prof. W.E. Acree, Jr. (Department of Chemistry, University of North Texas, Denton) who provided us with numerous reprints of his papers which have been helpful in the course of this work.

References

Acree, W.E. 1984. Thermodynamic Properties of Nonelectrolyte Solutions. Academic Press, Orlando.

Adjei, A., Newburger, J., Martin, A., 1980. Extended Hildebrand approach: solubility of caffeine in dioxane–water mixtures. J. Pharm. Sci. 69, 659–661.

Barzegar-Jalali, M., Jouyban-Gharamaleki, A., 1996. Models for calculating solubility in binary solvent systems. Int. J. Pharm. 140, 237–246.

Ben-Naim, A., 1977. Inversion of the Kirkwood–Buff theory of solutions: application to the water–ethanol system. J. Chem. Phys. 67, 4884–4890.

Bustamante, C., Bustamante, P., 1996. Nonlinear enthalpy–entropy compensation for the solubility of phenacetin in dioxane–water solvent mixtures. J. Pharm. Sci. 85, 1109–1111.

Bustamante, P., Escalera, B., Martin, A., Selles, E., 1993. A modification of the extended Hildebrand approach to predict the solubility of structurally related drugs in solvent mixtures. J. Pharm. Pharmacol. 45, 253–257.

Bustamante, P., Ochoa, R., Reillo, A., Escalera, J.-B., 1994. Chameleonic effect of sulfanilamide and sulfamethazine in solvent mixtures—solubility curves with 2 maxima. Chem. Pharm. Bull. 42, 1129–1133.

Domanska, U., 1990. Solubility of acetyl-substituted naphthols in binary solvent mixtures. Fluid Phase Equilibria. 55, 125–145.

Fan, C.H., Jafvert, C.T., 1997. Margules equations applied to PAH solubilities in alcohol–water mixtures. Environ. Sci. Technol. 31, 3516–3522.

Gonzalez, A.G., Herrador, M.A., Asuero, A.G., 1994. Solubility of theophylline in aqueous N,N-dimethylformamide mixtures. Int. J. Pharm. 108, 149–154.

Herrador, M.A., Gonzalez, A.G., 1997. Solubility prediction of caffeine in aqueous N,N-dimethylformamide mixtures using the extended Hildebrand solubility approach. Int. J. Pharm. 156, 239–244.

Jouyban, A., Romero, S., Chan, H.K., Clark, B.J., Bustamante, P., 2002. A cosolvency model to predict solubility of drugs at several temperatures from a limited number of solubility measurements. Chem. Pharm. Bull. 50, 594–599.

Jouyban-Gharamaleki, A., Valaee, L., Barzegar-Jalali, M., Clark, B.J., Acree, W.E., 1999. Comparison of various cosolvency models for calculating solute solubility in water–cosolvent mixtures. Int. J. Pharm. 177, 93–101.

Kirkwood, J.G., Buff, F.P., 1951. Statistical mechanical theory of solutions. 1. J. Chem. Phys. 19, 774–782.

Khossravi, D., Connors, K.A., 1992. Solvent effects on chemical processes. 1. Solubility of aromatic and heterocyclic compounds in binary aqueous organic solvents. J. Pharm. Sci. 81, 371–379.

Martin, A., Newburger, J., Adjei, A., 1980. Extended Hildebrand approach: solubility of theophylline in polar binary solvents. J. Pharm. Sci. 69, 487–491.

Martin, A., Paruta, A.N., Adjei, A., 1981. Extended Hildebrand solubility approach—methylxanthines in mixed-solvents. J. Pharm. Sci. 70, 1115–1120.

Martin, A., Wu, P.L., Adjei, A., Lindstrom, R.E., Elworthy, P.H., 1982. Extended Hildebrand solubility approach and the log linear solubility equation. J. Pharm. Sci. 71, 849–856.

Martin, A., Wu, P.L., Velasquez, T., 1985. Extended Hildebrand solubility approach: sulfonamides in binary and ternary solvents. Int. J. Pharm. 74, 277–282.

Ochsner, A.B., Belloto, R.J., Sokoloski, T.D., 1985. Prediction of xanthine solubilities using statistical techniques. J. Pharm. Sci. 74, 132–135.

Prausnitz, J.M., Lichtenthaler, R.N., Gomes de Azevedo, E. 1986. Molecular Thermodynamics of Fluid-phase Equilibria, second ed. Prentice-Hall, Englewood Cliffs, NJ.

Reillo, A., Escalera, B., Selles, E., 1993. Prediction of sulfanilamide solubility in dioxane–water mixtures. Pharmazie 48, 904–907.

Reillo, A., Bustamante, P., Escalera, B., Jimenez, M.M., Selles, E., 1995a. Solubility parameter-based methods for predicting the solubility of sulfapyridine in solvent mixtures. Drug Dev. Ind. Pharm. 21, 2073–2084.

Reillo, A., Cordoba, M., Escalera, B., Selles, E., Cordoba, M., 1995b. Prediction of sulfamethizole solubility in dioxane–water mixtures. Pharmazie 50, 472–475.

Romero, S., Reillo, A., Escalera, J.-B., Bustamante, P., 1996. The behavior of paracetamol in mixtures of amphiprotic and amphiprotic–aprotic solvents. Relationship of solubility curves to specific and nonspecific interactions. Chem. Pharm. Bull. 44, 1061–1064.

Ruckenstein, E., Shulgin, I., 2001. Entrainer effect in supercritical mixtures. Fluid Phase Equilibria 180, 345–359.

Ruckenstein, E., Shulgin, I., 2002. Salting-out or -in by fluctuation theory. Ind. Eng. Chem. Res. 41, 4674–4680.

Rubino, J.T., Obeng, E.K., 1991. Influence of solute structure on deviations from the log-linear solubility equation in propylene-glycol:water mixtures. J. Pharm. Sci. 80, 479–483.

Shulgin, I., Ruckenstein, E., 2002. Henry's constant in mixed solvents from binary data. Ind. Eng. Chem. Res. 41, 1689–1694.

Wu, P.L., Martin, A., 1983. Extended Hildebrand solubility approach: p-hydroxybenzoic acid in mixtures of dioxane and water. J. Pharm. Sci. 72, 587–592.

Yalkowsky, S.H., Roseman, T.J. 1981. Solubilization of drugs by cosolvents. In: Yalkowsky, S.H. (Ed.), Techniques of Solubilization of Drugs. Dekker, New York (Chapter 3).

Solubility of drugs in aqueous solutions
Part 2: Binary nonideal mixed solvent

E. Ruckenstein*, I. Shulgin

Department of Chemical Engineering, State University of New York at Buffalo, Amherst, NY 14260, USA

Received 27 February 2003; received in revised form 28 April 2003; accepted 28 April 2003

Abstract

As in a previous paper [Int. J. Pharm. 258 (2003) 193–201], the Kirkwood–Buff theory of solutions was employed to calculate the solubility of a solid in mixed solvents. Whereas in the former paper the binary solvent was assumed ideal, in the present one it was considered nonideal. A rigorous expression for the activity coefficient of a solute at infinite dilution in a mixed solvent [Int. J. Pharm. 258 (2003) 193–201] was used to obtain an equation for the solubility of a poorly soluble solid in a nonideal mixed solvent in terms of the solubilities of the solute in the individual solvents, the molar volumes of those solvents, and the activity coefficients of the components of the mixed solvent.

The Flory–Huggins and Wilson equations for the activity coefficients of the components of the mixed solvent were employed to correlate 32 experimental data sets regarding the solubility of drugs in aqueous mixed solvents. The results were compared with the models available in literature. It was found that the suggested equation can be used for an accurate and reliable correlation of the solubilities of drugs in aqueous mixed binary solvents. It provided slightly better results than the best literature models but has also the advantage of a theoretical basis.
© 2003 Elsevier Science B.V. All rights reserved.

Keywords: Solubility of drugs; Nonideal binary aqueous solvents; Fluctuation theory

1. Introduction

The solubility of drugs in aqueous mixed solvents often exhibits a maximum as a function of the mixed solvent composition. The higher solubility of a solid solute in a mixed solvent than in either of the pure solvents, was frequently observed (Acree, 1984; Prausnitz et al., 1986) and is not an exception as it seemed several decades ago.

Gordon and Scott (1952) observed an enhanced solubility of phenanthrene in the mixture of cyclohexane and methylene iodine, while Smith et al. (1959) noted such an enhancement in the solubility of iodine in the C_7F_{16}/CCl_4 mixture. In their book, Hildebrand and Scott (1962) pointed out that such enhancements can be predicted in the framework of the regular solution theory (the Scatchard–Hildebrand solubility parameter model), when the solubility parameter of the solid solute lies between those of the two solvents.

However, when the solute or either of the pure solvents is polar, the regular solution theory could no longer provide quantitative agreement regarding the solubility of a solid solute in a mixed solvent (Acree, 1984; Walas, 1985).

Many models, including various modifications of the Scatchard–Hildebrand solubility parameter model, were suggested for the correlation and prediction of

---

* Corresponding author. Tel.: +1-716-645-2911x2214; fax: +1-716-645-3822.
  E-mail addresses: feaeliru@acsu.buffalo.edu (E. Ruckenstein), ishulgin@eng.buffalo.edu (I. Shulgin).

0378-5173/03/$ – see front matter © 2003 Elsevier Science B.V. All rights reserved.

the solubility of solids in mixed solvents. Details regarding these methods and their comparison with experiment have been summarized in books and recent publications (Acree, 1984; Prausnitz et al., 1986; Barzegar-Jalali and Jouyban-Gharamaleki, 1996; Jouyban-Gharamaleki and Acree, 1998; Jouyban-Gharamaleki et al., 1999).

In a previous paper (Ruckenstein and Shulgin, 2003), the Kirkwood–Buff theory of solutions (Kirkwood and Buff, 1951) was employed to obtain an expression for the solubility of a solid (particularly a drug) in binary mixed (mainly aqueous) solvents. A rigorous expression for the composition derivative of the activity coefficient of a solute in a ternary solution (Ruckenstein and Shulgin, 2001) was used to derive an equation for the activity coefficient of the solute at infinite dilution in an ideal binary mixed solvent and further for the solubility of a poorly soluble solid. By considering that the excess volume of the mixed solvent depends on composition, the above equation was modified empirically by including one adjustable parameter. The modified equation was compared with the other three-parameter equations available in the literature to conclude that it provided a better agreement.

In the present paper, an equation for the activity coefficient of a solute at infinite dilution in a nonideal mixed solvent is used to derive expressions for its solubility in a nonideal binary mixed solvent.

The paper is organized as follows: first, an equation for the activity coefficient of a solute at infinite dilution in a binary nonideal mixed solvent (Ruckenstein and Shulgin, 2003) is employed to derive an expression for its solubility in terms of the properties of the mixed solvent. Second, various expressions for the activity coefficients of the cosolvents are inserted into the above equation. Finally, the obtained equations are used to correlate the drug solubilities in binary aqueous mixed solvents and the results are compared with experimental data and other models available in the literature.

2. Theory and formulas

The following rigorous expression for the activity coefficient ($\gamma_2^{t,\infty}$) of a solid solute (the designation of the components in this paper is as follows: the solid solute is component 2, the water is component 3 and the other cosolvent component 1) in a binary mixed solvent at infinite dilution can be written as (Ruckenstein and Shulgin, 2003):

$$\ln \gamma_2^{t,\infty} = -\int \frac{B}{V}\left[1+x_3^{b,1-3}\left(\frac{\partial \ln \gamma_3^{b,1-3}}{\partial x_3^{b,1-3}}\right)_{P,T}\right]dx_3^{b,1-3}$$
$$+ \frac{1}{2}\int \frac{(x_1^{b,1-3}-x_3^{b,1-3})}{x_1^{b,1-3}}$$
$$\times \left(\frac{\partial \ln \gamma_3^{b,1-3}}{\partial x_3^{b,1-3}}\right)_{P,T} dx_3^{b,1-3} + A \qquad (1)$$

where P and T are the pressure and temperature, $x_i^{b,1-3}$ and $\gamma_i^{b,1-3}$ ($i = 1, 3$) are the mole fraction and the activity coefficient of component i in the binary solvent 1–3, V is the molar volume of the binary 1–3 solvent, $A(P, T)$ is a composition-independent constant of integration, and B is a function of the Kirkwood–Buff integrals (Ruckenstein and Shulgin, 2003). If B is considered independent of the composition of the binary mixed solvent 1–3, Eq. (1) can be rewritten in the form:

$$\ln \gamma_2^{t,\infty} = -BI_1 + \frac{I_2}{2} + A \qquad (2)$$

where

$$I_1 = \int \frac{[1+x_3^{b,1-3}((\partial \ln \gamma_3^{b,1-3})/(\partial x_3^{b,1-3}))_{P,T}]}{V}dx_3^{b,1-3} \qquad (3)$$

and

$$I_2 = \int \frac{(x_1^{b,1-3}-x_3^{b,1-3})}{x_1^{b,1-3}}\left(\frac{\partial \ln \gamma_3^{b,1-3}}{\partial x_3^{b,1-3}}\right)_{P,T} dx_3^{b,1-3} \qquad (4)$$

On the other hand, the solubility of a poorly soluble solute in a mixed solvent is given by the expression (Prausnitz et al., 1986):

$$\frac{f_2^S}{f_2^L(T,P)} = x_2^t \gamma_2^{t,\infty} \qquad (5)$$

where $f_2^L(T, P)$ is a hypothetical fugacity of component 2 as a (subcooled) liquid at a given pressure (P) and temperature (T), f_2^S is the fugacity of the pure solid component 2 and x_2^t is the solubility of the solid solute. The combination of Eqs. (2) and (5) provides an expression for the solubility of a poorly soluble solid in a mixed binary solvent:

$$\ln x_2^t = BI_1 - \frac{I_2}{2} + \bar{A}(P, T) \quad (6)$$

where $\bar{A}(P, T) = -A(P, T) + \ln \left[f_2^S / f_2^L(T, P) \right]$.

The integrals I_1 and I_2 can be calculated if the composition dependencies of the activity coefficients and of the molar volume V of the binary mixed solvent 1–3 are known. The solubilities of the solute in the individual solvents are also needed for the calculation of the composition-independent constants $\bar{A}(P, T)$ and $B(P, T)$.

Because it contains the derivative of the activity coefficient and this gives rise to numerical errors, Eq. (6) is not entirely suitable for numerical calculations. Therefore, Eq. (6) was first modified by replacing the derivative of the activity coefficient with a less error-prone quantity.

The integral I_2 can be transformed by using the Gibbs–Duhem Eq. (7) for a binary system at constant temperature and pressure:

$$x_1^{b,1-3} d\ln \gamma_1^{b,1-3} = -x_3^{b,1-3} d\ln \gamma_3^{b,1-3} \quad (7)$$

Consequently,

$$\frac{(x_1^{b,1-3} - x_3^{b,1-3})}{x_1^{b,1-3}} \left(\frac{\partial \ln \gamma_3^{b,1-3}}{\partial x_3^{b,1-3}} \right)_{P,T}$$
$$= \left(\frac{\partial \ln \gamma_1^{b,1-3}}{\partial x_3^{b,1-3}} \right)_{P,T} + \left(\frac{\partial \ln \gamma_3^{b,1-3}}{\partial x_3^{b,1-3}} \right)_{P,T} \quad (8)$$

and the integral I_2 becomes:

$$I_2 = \ln \gamma_1^{b,1-3} + \ln \gamma_3^{b,1-3} \quad (9)$$

The integral I_1 will be modified by assuming that V can be still considered as ideal:

$$V = x_1^{b,1-3} V_1^0 + x_3^{b,1-3} V_3^0 \quad (10)$$

where V_i^0 is the molar volume of the pure component i ($i = 1, 3$).
Then,

$$I_1 = \frac{\ln(x_1^{b,1-3} V_1^0 + x_3^{b,1-3} V_3^0)}{V_3^0 - V_1^0} + \frac{\ln \gamma_3^{b,1-3}}{V_3^0 - V_1^0} - \frac{V_1^0}{V_3^0 - V_1^0}$$
$$\times \int \frac{1}{x_1^{b,1-3} V_1^0 + x_3^{b,1-3} V_3^0} d\ln \gamma_3^{b,1-3} \quad (11)$$

Integrating by parts, Eq. (11) becomes:

$$I_1 = \frac{\ln(x_1^{b,1-3} V_1^0 + x_3^{b,1-3} V_3^0)}{V_3^0 - V_1^0} + \frac{\ln \gamma_3^{b,1-3}}{V_3^0 - V_1^0}$$
$$- \frac{V_1^0}{V_3^0 - V_1^0} \frac{\ln \gamma_3^{b,1-3}}{x_1^{b,1-3} V_1^0 + x_3^{b,1-3} V_3^0}$$
$$- V_1^0 \int \frac{\ln \gamma_3^{b,1-3}}{(x_1^{b,1-3} V_1^0 + x_3^{b,1-3} V_3^0)^2} dx_3^{b,1-3} \quad (12)$$

Eqs. (9) and (12) for the integrals I_1 and I_2 no longer contain the derivative of the activity coefficient. The simplifying assumption that the molar volume can be treated as ideal (Eq. (10)) does not introduce major errors (see Appendix A).

Eq. (6) combined with Eqs. (9) and (12) allows one to calculate the solubility of a poorly soluble solid in a mixed solvent in terms of the solubilities in the individual solvents (which provide the values of $\bar{A}(P, T)$ and $B(P, T)$, their molar volumes and the activity coefficients of the components of the mixed solvent.

There are numerous expressions for the activity coefficients, which can be employed to calculate the solubility of a poorly soluble solid in a mixed binary solvent. In this paper two expressions for the activity coefficients will be used.

(1) The Flory–Huggins equation (Walas, 1985):

$$\ln \gamma_1^{b,1-3} = \ln \left[\frac{\varphi_1}{x_1^{b,1-3}} \right] + \left(1 - \frac{1}{r} \right) \varphi_3 + \chi \varphi_3^2 \quad (13)$$

and

$$\ln \gamma_3^{b,1-3} = \ln \left[\frac{\varphi_3}{x_3^{b,1-3}} \right] + (r - 1)\varphi_1 + \chi \varphi_1^2 \quad (14)$$

(2) The Wilson equation (Wilson, 1964):

$$\ln \gamma_1^{b,1-3}$$
$$= -\ln(x_1^{b,1-3} + x_3^{b,1-3} L_{13}) + x_3^{b,1-3}$$
$$\times \left[\frac{L_{13}}{x_1^{b,1-3} + x_3^{b,1-3} L_{13}} - \frac{L_{31}}{x_3^{b,1-3} + x_1^{b,1-3} L_{31}} \right] \quad (15)$$

and

$$\ln \gamma_3^{b,1-3} = -\ln(x_3^{b,1-3} + x_1^{b,1-3} L_{31}) - x_1^{b,1-3}$$
$$\times \left[\frac{L_{13}}{x_1^{b,1-3} + x_3^{b,1-3} L_{13}} - \frac{L_{31}}{x_3^{b,1-3} + x_1^{b,1-3} L_{31}} \right] \quad (16)$$

In Eqs. (13)–(16), $\varphi_i = (V_i^0 x_i^{b,1-3})/(V_1^0 x_1^{b,1-3} + V_3^0 x_3^{b,1-3})$ is the volume fraction of component i ($i = 1, 3$) in the mixed solvent 1–3, $r = (V_3^0)/(V_1^0)$, χ is the Flory–Huggins interaction parameter and L_{13} and L_{31} are the Wilson parameters. All these parameters are assumed to be composition independent.

Eq. (6) combined with Eqs. (9) and (12) leads to a slightly cumbersome expression. The expression for the integral in Eq. (12) when the Wilson Eq. (16) is used for the activity coefficient $\gamma_3^{b,1-3}$ is given in Appendix B.

3. Calculations and results

For the sake of comparison, the same 32 sets of experimental data regarding the solubility of drugs in aqueous mixed solvents correlated in a previous paper (Ruckenstein and Shulgin, 2003) are used here.

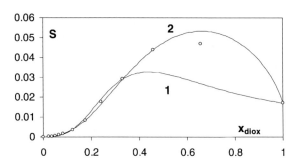

Fig. 1. Comparison between experimental (○) (Bustamante and Bustamante, 1996) and calculated (solid lines) solubilities of phenacetin (S is the mole fraction of phenacetin) in the mixed solvent water/dioxane (x_{diox} is the mole fraction of dioxane) at room temperature. The solubility was calculated using Eq. (6) combined with Eqs. (9) and (12): (1) the activity coefficients expressed via the Flory–Huggins equation; (2) the activity coefficients expressed via the Wilson equation.

Each data set is treated using Eq. (6) combined with Eqs. (9) and (12), with the activity coefficients expressed via the Flory–Huggins or Wilson equations.

The Flory–Huggins interaction parameter (χ) and the Wilson parameters (L_{13} and L_{31}) are considered here adjustable parameters and are calculated from the experimental data regarding the solubility of drugs in aqueous mixed solvents.

The results of the calculations are listed in Tables 1 and 2, and some details are provided for illustration in Figs. 1 and 2.

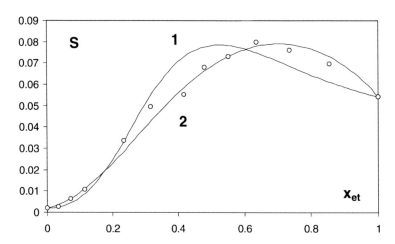

Fig. 2. Comparison between experimental (○) (Romero et al., 1996) and calculated (solid lines) solubilities of paracetamol (S is the mole fraction of paracetamol) in the mixed solvent water/ethanol (x_{et} is the mole fraction of ethanol) at room temperature. The solubility was calculated using Eq. (6) combined with Eqs. (9) and (12): (1) the activity coefficients expressed via the Flory–Huggins equation, (2) the activity coefficients expressed via the Wilson equation.

Table 1
The experimental data[a] regarding the solubilities (at room temperature) of drugs in aqueous mixed solvents and comparison with the predictions

System number	Cosolvent	Solute	Reference	Deviation from experimental data[b]	
				Flory–Huggins activity coefficients	Wilson activity coefficients
1	N,N-Dimethylformamide	Sulfadiazine	Martin et al. (1982)	30.5	13.7
2	N,N-Dimethylformamide	Theophyllene	Gonzalez et al. (1994)	13.4	7.8
3	N,N-Dimethylformamide	Caffeine	Herrador and Gonzalez (1997)	11.1	6.5
4	Dioxane	Caffeine	Adjei et al. (1980)	10.4	9.6
5	Dioxane	p-Hydroxybenzoic acid	Wu and Martin (1983)	15.3	3.5
6	Dioxane	Paracetamol	Romero et al. (1996)	16.5	8.6
7	Dioxane	Phenacetin	Bustamante and Bustamante (1996)	16.2	3.6
8	Dioxane	Sulfadiazine	Bustamante et al. (1993)	12.4	3.5
9	Dioxane	Sulfadimidine	Bustamante et al. (1993)	12.7	3.0
10	Dioxane	Sulfamethizole	Reillo et al. (1995a)	34.3	18.9
11	Dioxane	Sulfamethoxazole	Bustamante et al. (1993)	9.5	2.9
12	Dioxane	Sulfapyridine	Reillo et al. (1995b)	25.9	7.1
13	Dioxane	Sulfamethoxypyridazine	Bustamante et al. (1993)	10.6	2.6
14	Dioxane	Sulfanilamide	Reillo et al. (1993)	19.9	4.0
15	Dioxane	Sulfisomidine	Martin et al. (1985)	23.2	10.6
16	Dioxane	Theobromine	Martin et al. (1981)	3.2	8.4
17	Dioxane	Theophyllene	Martin et al. (1980)	15.5	6.3
18	Ethanol	Paracetamol	Romero et al. (1996)	11.4	6.2
19	Ethanol	Sulfamethazine	Bustamante et al. (1994)	12.3	5.8
20	Ethanol	Sulfanilamide	Bustamante et al. (1994)	14.6	3.2
21	Ethanol	Oxolinic acid	Jouyban et al. (2002)	14.6	5.2
22	Ethylene glycol	Naphthalene	Khossravi and Connors (1992)	13.9	9.8
23	Ethylene glycol	Theophyllene	Khossravi and Connors (1992)	4.8	3.4
24	Methanol	Theophyllene	Khossravi and Connors (1992)	5.2	8.9
25	Propylene glycol	Butyl p-aminobenzoate	Rubino and Obeng (1992)	10.2	9.0
26	Propylene glycol	Butyl p-hydroxybenzoate	Rubino and Obeng (1992)	17.1	24.0
27	Propylene glycol	Ethyl p-aminobenzoate	Rubino and Obeng (1992)	7.2	8.5
28	Propylene glycol	Ethyl p-hydroxybenzoate	Rubino and Obeng (1992)	17.9	2.3
29	Propylene glycol	Methyl p-aminobenzoate	Rubino and Obeng (1992)	4.3	6.6
30	Propylene glycol	Methyl p-hydroxybenzoate	Rubino and Obeng (1992)	7.2	12.4
31	Propylene glycol	Propyl p-aminobenzoate	Rubino and Obeng (1992)	10.3	9.7
32	Propylene glycol	Propyl p-hydroxybenzoate	Rubino and Obeng (1992)	5.2	16.1

[a] The same experimental data regarding the solubility of drugs in aqueous mixed solvents were used in a previous paper (Ruckenstein and Shulgin, 2003).
[b] Deviation from experimental data calculated as MPD (%) (the mean percentage deviation) defined as $(100\sum_{i=1}^{N_j}|(x_i^{exp}-x_i^{calc})/(x_i^{exp})|)/(N_j)$, where x_i^{exp} and x_i^{calc} are experimental and calculated (using Eq. (6) combined with Eqs. (9) and (12)) solubilities (mole fractions). N_j is the number of experimental points in the data set j.

Table 2
Comparison between the results of calculation of the drug solubilities using the present method (Eq. (6)) combined with Eqs. (9) and (12) and various literature models

Number of parameters	MPD (%)[a]				
	Eq. (6) combined with Eqs. (9) and (12)		Literature models		
	Flory–Huggins activity coefficients	Wilson activity coefficients	MRS[b]	GSM[c]	CNIBS/R-K[d]
3	14.4	–	15.9	15.9	22.3
4	–	7.7	18.7	9.1	10.7

[a] MPD (%) is the mean percentage deviation defined as $(100\sum_{j=1}^{M}\sum_{i=1}^{N_j}|(x_i^{exp} - x_i^{calc})/x_i^{exp}|)/(\sum_{j=1}^{M} N_j)$ where x_i^{exp} and x_i^{calc} are experimental and calculated solubilities (mole fractions), N_j is the number of experimental points in the data set (see Table 1) and M is the number of experimental data sets (here 32).

[b] MRS is the mixture response surface method (Ochsner et al., 1985). The value of MPD was taken from Table 2 of the Jouyban-Gharamaleki et al. (1999) paper.

[c] GSM is the general single model (Barzegar-Jalali and Jouyban-Gharamaleki, 1996). The value of MPD was taken from Table 2 of the Jouyban-Gharamaleki et al., 1999) paper.

[d] CNIBS/R-K is the combined nearly ideal binary solvent/Redlich–Kister equations (Acree et al., 1991). The value of MPD was taken from Table 2 of the Jouyban-Gharamaleki et al. (1999) paper.

When the activity coefficients are expressed via the Flory–Huggins equation, one adjustable parameter is introduced. When, however, the Wilson expressions are employed, two adjustable parameters are needed. The solubilities of the solute in the individual solvents are also necessary to calculate the composition-independent constants $\bar{A}(P, T)$ and $B(P, T)$. Therefore, our method can be considered a three-parameter method, when based on the Flory–Huggins equations, and a four-parameter one, when based on the Wilson equations (see Table 2).

4. Discussion

One can see from Tables 1 and 2 that our methods for the correlation of the solubility of drugs in aqueous mixed solvents provide accurate and reliable results. A comparison with the models available in the literature (Table 2) demonstrates that our Eq. (6) provides slightly better results than the best literature models with the same number of parameters.

Only one- and two-parameter activity coefficient expressions were employed in this paper. However,

Table 3
The Wilson parameters (L_{13} and L_{31}) determined from the solubilities of sulfadiazine, sulfadimidine, sulfamethizole, sulfamethoxazole, sulfapyridine, sulfamethoxypyridazine, sulfanilamide and sulfisomidine in water (1)/dioxane (3) mixtures

Solute	The Wilson parameters		Deviation (%) from experimental data, when the average values of the Wilson parameters are used
	L_{13}	L_{31}	
Sulfadiazine	0.11	0.10	19.4
Sulfadimidine	0.27	0.12	35.4
Sulfamethoxazole	0.19	0.12	31.1
Sulfamethoxypyridazine	0.19	0.10	22.0
Sulfamethizole	0.20	0.01	57.3
Sulfapyridine	0.31	0.04	20.7
Sulfanilamide	0.17	0.04	25.4
Sulfisomidine	0.12	0.02	48.6
Average	0.195	0.069	32.4

expressions for the activity coefficients with any number of parameters can be used.

It should be emphasized that the parameters involved in the activity coefficients are adjustable parameters which cannot be obtained easily from the properties of the mixed solvents, for instance the vapor–liquid equilibria. However, for the solubilities of structurally related caffeine and theophyllene in water/N,N-dimethylformamide, the values of the Wilson parameters are close to each other (1.96 and 0.12 for caffeine and 1.81 and 0.10 for theophyllene). If the Wilson parameters for theophyllene are used to predict the solubility of caffeine in water/N,N-dimethylformamide, a deviation of 8.8% from experimental data is obtained. The deviation was, however, 6.5% when the Wilson parameters were determined by fitting the experimental solubility data (Table 1). The values of the Wilson parameters determined from the solubilities of the structurally more different sulfonamides (sulfadiazine, sulfadimidine, sulfamethizole, sulfamethoxazole, sulfapyridine, sulfamethoxypyridazine, sulfanilamide and sulfisomidine) in water/dioxane mixtures are listed in Table 3. Even for such cases, the average values of the Wilson parameters can be used for a first estimation of the solubilities of the above group of drugs (Table 3).

Eq. (6) is a rigorous equation for the solubility of poorly soluble solids in a mixed solvent. The only approximation involved is that the solubilities of the solid in either of the pure solvents and in the mixed solvent are very small (infinite dilution approximation). It is not applicable when at least one of these solubilities has an appreciable value. Indeed (see Table 1), when the solubility of a solute in a nonaqueous solvent exceeds about 5 mol%, such as the solubilities of drugs in propylene glycol (Rubino and Obeng, 1992), the deviation from the experimental data is about 11.5% for the Wilson equation, whereas the average deviation for all 32 mixtures of Table 1 is only 7.7%.

5. Conclusion

In this paper, the fluctuation theory of solutions was applied to the solubility of drugs in aqueous mixed solvents. A rigorous expression for the activity coefficient of a solute at infinite dilution in a real mixed solvent was used to derive an equation for the solubility of poorly soluble solutes, such as drugs, in mixed solvents. The latter solubility is expressed in terms of the solubilities in the individual solvents, their molar volumes, and the activity coefficients of the constituents of the binary solvent. For illustration purposes, the one-parameter (Flory–Huggins) and the two-parameter (Wilson) expressions were employed for the activity coefficients of the constituents of the solvent.

Thirty-two experimental data sets were selected and used to test the equation suggested. The results were compared with the models available in literature. It was found that the suggested equation provides an accurate and reliable correlation of the solubility of drugs in aqueous mixed solvents with slightly better results than the best of the literature models.

Appendix A

The aim of this appendix is to evaluate the sensitivity of the integral in Eq. (12) to the ideality assumption of the molar volume. For this purpose, the composition dependence of $(\ln \gamma_3^{b,1-3})/(V^2)$ for the mixture water/1,4-dioxane at 25 °C was calculated for two cases: (1) $V^E = 0$ (V^E being the excess molar volume); and (2) $V^E \neq 0$ (the mixture water/1,4-dioxane was selected because it is the most frequently used mixed solvent considered in the present paper). The activity coefficient of water in water/1,4-dioxane mixture was calculated using the Wilson equation with the parameters provided by the Gmehling VLE compilation (Gmehling and Onken, 1977). The molar volume of the mixed solvent was calculated using the expression:

$$V = x_1^{b,1-3} V_1^0 + x_3^{b,1-3} V_3^0 \qquad (A.1)$$

or

$$V = x_1^{b,1-3} V_1^0 + x_3^{b,1-3} V_3^0 + V^E \qquad (A.2)$$

The composition dependence of the excess molar volume of the mixture water/1,4-dioxane at 25 °C was found in a paper by Aminabhavi and Gopalakrishna (1995).

The composition dependence of the integrant $(\ln \gamma_3^{b,1-3})/(V^2)$ is presented in Fig. 3, which demonstrates that the numerical values of the integrand are

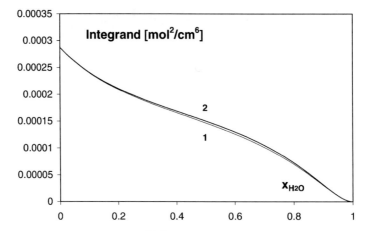

Fig. 3. The composition dependence of the integrand $(\ln \gamma_3^{b,1-3})/(V^2)$ of the integral in Eq. (12). (1) The molar volume of the mixed solvent was calculated using Eq. (A.1), (2) the molar volume of the mixed solvent was calculated using Eq. (A.2).

almost the same for the molar volumes expressed via both Eq. (A.1) or Eq. (A.2).

Appendix B

The aim of this appendix is to derive an analytical expression for the integral:

$$I = \int \frac{\ln \gamma_3^{b,1-3}}{(x_1^{b,1-3} V_1^0 + x_3^{b,1-3} V_3^0)^2} dx_3^{b,1-3} \quad (B.1)$$

in Eq. (12), when the activity coefficient is expressed via the Wilson Eq. (16).

The integration leads to:

$$I = \frac{L_{13}^2 (\ln V - \ln(x_1^{b,1-3} + x_3^{b,1-3} L_{13}))}{(V_3^0 - L_{13} V_1^0)^2} + \frac{L_{31}(-\ln V + \ln(x_3^{b,1-3} + x_1^{b,1-3} L_{31}))}{(V_1^0 - L_{31} V_3^0)^2}$$
$$+ \frac{V_3^0 (L_{13} L_{31} V_3^0 - L_{13} V_1^0 + L_{13} L_{31} V_1^0 - L_{31} V_3^0)}{(V_3^0 - L_{13} V_1^0)(L_{31} V_3^0 - V_1^0)(V_3^0 - V_1^0) V} + \frac{(-1 + L_{31}) \ln((1 - L_{31})V)}{(V_3^0 L_{31} - V_1^0)(V_3^0 - V_1^0)}$$
$$+ \frac{(x_3^{b,1-3} + x_1^{b,1-3} L_{31}) \ln(x_3^{b,1-3} + x_1^{b,1-3} L_{31})}{(L_{31} V_3^0 - V_1^0) V} \quad (B.2)$$

where $V = x_1^{b,1-3} V_1^0 + x_3^{b,1-3} V_3^0$.

References

Acree, W.E., 1984. Thermodynamic Properties of Nonelectrolyte Solutions. Academic Press, Orlando.

Acree, W.E., McCargar, J.W., Zvaigzne, A.I., Teng, I.L., 1991. Mathematical representation of thermodynamic properties. Carbazole solubilities in binary alkane + dibutyl ether and alkane + tetrahydropyran solvent mixtures. Phys. Chem. Liq. 23, 27–35.

Adjei, A., Newburger, J., Martin, A., 1980. Extended Hildebrand approach: solubility of caffeine in dioxane–water mixtures. J. Pharm. Sci. 69, 659–661.

Aminabhavi, T.M., Gopalakrishna, B., 1995. Density, viscosity, refractive index, and speed of sound in aqueous mixtures of N,N-dimethylformamide, dimethylsulfoxide, N,N-dimethylacetamide, acetonitrile, ethylene glycol, diethylene glycol, 1,4-dioxane, tetrahydrofuran, 2-methoxyethanol, and 2-ethoxyethanol at 298.15 K. J. Chem. Eng. Data 40, 856–861.

Barzegar-Jalali, M., Jouyban-Gharamaleki, A., 1996. Models for calculating solubility in binary solvent systems. Int. J. Pharm. 140, 237–246.

Bustamante, C., Bustamante, P., 1996. Nonlinear enthalpy–entropy compensation for the solubility of phenacetin in dioxane–water solvent mixtures. J. Pharm. Sci. 85, 1109–1111.

Bustamante, P., Escalera, B., Martin, A., Selles, E., 1993. A modification of the extended Hildebrand approach to predict the solubility of structurally related drugs in solvent mixtures. J. Pharm. Pharmacol. 45, 253–257.

Bustamante, P., Ochoa, R., Reillo, A., Escalera, J.-B., 1994. Chameleonic effect of sulfanilamide and sulfamethazine in solvent mixtures—solubility curves with 2 maxima. Chem. Pharm. Bull. 42, 1129–1133.

Gmehling, J., Onken, U., 1977. Vapor–Liquid Equilibrium Data Collection; DECHEMA Chemistry Data Series, vol. I. Part 1. DECHEMA, Frankfurt, Germany.

Gonzalez, A.G., Herrador, M.A., Asuero, A.G., 1994. Solubility of theophylline in aqueous N,N-dimethylformamide mixtures. Int. J. Pharm. 108, 149–154.

Gordon, L.J., Scott, R.L., 1952. Enhanced solubility in solvent mixtures. I. The system phenanthrene–cyclohexane–methylene iodide. J. Am. Chem. Soc. 74, 4138–4140.

Herrador, M.A., Gonzalez, A.G., 1997. Solubility prediction of caffeine in aqueous N,N-dimethylformamide mixtures using the extended Hildebrand solubility approach. Int. J. Pharm. 156, 239–244.

Hildebrand, J.H., Scott, R.L., 1962. Regular Solution. Prentice-Hall, Englewood Cliffs, NJ.

Jouyban-Gharamaleki, A., Acree, W.E., 1998. Comparison of models for describing multiple peaks in solubility profiles. Int. J. Pharm. 167, 177–182.

Jouyban-Gharamaleki, A., Valaee, L., Barzegar-Jalali, M., Clark, B.J., Acree, W.E., 1999. Comparison of various cosolvency models for calculating solute solubility in water–cosolvent mixtures. Int. J. Pharm. 177, 93–101.

Jouyban, A., Romero, S., Chan, H.K., Clark, B.J., Bustamante, P., 2002. A cosolvency model to predict solubility of drugs at several temperatures from a limited number of solubility measurements. Chem. Pharm. Bull. 50, 594–599.

Kirkwood, J.G., Buff, F.P., 1951. Statistical mechanical theory of solutions. 1. J. Chem. Phys. 19, 774–782.

Khossravi, D., Connors, K.A., 1992. Solvent effects on chemical processes. 1. Solubility of aromatic and heterocyclic-compounds in binary aqueous organic-solvents. J. Pharm. Sci. 81, 371–379.

Martin, A., Newburger, J., Adjei, A., 1980. Extended Hildebrand approach: solubility of theophylline in polar binary solvents. J. Pharm. Sci. 69, 487–491.

Martin, A., Paruta, A.N., Adjei, A., 1981. Extended Hildebrand solubility approach—methylxanthines in mixed-solvents. J. Pharm. Sci. 70, 1115–1120.

Martin, A., Wu, P.L., Adjei, A., Lindstrom, R.E., Elworthy, P.H., 1982. Extended Hildebrand solubility approach and the log linear solubility equation. J. Pharm. Sci. 71, 849–856.

Martin, A., Wu, P.L., Velasquez, T., 1985. Extended Hildebrand solubility approach: sulfonamides in binary and ternary solvents. Int. J. Pharm. 74, 277–282.

Ochsner, A.B., Belloto, R.J., Sokoloski, T.D., 1985. Prediction of xanthine solubilities using statistical techniques. J. Pharm. Sci. 74, 132–135.

Prausnitz, J.M., Lichtenthaler, R.N., Gomes de Azevedo, E., 1986. Molecular Thermodynamics of Fluid—Phase Equilibria, 2nd ed. Prentice-Hall, Englewood Cliffs, NJ.

Reillo, A., Escalera, B., Selles, E., 1993. Prediction of sulfanilamide solubility in dioxane–water mixtures. Pharmazie 48, 904–907.

Reillo, A., Bustamante, P., Escalera, B., Jimenez, M.M., Selles, E., 1995a. Solubility parameter-based methods for predicting the solubility of sulfapyridine in solvent mixtures. Drug Dev. Ind. Pharm. 21, 2073–2084.

Reillo, A., Cordoba, M., Escalera, B., Selles, E., Cordoba, M., 1995b. Prediction of sulfamethizole solubility in dioxane–water mixtures. Pharmazie 50, 472–475.

Romero, S., Reillo, A., Escalera, J.-B., Bustamante, P., 1996. The behavior of paracetamol in mixtures of amphiprotic and amphiprotic-aprotic solvents. Relationship of solubility curves to specific and nonspecific interactions. Chem. Pharm. Bull. 44, 1061–1064.

Ruckenstein, E., Shulgin, I., 2001. Entrainer effect in supercritical mixtures. Fluid Phase Equilibria 180, 345–359.

Ruckenstein, E., Shulgin, I., 2003. Solubility of drugs in aqueous solutions. Part 1. Ideal mixed solvent approximation. Int. J. Pharm. 258, 193–201.

Rubino, J.T., Obeng, E.K., 1992. Influence of solute structure on deviations from the log-linear solubility equation in propylene-glycol:water mixtures. J. Pharm. Sci. 80, 479–483.

Smith, E.B., Walkley, J., Hildebrand, J.H., 1959. Intermolecular forces involving chlorofluorocarbons. J. Phys. Chem. 63, 703–704.

Walas, S.M., 1985. Phase Equilibria in Chemical Engineering. Butterworth, Boston.

Wilson, G.M., 1964. Vapor–liquid equilibrium. XI. A new expression for the excess free energy of mixing. J. Am. Chem. Soc. 86, 127–130.

Wu, P.L., Martin, A., 1983. Extended Hildebrand solubility approach: p-hydroxybenzoic acid in mixtures of dioxane and water. J. Pharm. Sci. 72, 587–592.

Solubility of drugs in aqueous solutions
Part 3: Multicomponent mixed solvent

E. Ruckenstein*, I. Shulgin

Department of Chemical Engineering, State University of New York at Buffalo, Amherst, NY 14260, USA

Received 17 June 2003; received in revised form 21 August 2003; accepted 21 August 2003

Abstract

The results obtained previously by Ruckenstein and Shulgin [Int. J. Pharm. 258 (2003a) 193; Int. J. Pharm. 260 (2003b) 283] via the fluctuation theory of solutions regarding the solubility of drugs in binary aqueous mixed solvents were extended in the present paper to multicomponent aqueous solvents. The multicomponent mixed solvent was considered to behave as an ideal solution and the solubility of the drug was assumed small enough to satisfy the infinite dilution approximation.

An expression derived for the activity coefficient of a solid solute in a multicomponent solvent was used to obtain an equation for the solubility of a drug in terms of its solubilities in two subsystems of the multicomponent solvent and their molar volumes. Ultimately the solubility can be expressed in terms of those in binary or even in individual solvents and their molar volumes.

The method was applied to the solubility of tioconazole and 19-Nor-1α,25-dihydrovitamin D_2 in several ternary and in a quaternary aqueous mixed solvents. The predicted solubilities were compared with experimental data and good agreement was found.

© 2003 Elsevier B.V. All rights reserved.

Keywords: Solubility of drugs; Multicomponent mixed solvent; Fluctuation theory

1. Introduction

The two previous papers (Ruckenstein and Shulgin, 2003a, b) of this series were focused on the solubility of a solid (particularly a drug) in binary mixed (mainly aqueous) solvents. The present paper extends the method suggested in the above publications to the solubility of drugs in ternary and multicomponent mixed solvents.

While the binary aqueous mixed solvents usually increase the solubility of a poorly soluble drug compared to that in pure water, they could also increase the risk of toxicity. The right selection of a ternary and multicomponent aqueous mixed solvent can, however, improve the solubility of the drug with minimal toxic effects (Lachman et al., 1976).

The pharmaceutical practice has shown that many marketed liquid formulations, which utilize cosolvents, involve multiple solvents (Yalkowsky and Roseman, 1981). However, the experimental determinations of the solubilities in multicomponent solutions are time-consuming because of the large number of compositions needed to cover the concentration ranges of interest and can be very expensive because of the high prices of some modern drugs. For this reason, it is important to provide a reliable method for predicting the solubility of drugs in multicomponent

---

* Corresponding author. Tel.: +1-716-645-2911x2214; fax: +1-716-645-3822.
  E-mail addresses: feaeliru@acsu.buffalo.edu (E. Ruckenstein), ishulgin@eng.buffalo.edu (I. Shulgin).

0378-5173/$ – see front matter © 2003 Elsevier B.V. All rights reserved.

mixed solvents from available experimental solubilities in subsystems such as pure solvents, binary mixed solvents, etc.

For the solubility of a solid (solute, component 2) in a $(n-1)$ multicomponent mixed solvent one can write the following equation (Acree, 1984; Prausnitz et al., 1986):

$$\ln(x_2^n) = \ln\left(\frac{f_2^S}{f_2^L(T, P)}\right) - \ln(\gamma_2^n(T, P, \{x\})) \quad (1)$$

where x_2^n is the solubility (mole fraction) of the solid component 2 in a $(n-1)$-component mixed solvent, γ_2^n is the activity coefficients of the solid in its saturated solutions (n-component mixture composed of solute $+ (n-1)$-component mixed solvent), $f_2^L(T, P)$ is the hypothetical fugacity of the solid as a (subcooled) liquid at a given pressure (P) and temperature (T), f_2^S is the fugacity of a pure solid component 2, and $\{x\}$ indicates that the activity coefficient of the solid solute depends on composition. If the solubility of a $(n-1)$-component mixed solvent in the solid phase is negligible, then the right hand side of Eq. (1) depends only on the properties of the solute and its activity coefficient in the saturated solution of the n-component mixture.

The calculation of the activity coefficient of a solid in a saturated solution of a n-component mixture constitutes the main difficulty in predicting the solid solubility. Generally speaking, the activity coefficient of a solid in a saturated solution of a n-component mixture can be predicted using either group-contribution methods, such as UNIFAC and ASOG, or the experimental solubilities of the solid in subsystems of the multi-component mixed solvent combined with the Wilson, NRTL, etc. equation (Acree, 1984; Prausnitz et al., 1986).

The application of UNIFAC to the solubility of naphthalene in nonaqueous mixed solvents provided satisfactory results when compared to experimental data (Acree, 1984). However, the UNIFAC was inaccurate in predicting the solubilities of solids in aqueous solutions (Fan and Jafvert, 1997). Furthermore, the application of the traditional UNIFAC to mixtures containing a polymer or another large molecule, such as a drug, and low molecular weight solvents is debatable (Fredenslund and Sørensen, 1994). The reason is that the UNIFAC parameters were determined mostly from equilibrium properties of mixtures formed of low molecular weight compounds.

The prediction of the activity coefficient of a solid in its saturated solution in a n-component mixture from the experimental solubilities of the solid in subsystems, such as binary mixed solvents or even individual solvents, is very attractive, because the solubilities in many of the binary mixed solvents and individual solvents are known or can be determined rapidly and their determinations are cheaper than for multicomponent mixed solvents. The method most often used for the solubility of a solid in ternary and multicomponent mixed solvents is the combined nearly ideal binary solvent/Redlich–Kister equation (Acree et al., 1991). This method was applied to the solubility of a solid in ternary nonaqueous mixed solvents and even to the solubility of a solid in a 7-component nonaqueous mixed solvent (Jouyban-Gharamaleki et al., 2000a; Deng et al., 1999). Jouyban-Gharamaleki et al. (2000b) suggested to apply this method also to the solubility of drugs in multicomponent aqueous mixed solvents.

Recently (Ruckenstein and Shulgin, 2003c), a method was suggested to calculate the activity coefficient of a poorly soluble solid in an ideal multicomponent solvent in terms of its activity coefficients at infinite dilution in some subsystems of the multicomponent solvent. The method, based on the fluctuation theory of solutions (Kirkwood and Buff, 1951), provided the following expression for the activity coefficient of a poorly soluble solid solute in an ideal multicomponent solvent:

$$(\ln \gamma_2^{n,\infty})_{x_{i\neq 1,3}^n} = -\left(\frac{B \ln V}{(V_3^0 - V_1^0)}\right)_{x_{i\neq 1,3}^n} + A \quad (2)$$

where $\gamma_2^{n,\infty}$ is the activity coefficient of the solid solute (denoted 2) in a n-component mixture (solute $+ (n-1)$-component solvent), V is the molar volume of an ideal $(n-1)$-component solvent, V_i^0 is the molar volume of the individual i-solvent, x_i^n is the mole fraction of component i in the n-component mixture, and A and B are composition independent constants. The constants A and B can be determined from the activity coefficients of the solid solute in two $(n-1)$-component mixtures with the mole fraction of component 1 zero in one of them and the mole fraction of component 3 zero in the other one. Expression

(2) was used to predict the gas solubilities and the solubilities of solid nonelectrolytes in multicomponent mixed solvents (Ruckenstein and Shulgin, 2003c).

Expression (2) implies that $V_1^0 \neq V_3^0$. When $V_1^0 = V_3^0$, another expression for the activity coefficient of a poorly soluble solid solute in an ideal multicomponent solvent was obtained (Ruckenstein and Shulgin, 2003a):

$$(\ln \gamma_2^{n,\infty})_{x_{i\neq1,3}^n} = -\left(\frac{Bx_3^n}{V}\right)_{x_{i\neq1,3}^n} + A \qquad (2A)$$

Details regarding such cases are provided in the above cited paper. In the present paper, only expression (2) will be employed to predict the solubility of drugs in ternary and quaternary aqueous mixed solvents. It should be emphasized that Eq. (2) remains valid even for small differences between V_1^0 and V_3^0; it is not valid only when V_1^0 is mathematically equal to V_3^0 (very rare case).

2. Solubility of drugs in a multicomponent mixed solvent

In order to apply Eq. (2) to the solubility of a solid solute in a $(n-1)$-component solvent, one must calculate the constants A and B. For this purpose, we consider a $(n-1)$-component solvent with mole fractions $x_1^n, x_3^n, \ldots, x_n^n$, among which, as required by Eq. (2), all mole fractions with the exception of x_1^n and x_3^n are constant. Because $x_1^n + \sum_{i=3}^n x_i^n = 1$, it is clear that the sum of the mole fractions of components 1 and 3 must be constant. Consequently, the composition of the $(n-1)$-component solvent can be changed along the line $x_1^n + x_3^n = $ const. To determine the constants A and B one can use two limiting $(n-2)$-component solvents (along the line $x_1^n + x_3^n = $ const); the mole fraction of component i in one of them will be denoted y_i^{n-1} and in the other z_i^{n-1}. In the first, the mole fraction of component 3, y_3^{n-1}, and in the other one the mole fraction of component 1, z_1^{n-1}, is taken zero. Because $y_1^{n-1} + y_3^{n-1} = z_1^{n-1} + z_3^{n-1} = x_1^n + x_3^n = $ const, one obtains that $y_1^{n-1} = x_1^n + x_3^n$ and $z_3^{n-1} = x_1^n + x_3^n$.

Consequently,

- In the first limiting case, denoted I, the mole fractions are $y_1^{n-1} = x_1^n + x_3^n$, $y_3^{n-1} = 0$, $y_4^{n-1} = x_4^n, \ldots, y_n^{n-1} = x_n^n$ with $y_1^{n-1} + \sum_{i=3}^n y_i^{n-1} = 1$ and the mole fraction of the solute is y_2^{n-1}.
- In the second limiting case, denoted II, the mole fractions are $z_1^{n-1} = 0$, $z_3^{n-1} = x_1^n + x_3^n$, $z_4^{n-1} = x_4^n, \ldots, z_n^{n-1} = x_n^n$ with $\sum_{i=3}^n z_i^{n-1} = 1$ and the mole fraction of the solute is z_2^{n-1}.
- In the limiting cases I and II, Eq. (2) acquires the form:

$$\ln(\gamma_2^{n-1(I),\infty}) = -\frac{B \ln V^{(I)}}{(V_3^0 - V_1^0)} + A \qquad (3)$$

$$\ln(\gamma_2^{n-1(II),\infty}) = -\frac{B \ln V^{(II)}}{(V_3^0 - V_1^0)} + A \qquad (4)$$

where $V^{(I)}$ and $V^{(II)}$ are the molar volumes of the mixtures composed of $(n-2)$-component solvents I and II and the solid solute, respectively. Furthermore, for a poorly soluble solid, the molar volumes of the mixtures can be taken equal to the molar volumes of the solvents.

When the solubility of the solute is small (which is typical for drugs in aqueous mixed solvents), one can write the following expressions (see Eq. (1)) for the solubility of a solute in the above multicomponent mixed solvents:

$$\ln(x_2^n) = \ln\left(\frac{f_2^S}{f_2^L(T,P)}\right) - \ln(\gamma_2^{n,\infty}) \qquad (5)$$

$$\ln(y_2^{n-1}) = \ln\left(\frac{f_2^S}{f_2^L(T,P)}\right) - \ln(\gamma_2^{n-1(I),\infty}) \qquad (6)$$

and

$$\ln(z_2^{n-1}) = \ln\left(\frac{f_2^S}{f_2^L(T,P)}\right) - \ln(\gamma_2^{n-1(II),\infty}) \qquad (7)$$

where $\gamma_2^{n-1(I),\infty}$ and $\gamma_2^{n-1(II),\infty}$ are the activity coefficients of the solid solute at infinite dilution in the $(n-2)$-component solvents I and II, respectively.

Taking into account Eqs. (3) and (4), Eq. (2) can be recast as:

$$(\ln \gamma_2^{n,\infty})_{x_{i\neq1,3}^n} = \frac{(\ln V - \ln V^{(II)})\ln(\gamma_2^{n-1(I),\infty}) + (\ln V^{(I)} - \ln V)\ln(\gamma_2^{n-1(II),\infty})}{\ln V^{(I)} - \ln V^{(II)}} \qquad (8)$$

Eq. (8) provides an expression for the activity coefficient of a poorly soluble solid at infinite dilution in an ideal $(n-1)$-component mixed solvent in terms of its molar volume and the activity coefficients at infinite dilution in the two limiting cases I and II and their molar volumes.

The combination of Eq. (8) with Eqs. (5)–(7) yields an expression for the solubility of a poorly soluble solid in an ideal $(n-1)$-component mixed solvent in terms of its solubilities in the ideal $(n-2)$-component mixed solvents I and II and their molar volumes.

$$\ln(x_2^n) = \frac{(\ln V - \ln V^{(\text{II})})\ln(y_2^{n-1}) + (\ln V^{(\text{I})} - \ln V)\ln(z_2^{n-1})}{\ln V^{(\text{I})} - \ln V^{(\text{II})}} \quad (9)$$

Furthermore, the solubilities of a poorly soluble solid in ideal $(n-2)$-component mixed solvents I and II can be expressed through those in the ideal $(n-3)$-component mixed solvents and so on. Therefore, the suggested procedure allows one to predict the solubility of a poorly soluble solid in an ideal $(n-1)$-component mixed solvent from the solubilities in binary mixed solvents or even from the solubilities in the individual solvents.

3. Comparison with experiment

3.1. Ternary mixed solvents

The experimental solubility of tioconazole (Gould et al., 1984) in the following mixed solvents:

(1) ethanol–propylene glycol–water,
(2) ethanol–polyethylene glycol 400 (PEG 400)–water,
(3) propylene glycol–PEG 400–water,
   and the solubility of 19-Nor-1α,25-dihydrovitamin D_2 (an analog of vitamin D_2) (Stephens et al., 1999) in
(4) ethanol–propylene glycol–water

were selected for comparison of the developed method with experiment.

The above systems were selected because the experimental solubilities of tioconazole in the binary mixed solvents: ethanol–water, propylene glycol–water and PEG 400–water, and the solubilities of 19-Nor-1α,25-dihydrovitamin D_2 in the binary mixed solvents: ethanol–water and propylene glycol–water are available (Gould et al., 1984; Stephens et al., 1999).

The solubilities of the drugs in ternary aqueous mixed solvents were calculated from those in binary aqueous mixed solvents using Eq. (9). The solubilities in the limiting binary aqueous mixed solvents (y and z) were evaluated using two different procedures:

(1) The experimental solubility data were correlated using the following relation (Ruckenstein and Shulgin, 2003a):

$$\ln(x_2^{(b)}) = \frac{(\ln V^{(b)} - \ln V^{(\text{H}_2\text{O})})\ln(x_2^{(\text{co})}) + (\ln V^{(\text{co})} - \ln V^{(b)})\ln(x_2^{(\text{H}_2\text{O})})}{\ln V^{(\text{co})} - \ln V^{(\text{H}_2\text{O})}} \quad (10)$$

where $x_2^{(b)}$ is the drug solubility in the binary solvent: water + cosolvent (co), $x_2^{(\text{H}_2\text{O})}$ and $x_2^{(\text{co})}$ are the drug solubilities in water and cosolvent, respectively, $V^{(\text{H}_2\text{O})}$ and $V^{(\text{co})}$ are the molar volumes of water and cosolvent at 25 °C, respectively, and $V^{(b)} = x_{\text{co}}^b V^{(\text{co})} + x_{\text{H}_2\text{O}}^b V^{(\text{H}_2\text{O})} + e x_{\text{co}}^b x_{\text{H}_2\text{O}}^b$, where x_{co}^b and $x_{\text{H}_2\text{O}}^b$ are the mole fractions of the cosolvent and water, respectively, in the mixed solvent: water + cosolvent and e is an adjustable parameter introduced in a previous paper (Ruckenstein and Shulgin, 2003a).

Finally, the solubility of the drug for the compositions of the mixed solvents corresponding to the limiting binary mixtures I and II were calculated using Eq. (10).

(2) The solubilities in binary aqueous mixed solvents (y and z) were evaluated graphically from experimental data.

A comparison between predicted and experimental drug solubilities in ternary aqueous mixed solvents is made in Table 1.

It is worth mentioning that all the predictions listed in Table 1 were obtained on the basis of experimental drug solubilities in binary aqueous mixed solvents, without using any experimental drug solubilities in ternary aqueous mixed solvents.

One can see from Table 1 that the drug solubilities in ternary aqueous mixed solvents could be accurately predicted using the experimental drug solubilities in binary aqueous mixed solvents.

Table 1
Comparison between predicted and experimental drug solubilities in ternary solvents

Solute	Mixed solvent	Reference	Deviation (%) between experimental and predicted (Eq. (9)) solubilities[a]	
			The solubilities in binary solvents calculated using Eq. (10)[b]	The solubilities in binary solvents evaluated graphically from experimental data
Tioconazole	Ethanol–propylene glycol–water	Gould et al. (1984)	10.4	6.8
	Ethanol–PEG 400–water		19.6	15.4
	Propylene glycol–PEG 400–water		39.1	15.2
19-Nor-1α,25-dihydrovitamin D_2	Ethanol–propylene glycol–water	Stephens et al. (1999)	55.4	15.0

[a] Deviation from experimental data calculated as MPD (%) (mean percentage deviation) defined as $[100\sum_{i=1}^{N}|(x_i^{exp} - x_i^{calc})/x_i^{exp}|]/N$, where x_i^{exp} and x_i^{calc} are experimental and calculated (using Eq. (9)) solubilities (mole fractions) and N is the number of experimental points.

[b] Because we could not find in literature the solubilities of 19-Nor-1α,25-dihydrovitamin D_2 in ethanol and propylene glycol, they were taken equal to the solubility of vitamin D_2 in ethanol (Penau and Hagemann, 1946).

The difference in predicted solubilities when the solubilities in binary aqueous mixed solvents (y and z) were evaluated using Eq. (10) or obtained graphically from experimental data is understandable. The accuracy of Eq. (10) for predicting the drug solubility in binary aqueous mixed solvents is about 14% (the mean percentage deviation) (Ruckenstein and Shulgin, 2003a) and this inaccuracy plays a role in the prediction of the drug solubilities in ternary aqueous mixed solvents (see Table 2 for details).

3.2. Quaternary mixed solvent

We found in literature only one example regarding the drug solubilities in quaternary aqueous mixed solvents: the solubility of tioconazole in ethanol–propylene glycol–PEG 400–water (Gould et al., 1984).

The prediction of the solubility of tioconazole in ethanol–propylene glycol–PEG 400–water was carried out using the following steps:

(1) Two ternary solvents: I (ethanol–propylene glycol–water) and II (ethanol–PEG 400–water) were selected,
(2) The solubilities of tioconazole in the above ternary solvents were calculated as described in the previous section (Eq. (10) was used to evaluate the solubility of tioconazole in binary aqueous mixed solvents, see Table 2 for details),
(3) The solubilities of tioconazole in ethanol–propylene glycol–PEG 400–water mixed solvent

Table 2
Comparison between calculated (using Eq. (10)) and experimental drug solubilities in aqueous binary solvents

Solute	Cosolvent	Deviation from experimental data[a]	Value of e (cm³/mol)[b]
Tioconazole	Ethanol	4.12	40.87
Tioconazole	Propylene glycol	7.74	37.75
Tioconazole	PEG 400	18.60	507.09
19-Nor-1α,25-dihydrovitamin D_2[c]	Ethanol	27.56	−34.71
19-Nor-1α,25-dihydrovitamin D_2[c]	Propylene glycol	8.72	−78.63

[a] Deviation from experimental data calculated as MPD (%) (mean percentage deviation) defined as $[100\sum_{i=1}^{N}|(x_i^{exp} - x_i^{calc})/x_i^{exp}|]/N$, where x_i^{exp} and x_i^{calc} are experimental and calculated (using Eq. (10)) solubilities (mole fractions) and N is the number of experimental points.

[b] Parameter e was used in the following equation for molar volume of binary mixed solvent (see Ruckenstein and Shulgin, 2003a) $V^{(b)} = x_{co}^b V^{(co)} + x_{H_2O}^b V^{(H_2O)} + e x_{co}^b x_{H_2O}^b$.

[c] Because we could not find in literature the solubilities of 19-Nor-1α,25-dihydrovitamin D_2 in ethanol and propylene glycol, they were taken equal to the solubility of vitamin D_2 in ethanol (Penau and Hagemann, 1946).

Table 3
Comparison between predicted and experimental tioconazole solubilities in quaternary solvent

Solute	Mixed solvent	Reference	Deviation (%) between experimental and predicted (Eq. (9)) solubilities[a]
Tioconazole	Ethanol–propylene glycol–PEG 400–water	Gould et al. (1984)	10.6

[a] Deviation from experimental data calculated as MPD (%) (mean percentage deviation) defined as $[100\sum_{i=1}^{N}|(x_i^{exp} - x_i^{calc})/x_i^{exp}|]/N$, where x_i^{exp} and x_i^{calc} are experimental and calculated (using Eq. (9)) solubilities (mole fractions) and N is the number of experimental points.

were calculated with Eq. (9), using the solubilities of tioconazole in the ternary solvents obtained in the previous step.

The results of the predictions are listed in Table 3, which show that there is an excellent agreement between the experimental and predicted solubilities.

It is also noteworthy to emphasize that all the predictions listed in Table 3 were made on the basis of experimental drug solubilities in binary aqueous mixed solvents, without using any experimental drug solubilities in ternary and quaternary aqueous mixed solvents.

4. Discussion and conclusion

As in our previous publications regarding the solubility of drugs in aqueous mixed solvents (Ruckenstein and Shulgin, 2003a, b), the fluctuation theory of solutions was used as a theoretical tool. However, whereas the above publications were devoted to binary mixed solvents, the present one provides a predictive method for the solubility of drugs in multicomponent aqueous mixed solvents.

First, a rigorous expression for the activity coefficient of a solid solute at infinite dilution in an ideal multicomponent solvent was derived using the fluctuation theory of solution. Second, the obtained expression was used to express the solubility of a poorly soluble solid in an ideal multicomponent solvent in terms of the solubilities of this solid in two subsystems of the multicomponent solvent and their molar volumes. Finally, the developed procedure was used to predict the drug solubilities in ternary and quaternary aqueous mixed solvents using the drug solubilities in the constituent binary aqueous mixed solvents. The predicted solubilities were compared with the experimental ones and good agreement was found.

It is worth noting that good agreement was found despite two important limitations imposed on our method: (a) the multicomponent solvent was considered ideal, and (b) the drug solubility in a mixed solvent was supposed to be small enough to satisfy the infinite dilution approximation.

The developed predictive method can be applied not only to ternary and quaternary mixed solvents, but also to any multicomponent solvent.

Acknowledgements

We are indebted to Professor W.E. Acree, Jr. (Department of Chemistry, University of North Texas, Denton) for helpful discussions regarding some theoretical results of this paper. We like to thank Dr. L.C. Li (Abbott Laboratories, Abbott Park, IL) for providing us with data regarding the solubility of 19-Nor-1α,25-dihydrovitamin D_2 in water.

References

Acree, W.E., 1984. Thermodynamic Properties of Nonelectrolyte Solutions. Academic Press, Orlando.

Acree, W.E., McCargar, J.W., Zvaigzne, A.I., Teng, I.L., 1991. Mathematical representation of thermodynamic properties. Carbazole solubilities in binary alkane + dibutyl ether and alkane + tetrahydropyran solvent mixtures. Phys. Chem. Liq. 23, 27–35.

Deng, T.H., Horiuchi, S., De Fina, K.M., Hernandez, C.E., Acree, W.E., 1999. Solubility of anthracene in multicomponent solvent mixtures containing propanol, butanol, and alkanes. J. Chem. Eng. Data 44, 798–802.

Fan, C.H., Jafvert, C.T., 1997. Margules equations applied to PAH solubilities in alcohol–water mixtures. Environ. Sci. Technol. 31, 3516–3522.

Fredenslund, Aa., Sørensen, J.M., 1994. Group contribution estimation methods. In: Sandler, S.I. (Ed.), Models for Thermodynamic and Phase Equilibria Calculations. Dekker, New York (Chapter 4).

Gould, P.L., Goodman, M., Hanson, P.A., 1984. Investigation of the solubility relationships of polar, semi-polar and non-polar drugs in mixed co-solvent systems. Int. J. Pharm. 19, 149–159.

Jouyban-Gharamaleki, A., Clark, B.J., Acree, W.E., 2000a. Models to predict solubility in ternary solvents based on sub-binary experimental data. Chem. Pharm. Bull. 48, 1866–1871.

Jouyban-Gharamaleki, A., Clark, B.J., Acree, W.E., 2000b. Prediction of drug solubility in ternary solvent mixture. Drug Dev. Ind. Pharm. 26, 971–973.

Kirkwood, J.G., Buff, F.P., 1951. Statistical mechanical theory of solutions. 1. J. Chem. Phys. 19, 774–782.

Lachman, L., Lieberman, H.A., Kanig, J.L. (Eds.), 1976. The Theory and Practice of Industrial Pharmacy. Lea & Febiger, Philadelphia.

Prausnitz, J.M., Lichtenthaler, R.N., Gomes de Azevedo, E., 1986. Molecular Thermodynamics of Fluid—Phase Equilibria, 2nd ed. Prentice-Hall, Englewood Cliffs, NJ.

Penau, H., Hagemann, G., 1946. The purification and determination of Vitamin D_2 in pharmaceutical preparations and in several esters of calciferol. Helv. Chim. Acta 29, 1366–1371.

Ruckenstein, E., Shulgin, I., 2003a. Solubility of drugs in aqueous solutions. Part 1. Ideal mixed solvent approximation. Int. J. Pharm. 258, 193–201.

Ruckenstein, E., Shulgin, I., 2003b. Solubility of drugs in aqueous solutions. Part 2. Ideal mixed solvent approximation. Int. J. Pharm. 260, 283–291.

Ruckenstein, E., Shulgin, I., 2003c. Ideal multicomponent liquid solution as a mixed solvent. Ind. Eng. Chem. Res. 42, 4406–4413.

Stephens, D., Li, L.C., Pec, E., Robinson, D., 1999. A statistical experimental approach to cosolvent formulation of a water-insoluble drug. Drug Dev. Ind. Pharm. 25, 961–965.

Yalkowsky, S.H., Roseman, T.J., 1981. Solubilization of drugs by cosolvents. In: Yalkowsky, S.H. (Ed.), Techniques of Solubilization of Drugs. Dekker, New York (Chapter 3).

Solubility of drugs in aqueous solutions
Part 4. Drug solubility by the dilute approximation

E. Ruckenstein*, I. Shulgin

Department of Chemical Engineering, State University of New York at Buffalo, Amherst, NY 14260, USA

Received 20 October 2003; received in revised form 13 February 2004; accepted 2 March 2004

Available online 19 May 2004

Abstract

As in our previous publications in this journal [Int. J. Pharm. 258 (2003a) 193; Int. J. Pharm. 260 (2003b) 283; Int. J. Pharm. 267 (2003c) 121], this paper is concerned with the solubility of poorly soluble drugs in aqueous mixed solvents. In the previous publications, the solubilities of drugs were assumed to be low enough for the so-called infinite dilution approximation to be applicable. In contrast, in the present paper, the solubilities are considered to be finite and the dilute solution approximation is employed. As before, the fluctuation theory of solutions is used to express the derivatives of the activity coefficient of a solute in a ternary solution (dilute solute concentrations in a binary solvent) with respect to the concentrations of the solvent and cosolvent. The expressions obtained are combined with a theoretical equation for the activity coefficient of the solute. As a result, the activity coefficient of the solute was expressed through the activity coefficients of the solute at infinite dilution, solute mole fraction, some properties of the binary solvent (composition, molar volume and activity coefficients of the components) and parameters reflecting the nonidealities of binary species. The expression thus obtained was used to derive an equation for the solubility of poorly soluble drugs in aqueous binary solvents which was applied in two different ways. First, the nonideality parameters were considered as adjustable parameters, determined from experimental solubility data. Second, the obtained equation was used to correct the solubilities of drugs calculated via the infinite dilution approximation. It was shown that both procedures provide accurate correlations for the drug solubility.
© 2004 Elsevier B.V. All rights reserved.

Keywords: Solubility; Drugs; Dilute approximation

1. Introduction

In our previous papers regarding the solubility of poorly soluble drugs in aqueous mixed solvents (Ruckenstein and Shulgin, 2003a–c), the fluctuation theory of solutions (Kirkwood and Buff, 1951) was used for their correlation and prediction. Such information is useful because poor aqueous solubility can often affect the drug efficiency.

Whereas the first two publications of this series (Ruckenstein and Shulgin, 2003a,b) were concerned with binary mixed solvents, the third one (Ruckenstein and Shulgin, 2003c) was devoted to the solubility of drugs in multicomponent solvents.

In the above papers, the solubility of drugs in mixed solvents was assumed to be low enough for the infinite dilution approximation to be applicable. Let us examine this approximation in more detail. The solubility of solid substances in pure and mixed solvents

* Corresponding author. Tel.: +1-716-645-2911x2214; fax: +1-716-645-3822.

E-mail addresses: feaeliru@acsu.buffalo.edu (E. Ruckenstein), ishulgin@eng.buffalo.edu (I. Shulgin).

0378-5173/$ – see front matter © 2004 Elsevier B.V. All rights reserved.

can be described by the classical solid–liquid equilibrium equations (Acree, 1984; Prausnitz et al., 1986). For the solubilities of a solid solute (component 2) in water (component 3), cosolvent (component 1), and their mixture (mixed solvents 1–3), one can write the following equations

$$\frac{f_2^S}{f_2^L(T,P)} = y_2^{b_1} \gamma_2^{b_1}(T, P, \{y\}) \tag{1}$$

$$\frac{f_2^S}{f_2^L(T,P)} = y_2^{b_3} \gamma_2^{b_3}(T, P, \{y\}) \tag{2}$$

$$\frac{f_2^S}{f_2^L(T,P)} = y_2^t \gamma_2^t(T, P, \{y\}) \tag{3}$$

where $y_2^{b_1}$, $y_2^{b_3}$, and y_2^t are the solubilities (mole fractions) of the solid component 2 in the cosolvent, water, and their mixture, respectively; $\gamma_2^{b_1}$, $\gamma_2^{b_3}$, and γ_2^t are the activity coefficients of the solid solute in its saturated solutions in the cosolvent, water, and mixed solvent, respectively; $f_2^L(T, P)$ is the hypothetical fugacity of a solid as a (subcooled) liquid at a given pressure (P) and temperature (T); f_2^S is the fugacity of the pure solid component 2; and $\{y\}$ indicates that the activity coefficients of the solute depend on composition. If the solubilities of the pure and mixed solvents in the solid phase are negligible, then the left hand sides of Eqs. (1)–(3) depend only on the properties of the solute.

The infinite dilution approximation implies that the activity coefficients in Eqs. (1)–(3) can be replaced by their values at infinite dilution of the solute ($\gamma_2^{b_1,\infty}$, $\gamma_2^{b_3,\infty}$, and $\gamma_2^{t,\infty}$). However, the solubilities of drugs in aqueous mixed solvents are not always very low. While the solubilities of various drugs in water (only poorly soluble drugs are considered in the present paper) do not exceed 1–2 mol%, the solubilities of the same drugs in the popular cosolvents ethanol and 1,4-dioxane can reach 5–20 mol%, and the solubilities in the water/1,4-dioxane and water/ethanol mixtures are often appreciable and can reach 8–30 mol%. Therefore, the effect of the infinite dilution approximation on the accuracy of the predictions of the solubilities of poorly soluble drugs deserves to be examined.

In the present paper, dilute binary and ternary solutions (drug + water, drug + cosolvent, and drug + water + cosolvent) will be considered, hence the infinite dilution approximation will be replaced by the dilute solution approximation. The range in which the infinite dilution approximation is valid and the range in which the dilute approximation can be used were discussed by Kojima et al. (1997). They pointed out that the above composition ranges depend on the nature of the solute and solvent and on the types of intermolecular interactions in the mixtures involved. For example, mixtures with self-association of one of the components have a narrower range in which the dilute approximation is valid.

As for infinite dilution, the main difficulty in predicting the solid solute solubility in a mixed solvent for a dilute solution is provided by the calculation of the activity coefficient of the solute in a ternary mixture. To obtain an expression for the activity coefficient of a low concentration solute in a ternary mixture, the fluctuation theory of solution will be combined with the assumption that the system is dilute with respect to the solute.

The paper is organized as follows: first, an equation for the activity coefficient of a low concentration solute in individual and binary solvents will be written. This equation will be combined with the fluctuation theory of solutions and with Eqs. (1)–(3) to derive an expression for the drug solubility. Further, the expression obtained will be compared with experimental data and with the infinite dilution approximation (Ruckenstein and Shulgin, 2003a,b).

2. Theory

2.1. The activity coefficient of a solute in its dilute range in binary solvents

For a binary dilute mixture, Debenedetti and Kumar (1986) suggested the following series expansion for the fugacity coefficient of a solute (ϕ_2^b)

$$\ln \phi_2^b = \ln \phi_2^{b,\infty} - k_{22} x_2^b \tag{4}$$

where $\phi_2^{b,\infty}$ is the fugacity coefficient at infinite dilution, x_2^b is the mole fraction of the solute, and

$$k_{22} = -\left(\frac{\partial \ln \gamma_2^b}{\partial x_2^b}\right)_{P,T,x_2^b \to 0} = -\left(\frac{\partial \ln \phi_2^b}{\partial x_2^b}\right)_{P,T,x_2^b \to 0} \tag{5}$$

γ_2^b being the activity coefficient of the solute in the binary mixture.

The above expression was extended to ternary mixtures, containing a solute and a cosolvent in low concentrations by Chialvo (1993), Jonah and Cochran (1994), and Munoz et al. (1995).

In this paper we consider the case in which only the solute concentration is small (Ruckenstein and Shulgin, 2002). Let us consider those compositions (mole fraction) of the ternary mixture (x_1^t, x_2^t, x_3^t) which are located on the line connecting the points ($x_1^t = 0$, $x_2^t = 1$, $x_3^t = 0$) and ($x_1^t = x_1^0$, $x_2^t = 0$, $x_3^t = x_3^0$) in the Gibbs triangle (Fig. 1). This line connects the pure component 2 (a solute) and the binary mixtures 1–3 (cosolvent + solvent) with a mole fraction of component 1 equal to x_1^0. Physically speaking, this line represents the locus of the compositions of ternary mixtures formed by adding a solute (2) to a binary mixture of a solvent (3) and a cosolvent (1).

On the above line, the following relation holds

$$\left(\frac{x_1^t}{x_3^t}\right) = \left(\frac{x_1^0}{x_3^0}\right) = \alpha \qquad (6)$$

Because $x_1^t + x_2^t + x_3^t = 1$, one can write that

$$x_1^t = \alpha \frac{1 - x_2^t}{1 + \alpha} \qquad (7)$$

and

$$x_3^t = \frac{1 - x_2^t}{1 + \alpha} \qquad (8)$$

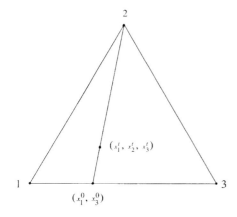

Fig. 1. The change of composition in a ternary mixture solute + binary solvent, when a solute (2) is added to a binary solvents (1–3) of composition (mole fractions) (x_1^0, x_3^0).

For the fugacity coefficient of a solute in a ternary dilute solution, one can write, at constant temperature and pressure, near the composition $x_1^t = x_1^0$, $x_2^t = 0$, $x_3^t = x_3^0$, the following expression

$$\ln \phi_2^t = \ln \phi_2^t(x_1^0, 0, x_3^0) + x_2^t \left(\frac{\partial \ln \phi_2^t}{\partial x_2^t}\right)_{P,T,\alpha,(x_1^0,0,x_3^0)} \qquad (9)$$

where ϕ_2^t is the fugacity coefficient of the solute in a ternary mixture and $\phi_2^t(x_1^0, 0, x_3^0) = \phi_2^{t,\infty}$ is its value at infinite dilution of the solute.

If, at a given pressure and temperature, the mole fractions of components 1 and 3 are taken as independent variables, one can rewrite Eq. (9) under the form

$$\ln \phi_2^t = \ln \phi_2^t(x_1^0, 0, x_3^0)$$
$$+ x_2^t \left[\left(\frac{\partial \ln \phi_2^t}{\partial x_1^t}\right)_{P,T,x_3^t,(x_1^0,0,x_3^0)} \left(\frac{\partial x_1^t}{\partial x_2^t}\right)_\alpha\right.$$
$$\left.+ \left(\frac{\partial \ln \phi_2^t}{\partial x_3^t}\right)_{P,T,x_1^t,(x_1^0,0,x_3^0)} \left(\frac{\partial x_3^t}{\partial x_2^t}\right)_\alpha\right] \qquad (10)$$

which, taking into account Eqs. (7) and (8), becomes

$$\ln \phi_2^t = \ln \phi_2^t(x_1^0, 0, x_3^0)$$
$$- \frac{x_2^t}{1+\alpha}\left[\alpha\left(\frac{\partial \ln \phi_2^t}{\partial x_1^t}\right)_{P,T,x_3^t,(x_1^0,0,x_3^0)}\right.$$
$$\left.+ \left(\frac{\partial \ln \phi_2^t}{\partial x_3^t}\right)_{P,T,x_1^t,(x_1^0,0,x_3^0)}\right] \qquad (11)$$

or equivalently,

$$\ln \phi_2^t = \ln \phi_2^t(x_1^0, 0, x_3^0)$$
$$- x_2^t\left[x_1^0\left(\frac{\partial \ln \phi_2^t}{\partial x_1^t}\right)_{P,T,x_3^t,(x_1^0,0,x_3^0)}\right.$$
$$\left.+ x_3^0\left(\frac{\partial \ln \phi_2^t}{\partial x_3^t}\right)_{P,T,x_1^t,(x_1^0,0,x_3^0)}\right] \qquad (12)$$

A similar equation can be written for the activity coefficient of a low concentration solute in a ternary mixture

$$\ln \gamma_2^t = \ln \gamma_2^t(x_1^0, 0, x_3^0)$$
$$- x_2^t \left[x_1^0 \left(\frac{\partial \ln \gamma_2^t}{\partial x_1^t} \right)_{P,T,x_3^t,(x_1^0,0,x_3^0)} \right.$$
$$\left. + x_3^0 \left(\frac{\partial \ln \gamma_2^t}{\partial x_3^t} \right)_{P,T,x_1^t,(x_1^0,0,x_3^0)} \right] \quad (13)$$

Eq. (13) will be used for the drug solubility when its saturated solution in a binary solvent can be considered dilute. First, expressions for the two partial derivatives in Eq. (13) will be derived on the basis of the fluctuation theory of solutions (Kirkwood and Buff, 1951).

2.2. Expressions for the derivatives $(\partial \ln \gamma_2^t/\partial x_1^t)_{P,T,x_3^t}$ and $(\partial \ln \gamma_2^t/\partial x_3^t)_{P,T,x_1^t}$

It was shown previously, that, for the derivatives of the activity coefficient (γ_2^t) one can write the following relations (Ruckenstein and Shulgin, 2001)

$$\left(\frac{\partial \ln \gamma_2^t}{\partial x_1^t} \right)_{T,P,x_3^t}$$
$$= \frac{-c_2 c_3 (G_{12} + G_{33} - G_{13} - G_{23}) + c_1 c_2 \Delta_{12} + c_2 c_3 \Delta_{23} + c_1 c_2 c_3 \Delta_{123}}{x_2^t (c_1 + c_2 + c_3 + c_1 c_2 \Delta_{12} + c_1 c_3 \Delta_{13} + c_2 c_3 \Delta_{23} + c_1 c_2 c_3 \Delta_{123})} \quad (14)$$

and

$$\left(\frac{\partial \ln \gamma_2^t}{\partial x_3^t} \right)_{T,P,x_1^t}$$
$$= \frac{-c_1 c_2 (G_{11} + G_{23} - G_{12} - G_{13}) + c_1 c_2 \Delta_{12} + c_2 c_3 \Delta_{23} + c_1 c_2 c_3 \Delta_{123}}{x_2^t (c_1 + c_2 + c_3 + c_1 c_2 \Delta_{12} + c_1 c_3 \Delta_{13} + c_2 c_3 \Delta_{23} + c_1 c_2 c_3 \Delta_{123})} \quad (15)$$

where c_k is the bulk molecular concentration of component k and $G_{\alpha\beta}$ is the Kirkwood–Buff integral given by

$$G_{\alpha\beta} = \int_0^\infty (g_{\alpha\beta} - 1) 4\pi r^2 \, dr \quad (16)$$

In the above expressions, $g_{\alpha\beta}$ is the radial distribution function between species α and β, r is the distance between the centers of molecules α and β, and $\Delta_{\alpha\beta}$ and Δ_{123} are defined as follows

$$\Delta_{\alpha\beta} = G_{\alpha\alpha} + G_{\beta\beta} - 2G_{\alpha\beta}, \quad \alpha \neq \beta \quad (17)$$

and

$$\Delta_{123} = G_{11}G_{22} + G_{11}G_{33} + G_{22}G_{33} + 2G_{12}G_{13}$$
$$+ 2G_{12}G_{23} + 2G_{13}G_{23} - G_{12}^2 - G_{13}^2 - G_{23}^2$$
$$- 2G_{11}G_{23} - 2G_{22}G_{13} - 2G_{33}G_{12} \quad (18)$$

It was shown that the expressions in the brackets in the numerators of Eqs. (14) and (15) and Δ_{123} can be expressed in terms of $\Delta_{\alpha\beta}$ as follows (Ruckenstein and Shulgin, 2001)

$$G_{12} + G_{33} - G_{13} - G_{23} = \frac{\Delta_{13} + \Delta_{23} - \Delta_{12}}{2} \quad (19)$$

$$G_{11} + G_{23} - G_{12} - G_{13} = \frac{\Delta_{12} + \Delta_{13} - \Delta_{23}}{2} \quad (20)$$

and

$$\Delta_{123} = -\frac{(\Delta_{12})^2 + (\Delta_{13})^2 + (\Delta_{23})^2 - 2\Delta_{12}\Delta_{13} - 2\Delta_{12}\Delta_{23} - 2\Delta_{13}\Delta_{23}}{4} \quad (21)$$

The insertion of Eqs. (19)–(21) into Eqs. (14) and (15) provides the following expressions for the derivatives $(\partial \ln \gamma_2^t/\partial x_1^t)_{P,T,x_3^t}$ and $(\partial \ln \gamma_2^t/\partial x_3^t)_{P,T,x_1^t}$ in terms of $\Delta_{\alpha\beta}$ and the concentrations of the solute-free mixed solvent

$$\lim_{x_2 \to 0} \left(\frac{\partial \ln \gamma_2^t}{\partial x_1^t} \right)_{T,P,x_3^t}$$
$$= \frac{(c_1^0 + c_3^0)\{(c_1^0 + 0.5c_3^0)\Delta_{12} + 0.5c_3^0 \Delta_{23} - 0.5c_3^0 \Delta_{13}\}_{x_2^t=0}}{c_1^0 + c_3^0 + c_1^0 c_3^0 \Delta_{13}}$$
$$- \frac{c_1^0 c_3^0 (c_1^0 + c_3^0)((\Delta_{12})^2 + (\Delta_{13})^2 + (\Delta_{23})^2 - 2\Delta_{12}\Delta_{13} - 2\Delta_{12}\Delta_{23} - 2\Delta_{13}\Delta_{23})_{x_2^t=0}}{4(c_1^0 + c_3^0 + c_1^0 c_3^0 \Delta_{13})}$$
$$\quad (22)$$

and

$$\lim_{x_2 \to 0} \left(\frac{\partial \ln \gamma_2^t}{\partial x_3^t}\right)_{T,P,x_1^t}$$

$$= \frac{(c_1^0 + c_3^0)\{0.5c_1^0 \Delta_{12} + (0.5c_1^0 + c_3^0)\Delta_{23} - 0.5c_1^0 \Delta_{13}\}_{x_2^t=0}}{c_1^0 + c_3^0 + c_1^0 c_3^0 \Delta_{13}}$$

$$- \frac{c_1^0 c_3^0 (c_1^0 + c_3^0)((\Delta_{12})^2 + (\Delta_{13})^2 + (\Delta_{23})^2 - 2\Delta_{12}\Delta_{13} - 2\Delta_{12}\Delta_{23} - 2\Delta_{13}\Delta_{23})_{x_2^t=0}}{4(c_1^0 + c_3^0 + c_1^0 c_3^0 \Delta_{13})}$$

(23)

where c_1^0 and c_3^0 are the bulk molecular concentrations of components 1 and 3 in the solute-free binary 1–3 solvent.

The derivatives $(\partial \ln \gamma_2^t / \partial x_1^t)_{P,T,x_3^t}$ and $(\partial \ln \gamma_2^t / \partial x_3^t)_{P,T,x_1^t}$ are expressed in Eqs. (22) and (23) in terms of $\Delta_{\alpha\beta}$ and the concentrations of the solute-free mixed solvent. It is worth noting that $\Delta_{\alpha\beta}$ is a measure of nonideality (Ben-Naim, 1977) of the binary mixture $\alpha - \beta$, because for an ideal mixture $\Delta_{\alpha\beta} = 0$. Furthermore, being measures of nonideality, the parameters $\Delta_{\alpha\beta}$ have a clear physical meaning and this fact is useful in the thermodynamic analysis of multicomponent mixtures.

2.3. Equations for the solubility of a solid in a binary solvent

Insertion of Eqs. (22)–(23) into Eq. (12) leads to

$$\ln \gamma_2^t = \ln \gamma_2^t(x_1^0, 0, x_3^0)$$

$$- x_2^t \frac{(x_1^0 \Delta_{12} + x_3^0 \Delta_{23} - x_1^0 x_3^0 \Delta_{13})_{x_2^t=0}}{V + x_1^0 x_3^0 \Delta_{13}}$$

$$+ \frac{x_1^0 x_3^0 x_2^t ((\Delta_{12})^2 + (\Delta_{13})^2 + (\Delta_{23})^2 - 2\Delta_{12}\Delta_{13} - 2\Delta_{12}\Delta_{23} - 2\Delta_{13}\Delta_{23})_{x_2^t=0}}{4V(V + x_1^0 x_3^0 \Delta_{13})}$$

(24)

where V is the molar volume of the solute-free binary solvent.

An expression for the activity coefficient of a solute at infinite dilution in a ternary mixture $\gamma_2^t(x_1^0, 0, x_3^0)$ was obtained elsewhere (Ruckenstein and Shulgin, 2003b) and has the form

$$\ln \gamma_2^{t,\infty} = \ln \gamma_2^t(x_1^0, 0, x_3^0)$$

$$= -(\Delta_{12} - \Delta_{23})_{x_2^t=0} \left(\frac{I_1}{2}\right) + \left(\frac{I_2}{2}\right) + A$$

(25)

where A is a composition independent constant

$$I_1 = \int \frac{1 + x_3^0 (\partial \ln \gamma_3^b / \partial x_3^0)_{P,T}}{V} dx_3^0$$

(26)

and

$$I_2 = \ln \gamma_1^b + \ln \gamma_3^b$$

(27)

In Eqs. (26) and (27), γ_1^b and γ_3^b are the activity coefficients of the cosolvent and solvent in a solute-free binary solvent.

The combination of Eqs. (24)–(27) with the equation for the solid–liquid equilibrium provides a relation for the solubility of a solute forming a dilute solution in a ternary mixture.

$$\ln y_2^t = (\Delta_{12} - \Delta_{23})_{x_2^t=0} \left(\frac{I_1}{2}\right) - \left(\frac{I_2}{2}\right)$$

$$+ y_2^t \frac{(x_1^0 \Delta_{12} + x_3^0 \Delta_{23} - x_1^0 x_3^0 \Delta_{13})_{x_2^t=0}}{V + x_1^0 x_3^0 \Delta_{13}} + \bar{A}$$

$$- y_2^t \frac{x_1^0 x_3^0 ((\Delta_{12})^2 + (\Delta_{13})^2 + (\Delta_{23})^2 - 2\Delta_{12}\Delta_{13} - 2\Delta_{12}\Delta_{23} - 2\Delta_{13}\Delta_{23})_{x_2^t=0}}{4V(V + x_1^0 x_3^0 \Delta_{13})}$$

(28)

where $\bar{A}(P,T) = -A(P,T) + \ln[f_2^S / f_2^L(T,P)]$ is a composition-independent constant.

Eq. (28) allows one to calculate the solubility of a solute in a binary mixed solvent if the composition dependence of the activity coefficients, the molar volume V, the nonideality parameters Δ_{12}, Δ_{23} and the constant $\bar{A}$ are known. The nonideality parameters $\Delta_{\alpha\beta}$ for a binary mixture $\alpha - \beta$ can be obtained from the composition dependence of the activity coefficients in the above mixture using the expression (Kirkwood and Buff, 1951)

$$\Delta_{\alpha\beta} = -\frac{V(\partial \ln \gamma_\beta^b / \partial x_\beta^0)_{P,T}}{x_\alpha^0 - x_\alpha^0 x_\beta^0 (\partial \ln \gamma_\beta^b / \partial x_\beta^0)_{P,T}}$$

(29)

Table 1
The solubility of drugs in binary solvents calculated with Eq. (28)

Drug	Mixed solvent	Deviation from experimental data[a]	
		Eq. (28) combined with Wilson's equation	Infinite dilution approximation combined with Wilson's equation[b]
Caffeine	Water/N,N-dimethylformamide	2.8	6.5
Caffeine	Water/1,4-dioxane	5.3	9.6
Sulfamethizole	Water/1,4-dioxane	16.8	18.9
Methyl p-hydroxybenzoate	Water/propylene glycol	12.8	12.4
Methyl p-aminobenzoate	Water/propylene glycol	6.5	6.6
Ethyl p-aminobenzoate	Water/propylene glycol	8.1	8.5
Propyl p-hydroxybenzoate	Water/propylene glycol	13.5	16.1
Butyl p-hydroxybenzoate	Water/propylene glycol	22.4	24.0

[a] Deviation from experimental data calculated as MPD (%) (the mean percentage deviation) defined as $100 \times \sum_{i=1}^{N} |(x_i^{\exp} - x_i^{\text{calc}})/x_i^{\exp}|/N$, where $x_i^{\exp}$ and x_i^{calc} are the experimental and calculated solubilities, and N is the number of experimental points.
[b] These results were taken from our previous publication (Ruckenstein and Shulgin, 2003b).

Eq. (29) can be used to calculate the parameter $\Delta_{\alpha\beta}$ from vapor–liquid equilibrium data for mixed binary solvents. Unfortunately, for most solute + individual solvent pairs such data are not available.

3. Application of Eq. (28) to the solubility of drugs in a binary solvent

Being a transcendent equation, Eq. (28) cannot provide an explicit expression for the solubility of a drug (y_2^t), but has to be solved numerically for every set of parameters.

In order to check Eq. (28), the solubilities of caffeine in the water/N,N-dimethylformamide (Herrador and Gonzalez, 1997) and water/1,4-dioxane mixtures (Adjei et al., 1980), as well as the solubilities of sulfamethizole in the mixture water/1,4-dioxane (Reillo et al., 1995) and of five solutes in water/propylene glycol (Rubino and Obeng, 1991) were employed.

First, Δ_{12}, Δ_{23}, and $\bar{A}$ were considered adjustable parameters which were determined by fitting Eq. (28) to the experimental solubility data. The activity coefficients of the components in binary solvents were expressed via the Wilson equation (Wilson, 1964) (of course, any other expressions for the activity coefficients can be used)

$$\ln \gamma_1^b = -\ln(x_1^0 + x_3^0 L_{13})$$
$$+ x_3^0 \left(\frac{L_{13}}{x_1^0 + x_3^0 L_{13}} - \frac{L_{31}}{x_3^0 + x_1^0 L_{31}} \right) \quad (30)$$

and

$$\ln \gamma_3^b = -\ln(x_3^0 + x_1^0 L_{31})$$
$$- x_1^0 \left(\frac{L_{13}}{x_1^0 + x_3^0 L_{13}} - \frac{L_{31}}{x_3^0 + x_1^0 L_{31}} \right) \quad (31)$$

where L_{13} and L_{31} are the Wilson parameters.

The parameters L_{13} and L_{31} were also determined from the experimental solubility data. Therefore, Eq. (28) can be considered as a five parameters equation. The results of the calculations as well as a comparison with those obtained under the infinite dilution approximation are listed in Table 1.

Table 1 shows that Eq. (28) provides slightly better results that the correlation based on the infinite dilution approximation. However, it is not clear whether this improvement was caused by the use of the more realistic dilute approximation, or of a larger number of adjustable parameters (five in the present case instead of four in the equation based on the infinite dilution approximation).

The new equation can be consider as a correction to the infinite dilution approximation. Indeed, combining Eq. (24) with Eq. (3) and with the following equation involving the infinite dilution approximation

$$\frac{f_2^S}{f_2^L(T, P)} = z_2^t \gamma_2^t(x_1^0, 0, x_3^0) \quad (32)$$

Table 2
Comparison between the drug solubilities in aqueous binary solvents calculated using Eq. (34) and the infinite dilution approximation (Ruckenstein and Shulgin, 2003a)

System number	Cosolvent	Solute	MPD (%)[a]	
			Eq. (34)[b]	The infinite dilution approximation (Ruckenstein and Shulgin, 2003a)[c]
1	N,N-dimethylformamide	Sulfadiazine	11.8	11.4
2	N,N-dimethylformamide	Theophyllene	14.1	14.1
3	N,N-dimethylformamide	Caffeine	11.9	11.9
4	Dioxane	Caffeine	10.2	12.8
5	Dioxane	p-Hydroxybenzoic acid	21.7	28.1
6	Dioxane	Paracetamol	7.3	15.4
7	Dioxane	Phenacetin	6.2	6.9
8	Dioxane	Sulfadiazine	5.0	7.6
9	Dioxane	Sulfadimidine	7.4	5.4
10	Dioxane	Sulfamethizole	12.0	12.7
11	Dioxane	Sulfamethoxazole	9.1	10.3
12	Dioxane	Sulfapyridine	7.6	9.0
13	Dioxane	Sulfamethoxypyridazine	6.6	7.8
14	Dioxane	Sulfanilamide	9.1	14.6
15	Dioxane	Sulfisomidine	12.0	13.0
16	Dioxane	Theobromine	23.6	23.7
17	Dioxane	Theophyllene	13.7	16.6
18	Ethanol	Paracetamol	7.3	15.4
19	Ethanol	Sulfamethazine	7.5	7.6
20	Ethanol	Sulfanilamide	22.2	22.5
21	Ethanol	Oxolinic acid	9.5	9.5
22	Ethylene glycol	Naphthalene	9.1	9.3
23	Ethylene glycol	Theophyllene	4.6	4.6
24	Methanol	Theophyllene	11.1	11.1
25	Propylene glycol	Butyl p-aminobenzoate	19.6	19.7
26	Propylene glycol	Butyl p-hydroxybenzoate	36.3	36.4
27	Propylene glycol	Ethyl p-aminobenzoate	10.7	10.7
28	Propylene glycol	Ethyl p-hydroxybenzoate	4.0	4.6
29	Propylene glycol	Methyl p-aminobenzoate	9.3	9.3
30	Propylene glycol	Methyl p-hydroxybenzoate	17.8	18.4
31	Propylene glycol	Propyl p-aminobenzoate	13.9	14.2
32	Propylene glycol	Propyl p-hydroxybenzoate	26.8	27.1
Average[d]			11.8	13.3

[a] Deviation from experimental data calculated as MPD (%) (the mean percentage deviation) defined as $100 \times \sum_{i=1}^{N_j} |(x_i^{\exp} - x_i^{\text{calc}})/x_i^{\exp}|/N_j$, where $x_i^{\exp}$ and x_i^{calc} are experimental and calculated solubilities (mole fractions), and N_j is the number of experimental points in the data set j.

[b] The parameter Δ_{13} was calculated from vapor–liquid equilibrium data for binary solvents using Eq. (29). The activity coefficients of the components in the binary solvents were expressed via the Wilson equation (Wilson, 1964) and the Wilson parameters L_{13} and L_{31} were taken from Gmehling's vapor–liquid equilibrium data compilation (Gmehling et al., 1977–2003).

[c] The values of MPD were calculated in a previous paper (Ruckenstein and Shulgin, 2003a).

[d] The average was calculated as $100 \times \sum_{j=1}^{M} \sum_{i=1}^{N_j} |(x_i^{\exp} - x_i^{\text{calc}})/x_i^{\exp}|/\sum_{j=1}^{M} N_j$ where $x_i^{\exp}$ and x_i^{calc} are the experimental and calculated solubilities (mole fractions), N_j is the number of experimental points in the data set j, and M is the number of experimental data sets (here 32).

one obtains

$$\ln y_2^t = \ln z_2^t + y_2^t \frac{\{x_1^0 \Delta_{12} + x_3^0 \Delta_{23} - x_1^0 x_3^0 \Delta_{13}\}_{x_2^t=0}}{V + x_1^0 x_3^0 \Delta_{13}}$$

$$- y_2^t \frac{\begin{array}{c} x_1^0 x_3^0((\Delta_{12})^2 + (\Delta_{13})^2 + (\Delta_{23})^2 \\ -2\Delta_{12}\Delta_{13} - 2\Delta_{12}\Delta_{23} \\ -2\Delta_{13}\Delta_{23})_{x_2^t=0} \end{array}}{4V(V + x_1^0 x_3^0 \Delta_{13})} \quad (33)$$

where z_2^t is the solubility of the solute under the infinite dilution approximation.

Because the infinite dilution approximation provides in many cases accurate results (Ruckenstein and Shulgin, 2003a), the difference between y_2^t and z_2^t is expected to be small. Consequently, one can expand $\ln(y_2^t/z_2^t)$ in a Taylor series to obtain for the solute solubility in the dilute approximation, the expression

$$y_2^t = \frac{z_2^t}{1 - z_2^t \Phi} \quad (34)$$

where

$$\Phi = \frac{(x_1^0 \Delta_{12} + x_3^0 \Delta_{23} - x_1^0 x_3^0 \Delta_{13})_{x_2^t=0}}{V + x_1^0 x_3^0 \Delta_{13}}$$

$$- \frac{x_1^0 x_3^0((\Delta_{12})^2 + (\Delta_{13})^2 + (\Delta_{23})^2 - 2\Delta_{12}\Delta_{13} - 2\Delta_{12}\Delta_{23} - 2\Delta_{13}\Delta_{23})_{x_2^t=0}}{4V(V + x_1^0 x_3^0 \Delta_{13})}$$

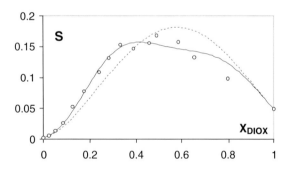

Fig. 2. Comparison between experimental (○) (Romero et al., 1996) and predicted (the solid line is based on Eq. (34), while the dashed line is based on the infinite dilution approximation (Ruckenstein and Shulgin, 2003a)) solubilities of the paracetomol (S is the mole fraction of paracetomol) in the binary solvent water/1,4-dioxane (x_{DIOX} is the mole fraction of dioxane) at room temperature.

Eq. (34) allows one to correct the solubility of a solute under the infinite dilution approximation if the properties of the binary solvent and the nonideality parameters Δ_{12} and Δ_{23} are known. Any of the methods available can be used to calculate the solubility of a solute under the infinite dilution approximation. For illustration purposes we selected a method suggested by us previously, and use Eq. (34) for the same 32 experimental sets, which were utilized there (Ruckenstein and Shulgin, 2003a). The results of the calculations are given in Table 2. Fig. 2 provides details for a particular case.

4. Discussion and conclusion

In contrast to previous papers (Ruckenstein and Shulgin, 2003a–d), the solubility of the drug in a binary solvent is considered to be finite, and the infinite dilution approximation is replaced by a more realistic one, the dilute solution approximation. An expression for the activity coefficient of a solute at low concentrations in a binary solvent was derived by combining the fluctuation theory of solutions (Kirkwood and Buff, 1951) with the dilute approximation. This procedure allowed one to relate the activity coefficient of a solute forming a dilute solution in a binary solvent to the solvent properties and some parameters characterizing the nonidealities of the various pairs of the ternary mixture.

Eq. (28) thus obtained can be used to represent the solubility of poorly soluble drugs in aqueous mixed solvents if information about the properties of the binary solvent (composition, phase equilibria and molar volume), the nonideality parameters and the constant $\bar{A}$ is available. These parameters can be considered as adjustable, and determined by fitting the experimental solubilities in the binary solvent. We applied such a procedure to the solubilities of caffeine in water/N,N-dimethylformamide (Herrador and Gonzalez, 1997) and water/1,4-dioxane (Adjei et al., 1980), of sulfamethizole in water/1,4-dioxane (Reillo et al., 1995) as well as of five solutes in water/propylene glycol (Rubino and Obeng, 1991). It was shown that Eq. (28) provides accurate correlations of the experimental data.

In essence, the developed computational scheme is a first order perturbation to the infinite dilution approximation. Therefore, the results regarding the solubility of poorly soluble drugs in aqueous mixed solvents obtained from the equations based on the infinite dilution approximation, can be slightly improved by the suggested method. The procedure was applied to 32 experimental data sets to show that the infinite dilution approximation is improved by the dilute solution approximation.

References

Acree, W.E., 1984. Thermodynamic Properties of Nonelectrolyte Solutions. Academic Press, Orlando.

Adjei, A., Newburger, J., Martin, A., 1980. Extended Hildebrand approach: solubility of caffeine in dioxane–water mixtures. J. Pharm. Sci. 69, 659–661.

Ben-Naim, A., 1977. Inversion of the Kirkwood–Buff theory of solutions: application to the water–ethanol system. J. Chem. Phys. 67, 4884–4890.

Chialvo, A., 1993. Solute–solute and solute–solvent correlations in dilute near-critical ternary mixtures: mixed-solute and entrainer effects. J. Phys. Chem. 97, 2740–2744.

Debenedetti, P.G., Kumar, S.K., 1986. Infinite dilution fugacity coefficients and the general behavior of dilute binary systems. AIChE J. 32, 1253–1262.

Gmehling, J., et al. 1977–2003. Vapor–Liquid Equilibrium Data Collection: DECHEMA Chemistry Data Series, vol. I, 19 Parts. DECHEMA, Frankfurt.

Herrador, M.A., Gonzalez, A.G., 1997. Solubility prediction of caffeine in aqueous N,N-dimethylformamide mixtures using the extended Hildebrand solubility approach. Int. J. Pharm. 156, 239–244.

Jonah, D.A., Cochran, H.D., 1994. Chemical potentials in dilute, multicomponent solutions. Fluid Phase Equilib. 92, 107–137.

Kirkwood, J.G., Buff, F.P., 1951. Statistical mechanical theory of solutions. Part 1. J. Chem. Phys. 19, 774–782.

Kojima, K., Zhang, S., Hiaki, T., 1997. Measuring methods of infinite dilution activity coefficients and a database for systems including water. Fluid Phase Equilib. 131, 145–179.

Munoz, F., Li, T.W., Chimowitz, E.H., 1995. Henry's law and synergism in dilute near-critical solutions: theory and simulation. AIChE J. 41, 389–401.

Prausnitz, J.M., Lichtenthaler, R.N., Gomes de Azevedo, E., 1986. Molecular Thermodynamics of Fluid-Phase Equilibria, second ed. Prentice-Hall, Englewood Cliffs, NJ.

Reillo, A., Bustamante, P., Escalera, B., Jimenez, M.M., Selles, E., 1995. Solubility parameter-based methods for predicting the solubility of sulfapyridine in solvent mixtures. Drug Dev. Ind. Pharm. 21, 2073–2084.

Romero, S., Reillo, A., Escalera, J.-B., Bustamante, P., 1996. The behavior of paracetamol in mixtures of amphiprotic and amphiprotic–aprotic solvents. Relationship of solubility curves to specific and nonspecific interactions. Chem. Pharm. Bull. 44, 1061–1064.

Ruckenstein, E., Shulgin, I., 2001. Entrainer effect in supercritical mixtures. Fluid Phase Equilib. 180, 345–359.

Ruckenstein, E., Shulgin, I., 2002. The solubility of solids in mixtures composed of a supercritical fluid and an entrainer. Fluid Phase Equilib. 200, 53–67.

Ruckenstein, E., Shulgin, I., 2003a. Solubility of drugs in aqueous solutions. Part 1. Ideal mixed solvent approximation. Int. J. Pharm. 258, 193–201.

Ruckenstein, E., Shulgin, I., 2003b. Solubility of drugs in aqueous solutions. Part 2. Ideal mixed solvent approximation. Int. J. Pharm. 260, 283–291.

Ruckenstein, E., Shulgin, I., 2003c. Solubility of drugs in aqueous solutions. Part 3. Multicomponent mixed solvent. Int. J. Pharm. 267, 121–127.

Ruckenstein, E., Shulgin, I., 2003d. Ideal multicomponent liquid solution as a mixed solvent. Ind. Eng. Chem. Res. 42, 4406–4413.

Rubino, J.T., Obeng, E.K., 1991. Influence of solute structure on deviations from the log–linear solubility equation in propylene–glycol: water mixtures. J. Pharm. Sci. 80, 479–483.

Wilson, G.M., 1964. Vapor–liquid equilibrium. Part XI. A new expression for the excess free energy of mixing. J. Am. Chem. Soc. 86, 127–130.

Solubility of drugs in aqueous solutions
Part 5. Thermodynamic consistency test for the solubility data

E. Ruckenstein*, I. Shulgin

Department of Chemical and Biological Engineering, State University of New York at Buffalo, Amherst, NY 14260, USA

Received 2 August 2004; received in revised form 12 October 2004; accepted 17 November 2004
Available online 24 January 2005

Abstract

This paper is devoted to the verification of the quality of experimental data regarding the solubility of sparingly soluble solids, such as drugs, environmentally important substances, etc. in mixed solvents. A thermodynamic consistency test based on the Gibbs–Duhem equation for ternary mixtures is suggested. This test has the form of an equation, which connects the solubilities of the solid, and the activity coefficients of the constituents of the solute-free mixed solvent in two mixed solvents of close compositions.

The experimental data regarding the solubility of sparingly soluble substances can be verified with the suggested test if accurate data for the activity coefficients of the constituents of the solute-free mixed solvent are available.

The test was applied to a number of systems representing the solubilities of sparingly soluble substances in mixed solvents. First, the test was scrutinized for four nonaqueous systems for which accurate solubility data were available. Second, the suggested test was applied to a number of systems representing experimental data regarding the solubility of sparingly soluble substances in aqueous mixed solvents.
© 2005 Published by Elsevier B.V.

Keywords: Drug solubility; Mixed solvent; Thermodynamic consistency test

1. Introduction

The solubility of drugs in water and aqueous mixed solvents is one of the important topics in pharmaceutical science and industry. However, the literature data regarding the aqueous solubility are not always reliable and large discrepancies between the data from different authors are typical. Indeed, according to a recently published compilation of aqueous solubilities (Yalkowsky and He, 2003), the aqueous solubility of naphthalene at room temperature measured by different authors varies from 0.0125 to 0.04 g/L, for anthracene from 3×10^{-4} to 7.3×10^{-4} g/L, and so on. The same or even worse situation could be observed for the solubilities of drugs in aqueous mixed solvents. Consequently, it is difficult

---

* Corresponding author. Tel.: +1 716 645 2911x2214; fax: +1 716 645 3822.
E-mail addresses: feaeliru@acsu.buffalo.edu (E. Ruckenstein), ishulgin@eng.buffalo.edu (I. Shulgin).

0378-5173/$ – see front matter © 2005 Published by Elsevier B.V.

to judge whether the solubility data are accurate or not, and it is important to have a rigorous test for checking the experimental solubility data and selecting the correct ones. Because we could not find such a method in the literature, the purpose of the present paper is to suggest a thermodynamic method for testing the accuracy of the experimental data regarding the solubility of drugs in aqueous mixed solvents.

Thermodynamic consistency tests are well known, and have been frequently used for vapour–liquid equilibrium data in binary mixtures (for reviews one can see Gmehling and Onken, 1977; Acree, 1984; Prausnitz et al., 1986). These tests are based on the Gibbs–Duhem equation and allow one to grade the experimental data for vapor–liquid equilibrium in binary mixtures. A more difficult problem is the consistency of data regarding vapor–liquid equilibrium in ternary or multicomponent mixtures. However, several thermodynamic consistency tests, also based on the Gibbs–Duhem equation, were suggested for vapor–liquid equilibrium in ternary or multicomponent mixtures (Li and Lu, 1959; McDermott and Ellis, 1965).

2. General relations for multicomponent mixtures

The isothermal–isobaric Gibbs–Duhem equation for an N-component mixture ($N \geq 2$) can be written as follows

$$\sum_{i=1}^{N} x_i d(\ln \gamma_i) = 0 \tag{1}$$

where x_i and γ_i are the mole fraction and the activity coefficient of component i in the N-component mixture. Integrating Eq. (1) directly along a loop of points a, b, c, ..., y, z, ... by using the trapezoidal rule, one can obtain the following equation (Li and Lu, 1959)

$$\sum_{i=1}^{N} \left\{ \frac{x_i^{(a)} + x_i^{(b)}}{2} [\ln \gamma_i^{(b)} - \ln \gamma_i^{(a)}] \right.$$
$$+ \frac{x_i^{(b)} + x_i^{(c)}}{2} [\ln \gamma_i^{(c)} - \ln \gamma_i^{(b)}]$$
$$\left. + \cdots + \frac{x_i^{(y)} + x_i^{(z)}}{2} [\ln \gamma_i^{(z)} - \ln \gamma_i^{(y)}] + \cdots \right\} = 0 \tag{2}$$

McDermott and Ellis (1965) applied Eq. (2) to a pair of points c and d. In this case, Eq. (2) reduces to

$$\sum_{i=1}^{N} (x_i^{(c)} + x_i^{(d)})[\ln \gamma_i^{(d)} - \ln \gamma_i^{(c)}] = 0 \tag{3}$$

The McDermott and Ellis consistency test means that if the vapor–liquid equilibrium data for points c and d are correct, then Eq. (3) should be satisfied. Eq. (3) will be used to derive a thermodynamic consistency test for verifying the experimental data regarding the solubility of drugs in aqueous mixed solvents.

3. Thermodynamic consistency test regarding the solubility of drugs in binary aqueous mixed solvents

For the solubility of a solid substance (solute, component 2) in a mixed solvent 1–3, one can write the following equation (Prausnitz et al., 1986):

$$\frac{f_2^S}{f_2^L(T, P)} = x_{2,t} \gamma_{2,t}(T, P, \{x\}) \tag{4}$$

where $x_{2,t}$ and $\gamma_{2,t}$ are the solubility (mole fraction) and the activity coefficient of the solid in its saturated solution in a mixed solvent, $f_2^L(T, P)$ is the hypothetical fugacity of a solid as a (sub-cooled) liquid at a given pressure (P) and temperature (T), f_2^S is the fugacity of a pure solid component 2, and $\{x\}$ indicates that the activity coefficient of the solid depends on composition. If the solubility of the mixed solvent in the solid phase is negligible, then the left hand side of Eq. (4) depends only on the properties of the solute.

Rewriting of Eq. (3) for a ternary mixture yields the expression

$$(x_1^{(c)} + x_1^{(d)})[\ln \gamma_1^{(d)} - \ln \gamma_1^{(c)}] + (x_2^{(c)} + x_2^{(d)})$$
$$\times [\ln \gamma_2^{(d)} - \ln \gamma_2^{(c)}] + (x_3^{(c)} + x_3^{(d)})$$
$$\times [\ln \gamma_3^{(d)} - \ln \gamma_3^{(c)}] = 0 \tag{5}$$

Let us consider the solubilities of a poorly soluble solid in two mixed solvents of close compositions (points c and d). Because these solubilities satisfy Eq. (4), one can express the activity coefficients of the solid via Eq. (4), and Eq. (5) acquires the form

$$(x_1^{(c)} + x_1^{(d)})[\ln \gamma_1^{(d)} - \ln \gamma_1^{(c)}] + (x_2^{(c)} + x_2^{(d)})$$
$$\times [\ln x_2^{(c)} - \ln x_2^{(d)}] + (x_3^{(c)} + x_3^{(d)})$$
$$\times [\ln \gamma_3^{(d)} - \ln \gamma_3^{(c)}] = 0 \qquad (6)$$

Let us suppose that the solubility of the solid in the mixed solvent is so low, that one can consider the activity coefficients of the solvent and cosolvent equal to those in the solute-free binary solvent mixture ($\gamma_{1,0}$ and $\gamma_{3,0}$). In addition, the following relations for the mole fractions of the constituents of the solvent can be used

$$x_{1,t} = x_{1,0} - x_{1,0} x_{2,t} \qquad (7)$$

and

$$x_{3,t} = x_{3,0} - x_{3,0} x_{2,t} \qquad (8)$$

where $x_{1,0}$ and $x_{3,0}$ are the mole fractions of constituents 1 and 3 in a solute-free mixed solvent.

Consequently, Eq. (6) becomes

$$(x_{1,0}^{(c)} + x_{1,0}^{(d)} - x_{1,0}^{(c)} x_2^{(c)} - x_{1,0}^{(d)} x_2^{(d)})[\ln \gamma_{1,0}^{(d)} - \ln \gamma_{1,0}^{(c)}]$$
$$+ (x_2^{(c)} + x_2^{(d)})[\ln x_2^{(c)} - \ln x_2^{(d)}] + (x_{3,0}^{(c)} + x_{3,0}^{(d)}$$
$$- x_{3,0}^{(c)} x_2^{(c)} - x_{3,0}^{(d)} x_2^{(d)})[\ln \gamma_{3,0}^{(d)} - \ln \gamma_{3,0}^{(c)}] = 0 \qquad (9)$$

The last equation can be simplified by applying Eq. (5) to the pair of binary mixed solvent mixtures of compositions $(x_{1,0}^{(c)}, x_{3,0}^{(c)})$ and $(x_{1,0}^{(d)}, x_{3,0}^{(d)})$. For this pair, Eq. (5) becomes

$$(x_{1,0}^{(c)} + x_{1,0}^{(d)})[\ln \gamma_{1,0}^{(d)} - \ln \gamma_{1,0}^{(c)}]$$
$$+ (x_{3,0}^{(c)} + x_{3,0}^{(d)})[\ln \gamma_{3,0}^{(d)} - \ln \gamma_{3,0}^{(c)}] = 0 \qquad (10)$$

Subtracting Eq. (10) from Eq. (9) yields

$$(x_{1,0}^{(c)} x_2^{(c)} + x_{1,0}^{(d)} x_2^{(d)})[\ln \gamma_{1,0}^{(d)} - \ln \gamma_{1,0}^{(c)}] + (x_2^{(c)} + x_2^{(d)})$$
$$\times [\ln x_2^{(d)} - \ln x_2^{(c)}] + (x_{3,0}^{(c)} x_2^{(c)} + x_{3,0}^{(d)} x_2^{(d)})$$
$$\times [\ln \gamma_{3,0}^{(d)} - \ln \gamma_{3,0}^{(c)}] = 0 \qquad (11)$$

Eq. (11) provides a thermodynamic consistency test for the solubility of poorly soluble substances, such as drugs, environmentally important substances, etc. in mixed solvents in terms of the activity coefficients of the constituents of the binary solute-free mixed solvent and mixed solvent composition.

Two limitations are involved in the derivation of the above equation: (1) the compositions of mixed solvents (points c and d) should be close enough to each other for the trapezoidal rule used to integrate the Gibbs–Duhem equation to be valid, (2) the solubility of the solid should be low enough for the activity coefficients of the solvent and cosolvent to be taken equal to those in a solute-free binary solvent mixture. In addition, the fugacity of the solid phase in Eq. (4) should remain the same for all mixed solvent compositions considered.

4. Numerical estimations

Of course, for real mixtures the left hand side of Eq. (11) is not exactly equal to zero; it has certain finite values even for very accurate data. Let us denote that value with D. McDermott and Ellis (McDermott and Ellis, 1965) suggested that the vapor–liquid equilibrium data in a ternary mixture are thermodynamically consistent if $|D|$ for Eq. (6) is smaller than $D_{max} = 0.01$. Now we should find the value of D_{max} for the solubility of poorly soluble substances in mixed solvents for Eq. (11). Of course, this value should differ from that for the vapor–liquid equilibrium.

In order to find D_{max} the following procedure was employed:

(1) Several data sets for the solubilities of poorly soluble substances in mixed solvents were selected from Solubility Data Series (Acree, 1995);
(2) The selected data were correlated with reliable equations (Ruckenstein and Shulgin, 2003a,b);
(3) Using the above equations, the solubility of the solute was calculated for small changes in the mixed solvent composition (2.5 mol%);
(4) The value of D was calculated for each of the two neighboring points;
(5) Artificial deviations ("errors") were added to selected points and a criterium for thermodynamic consistency was identified.

Table 1
Correlation of the experimental data regarding the solubility (at room temperature) of anthracene in mixed solvents

Solvent + cosolvent	Solute	Reference	Deviation from experimental data[a]	
			3-Parameter equation (Ruckenstein and Shulgin, 2003a)	4-Parameter equation (Ruckenstein and Shulgin, 2003b)
1-Propanol-2-propanol	Anthracene	Acree, 1995	0.41	0.39
n-Hexane-cyclohexane	Anthracene	Acree, 1995	0.48	0.29

[a] Deviation from experimental data calculated as the mean percentage deviation (MPD) (%) defined as $\left(100\sum_{i=1}^{N_j}|(x_i^{exp} - x_i^{calc})/(x_i^{exp})|\right)/N_j$, where x_i^{exp} and x_i^{calc} are experimental and calculated solubilities (mole fractions), and N_j is the number of experimental points in the data set j.

5. The use of the solubilities of anthracene in 1-propanol-2-propanol and anthracene in n-hexane–cyclohexane mixtures for the determination of the D_{max} value

The experimental data regarding the solubility of anthracene in 1-propanol-2-propanol and anthracene in n-hexane–cyclohexane mixtures were taken from the Solubility Data Series (Acree, 1995) and correlated with equations based on the fluctuation theory of solutions (Ruckenstein and Shulgin, 2003a,b). The results of these correlations are presented in Table 1. The values of D, calculated using Eq. (11), are plotted in Fig. 1a and b (Throughout this paper, the activity coefficients of the constituents of a solute-free mixed solvent were calculated with the Wilson equation (Wilson, 1964), using the Wilson parameters the values listed in Gmehling's vapor–liquid compilation (Gmehling and Onken, 1977)). In order to understand how the errors affected the values of D, 20% "error" was added to every second point and the D values were again calculated via Eq. (11). The results of these calculations are presented in Fig. 2. Figs. 1 and 2 show that for thermodynamically consistent data $|D| < D_{max} = 10^{-4}$. One should note that the solubility of anthracene in 1-propanol-2-propanol varies in the range 4.1×10^{-4} to 5.9×10^{-4} mole fraction and the solubility of anthracene in n-hexane-cyclohexane varies in the range 1.3×10^{-3} to 1.6×10^{-3} mole fraction. It is of interest to calculate the D values for more soluble substances. We carried out such calculations (Fig. 3) for the solubility of pyrene in 1-propanol-2-propanol (for which the solubility varies in the range 3.9×10^{-3} to 4.3×10^{-3} mole fraction) and pyrene in n-hexane–cyclohexane mixtures (for which the solubility varies in the range 8.5×10^{-3} to 10.9×10^{-3} mole fraction) (Acree, 1995; Zvaigzne et al., 1995). Fig. 3 shows that the established limit ($|D| < D_{max} = 10^{-4}$) is valid when the mole fraction solubility is smaller than 1 mol%.

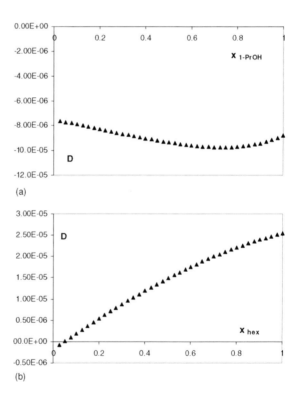

Fig. 1. D values obtained via Eq. (11) for the solubilities of anthracene in 1-propanol-2-propanol (a) and anthracene in n-hexane-cyclohexane (b).

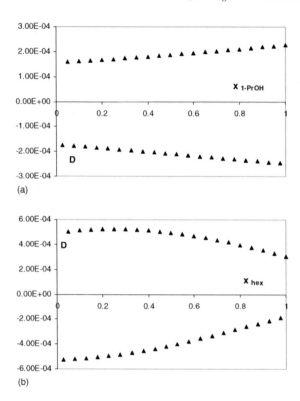

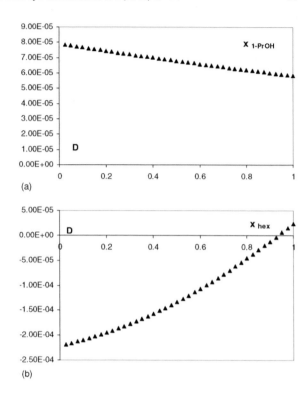

Fig. 2. D values obtained via Eq. (11) for the solubilities of anthracene in 1-propanol-2-propanol (a) and anthracene in n-hexane-cyclohexane (b) when 20% "errors" were added to every second point.

Fig. 3. D values obtained via Eq. (11) for the solubilities of pyrene in 1-propanol-2-propanol (a) and pyrene in n-hexane-cyclohexane (b).

6. Application of Eq. (11) to the solubility of poorly soluble solids in aqueous mixed solvents

6.1. Solubility of naphthalene in ethanol–water mixtures

There are several experimental determinations of the solubility of naphthalene in ethanol–water mixtures at room temperature (Bennett and Canady, 1984; Morris, 1988; Dickhut et al., 1989; LePree et al., 1994). These data deviate appreciably from each other (Fig. 4). The analysis of the above data with Eq. (11) (Table 2) indicated that those regarding the solubility of naphthalene in ethanol–water mixtures at room temperature, obtained by various authors, were thermodynamically consistent in the dilute region; however, the data of LePree et al. (1994), and Morris (1988) are thermodynamically inconsistent at high mole fractions of ethanol. Only the data for ethanol mole fractions less than 0.3 were analyzed by us, because the experimental determinations in the above publications were made with small changes in composition in that range only, and with large changes outside that range. For the latter cases, the trapezoidal rule for the integration of the Gibbs–Duhem equation might no longer be valid.

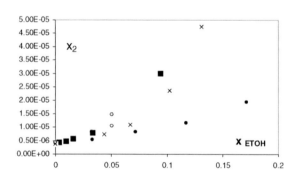

Fig. 4. The solubility of naphthalene (x_2) in ethanol–water mixtures at room temperature: (○) Bennett and Canady, 1984; (●) Morris, 1988; (■) Dickhut et al., 1989; (×) LePree et al., 1994). x_{ETOH} is the mole fraction of ethanol in the solute-free mixed solvent.

Table 2
D values obtained via Eq. (11) for data regarding the solubility of naphthalene in ethanol–water mixtures

Mole fraction of ethanol in the solute-free mixed solvent	Mole fraction of naphthalene solubility	Reference	D
0.0333	5.3E−06	A	
0.0720	8.2E−06	A	−5.9E−06
0.1173	1.2E−05	A	−6.6E−06
0.1713	1.9E−05	A	−1.6E−05
0.2367	2.3E−04	A	−6.2E−04
0.0159	5.9E−06	B	
0.0329	8.3E−06	B	−3.1E−06
0.0508	1.1E−05	B	−4.9E−06
0.0508	1.5E−05	B	−4.5E−06
0.0031	4.5E−06	C	
0.0095	4.8E−06	C	−5.5E−07
0.0161	5.8E−06	C	−2.0E−06
0.0333	8.0E−06	C	−4.4E−06
0.0937	3.0E−05	C	−5.1E−05
0.0438	7.3E−06	E	
0.0672	1.1E−05	E	−7.0E−06
0.1024	2.4E−05	E	−7.5E−06
0.1308	4.8E−05	E	−2.7E−05
0.1826	1.6E−04	E	−5.0E−05
0.2101	2.7E−04	E	−2.5E−04

A (Morris, 1988); B (Bennett and Canady, 1984); C (Dickhut et al., 1989); E (LePree et al., 1994).

In the present paper, Eq. (11) was used to analyze separately each of the sets of experimental data listed above. Therefore, each of the examinations was concerned with the internal consistency of a selected set.

6.2. Solubility of naphthalene in acetone–water mixtures

The analysis of the experimental solubilities of naphthalene in acetone–water mixtures at room

Table 3
D values obtained via Eq. (11) for data regarding the solubility of naphthalene in acetone–water mixtures

Mole fraction of ethanol in the solute-free mixed solvent	Mole fraction of naphthalene solubility	Reference	D
0.0176	7.4E−06	A	
0.0557	2.9E−05	A	−5.0E−05
0.0907	7.3E−05	A	−1.3E−04
0.1339	2.9E−04	A	−4.6E−04
0.1816	4.2E−04	A	−2.6E−04
0.2261	1.9E−03	A	−3.5E−03
0.0128	1.0E−05	B	
0.0266	1.7E−05	B	−1.3E−05
0.0580	6.9E−05	B	−1.2E−04
0.0954	2.0E−04	B	−2.8E−04
0.1410	7.2E−04	B	−1.2E−03
0.1975	2.3E−03	B	−3.5E−03
0.0266	9.3E−06	C	
0.0580	3.0E−05	C	−4.6E−05
0.0954	7.9E−05	C	−1.1E−04
0.1410	3.1E−04	C	−5.2E−04
0.1975	1.2E−03	C	−2.1E−03

A (LePree et al., 1994); B (Fu and Luthy, 1985); C (Morris, 1988).

temperature (LePree et al., 1994; Fu and Luthy, 1985; Morris, 1988), summarized in Table 3, shows that, as in the previous case, there is thermodynamic consistency in the diluted region. However, the data become increasingly inaccurate in more concentrated mixed solvents. Again, only the data for mole fractions of acetone less than 0.3 were considered.

6.3. Solubility of naphthalene in ethylene glycol–water mixtures

The analysis of the experimental solubilities of naphthalene in ethylene glycol–water mixtures at room temperature (Khossravi and Connors, 1992; Huot et al., 1991) showed that both experimental sets were accurate in a wide composition range with the exception of the points between $X_{ETD} \approx 0.5$ and 0.6 (Fig. 5).

6.4. Solubility of sulphamethoxypyridazine in ethanol–water mixtures

The solubility of sulphamethoxypyridazine in ethanol–water mixtures represents a rare kind of drug solubility in an aqueous mixed solvent, because it exhibits two solubility maxima on the curve solubility versus mixed solvent composition (Escalera et al., 1994). It is of interest to verify if such behavior satisfies the thermodynamic consistency criterion. The values of D were calculated using Eq. (11), and

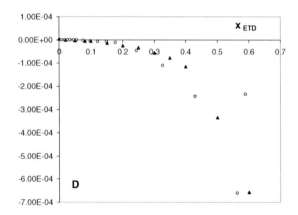

Fig. 5. D values calculated with Eq. (11) for the solubility of naphthalene in ethylene glycol–water mixtures at room temperature; (○) (Khossravi and Connors, 1992); (▲) (Huot et al., 1991). x_{ETD} is the mole fraction of ethylene glycol in a solute-free ethylene glycol–water mixture.

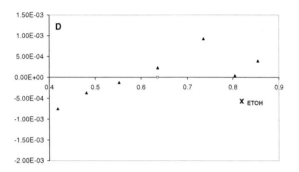

Fig. 6. D values (▲) calculated with Eq. (11) for the solubility of sulphamethoxypyridazine in ethanol–water mixture at room temperature. x_{EtOH} is the mole fraction of ethanol in a solute-free ethanol–water mixture.

the results are presented in Fig. 6. The latter figure shows that the second maximum (mole fraction of ethanol approximately 0.75) is thermodynamically less consistent than the first maximum (mole fraction of ethanol approximately 0.5).

7. Discussion and conclusion

The Gibbs–Duhem equation for ternary mixtures is used to analyze the quality of experimental data pertaining to the solubility of drugs and other poorly soluble solids in a binary mixed solvent. In order to test the quality of the data, a thermodynamic consistency test is suggested. This test is based on the thermodynamic relation between the solubilities of a solid in a binary mixed solvent at two different compositions and the activity coefficients of the constituents of the solute-free mixed solvent. The suggested test is applicable to all kinds of systems with the following limitations: (1) the solubility of the solid should be low, (2) the above two compositions of the mixed solvent should be close enough to each other.

The test was applied to a number of systems representing different types of solubilities of drugs and other poorly soluble substances in binary mixed solvents. It was shown that the suggested test could be helpful in the analysis of such experimental data.

References

Acree, W.E., 1984. Thermodynamic Properties of Nonelectrolyte Solutions. Academic Press, Orlando.

Acree, W.E. (Ed.), 1995. Polycyclic Aromatic Hydrocarbons: Binary Non-aqueous Systems. Solubility Data Series, vol. 58–59. Oxford University Press, Oxford.

Bennett, D., Canady, W.J., 1984. Thermodynamics of solution of naphthalene in various water ethanol mixtures. J. Am. Chem. Soc. 106, 910–915.

Dickhut, R.M., Andren, A.W., Armstrong, D.E., 1989. Naphthalene solubility in selected organic solvent-water mixtures. J. Chem. Eng. Data 34, 438–443.

Escalera, J.B., Bustamante, P., Martin, A., 1994. Predicting the solubility of drugs in solvent mixtures - multiple solubility maxima and the chameleonic effect. J. Pharm. Pharmacol. 46, 172–176.

Fu, J.-K., Luthy, R.G., 1985. Pollutant Sorption to Soils and Sediments in Organic/Aqueous Solvent Systems. U.S. Environmental Protection Agency, Environmental Research Laboratory, Athens, GA.

Gmehling, J., Onken, U., 1977. Vapor–Liquid Equilibrium Data Collection; DECHEMA Chemistry Data Series, vol. I, Part 1. DECHEMA: Frankfurt, Germany.

Huot, J.-Y., Page, M., Jolicoeur, C., 1991. Thermodynamic properties of naphthalene and uric acid in ethylene glycol-water mixtures. J. Sol. Chem. 20, 1093–1112.

Khossravi, D., Connors, K.A., 1992. Solvent effects on chemical processes. 1. Solubility of aromatic and heterocyclic-compounds in binary aqueous organic-solvents. J. Pharm. Sci. 81, 371–379.

LePree, J.M., Mulski, M.J., Connors, K.A., 1994. Solvent effects on chemical processes. 6. the phenomenological model applied to the solubility of naphthalene and 4-nitroaniline in binary aqueous–organic solvent mixtures. J. Chem. Soc., Perkin Trans. 2, 1491–1497.

Li, J.C.M., Lu, B.C.Y., 1959. A note on thermodynamic consistency of ternary vapor–liquid equilibrium data. Can. J. Chem. Eng. 37, 117–120.

McDermott, C., Ellis, S.R.M.A., 1965. A multicomponent consistency test. Chem. Eng. Sci. 20, 293–296.

Morris, K. R. Solubility of aromatic compounds in mixed solvents. Ph.D. Dissertation. University of Arizona, 1988.

Prausnitz, J.M., Lichtenthaler, R.N., Gomes de Azevedo, E., 1986. Molecular Thermodynamics of Fluid-Phase Equilibria, second ed. Prentice-Hall, Englewood Cliffs, NJ.

Ruckenstein, E., Shulgin, I., 2003a. Solubility of drugs in aqueous solutions. Part 1: Ideal mixed solvent approximation. Int. J. Pharm. 258, 193–201.

Ruckenstein, E., Shulgin, I., 2003b. Solubility of drugs in aqueous solutions. Part 2: Nonideal mixed solvent. Int. J. Pharm. 260, 283–291 (see also Errata: Ruckenstein, E., Shulgin, I., 2004. Int. J. Pharm. 278, 475).

Wilson, G.M., 1964. Vapor–liquid equilibrium. XI: A new expression for the excess free energy of mixing. J. Am. Chem. Soc. 86, 127–130.

Yalkowsky, S.H., He, Y., 2003. Handbook of Aqueous Solubility Data. CRC Press, Boca Raton, Fla.

Zvaigzne, A.I., Miller, B.J., Acree, W.E., 1995. Naphthalene solubility in selected organic solvent–water mixtures. J. Chem. Eng. Data 40, 1267–1269.

Environ. Sci. Technol. **2005,** *39,* 1623−1631

Solubility of Hydrophobic Organic Pollutants in Binary and Multicomponent Aqueous Solvents

E. RUCKENSTEIN* AND I. SHULGIN
Department of Chemical and Biological Engineering, State University of New York at Buffalo, Amherst, New York 14260

The present paper deals with the application of fluctuation theory of solutions to the solubility of poorly soluble substances of environmental significance in aqueous mixed solvents. The fluctuation theory of ternary solutions was first used to derive an expression for the activity coefficient of a solute at infinite dilution in a binary mixed solvent. This equation contains the activity coefficients of the constituents of the solute-free mixed solvent and the molar volume of the solute-free mixed solvent. Further, the derived expression for the activity coefficient of a solute at infinite dilution was used to generate a number of expressions for the solubility of solids in aqueous mixed solvents. Several expressions for the activity coefficients of the components were considered: first, the mixed solvent was considered an ideal mixture; second, the activity coefficients of the constituents of the binary solvent were expressed using the two-suffix Margules equations; third, the activity coefficients of the constituents of the binary solvent were expressed using the Wilson equations. The obtained expressions were applied to 25 experimental data sets pertaining to the solubilities of hydrophobic organic pollutants (HOP) in aqueous mixed solvents. It was found that the suggested equations can be used for an accurate and reliable correlation of the solubilities in aqueous mixed binary solvents. The best results were obtained by combining our expression for the activity coefficient of a solute at infinite dilution in a mixed solvent with the Wilson equations for the activity coefficients of the constituents of a solute-free mixed solvent. The derived equations can also be used for predicting the solubilities of poorly soluble environmentally important compounds in aqueous mixed solvents using for the Wilson parameters those obtained from vapor−liquid equilibrium data. A similar methodology was applied to the solubility of poorly soluble substances of environmental significance in multicomponent (ternary and higher) aqueous mixed solvents. The expression for the activity coefficient of a solute in an ideal multicomponent mixed solvent was used to derive an equation for the solubility of a poorly soluble solute in an ideal multicomponent mixed solvent in terms of its solubilities in two subsystems of the multicomponent solvent and their molar volumes. Ultimately the solubility could be expressed in terms of those in binary or even in the individual constituents of the solvent and their molar volumes. The computational method was applied to predict the solubilities of naphthalene and anthracene in ternary, quaternary and quinary aqueous mixed solvents. The results were compared with experiment and good agreement was obtained.

Introduction

The solubility of an organic compound in water is one of the key factors that affects its environmental behavior (*1−3*). The aqueous solubility is a fundamental parameter in assessing the extent of dissolution of environmentally important substances and their persistence in an aquatic environment. The extent to which aquatic biota is exposed to a toxicant is largely controlled by the aqueous solubility. In addition, these solubilities are of thermodynamic interest in elucidating the nature of these highly nonideal solutions (*1,2*).

As well-known, the solubility in water of a poorly soluble solid substance, including many of environmentally important substances, can be affected by the addition of cosolutes and cosolvents into water. As a rule the solubility of hydrophobic organic pollutants (HOP) is much greater in solutions containing organic cosolvents that in pure water (*2*). Moreover, the knowledge of the solubilities of environmentally important compounds in mixtures of water and organic cosolvents is important in the treatment of industrial wastewaters (*1*). One may encounter such cases in industrial wastewaters or at waste disposal sites where, because of careless dumping procedures, the leachates may contain a high fraction of organic solvent(s) (*1*).

While experimental data regarding the solubilities of many environmentally important substances in pure water are available in the literature (*4−6*), experimental data regarding the solubilities in aqueous mixed solvents (water + organic cosolvents) are scarce (*2, 7−10*). Therefore, there is a great need for a predictive method of the solubilities of environmentally important substances in both pure water and mixtures of water with one or several organic cosolvents. Because the majority of environmentally important substances are highly hydrophobic, they have a low solubility in water. Therefore, the thermodynamic quantity which governs their behavior in aqueous solutions is the activity coefficient at infinite dilution (*11, 12*). Consequently, the prediction of the activity coefficient of a solute at infinite dilution in water and in mixed aqueous solvents constitutes the main difficulty in the prediction of the solubilities of environmentally important substances in both pure water and mixtures of water with one or several organic cosolvent.

It seems promising to use for such predictions group-contribution methods, such as UNIFAC (*13*). The application of UNIFAC to the solubility of naphthalene in nonaqueous mixed solvents provided satisfactory results (*14*). Unfortunately, the accuracy of the UNIFAC regarding the solubility of solids in aqueous solutions is controversial (*15−17*). Large deviations from the experimental activity coefficients at infinite dilution and octanol/water partition coefficients have been reported (*16, 17*) when the classical old version of UNIFAC interaction parameters (*13*) was used. To improve the prediction of the activity coefficients at infinite dilution and of the octanol/water partition coefficients of environmentally important substances, special ad hoc sets of parameters for environmental applications were introduced (*16−18*).

The present paper is focused on the theoretical modeling of the solubility of HOP in binary and multicomponent aqueous solvents. Many methods, mostly empirical and

---

* Corresponding author phone: (716)645-2911/ext. 2214; fax: (716)645-3822; e-mail: feaeliru@acsu.buffalo.edu.

semiempirical, were suggested for the correlation and prediction of the solubility of solids in mixed solvents (14, 19). Among the methods used for the solubility of environmentally important components in aqueous solvents, which have some theoretical basis, one should point out the method based on the Margules equation for the activity coefficient of a solute (15) and the Combined Nearly Ideal Binary Solvent/ Redlich−Kister (NIBS/R−K) equation (20). The reader can find a review of the other methods in refs 1 and 21.

In this paper, a previously developed expression for the activity coefficient of a solute at infinite dilution in multicomponent solutions (22−24) will be applied to the solubility of environmentally significant compounds in aqueous solvent mixtures. The above expression for the activity coefficient of a solute at infinite dilution in multicomponent solutions (22−24) is based on the fluctuation theory of solutions (25). This model-free thermodynamic expression can be applied to both binary and multicomponent solvents.

The paper is organized as follows: first, the thermodynamic relations for the solubility of poorly soluble solids in pure and multicomponent mixed solvents are written. Second, an equation for the activity coefficient of a solute at infinite dilution in a binary nonideal mixed solvent (23) is employed to derive an expression for its solubility in terms of the properties of the mixed solvent. Third, various expressions for the activity coefficients of the cosolvents, such as Margules and Wilson equations (19), are inserted into the above equation for the solubility. The obtained equations are used to correlate the HOP solubilities in binary aqueous mixed solvents and the results are compared with experiment. Finally, the case of an ideal multicomponent solvent is considered and used for ternary and higher mixed solvents.

Thermodynamic Relations for the Solubility of Poorly Soluble Solids in Pure and Binary Mixed Solvents. For poorly soluble solids, such as the HOP, one can use the infinite dilution approximation and consider that the activity coefficient of a solute in pure and mixed solvents to be equal to those at infinite dilution. Therefore, for the solubilities of a solid substance (solute, component 2) in water (component 3), cosolvent (component 1) and their mixture (mixed solvent 1−3), one can write the following equations (14, 19)

$$f_2^S/f_2^L(T,P) = x_2^{b_1} \gamma_2^{b_1,\infty} \quad (1)$$

$$f_2^S/f_2^L(T,P) = x_2^{b_3} \gamma_2^{b_3,\infty} \quad (2)$$

$$f_2^S/f_2^L(T,P) = x_2^t \gamma_2^{t,\infty} \quad (3)$$

where $\gamma_2^{b_1,\infty}$ and $\gamma_2^{b_3,\infty}$ are the activity coefficients at infinite dilution of the solute in the individual solvents 1 and 3, $\gamma_2^{t,\infty}$ is the activity coefficient of a solute in a mixed solvent at infinite dilution, $x_2^{b_1}$, $x_2^{b_3}$ and x_2^t are the solubilities (mole fractions) of the solid component 2 in the cosolvent, water and their mixture, respectively, $f_2^L(T,P)$ is the hypothetical fugacity of a solid as a (subcooled) liquid at a given pressure (P) and temperature (T), and f_2^S is the fugacity of a pure solid component 2. If the solubilities of the pure and mixed solvents in the solid phase are negligible, then the left-hand sides of Equations 1−3 depend only on the properties of the solute. Equations 1−3 show that the solubilities of solid substances in pure and mixed solvents can be calculated if their activity coefficients at infinite dilution in the binary and ternary saturated solutions (1−2, 2−3 and 1−2−3) are known. Equation 3 is valid not only for binary, but also for ternary and higher mixed solvents.

Activity Coefficient of a Solute in a Binary solvent at Infinite Dilution via Fluctuation Theory. The following expression for the activity coefficient ($\gamma_2^{t,\infty}$) of a solid solute in a binary mixed solvent at infinite dilution can be written (23) on the basis of the fluctuation theory of ternary mixtures (see Appendix for details)

$$\ln \gamma_2^{t,\infty} = -B(P,T)I_1 + \frac{\ln \gamma_1^{b,1-3} + \ln \gamma_3^{b,1-3}}{2} + A(P,T) \quad (4)$$

with

$$I_1 = \frac{\ln(x_1^{b,1-3}V_1^0 + x_3^{b,1-3}V_3^0)}{V_3^0 - V_1^0} + \frac{\ln \gamma_3^{b,1-3}}{V_3^0 - V_1^0} - \frac{V_1^0}{V_3^0 - V_1^0} \frac{\ln \gamma_3^{b,1-3}}{x_1^{b,1-3}V_1^0 + x_3^{b,1-3}V_3^0} -$$

$$V_1^0 \int \frac{\ln \gamma_3^{b,1-3}}{(x_1^{b,1-3}V_1^0 + x_3^{b,1-3}V_3^0)^2} dx_3^{b,1-3} \quad (5)$$

where $x_i^{b,1-3}$ and $\gamma_i^{b,1-3}$ (i=1, 3) are the mole fraction and the activity coefficient of component i in the binary solvent 1−3, V_i^0 is the molar volume of the pure component i (i=1, 3), $A(P,T)$ is a composition-independent quantity, and $B(P,T)$, also a composition-independent quantity, is a function of the Kirkwood−Buff integrals (23). The molar volume (V) of the binary 1−3 solvent was approximated in Equation 5 by the ideal one

$$V = x_1^{b,1-3}V_1^0 + x_3^{b,1-3}V_3^0 \quad (6)$$

The combination of Equations 3 and 4 provides an expression for the solubility of a poorly soluble solid in a mixed binary solvent

$$\ln x_2^t = B(P,T)I_1 - \frac{\ln \gamma_1^{b,1-3} + \ln \gamma_3^{b,1-3}}{2} + \bar{A}(P,T) \quad (7)$$

where $\bar{A}(P,T) = -A(P,T) + \ln [f_2^S/f_2^L(T,P)]$.

The two composition-independent quantities $\bar{A}(P,T)$ and $B(P,T)$ can be expressed in terms of the solubilities of the solid in the individual solvents. Consequently, Equation 7 expresses the solubility of a poorly soluble solid in a mixed solvent in terms of the solubilities in the individual solvents, their molar volumes, and the activity coefficients of the constituents of the binary solvent. It is worth emphasizing that Equation 7 is an equation for the solubility of a poorly soluble solid in a mixed solvent. The only approximation involved is that the solubilities in either the pure solvents and in the mixed solvent are very small (infinite dilution approximation).

The Solubility of Poorly Soluble Solids in a Binary Solvent by Combining Equation 7 with Various Expressions for the Activity Coefficients of the Constituents of the Binary Solvent. 1) The mixed solvent is an ideal binary mixture

$$\gamma_i^{b,1-3} = 1 \ (i=1, 3) \quad (8)$$

This approximation leads to the following expression for the solubility of poorly soluble solids in a mixed solvent (22)

$$\ln x_2^t = \frac{(\ln V - \ln V_3^0)\ln x_2^{b_1} + (\ln V_1^0 - \ln V)\ln x_2^{b_3}}{\ln V_1^0 - \ln V_3^0},$$

$$\text{when } V_1^0 \neq V_3^0 \quad (9)$$

However, when $V_1^0 = V_3^0$, Equation 9 leads to a nondetermination 0/0. In this case, one obtains (22) the following expression

$$\ln x_2^t = x_1^{b,1-3}\ln x_2^{b_1} + x_3^{b,1-3}\ln x_2^{b_3} \quad (10)$$

It should be pointed out that Equations 9 and 10 cannot represent the composition dependence of the solubilities when the deviation from ideality is large. For example, they cannot provide a maximum in the solubility versus mixed solvent composition. However, such cases are frequently encountered (see the examples listed in refs 14, 22, 23, 27). To represent the large deviations from ideal behavior, such as a maximum, on the solubility curve, the ideal molar volume of the mixed solvent (Equation 6) will be replaced in Equation 9 by

$$V = x_1^{b,1-3}V_1^0 + x_3^{b,1-3}V_3^0 + ex_1^{b,1-3}x_3^{b,1-3} \quad (11)$$

where e is an empirical parameter which can be evaluated from the solubility data in a mixed solvent. Such a simple modification enabled Equation 9 to represent solubility curves exhibiting maxima. However, one should not expect Equation 11 to satisfactorily represent the molar volume of the binary solvent.

2) The activity coefficients of the constituents of the binary solvent are expressed through the two-suffix Margules equations (19)

$$\ln \gamma_1^{b,1-3} = F(x_3^{b,1-3})^2 \quad (12)$$

$$\ln \gamma_3^{b,1-3} = F(x_1^{b,1-3})^2 \quad (13)$$

where F is a temperature-dependent constant.

In this case, the integral in Equation 5 becomes

$$\int \frac{\ln \gamma_3^{b,1-3}}{(x_1^{b,1-3}V_1^0 + x_3^{b,1-3}V_3^0)^2}dx_3^{b,1-3} = \left(\frac{Fx_3^{b,1-3}}{(V_3^0 - V_1^0)^2} - \frac{F(V_3^0)^2}{(V_3^0 - V_1^0)^3 V} - \frac{2FV_3^0\ln(V)}{(V_3^0 - V_1^0)^3}\right) \quad (14)$$

3) The activity coefficients of the constituents of the binary solvent are expressed through the Wilson equations (26)

$$\ln \gamma_1^{b,1-3} = -\ln(x_1^{b,1-3} + x_3^{b,1-3}L_{13}) + x_3^{b,1-3} \times \left[\frac{L_{13}}{x_1^{b,1-3} + x_3^{b,1-3}L_{13}} - \frac{L_{31}}{x_3^{b,1-3} + x_1^{b,1-3}L_{31}}\right] \quad (15)$$

and

$$\ln \gamma_3^{b,1-3} = -\ln(x_3^{b,1-3} + x_1^{b,1-3}L_{31}) - x_1^{b,1-3} \times \left[\frac{L_{13}}{x_1^{b,1-3} + x_3^{b,1-3}L_{13}} - \frac{L_{31}}{x_3^{b,1-3} + x_1^{b,1-3}L_{31}}\right] \quad (16)$$

where L_{13} and L_{31} are the Wilson parameters, which are composition independent.

In this case, the integral in Equation 5 becomes (23)

$$\int \frac{\ln \gamma_3^{b,1-3}}{(x_1^{b,1-3}V_1^0 + x_3^{b,1-3}V_3^0)^2}dx_3^{b,1-3} = \frac{L_{13}^2(\ln V - \ln(x_1^{b,1-3} + x_3^{b,1-3}L_{13}))}{(V_3^0 - L_{13}V_1^0)^2} + \frac{L_{31}(-\ln V + \ln(x_3^{b,1-3} + x_1^{b,1-3}L_{31}))}{(V_1^0 - L_{31}V_3^0)^2} + \frac{V_3^0(L_{13}L_{31}V_3^0 - L_{13}V_1^0 + L_{13}L_{31}V_1^0 - L_{31}V_3^0)}{(V_3^0 - L_{13}V_1^0)(L_{31}V_3^0 - V_1^0)(V_3^0 - V_1^0)V} + \frac{(-1 + L_{31})\ln((1 - L_{31})V)}{(V_3^0 L_{31} - V_1^0)(V_3^0 - V_1^0)} + \frac{(x_3^{b,1-3} + x_1^{b,1-3}L_{31})\ln(x_3^{b,1-3} + x_1^{b,1-3}L_{31})}{(L_{31}V_3^0 - V_1^0)V} \quad (17)$$

The following equations: M1) Equation 9 with ideal molar volume, M2) Equation 9 with the molar volume expressed via Equation 11, M3) Equation 7 combined with Equations 5, 12–14, and M4) Equation 7 combined with Equations 5, 15–17 will be tested for the solubilities of the HOP in aqueous mixed solvents.

As one can see, M1 contains no adjustable parameter, M2 contains one adjustable parameter, M3-one and M4-two adjustable parameters. However, in addition, all equations (M1-M4) require the solubilities in individual solvents, pure water and organic cosolvent. By considering them as parameters, M1 becomes a two-parameter equation ($x_2^{b_1}$ and $x_2^{b_3}$); M2 — a three-parameter equation ($x_2^{b_1}, x_2^{b_3}$ and e); M3 — a three-parameter equation ($x_2^{b_1}, x_2^{b_3}$ and F); and M4 — a four-parameter equation ($x_2^{b_1}, x_2^{b_3}, L_{13}$ and L_{31}). Of course, expressions for the activity coefficients with any number of parameters can be used.

Activity Coefficient of a Solute in Multicomponent (Ternary and Higher) Mixed Solvents. Let us consider a n-component mixture containing a solute (component 2), water and (n-2) organic cosolvents. If one considers the (n-1) mixed solvent as an ideal mixture, one can rewrite Equation 4 for the activity coefficient of a solid solute at infinite dilution in a multicomponent (ternary and higher) solvent (24)

$$(\ln \gamma_2^{n,\infty})_{x_{i\neq 1,3}^n} = -\left(\frac{B \ln W}{(V_3^0 - V_1^0)}\right)_{x_{i\neq 1,3}^n} + A \quad (18)$$

where $\gamma_2^{n,\infty}$ is the activity coefficient of the solid solute at infinite dilution in a n-component mixture (solute+ (n-1) component solvent), W is the molar volume of an ideal (n-1) — component solvent, V_i^0 is the molar volume of the individual cosolvent i, x_i^n is the mole fraction of component i in the n-component mixture, and A and B are composition independent constants. The constants A and B can be determined from the activity coefficients of the solid solute in two (n-1) — component mixtures with the mole fraction of component 1 equal to zero in one of them and the mole fraction of component 3 equal to zero in the other one. It should be noted that expression 18 is valid on the line on which the sum of the mole fractions of components 1 and 3 is constant. Of course, a similar expression can be written for any pair of components of the mixed solvent.

To apply Equation 18 to the solubility of a solid solute in a (n-1)-component solvent, one must calculate the constants A and B. As already noted, Equation 18 is valid along the line for which

$$x_1^n + x_3^n = const \quad (19)$$

In particular, it is valid in the two limiting cases for which (I) $x_3^n = 0$ and (II) $x_1^n = 0$. These two limiting cases represent, two different (n-2) − component mixed solvents with the following compositions: In the first limiting case (I), the mole fractions are $y_1^{n-1} = x_1^n + x_3^n$, $y_3^{n-1} = 0$, $y_4^{n-1} = x_4^n$, ..., $y_n^{n-1} = x_n^n$ with $y_1^{n-1} + \sum_{i=3}^{n} y_i^{n-1} = 1$ and the mole fraction of the solute is y_2^{n-1}. In the second limiting case (II), the mole fractions are $z_1^{n-1} = 0$, $z_3^{n-1} = x_1^n + x_3^n$, $z_4^{n-1} = x_4^n$, ..., $z_n^{n-1} = x_n^n$ with $\sum_{i=3}^{n} z_i^{n-1} = 1$ and the mole fraction of the solute is z_2^{n-1}.

In the limiting cases I and II, Equation 18 acquires the forms

$$\ln \gamma_2^{n-1(I),\infty} = -\frac{B \ln V^{(I)}}{(V_3^0 - V_1^0)} + A \quad (20)$$

$$\ln \gamma_2^{n-1(II),\infty} = -\frac{B \ln V^{(II)}}{(V_3^0 - V_1^0)} + A \quad (21)$$

where $\gamma_2^{n-1(I),\infty}$ and $\gamma_2^{n-1(II),\infty}$ are the activity coefficients of the solid solute at infinite dilution in the (n-2) − component solvents I and II, respectively; $V^{(I)}$ and $V^{(II)}$ are the molar volumes of the mixtures composed of (n-2) − component solvents I and II and the solid solute, respectively. Furthermore, for a poorly soluble solid, the molar volumes of the mixtures can be taken equal to the molar volumes of the solvents.

The combination of Equation 18 with Equations 20 and 21 provides the following expression for the activity coefficient of a poorly soluble solid at infinite dilution in an ideal (n-1)-component solvent mixture

$$(\ln \gamma_2^{n,\infty})_{x_{i\neq 1,3}^n} =$$
$$\frac{(\ln W - \ln V^{(II)})\ln \gamma_2^{n-1(I),\infty} + (\ln V^{(I)} - \ln W)\ln \gamma_2^{n-1(II),\infty}}{\ln V^{(I)} - \ln V^{(II)}} \quad (22)$$

Equation 22 relates the activity coefficient of a poorly soluble solid at infinite dilution in an ideal (n-1)-component mixed solvent to the molar volume W and the activity coefficients at infinite dilution in the two limiting cases I and II and their molar volumes. The same procedure can be applied to the activity coefficient of a poorly soluble solid at infinite dilution in two ideal (n-2) − component solvents I and II and so on. Ultimately, $\gamma_2^{n,\infty}$ in Equation 18 can be predicted from the activity coefficients of a poorly soluble solid at infinite dilution in binary or even in the individual constituents of the solvent and their molar volumes.

Expression for the Solubility of a Poorly Soluble Solid in a Multicomponent Mixed Solvent. By inserting into Equation 22 the expressions of $\gamma_2^{n,\infty}$, $\gamma_2^{n-1(I),\infty}$ and $\gamma_2^{n-1(II),\infty}$ from Equations 1−3 written for the solubilities of a solid in an ideal (n-1)-component solvent mixture and in two ideal (n-2) − component solvents I and II, one can obtain the following final expression for the solubility of a poorly soluble solid in an ideal (n-1)-component solvent in terms of its solubilities in the ideal (n-2)-component mixed solvents I and II and their molar volumes:

$$\ln x_2^n = \frac{(\ln W - \ln V^{(II)})\ln y_2^{n-1} + (\ln V^{(I)} - \ln W)\ln z_2^{n-1}}{\ln V^{(I)} - \ln V^{(II)}} \quad (23)$$

Furthermore, the solubilities of a poorly soluble solid in the ideal (n-2)-component mixed solvents I and II can be expressed through those in the ideal (n-3)-component mixed solvents, and so on. Therefore, the suggested procedure

TABLE 1. Experimental Data Regarding the Solubilities (under Ambient Conditions) of the HOP in Aqueous Mixed Binary Solvents Used in Calculations

System number	Cosolvent	Solute	Number of experimental points[a]	Ref
1	methanol	Naphthalene	12	(2)
2	ethanol	Naphthalene	12	(2)
3	1-propanol	Naphthalene	10	(2)
4	acetone	Naphthalene	8	(2)
5	methanol	Phenanthrene	15	(2)
6	ethanol	Phenanthrene	15	(2)
7[b]	acetone	Naphthol	9	(2)
8	ethanol	Naphthalene	9	(9)
9	methanol	4-chlorobiphenyl	8	(10)
10	ethanol	4-chlorobiphenyl	8	(10)
11	1-propanol	4-chlorobiphenyl	8	(10)
12	methanol	2, 4, 6-trichloro-biphenyl	8	(10)
13	methanol	2, 2′, 4, 4′, 6, 6′-hexachloro-biphenyl	6	(10)
14	methanol	Anthracene	22	(8)
15	ethanol	Anthracene	22	(8)
16	2-propanol	Anthracene	22	(8)
17	acetone	Anthracene	22	(8)
18	acetonitrile	Anthracene	22	(8)
19	methanol	Atrazine	22	(8)
20	acetone	Atrazine	22	(8)
21	methanol	Naphthalene	22	(8)
22	ethanol	Naphthalene	22	(8)
23	2-propanol	Naphthalene	22	(8)
24	acetone	Naphthalene	22	(8)
25	acetonitrile	Naphthalene	22	(8)

[a] The solubilities in the individual solvents are included in the total number of experimental points. [b] The solubility of naphthol in acetone was taken from reference (28).

allows one to predict the solubility of a poorly soluble solid in an ideal (n-1)-component mixed solvent from the solubilities in binary constituents of the solvents or even from those in the individual constituents of the solvent.

Experimental Data

25 experimental data sets pertaining to the solubilities of HOP in binary aqueous solvents were selected (Table 1).

The main difficulties in this selection were the following: 1) the total number of experimental data regarding the HOP in aqueous mixed solvents is small, much smaller that the number of experimental data regarding the solubilities of drugs in aqueous mixed solvents (1). 2) There is no thermodynamic consistency test, such as those for vapor−liquid equilibrium (29), for checking the self-consistency of the data regarding the solubility of a solid in a mixed solvent. Therefore, it is difficult to evaluate whether the solubility data are accurate or contain errors.

Jorgensen and Duffy examined the accuracy of the log of the experimental solubilities of drugs in water (30) and found that they have standard deviations of about 0.58 in log units. In the present paper, the experimental data from literature with higher deviations in solubilities in pure water (>0.5 in log units) were not selected for comparison.

One should emphasize that the experimental data of different research groups exhibit large deviations from each other. This is illustrated in Figure 1 for the solubility of naphthalene in the ethanol + water mixture. Figure 1 shows that there are large differences between the experimental solubilities of naphthalene (2, 8 and 9) in the ethanol + water mixture, which are particularly large for mole fractions of ethanol between 0.05 and 0.25. Therefore, the inaccuracy of the experimental data should be taken into account, since

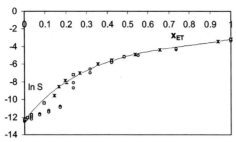

FIGURE 1. Comparison between experimental (o, ref *8*), (∗, ref *2*), and (□, ref *9*), as well as calculated (solid lines) solubilities of naphthalene (S is the mole fraction of naphthalene) in the mixed solvent ethanol/water (x_{ET} is the mole fraction of ethanol) at room temperature. The solubility was calculated using Equation M4. The adjustable constants were found using the experimental data from (*2*).

TABLE 2. Information about the Experimental Data Regarding the Solubilities of Naphthalene and Anthracene in Ternary, Quaternary and Quinary Aqueous Mixed Solvents Used in the Calculations

Number	Solute	Multicomponent solvent	Ref
		Ternary mixed solvent	
1	naphthalene	water/methanol/1-propanol	9
2	naphthalene	water/methanol/1-butanol	9
3	anthracene	water/methanol/acetone	8
4	naphthalene	water/methanol/acetonitrile	8
		Quaternary mixed solvent	
5	naphthalene	water/methanol/1-propanol/1-butanol	9
		Quinary mixed solvent	
6	naphthalene	water/methanol/ethanol/isopropanol/acetone	8
7	anthracene	water/methanol/ethanol/isopropanol/acetone	8

TABLE 3. Comparison between Experimental Solubilities in Binary Mixed Solvents and Solubilities Calculated Using Different Equations[A]

Deviations (%) between experimental and calculated solubilities

System no.[B]	Equation M1		Equation M2		Equation M3		Equation M4	
	MPD_1[C]	MPD_2[D]	MPD_1	MPD_2	MPD_1	MPD_2	MPD_1	MPD_2
1	15.6	1.8	7.9	0.78	7.5	0.7	7.0	0.6
2	54.5	15.7	17.5	2.6	25.0	4.7	12.7	1.4
3	69.6	32.5	30.4	6.4	40.5	11.0	14.8	5.5
4	55.6	15.8	5.4	0.6	15.9	1.9	5.8	0.6
5	30.1	4.4	3.3	0.3	8.7	1.0	4.3	0.5
6	44.3	8.5	13.9	1.6	28.1	3.4	8.0	1.1
7	39.6	8.8	12.2	1.8	14.5	2.2	11.2	1.6
8	24.9	6.8	15.4	3.2	9.6	1.4	8.6	1.5
9	22.3	1.9	19.3	1.2	17.2	1.0	11.9	0.6
10	34.6	5.9	31.7	4.8	24.3	2.3	20.7	2.4
11	51.5	13.0	30.6	4.1	30.0	5.4	27.7	3.0
12	22.9	2.2	18.2	1.0	14.0	0.7	10.0	0.4
13	33.5	4.7	16.4	1.1	11.1	0.7	9.2	0.4
14	28.9	3.1	16.2	1.5	18.2	1.7	14.3	1.4
15	56.5	10.4	30.8	3.4	31.6	3.3	25.8	1.9
16	67.5	17.8	37.5	5.0	45.0	7.3	29.7	1.8
17	71.8	18.0	23.8	2.2	39.5	5.1	19.3	2.2
18	67.6	16.0	28.4	3.1	42.4	6.1	16.1	1.6
19	65.4	24.5	49.0	16.6	49.5	17.7	18.7	2.7
20	71.6	25.1	39.1	9.7	47.7	13.0	24.1	6.0
21	21.6	3.9	17.0	2.5	13.1	1.6	10.7	1.3
22	56.7	10.9	41.7	15.7	46.8	14.9	38.3	12.0
23	55.5	20.5	32.0	4.0	35.5	6.6	25.6	2.4
24	59.3	24.6	15.5	2.9	24.6	5.7	11.5	2.0
25	59.3	21.9	21.7	3.9	33.8	7.4	13.8	1.9

[A] Equations M1-M4 are described in text after eq 17. [B] The systems are in the same order as in Table 1. [C] Deviation from experimental data calculated as MPD_1 (%) (the mean percentage deviation) defined as

$$\frac{100 \sum_{i=1}^{N_j} \left| \frac{x_i^{exp} - x_i^{calc}}{x_i^{exp}} \right|}{N_j}$$

where x_i^{exp} and x_i^{calc} are the experimental and calculated solubilities, N_j is the number of experimental points in the data set j. [D] Deviation from experimental data calculated as MPD_2 (%) defined as

$$\frac{100 \sum_{i=1}^{N_j} \left| \frac{\ln(x_i^{exp}) - \ln(x_i^{calc})}{\ln(x_i^{exp})} \right|}{N_j}.$$

the accuracy of any prediction "cannot exceed the accuracy of the experimental data" (*30*).

Much less data is available for the solubilities of HOP in multicomponent aqueous solvents. The literature provides the solubilities of naphthalene and anthracene in ternary, quaternary and quinary aqueous mixed solvents. Detailed information about the experimental data used in our calculations is listed in Table 2. So far there is no method for testing the self-consistency of the experimental data regarding the solubility of a poorly soluble solid in mixed solvents and the accuracy of the data in Table 2 could not be verified.

For our computational scheme, it is important to have the solubilities of naphthalene and anthracene in binary aqueous mixed solvents, which are subsystems of the ternary, quaternary and quinary aqueous mixed solvents listed in Table 2 and they are available (*8*, *9*). One can therefore compare the predictions from the solubilities in the individual constituents of the solvent to those obtained on the basis of the solubilities in binary mixed solvents.

Results of the Calculations

Binary Mixed Solvent. Calculations involving 25 experimental data sets (Table 1) were carried out using the equations listed above (M1, M2, M3 and M4), and the results are summarized in Tables 3 and 4. An additional Table containing the values of the adjustable parameters obtained for Equations M2-M4 as well as the values of $\bar{A}$ and B is given as Supporting Information.

Ternary Mixed Solvent. The solubilities of naphthalene and anthracene in the ternary aqueous mixed solvents (see Table 2) were calculated using Equation 23. The prediction of the solubility of naphthalene in water/methanol/1-propanol mixed solvent (system 1 of Table 2) is used as an example. As mentioned above the solubility can be calculated in two different ways:

1) From the solubilities in the individual constituents of the solvent and their molar volumes. The molar volume of a ternary mixed solvent (W) in Equation 23 can be obtained from the molar volumes of water, methanol and 1-propanol as an ideal molar volume

$$W = x_{H_2O}^t V_{H_2O}^0 + x_{MeOH}^t V_{MeOH}^0 + x_{PrOH}^t V_{PrOH}^0 \quad (24)$$

where $x_{H_2O}^t$, x_{MeOH}^t and x_{PrOH}^t are the mole fractions of water, methanol and 1-propanol in the mixed solvent and $V_{H_2O}^0$, V_{MeOH}^0 and V_{PrOH}^0 are the molar volumes of the pure water, methanol and 1-propanol. There are 3 options for selecting the binary subsystems I and II: 1) water/methanol and water/1-propanol, 2) water/methanol and methanol/1-propanol, or 3) water/1-propanol and methanol/1-propanol. One can show that all these selections lead to the same result. However, in our calculations we will select the aqueous binary subsystems (water/methanol and water/1-propanol), because the solubilities of naphthalene in binary aqueous mixed solvents are available in the same references as for the ternary data (*8*, *9*). Hence, the binary subsystem I is the binary system water/1-propanol with mole fractions of components: $y_{PrOH}^{b,I}$

TABLE 4. Comparison between Average Deviations of the Solubilities in Binary Solvents Calculated Using Equations M1-M4 and Experimental Data

	Average deviation (%)[a]							
	M1		M2		M3		M4	
Experimental data	MPD$_1$	MPD$_2$	MPD$_1$	MPD$_2$	MPD$_1$	MPD$_2$	MPD$_1$	MPD$_2$
All 25 experimental data sets from Table 1	50.8	14.1	25.0	4.7	30.2	6.0	17.5	2.6
13 experimental data sets from references (2, 9 and 10)	38.5	9.4	16.0	2.2	19.1	2.9	11.0	1.5
12 experimental data sets from reference (8)	56.8	15.8	29.4	5.7	35.6	7.2	20.7	3.0

[a] Average deviation (%) is the mean percentage deviation defined in case MPD$_1$ as

$$\frac{100 \sum_{j=1}^{M} \sum_{i=1}^{N_j} \left| \frac{x_i^{exp} - x_i^{calc}}{x_i^{exp}} \right|}{\sum_{j=1}^{M} N_j}$$

and in case MPD$_2$ as

$$\frac{100 \sum_{j=1}^{M} \sum_{i=1}^{N_j} \left| \frac{\ln(x_i^{exp}) - \ln(x_i^{calc})}{\ln(x_i^{exp})} \right|}{\sum_{j=1}^{M} N_j}$$

where x_i^{exp} and x_i^{calc} are experimental and calculated solubilities (mole fractions), N_j is the number of experimental points in the data set j (see Table 1), M is the number of data sets.

$= x_{MeOH}^t + x_{PrOH}^t$ and $y_{H_2O}^{b,I} = 1 - y_{PrOH}^{b,I}$ and molar volume

$$V^{(I)} = y_{H_2O}^{b,I} V_{H_2O}^0 + y_{PrOH}^{b,I} V_{PrOH}^0 \quad (25)$$

The binary subsystem II is water/methanol with mole fractions of components: $z_{MeOH}^{b,II} = x_{MeOH}^t + x_{PrOH}^t$ and $z_{H_2O}^{b,II} = 1 - y_{MeOH}^{b,II}$ and molar volume

$$V^{(II)} = z_{H_2O}^{b,II} V_{H_2O}^0 + z_{MeOH}^{b,II} V_{MeOH}^0 \quad (26)$$

The solubilities of naphthalene in the binary subsystems I and II can be calculated from Equation 23 for binary mixed solvents, which requires only the solubilities in the individual constituents of the solvent.

2) From the experimental solubilities in binary solvents and molar volumes of the individual constituents. The calculation procedure is the same as above, with the exception that the solubilities of naphthalene in the binary subsystems I and II were obtained from experiment. The calculation of the solubilities in binary aqueous solvents was described in detail in previous sections. It should be noted that the water/methanol/1-butanol mixed solvent and the binary subsystem water/1-butanol are not completely miscible. Only the homogeneous regions of mixed solvents were considered in this paper.

The solubilities of naphthalene and anthracene in ternary aqueous mixed solvents are predicted and compared with experiment in Table 5.

Quaternary Mixed Solvent. The solubilities of naphthalene in the water/methanol/1-propanol/1-butanol solvent mixture are predicted using Equation 23. The molar volume of an ideal quaternary mixed solvent was calculated as

$$W = x_{H_2O}^q V_{H_2O}^0 + x_{MeOH}^q V_{MeOH}^0 + x_{PrOH}^q V_{PrOH}^0 + x_{BuOH}^q V_{BuOH}^0 \quad (27)$$

where x_i^q is the molar fraction of component i in the quaternary mixed solvent. The two selected ternary subsystems I and II were water/methanol/1-propanol and water/methanol/1-butanol with compositions: I ($y_{PrOH}^{t,I} = x_{BuOH}^q + x_{PrOH}^q$, $y_{H_2O}^{t,I} = x_{H_2O}^q$ and $y_{MeOH}^{t,I} = x_{MeOH}^q$) and II ($z_{BuOH}^{t,II} = x_{BuOH}^q + x_{PrOH}^q$, $z_{H_2O}^{t,II} = x_{H_2O}^q$ and $z_{MeOH}^{t,II} = x_{MeOH}^q$). The solubilities of naphthalene in the ternary subsystems I and II were calculated as described in the section "Ternary mixed solvent".

It should be noted that the water/methanol/1-propanol/1-butanol solvent is not completely miscible. Only the homogeneous regions of mixed solvents were considered in this paper. The solubility predictions for the quaternary mixed solvent are listed and compared with experiment in Table 5.

Quinary Mixed Solvent. The solubilities of naphthalene and anthracene in quinary mixed solvents (water/methanol/ethanol/2-propanol/acetone) were predicted using Equation 23. The molar volume of an ideal quinary mixed solvent was calculated as

$$W = x_{H_2O}^{qu} V_{H_2O}^0 + x_{MeOH}^{qu} V_{MeOH}^0 + x_{EtOH}^{qu} V_{EtOH}^0 + x_{i-PrOH}^{qu} V_{i-PrOH}^0 + x_{Acet}^{qu} V_{Acet}^0 \quad (28)$$

where x_i^{qu} is the molar fraction of component i in the quinary mixed solvent. The two selected quaternary subsystems I and II were water/methanol/ethanol/2-propanol and water/methanol/ethanol/acetone with compositions: I ($y_{i-PrOH}^{q,I} = x_{Acet}^{qu} + x_{i-PrOH}^{qu}$, $y_{H_2O}^{q,I} = x_{H_2O}^{qu}$, $y_{MeOH}^{q,I} = x_{MeOH}^{qu}$ and $y_{EtOH}^{q,I} = x_{EtOH}^{qu}$) and II ($z_{Acet}^{q,I} = x_{Acet}^{qu} + x_{i-PrOH}^{qu}$, $z_{H_2O}^{q,I} = x_{H_2O}^{qu}$, $z_{MeOH}^{q,I} = x_{MeOH}^{qu}$ and $z_{EtOH}^{q,I} = x_{EtOH}^{qu}$). The solubilities of naphthalene and anthracene in quaternary subsystems I and II were calculated as described in the section "Quaternary mixed solvent". The results of the solubility prediction for quinary mixed solvent are listed and compared with experiment in Table 5.

Discussion

The results listed in Tables 3 and 4 demonstrate that equation M4, which combines eq 7 with the Wilson equation, provides the best results for all 25 experimental sets analyzed. Equation M2, which contains an empirical parameter performed surprisingly well, exhibiting an accuracy comparable to those of the theoretical equation M3 (eq 7 combined with the Margules equation). Equation 9 (M1) based on an ideal mixed

TABLE 5. Comparison between the Experimental Solubilities of Naphthalene and Anthracene in Multicomponent (Ternary, Quaternary and Quinary) Aqueous Mixed Solvents and the Solubilities Predicted with Equation 23

Solute	Multicomponent mixed solvent	Deviation (%) from experimental data			
		Using solubilities in individual solvents		Using solubilities in binaries	
		MPD$_1$[A]	MPD$_2$[B]	MPD$_1$	MPD$_2$
	Ternary mixed solvent				
naphthalene	water/methanol/1-propanol	7.8	0.7	3.3	0.3
naphthalene	water/methanol/1-butanol	11.9	1.1	4.8	0.4
anthracene	water/methanol/acetone	76.4	16.0	36.1	4.8
naphthalene	water/methanol/acetonitrile	62.2	18.9	14.8	2.6
	Quaternary mixed solvent				
naphthalene	water/methanol/1-propanol/1-butanol	7.7	0.7	2.6	0.2
	Quinary mixed solvent				
naphthalene	water/methanol/ethanol/2-propanol/acetone	54.3	14.7	28.8	3.8
anthracene	water/methanol/ethanol/2-propanol/acetone	76.1	14.7	30.6	2.7

[A] Deviation from experimental data calculated as MPD$_1$ (%) (the mean percentage deviation) defined as

$$\frac{100 \sum_{i=1}^{N_j} \left| \frac{x_i^{exp} - x_i^{calc}}{x_i^{exp}} \right|}{N_j}$$

where x_i^{exp} and x_i^{calc} are experimental and calculated solubilities, N_j is the number of experimental points in the data set j. [B] Deviation from experimental data calculated as MPD$_2$ defined as

$$\frac{100 \sum_{i=1}^{N_j} \left| \frac{\ln(x_i^{exp}) - \ln(x_i^{calc})}{\ln(x_i^{exp})} \right|}{N_j}$$

where x_i^{exp} and x_i^{calc} are experimental and calculated solubilities, N_j is the number of experimental points in the data set j.

solvent approximation provided the least accurate results among the equations M1-M4, but can be used as a first approximation for predictions. However, one should again emphasize that Equation M1 cannot predict a maximum in the solubility versus mixed solvent composition, which was frequently observed for the solubility of poorly soluble drugs and environmentally important substances in aqueous mixed solvents (see the examples listed in refs 14, 22, 23, and 27).

The main advantage of the proposed computational scheme is its theoretical basis. The parameters of the suggested equations M2-M4 (with the exception of the parameter e in Equation 11) have a clear physical meaning, being the parameters of the activity coefficients of the constituents of the mixed solvent.

Generally speaking, one can use for the values of the parameters those obtained from vapor−liquid or solid−liquid equilibria. Figure 2 presents such predictions for the solubilities of naphthalene in methanol + water and ethanol + water mixtures. One can see that while the agreement for the solubilities of naphthalene in methanol + water is excellent, the prediction of the solubilities of naphthalene in ethanol + water is only fair.

One should mention that the experimental data taken from different references have different qualities. Indeed, Table 4 shows that the average deviations of the experimental data taken from refs 2, 9 and 10 are much smaller than those from ref 8. Figure 1 also shows that the experimental data taken from refs 2 and 9 differ from those taken from ref 8.

The prediction of the solubility of poorly soluble substances of environmental significance in multicomponent (ternary and higher) aqueous mixed solvents is a difficult task because it requires the knowledge of the activity coefficient of a solute in a multicomponent mixed solvent. The method most often used for the solubility of a solid in ternary and multicomponent mixed solvents is the combined nearly ideal binary solvent/Redlich − Kister equation (33). That equation was applied to the solubility of a solid in ternary

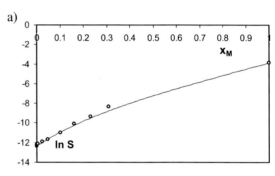
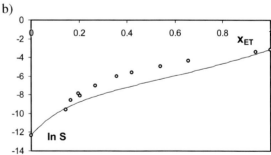

FIGURE 2. Solubilities of naphthalene (S is the mole fraction of naphthalene) in the mixtures: a) methanol + water and b) ethanol + water. The experimental data (o) were taken from Ref. (2). The solid lines represent the solubilities of naphthalene predicted using equation M4. The Wilson constants were taken from Gmeling's vapor−liquid equilibrium compilation (29). Thus, the only solubilities in pure water and cosolvents were used for prediction.

nonaqueous mixed solvents and even to the solubility of a solid in a 7-component nonaqueous mixed solvent (34, 35). However, we could not find in the literature any example of applications of theoretical methods to the solubility of environmentally important substances in multicomponent aqueous mixed solvents. In this paper, the ideal solution approximation (Equation 18) was used for a mixed solvent. It was supposed that the main contribution to nonideality of the very dilute mixture solute + multicomponent solvent stems from the interactions between the solute molecules and the molecules of the mixed solvent and not from the nonideality of the mixed solvent. Indeed, the experimental data regarding the activity coefficients indicated that such an assumption is quite reasonable for aqueous solutions of large molecules containing various functional groups, such as the drugs and the substances of environmental significance. Indeed, whereas the activity coefficient of the components of binary mixtures of water and typical organic cosolvents such as alcohols, acetone, acetonitrile, etc. are usually between 1 and 10 and they rarely reach few dozens, the activity coefficients at infinite dilution of large molecules, such as the polycyclic aromatics are usually greater than thousands or tens of thousands, or even larger (11, 12, 36).

Examination of the solubility prediction results (Table 5) reveals that the method suggested provides reasonable predictions for the solubility of environmentally important compounds in ternary, quaternary and quinary mixed solvents. Only the prediction of the solubilities of anthracene in water/methanol/acetone mixed solvent are not very accurate. However, they might have been caused by the inaccuracies of the experimental data (8) for this system. It should be noted, that the calculated solubilities of anthracene in the binaries water/methanol and water/acetone (9) were also in disagreement with the data of ref 8.

The predictions of the solubilities in ternary, quaternary and quinary mixed solvents were made from the molar volumes of water and cosolvents and a) the solubilities in the individual constituents of the solvent or b) solubilities in binary constituents. The comparison between a) and b) predictions provided an expected result, namely that the predictions involving the solubilities in binary constituents of the solvent are more accurate than those obtained in terms of the solubilities in the individual constituents of the solvent. However, the latter provides also sufficiently accurate results.

It is worth noting that good agreement was obtained despite of two important limitations imposed on the method: a) the multicomponent solvent was considered ideal, b) the solid solubility in a mixed solvent was supposed to be small enough for the infinite dilution approximation to be valid. This good agreement proves that indeed the main contribution to nonideality of a mixture composed of a large solute, such as proteins, polymers, drugs or substances of environmental significance, and multicomponent aqueous solvents stems from the interactions between the solute molecules and the molecules of the mixed solvent, and not from the interactions between the latter molecules.

Appendix

The aim of this Appendix is to provide some details regarding the derivation of an equation for the activity coefficient ($\gamma_2^{t,\infty}$) of a solid solute in a binary mixed solvent at infinite dilution.

The Kirkwood–Buff theory of solutions (25) for ternary mixtures (31) provides the following expression for the composition derivative of $\gamma_{2,t}$ at infinite dilution

$$\lim_{x_2^t \to 0} \left(\frac{\partial \ln \gamma_{2,t}}{\partial x_3^t}\right)_{T,P,x_2^t} = -[(c_1^0 + c_3^0)((c_1^0 + c_3^0)(\Delta_{12} - \Delta_{23})_{x_2^t = 0} + (c_1^0 - c_3^0)(\Delta_{13})_{x_2^t = 0})]/[2(c_1^0 + c_3^0) + c_1^0 c_3^0 (\Delta_{13})_{x_2^t = 0}] \quad \text{(A-1)}$$

where x_3^t is the mole fraction of component 3 in the ternary mixture, c_k^0 ($k=1, 3$) is the molecular concentration of component k in the binary mixture 1–3 and $G_{\alpha\beta}$ is the Kirkwood–Buff integral given by

$$G_{\alpha\beta} = \int_0^\infty (g_{\alpha\beta} - 1) 4\pi r^2 dr \quad \text{(A-2)}$$

In the above expressions, $g_{\alpha\beta}$ is the radial distribution function between species α and β, r is the distance between the centers of molecules α and β, and $\Delta_{\alpha\beta}$ are defined as

$$\Delta_{\alpha\beta} = G_{\alpha\alpha} + G_{\beta\beta} - 2G_{\alpha\beta}, \alpha \neq \beta \quad \text{(A-3)}$$

It should be noted that $\Delta_{\alpha\beta}$ is a measure of the nonideality (32) of the binary mixture $\alpha - \beta$, because for an ideal mixture $\Delta_{\alpha\beta} = 0$.

Kirkwood and Buff (25) obtained the following expression for the concentration derivative of the activity coefficient of component α in a binary mixture $\alpha-\beta$

$$\left(\frac{\partial \ln \gamma_\alpha}{\partial x_\alpha}\right)_{P,T} = \frac{c_\beta^0 \Delta_{\alpha\beta}}{1 + c_\alpha^0 x_\beta \Delta_{\alpha\beta}} \quad \text{(A-4)}$$

where x_i and γ_i are the mole fraction and the activity coefficient of component i in the binary mixture $\alpha-\beta$.

Introducing Δ_{13} from Equation (A-4) in Equation (A-1) and integrating yields

$$\ln \gamma_2^{t,\infty} = -\int (c_1^0 + c_3^0) \frac{(\Delta_{12} - \Delta_{23})_{x_2^t = 0}}{2} \left[1 + x_3^{b,1-3}\left(\frac{\partial \ln \gamma_3^{b,1-3}}{\partial x_3^{b,1-3}}\right)_{P,T}\right] dx_3^{b,1-3} + \frac{1}{2}\int \frac{(x_1^{b,1-3} - x_3^{b,1-3})}{x_1^{b,1-3}}\left(\frac{\partial \ln \gamma_3^{b,1-3}}{\partial x_3^{b,1-3}}\right)_{P,T} dx_3^{b,1-3} + U \quad \text{(A-5)}$$

where $x_i^{b,1-3}$ ($i=1, 3$) is the mole fraction of component i in the mixed solvent, $\gamma_2^{t,\infty}$ is the activity coefficient of a solute in a mixed solvent at infinite dilution and U is a constant of integration. It was supposed that $(\Delta_{12} - \Delta_{23})_{x_2^t = 0}$ is a composition independent quantity. This assumption was suggested by the fact that $(\Delta_{i2})_{x_2^t = 0}$ is a measure of the nonideality between component i ($i=1$ or 3) and 2 at infinite dilution of component 2. In the first approximation, we suppose that $(\Delta_{i2})_{x_2^t = 0}$ is the same as in the binary mixture i-2 at infinite dilution of component 2. Therefore the difference between $(\Delta_{12})_{x_2^t = 0}$ and $(\Delta_{23})_{x_2^t = 0}$ is in a first approximation a composition independent quantity.

Equation (A-5) can be transformed into Equation 4 of the text (23). However, one should note that Equation 4 was derived assuming that $(\Delta_{12} - \Delta_{23})_{x_2^t = 0}$ is a composition independent quantity.

Supplementary Material

Table

The values of the adjustable parameters obtained for Equations M2-M4 [A]

System number [B]	Values of adjustable parameters		
	Equation M2	Equation M3	Equation M4
	e, [cm^3/mol]	F, ($\bar{B}$ and $\bar{A}$ [cm^3/mol])	L_{13} and L_{31} ($\bar{B}$ and $\bar{A}$ [cm^3/mol])
1	4.348	-0.0992 (-402.166 and -46.725)	3.2656 and 0.0067 (-265.612 and -41.266)
2	30.322	-0.2232 (-611.236 and -41.429)	0.7376 and 0.2182 (-535.468 and -121.325)
3	69.805	-0.3495 (-825.609 and -40.138)	0.2187 and 0.1587 (-1178.561 and -5951.405)
4	77.044	-0.4303 (-1321.4 and -51.426)	0.3830 and 0.1778 (-980.696 and -463.206)
5	7.271	-0.1017 (-612.342 and -68.963)	1.3589 and 0.2544 (-443.2819 and -107.5740)
6	26.630	-0.2078 (-767.151 and -55.081)	0.7387 and 0.2248 (-704.185 and -160.418)
7	39.425	-0.3434 (-724.323 and -33.966)	0.4575 and 0.4183 (-570.078 and -161.987)
8	17.310	-0.2427 (-662.576 and -42.501)	1.6386 and 0.0670 (-426.512 and -51.891)
9	3.203	-0.0689 (-453.914 and -65.983)	2.2359 and 0.1317 (-352.070 and -68.770)
10	5.535	-0.2288 (-846.573 and -60.833)	2.9568 and 0.0048 (-487.184 and -53.557)
11	60.679	-0.3386 (-1062.06 and -57.312)	0.4798 and 0.1398 (-1099.484 and -272.758)

12	5.163	-0.0936 (-605.598 and -76.783)	1.9002 and 0.1442 (-429.401 and -88.826)
13	8.701	-0.1171 (-901.869 and -93.832)	1.4240 and 0.1615 (-605.136 and -147.416)
14	4.234	-0.0807 (-439.148 and -59.832)	1.1323 and 0.4561 (-350.598 and -103.732)
15	25.258	-0.2388 (-790.248 and -54.856)	0.7629 and 0.1659 (-665.841 and -147.925)
16	66.167	-0.3534 (-948.369 and -50.290)	0.3508 and 0.1082 (-1172.959 and -648.760)
17	56.912	-0.3098 (-1037.19 and -56.052)	0.4494 and 0.1765 (-1133.835 and -341.156)
18	35.763	-0.22 (-864.699 and -61.708)	0.5393 and 0.1343 (-812.0732 and -511.146)
19	14.251	-0.1370 (-414.149 and -40.562)	0.0248 and 0.0370 (-2994.097 and -3.343)
20	65.088	-0.3441 (-648.405 and -34.288)	0.0303 and 0.0911 (-2781.0035 and -7.1439)
21	2.791	5.1632 (-10.601 and -35.574)	2.4368 and 0.1216 (-253.294 and -44.967)
22	-15.994	1.3383 (-81.154 and -29.628)	11.8168 and 0.000001 (-236.6571 and -21.1199)
23	65.196	-0.3509 (-780.332 and -38.465)	0.5418 and 0.1532 (-747.381 and -142.115)
24	49.941	-0.3279 (-909.602 and -44.382)	0.7510 and 0.1582 (-757.055 and -104.408)
25	32.397	-0.2122 (-708.843 and -48.489)	0.6195 and 0.1600 (-633.894 and -264.043)

[A] Equations M2-M4 are described in text after Eq. 17,

[B] The systems are in the same order as in Table 1.

Literature Cited

(1) Schwarzenbach, R. P.; Gschwend, P. M.; Imboden, D. M. *Environmental organic chemistry*; J. Wiley: New York, 1993.

(2) Fu, J.-K.; Luthy, R. G. *Pollutant sorption to soils and sediments in organic/aqueous solvent systems*; Athens, GA: U.S. Environmental Protection Agency, Environmental Research Laboratory, 1985.

(3) Lee, L. S.; Bellin, C. A.; Pinal, R.; Rao, P. S. C. Cosolvent effects on sorption of organic acids by soils from mixed solvents. *Environ. Sci. Technol.* **1993**, *27*, 165−171.

(4) Stephen, H.; Stephen, T. *Solubilities of Inorganic and Organic Compounds*; Pergamon Press Ltd.: Oxford, 1963.

(5) Mackay, D.; Shiu, W. Y. Aqueous solubility of polynuclear aromatic hydrocarbons. *J. Chem. Eng. Data* **1977**, *22*, 399−402.

(6) *Polycyclic aromatic hydrocarbons in pure and binary solvents/Solubility Data Series*; Vol. 54, vol. editor, William E. Acree; Pergamon: 1982.

(7) Shiu, W. Y.; Maijanen, A.; Ng, A. L. Y.; Mackay, D. Preparation of aqueous solutions of sparingly soluble organic substances. 2. Multicomponent systems − hydrocarbon mixtures and petroleum-products. *Environ. Toxicol. Chem.* **1988**, *7*, 125−137.

(8) Morris, K. R. *Solubility of aromatic compounds in mixed solvents*; University of Arizona: Ph.D. Dissertation, 1988.

(9) Dickhut, R. M.; Andren, A. W.; Armstrong, D. E. Naphthalene solubility in selected organic solvent−water mixtures. *J. Chem. Eng. Data* **1989**, *34*, 438−443.

(10) Li, A.; Andren, A. W. Solubility of polychlorinated biphenyls in water-alcohol mixtures. 1. Experimental data. *Environ. Sci. Technol.* **1994**, *28*, 47−52.

(11) Sandler, S. I. Infinite dilution activity coefficients in chemical, environmental and biochemical engineering. *Fluid Phase Equilibr.* **1996**, *116*, 343−353.

(12) Sandler, S. I. Unusual chemical thermodynamics. *J. Chem. Thermodyn.* **1999**, *31*, 3−25.

(13) Fredenslund, A.; Gmehling, J.; Rasmussen, P. *Vapor-Liquid Equilibria Using UNIFAC*, Elsevier: Amsterdam, 1977.

(14) Acree, W. E. *Thermodynamic properties of nonelectrolyte solutions*, Academic Press: Orlando, 1984.

(15) Fan, C. H.; Jafvert, C. T. Margules equations applied to PAH solubilities in alcohol-water mixtures. *Environ. Sci. Technol.* **1997**, *31*, 3516−3522.

(16) Kan, A. T.; Tomson, M. B. UNIFAC prediction of aqueous and nonaqueous solubilities of chemicals with environmental interest. *Environ. Sci. Technol.* **1996**, *30*, 1369−1376.

(17) Wienke, G.; Gmehling, J. Prediction of octanol − water − partition coefficients, Henry-coefficients and water solubilities using UNIFAC. *Toxicol. Environ. Chem.* **1998**, *65*, 57−86.

(18) Li, A.; Doucette, W. J.; Andren, A. W. Estimation of aqueous solubility, octanol/water partition-coefficient, and henrys law constant for polychlorinated biphenyls using UNIFAC. *Chemosphere* **1994**, *29*, 657−669.

(19) Prausnitz, J. M.; Lichtenthaler, R. N.; Gomes de Azevedo, E. *Molecular thermodynamics of fluid − Phase equilibria*, 2nd ed.; Prentice − Hall: Englewood Cliffs, NJ, 1986.

(20) Jouyban-Gharamaleki, A.; Acree, W. E.; Clark, B. J. Comment on "Margules equations applied to PAH solubilities in alcohol-water mixtures". *Environ. Sci. Technol.* **1999**, *33*, 1953−1954.

(21) Dickhut, R. M.; Armstrong, D. E.; Andren, A. W. The solubility of hydrophobic aromatic chemicals in organic solvent water mixtures − evaluation of 4 mixed solvent solubility estimation methods. *J. Environ. Toxicol. Chem.* **1991**, *10*, 881−889.

(22) Ruckenstein, E.; Shulgin, I. Solubility of drugs in aqueous solutions. Part 1: Ideal mixed solvent approximation. *Int. J. Pharmaceut.* **2003**, *258*, 193−201.

(23) Ruckenstein, E.; Shulgin, I. Solubility of drugs in aqueous solutions. Part 2: Nonideal mixed solvent. *Int. J. Pharmaceut.* **2003**, *260*, 283−291. (see also Errata: Ruckenstein, E.; Shulgin, I. *Int. J. Pharmaceut.* **2004**, *278*, 475).

(24) Ruckenstein, E.; Shulgin, I. Ideal multicomponent liquid solution as a mixed solvent. *Ind. Eng. Chem. Res.* **2003**, *42*, 4406−4413.

(25) Kirkwood, J. G.; Buff, F. P. Statistical mechanical theory of solutions. I. *J. Chem. Phys.* **1951**, *19*, 774−782.

(26) Wilson, G. M. Vapor-liquid equilibrium. XI: A new expression for the excess free energy of mixing. *J. Am. Chem. Soc.* **1964**, *86*, 127−130.

(27) Pinal, R.; Lee, L. S.; Rao, P. S. C. Prediction of the solubility of hydrophobic compounds in nonideal solvent mixtures. *Chemosphere* **1991**, *22*, 939−951.

(28) Azizian, S.; Pour, A. H. Solubility of 2-naphthol in organic nonelectrolyte solvents. Comparison of observed versus predicted values based upon mobile order and regular solution theories. *J. Chem. Res.-S.* **2003**, *7*, 402−404.

(29) Gmehling, J.; Onken, U. *Vapor-Liquid Equilibrium Data Collection*; DECHEMA Chemistry Data Series; DECHEMA: Frankfurt, Germany, Vol. I, Part 1, 1977.

(30) Jorgensen, W. L.; Duffy, E. M. Prediction of drug solubility from structure. *Adv. Drug Deliver. Rev.* **2002**, *54*, 355−366.

(31) Ruckenstein, E.; Shulgin, I. Entrainer effect in supercritical mixtures. *Fluid Phase Equilibria* **2001**, *180*, 345−359.

(32) Ben-Naim, A. Inversion of the Kirkwood−Buff theory of solutions: application to the water-ethanol system. *J. Chem. Phys.* **1977**, *67*, 4884−4890.

(33) Acree, W. E.; McCargar, J. W.; Zvaigzne, A. I.; Teng, I. L. Mathematical representation of thermodynamic properties − carbazole solubilities in binary alkane + dibutyl ether and alkane + tetrahydropyran solvent mixtures. *Phys. Chem. Liq.* **1991**, *23*, 27−35.

(34) Deng, T. H.; Horiuchi, S.; De Fina, K. M.; Hernandez, C. E.; Acree, W. E. Solubility of anthracene in multicomponent solvent mixtures containing propanol, butanol, and alkanes. *J. Chem. Eng. Data* **1999**, *44*, 798−802.

(35) Jouyban-Gharamaleki, A.; Clark, B. J.; Acree, W. E. Models to predict solubility in ternary solvents based on sub-binary experimental data. *Chem. Pharm. Bull.* **2000**, *48*, 1866−1871.

(36) Kojima, K.; Zhang, S.; Hiaki, T. Measuring methods of infinite dilution activity coefficients and a database for systems including water. *Fluid Phase Equilibr.* **1997**, *131*, 145−179.

Received for review June 29, 2004. Revised manuscript received November 29, 2004. Accepted December 2, 2004.

ES0490166

Chapter 5

Aqueous solutions of biomolecules

5.1 A protein molecule in an aqueous mixed solvent: Fluctuation theory outlook.
5.2 Relationship between preferential interaction of a protein in an aqueous mixed solvent and its solubility.
5.3 A protein molecule in a mixed solvent: the preferential binding parameter via the Kirkwood – Buff theory.
5.4 Preferential hydration and solubility of proteins in aqueous solutions of polyethylene glycol.
5.5 Effect of salts and organic additives on the solubility of proteins in aqueous solutions.
5.6 Local composition in the vicinity of a protein molecule in an aqueous mixed solvent.
5.7 Local composition in solvent + polymer or biopolymer systems.
5.8 Various contributions to the osmotic second virial coefficient in protein-water-cosolvent solutions.

Introduction to Chapter 5

Chapter 5 is concerned with aqueous solutions of biomolecules, particularly proteins. The emphasis is on the structural features of such solutions at the molecular (or nanometer) level and on their thermodynamic properties.

Aqueous solutions of proteins are of paramount importance in modern science. One can mention, for example, that numerous human diseases are connected with the behavior of proteins in the human body (which is, in essence, an aqueous biological solution). A better understanding of the properties of such systems at both micro- and macro-levels, is relevant in medicine, biology, biochemistry, and other life sciences.

However, it is difficult to apply the methods usually used for low molecular weight systems to these solutions. Protein molecule have hydrophobic and hydrophilic moieties and are charged. In addition, they form H-bonds with other protein molecules, with water, with nonaqueous cosolvents and even with themselves. It is understandable that the classical thermodynamics of small molecules cannot provide insight into the properties of such mixtures.

This generated interest in the rigorous statistical thermodynamic theories of Kirkwood-Buff and McMillan-Mayer. In Chapter 5, the emphasis is on the application of the Kirkwood-Buff theory to aqueous solutions of proteins.

Two types of mixtures: (water / protein (5.7) and water / protein / cosolvent (5.1–5.6 and 5.8), are considered.

First, for water / protein / cosolvent mixtures, relations between measurable properties (such as the preferential binding parameter) and the Kirkwood-Buff integrals are derived (5.1 and 5.3). Further, the established relations are used to examine the local composition around a dilute protein molecule in water / protein / cosolvent mixtures (5.1, 5.3–5.4 and 5.6). Such analysis allows one to explain whether a protein molecule is preferentially hydrated or preferentially solvated in a water / protein / cosolvent mixture. The derived equations were used to establish a relation between the preferential binding parameter and the protein solubility. This expression was used in two different ways: (1) to propose a simple criterion for the salting-in or salting-out by various cosolvents of the protein solubility in water, (2) to derive equations which predict the solubility of a protein in a binary aqueous solution in terms of the preferential binding parameter (5.2, 5.4–5.6). The mixture water / protein / polyethylene glycol (PEG) was particularly scrutinized (5.4) because PEG is the most successful precipitant for protein crystallization and is widely used by the protein chemists. The correlations between the aqueous protein solubility and the osmotic second virial coefficient or the preferential binding parameter were reviewed (5.5). Whereas the preferential binding parameter reflects the protein / water (and protein / cosolvent) interactions, the osmotic second virial coefficients is a measure of the protein / protein interactions. A detailed analysis of the various contributions (protein / water, protein / cosolvent and protein / protein) to the osmotic second virial coefficient in protein-water-cosolvent solutions is presented (5.8).

Water / protein (polymer) mixtures are also considered (5.7). The local composition around a protein (or polymer) molecule is examined, and its connection to the protein / water and protein / protein intermolecular interactions is scrutinized.

THE JOURNAL OF CHEMICAL PHYSICS **123**, 054909 (2005)

A protein molecule in an aqueous mixed solvent: Fluctuation theory outlook

Ivan L. Shulgin[a] and Eli Ruckenstein[b]
Department of Chemical & Biological Engineering, State University of New York at Buffalo, Amherst, New York 14260

(Received 14 April 2005; accepted 7 July 2005; published online 11 August 2005)

In the present paper a procedure to calculate the properties of proteins in aqueous mixed solvents, particularly the excesses of the constituents of the mixed solvent near the protein molecule and the preferential binding parameters, is suggested. Expressions for the Kirkwood-Buff integrals in ternary mixtures and for the preferential binding parameter were derived and used to calculate various properties of infinitely dilute proteins in aqueous mixed solvents. The derived expressions and experimental information regarding the partial molar volumes and the preferential binding parameters were used to calculate the excesses (deficits) of water and cosolvent (in comparison with the bulk concentrations of protein-free mixed solvent) in the vicinity of ribonuclease A, ribonuclease T1, and lysozyme molecules. The calculations showed that water was in excess in the vicinity of ribonuclease A for water/glycerol and water/trehalose mixtures, and the cosolvent urea was in excess in the vicinity of ribonuclease T1 and lysozyme. The derivative of the activity coefficient of the protein with respect to the mole fraction of water was also calculated. This derivative was negative for the water/glycerol and water/trehalose mixed solvents and positive for the water/urea mixture. The mixture of lysozyme in the water/urea solvent is of particular interest, because the lysozyme at pH 7.0 is in its native state up to 9.3M urea, while at pH 2.0 it is denatured between 2.5 and 5M and higher concentrations of urea. Our results demonstrated a striking similarity in the hydration of lysozyme at both pHs. It is worthwhile to note that the excesses of urea were only weakly composition dependent on both cases. © *2005 American Institute of Physics.*
[DOI: 10.1063/1.2011388]

INTRODUCTION

The addition of one more component (a cosolvent) to aqueous solutions of proteins can dramatically change the properties of those solutions, such as the protein solubility, protein self-assembling, and protein stability.[1–6] Indeed, the solubility of proteins can be essentially changed by the addition of a third component.[1–6] It is well known for a long time that the addition of certain compounds (such as urea) can cause protein denaturation, and that other cosolvents, such as glycerol, sucrose, etc., can stabilize at high concentrations the protein structure and preserve its enzymatic activity.[1–6]

The most plausible explanation of these observations could be connected with the preferential interactions of some cosolvents with the proteins. How the water and cosolvent interact with the protein, how the water interacts with the cosolvent, and how much the local composition in the vicinity of the protein surface differs from that in the protein-free solution are natural questions to be asked.

Timasheff subdivided the cosolvents into several groups:[7] "When a protein molecule is immersed into a solvent consisting of water and another chemical species (a cosolvent), the interactions between the protein and the solvent components may lead to three possible situations: (1) the cosolvent is present at the protein surface in excess over its concentration in the bulk (this is what constitutes binding); (2) the water is present in excess at the protein surface; this means that the protein has a higher affinity for water than for the cosolvent (this situation is referred to as preferential hydration, or preferential exclusion of the cosolvent); (3) the protein is indifferent to the nature of molecules (water or cosolvent) with which it comes in contact, so that no solvent concentration perturbation occurs at the protein surface."

One of the characteristics of the effect of a cosolvent on the behavior of a protein is the preferential binding parameter.[7] It can be defined using various concentration scales as follows (component 1 is water, component 2 is a protein, and component 3 is a cosolvent):

(1) in molal concentrations

$$\Gamma_{23}^{(m)} \equiv \lim_{m_2 \to 0} (\partial m_3/\partial m_2)_{T,P,\mu_3}, \quad (1)$$

where m_i is the molality of component i, P is the pressure, T the temperature, and μ_i is the chemical potential of component i.

(2) in molar concentrations

$$\Gamma_{23}^{(c)} \equiv \lim_{c_2 \to 0} (\partial c_3/\partial c_2)_{T,P,\mu_3}, \quad (2)$$

where c_i is the molar concentration of component i. It should be noted that $\Gamma_{23}^{(m)}$ and $\Gamma_{23}^{(c)}$ are defined at infinite dilution of

---
[a] Electronic mail: ishulgin@eng.buffalo.edu
[b] Author to whom correspondence should be addressed; Fax: (716) 645-3822. Electronic mail: feaeliru@acsu.buffalo.edu

protein (throughout this paper, only infinitely dilute protein solutions are considered).

The preferential binding parameter $\Gamma_{23}^{(m)}$ can be determined experimentally using various methods such as dialysis equilibrium,[8(a)] vapor pressure osmometry,[8(b)] etc. The preferential binding parameters expressed by Eqs. (1) and (2) can be calculated in terms of each other and they can have different signs[4] (for details see Appendix A). The theoretical investigation of the preferential binding parameters is complicated because it involves the thermodynamics of ternary and, for a number of cosolvents larger than one, multicomponent mixtures. In what follows it will be shown that the Kirkwood-Buff theory of solution[9] can shed some light about the effect of the addition of one more component to an aqueous protein solution. So far there are several publications[10–15] in which the Kirkwood-Buff theory of solution (or fluctuation theory) was applied to the analysis of the protein behavior in the presence of a cosolvent. The most important expression derived in the present paper, which connects the Kirkwood-Buff theory of solution with the present particular problem, is that for the preferential binding parameter $\Gamma_{23}^{(c)}$ (Appendix B)

$$\Gamma_{23}^{(c)} = c_3(G_{23} - G_{13}), \quad (3)$$

where G_{13} and G_{23} are the Kirkwood-Buff integrals defined as[9]

$$G_{\alpha\beta} = \int_0^\infty (g_{\alpha\beta} - 1) 4\pi r^2 dr. \quad (4)$$

$g_{\alpha\beta}$ is the radial distribution function between species α and β, and r is the distance between the centers of molecules α and β.

One can also write the following expression for the partial molar volume of a protein at infinite dilution in a mixed solvent (V_2^∞) in terms of the Kirkwood-Buff theory of solution:[16]

$$V_2^\infty = -c_1 V_1 G_{12} - c_3 V_3 G_{23} + kTk_T$$
$$\cong -c_1 V_1 G_{12} - c_3 V_3 G_{23}, \quad (5)$$

where V_i is the partial molar volume of component i, k is the Boltzmann constant, and k_T is the isothermal compressibility of the mixed solvent.

The Kirkwood-Buff integrals G_{12} and G_{23} can be calculated by combining Eqs. (3) and (5) and using the experimental data for $\Gamma_{23}^{(m)}$, V_2^∞, V_1 and V_3, and the values of G_{13} calculated from the properties of protein-free mixed solvent. The Kirkwood-Buff integrals G_{12} and G_{23} are of prime interest because they can provide information regarding the behavior of a mixed solvent in the vicinity of a protein molecule.

The knowledge of the Kirkwood-Buff integrals for dilute mixtures can be very helpful in the analysis of the local water/cosolvent composition[17–21] in the vicinity of a solute molecule. Ultimately, it can provide information about the effect of various cosolvents on the protein behavior in aqueous solutions.

In the present paper, the Kirkwood-Buff theory of solutions will be used to examine dilute mixtures of proteins in a mixed solvent as follows: first, explicit expressions for the Kirkwood-Buff integrals $G_{\alpha\beta}$ in ternary mixtures will be obtained and used to calculate the preferential binding parameters of infinitely dilute mixtures of proteins in aqueous mixed solvents. Various solution behaviors starting with the ideal mixture will be considered. The final goal of the paper is to obtain information about the local composition near a protein molecule and to predict the preferential binding parameter in various systems.

THE KIRKWOOD-BUFF INTEGRALS IN TERNARY MIXTURES

Explicit expressions for the Kirkwood-Buff integrals $G_{\alpha\beta}$ in ternary mixtures can be derived as described in the literature.[9,22,23] For an infinitely dilute solute, one can derive the following expressions (for details see Appendix C):

$$G_{12} = kTk_T - \frac{J_{21}V_3c_3 + J_{11}V_2^\infty c_1}{(c_1 + c_1 J_{11} + c_3)}$$
$$- \frac{V_3 c_3(c_1+c_3)(V_1-V_3) + V_2^\infty(c_1+c_3)}{(c_1+c_1 J_{11}+c_3)}, \quad (6)$$

$$G_{23} = kTk_T + \frac{J_{21}V_1 c_1 - J_{11}c_1 V_2^\infty}{(c_1+c_1 J_{11}+c_3)}$$
$$+ \frac{c_1 V_1(c_1+c_3)(V_1-V_3) - V_2^\infty(c_1+c_3)}{(c_1+c_1 J_{11}+c_3)}, \quad (7)$$

and

$$G_{13} = kTk_T - \frac{(c_1+c_3)^2 V_1 V_3}{(c_1+c_1 J_{11}+c_3)}, \quad (8)$$

where $J_{11} = \lim_{x_2 \to 0}(\partial \ln \gamma_1 / \partial x_1)_{x_2}$, $J_{21} = \lim_{x_2 \to 0}(\partial \ln \gamma_2 / \partial x_1)_{x_2}$, x_i is the mole fraction of component i, and γ_i is the activity coefficient of component i in a mole fraction scale.

Combining Eqs. (3), (6), and (8) yields the following expression for $\Gamma_{23}^{(c)}$:

$$\Gamma_{23}^{(c)} = \frac{c_1 c_3 (J_{21}V_1 - J_{11}V_2^\infty)}{(c_1+c_1 J_{11}+c_3)} + \frac{c_3(c_1+c_3)(V_1-V_2^\infty)}{(c_1+c_1 J_{11}+c_3)}. \quad (9)$$

Equations (6)–(9) are new equations free of any approximations.

It should be noted that only the derivative J_{21} and the partial molar volume of the protein V_2^∞ depend on the protein characteristics; all the other quantities in Eq. (9) can be determined from the characteristics of the protein-free mixed solvent. Equation (9) shows that the preferential binding parameter $\Gamma_{23}^{(c)}$ can be decomposed into the sum of two terms. One of them, depends on the protein nature, reflected in J_{21} and V_2^∞, and the other one depends only on the properties of the protein-free mixed solvent.

Equations (6)–(9) allows one to derive expressions for the Kirkwood-Buff integrals G_{12} and G_{23}, and the preferential binding parameter $\Gamma_{23}^{(c)}$ for various kinds of ternary mixtures.

(1) *Ideal ternary mixture* [superscript "(id)"]. In this case

all activity coefficients are equal to unity, all partial molar volumes are equal to the molar volumes of the pure components,[24] and Eqs. (6)–(9) become

$$G_{12}^{(id)} = kTk_T^{(id)} - V_3^0 c_3 (V_1^0 - V_3^0) - V_2^0, \quad (10)$$

$$G_{23}^{(id)} = kTk_T^{(id)} + V_1^0 c_1 (V_1^0 - V_3^0) - V_2^0, \quad (11)$$

$$G_{13}^{(id)} = kTk_T^{(id)} - (c_1 + c_3) V_1^0 V_3^0, \quad (12)$$

and

$$\Gamma_{23}^{(c,id)} = c_3 (V_1^0 - V_2^0), \quad (13)$$

where V_i^0 is the molar volume of the pure component i.

Equation (13) shows that the preferential binding parameter $\Gamma_{23}^{(c,id)}$ is not zero for an ideal mixture, fact also noted a long time ago by Hade and Tanford.[25]

(2) *Ideal mixed solvent* [superscript "(IS)"] *approximation*.[26,27] In this case, $J_{11}=0$ and Eqs. (6)–(9) acquire the forms

$$G_{12}^{(IS)} = kTk_T - \frac{J_{21} V_3^0 c_3}{c_1 + c_3} - V_3^0 c_3 (V_1^0 - V_3^0) - V_2^\infty, \quad (14)$$

$$G_{23}^{(IS)} = kTk_T + \frac{J_{21} V_1^0 c_1}{c_1 + c_3} + c_1 V_1^0 (V_1^0 - V_3^0) - V_2^\infty, \quad (15)$$

$$G_{13}^{(IS)} = kTk_T - (c_1 + c_3) V_1^0 V_3^0, \quad (16)$$

and

$$\Gamma_{23}^{(c,IS)} = \frac{c_1 c_3 J_{21} V_1^0}{(c_1 + c_3)} + c_3 (V_1^0 - V_2^\infty). \quad (17)$$

This approximation implies that the interaction between the solute and the constituents of the mixed solvent is much stronger than those between the constituents of the mixed solvent. In other words, the main contribution to the nonideality of the very dilute mixture protein+mixed solvent stems from the nonideality due to the interactions of the solute with the mixed solvent and not from the nonideality of the mixed solvent. Indeed, the experimental data regarding the activity coefficients indicate that such an assumption is quite plausible for aqueous solutions of large molecules containing various functional groups, such as proteins, drugs, etc. Indeed, whereas the activity coefficients of the components of binary mixtures of water and typical organic cosolvents, such as alcohols, acetone, acetonitrile, etc., are usually between 1 and 10 and rarely reach few dozens, the activity coefficients at infinite dilution of large molecules, such as proteins, drugs, etc., are usually greater than thousand or tens of thousands, or even larger.[28–30]

THE EXCESS AND DEFICIT NUMBERS OF MOLECULES OF WATER AND COSOLVENT AROUND A PROTEIN MOLECULE

The conventional way to calculate the excess (or deficit) number of i molecules around a central molecule j is based on the following relation:[31]

$$\Delta n'_{ij} = c_i G_{ij} \quad (18)$$

However, Matteoli and Lepori[32] and Mateolli[33] observed that $\Delta n'_{ij}$ calculated with Eq. (18) has nonzero values for ideal binary systems, even though they are expected to vanish. It was also noted[34] that there are many systems for which all the Kirkwood-Buff integrals in binary systems are negative in certain ranges of composition. As a result, in such cases all $\Delta n'_{ij}$ would be negative, and this is not plausible.

For the above reasons, it was suggested[32–34] to calculate the excess (or deficit) number of molecules i around a central molecule j with the following relation:

$$\Delta n_{ij} = c_i (G_{ij} - G_{ij}^R), \quad (19)$$

where G_{ij}^R is the Kirkwood-Buff integral of a reference state. Matteoli and Lepori[32] and Mateolli[33] suggested the ideal solution [(id)] as the reference state because then Δn_{ij} becomes zero for an ideal solution. However, intuition suggests that the excesses and deficits should satisfy the so-called volume conservation condition[32,33] which for a binary mixture can be formulated as follows: "the volume occupied by the excess i molecules around a j molecule must be equal to the volume left free by the j molecules around the same j molecule." One can show that the excesses and deficits calculated with the ideal mixture as reference state do not satisfy the above volume conservation condition. For this reason, a new reference state was suggested by Shulgin and Ruckenstein[34] in which all the activity coefficients are taken equal to unity but there are no constraints on the partial molar volumes of the components (superscript "(SR)"). This reference state satisfies the volume conservation condition and provides zero excesses for ideal mixtures for both binary and ternary mixtures.[23,34] However,[34,35] far from critical conditions, the above two reference states provide almost the same results. In this paper, we consider $G_{ij}^R = G_{ij}^{(SR)}$.

For the [(SR)] reference state, Eqs. (6) and (7) for G_{12} and G_{23} for an infinitely dilute protein in a mixed solvent can be recast as follows:

$$G_{12}^{(SR)} = kTk_T^{(SR)} - V_3 c_3 (V_1 - V_3) - V_2^\infty \quad (20)$$

and

$$G_{23}^{(SR)} = kTk_T^{(SR)} - V_1 c_1 (V_1 - V_3) - V_2^\infty. \quad (21)$$

NUMERICAL ESTIMATIONS FOR VARIOUS SYSTEMS

In order to obtain information regarding the behavior of various cosolvents in the vicinity of a protein surface, the following quantities were calculated: (1) the Kirkwood-Buff integrals G_{12} and G_{23} for an infinitely dilute protein in a mixed solvent, (2) the Kirkwood-Buff integrals $G_{12}^{(SR)}$ and $G_{23}^{(SR)}$ for the reference mixture, (3) the excess (or deficit) number of molecules of water and cosolvent around a protein molecule, and (4) the preferential binding parameter $\Gamma_{23}^{(c)}$.

The Kirkwood-Buff integrals G_{12} and G_{23} were calculated by solving Eqs. (3) and (5). First, the Kirkwood-Buff integral G_{23} was calculated from Eq. (3) by assuming that for dilute solutions of proteins $|G_{23}| \gg |G_{13}|$. Indeed, $|G_{13}| \leq 70$ [cm^3/mol] for aqueous mixtures of glycerol and

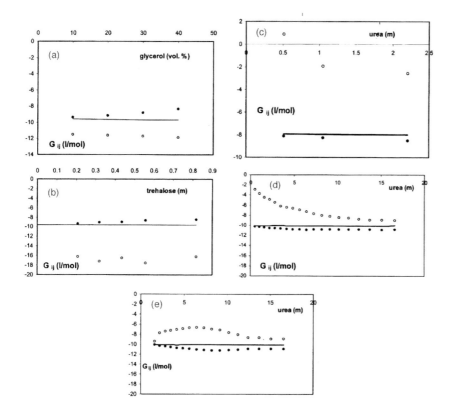

FIG. 1. The Kirkwood-Buff integrals G_{12} and G_{23} for an infinitely dilute protein (2) in water (1)+cosolvent (3) mixture. The solid line represents $G_{12}^{(SR)}$ and $G_{23}^{(SR)}$ for the reference mixture calculated using Eqs. (20) and (21). The numerical values of $G_{12}^{(SR)}$ and $G_{23}^{(SR)}$ are so close to each other that at the scale of the figure they superpose on a single curve. The symbol (●) represents G_{12} and the symbol (○) represents G_{23}. The values of G_{12} and G_{23} were calculated by solving Eqs. (3) and (5).

(a) Ribonuclease A (2) in water (1)+glycerol (3) mixture.
(b) Ribonuclease A (2) in water (1)+trehalose (3) mixture.
(c) Ribonuclease T1 (2) in water (1)+urea (3) mixture.
(d) Lysozyme (2) in water (1)+urea (3) mixture ($pH=7.0$).
(e) Lysozyme (2) in water (1)+urea (3) mixture ($pH=2.0$).

urea,[35,36] while the Kirkwood-Buff integrals for the pairs containing proteins ($|G_{23}|$ and $|G_{12}|$) have values of the order of 10^4 [cm^3/mol] or larger. Second, G_{12} was calculated using Eq. (5).

The Kirkwood-Buff integrals $G_{12}^{(SR)}$ and $G_{23}^{(SR)}$ were calculated from Eqs. (20) and (21) (the term involving the isothermal compressibility was omitted because it is much smaller than the other terms).

The preferential binding parameter $\Gamma_{23}^{(c)}$ was calculated with Eq. (17), in which J_{21} was considered as an adjustable composition-independent parameter determined by fitting the experimental data regarding $\Gamma_{23}^{(c)}$. Equation (9) can be also used to calculate $\Gamma_{23}^{(c)}$ if accurate data regarding the activity coefficients in protein-free mixed solvents are available.

All the above quantities were calculated using experimental information[36–40] about the partial molar volumes (V_1, V_3, and V_2^∞) and the preferential binding parameter $\Gamma_{23}^{(m)}$ for the following mixtures.

(1) Ribonuclease A (2) in water (1)+glycerol (3) mixture ($pH=2.9$ and 20 °C).
(2) Ribonuclease A (2) in water (1)+trehalose (3) mixture (pH 5.5, 20 °C).
(3) Ribonuclease T1 (2) in water (1)+urea (3) mixture (pH 7.0, 25 °C).
(4) Lysozyme (2) in water (1)+urea (3) mixture (pH 7.0, 20 °C).
(5) Lysozyme (2) in water (1)+urea (3) mixture (pH 2.0, 20 °C).

RESULTS AND DISCUSSION

The results of the calculations are presented in Figs. 1–3 and Table I. Fig. 1(a) presents the Kirkwood-Buff integrals G_{12} and G_{23} for an infinitely dilute ribonuclease A (2) in water (1)+glycerol (3) mixtures. All the Kirkwood-Buff integrals have negative values. However, G_{12} and G_{23} have different sign deviations from $G_{12}^{(SR)}$ and $G_{23}^{(SR)}$. The same observation is valid for all the mixtures investigated. The calculated excesses and deficits in the vicinity of a molecule of ribonuclease A are presented in Fig. 2(a) and it should be emphasized that their values satisfy the volume conservation condition. Indeed, for a volume fraction of glycerol of 30% we have the following excesses (or deficits): $\Delta n_{12} \cong 33.3$ moles of water which have a volume of 0.602 liters and $\Delta n_{23} \cong -8.2$ moles of glycerol which have a volume of 0.602 liters. In contrast, the calculations of excesses (or deficits) with the expression ($c_i G_{ij}$) provided the following values for a volume fraction of glycerol of 30%: $c_1 G_{12} = -341.8$ moles of water which have a volume equal -6.18 liters and $c_3 G_{23} = -48.1$ moles of water which have a volume equal -3.52 liters. A deficit of glycerol in the vicinity of a protein was also found by Gekko and Timasheff.[37]

Similar results regarding the Kirkwood-Buff integrals G_{12} and G_{23}, and excesses or deficits of water and cosolvent in the vicinity of a protein molecule were obtained for an infinitely dilute ribonuclease A (2) in water (1)+trehalose (3) mixture [See Figs. 1(b) and 2(b)]. Again, our calculations demonstrate that in the vicinity of ribonuclease A at high

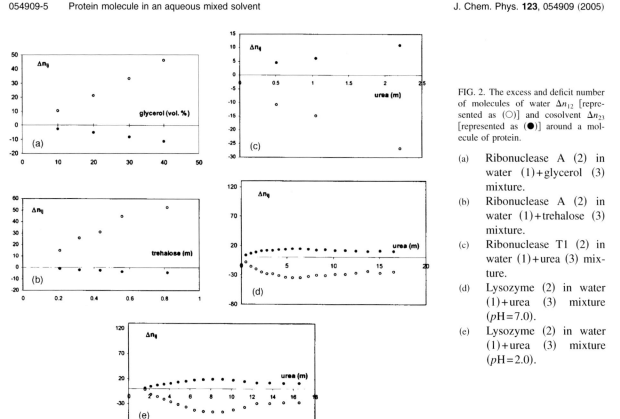

FIG. 2. The excess and deficit number of molecules of water Δn_{12} [represented as (○)] and cosolvent Δn_{23} [represented as (●)] around a molecule of protein.

(a) Ribonuclease A (2) in water (1)+glycerol (3) mixture.
(b) Ribonuclease A (2) in water (1)+trehalose (3) mixture.
(c) Ribonuclease T1 (2) in water (1)+urea (3) mixture.
(d) Lysozyme (2) in water (1)+urea (3) mixture (pH=7.0).
(e) Lysozyme (2) in water (1)+urea (3) mixture (pH=2.0).

dilution of protein, the water is in excess and trehalose in deficit compared to the bulk concentrations of protein-free mixed solvent.

The results regarding the infinitely dilute mixtures of ribonuclease T1 and lysozyme in water/urea mixtures are presented in Figs. 1(c)–1(e), 2(c)–2(e), and 3(c)–3(e), and Table I. Urea is a cosolvent of particular interest because it can cause protein denaturation, in contrast to glycerol and trehalose which increase the protein stability. Figures 1(c), 2(c), and 3(c) present the results obtained for ribonuclease T1 (2) in water (1)+urea (3) mixture (pH 7.0, 25 °C). Figures 1(d), 2(d), and 3(d) present the results obtained for lysozyme (2) in water (1)+urea (3) mixture (pH 7.0, 20 °C) and Figs. 1(e), 2(e), and 3(e) present the results obtained for lysozyme (2) in water (1)+urea (3) mixture (pH 2.0, 20 °C). Whereas the ribonuclease T1 and the lysozyme at pH 7.0 and up to 9.3M urea are in their native state, the lysozyme at pH 2.0 becomes denatured between 2.5 and 5.0M and higher concentrations of urea. Our calculations showed that urea is in excess at all compositions.

It is worth recalling here the unpublished 1948 opinion of W. Kauzmann (as quoted by Timasheff and Xie[40]) that "the bulkiness of the cosolvent molecules creates around a protein molecule a zone that is impenetrable to cosolvent, the thickness of which is determined by the distance of closest approach between protein and ligand molecules. This region can be penetrated by the smaller water molecules. Hence, it is enriched in water relative to bulk solvent." However, it is not clear to what extent this opinion is applicable to urea as a cosolvent, because urea is also a small molecule (the partial molar volume of urea at infinite dilution in aqueous mixture is about 43–44 [cm^3/mol]). The literature provides sometimes conflicting viewpoints regarding the local compositions in dilute mixtures of proteins in the water/urea solvent. Timasheff[8(a)] pointed out that urea can be both preferentially bound to and preferentially excluded from the surface of different proteins. Recent molecular-dynamics simulation[41] predicted that most of the constituent groups of ribonuclease T1 either preferentially bind urea or are indifferent to urea or water. Another recent study[42] found that urea is moderately accumulated in the vicinity of the polar amide surfaces of proteins and is neither accumulated nor significantly excluded from the anionic areas of the protein. It is worth noting that the methodology suggested in the present paper allows one to conclude not only whether there is a preferential hydration or not, but it additionally provides values for the excesses (or deficits) of water and cosolvent in the vicinity of a protein molecule.

One should note that the results obtained regarding the excess (or deficit) of water and cosolvent in the vicinity of a protein in a multiple solvent are valid for the whole protein molecule and not for particular functional groups of the protein. It is perfectly possible for urea to be in excess around the whole protein molecule but in deficit in the vicinity of certain functional groups.

It is of interest to compare the excesses (or deficits) of water and cosolvent in the vicinity of lysozyme (Δn_{12} and Δn_{23}) for water/urea mixed solvent at different pHs (2.0 and 7.0), because the lysozyme molecule at pH=7.0 and for a concentration of urea up to 9.3M is in a native state and at

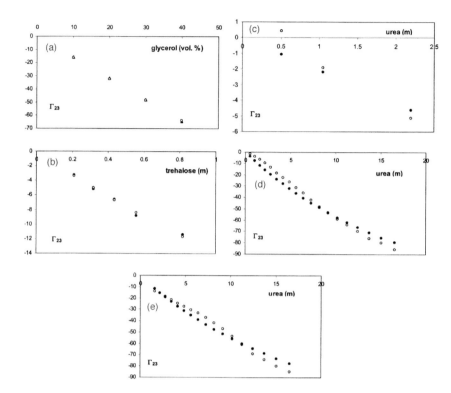

FIG. 3. The preferential binding parameter $\Gamma_{23}^{(c)}$ for an infinitely dilute protein (2) in water (1)+cosolvent (3) mixture: comparison between experiment (○) and calculation (●) [Eq. (14)]. The preferential binding parameter $\Gamma_{23}^{(c)}$ was calculated using the experimental values from Refs. 37–40 for $\Gamma_{23}^{(m)}$.

(a) Ribonuclease A (2) in water (1)+glycerol (3) mixture.
(b) Ribonuclease A (2) in water (1)+trehalose (3) mixture.
(c) Ribonuclease T1 (2) in water (1)+urea (3) mixture.
(d) Lysozyme (2) in water (1)+urea (3) mixture (pH=7.0).
(e) Lysozyme (2) in water (1)+urea (3) mixture (pH=2.0).

pH=2.0 and for an urea concentration starting between 2.5 and 5.0M and larger in a denaturated one. Figures 2(d) and 2(e) exhibit a striking similarity regarding the excesses (or deficits) of water and cosolvent in the vicinity of a lysozyme molecule at both pHs. One should mention that the excess of urea (deficit of water) has a weak composition dependence compared to the cases with preferential hydration (ribonuclease A in water+glycerol and water+trehalose mixtures). It can be caused by a strong preferential interaction of some functional groups (peptide groups?) of the lysozyme with urea, the remaining part of lysozyme molecule being surrounded by the "bulk" water/urea mixture. The suggestion about the remaining part of the lysozyme is in agreement with the observation[35] that the water/urea mixture can be considered an almost ideal one.

Table I shows that J_{21} for water/glycerol and water/trehalose mixtures are negative and J_{21} for water/urea mixture is positive. Therefore, one can notice a correspondence between the preferential hydration and the sign of J_{21}. If water is in excess in the vicinity of a protein molecule J_{21} is positive and when the cosolvent is in excess J_{21} is negative. J_{21} depends on the nature of the protein and is expected to depend on the mixed solvent composition as well. However, our calculations based on Eq. (17) and the experimental values of the preferential binding parameter have revealed a weak composition dependence of this parameter. One can see from Fig. 3 that the one-parameter (J_{21}) Eq. (17) can accurately represent the preferential binding parameter $\Gamma_{23}^{(c)}$ by considering J_{21} as composition independent.

It is of interest to establish a connection between the preferential binding parameters $\Gamma_{23}^{(c)}$ and $\Gamma_{23}^{(m)}$ and the excess (deficit) of water or cosolvent near a protein molecule. Does a positive value of $\Gamma_{23}^{(c)}$ [$\Gamma_{23}^{(m)}$] mean an excess of cosolvent? In order to understand this connection one should remember that $\Gamma_{23}^{(c)}$ and $\Gamma_{23}^{(m)}$ are connected via the relation[4] (see Appendix A),

$$\Gamma_{23}^{(c)} = (1 - c_3 V_3)\Gamma_{23}^{(m)} - c_3 V_2^\infty \qquad (22)$$

which shows[4,43] that $\Gamma_{23}^{(c)}$ and $\Gamma_{23}^{(m)}$ can have different signs. Some of the mixtures considered in the present papers (ribonuclease T1 in water+urea and lysozyme in water+urea) exhibit this feature. While for almost all compositions of the mixed solvent $\Gamma_{23}^{(m)}$ are in above cases positive, $\Gamma_{23}^{(c)}$ is negative. Our calculations have shown that the sign of $\Gamma_{23}^{(m)}$ indicates preferential hydration when it is positive and preferential exclusion of water when it is negative.

TABLE I. The derivative J_{21} of the activity coefficient for the systems investigated.

System	J_{21}
Ribonuclease A (2) in water (1)+ glycerol (3) mixture	−129.85
Ribonuclease A (2) in water (1)+ trehalose (3) mixture	−401.10
Ribonuclease T1 (2) in water (1) +urea (3) mixture	326.51
Lysozyme (2) in water (1)+urea (3) mixture (pH 7.0, 20 °C)	119.42
Lysozyme (2) in water (1)+urea (3) mixture (pH 2.0, 20 °C)	136.59

CONCLUSION

In the present paper the Kirkwood-Buff theory of ternary solutions was applied to infinitely dilute proteins in aqueous mixed solvents. Novel expressions for the Kirkwood-Buff integrals G_{12}, G_{23}, and G_{13}, and the preferential binding parameter $\Gamma_{23}^{(c)}$ have been derived and used to calculate the various properties of infinitely dilute proteins in aqueous mixed solvents. In particular, the Kirkwood-Buff integrals G_{12} and G_{23}, the excess (or deficit) of water and cosolvent, and the derivatives of the activity coefficients of a protein and cosolvent were calculated for five different mixtures involving infinitely dilute proteins in various aqueous mixed solvents.

The results demonstrated that water was in excess in the vicinity of ribonuclease A for water/glycerol and water/trehalose mixtures and the cosolvent (urea) was in excess in the vicinity of ribonuclease T1 and lysozyme.

Other noteworthy results were obtained regarding the excesses (or deficits) of water and urea for denaturated and native lysozyme at pH=2.0 and pH=7.0, respectively. The results demonstrate a striking similarity at both pHs. It should be mentioned that the urea excesses are in both cases weak composition dependent.

APPENDIX A: RELATION BETWEEN $\Gamma_{23}^{(m)}$ AND $\Gamma_{23}^{(c)}$

Because

$$c_i = \frac{x_i}{V} \text{ and } dc_i = \frac{dx_i}{V} - \frac{x_i dV}{V^2},$$

where x_i is the mole fraction of component i and V is the molar volume,

$$dc_2 = \frac{dx_2}{V} - \frac{x_2}{V^2}\left(\left(\frac{\partial V}{\partial x_2}\right)_{x_3} dx_2 + \left(\frac{\partial V}{\partial x_3}\right)_{x_2} dx_3\right)$$

and

$$dc_3 = \frac{dx_3}{V} - \frac{x_3}{V^2}\left(\left(\frac{\partial V}{\partial x_2}\right)_{x_3} dx_2 + \left(\frac{\partial V}{\partial x_3}\right)_{x_2} dx_3\right).$$

By dividing the latter two equations one obtains

$$\left(\frac{\partial c_3}{\partial c_2}\right) = \left(\frac{-x_3\left(\frac{\partial V}{\partial x_2}\right)_{x_3} + \left(V - x_3\left(\frac{\partial V}{\partial x_3}\right)_{x_2}\right)\frac{dx_3}{dx_2}}{\left(V - x_2\left(\frac{\partial V}{\partial x_2}\right)_{x_3}\right) - x_2\left(\frac{\partial V}{\partial x_3}\right)_{x_2}\frac{dx_3}{dx_2}}\right). \quad (A1)$$

In terms of molalities,

$$m_2 = \frac{1000 x_2}{M_1 x_1} \text{ and } m_3 = \frac{1000 x_3}{M_1 x_1}$$

from which one obtains

$$dm_2 = \frac{1000 dx_2}{M_1 x_1} + \frac{1000 x_2 d(x_2 + x_3)}{M_1 (x_1)^2},$$

$$dm_3 = \frac{1000 dx_3}{M_1 x_1} + \frac{1000 x_3 d(x_2 + x_3)}{M_1 (x_1)^2},$$

and

$$\left(\frac{dm_3}{dm_2}\right) = \left(\frac{x_3 + (x_1 + x_3) dx_3/dx_2}{(x_1 + x_2) + x_2(dx_3/dx_2)}\right). \quad (A2)$$

In terms of mole fractions, the preferential binding parameter $\Gamma_{23}^{(x)}$ is defined as

$$\Gamma_{23}^{(x)} \equiv \lim_{x_2 \to 0}(\partial x_3/\partial x_2)_{T,P,\mu_3}. \quad (A3)$$

At infinite dilution of the protein, Eqs. (A1) and (A3) lead to

$$\Gamma_{23}^{(c)} = \frac{-x_3(V_2^\infty - V_1) + V_1 \Gamma_{23}^{(x)}}{x_1 V_1 + x_3 V_3} \quad (A4)$$

and Eqs. (A2) and (A3) to

$$\Gamma_{23}^{(m)} = \frac{x_3 + \Gamma_{23}^{(x)}}{x_1}. \quad (A5)$$

Consequently

$$\Gamma_{23}^{(c)} = (1 - c_3 V_3)\Gamma_{23}^{(m)} - c_3 V_2^\infty. \quad (A6)$$

Equation (A6) was derived in a different manner in Ref. 4.

APPENDIX B: DERIVATION OF EQ. (3)

Because

$$d\mu_3 = \left(\frac{\partial \mu_3}{\partial T}\right)_{P,x} dT + \left(\frac{\partial \mu_3}{\partial P}\right)_{T,x} dP + \left(\frac{\partial \mu_3}{\partial x_2}\right)_{T,P,x_3} dx_2 + \left(\frac{\partial \mu_3}{\partial x_3}\right)_{T,P,x_2} dx_3,$$

one can write that

$$(\partial x_3/\partial x_2)_{T,P,\mu_3} = -\frac{(\partial \mu_3/\partial x_2)_{T,P,x_3}}{(\partial \mu_3/\partial x_3)_{T,P,x_2}}. \quad (B1)$$

Expressing $\lim_{x_2 \to 0}(\partial \mu_3/\partial x_3)_{T,P,x_2}$ and $\lim_{x_2 \to 0}(\partial \mu_3/\partial x_2)_{T,P,x_3}$ in terms of the Kirkwood-Buff integrals,[23] one obtains

$$\Gamma_{23}^{(x)} = \frac{c_1 c_3(G_{23} - G_{13} + G_{11} - G_{12})}{c_1 + c_3}, \quad (B2)$$

which combined with Eq. (A4) and relations for V_2^∞, V_1, and V_3 in terms of the Kirkwood-Buff integrals[23] yields Eq. (3) of the text. It should be mentioned that Eq. (3) for $\Gamma_{23}^{(c)}$ differs from that used in the literature.[10–13]

APPENDIX C: ANALYTICAL EXPRESSIONS FOR THE KIRKWOOD-BUFF INTEGRALS IN TERNARY MIXTURE

The Kirkwood-Buff integrals in an n-component mixture can be obtained from the following relation:[9]

$$G_{\alpha\beta} = \frac{v|A|_{\alpha\beta}}{\langle N_\alpha\rangle\langle N_\beta\rangle|A|} - \frac{\delta_{\alpha\beta}}{c_\beta}, \quad (C1)$$

where $\langle N_\alpha\rangle$ and $\langle N_\beta\rangle$ are the average numbers of molecules of α and β, respectively, in the volume v, $\delta_{\alpha\beta}$ is the Kronecker symbol ($\delta_{\alpha\beta}=1$ for $\alpha=\beta$ and $\delta_{\alpha\beta}=0$ for $\alpha\neq\beta$), c_β is the bulk molecular concentration of species β ($c_\beta=\langle N_\beta\rangle/v$), $|A|_{\alpha\beta}$ represents the cofactor of $A_{\alpha\beta}$ in the determinant $|A|$, and $A_{\alpha\beta}$ is given by[9]

$$A_{\alpha\beta} = \frac{1}{kT}\left(\left(\frac{\partial\mu_\alpha}{\partial N_\beta}\right)_{T,P,N_\gamma} + \frac{v_\alpha v_\beta}{k_T v}\right). \quad (C2)$$

In Eq. (C2) k is the Boltzmann constant, T the absolute temperature, P the pressure, μ_α the chemical potential per molecule of species α, v_α and v_β are the partial molar volumes per molecule of species α and β, respectively, and k_T is the isothermal compressibility.

Equation (C1) can be recast in the following form:

$$G_{\alpha\beta} = \frac{|A|_{\alpha\beta}}{v c_\alpha c_\beta |A|} - \frac{\delta_{\alpha\beta}}{c_\beta}. \quad (C3)$$

The above equations are valid for any n-component system. For ternary mixtures, the Kirkwood-Buff integrals can be obtained from (C3) using for $\mu_{\alpha\beta}=(\partial\mu_\alpha/\partial N_\beta)_{T,P,N_\gamma}$ the expressions[22]

$$\bar{\mu}_{11} = (1-x_1)\left(\frac{\partial\mu_1}{\partial x_1}\right)_{x_2} - x_2\left(\frac{\partial\mu_1}{\partial x_2}\right)_{x_1}, \quad (C4)$$

$$\bar{\mu}_{22} = (1-x_2)\left(\frac{\partial\mu_2}{\partial x_2}\right)_{x_1} - x_1\left(\frac{\partial\mu_2}{\partial x_1}\right)_{x_2}, \quad (C5)$$

$$\bar{\mu}_{33} = \frac{1}{x_3}\left[x_1^2\left(\frac{\partial\mu_1}{\partial x_1}\right)_{x_2} + x_1 x_2\left\{\left(\frac{\partial\mu_1}{\partial x_2}\right)_{x_1} + \left(\frac{\partial\mu_2}{\partial x_1}\right)_{x_2}\right\} + x_2^2\left(\frac{\partial\mu_2}{\partial x_2}\right)_{x_1}\right], \quad (C6)$$

where $\bar{\mu}_{\alpha\beta} = (N_1+N_2+N_3)\mu_{\alpha\beta}$.

For $\mu_{\alpha\beta}=(\partial\mu_\alpha/\partial N_\beta)_{T,P,N_\gamma}$ with $\alpha\neq\beta$, the following expression were obtained using the Gibbs-Duhem equation:[22]

$$\mu_{\alpha\beta} = \frac{c_\gamma^2 \mu_{\gamma\gamma} - c_\alpha^2 \mu_{\alpha\alpha} - c_\beta^2 \mu_{\beta\beta}}{2c_\alpha c_\beta} \text{ with } \gamma\neq\alpha,\beta. \quad (C7)$$

The derivatives of the chemical potentials at constant pressure and temperature can be expressed in terms of the activity coefficients γ_i as follows:[22]

$$\left(\frac{\partial\mu_1}{\partial x_1}\right)_{x_2} = RT\left[\frac{1}{x_1} + \left(\frac{\partial\ln\gamma_1}{\partial x_1}\right)_{x_2}\right], \quad (C8)$$

$$\left(\frac{\partial\mu_2}{\partial x_2}\right)_{x_1} = RT\left[\frac{1}{x_2} + \left(\frac{\partial\ln\gamma_2}{\partial x_2}\right)_{x_1}\right], \quad (C9)$$

$$\left(\frac{\partial\mu_1}{\partial x_2}\right)_{x_1} = RT\left(\frac{\partial\ln\gamma_1}{\partial x_2}\right)_{x_1}, \quad (C10)$$

and

$$\left(\frac{\partial\mu_2}{\partial x_1}\right)_{x_2} = RT\left(\frac{\partial\ln\gamma_2}{\partial x_1}\right)_{x_2}. \quad (C11)$$

where R is the universal gas constant.

Combining Eqs. (C4)–(C11) with Eqs. (C2) and (C3) allows one to obtain the following Kirkwood-Buff integrals of ternary mixtures for an infinitely dilute solute (of course, the numerous algebraic transformations necessary could be carried out by using an algebraic software such as MATHEMATICA or MAPLE),

$$G_{12} = kTk_T - J_{12}\frac{V_3(c_1+c_3)}{2(c_1+c_1 J_{11}+c_3)} - J_{21}\frac{V_3 c_3}{2(c_1+c_1 J_{11}+c_3)} + J_{11}\frac{V_3 c_1 - 2V_2^\infty c_1}{2(c_1+c_1 J_{11}+c_3)} - \frac{V_3 c_3(c_1+c_3)(V_1-V_3)}{(c_1+c_1 J_{11}+c_3)} - \frac{V_2^\infty(c_1+c_3)}{(c_1+c_1 J_{11}+c_3)}, \quad (C12)$$

$$G_{23} = kTk_T + J_{12}\frac{V_1 c_1(c_1+c_3)}{2c_3(c_1+c_1 J_{11}+c_3)} + J_{21}\frac{V_1 c_1}{2(c_1+c_1 J_{11}+c_3)} - J_{11}\frac{c_1(V_1 c_1 + 2c_3 V_2^\infty)}{2c_3(c_1+c_1 J_{11}+c_3)} + \frac{c_1 V_1(c_1+c_3)(V_1-V_3)}{(c_1+c_1 J_{11}+c_3)} - \frac{V_2^\infty(c_1+c_3)}{(c_1+c_1 J_{11}+c_3)}, \quad (C13)$$

and

$$G_{13} = kTk_T - \frac{(c_1+c_3)^2 V_1 V_3}{(c_1+c_1 J_{11}+c_3)}, \quad (C14)$$

where $J_{11}=\lim_{x_2\to 0}(\partial\ln\gamma_1/\partial x_1)_{x_2}$, $J_{12}=\lim_{x_2\to 0}(\partial\ln\gamma_1/\partial x_2)_{x_1}$, and $J_{21}=\lim_{x_2\to 0}(\partial\ln\gamma_2/\partial x_1)_{x_2}$. However, J_{11}, J_{12}, and J_{21} are not independent quantities. Indeed, it was shown that[9,44]

$$J_{11} = -\frac{c_3(c_1+c_3)[G_{11}+G_{33}-2G_{13}]}{c_1+c_3+c_3 c_1[G_{11}+G_{33}-2G_{13}]}, \quad (C15)$$

$$J_{12} = -\frac{c_3(c_1+c_3)[G_{33}+G_{12}-G_{13}-G_{23}]}{c_1+c_3+c_3 c_1[G_{11}+G_{33}-2G_{13}]}, \quad (C16)$$

and

$$J_{21} = \frac{(c_1+c_3)\{c_1[G_{11}+G_{23}-G_{12}-G_{13}]+c_3[-G_{12}-G_{33}+G_{13}+G_{23}]\}}{c_1+c_3+c_3c_1[G_{11}+G_{33}-2G_{13}]}. \tag{C17}$$

After some algebraic manipulations of Eqs. (C15)–(C17), the following relation between the derivatives J_{11}, J_{12}, and J_{21} was obtained

$$J_{12} = (c_1 J_{11} + c_3 J_{21})/(c_1+c_3). \tag{C18}$$

The insertion of Eq. (C18) into Eqs. (C12) and (C13) provides the following expressions for G_{12} and G_{23} at infinite protein dilution:

$$G_{12} = kTk_T - \frac{J_{21}V_3 c_3 + J_{11}V_2^\infty c_1}{(c_1+c_1 J_{11}+c_3)}$$

$$- \frac{V_3 c_3(c_1+c_3)(V_1-V_3)+V_2^\infty(c_1+c_3)}{(c_1+c_1 J_{11}+c_3)}, \tag{C19}$$

$$G_{23} = kTk_T + \frac{J_{21}V_1 c_1 - J_{11}c_1 V_2^\infty}{(c_1+c_1 J_{11}+c_3)}$$

$$+ \frac{c_1 V_1(c_1+c_3)(V_1-V_3)-V_2^\infty(c_1+c_3)}{(c_1+c_1 J_{11}+c_3)}. \tag{C20}$$

[1] H. R. Mahler and E. H. Cordes, *Biological Chemistry* (Harper & Row, New York, 1966).
[2] W. Kauzmann, *Denaturation of Proteins and Enzymes. In the Mechanism of Enzyme Action*, edited by W. D. McElroy and B. Glass (Johns Hopkins, Baltimore, 1954), pp. 71–110.
[3] W. Kauzmann, Adv. Protein Chem. **14**, 1 (1959).
[4] E. F. Casassa and H. Eisenberg, Adv. Protein Chem. **19**, 287 (1964).
[5] J. Wyman, Adv. Protein Chem. **19**, 223 (1964).
[6] S. N. Timasheff, Annu. Rev. Biophys. Biomol. Struct. **22**, 67 (1993).
[7] S. N. Timasheff, *A Physicochemical Basis for the Selection of Osmolytes by Nature. In Water and Life: Comparative Analysis of Water Relationships at the Organismic, Cellular, and Molecular levels*, edited by G. N. Somero, C. B. Osmond, C. L. Bolis (Springer, Berlin, 1992), pp. 70–84.
[8] (a) S. N. Timasheff, Adv. Protein Chem. **51**, 355 (1998). (b) W. T. Zhang, M. W. Capp, J. P. Bond, C. F. Anderson, and M. T. Record, Jr., Biochemistry **35**, 10506 (1996).
[9] J. G. Kirkwood and F. P. Buff, J. Chem. Phys. **19**, 774 (1951).
[10] S. Shimizu, Proc. Natl. Acad. Sci. U.S.A. **101**, 1195 (2004).
[11] S. Shimizu, J. Chem. Phys. **120**, 4989 (2004).
[12] S. Shimizu and D. Smith, J. Chem. Phys. **121**, 1148 (2004).
[13] S. Shimizu and C. L. Boon, J. Chem. Phys. **121**, 9147 (2004).
[14] P. E. Smith, J. Phys. Chem. B **108**, 16271 (2004).
[15] P. E. Smith, J. Phys. Chem. B **108**, 18716 (2004).
[16] A. Ben-Naim, *Statistical Thermodynamics for Chemists and Biochemists* (Plenum, New York, 1992).
[17] R. G. Rubio, M. G. Prolongo, M. D. Pena, and J. A. R. Renuncio, J. Phys. Chem. **91**, 1177 (1987).
[18] A. Ben-Naim, Cell Biophys. **12**, 255 (1988).
[19] A. Ben-Naim, Pure Appl. Chem. **62**, 25 (1990).
[20] I. Shulgin and E. Ruckenstein, J. Phys. Chem. **103**, 872 (1999).
[21] I. Shulgin and E. Ruckenstein, J. Phys. Chem. **103**, 4900 (1999).
[22] E. Matteoli and L. Lepori, *Interaction in Mixtures Studied through the Kirkwood-Buff Theory. In Fluctuation Theory of Mixtures*, edited by E. Matteoli and G. A. Mansoori (Taylor & Francis, New York, 1990), pp. 259–271.
[23] E. Ruckenstein and I. Shulgin, Fluid Phase Equilib. **180**, 281 (2001).
[24] J. M. Prausnitz, R. N. Lichtenthaler, and E. Gomes de Azevedo, *Molecular Thermodynamics of Fluid: Phase Equilibria*, 2nd ed. (Prentice-Hall, Englewood Cliffs, NJ, 1986).
[25] E. P. K. Hade and C. Tanford, J. Am. Chem. Soc. **89**, 5034 (1967).
[26] E. Ruckenstein and I. Shulgin, Int. J. Pharm. **258**, 193 (2003).
[27] E. Ruckenstein and I. Shulgin, Environ. Sci. Technol. **39**, 1623 (2005).
[28] S. I. Sandler, Fluid Phase Equilib. **116**, 343 (1996).
[29] S. I. Sandler, J. Chem. Thermodyn. **31**, 3 (1999).
[30] K. Kojima, S. Zhang, and T. Hiaki, Fluid Phase Equilib. **131**, 145 (1997).
[31] A. Ben-Naim, J. Chem. Phys. **67**, 4884 (1977).
[32] E. Matteoli and L. Lepori, J. Chem. Soc., Faraday Trans. **91**, 431 (1995).
[33] E. Matteoli, J. Phys. Chem. B **101**, 9800 (1997).
[34] I. Shulgin and E. Ruckenstein, J. Phys. Chem. B **103**, 2496 (1999).
[35] R. Chitra and P. E. Smith, J. Phys. Chem. B **106**, 1491 (2002).
[36] Y. Marcus, Phys. Chem. Chem. Phys. **2**, 4891 (2000).
[37] K. Gekko and S. N. Timasheff, Biochemistry **20**, 4667 (1981).
[38] G. Xie and S. N. Timasheff, Biophys. Chem. **64**, 25 (1997).
[39] T.-Y. Lin and S. N. Timasheff, Biochemistry **33**, 12695 (1994).
[40] S. N. Timasheff and G. Xie, Biophys. Chem. **105**, 421 (2003).
[41] B. M. Baynes and B. L. Trout, J. Phys. Chem. B **107**, 14058 (2003).
[42] J. Hong, M. W. Capp, C. F. Anderson, R. M. Soecker, D. J. Felitsky, M. W. Anderson, and M. T. Record, Jr., Biochemistry **43**, 14744 (2004).
[43] M. E. Noelken and S. N. Timasheff, J. Biol. Chem. **242**, 5080 (1967).
[44] I. Shulgin and E. Ruckenstein, Ind. Eng. Chem. Res. **41**, 1689 (2002).

Biophysical Chemistry

Biophysical Chemistry 118 (2005) 128–134

http://www.elsevier.com/locate/biophyschem

Relationship between preferential interaction of a protein in an aqueous mixed solvent and its solubility

Ivan L. Shulgin, Eli Ruckenstein *

Department of Chemical and Biological Engineering, State University of New York at Buffalo, Amherst, NY 14260, USA

Received 29 June 2005; received in revised form 22 July 2005; accepted 25 July 2005
Available online 2 November 2005

Abstract

The present paper is devoted to the derivation of a relation between the preferential solvation of a protein in a binary aqueous solution and its solubility. The preferential binding parameter, which is a measure of the preferential solvation (or preferential hydration) is expressed in terms of the derivative of the protein activity coefficient with respect to the water mole fraction, the partial molar volume of protein at infinite dilution and some characteristics of the protein-free mixed solvent. This expression is used as the starting point in the derivation of a relationship between the preferential binding parameter and the solubility of a protein in a binary aqueous solution.

The obtained expression is used in two different ways: (1) to produce a simple criterion for the salting-in or salting-out by various cosolvents on the protein solubility in water, (2) to derive equations which predict the solubility of a protein in a binary aqueous solution in terms of the preferential binding parameter. The solubilities of lysozyme in aqueous sodium chloride solutions (pH=4.5 and 7.0), in aqueous sodium acetate (pH=8.3) and in aqueous magnesium chloride (pH=4.1) solutions are predicted in terms of the preferential binding parameter without any adjustable parameter. The results are compared with experiment, and for aqueous sodium chloride mixtures the agreement is excellent, for aqueous sodium acetate and magnesium chloride mixtures the agreement is only satisfactory.
© 2005 Elsevier B.V. All rights reserved.

Keywords: Protein; Aqueous mixed solvent; Preferential binding parameter; Solubility

1. Introduction

The solvation behavior of a macromolecule such as a protein in a binary aqueous solvent is important in the understanding of such solutions [1–5]. A macromolecule can be preferentially hydrated when the concentration of water in the vicinity of the macromolecule (local concentration of water) is higher than the bulk concentration. The macromolecule can be preferentially solvated when the concentration of the cosolvent in the vicinity of the macromolecule is higher than the bulk cosolvent concentration. A measure of the solvation (or hydration) is the preferential binding parameter [2–6], which can be defined using various concentration scales (component 1 is water, component 2 is a protein and component 3 is a cosolvent):

(1) in molal concentrations

$$\Gamma_{23}^{(m)} \equiv \lim_{m_2 \to 0} (\partial m_3/\partial m_2)_{T,P,\mu_3} \qquad (1)$$

where m_i is the molality of component i, P is the pressure, T the temperature (throughout this paper only isothermal–isobaric conditions are considered), and μ_i is the chemical potential of component i.

(2) in molar concentrations

$$\Gamma_{23}^{(c)} \equiv \lim_{c_2 \to 0} (\partial c_3/\partial c_2)_{T,P,\mu_3} \qquad (2)$$

where c_i is the molar concentration of component i. It should be noted that $\Gamma_{23}^{(m)}$ and $\Gamma_{23}^{(c)}$ are defined at infinite dilution of the protein.

---

* Corresponding author. Tel.: +1 716 645 2911x2214; fax: +1 716 645 3822.
  E-mail addresses: ishulgin@eng.buffalo.edu (I.L. Shulgin), feaeliru@acsu.buffalo.edu (E. Ruckenstein).

Many characteristics of a protein in aqueous solvents are connected to its preferential solvation (or preferential hydration). The protein stability is a well-known example. Indeed, the addition of certain compounds (such as urea) can cause protein denaturation, whereas the addition of other cosolvents, such as glycerol, sucrose, etc. can stabilize at high concentrations the protein structure and preserve its enzymatic activity [4–7]. The analysis of literature data shows that as a rule $\Gamma_{23}^{(m)}>0$ for the former and $\Gamma_{23}^{(m)}<0$ for the latter compounds. Recently, the authors of the present paper showed how the excess (or deficit) number of water (or cosolvent) molecules in the vicinity of a protein molecule can be calculated in terms of $\Gamma_{23}^{(m)}$, the molar volume of the protein at infinite dilution and the properties of the protein-free mixed solvent [8]. The protein solubility in an aqueous mixed solvent is another important quantity which can be connected to the preferential solvation (or hydration) [9–13] and can help to understand the protein behavior [9–17].

The aim of the present paper is to establish a relation between: (1) the preferential solvation (or hydration) of a protein and (2) the protein solubility in an aqueous mixed solvent. The obtained relation will be used to predict the protein solubility in an aqueous solvent in terms of the preferential binding parameter.

The preferential binding parameter $\Gamma_{23}^{(m)}$ can be measured experimentally using various methods such as sedimentation [4], dialysis equilibrium [7], vapor pressure osmometry [14], etc. and has been determined for numerous systems [2–7,9–13,18–22]. It is of interest to use these experimental results for the evaluation of protein solubility.

The results obtained will be presented as follows: (1) firstly a relation between the protein solubility and the preferential binding parameter in a binary solvent will be established; (2) secondly the established relation will be used to derive criteria for the effect of cosolvents (salting-in or salting-out), (3) thirdly the experimental data for the preferential binding parameter $\Gamma_{23}^{(m)}$ will be used to predict the protein solubility and the obtained results will be compared with available experimental data.

2. Theoretical part

In a previous paper [8], the following expression for the preferential binding parameter $\Gamma_{23}^{(c)}$ was derived on the basis of the Kirkwood–Buff theory of ternary solutions:

$$\Gamma_{23}^{(c)} = \frac{c_1 c_3 (J_{21} V_1 - J_{11} V_2^\infty)}{(c_1 + c_1 J_{11} + c_3)} + \frac{c_3 (c_1 + c_3)(V_1 - V_2^\infty)}{(c_1 + c_1 J_{11} + c_3)} \quad (3)$$

where V_i is the partial molar volume of component i, V_2^∞ is the partial molar volume of a protein at infinite dilution in a mixed solvent,

$$J_{11} = \lim_{x_2 \to 0} \left(\frac{\partial \ln \gamma_1}{\partial x_1}\right)_{x_2},$$

$$J_{21} = \lim_{x_2 \to 0} \left(\frac{\partial \ln \gamma_2}{\partial x_1}\right)_{x_2},$$

x_i is the mole fraction of component i, and γ_i is the activity coefficient of component i at a mole fraction scale.

It should be noted that the quantities $\Gamma_{23}^{(c)}$, V_2^∞ and J_{21} of Eq. (3) depend on the nature of the protein, while all the other ones are related to the properties of the protein-free mixed solvent.

Eq. (3) can be rewritten as

$$J_{21} = \left(\frac{\partial \ln \gamma_2}{\partial x_1}\right)_{x_2=0}$$
$$= -\frac{c_3(c_1+c_3)V_1 - \left(\Gamma_{23}^{(c)} + c_3 V_2^\infty\right)(c_1 + c_1 J_{11} + c_3)}{c_1 c_3 V_1} \quad (4)$$

Because [2,8,23]

$$\Gamma_{23}^{(c)} = (1 - c_3 V_3)\Gamma_{23}^{(m)} - c_3 V_2^\infty \quad (5)$$

and experiment provides $\Gamma_{23}^{(m)}$, Eq. (4) can be recast in the form

$$\left(\frac{\partial \ln \gamma_2}{\partial x_1}\right)_{x_2=0}$$
$$= -\frac{c_3(c_1+c_3)V_1 - \Gamma_{23}^{(m)}(1 - c_3 V_3)(c_1 + c_1 J_{11} + c_3)}{c_1 c_3 V_1} \quad (6)$$

For poorly soluble solids, such as the proteins, one can use the infinite dilution approximation and consider that the activity coefficient of the protein in a mixed solvent is equal to that at infinite dilution. Therefore, for the solubility y_2 of a protein (solute, component 2) in a mixed solvent 1–3, one can write the following equation [24]:

$$f_2^S / f_2^L(T,P) = y_2 \gamma_2^\infty \quad (7)$$

where γ_2^∞ is the activity coefficient of a protein in a mixed solvent at infinite dilution, $f_2^L(T,P)$ is the hypothetical fugacity of a solid as a (subcooled) liquid at a given pressure (P) and temperature (T), and f_2^S is the fugacity of the pure solid component 2. If the solubility of the mixed solvent in the solid phase is negligible, then the left hand side of Eq. (7) depends only on the properties of the solute.

The combination of Eqs. (6) and (7) yields the following relation for the solubility of a protein in a mixed solvent

$$\left(\frac{\partial \ln y_2}{\partial x_1}\right) = \frac{c_3(c_1+c_3)V_1 - \Gamma_{23}^{(m)}(1-c_3V_3)(c_1+c_1J_{11}+c_3)}{c_1c_3V_1} \quad (8)$$

2.1. Salting-in or salting-out?

Eq. (8) allows one to derive a criterium for salting-in or salting-out for small cosolvent concentrations. Starting from the Gibbs–Duhem equation for a binary mixture

$$x_1 \frac{d\ln\gamma_1}{dx_1} + x_3 \frac{d\ln\gamma_3}{dx_1} = 0 \quad (9a)$$

one can conclude that

$$\lim_{x_3\to 0} J_{11} = 0 \quad (9b)$$

Eq. (8) can be therefore written for $c_3 \to 0$ in the form

$$\left(\frac{\partial \ln y_2}{\partial x_3}\right) = -\left(\frac{\partial \ln y_2}{\partial x_1}\right) = \frac{\alpha}{V_1^0} - 1 \quad (10)$$

where $\alpha = \lim_{c_3\to 0} \frac{\Gamma_{23}^{(m)}}{c_3}$ and V_1^0 is the molar volume of pure water. Salting-in occurs when

$$\left(\frac{\partial \ln y_2}{\partial x_3}\right) > 0, \quad \text{hence when } \alpha > V_1^0 \quad (11)$$

and salting-out occurs when

$$\left(\frac{\partial \ln y_2}{\partial x_3}\right) < 0, \quad \text{hence when } \alpha < V_1^0 \quad (12)$$

It is well-known [8,19,25,26] that the preferential binding parameter $\Gamma_{23}^{(m)}$ is proportional to the concentration of the cosolvent at least at low concentrations. Consequently the salting-in or salting-out depends on the slope of the curve $\Gamma_{23}^{(m)}$ versus concentration for small c_3. The application of the established criteria to salting-in or salting-out in real systems is illustrated in Table 1.

The above criteria (Eqs. (11) and (12)) are valid:

(1) for $c_3 \to 0$, hence when a small amount of cosolvent is added to the pure water;
(2) for ternary mixtures (water (1)–protein (2)–cosolvent (3)) (the experimental results regarding the preferential binding parameter) $\Gamma_{23}^{(m)}$ and the solubilities were obtained for mixtures which involve in addition a buffer, and the effect of the buffer is taken into account only indirectly via the preferential binding parameter $\Gamma_{23}^{(m)}$);
(3) for infinite dilution (this means that the protein solubility is supposed to be small enough to satisfy the infinite dilution approximation ($\gamma_2 \cong \gamma_2^\infty$));
(4) for experimental preferential binding parameters) $\Gamma_{23}^{(m)}$ and solubilities determined at low cosolvent concentrations (however, the preferential binding parameter $\Gamma_{23}^{(m)}$ and the solubilities were usually determined for molalities larger than 0.5 and those values had to be used for the cases listed in Table 1 because no other experimental data are available).

2.2. Simple equation for the protein solubility in a mixed solvent

The combination of Eqs. (4) and (7) leads to the following expression for the solubility of a protein in a mixed solvent

$$\left(\frac{\partial \ln y_2}{\partial x_1}\right) = \frac{c_3(c_1+c_3)V_1 - \left(\Gamma_{23}^{(c)} + c_3 V_2^\infty\right)(c_1+c_1J_{11}+c_3)}{c_1c_3V_1} \quad (13)$$

The integration of Eq. (13) yields for the solubility y_2 of the protein in a mixed solvent for a water mole fraction x_1 the expression

$$\ln\frac{y_2}{y_2^w} = \int_1^{x_1} \frac{\left(V_1 - V_2^\infty - \Gamma_{23}^{(c)}/c_3\right)dx_1}{x_1 V_1}$$
$$- \int_1^{x_1} \frac{\left(\Gamma_{23}^{(c)}/c_3 + V_2^\infty\right)J_{11}}{V_1}dx_1 \quad (14)$$

where y_2^w is the protein solubility in cosolvent-free water plus buffer.

Eq. (14) allows one to calculate the protein solubility if the composition dependencies of J_{11}, $\Gamma_{23}^{(c)}$ (or $\Gamma_{23}^{(m)}$) and partial molar volumes are available.

Table 1
Application of criteria (Eqs. (11),(12)) for salting-in or salting-out to aqueous solutions of proteins

Protein	Cosolvent[a]	Experimental data used		Do the criteria (Eqs. (11) (12)) work?
		Solubility (salting-in or salting-out, conditions, references)	Preferential binding parameter $\Gamma_{23}^{(m)}$ (conditions, references)	
Lysozyme	NaCl	Salting-out, $T=0–40$ °C, pH=3–10 [27–31]	pH=4.5 [32], pH=3–7 [12]	Yes
Lysozyme	MgCl$_2$	Salting-out, $T=18$ °C, pH=4.5 [27]	pH=3.0, 4.5 [13]	Yes
Lysozyme	NaAcO	Salting-out, $T=18$ °C, pH=4.5, 8.3 [27]	pH=4.5–4.71 [32]	Yes
Ribonuclease Sa	Urea	Salting-in, $T=25$ °C, pH=3.5, 4.0 [16]	pH=2.0, 4.0, 5.8 [33][b]	Yes
Lysozyme	Glycerol	Salting-in, $T=25$ °C, pH=4.6 [34]	pH=2.0, 5.8 [35]	No
β-Lactoglobulin	NaCl	Salting-in, $T=25$ °C, pH=5.15–5.3 [36]	pH=1.55–10 [12]	No

[a] The term "cosolvent" is also used here for electrolytes.
[b] The preferential binding parameters were determined for ribonuclease A in 30 vol.% glycerol solution.

Eq. (14) can be simplified if one takes into account that at least at low cosolvent concentrations $\Gamma_{23}^{(c)}$ is proportional to the concentration $c_3 (\Gamma_{23}^{(c)} = \beta c_3)$ [8,19,25,26] and by assuming in addition that the partial molar volumes V_2^∞ and V_1 are composition independent. With these two approximations, Eq. (14) becomes

$$\ln \frac{y_2}{y_2^w} = \frac{(V_1 - V_2^\infty - \beta)}{V_1} \ln x_1 - \frac{(\beta + V_2^\infty)}{V_1} (\ln \gamma_1)_{x_2=0} \quad (15a)$$

and hence

$$\ln \frac{y_2}{y_2^w} = -\frac{(V_2^\infty + \beta) \ln a_w}{V_1} + \ln x_1 \quad (15b)$$

where a_w is the water activity in the protein-free mixed solvent. Taking into account Eq. (5) and the relation $\frac{\Gamma_{23}^{(m)}}{c_3} = \alpha$, Eq. (15b) can be recast as follows

$$\ln \frac{y_2}{y_2^w} = -\frac{\left(\alpha - V_3 \Gamma_{23}^{(m)}\right) \ln a_w}{V_1} + \ln x_1 \approx -\frac{\left(\alpha - V_3 \Gamma_{23}^{(m)}\right) \ln a_w}{V_1} \quad (16)$$

Because as noted a long time ago [4] $V_3 \Gamma_{23}^{(m)}$ "is two order of magnitude smaller" than α, $\alpha \gg V_3 \Gamma_{23}^{(m)}$, and Eq. (16) can be further simplified to

$$\ln \frac{y_2}{y_2^w} = -\frac{\alpha \ln a_w}{V_1} \quad (17)$$

Eqs. (14) (15a) (15b) (16) (17) provide interrelations between the preferential binding parameter $\Gamma_{23}^{(c)}$ (or $\Gamma_{23}^{(m)}$) and the protein solubility in a mixed solvent.

3. Calculations

In order to illustrate the results obtained regarding the solubility, several systems, for which experimental data regarding both the preferential binding parameter and the protein solubility in a mixed solvent were available, were selected. The solubilities of proteins were calculated with Eq. (17). In order to predict the solubility of a protein as a function of composition one should have information about y_2^w, V^1, α and the composition dependence of the activity of water a_w in the protein-free mixed solvent. The values of y_2^w were taken from the original references regarding the solubilities [28,29], V_1 was taken equal to the molar volume of pure water at a given temperature, and α was calculated from the original references regarding the preferential binding parameters [12,32]. The concentration dependence of the activity of water a_w in protein-free mixed solvents were calculated from the experimental data for the osmotic coefficient φ [37–39] using the expression [24]

$$\ln a_w = -\varphi M_w m_3 v \quad (18)$$

where M_w is the molar weight of water, m_3 is the molality of the cosolvent in the protein-free mixed solvent, and v is the number of ions formed through complete dissociation of the electrolyte.

3.1. Water (1)–Lysozyme (2)–Sodium Chloride (3)

The lysozyme solubilities in aqueous solutions of sodium chloride are predicted for pH=4.5 and pH=6.5. In these predictions only the values of the preferential binding parameter were used and no additional (or adjustable) parameters were involved. The results are presented in Figs. 1 and 2 and the experimental preferential binding parameters used are listed in Table 2. The solubilities at pH=6.5 were predicted from the preferential binding parameter determined at pH=7.0 because the values for pH=6.5 were not available. The concentration dependence of the water activity in solutions of sodium chloride was obtained from Eq. (18) using an accurate semiempirical equation for the osmotic coefficient [37].

3.2. Water (1)–Lysozyme (2)–Sodium Acetate (3)

The lysozyme solubilities in aqueous solutions of sodium acetate were calculated for pH=8.3 and the results are presented in Fig. 3. The experimental preferential binding parameters are listed in Table 2 (the values for pH=4.68–4.7 were, however, used because those for pH=8.3 were not available). The concentration dependence of the water activity in solutions of sodium chloride was obtained from Eq. (18) using the Pitzer equation for the osmotic coefficient [38].

3.3. Water (1)–Lysozyme (2)–Magnesium Chloride (3)

The lysozyme solubilities in aqueous solutions of magnesium chloride were calculated for a pH=4.1 and the results are presented in Fig. 4. The experimental preferential binding parameter is listed in Table 2 (the value for pH=4.5

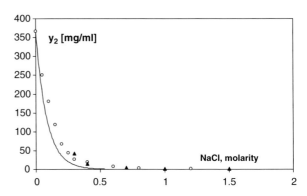

Fig. 1. Lysozyme solubility in aqueous solutions of sodium chloride at pH=4.5. The solid line represents the prediction based on Eq. (17), (o) and (▲) are the experimental data from Refs. [27,28], respectively.

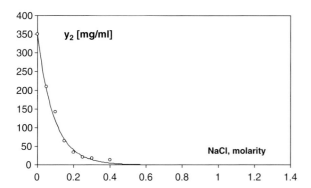

Fig. 2. Lysozyme solubility in aqueous solutions of sodium chloride at pH=6.5. The solid line represents the prediction based on Eq. (17), (o) are the experimental data from Ref. [28].

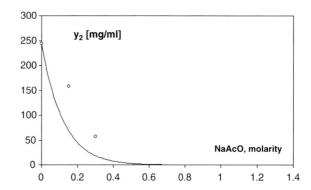

Fig. 3. Lysozyme solubility in aqueous solutions of sodium acetate at pH=8.3. The solid line represents the prediction based on Eq. (17), (o) are the experimental data from Ref. [29].

was used because that for pH=4.1 was not available). The concentration dependence of the water activity in solutions of magnesium chloride was obtained from Eq. (18) using the Pitzer equation for the osmotic coefficient [39]. For this system, the solubility data for y_2^w was not available and therefore the experimental solubility [27] for $m_3=0.4$ (5.8 [mg/ml]) was employed as the lower limit of integration in Eq. (14). Using the approximations involved in the derivation of Eq. (17), one obtains

$$\ln\frac{y_2}{y_2^*} = -\frac{\alpha\ln(a_w/a_w^*)}{V_1} \quad (19)$$

where y_2^* and a_w^* are the molar fraction solubility and the water activity, respectively, both at $m_3=0.4$.

3.4. Comments regarding the solubility predictions

The scheme employed to predict the solubility of a protein in a mixed solvent involves a number of simplifications:

(1) The derived equations (Eqs. (14) (15a) (15b) (16) (17)) involve the infinite dilution approximation ($\gamma_2 \cong \gamma_2^\infty$).
2) The equations are established for a ternary mixture (water (1)–protein (2)–cosolvent (3)). However, all the experimental results regarding the preferential binding parameter $\Gamma_{23}^{(m)}$ and the solubility involve in addition to the above three components also a buffer. The effect of the buffer is taken into account only indirectly via the preferential binding parameter $\Gamma_{23}^{(m)}$.
3) The parameter α was determined as the slope of the composition dependence of the preferential binding parameter $\Gamma_{23}^{(m)}$, assuming that the latter is proportional to the concentration.

4. Discussion

In the present paper, a connection between the preferential binding parameter of a protein and its solubility in an aqueous solvent was established. The preferential binding parameter is a measure of the protein / water and protein / cosolvent interaction at molecular level [6,19]. Regarding the preferential binding parameter, Timasheff subdivided the cosolvents into several groups [6]: "When a protein molecule is immersed into a solvent consisting of water and another chemical species (a cosolvent), the interactions between the protein and the

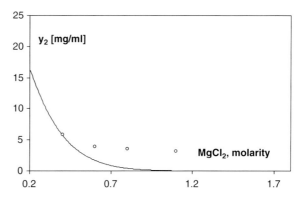

Fig. 4. Lysozyme solubility in aqueous solutions of magnesium chloride at pH=4.1. The solid line represents the prediction based on Eq. (17), (o) are the experimental data from Ref. [27].

Table 2
Experimental preferential binding parameters used for solubility predictions

Protein	Cosolvent	Molality	pH	Preferential binding parameter $\Gamma_{23}^{(m)}$ [mol/mol]	Reference
Lysozyme	NaCl	1	4.5	−6.2	[12]
Lysozyme	NaCl	1	7.0	−5.8	[12]
Lysozyme	NaAcO	0.5 and 1	4.68 and 4.71	−5.14 and −7.5	[32]
Lysozyme	MgCl$_2$	1	4.5	−1.79	[32]

solvent components may lead to three possible situations: (1) the cosolvent is present at the protein surface in excess over its concentration in the bulk (this is what constitutes binding); (2) the water is present in excess at the protein surface; this means that the protein has a higher affinity for water than for the cosolvent (this situation is referred to as preferential hydration, or preferential exclusion of the cosolvent); (3) the protein is indifferent to the nature of molecules (water or cosolvent) with which it comes in contact, so that no solvent concentration perturbation occurs at the protein surface".

The present analysis shows that the same classification can be made with respect to the effect of a small amount of a cosolvent on the protein solubility in an aqueous solvent (see Eqs. (11) and (12)). Namely, the cosolvents of the first Timasheff's group (e. g. urea) increase the protein solubility compared to the solubility in water when a small amount is added to water. Compounds of the second group (e. g. salts) decrease the solubility and they are well-known salting-out agents [14]. Substances of the third group (we have no example) do not essentially change the solubility compared to the solubility in pure water.

The present paper emphasizes how the preferential binding parameter is related to the solubility and how the preferential binding parameter can be used to predict the solubility. Eq. (13) (or its equivalent Eq. (8)) provides the most general equation that connects the preferential binding parameter and the solubility. The integration of this equation leads to Eq. (14) which allows one to predict the protein solubility in a mixed solvent if the composition dependencies of J_{11}, $\Gamma_{23}^{(c)}$ (or $\Gamma_{23}^{(m)}$) and partial molar volumes are available. A simplified form of Eqs. (14) and (17), can predict the solubility if information about y_2^w, V_1, α and the composition dependence of the activity of water a_w in a protein-free mixed solvent is available. Eq. (17) was used in this paper to predict the protein solubility in an aqueous mixed solvent. The results of predictions (Figs. 1–4) demonstrate that the experimental data regarding the preferential binding parameter could be successfully used to predict the solubility of proteins in aqueous mixed solvents. It should be pointed out that no additional parameters (adjustable parameters) were used. However, the present approach involves a number of approximations, among which the infinite dilution approximation deserves an additional comment, because the solubility of some proteins can be relatively large. For example, the solubility of lysozyme in aqueous solutions of sodium chloride at pH=4.5 can be as high as 365 mg/ml. However, in the mole fraction scale, this solubility is smaller than 5 10^{-4} (2000 molecules of water per molecule of lysozyme), value which seems to be sufficiently low for the system to be considered dilute. The accuracy of the predictions are highly dependent on the quality of experimental data regarding the preferential binding parameter, the solubility and the water activity in protein-free mixed solvents.

5. Conclusion

A relationship between the derivative of the activity coefficient of the protein with respect to the mole fraction of water at infinite dilution of protein and the preferential binding parameter was used to connect the solubility of a protein in an aqueous mixed solvent to the preferential binding parameter. This relation was used to examine the salting-in and salting-out effect of various compounds on the protein solubility in water and to predict the protein solubility.

References

[1] H.K. Schachman, M.A. Lauffer, The hydration, size and shape of tobacco mosaic virus, J. Am. Chem. Soc. 71 (1949) 536–541.
[2] E.F. Casassa, H. Eisenberg, Thermodynamic analysis of multi-component solutions, Adv. Protein Chem. 19 (1964) 287–395.
[3] J. Wyman Jr., Linked functions and reciprocal effects in hemoglobin: a second look, Adv. Protein Chem. 19 (1964) 223–286.
[4] I.D. Kuntz, W. Kauzmann, Hydration of proteins and polypeptides, Adv. Protein Chem. 28 (1974) 239–343.
[5] S.N. Timasheff, The control of protein stability and association by weak interactions with water: how do solvents affect these processes? Annu. Rev. Biophys. Biomol. Struct. 22 (1993) 67–97.
[6] S.N. Timasheff, A physicochemical basis for the selection of osmolytes by nature, in: G.N. Somero, C.B. Osmond, C.L. Bolis (Eds.), Water and Life: Comparative Analysis of Water Relationships at the Organismic, Cellular, and Molecular Levels, Springer-Verlag, Berlin, 1992, pp. 70–84.
[7] S.N. Timasheff, Control of protein stability and reactions by weakly interacting cosolvents: the simplicity of the complicated, Adv. Protein Chem. 51 (1998) 355–432.
[8] I.L. Shulgin, E. Ruckenstein, A protein molecule in an aqueous mixed solvent: fluctuation theory outlook, J. Chem. Phys. 123 (2005) 054909.
[9] T. Arakawa, S.N. Timasheff, Mechanism of protein salting in and salting out by divalent cation salts: balance between hydration and salt binding, Biochemistry 23 (1984) 5912–5923.
[10] T. Arakawa, S.N. Timasheff, Theory of protein solubility, Methods Enzymol. 114 (1985) 49–77.
[11] S.N. Timasheff, T. Arakawa, Mechanism of protein precipitation and stabilization by co-solvents, J. Cryst. Growth 90 (1988) 39–46.
[12] T. Arakawa, S.N. Timasheff, Abnormal solubility behavior of beta-lactoglobulin — salting-in by glycine and NaCl, Biochemistry 26 (1987) 5147–5153.
[13] T. Arakawa, R. Bhat, S.N. Timasheff, Preferential interactions determine protein solubility in 3-component solutions — the MgCl2 system, Biochemistry 29 (1990) 1914–1923.
[14] E.J. Cohn, J.T. Edsall, Proteins, Amino Acids and Peptides, Reinhold, New York, 1943.
[15] Y. Qu, C.L. Bolen, D.W. Bolen, Osmolyte-driven contraction of a random coil protein, Proc. Natl. Acad. Sci. U. S. A. 95 (1998) 9268–9273.
[16] C.N. Pace, S. Trevino, E. Prabhakaran, et al., Protein structure, stability and solubility in water and other solvents, Philos. Trans. R. Soc. Lond., B Biol. Sci. 359 (2004) 1225–1235.
[17] D.W. Bolen, Effects of naturally occurring osmolytes on protein stability and solubility: issues important in protein crystallization, Methods 34 (2004) 312–322.
[18] W.T. Zhang, M.W. Capp, J.P. Bond, C.F. Anderson, M.T. Record Jr., Thermodynamic characterization of interactions of native bovine serum albumin with highly excluded (glycine betaine) and moderately

accumulated (urea) solutes by a novel application of vapor pressure osmometry, Biochemistry 35 (1996) 10506–10516.
[19] M.T. Record Jr., W.T. Zhang, C.F. Anderson, Analysis of effects of salts and uncharged solutes on protein and nucleic acid equilibria and processes: a practical guide to recognizing and interpreting polyelectrolyte effects, Hofmeister effects and osmotic effects of salts, Adv. Protein Chem. 51 (1998) 281–353.
[20] D.J. Felitsky, M.T. Record Jr., Application of the local-bulk partitioning and competitive binding models to interpret preferential interactions of glycine betaine and urea with protein surface, Biochemistry 43 (2004) 9276–9288.
[21] C.F. Anderson, E.S. Courtenay, M.T. Record Jr., Thermodynamic expressions relating different types of preferential interaction coefficients in solutions containing two solute components, J. Phys. Chem., B 106 (2002) 418–433.
[22] E.S. Courtenay, M.W. Capp, M.T. Record Jr., Thermodynamics of interactions of urea and guanidinium salts with protein surface: relationship between solute effects on protein processes and changes in water-accessible surface area, Protein Sci. 10 (2001) 2485–2497.
[23] H. Eisenberg, Biological Macromolecules and Polyelectrolytes in Solution, Clarendon Press, Oxford, 1976.
[24] J.M. Prausnitz, R.N. Lichtenthaler, E. Gomes de Azevedo, Molecular Thermodynamics of Fluid — Phase Equilibria, 2nd ed., Prentice-Hall, Englewood Cliffs, NJ, 1986.
[25] E.S. Courtenay, M.W. Capp, C.F. Anderson, M.T. Record Jr., Vapor pressure osmometry studies of osmolyte-protein interactions: implications for the action of osmoprotectants in vivo and for the interpretation of 'osmotic stress' experiments in vitro, Biochemistry 39 (2000) 4455–4471.
[26] B.M. Baynes, B.L. Trout, Proteins in mixed solvents: a molecular-level perspective, J. Phys. Chem., B 107 (2003) 14058–14067.
[27] M.M. Ries-Kautt, A.F. Ducruix, Relative effectiveness of various ions on the solubility and crystal growth of lysozyme, J. Biol. Chem. 264 (1989) 745–748.
[28] P. Retailleau, M. Ries-Kautt, A. Ducruix, No salting-in of lysozyme chloride observed at low ionic strength over a large range of pH, Biophys. J. 72 (1997) 2156–2163.
[29] P. Retailleau, A.F. Ducruix, M. Ries-Kautt, Importance of the nature of anions in lysozyme crystallization correlated with protein net charge variation, Acta Crystallogr., D Biol. Crystallogr. 58 (2002) 1576–1581.
[30] B. Guo, S. Kao, H. McDonald, A. Asanov, L.L. Combs, W.W. Wilson, Correlation of second viral coefficients and solubilities useful in protein crystal growth, J. Cryst. Growth 196 (1999) 424–433.
[31] E.L. Forsythe, R.A. Judge, M.L. Pusey, Tetragonal chicken egg white lysozyme solubility in sodium chloride solutions, J. Chem. Eng. Data 44 (1999) 637–640.
[32] T. Arakawa, S.N. Timasheff, Preferential interactions of proteins with salts in concentrated solutions, Biochemistry 21 (1982) 6545–6552.
[33] T.Y. Lin, S.N. Timasheff, Why do some organisms use a urea–methylamine mixture as osmolytes?, Biochemistry 33 (1994) 12695–12701.
[34] A.M. Kulkarni, C.F. Zukoski, Nanoparticle crystal nucleation: influence of solution conditions, Langmuir 18 (2002) 3090–3099.
[35] K. Gekko, S.N. Timasheff, Thermodynamic and kinetic examination of protein stabilisation by glycerol, Biochemistry 20 (1981) 4677–4686.
[36] J.M. Treece, R.S. Sheinson, T.L. McMeekin, The solubilities of -lactoglobulins A, B and AB, Arch. Biochem. Biophys. 108 (1964) 99–108.
[37] W.J. Hamer, Y.-C. Wu, Osmotic coefficients and mean activity coefficients of uni-univalent electrolytes in water at 25 °C, J. Phys. Chem. Ref. Data 1 (1972) 1047–1099.
[38] R. Beyer, M. Steiger, Vapor pressure measurements and thermodynamic properties of aqueous solutions of sodium acetate, J. Chem. Thermodyn. 34 (2002) 1057–1071.
[39] J.A. Rard, D.G. Miller, Isopiestic determination of the osmotic and activity coefficients of aqueous magnesium chloride solutions at 25 °C, J. Chem. Eng. Data 26 (1981) 38–43.

Comment to the Editor

A Protein Molecule in a Mixed Solvent: The Preferential Binding Parameter via the Kirkwood-Buff Theory

In a recent article (Schurr, J. M., D. P. Rangel, and S. R. Aragon. 2005. A contribution to the theory of preferential interaction coefficients. *Biophys. J.* 89:2258–2276), a detailed derivation of an expression for the preferential binding coefficient via the Kirkwood-Buff theory of solutions was presented. The authors of this Comment (Shulgin, I. L., and E. Ruckenstein. 2005. A protein molecule in an aqueous mixed solvent: fluctuation theory outlook. *J. Chem. Phys.* 123:054909) also recently established on the basis of the Kirkwood-Buff theory of solutions an equation for the preferential binding of a cosolvent to a protein. There are other publications that relate the preferential binding parameter to the Kirkwood-Buff theory of solutions for protein + binary mixed solvents. The expressions derived in the two articles mentioned above are different because the definitions of the preferential binding parameter are different. However, there are articles in which the definitions of the preferential binding parameter are the same, but the derived equations that relate the preferential binding parameter to the Kirkwood-Buff integrals are different. The goal of this Comment is to examine the various expressions that relate the preferential binding parameter to the Kirkwood-Buff theory.

INTRODUCTION

An important characteristic of a solution of a protein (component 2) in a mixture water (1) + cosolvent (3) is the preferential binding parameter $\Gamma_{23}^{(m)}$ (1–6)

$$\Gamma_{23}^{(m)} \equiv \lim_{m_2 \to 0} (\partial m_3 / \partial m_2)_{T,P,\mu_3}, \tag{1}$$

where m_i is the molality of component i, P is the pressure, T is the absolute temperature, and μ_i is the chemical potential of component i. The preferential binding parameter can be also defined at a molarity scale by

$$\Gamma_{23}^{(c)} \equiv \lim_{c_2 \to 0} (\partial c_3 / \partial c_2)_{T,P,\mu_3}, \tag{2}$$

where c_i is the molar concentration of component i. It should be emphasized that $\Gamma_{23}^{(m)}$ and $\Gamma_{23}^{(c)}$ are defined at infinite protein dilution.

The preferential binding parameter $\Gamma_{23}^{(m)}$ was determined experimentally (5–7) and provides information regarding the interactions between a protein and the components of the mixed solvent. As a rule (1–5), $\Gamma_{23}^{(m)} < 0$, the protein is preferentially hydrated, for cosolvents such as glycerol, sucrose, etc., which can stabilize at high concentrations the protein structure and preserve its enzymatic activity (3–5), and $\Gamma_{23}^{(m)} > 0$, the protein is preferentially solvated by cosolvents (such as urea), which can cause protein denaturation.

In literature (8) a number of different definitions of the preferential binding parameter (coefficient) have been employed. They can be connected by thermodynamic relations for ternary mixtures (8). In this Comment the preferential binding parameter will be mostly defined by Eqs. 1 and 2.

Because the preferential binding parameter is a meaningful physical quantity, attempts have been made to relate it to a general theory of solutions, such as the Kirkwood-Buff theory of solutions (9). Several authors reported results in this direction (10–17). The authors of this Comment derived the following equation for $\Gamma_{23}^{(c)}$ (16):

$$\Gamma_{23}^{(c)} = c_3(G_{23} - G_{13}), \tag{3}$$

where G_{13} and G_{23} are the Kirkwood-Buff integrals defined as (9)

$$G_{\alpha\beta} = \int_0^\infty (g_{\alpha\beta} - 1) 4\pi r^2 dr, \tag{4}$$

where $g_{\alpha\beta}$ is the radial distribution function between species α and β, and r is the distance between the centers of molecules α and β.

Equation 3 differs from the expression of $\Gamma_{23}^{(c)}$ employed in Shimizu (10,11):

$$\Gamma_{23}^{(c)} = c_3(G_{23} - G_{12}). \tag{5}$$

In a recent article in this journal (17), the Kirkwood-Buff theory of solutions was used to express the preferential binding coefficient $\Gamma_3(2)$, defined as

$$\Gamma_3(2) \equiv -\lim_{c_2 \to 0} (\partial \mu_2 / \partial \mu_3)_{T,P,c_2}, \tag{6}$$

in terms of the Kirkwood-Buff integrals. It was found (17) that

$$\Gamma_3(2) = c_3(G_{23} - G_{12}). \tag{7}$$

As noted in Schurr et al. (17) the preferential binding coefficient $\Gamma_3(2)$ defined by Eq. 6 differs from the preferential binding parameter $\Gamma_{23}^{(c)}$ defined by Eq. 2.

Submitted September 7, 2005, and accepted for publication October 12, 2005.

Address reprint requests to Eli Ruckenstein, Tel.: 716-645-2911, ext. 2214; Fax: 716-645-3822; E-mail: feaeliru@acsu.buffalo.edu.

© 2006 by the Biophysical Society

0006-3495/06/01/704/04 $2.00

However, Eqs. 3 and 5 are different equations even though they are based on the same definition of the preferential binding parameter and have the same theoretical basis: the Kirkwood-Buff theory of solutions. To make a selection between Eqs. 3 and 5 a simple limiting case, the ideal ternary mixture, will be examined using the traditional thermodynamics, and the results will be compared to those provided by Eqs. 3 and 5.

IDEAL TERNARY MIXTURE

Let us consider an ideal ternary mixture. According to the definition of an ideal mixture (18), the activities of the components (a_i) are equal to their mol fractions (x_i) and their partial molar volumes are equal to those of the pure components ($V_i = V_i^0$).

Because

$$d\mu_3 = \left(\frac{\partial \mu_3}{\partial T}\right)_{P,c_2,c_3} dT + \left(\frac{\partial \mu_3}{\partial P}\right)_{T,c_2,c_3} dP + \left(\frac{\partial \mu_3}{\partial c_2}\right)_{T,P,c_3} dc_2 + \left(\frac{\partial \mu_3}{\partial c_3}\right)_{T,P,c_2} dc_3, \quad (8)$$

one can write for an ideal mixture

$$\Gamma_{23}^{(c)} \equiv \lim_{c_2 \to 0} (\partial c_3/\partial c_2)_{T,P,\mu_3} = -\lim_{c_2 \to 0} \frac{\left(\frac{\partial \mu_3}{\partial c_2}\right)_{T,P,c_3}}{\left(\frac{\partial \mu_3}{\partial c_3}\right)_{T,P,c_2}}$$

$$= -\lim_{c_2 \to 0} \frac{\left(\frac{\partial \ln x_3}{\partial c_2}\right)_{T,P,c_3}}{\left(\frac{\partial \ln x_3}{\partial c_3}\right)_{T,P,c_2}}. \quad (9)$$

For isothermal-isobaric conditions

$$dc_2 = \frac{dx_2}{V} - \frac{x_2}{V^2}\left(\left(\frac{\partial V}{\partial x_2}\right)_{x_3} dx_2 + \left(\frac{\partial V}{\partial x_3}\right)_{x_2} dx_3\right), \quad (10)$$

and

$$dc_3 = \frac{dx_3}{V} - \frac{x_3}{V^2}\left(\left(\frac{\partial V}{\partial x_2}\right)_{x_3} dx_2 + \left(\frac{\partial V}{\partial x_3}\right)_{x_2} dx_3\right), \quad (11)$$

where V is the molar volume of the ternary mixture.

When c_3 is a constant, Eqs. 10 and 11 lead to

$$\left(\frac{\partial c_2}{\partial x_3}\right)_{c_3} = \frac{V - x_2\left(\frac{\partial V}{\partial x_2}\right)_{x_3} - x_3\left(\frac{\partial V}{\partial x_3}\right)_{x_2}}{x_3 V \left(\frac{\partial V}{\partial x_2}\right)_{x_3}}, \quad (12)$$

and when c_2 is a constant, Eqs. 10 and 11 lead to

$$\left(\frac{\partial c_3}{\partial x_3}\right)_{c_2} = \frac{V - x_2\left(\frac{\partial V}{\partial x_2}\right)_{x_3} - x_3\left(\frac{\partial V}{\partial x_3}\right)_{x_2}}{V\left(V - x_2\left(\frac{\partial V}{\partial x_2}\right)_{x_3}\right)}. \quad (13)$$

By inserting Eqs. 12 and 13 into Eq. 9 at infinite dilution of component 2, one obtains the following expression for $\Gamma_{23}^{(c)}$ of an ideal ternary mixture:

$$\Gamma_{23}^{(c)}(ideal) = c_3(V_1^0 - V_2^0). \quad (14)$$

On the other hand, expressions for $\Gamma_{23}^{(c)}$ for an ideal ternary solution can be also derived by combining Eq. 3 or Eq. 5 with the following Kirkwood-Buff integrals for ideal ternary mixtures (16):

$$G_{12}^{(id)} = kTk_T^{(id)} - V_3^0 c_3(V_1^0 - V_3^0) - V_2^0 \quad (15)$$

$$G_{23}^{(id)} = kTk_T^{(id)} + V_1^0 c_1(V_1^0 - V_3^0) - V_2^0 \quad (16)$$

$$G_{13}^{(id)} = kTk_T^{(id)} - (c_1 + c_3)V_1^0 V_3^0, \quad (17)$$

where k is the Boltzmann constant and k_T is the isothermal compressibility.

Equation 3 leads to

$$\Gamma_{23}^{(c)}(ideal) = c_3(V_1^0 - V_2^0), \quad (18)$$

whereas Eq. 5 to

$$\Gamma_{23}^{(c)}(ideal) = c_3(V_1^0 - V_3^0). \quad (19)$$

DISCUSSION

One can see that the result obtained on the basis of Eq. 3 (Eq. 18) coincides with Eq. 14 derived from general thermodynamic considerations, whereas that based on Eq. 5 does not. The numerical difference between the two expressions is very large because the molar volume of a protein is, usually, much larger than the molar volume of the cosolvent.

Whereas the above discussion involves $\Gamma_{23}^{(c)}$, the quantity $\Gamma_{23}^{(m)}$, which is usually determined experimentally (2–7), is related to $\Gamma_{23}^{(c)}$ through the equation (1,16)

$$\Gamma_{23}^{(c)} = (1 - c_3 V_3)\Gamma_{23}^{(m)} - c_3 V_2^\infty, \quad (20)$$

where V_2^∞ is the partial molar volume of the protein at infinite dilution. V_2^∞ and V_3 can be expressed at infinite dilution of component 2 in terms of the Kirkwood-Buff integrals as follows (19):

$$V_2^\infty = \frac{1+c_1(G_{11}-G_{12})+c_3(G_{33}-G_{23})+c_1c_3(-G_{12}G_{33}+G_{12}G_{13}-G_{13}^2+G_{13}G_{23}+G_{11}G_{33}-G_{11}G_{23})}{c_1+c_3+c_1c_3(G_{11}+G_{33}-2G_{13})}, \quad (21)$$

and (9)

$$V_3 = \frac{1+(G_{11}-G_{13})c_1}{c_1+c_3+c_1c_3(G_{11}+G_{33}-2G_{13})}. \quad (22)$$

By combining Eqs. 3, 20, 21, and 22, one obtains after some algebra the following simple expression:

$$\Gamma_{23}^{(m)} = \frac{c_3}{c_1} + c_3(G_{23}-G_{12}+G_{11}-G_{13}). \quad (23)$$

Whereas G_{12} and G_{23} depend on the protein characteristics, G_{11} and G_{13} depend only on the characteristics of the protein-free mixed solvent.

For usual cosolvents (organic solvents, salts, etc.), one can use the following approximation of Eq. 23 in the dilute cosolvent range:

$$\Gamma_{23}^{(m)} \approx c_3(G_{23}-G_{12}). \quad (24)$$

Indeed, $|G_{11}|$ and $|G_{13}|$ are much smaller than the Kirkwood-Buff integrals for the pairs involving the protein ($|G_{12}|$ and $|G_{23}|$). Table 1 provides their values for the system water (1) + lysozyme (2) + urea (3) (pH 7.0, 20°C). However, when $|G_{11}|$ and $|G_{13}|$ are large, and this occurs when the cosolvent is, for example, a polymer ($G_{13} \approx -1000$ (cm^3/mol) for the system water/polyethylene glycol 2000 at a weight fraction of polyethylene glycol of 0.02 (21)), the complete Eq. 23 should be used. This conclusion is valid for all large cosolvent molecules (polymers, biomolecules, etc.).

Let us consider the biochemical equilibrium between infinitely dilute native (N) and denaturated (D) states of a protein in a mixed solvent. The changes of the preferential binding parameters $\Gamma_{23}^{(c)}$, $\Gamma_{23}^{(m)}$, and $\Gamma_3(2)$ in this process are given by

$$\Delta\Gamma_{23}^{(c)} = c_3(G_{23}(D)-G_{23}(N)) = c_3\Delta G_{23} \quad (25)$$

$$\Delta\Gamma_{23}^{(m)} = c_3(G_{23}(D)-G_{23}(N)-G_{12}(D)+G_{12}(N))$$
$$= c_3(\Delta G_{23}-\Delta G_{12}), \quad (26)$$

TABLE 1 Numerical values of the Kirkwood-Buff integrals for the water (1) + lysozyme (2) + urea (3) (pH 7.0, 20°C) system

c_3 (mol/l)	$\|G_{12}\|$ (cm^3/mol) (16)	$\|G_{23}\|$ (cm^3/mol) (16)	$\|G_{11}\|$ (cm^3/mol) (20)	$\|G_{13}\|$ (cm^3/mol) (20)
1	10,350	3700	~16	~42
3	10,630	6180	~12	~45

and

$$\Delta\Gamma_3(2) = c_3(G_{23}(D)-G_{23}(N)-G_{12}(D)+G_{12}(N))$$
$$= c_3(\Delta G_{23}-\Delta G_{12}) = \Delta\Gamma_{23}^{(m)}. \quad (27)$$

Equations 25 and 26 follow from Eqs. 3 and 23 by taking into account that G_{11} and G_{13} are characteristics of the protein-free mixed solvent at infinite protein dilution.

The equilibrium constant K of biochemical equilibrium between infinitely dilute native (N) and denaturated (D) states of a protein in a mixed solvent can be expressed in terms of $\Delta\Gamma_{23}^{(m)}$ (22)

$$\left(\frac{\partial \ln K}{\partial \ln a_3}\right)_{T,P,m_2} = \frac{\left(\frac{\partial \mu_3}{\partial m_2}\right)^N_{T,P,m_3} - \left(\frac{\partial \mu_3}{\partial m_2}\right)^D_{T,P,m_3}}{\left(\frac{\partial \mu_3}{\partial m_3}\right)_{T,P,m_2}} = \Delta\Gamma_{23}^{(m)}, \quad (28)$$

where $\Delta\Gamma_{23}^{(m)}$ can be provided by experiment (23).

Using for G_{12} and G_{23} expressions from Shulgin and Ruckenstein (16), Eq. 28 can be also rewritten in the form

$$\left(\frac{\partial \ln K}{\partial \ln a_3}\right)_{T,P,m_2} = \frac{c_3\Delta J_{21}}{c_1+c_1J_{11}+c_3}, \quad (29)$$

where $J_{11} = \lim_{x_2 \to 0}(\partial \ln \gamma_1/\partial x_1)_{x_2}$, $J_{21} = \lim_{x_2 \to 0}(\partial \ln \gamma_2/\partial x_1)_{x_2}$, and γ_i is the activity coefficient of component i at a mol fraction scale. Let us note that J_{11} is characteristic of the protein-free mixed solvent at infinite protein dilution.

We are indebted to Prof. J. Michael Schurr (Dept. of Chemistry, University of Washington, Seattle, WA) for helpful comments regarding this manuscript and for drawing our attention to the fact that the coefficients $\Gamma_3(2)$ and $\Gamma_{23}^{(c)}$ are different.

REFERENCES

1. Casassa, E. F., and H. Eisenberg. 1964. Thermodynamic analysis of multi-component solutions. *Adv. Protein Chem.* 19:287–395.
2. Noelken, M. E., and S. N. Timasheff. 1967. Preferential solvation of bovine serum albumin in aqueous guanidine hydrochloride. *J. Biol. Chem.* 1967:5080–5085.
3. Kuntz, I. D., and W. Kauzmann. 1974. Hydration of proteins and polypeptides. *Adv. Protein Chem.* 28:239–345.
4. Timasheff, S. N. 1993. The control of protein stability and association by weak interactions with water: how do solvents affect these processes? *Annu. Rev. Biophys. Biomol. Struct.* 22:67–97.
5. Timasheff, S. N. 1998. Control of protein stability and reactions by weakly interacting cosolvents: the simplicity of the complicated. *Adv. Protein Chem.* 51:355–432.
6. Record, M. T., W. T. Zhang, and C. F. Anderson. 1998. Analysis of effects of salts and uncharged solutes on protein and nucleic acid

equilibria and processes: a practical guide to recognizing and interpreting polyelectrolyte effects, Hofmeister effects and osmotic effects of salts. *Adv. Protein Chem.* 51:281–353.

7. Zhang, W. T., M. W. Capp, J. P. Bond, C. F. Anderson, and M. T. Record, Jr. 1996. Thermodynamic characterization of interactions of native bovine serum albumin with highly excluded (glycine betaine) and moderately accumulated (urea) solutes by a novel application of vapor pressure osmometry. *Biochemistry.* 35:10506–10516.

8. Anderson, C. F., E. S. Courtenay, and M. T. Record, Jr. 2002. Thermodynamic expressions relating different types of preferential interaction coefficients in solutions containing two solute components. *J. Phys. Chem. B.* 106:418–433.

9. Kirkwood, J. G., and F. P. Buff. 1951. The statistical mechanical theory of solutions. *J. Chem. Phys.* 19:774–777.

10. Shimizu, S. 2004. Estimating hydration changes upon biomolecular reactions from osmotic stress, high pressure, and preferential interaction coefficients. *Proc. Natl. Acad. Sci. USA.* 101:1195–1199.

11. Shimizu, S. 2004. Estimation of excess solvation numbers of water and cosolvents from preferential interaction and volumetric experiments. *J. Chem. Phys.* 120:4989–4990.

12. Shimizu, S., and D. J. Smith. 2004. Preferential hydration and the exclusion of cosolvents from protein surfaces. *J. Chem. Phys.* 121:1148–1154.

13. Shimizu, S., and C. L. Boon. 2004. The Kirkwood-Buff theory and the effect of cosolvents on biochemical reactions. *J. Chem. Phys.* 121:9147–9155.

14. Smith, P. E. 2004. Local chemical potential equalization model for cosolvent effects on biomolecular equilibria. *J. Phys. Chem. B.* 108:16271–16278.

15. Smith, P. E. 2004. Cosolvent interactions with biomolecules: relating computer simulation data to experimental thermodynamic data. *J. Phys. Chem. B.* 108:18716–18724.

16. Shulgin, I. L., and E. Ruckenstein. 2005. A protein molecule in an aqueous mixed solvent: fluctuation theory outlook. *J. Chem. Phys.* 123:054909.

17. Schurr, J. M., D. P. Rangel, and S. R. Aragon. 2005. A contribution to the theory of preferential interaction coefficients. *Biophys. J.* 89:2258–2276.

18. Prausnitz, J. M., R. N. Lichtenthaler, and E. Gomes de Azevedo. 1986. Molecular Thermodynamics of Fluid-Phase Equilibria, 2nd Ed. Prentice-Hall, Englewood Cliffs, NJ.

19. Ruckenstein, E., and I. Shulgin. 2001. Entrainer effect in supercritical mixtures. *Fluid Phase Equilib.* 180:345–359.

20. Chitra, R., and P. E. Smith. 2002. Molecular association in solution: a Kirkwood-Buff analysis of sodium chloride, ammonium sulfate, guanidinium chloride, urea, and 2,2,2-trifluoroethanol in water. *J. Phys. Chem. B.* 106:1491–1500.

21. Vergara, A., L. Paduano, and R. Sartorio. 2002. Kirkwood-Buff integrals for polymer solvent mixtures. Preferential solvation and volumetric analysis in aqueous PEG solutions. *Phys. Chem. Chem. Phys.* 4:4716–4723.

22. Timasheff, S. N. 1992. Water as ligand: preferential binding and exclusion of denaturants in protein unfolding. *Biochemistry.* 31:9857–9864.

23. Timasheff, S. N., and G. Xie. 2003. Preferential interactions of urea with lysozyme and their linkage to protein denaturation. *Biophys. Chem.* 105:421–448.

Ivan L. Shulgin and Eli Ruckenstein

Department of Chemical & Biological Engineering
State University of New York at Buffalo
Amherst, New York

Preferential hydration and solubility of proteins in aqueous solutions of polyethylene glycol

Ivan L. Shulgin, Eli Ruckenstein *

Department of Chemical and Biological Engineering, State University of New York at Buffalo, Amherst, NY 14260, USA

Received 19 October 2005; received in revised form 29 November 2005; accepted 29 November 2005
Available online 22 December 2005

Abstract

This paper is focused on the local composition around a protein molecule in aqueous mixtures containing polyethylene glycol (PEG) and the solubility of proteins in water+PEG mixed solvents. Experimental data from literature regarding the preferential binding parameter were used to calculate the excesses (or deficits) of water and PEG in the vicinity of β-lactoglobulin, bovine serum albumin, lysozyme, chymotrypsinogen and ribonuclease A. It was concluded that the protein molecule is preferentially hydrated in all cases (for all proteins and PEGs investigated). The excesses of water and deficits of PEG in the vicinity of a protein molecule could be explained by a steric exclusion mechanism, i.e. the large difference in the sizes of water and PEG molecules.

The solubility of different proteins in water+PEG mixed solvent was expressed in terms of the preferential binding parameter. The slope of the logarithm of protein (lysozyme, β-lactoglobulin and bovine serum albumin) solubility versus the PEG concentration could be predicted on the basis of experimental data regarding the preferential binding parameter. For all the cases considered (various proteins, various PEGs molecular weights and various pHs), our theory predicted that PEG acts as a salting-out agent, conclusion in full agreement with experimental observations. The predicted slopes were compared with experimental values and while in some cases good agreement was found, in other cases the agreement was less satisfactory. Because the established equation is a rigorous thermodynamic one, the disagreement might occur because the experimental results used for the solubility and/or the preferential binding parameter do not correspond to thermodynamic equilibrium.
© 2005 Elsevier B.V. All rights reserved.

Keywords: Protein; Aqueous polyethylene glycol; Preferential binding parameter; Solubility; Fluctuation theory

1. Introduction

Polyethylene glycol (PEG), one of the most useful protein salting-out agents, is considered to be the most successful precipitant for protein crystallization [1]. Unlike ethanol and other organic precipitating agents, PEG has little tendency to denature or to specifically interact with the proteins even when present in high concentrations and at elevated temperatures [2]. One can also vary its molecular weight in order to select the best choice for the precipitation of a particular protein. McPherson considers that PEG may be the best reagent for crystallizing proteins and that the optimum PEG molecular weight for this purpose is between 2000 and 6000 [2]. The optimum PEG molecular weight was selected on the basis of the viscosity of the solution, the denaturation action of aqueous PEG and protein solubility. It should also be mentioned that PEG is neither corrosive nor toxic, is not inflammable and has a very low vapor pressure [1]. In addition, polyethylene glycol is available commercially in good quality at reasonable prices.

However, the mechanism of PEG induced precipitation is almost unknown. The interactions between the protein (component 2), PEG (3) and water (1) as well as the local properties of PEG+water mixed solvent near the protein surface are of particular interest from this point of view.

One of the most informative experimental quantity for the understanding of the above issues is the preferential binding parameter [3–9]. The preferential binding parameter can be expressed at different concentration scales:

1) in molal concentrations

$$\Gamma_{23}^{(m)} \equiv \lim_{m_2 \to 0} (\partial m_3 / \partial m_2)_{T,P,\mu_3} \tag{1}$$

---

* Corresponding author. Tel.: +1 716 645 2911x2214; fax: +1 716 645 3822.
E-mail addresses: ishulgin@eng.buffalo.edu (I.L. Shulgin), feaeliru@acsu.buffalo.edu (E. Ruckenstein).

where m_i is the molality of component i, P is the pressure, T the temperature (throughout this paper only isothermal–isobaric conditions are considered), and μ_i is the chemical potential of component i.

2) in molar concentrations

$$\Gamma_{23}^{(c)} \equiv \lim_{c_2 \to 0} (\partial c_3 / \partial c_2)_{T,P,\mu_3} \qquad (2)$$

where c_i is the molar concentration of component i. It should be noted that $\Gamma_{23}^{(m)}$ and $\Gamma_{23}^{(c)}$ are defined at the infinite dilution of the protein.

Several authors reported measurements of the preferential binding parameter in the system water (1)/protein (2)/PEG (3) [10–14]. It was found that for various proteins, various PEGs molecular weights, and various PEG concentrations, the protein is preferentially hydrated and the PEG is excluded from the vicinity of the protein molecule. The prevalent viewpoint which explains such a behavior is based on the steric exclusion mechanism suggested by Kauzmann and cited in Ref. [15]. According to this mechanism [12,14], the deficit of PEG and the excess of water (in comparison with the bulk concentrations) are located in the shell (volume of exclusion) between the protein surface and a sphere of radius R (see Fig. 1) [12,14]. However, Lee and Lee [10,11] suggested that the preferential exclusion of the PEG from the protein surface also involves the protein hydrophobicity and charge.

In our recent publication [16], it was shown how the experimental data regarding the preferential binding parameter can be used to calculate the excess (or deficit) of water and cosolvent in the vicinity of a protein molecule. The methodology, based on the Kirkwood–Buff theory of solutions [17], allowed one to compare the concentrations of water and cosolvent molecules in the vicinity of the protein molecule with the bulk solution concentrations (water+cosolvent mixture) in absence of the protein molecule and ultimately to draw a conclusion about the preferential hydration or preferential solvation. Furthermore, such data allowed one to analyze the ability of a cosolvent to stabilize a protein, because the preferential hydration of a protein in an aqueous solution containing an organic compound is related to the ability of the latter to stabilize the structure of the protein [18–20].

Another important use of the preferential binding parameter is its connection to the protein solubility. The authors of the present paper [21] showed that the preferential binding parameter is closely related to the solubility of a protein in a mixed solvent and that the experimental data regarding the preferential binding parameter can be used to predict how a cosolvent changes the solubility (salting-in or salting-out) or even to predict the solubility in a wide range of cosolvent concentrations.

Consequently, three important characteristics of the protein behavior in aqueous solutions (stability, preferential hydration (or solvation) and solubility) can be related to the preferential binding parameter. For instance, the addition of glycerol leads to an excess of water in the vicinity of the protein [16], i.e. the protein is preferentially hydrated; in addition, glycerol can be used to stabilize the native structure of the protein [18]. Glycerol also decreases the solubility of the protein, i.e. glycerol is a salting-out agent [21]. In contrast, the addition of urea leads to an excess of urea in the vicinity of the protein [22], i.e. the protein is preferentially solvated [18]. Urea increases the solubility of the protein, i.e. urea induces a salting-in effect [21]. In addition, it is well-known that urea can cause protein denaturation [4–6,22]. A similar analysis can be carried out for such an important cosolvent as the polyethylene glycol.

In the present paper, the Kirkwood–Buff theory of solutions will be used to examine dilute mixtures of various proteins in aqueous solutions containing PEGs of different molecular weights in terms of the preferential binding parameter. As already mentioned, extensive experimental data regarding the preferential binding parameter in the systems water/protein/PEG are available in the literature [10–14].

The purpose of the present analysis is two-fold: (i) to examine the local composition of a mixed solvent around different proteins in dilute solutions of proteins in water/PEG mixed solvents as a function of the PEG concentration and molecular weight and to use the obtained results to identify the mechanism of protein hydration in the presence of PEG, and (ii) to predict the solubility of proteins in water/PEG mixed solvents. Finally, the obtained results are compared with experiments available in the literature.

From a theoretical viewpoint, the present analysis can allow one to better understand the interactions between a protein and the constituents of a mixed solvent in the system water/protein/PEG and how these interactions differ from those between a protein and the constituents of a mixed solvent containing "regular" and not polymeric cosolvents. From a practical perspective, the results of this paper could allow one to better understand and improve the design of protein precipitation techniques with polyethylene glycol.

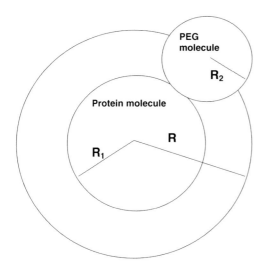

Fig. 1. The excess of water (in comparison with the bulk concentration) is located in the shell (volume of exclusion) between the protein surface and a sphere with an effective radius $R=R_1+R_2$, where R_1 is the radius of the protein molecule and R_2 is the radius of the PEG molecule (it is supposed that both the protein and the PEG molecules have spherical shapes). This figure is adapted from Refs. [12,14].

2. Theoretical part

2.1. Expressions for the preferential binding parameter via the Kirkwood–Buff integrals

It was demonstrated that the preferential binding parameters $\Gamma_{23}^{(m)}$ and $\Gamma_{23}^{(c)}$ can be expressed via the Kirkwood–Buff integrals as follows [23]:

$$\Gamma_{23}^{(m)} = \frac{c_3}{c_1} + c_3(G_{23}-G_{12}+G_{11}-G_{13}) \quad (3)$$

and [16]

$$\Gamma_{23}^{(c)} = c_3(G_{23}-G_{13}) \quad (4)$$

where $G_{\alpha\beta}$ are the Kirkwood–Buff integrals defined as [17]

$$G_{\alpha\beta} = \int_0^\infty (g_{\alpha\beta}-1)4\pi r^2 dr \quad (5)$$

$g_{\alpha\beta}$ is the radial distribution function between species α and β, and r is the distance between the centers of molecules α and β.

The Kirkwood–Buff integrals G_{11} and G_{13} can be calculated from the characteristics of the protein-free mixed solvents, whereas G_{12} and G_{23} depend on the properties of the infinitely dilute protein solutions [16].

One can also write the following expression for the partial molar volume of a protein at infinite dilution in a mixed solvent (V_2^∞) in terms of the Kirkwood–Buff theory of solution [24]

$$V_2^\infty = -c_1 V_1 G_{12} - c_3 V_3 G_{23} + kTk_T \cong -c_1 V_1 G_{12} - c_3 V_3 G_{23} \quad (6)$$

where V_i is the partial molar volume of component i, k is the Boltzmann constant, and k_T is the isothermal compressibility of the mixed solvent. The partial molar volumes V_1 and V_3 are those for the protein-free mixed solvent, and the contribution of kTk_T is usually negligible in comparison with $c_1 V_1 G_{12}$ and $c_3 V_3 G_{23}$.

Experimental data regarding $\Gamma_{23}^{(m)}$ and V_2^∞ are available in the literature for many water/protein/cosolvent systems [4–14,18–20,22]. The Kirkwood–Buff integrals G_{11} and G_{13} are for the binary mixture water/cosolvent and can be calculated as described in the literature [24–28]. The Kirkwood–Buff integrals G_{12} and G_{23} can be calculated from Eqs. (3) and (6) using experimental data for $\Gamma_{23}^{(m)}$, V_2^∞, V_1 and V_3.

2.2. Analytical expressions for the Kirkwood–Buff integrals G_{12}, G_{23}, G_{11} and G_{13} at infinite dilution of a protein

The analytical expressions for the Kirkwood–Buff integrals G_{12}, G_{23}, G_{11} and G_{13} at infinite dilution of a protein are helpful in analyzing the different factors that affect the preferential binding parameter and ultimately the preferential hydration or solvation. The following analytical expressions for the Kirkwood–Buff integrals G_{12} and G_{23} can be written [16]:

$$G_{12} = kTk_T - \frac{J_{21}V_3 c_3 + J_{11}V_2^\infty c_1}{(c_1 + c_1 J_{11} + c_3)} - \frac{V_3 c_3(c_1+c_3)(V_1-V_3) + V_2^\infty(c_1+c_3)}{(c_1 + c_1 J_{11} + c_3)} \quad (7)$$

and

$$G_{23} = kTk_T + \frac{J_{21}V_1 c_1 - J_{11}c_1 V_2^\infty}{(c_1 + c_1 J_{11} + c_3)} + \frac{c_1 V_1(c_1+c_3)(V_1-V_3) - V_2^\infty(c_1+c_3)}{(c_1 + c_1 J_{11} + c_3)} \quad (8)$$

where $J_{11} = \lim_{x_2 \to 0}\left(\frac{\partial \ln \gamma_1}{\partial x_1}\right)_{x_2}$, $J_{21} = \lim_{x_2 \to 0}\left(\frac{\partial \ln \gamma_2}{\partial x_1}\right)_{x_2}$, x_i is the mole fraction of component i, and γ_i is the activity coefficient of component i in a mole fraction scale. The expressions for the Kirkwood–Buff integrals G_{11} and G_{13} are also well-known from the literature (see for example [26,27])

$$G_{11} = kTk_T - \frac{(c_1+c_3)^2 V_1 V_3}{(c_1 + c_1 J_{11} + c_3)} + \frac{(c_1+c_3)(V_3-V_1)-J_{11}}{(c_1 + c_1 J_{11} + c_3)} \quad (9)$$

and

$$G_{13} = kTk_T - \frac{(c_1+c_3)^2 V_1 V_3}{(c_1 + c_1 J_{11} + c_3)}. \quad (10)$$

Expressions for G_{12}, G_{23}, G_{11} and G_{13} for an ideal ternary mixture at infinite dilution of a protein can be obtained from Eqs. (7)–(10) by taking into account that according to the definition of an ideal mixture [29], the activities of the components are equal to their mole fractions (x_i) and their partial molar volumes are equal to those of the pure components ($V_i = V_i^0$). These expressions are [16]

$$G_{12}^{(id)} = kTk_T^{(id)} - V_3^0 c_3(V_1^0 - V_3^0) - V_2^0 \quad (11)$$

$$G_{23}^{(id)} = kTk_T^{(id)} + V_1^0 c_1(V_1^0 - V_3^0) - V_2^0 \quad (12)$$

$$G_{11}^{(id)} = kTk_T^{(id)} - (c_1+c_3)V_1^0 V_3^0 + V_3^0 - V_1^0 \quad (13)$$

and

$$G_{13}^{(id)} = kTk_T^{(id)} - (c_1+c_3)V_1^0 V_3^0. \quad (14)$$

All the above equations will be later used in the calculations of the excesses and deficits of the constituents of a mixed solvent in the vicinity of a protein surface.

2.3. Excess and deficit numbers of molecules of water and cosolvent around a protein molecule

The conventional way to calculate the excess number of molecules i around a molecule j ($\Delta n'_{ij}$) is provided by the relation [25]:

$$\Delta n'_{ij} = c_i G_{ij}. \quad (15)$$

However, as noted by Matteoli and Lepori [30] and Matteoli [31], the above expression leads for an ideal binary mixture to non-zero values, even though they are expected to vanish. For the above reasons, Eq. (15) was replaced by [27,30,31]:

$$\Delta n_{ij} = c_i(G_{ij} - G_{ij}^R) \quad (16)$$

where G_{ij}^R is the Kirkwood–Buff integral of a reference state. Matteoli and Lepori [30] and Matteoli [31] suggested the ideal solution ($^{(id)}$) as the reference state because then Δn_{ij} becomes zero for an ideal solution, as intuition suggests that it should be. Shulgin and Ruckenstein [27] suggested a reference state in which all the activity coefficients are equal to unity but there are no constraints on the partial molar volumes of the components (*superscript* $^{(SR)}$). This reference state provides zero excesses for ideal mixtures for both binary and ternary mixtures [27,32]. It also satisfies the volume conservation condition, which for a binary mixture can be expressed as "the volume occupied by the excess i molecules around a j molecule must be equal to the volume left free by the j molecules around the same j molecule" [30,31]. Far from critical conditions, the above two reference states provide almost the same results [27,33]. In this paper, we consider $G_{ij}^R = G_{ij}^{(SR)}$.

For the ($^{(SR)}$) reference state, Eqs. (7) and (8) for G_{12} and G_{23} for an infinitely dilute protein in a mixed solvent can be recast as follows [16]:

$$G_{12}^{(SR)} = kTk_T^{(SR)} - V_3 c_3(V_1 - V_3) - V_2^\infty \quad (17)$$

and

$$G_{23}^{(SR)} = kTk_T^{(SR)} - V_1 c_1(V_1 - V_3) - V_2^\infty. \quad (18)$$

G_{12} and G_{23} will be first calculated by combining Eqs. (3) and (6) with experimental data regarding the preferential binding parameter $\Gamma_{23}^{(m)}$ and the partial molar volumes V_2^∞, V_1 and V_3. Furthermore the excess (or deficit) number of molecules of water and PEG around a protein molecule will be calculated using Eq. (16).

2.4. Relationship between preferential interaction of a protein in an aqueous PEG solution and protein solubility

The solubility of a protein in a water+cosolvent mixture depends on many factors such as temperature, cosolvent concentration, pH, type of buffer used, etc. The solubilities of proteins in aqueous PEGs solutions have been investigated both experimentally [34–42] and theoretically [38,41,43–48].

The experimental results showed that: (i) the addition of PEG decreases the protein solubility; (ii) the low molecular weight PEG is a less effective precipitant than the high molecular weight PEG, and (iii) the log of protein solubility versus PEG concentration is linear. The authors of the present paper [21] recently derived the following relation for the solubility of a protein in a mixed solvent as a function of the cosolvent mole fraction

$$\left(\frac{\partial \ln y_2}{\partial x_3}\right) = -\frac{c_3(c_1 + c_3)V_1 - \Gamma_{23}^{(m)}(1 - c_3 V_3)(c_1 + c_1 J_{11} + c_3)}{c_1 c_3 V_1} \quad (19)$$

where y_2 is the protein solubility in mole fraction.

Eq. (19) is a rigorous thermodynamic equation at infinite protein dilution. It allows one to derive a simple criteria for salting-in or salting-out for low cosolvent concentrations. At low cosolvent concentrations ($c_3 \to 0$)

$$\left(\frac{\partial \ln y_2}{\partial x_3}\right) = -\left(\frac{\partial \ln y_2}{\partial x_1}\right) = \frac{\alpha}{V_1^0} - 1 \quad (20)$$

where $\alpha = \lim_{c_3 \to 0} \frac{\Gamma_{23}^{(m)}}{c_3}$ and V_1^0 is the molar volume of pure water. Consequently, one can conclude that for low cosolvent concentrations salting-in occurs when

$$\left(\frac{\partial \ln y_2}{\partial x_3}\right) > 0, \quad \text{hence when } \alpha > V_1^0 \quad (21)$$

and salting-out occurs when

$$\left(\frac{\partial \ln y_2}{\partial x_3}\right) < 0, \quad \text{hence when } \alpha < V_1^0. \quad (22)$$

It is well-known [8,49,50] that the preferential binding parameter $\Gamma_{23}^{(m)}$ is, at least at low cosolvent concentrations, proportional to the cosolvent concentration. Consequently salting-in or salting-out depends on the slope of the curve $\Gamma_{23}^{(m)}$ versus concentration for small values of c_3.

The above criteria for salting-in or salting-out (Eqs. (21), (22)) are valid [21]: (i) for $c_3 \to 0$, hence when a small amount of cosolvent is added to pure water; (ii) for ternary mixtures (water (1)–protein (2)–cosolvent (3)) (it should be emphasized that the experimental results regarding the preferential binding parameter $\Gamma_{23}^{(m)}$ and the solubilities are usually for mixtures which involve in addition a buffer, and the effect of the buffer is taken into account only indirectly via the preferential binding parameter $\Gamma_{23}^{(m)}$); (iii) for infinite dilution (this means that the protein solubility is supposed to be small enough to satisfy the infinite dilution approximation ($\gamma_2 \cong \gamma_2^\infty$, where γ_2^∞ is the activity coefficient of a protein at infinite dilution)); (iiii) for experimental preferential binding parameters $\Gamma_{23}^{(m)}$ and solubilities determined at low cosolvent concentrations.

Physically speaking, Eq. (20) provides the slope of a curve representing the dependence of log mole fraction of the solubility versus the cosolvent mole fraction.

The following expression for protein solubility in a dilute cosolvent solution can be derived from Eq. (19), when $\Gamma_{23}^{(m)}$ is proportional to c_3 [21]:

$$\ln \frac{y_2}{y_2^w} = -\frac{(\alpha - V_3 \Gamma_{23}^{(m)}) \ln a_w}{V_1} + \ln x_1 \approx -\frac{(\alpha - V_3 \Gamma_{23}^{(m)}) \ln a_w}{V_1}$$
$$= -\frac{(1 - V_3 c_3)\alpha \ln a_w}{V_1} = -c_1 \alpha \ln a_w \quad (23)$$

where y_2^w is the protein solubility in the cosolvent-free water plus buffer and a_w is the water activity in the protein-free mixed solvent. Eq. (23) allows one to calculate the protein solubility if the composition dependence of $\Gamma_{23}^{(m)}$ is available (a_w and the partial molar volumes V_1 and V_3 are characteristics of the protein-free mixed solvent).

3. Numerical estimations for various water/protein/PEG systems

3.1. Kirkwood–Buff integrals and the excess (or deficit) number of molecules of water and PEG around a protein molecule

The Kirkwood–Buff integrals and the excess (or deficit) number of molecules of water and PEG around a protein molecule were calculated for β-lactoglobulin (β-LG), bovine serum albumin (BSA), lysozyme, chymotrypsinogen and ribonuclease A (RNase A). Experimental data regarding $\Gamma_{23}^{(m)}$ and V_2^∞ for these systems are available in the literature [11,12,14]. The partial molar volumes V_1 and V_3 in the aqueous solutions of PEGs were calculated using the experimental data and the correlations suggested in Ref. [51].

The Kirkwood–Buff integrals G_{12} and G_{23} were calculated from Eqs. (3) and (6) using experimental data for $\Gamma_{23}^{(m)}$, V_2^∞, V_1 and V_3. The difference between G_{11} and G_{13} in the right hand side of Eq. (3) was calculated using the expression (see Eqs. (9) and (10)):

$$G_{11}-G_{13}=\frac{(c_1+c_3)(V_3-V_1)-J_{11}}{(c_1+c_1J_{11}+c_3)}\approx V_3-V_1. \quad (24)$$

The above approximation simplifies the calculations without affecting much the accuracy (see Appendix).

The calculated Kirkwood–Buff integrals G_{12} and G_{23} can be found in the Supplementary Material.

The excess (or deficit) numbers of molecules of water and PEG (Δn_{12} and Δn_{23}) around a protein molecule were calculated using Eq. (16) and are listed in Table 1. The results demonstrate that in all cases there is preferential exclusion of PEG from the surface of the protein, conclusion in agreement with previous observations [10–14]. There is only one exception (water/lysozyme/PEG 200, at a concentration of 10 g PEG/100 ml solution). However, this is probably caused by the inaccuracy in the experimental value of $\Gamma_{23}^{(m)}$ (according to the authors of Ref. [12], for this composition, $\Gamma_{23}^{(m)}=0.66\pm1.32$ [mol/mol]).

In order to better understand why PEG is preferentially excluded from the vicinity of a protein molecule, we calculated the excess (or deficit) number of molecules of water and PEG around a protein molecule as a function of the PEG molecular weight at a constant PEG weight concentration. The results of the calculations are presented in Fig. 2 (only the Δn_{12}s are plotted; Δn_{23} can be calculated using the balance relation $V_3\Delta n_{23}=-V_1\Delta n_{12}$). One can see from Fig. 2 that the excess number of molecules of water around a protein increases monotonically with increasing PEG molecular weight (and, respectively, molar volume) in agreement with the above mentioned steric exclusion mechanism.

Table 1
The excess (or deficit) number of molecules of water and PEG around a protein molecule as a function of cosolvent concentration

System	g of PEG/100 ml of solution	Δn_{12} [mol/mol]	Δn_{23} [mol/mol]	ΔV [l/mol protein][A]
Water/lysozyme/PEG 400 (pH=3.0) [11][B]	2.8	109.9	−5.9	2.0
	5.6	160.6	−8.6	2.9
	11.2	186.1	−10.0	3.4
	22.4	211.6	−11.4	3.8
	33.6	240.8	−12.9	4.4
	44.8	211.1	−11.1	3.8
Water/lysozyme/PEG 1000 (pH=3.0) [11]	2.5	45.6	−1.0	0.8
	5	44.6	−1.0	0.8
	10	85.1	−1.8	1.5
	20	115.8	−2.5	2.1
	30	138.8	−3.9	2.5
Water/lysozyme/PEG 4000 (pH=3.0) [11]	0.5	45.5	−0.2	0.8
	1.25	135.6	−0.7	2.5
	2.5	166.1	−0.9	3.0
	3.75	170.5	−0.9	3.1
Water/β-LG/PEG 200 (pH=2.0) [12]	10	−5.0	0.5	−0.1
	20	18.1	−2.0	0.3
	30	18.8	−2.0	0.3
	40	14.2	−1.5	0.3
Water/1.2 β-LG/PEG 400 (pH=2.0) [12]	30	66.0	−3.5	1.2
	40	57.2	−3.0	1.0
Water/β-LG/PEG 600 (pH=2.0) [12]	10	53.7	−1.9	1.0
	20	79.4	−2.9	1.4
	30	59.6	−2.1	1.1
Water/β-LG/PEG 1000 (pH=2.0) [12]	10	104.3	−2.2	1.9
	20	109.6	−2.4	2.0
	30	105.5	−2.3	1.9

[A] ΔV is the volume occupied by the excess of water (or by the deficit of PEG) molecules around a protein molecule, [B] the source of experimental data regarding $\Gamma_{23}^{(m)}$ and V_2^∞ used in calculations.

In order to provide additional insight on the PEG exclusion, the dependence of the excess number of molecules of water in the vicinity of a protein molecule is plotted against the volume of exclusion (V_S), volume inaccessible to the PEG molecules in the vicinity of a protein molecule (see Fig. 1). The volume of exclusion (cm³/mol protein) was calculated using the expression [12,14]:

$$V_S = (4\pi N_A/3)[(R_1+R_2)^3-R_1^3]\cdot 10^{-24} \quad (25)$$

where N_A is Avogadro's number.

The radii of various protein molecules were taken from Ref. [14] where they were estimated from the partial specific volumes of the proteins and protein molecular weights. The radii of the PEG molecules were taken equal to their radii of gyration [14]. The dependence of the excess number of molecules of water in the vicinity of a protein molecule against the volume of exclusion is presented in Fig. 3. It shows that in all cases the excess number of molecules of water in the vicinity of a protein molecule is almost proportional to the exclusion volume. This proportionality constitutes an argument in the favor of the steric exclusion mechanism. However, one should note that the excess water molecules are assumed to be located in the volume of exclusion and it is supposed that both the protein and the PEG molecules have spherical shapes.

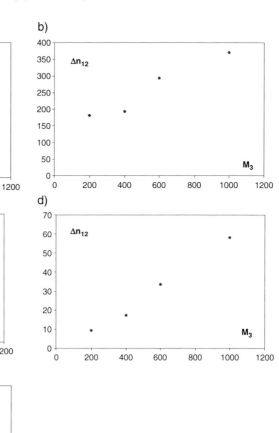

Fig. 2. Dependence of Δn_{12} [mol/mol] on molecular weight of PEG (M_3) for various proteins: a) BSA (pH=3.0), b) BSA (pH=7.0), c) chymotrypsinogen (pH=3.0), d) lysozyme (pH=7.0), e) RNase A (pH=2.0). Experimental data regarding $\Gamma_{23}^{(m)}$ and V_2^∞ for these systems were taken from Ref. [14].

A steric exclusion mechanism implies that a geometric factor and not an energetic one, such as the differences in the intermolecular interactions between the constituents of the water+protein+PEG mixtures is responsible for the local composition around a protein molecule. This constitutes the main difference between the preferential binding in water+protein+PEG mixtures and water+protein+low molecular weight cosolvents (such as urea, glycerol, alcohol, etc.) mixtures.

3.2. Solubility of different proteins in water+PEG mixed solvents

Generally the experimental data regarding the solubility of a protein in a water+PEG mixture are presented as a linear dependence of the logarithm of the solubility versus the PEG concentration. Eq. (23) can be used for solubility calculations when the values of y_2^w are available; unfortunately, such data could not be found in the literature. Eq. (20) can be, however, used to predict the slope of the solubility curve and this prediction can be compared with experiment. The calculated and measured results are listed in Table 2. Table 2 reveals that for various proteins and various PEGs molecular weights, Eq. (20) predicts a negative slope $\left(\frac{\partial \ln y_2}{\partial x_3}\right)$ at $c_3=0$, and hence a salting-out effect of PEG on protein solubility. Such a conclusion is in agreement with most experimental data regarding the protein solubility in aqueous PEG mixtures [34–42]. However, there are a few investigations [52,53] in which a salting-in effect of PEG on protein solubility was found. One can also see from Table 2 that there is no complete agreement between the experimental solubilities of various proteins in aqueous PEG mixtures obtained in different laboratories. For example, the value of $\left(\frac{\partial \ln y_2}{\partial x_3}\right) = -0.08$ at $c_3=0$ for the solubility of BSA in water+PEG 3350 at pH=4.6 [41] is very different from $\left(\frac{\partial \ln y_2}{\partial x_3}\right) = -0.23$ at $c_3=0$ for the solubility of HSA in water+PEG 4000 at pH=4.5 [38]. These two cases are, however, very similar and one should expect the slopes to be comparable.

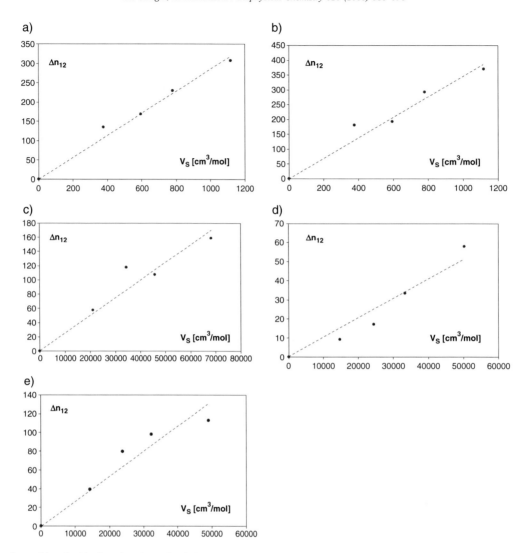

Fig. 3. Dependence of Δn_{12} [mol/mol] on the volume of exclusion (V_S) for various proteins: a) BSA (pH=3.0), b) BSA (pH=7.0), c) chymotrypsinogen (pH=3.0), d) lysozyme (pH=7.0), e) RNase A (pH=2.0). The dashed lines are shown for comparison. Experimental data regarding $\Gamma_{23}^{(m)}$ and V_2^∞ for these systems were taken from Ref. [14].

It should also be mentioned that there are large discrepancies between the experimental values for the preferential binding parameter $\Gamma_{23}^{(m)}$ obtained by different authors. For example, $\Gamma_{23}^{(m)} = -2.0$ [mol/mol] for BSA in water/PEG 1000 at 10% PEG (w/v) and pH=3 [11], and for the same concentration $\Gamma_{23}^{(m)} = -7.21$ [mol/mol] (at pH=2) [14]. Another example, $\Gamma_{23}^{(m)} = -2.45$ [mol/mol] for β-LG in water/PEG 1000 at 10% PEG (w/v) and pH=2 [12], and for the same concentration $\Gamma_{23}^{(m)} = -0.18$ [mol/mol] (at pH=3) [11]. Such large differences led to the scattering in predicted solubilities.

Our Eqs. (19) and (23) are rigorous thermodynamic relations which provide a relation between the protein solubility and the preferential binding parameter $\Gamma_{23}^{(m)}$. These thermodynamic equations provide a consistency test between the protein solubility $\left(\frac{\partial \ln y_2}{\partial x_3}\right)$ and the preferential binding parameter $\Gamma_{23}^{(m)}$. If either the protein solubility and/or the preferential binding parameter do not correspond to thermodynamic equilibrium, then Eqs. (19) and (23) cannot be satisfied. We employed all available experimental data regarding the preferential binding parameter for the systems water (1)+protein (2)+PEG (3) to calculate the slope of the protein solubility $\left(\frac{\partial \ln y_2}{\partial x_3}\right)$ at $c_3=0$ and compared the calculated values with the experimental ones. Table 2 shows that there are cases in which there is a reasonable agreement between the experimental and the predicted values. This occurs, for instance, for the solubility of BSA (HSA) in water+high (larger or equal to 4000) molecular weight PEG. There are, however, also cases in which there are differences as large as a factor of six. As already emphasized, in the latter cases either the solubility and/or the preferential binding parameter determined experimentally do not correspond to thermodynamic equilibrium.

Table 2
Comparison between the experimental slopes of the solubility versus PEG mole fraction curve $\left(\frac{\partial \ln y_2}{\partial x_3}\right)$ with the results predicted by Eq. (20)

Experiment			Prediction		
Protein+PEG	pH, reference[A]	Slope	Protein+PEG	pH, reference[B]	Slope
Lysozyme+PEG 4000	pH=7.0, [38]	−0.02	Lysozyme+PEG 200	pH=7.0, [14]	−0.04
			Lysozyme+PEG 400	pH=7.0, [14]	−0.04
			Lysozyme+PEG 400	pH=3.0, [11]	−0.16
			Lysozyme+PEG 600	pH=7.0, [14]	−0.04
			Lysozyme+PEG 1000	pH=7.0, [14]	−0.05
			Lysozyme+PEG 1000	pH=3.0, [11]	−0.01
			Lysozyme+PEG 2000	pH=7.0, [14]	−0.07
			Lysozyme+PEG 3000	pH=7.0, [14]	−0.05
			Lysozyme+PEG 4000	pH=7.0, [14]	−0.11
			Lysozyme+PEG 4000	pH=3.0, [11]	−0.13
			Lysozyme+PEG 6000	pH=7.0, [14]	−0.09
β-LG+PEG 20,000	pH=5.0, [36]	−0.04 and −0.05[C]	β-LG+PEG 200	pH=2.0, [12]	−0.02
			β-LG+PEG 400	pH=2.0, [12]	−0.04
			β-LG+PEG 600	pH=2.0, [12]	−0.04
			β-LG+PEG 1000	pH=2.0, [12]	−0.04
			β-LG+PEG 1000	pH=3.0, [11]	−0.01
			β-LG+PEG 2000	pH=2.0, [14]	−0.21
			β-LG+PEG 3000	pH=2.0, [14]	−0.18
			β-LG+PEG 4000	pH=2.0, [14]	−0.17
			β-LG+PEG 6000	pH=2.0, [14]	−0.13
HSA[D]+PEG 400	pH=4.5, [38]	−0.09	BSA+PEG 200	pH=7.0, [14]	−0.84
HSA+PEG 600	pH=4.5, [38]	−0.11			
HSA+PEG 1000	pH=4.5, [38]	−0.14			
BSA+PEG 1450	pH=4.6, [41]	−0.09 and −0.10	BSA+PEG 400	pH=7.0, [14]	−0.45
BSA+PEG 1450	pH=7.0, [41]	−0.09 and −0.10			
BSA+PEG 3350	pH=4.6, [41]	−0.08 and −0.08			
BSA+PEG 3350	pH=7.0, [41]	−0.11 and −0.10	BSA+PEG 600	pH=7.0, [14]	−0.46
BSA+PEG 3350	pH=8.0, [41]	−0.02 and −0.03			
HSA+PEG 4000	pH=3.8, [38]	−0.16	BSA+PEG 1000	pH=7.0, [14]	−0.29
HSA+PEG 4000	pH=4.5, [38]	−0.23			
HSA+PEG 4000	pH=5.2, [38]	−0.21			
HSA+PEG 4000	pH=4.5, [38]	−0.15	BSA+PEG 1000	pH=3.0, [11]	−0.08
HSA+PEG 6000	pH=4.5, [38]	−0.27			
BSA+PEG 6000	pH=4.0, [34]	−0.29			
BSA+PEG 6000	pH=5.1, [34]	−0.27	BSA+PEG 2000	pH=7.0, [14]	−0.33
BSA+PEG 6000	pH=5.8, [34]	−0.26			
BSA+PEG 8000	pH=4.6, [41]	−0.13 and −0.12	BSA+PEG 3000	pH=7.0, [14]	−0.20
BSA+PEG 8000	pH=7.0, [41]	−0.20 and −0.12			
BSA+PEG 8000	pH=8.0, [41]	−0.07 and −0.07			
BSA+PEG 10,000	pH=4.6, [41]	−0.13 and −0.12	BSA+PEG 4000	pH=7.0, [14]	−0.21
BSA+PEG 10,000	pH=7.0, [41]	−0.08 and −0.20			
BSA+PEG 10,000	pH=8.0, [41]	−0.04 and −0.06			
HSA+PEG 20,000	pH=4.5, [38]	−0.27	BSA+PEG 6000	pH=7.0, [14]	−0.19
BSA+PEG 20,000	pH=5.0, [34]	−0.22 and −0.24			

[A]References regarding the experimental values of solubility data, [B]references regarding the experimental values of the preferential binding parameter, [C]two experimental data sets are available and [D]HSA designates human serum albumin.

According to experimental observations [34–41], the logarithm of the protein solubility versus PEG concentration exhibits linearity over a wide range of PEG concentrations. Eq. (23) allows one to examine this issue. For the sake of simplicity only the dilute region ($m_3 < 0.5$) was considered. It is worth noting that most of the experimental measurements of the preferential binding parameter $\Gamma_{23}^{(m)}$ and protein solubility for the system water+protein+PEG were carried out in this composition range [11–14,34–42]. In this composition range the preferential binding parameter $\Gamma_{23}^{(m)}$ is proportional to the concentration of the cosolvent [8,49,50] and Eq. (23) becomes a rigorous one. Eq. (23) reveals that the linearity (or nonlinearity) of $\ln y_2$ versus the PEG concentration depends entirely on the characteristics of the protein-free mixed solvent water/PEG. The dependence of the product $c_1 \ln a_w$ on PEG concentration is plotted in Fig. 4 for various PEG molecular weights. It shows an almost linear behavior of the logarithm of the protein solubility versus PEG concentration for PEG 1000 and PEG 4000. However, the logarithm of protein solubility versus the PEG concentration for PEG 6000 is nonlinear and one can expect the same nonlinear behavior to occur for higher molecular weight PEGs.

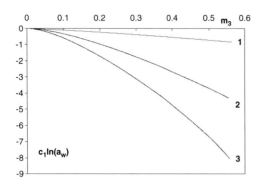

Fig. 4. Dependence of the product $c_1 \ln a_w$ in protein-free mixed solvent water/PEG on the PEG molality: 1) PEG 1000, 2) PEG 4000, 3) PEG 6000. Water activities in water/PEG mixtures were calculated as indicated in Ref. [54].

Another important issue is the salting-out strength of various molecular weight PEGs. It is well-known from literature that the low molecular weight PEGs are less effective precipitants than the high molecular weight PEG [11–14,34–42]. Eq. (20) provides the same result, because the salting-out effectiveness is proportional to the slope $\left(\frac{\partial \ln \gamma_2}{\partial x_3}\right)$ at $c_3 = 0$ and hence to $\alpha = \lim_{c_3 \to 0} \frac{\Gamma_{23}^{(m)}}{c_3}$ which, according to the $\Gamma_{23}^{(m)}$ experimental data [11–14], increases with increasing molecular weight of PEG [11–14].

4. Conclusion

In this paper, the Kirkwood–Buff theory of solutions is used to examine the effect of PEG on aqueous protein solutions, the focus being on the local composition of the mixed solvent in the vicinity of the protein molecule and on the protein solubility. The theoretical considerations led to equations that connect the experimental preferential binding parameter with the excess (or deficit) numbers of water and cosolvent molecules around a protein molecule. Calculations were carried out for various proteins in various PEG solutions. The results showed that in all cases the proteins were preferentially hydrated. Evidence was also brought that the hydration is a result of steric exclusion.

In addition, the solubility of a protein in water+PEG mixed solvent was examined. For this purpose, a previously [21] derived relationship between the preferential binding parameter and the solubility of a protein in a binary aqueous solution was used to predict the slope of the logarithm of the protein solubility versus the PEG concentration in terms of the experimental preferential binding parameter. Slopes were predicted for the solubilities of lysozyme, β-lactoglobulin and bovine serum albumin in water+PEG mixtures for various pHs and various PEG molecular weights and compared with experiment. For all considered cases (various proteins, various PEG molecular weights and various pHs), the theory predicts that the PEG acts as a salting-out agent, conclusion in agreement with experimental observations. Numerical comparison between the predicted and the experimental slopes showed good agreement in some cases (the solubility of BSA (HSA) in water+high (larger or equal to 4000) molecular weight PEG). In other cases, such as the case of the solubility of BSA in water+low molecular weight PEG the agreement was not satisfactory probably because the experimental data regarding the preferential binding parameter and/or the solubility do not correspond to the thermodynamic equilibrium. The equations were also used to shed some light on the linearity of the logarithm of protein solubility versus PEG concentration and on the salting-out effectiveness of PEG of various molecular weights.

It is noteworthy to point out that the preferential binding parameter provides an interconnection between the local and bulk properties in water+protein+cosolvent mixtures. Indeed, when the preferential binding parameter $\Gamma_{23}^{(m)}$ is negative, a protein is preferentially hydrated (water is in excess), the protein is additionally stabilized and its solubility is decreased by the cosolvent. It seems that there is no exception to this rule.

Appendix A

The purpose of this Appendix is to compare the Kirkwood–Buff integrals G_{12} and G_{23} calculated by combining Eqs. (3) and (6) with two expressions for $(G_{11} - G_{13})$:

1) a rigorous expression

$$G_{11} - G_{13} = \frac{(c_1 + c_3)(V_3 - V_1) - J_{11}}{(c_1 + c_1 J_{11} + c_3)} \quad (A-1)$$

2) a simplified expression in which $J_{11} = 0$

$$G_{11} - G_{13} = V_3 - V_1 \quad (A-2)$$

The results of the calculations of G_{12} and G_{23} for the system water (1)/β-lactoglobulin (2)/PEG 1000 (3) are listed in Table 3.

The partial molar volumes V_1 and V_3 for water/PEG 1000 were calculated using the experimental data and correlations provided in Ref. [51] and J_{11} was calculated using the concentration dependence of the water activity in water/PEG 1000 mixture [54].

Table 3 shows that the simplified expression for $(G_{11} - G_{13})$ can be used without essential change in accuracy.

Table 3
G_{12} and G_{23} for the system water (1)/β-lactoglobulin (2)/PEG 1000 (3)

g of PEG/100 ml of solution	G_{12} [cm³/mol]		G_{23} [cm³/mol]	
	Calculated using expression (A-1)	Calculated using expression (A-2)	Calculated using expression (A-1)	Calculated using expression (A-2)
10	−11,750	−11,770	−37,060	−36,840
20	−11,360	−11,430	−26,390	−26,050
30	−11,120	−11,250	−22,080	−21,690

Appendix B. Supplementary Materials

The calculated Kirkwood-Buff integrals G_{12} and G_{23} as a function of cosolvent concentration.

System	g of PEG / 100 ml of solution	G_{12} [l/mol]	G_{23} [l/mol]
Water / lysozyme / PEG 400 (pH=3.0) [11] [A]	2.8	-7.9	-94.2
	5.6	-7.2	-71.7
	11.2	-6.6	-46.2
	22.4	-5.6	-31.0
	33.6	-4.3	-26.1
	44.8	-4.3	-20.7
Water / lysozyme / PEG 1000 (pH=3.0) [11]	2.5	-9.1	-50.1
	5	-9.2	-30.0
	10	-8.5	-29.3
	20	-7.6	-23.4
	30	-6.8	-20.9
Water / lysozyme / PEG 4000 (pH=3.0) [11]	0.5	-9.2	-205.8
	1.25	-7.7	-246.6
	2.5	-7.2	-156.6
	3.75	-7.1	-111.2
Water / β-LG / PEG 200 (pH=2.0) [12]	10	-14.0	-12.8
	20	-13.5	-15.8
	30	-13.5	-15.3
	40	-13.5	-14.6
Water / β-LG / PEG 400 (pH=2.0) [12]	30	-12.2	-18.9
	40	-12.2	-17.1
Water / β-LG / PEG 600 (pH=2.0) [12]	10	-12.8	-25.9
	20	-12.1	-22.9
	30	-12.4	-18.6
Water / β-LG / PEG 1000 (pH=2.0) [12]	10	-11.7	-37.1
	20	-11.4	-26.4
	30	-11.1	-22.1

[A] the source of experimental data regarding $\Gamma_{23}^{(m)}$ and V_2^∞ used in calculations.

References

[1] A. McPherson, Crystallization of Biological Macromolecules, Cold Spring Harbor Laboratory Press, Cold Spring Harbor, NY, 1999.
[2] K.C. Ingham, Protein precipitation with polyethylene-glycol, Methods Enzymol. 104 (1984) 351–356.
[3] E.F. Casassa, H. Eisenberg, Thermodynamic analysis of multi-component solutions, Adv. Protein Chem. 19 (1964) 287–395.
[4] I.D. Kuntz, W. Kauzmann, Hydration of proteins and polypeptides, Adv. Protein Chem. 28 (1974) 239–343.
[5] S.N. Timasheff, The control of protein stability and association by weak interactions with water: how do solvents affect these processes? Annu. Rev. Biophys. Biomol. Struct. 22 (1993) 67–97.
[6] S.N. Timasheff, A physicochemical basis for the selection of osmolytes by nature, in: G.N. Somero, C.B. Osmond, C.L. Bolis (Eds.), Water and Life: Comparative Analysis of Water Relationships at the Organismic, Cellular, and Molecular Levels, Springer-Verlag, Berlin, 1992, pp. 70–84.
[7] S.N. Timasheff, Control of protein stability and reactions by weakly interacting cosolvents: the simplicity of the complicated, Adv. Protein Chem. 51 (1998) 355–432.
[8] M.T. Record Jr., T. Zhang, C.F. Anderson, Analysis of effects of salts and uncharged solutes on protein and nucleic acid equilibria and processes: a practical guide to recognizing and interpreting polyelectrolyte effects, Hofmeister effects and osmotic effects of salts, Adv. Protein Chem. 51 (1998) 281–353.
[9] H. Eisenberg, Biological Macromolecules and Polyelectrolytes in Solution, Clarendon Press, Oxford, 1976.
[10] J.C. Lee, L.L.Y. Lee, Interaction of calf brain tubulin with poly(ethyleneglycols), Biochemistry 18 (1979) 5518–5526.
[11] J.C. Lee, L.L.Y. Lee, Preferential solvent interactions between proteins and polyethylene glycol, J. Biol. Chem. 256 (1981) 625–631.
[12] T. Arakawa, S.N. Timasheff, Mechanism of poly(ethylene glycol) interaction with proteins, Biochemistry 24 (1985) 6756–6762.
[13] L.L.Y. Lee, J.C. Lee, Thermal stability of proteins in the presence of poly(ethylene glycols), Biochemistry 26 (1987) 7813–7819.
[14] R. Bhat, S.N. Timasheff, Steric exclusion is the principal source of the preferential hydration of proteins in the presence of polyethylene glycols, Protein Sci. 1 (1992) 1133–1143.
[15] H.K. Schachman, M.A. Lauffer, The hydration, size and shape of tobacco mosaic virus, J. Am. Chem. Soc. 71 (1949) 536–541.
[16] I.L. Shulgin, E. Ruckenstein, A protein molecule in an aqueous mixed solvent: fluctuation theory outlook, J. Chem. Phys. 123 (2005) 054909.
[17] J.G. Kirkwood, P. Buff, The statistical mechanical theory of solutions, J. Chem. Phys. 19 (1951) 774–777.
[18] K. Gekko, S.N. Timasheff, Thermodynamic and kinetic examination of protein stabilisation by glycerol, Biochemistry 20 (1981) 4677–4686.
[19] T. Arakawa, S.N. Timasheff, Stabilization of protein-structure by sugars, Biochemistry 21 (1982) 6536–6544.
[20] T. Arakawa, S.N. Timasheff, Preferential interactions of proteins with salts in concentrated solutions, Biochemistry 21 (1982) 6545–6552.
[21] I.L. Shulgin, E. Ruckenstein, Relationship between preferential interaction of a protein in an aqueous mixed solvent and its solubility, Biophys. Chem. 118 (2005) 128–134.
[22] S.N. Timasheff, G. Xie, Preferential interactions of urea with lysozyme and their linkage to protein denaturation, Biophys. Chem. 105 (2003) 421–448.
[23] I.L. Shulgin, E. Ruckenstwein, A protein molecule in a mixed solvent: the preferential binding parameter via the Kirkwood – Buff theory, Biophys. J. 90 (2006) 704–707.
[24] A. Ben-Naim, Statistical Thermodynamics for Chemists and Biochemists, Plenum, New York, 1992.
[25] A. Ben-Naim, Inversion of the Kirkwood–Buff theory of solutions: application to the water–ethanol system, J. Chem. Phys. 67 (1977) 4884–4890.
[26] E. Matteoli, L. Lepori, Solute–solute interactions in water: II. An analysis through the Kirkwood–Buff integrals for 14 organic solutes, J. Chem. Phys. 80 (1984) 2856–2863.
[27] I. Shulgin, E. Ruckenstein, Kirkwood–Buff integrals in aqueous alcohol systems: comparison between thermodynamic calculations and X-ray scattering experiments, J. Phys. Chem., B 103 (1999) 2496–2503.
[28] A. Vergara, L. Paduano, R. Sartorio, Kirkwood–Buff integrals for polymer solvent mixtures. Preferential solvation and volumetric analysis in aqueous PEG solutions, Phys. Chem. Chem. Phys. 4 (2002) 4716–4723.
[29] J.M. Prausnitz, R.N. Lichtenthaler, E. Gomes de Azevedo, Molecular Thermodynamics of Fluid-Phase Equilibria, 2nd ed., Prentice-Hall, Englewood Cliffs, NJ, 1986.
[30] E. Matteoli, L. Lepori, Kirkwood–Buff integrals and preferential solvation in ternary nonelectrolyte mixtures, J. Chem. Soc., Faraday Trans. 91 (1995) 431–436.
[31] E. Matteoli, A study on Kirkwood–Buff integrals and preferential solvation in mixtures with small deviations from ideality and/or with size mismatch of components. Importance of a proper reference system, J. Phys. Chem., B 101 (1997) 9800–9810.
[32] E. Ruckenstein, I. Shulgin, Effect of a third component on the interactions in a binary mixture determined from the fluctuation theory of solutions, Fluid Phase Equilib. 180 (2001) 281–297.
[33] R. Chitra, P.E. Smith, Molecular association in solution: a Kirkwood–Buff analysis of sodium chloride, ammonium sulfate, guanidinium chloride, urea, and 2,2,2-trifluoroethanol in water, J. Phys. Chem. B 106 (2002) 1491–1500.
[34] I.R.M. Juckes, Fractionation of proteins and viruses with polyethylene glycol, Biochim. Biophys. Acta 229 (1971) 535–546.
[35] P. Foster, P. Dunnill, M.D. Lilly, The precipitation of enzymes from cell extracts of Saccharomyces cerevisiae by polyethylene glycol, Biochim. Biophys. Acta 317 (1973) 505–516.
[36] C.R. Middaugh, W.A. Tisel, R.N. Haire, A. Rosenberg, Determination of the apparent thermodynamic activities of saturated protein solution, J. Biol. Chem. 254 (1979) 367–370.
[37] W.A. Tisel, R.N. Haire, J.G. White, Polyphasic linkage between protein solubility and ligand-binding in the hemoglobin–polyethylene glycol system, J. Biol. Chem. 255 (1980) 8975–8978.
[38] D.H. Atha, K.C. Ingham, Mechanism of precipitation of proteins by polyethylene glycols: analysis in terms of excluded volume, J. Biol. Chem. 256 (1981) 12108–12117.
[39] F. Haskó, R. Vaszileva, Solubility of plasma proteins in the presence of polyethylene glycol, Biotechnol. Bioeng. 24 (1982) 1931–1939.
[40] R.N. Haire, W.A. Tisel, J.C. White, A. Rosenberg, On the precipitation of proteins by polymers: the hemoglobulin–polyethylene, Biopolymers 23 (1984) 2761–2779.
[41] H. Mahadevan, C.K. Hall, Experimental analysis of protein precipitation by polyethylene glycol and comparison with theory, Fluid Phase Equilib. 78 (1992) 297–321.
[42] C.L. Stevenson, M.J. Hageman, Estimation of recombinant bovine somatotropin solubility by excluded-volume interaction with polyethylene glycols, J. Pharm. Sci. 12 (1995) 1671–1676.
[43] R. Guo, M. Guo, G. Narsimhan, Thermodynamics of precipitation of globular proteins by nonionic polymers, Ind. Eng. Chem. Res. 35 (1996) 3015–3026.
[44] E. Edmond, A.G. Ogston, An approach to the study of phase separation in ternary aqueous systems, Biochem. J. 109 (1968) 569–576.
[45] W. Melander, C. Horvath, Salt effect on hydrophobic interactions in precipitation and chromatography of proteins: an interpretation of the lyotropic series, Arch. Biochem. Biophys. 183 (1977) 200–221.
[46] T. Arakawa, S.N. Timasheff, Theory of protein solubility, Methods Enzymol. 114 (1985) 49–77.
[47] H. Mahadevan, C.K. Hall, Statistical–mechanical model of protein precipitation by non-ionic polymer, AIChE J. 36 (1990) 1517–1528.
[48] H. Mahadevan, C.K. Hall, Theory of precipitation of protein mixtures by nonionic polymer, AIChE J. 38 (1992) 573–591.
[49] E.S. Courtenay, M.W. Capp, C.F. Anderson, M.T. Record Jr., Vapor pressure osmometry studies of osmolyte–protein interactions: implications for the action of osmoprotectants in vivo and for the interpretation of 'osmotic stress' experiments in vitro, Biochemistry 39 (2000) 4455–4471.
[50] B.M. Baynes, B.L. Trout, Proteins in mixed solvents: a molecular-level perspective, J. Phys. Chem., B 107 (2003) 14058–14067.
[51] S. Kirincic, C.A. Klofutar, Volumetric study of aqueous solutions of poly(ethylene glycol)s at 298.15 K, Fluid Phase Equilib. 149 (1998) 233–247.
[52] A. Kulkarni, C. Zukoski, Depletion interactions and protein crystallization, J. Cryst. Growth 232 (2001) 156–164.
[53] A. Kulkarni, C. Zukoski, Nanoparticle crystal nucleation: influence of solution conditions, Langmuir 18 (2002) 3090–3099.
[54] D.-Q. Lin, Z.-Q. Zhu, L.-H. Mei, L.-R. Yang, Isopiestic determination of the water activities of poly(ethylene glycol)+salt+water systems at 25 °C, J. Chem. Eng. Data 41 (1996) 1040–1042.

Effect of salts and organic additives on the solubility of proteins in aqueous solutions

Eli Ruckenstein *, Ivan L. Shulgin

Department of Chemical and Biological Engineering, State University of New York at Buffalo, Amherst, NY 14260, USA

Available online 30 June 2006

Abstract

The goal of this review is to examine the effect of salts and organic additives on the solubility of proteins in aqueous mixed solvents. The focus is on the correlation between the aqueous protein solubility and the osmotic second virial coefficient or the preferential binding parameter. First, several approaches which connect the solubility and the osmotic second virial coefficient are presented. Most of the experimental and theoretical results correlate the solubility and the osmotic second virial coefficient in the presence of salts. The correlation of the aqueous protein solubility with the osmotic second virial coefficient when the cosolvent is an organic component requires additional research. Second, the aqueous protein solubility is correlated with the preferential binding parameter on the basis of a theory developed by the authors of the present review. This theory can predict (i) the salting-in or -out effect of a cosolvent and (ii) the initial slope of the solubility curve. Good agreement was obtained between theoretical predictions and experimental results.
© 2006 Elsevier B.V. All rights reserved.

Keywords: Protein solubility; Osmotic second virial coefficient; Preferential binding parameter; Mixed solvent; Salting-in and salting-out

Contents

1.	Introduction	97
2.	The aqueous protein solubility and the osmotic second virial coefficient	98
3.	The aqueous protein solubility and the preferential binding parameter	99
4.	Discussion	101
5.	Conclusion	101
	References	102

1. Introduction

Research regarding the solubility of proteins in water and aqueous solutions has spanned for more than a century and numerous reviews covering this topic are available [1–8]. It is well known that the solubility of a protein in a water+cosolvent mixture depends on many factors such as temperature, cosolvent concentration, pH, type of buffer used, etc. The focus of the present review is on the dependence of the protein solubility on the cosolvent concentration, at constant pH, temperature and pressure.

The effect of the addition of a cosolvent on the aqueous protein solubility was examined both experimentally and theoretically. Experiments have shown that the addition of organic substances reduces the aqueous protein solubility [5,6,8]. Therefore, the organic substances constitute salting-out agents. Arakawa and Timasheff [5] explained this salting-out effect as a result of (i) the decrease of the dielectric constant because the dielectric constants of organic substances are smaller than that of water, and (ii) the "redistribution of water

---

* Correspondence author. Tel.: +1 716 645 2911x2214; fax: +1 716 645 3822.
  E-mail addresses: feaeliru@acsu.buffalo.edu (E. Ruckenstein), ishulgin@eng.buffalo.edu (I.L. Shulgin).

0001-8686/$ - see front matter © 2006 Elsevier B.V. All rights reserved.

and organic molecules around the protein molecule, i.e., the preferential interactions of solvent components with the protein". However, not all organic substances decrease the aqueous protein solubility, for example, the addition of urea increases the aqueous solubility of ribonuclease Sa [9].

The addition of a salt to an aqueous protein solution leads to a more complex behavior of the aqueous protein solubility. Old solubility measurements [10,11] suggested that (i) a small amount of salt increases the aqueous protein solubility, and (ii) a large amount of salt decreases the aqueous protein solubility. At sufficiently large salt concentrations the solubility of a protein can be expressed by the empirical Cohn equation [1]:

$$\ln S_2 = \alpha - \beta c_3 \qquad (1)$$

where S_2 is the protein solubility (component 1 is water, component 2 is the protein and component 3 is the cosolvent), c_3 is the salt molarity, and α and β are empirical constants.

However, the measurements regarding the protein solubility in the presence of a salt carried out in the last two decades [12–20] revealed only salting-out effects. These measurements, carried out mostly for water–lysozyme–NaCl, can be considered to be reliable because there is agreement between the results obtained in different laboratories. In addition, these measurements showed that the Cohn equation cannot represent the dependence of the solubility on the cosolvent concentration. The data available, particularly for salts containing multivalent ions, are not sufficient to draw accurate conclusions about the dependence of the log solubility on cosolvent composition.

The theory of aqueous protein solubility has also attracted attention [20–39]. In this paper, two of the most promising approaches are reviewed. The approaches are:

1) correlation of the aqueous protein solubility with the osmotic second virial coefficient [31–39], and
2) correlation of the aqueous protein solubility with the preferential binding parameter [5,40–42].

2. The aqueous protein solubility and the osmotic second virial coefficient

The following expression can be written for the osmotic pressure [43]

$$\frac{\pi}{RT} = c_2/M_2 + B_2 c_2^2 + B_3 c_2^3 + \qquad (2)$$

where π is the osmotic pressure, R is the universal gas constant, T is the absolute temperature, M_2 is the protein molecular weight, B_2 and B_3 are the osmotic virial coefficients and c_2 is the molarity of the protein.

According to McMillan and Mayer [44], B_2 can be expressed in terms of the interaction between the protein molecules via the potential of mean force (W_{22})

$$B_2 = -\frac{N_A}{2M_2^2} \int_0^\infty \left[e^{-W_{22}/kT} - 1 \right] 4\pi r^2 dr \qquad (3)$$

where N_A is the Avogadro number, k is the Boltzmann constant and r is the center-to-center separation of two protein molecules.

The osmotic second virial coefficient was used to examine the crystallization of proteins and their solubility in water and in aqueous mixed solvents.

Firstly, George and Wilson [31] found empirically that the osmotic second virial coefficient could be correlated with the quality of crystallization of proteins from aqueous solutions. They found that good crystals could be obtained when the osmotic second virial coefficient was located in a window (crystallization slot) between -2×10^{-4} and -8×10^{-4} mL mol/g^2. For $B_2 > -2 \times 10^{-4}$, the protein–protein interactions are not strong enough for crystallization to occur and when $B_2 < -8 \times 10^{-4}$, the protein–protein interactions are too strong and amorphous precipitates are formed because the process is too rapid for the protein molecules to acquire crystalline structures. George and Wilson [31] findings constitute a useful screening criterion for protein crystallization because the osmotic second virial coefficient can be relatively easily obtained from static light scattering, small-angle X-ray and neutron scattering, osmometry etc. [31,32,34,45–49].

Secondly, it was found that the osmotic second virial coefficient and the aqueous solubility of a protein are not independent quantities and relations between them were established [33–35]. A simple relation between the osmotic second virial coefficient and the aqueous solubility S_2 (g/ml) of a protein was obtained from the condition of equilibrium between a protein in a solution and in a crystalline phase [34]. It has the following form

$$B_2 = \frac{-\Delta \mu_2}{RT} \frac{1}{2M_2 S_2} - \frac{\ln S_2}{2M_2 S_2} \qquad (4)$$

where $\Delta \mu_2 = \mu_2^0(s) - \mu_2(c)$, $\mu_2^0(s)$ is the standard chemical potential of the protein in a solution (mole fraction scale), which depends on temperature and pressure, and $\mu_2(c)$ is the chemical potential of the protein in the crystalline phase.

Eq. (4) provides a connection between the osmotic second virial coefficient B_2 and the aqueous solubility S_2. However, because $\Delta \mu_2$ is usually unknown, Eq. (4) cannot be used to predict the aqueous solubility from the osmotic second virial coefficient.

A square-well potential model for the interaction between protein molecules was used to derive a relation between the osmotic second virial coefficient B_2 and the aqueous solubility [33]. The following expression, which is valid at low solubilities, was obtained

$$B_2 = \frac{4}{\rho M_2} \left[1 - A \left\{ \left(\frac{\phi_S}{m} \right)^{-(2/z)} - 1 \right\} \right] \qquad (5)$$

where ρ is the density of the protein ($\rho \cong 1.36$ g/cm^3), $m = M_2/(18\rho)$ is the number of water molecules that can be placed in the volume of one protein molecule, z is the coordination number (the number of nearest-neighbor protein molecules in the protein crystal), ϕ_S is the protein solubility expressed in volume fraction of the protein, and A is a quantity, which depends on the anisotropy of the crystal and range of interaction between molecules, and which can be considered as an adjustable parameter. Eq. (5) provides an accurate correlation between the osmotic second virial coefficient and the aqueous solubility of lysozyme in the presence of various cosolvents (salts). This dependence has a monotonic character, i.e. the solubility of lysozyme increases with increasing osmotic second virial coefficient. According to experiments both the solubility of lysozyme and the osmotic second virial coefficient decrease with increasing cosolvent (salt) concentration [13–20,34]. Recent experiments [39] regarding the osmotic second virial coefficient of aqueous lysozyme in the presence of alcohols provide, however, a different picture. Indeed, when an alcohol is added to an aqueous lysozyme solution, the osmotic second virial coefficient increases [39], whereas the alcohols are well-known protein precipitants, i.e. the aqueous lysozyme solubility decreases with the addition of an alcohol [2,5,6,8,50].

Using classical thermodynamics, another relation between the osmotic second virial coefficient and the aqueous solubility was established [35]. In that paper, the aqueous mixed solvent is treated as a single component and the obtained relation contains two adjustable parameters. This equation was used to correlate the osmotic second virial coefficient and the aqueous protein solubility in the systems water–lysozyme–salt (NaCl) and water–ovalbumin–salt ($(NH_4)_2SO_4$).

3. The aqueous protein solubility and the preferential binding parameter

The preferential binding parameter [51–56] can be defined in various concentration scales (component 1 is water, component 2 is a protein and component 3 is a cosolvent):

1) in molal concentrations

$$\Gamma_{23}^{(m)} \equiv \lim_{m_2 \to 0} (\partial m_3 / \partial m_2)_{T,P,\mu_3} \qquad (6)$$

where m_i is the molality of component i, P is the pressure and μ_i is the chemical potential of component i.

2) in molar concentrations

$$\Gamma_{23}^{(c)} \equiv \lim_{c_2 \to 0} (\partial c_3 / \partial c_2)_{T,P,\mu_3} \qquad (7)$$

where c_i is the molar concentration of component i. One should notice that $\Gamma_{23}^{(m)}$ and $\Gamma_{23}^{(c)}$ are defined at infinite dilution of the protein.

The preferential binding parameter $\Gamma_{23}^{(m)}$ was measured experimentally by sedimentation [53], dialysis equilibrium [56], vapor pressure osmometry [57,58], etc. for numerous systems [51–68].

The preferential binding parameter $\Gamma_{23}^{(m)}$ provides information about the interactions between a protein and the components of a mixed solvent. $\Gamma_{23}^{(m)} < 0$ means that the protein is preferentially hydrated in the presence of a cosolvent such as glycerol, sucrose, etc. [5,51,53–56,59–63]. These cosolvents stabilize at high concentrations the protein structure and preserve its enzymatic activity [53–56,59–63]. $\Gamma_{23}^{(m)} > 0$ means that the protein is preferentially solvated by the cosolvent [5,51,53–56,59–63]. This occurs, for instance, for urea, which can cause protein denaturation.

Timasheff and coworkers [5,59–63] were the first to notice that there is a connection between the preferential binding parameter $\Gamma_{23}^{(m)}$ and the aqueous protein solubility. On the basis of their measurements and literature data regarding the preferential binding parameter and the aqueous protein solubility, they concluded that there is a general correlation between these quantities [5,59–63]. Particularly, they concluded that preferential hydration of a protein ($\Gamma_{23}^{(m)} < 0$) is equivalent to a salting-out behavior, i.e. the addition of a cosolvent decreases the protein solubility [5,65]. Thus, the local composition of the components of a mixed solvent is one of the most important factors affecting the aqueous protein solubility [5,40,59–63].

The authors of the present paper have developed a theory which connects the preferential binding parameter and the aqueous protein solubility [41,42]. The central element of the theory is the following relation (its derivation is provided in the Appendix) which relates the solubility of a protein to the mixed solvent composition and the preferential binding parameter [41,42]:

$$\left(\frac{\partial \ln y_2}{\partial x_3}\right) = -\frac{c_3(c_1+c_3)V_1 - \Gamma_{23}^{(m)}(1-c_3 V_3)(c_1 + c_1 J_{11} + c_3)}{c_1 c_3 V_1} \qquad (8)$$

In Eq. (8), y_2 is the protein solubility in mole fraction, $J_{11} = \lim_{x_2 \to 0} \left(\frac{\partial \ln \gamma_1}{\partial x_1}\right)_{x_2}$, x_i is the mole fraction of component i, γ_i is the activity coefficient of component i in a mole fraction scale, V_i is the partial molar volume of component i.

Eq. (8) is a rigorous thermodynamic equation at infinite protein dilution; it allows one to derive a simple criterion for salting-in or salting-out at low cosolvent concentrations. Indeed, at low cosolvent concentrations ($c_3 \to 0$) it becomes

$$\left(\frac{\partial \ln y_2}{\partial x_3}\right) = -\left(\frac{\partial \ln y_2}{\partial x_1}\right) = \frac{\alpha}{V_1^0} - 1 \qquad (9)$$

where $\alpha = \lim_{c_3 \to 0} \frac{\Gamma_{23}^{(m)}}{c_3}$ and V_1^0 is the molar volume of pure water. One can conclude that for low cosolvent concentrations salting-in occurs when

$$\left(\frac{\partial \ln y_2}{\partial x_3}\right) > 0, \qquad \text{hence when } \alpha > V_1^0 \qquad (10)$$

and salting-out occurs when

$$\left(\frac{\partial \ln y_2}{\partial x_3}\right) < 0, \qquad \text{hence when } \alpha < V_1^0 \qquad (11)$$

The above criteria for salting-in or salting-out (Eqs. (10), (11) are valid [41,42]: (i) for $c_3 \to 0$, hence when only a small amount of cosolvent is added to pure water; (ii) for ternary mixtures (water (1)–protein (2)–cosolvent (3)). It should be emphasized that those mixtures contain in addition a buffer, the effect of which is taken into account only indirectly through the preferential binding parameter $\Gamma_{23}^{(m)}$; (iii) for infinite dilution (this means that the protein solubility is supposed to be low enough to satisfy the infinite dilution approximation ($\gamma_2 \cong \gamma_2^\infty$, where γ_2^∞ is the activity coefficient of the protein at infinite dilution).

The following expression for protein solubility in a dilute cosolvent solution can be derived from Eq. (8) when $\Gamma_{23}^{(m)}$ is proportional to c_3 ($\Gamma_{23}^{(m)} = \alpha c_3$) [41,42]:

$$\ln\frac{y_2}{y_2^w} = -\frac{(\alpha - V_3 \Gamma_{23}^{(m)})\ln a_w}{V_1} + \ln x_1 \approx -\frac{(\alpha - V_3 \Gamma_{23}^{(m)})\ln a_w}{V_1}$$
$$= -\frac{(1 - V_3 c_3)\alpha \ln a_w}{V_1} = -c_1 \alpha \ln a_w \quad (12)$$

where y_2^w is the protein solubility in a cosolvent-free water plus buffer and a_w is the water activity in the protein-free mixed solvent. Eq. (12) allows one to calculate the protein solubility if the composition dependence of $\Gamma_{23}^{(m)}$ is available (a_w and the partial molar volumes V_1 and V_3 are characteristics of the protein-free mixed solvent). As already mentioned above, Eq. (12) was derived by assuming that $\Gamma_{23}^{(m)}$ was proportional to c_3. Indeed [58,64,69], the preferential binding parameter $\Gamma_{23}^{(m)}$ is, at least at low cosolvent concentrations, proportional to the cosolvent concentration.

Eqs. (8)–(12) demonstrate that the experimental preferential binding parameter $\Gamma_{23}^{(m)}$ can be used to predict the protein solubility. Of course, in turn the experimental protein solubility can be used to evaluate the preferential binding parameter.

Let us apply Eqs. (8)–(12) to real systems. First, Eqs. (10), (11) can be used to determine the type of cosolvent: salting-in or-out. Because in many cases $|\alpha| \gg V_1^0 \approx 18$ cm^3/mol, criteria (10)–(11) can be rewritten in the following simplified form (again, for $c_3 \to 0$). Salting-in occurs when

$$\left(\frac{\partial \ln y_2}{\partial x_3}\right) > 0, \quad \text{hence when } \Gamma_{23}^{(m)} > 0 \quad (13)$$

and salting-out occurs when

$$\left(\frac{\partial \ln y_2}{\partial x_3}\right) < 0, \quad \text{hence when } \Gamma_{23}^{(m)} < 0 \quad (14)$$

Criterion (14) was also suggested by Timasheff and coworkers [5,63].

The application of the established criteria (Eqs. (10), (11) or (13), (14)) provides a simple physical picture for the salting-in or the salting-out by a cosolvent: when a protein molecule is preferentially hydrated ($\Gamma_{23}^{(m)} < 0$), the addition of a small amount of cosolvent decreases the protein solubility; when the water is preferentially excluded from a protein surface ($\Gamma_{23}^{(m)} > 0$), the addition of a small amount of cosolvent increases the protein solubility.

The application of the established criteria (Eqs. (10), (11) or (13), (14)) to salting-in or-out in real systems is illustrated in Table 1 (a part of this Table was taken from our previous paper [41] but a part is new). Table 1 demonstrates that almost in all cases the established criteria (Eqs. (10), (11) or (13), (14)) provide types of salting effect by the cosolvents which coincide with the experimental solubility observations. However, in two cases (water–lysozyme–glycerol and water–β-lactoglobulin–NaCl) our criteria predict salting-out effects but the experimental data for solubility indicate salting-in. It seems that the experimental results [71] for water–lysozyme–glycerol (salting-in) are not correct, because glycerol is a well-known protein precipitant and hence the aqueous lysozyme solubility should decrease with the addition of glycerol to aqueous lysozyme solutions [2,5,6,8,50]. It is not yet clear why there is a discrepancy

Table 1
Application of criteria (Eqs. (10), (11) or (13), (14)) for salting-in or salting-out to aqueous solutions of proteins

Protein	Cosolvent [A]	Experimental data used		Do the criteria Eqs. (10), (11) or (13), (14) work?
		Solubility (salting-in or salting-out, conditions, references)	Preferential binding parameter $\Gamma_{23}^{(m)}$ (conditions, references)	
Lysozyme	NaCl	Salting-out, $T=0$–40 °C, pH=3–10 [15–20]	pH=4.5 [59], pH=3–7 [62]	Yes
Lysozyme	MgCl$_2$	Salting-out, $T=18$ °C, pH=4.5 [15]	pH=3.0, 4.5 [63]	Yes
Lysozyme	NaAcO	Salting-out, $T=18$ °C, pH=4.5, 8.3 [15]	pH=4.5–4.71 [59]	Yes
Ribonuclease Sa	Urea	Salting-in, $T=25$ °C, pH=3.5, 4.0 [9]	pH=2.0, 4.0, 5.8 [70] [B]	Yes
Lysozyme	Glycerol	Salting-in, $T=25$ °C, pH=4.6 [71]	pH=2.0, 5.8 [72]	No
β-Lactoglobulin	NaCl	Salting-in, $T=25$ °C, pH=5.15–5.3 [73]	pH=1.55–10 [62]	No
Ribonuclease A	MPD [C]	Salting-out, $T=25$ °C, pH=5.8 [5]	pH=5.8 [5]	Yes
Lysozyme	PEG 400 [C]	Salting-out, $T=25$ °C, pH=7.0 [74]	pH=7.0 [75]	Yes
β-Lactoglobulin	PEG 20,000	Salting-out, $T=25$ °C, pH=2.0 [76]	pH=2.0 [75, 77] [D]	Yes
BSA [C]	PEG 1450-PEG 20,000	Salting-out, $T=25$ °C, pH=4.5–8.0 [26,74,78]	pH=3.0 [79] [E] and pH=7.0 [75] [F]	Yes

[A] The term "cosolvent" is also used here for electrolytes.
[B] The preferential binding parameters were determined for ribonuclease A in 30 volume % glycerol solution.
[C] The MPD stands for 2-methyl-2, 4-pentanediol; PEG 400-polyethylene glycol with molecular weight 400, BSA-bovine serum albumin.
[D] The preferential binding parameters were determined for PEG 200–6000.
[E] The preferential binding parameters were determined for PEG 1000.
[F] The preferential binding parameters were determined for PEG 400–6000.

between the experimental solubility and the criteria employed for the case water–β-lactoglobulin–NaCl.

Eq. (9) can be also used to calculate the initial slope (at $c_3 \to 0$) of the dependence of the protein solubility against the cosolvent concentration. The slopes obtained through Eq. (9) can be compared with the experimental slopes. This comparison is made in Table 2. Table 2 shows that there is good agreement between the experimental slopes and those predicted by Eq. (9). There are, however, also cases in which there are large differences (for instance for the water+lysozyme+NaCl at pH=6.5). They can be caused by the experimental errors in the determination of either the solubility and/or the preferential binding parameter.

As noted in the Introduction, the Cohn equation, (Eq. (1)), considers that log of protein solubility is a linear function of the cosolvent molarity. In reality [5,14], the above dependence is not linear. Fig. 1 presents some accurate experimental data regarding the aqueous solubility of lysozyme and shows that linearity occurs only in the dilute region ($c_3 < 0.5$). Our Eq. (12) allows one to explain this behavior. Only the dilute region ($c_3 < 0.5$) is considered because in this composition range the preferential binding parameter $\Gamma_{23}^{(m)}$ is proportional to the concentration of the cosolvent [58,64,69] and Eq. (12) involves this approximation. Eq. (12) reveals that the linearity or nonlinearity of $\ln y_2$ versus cosolvent concentration depends on the water activity in the protein-free aqueous mixed solvent. Eq. (12) was used to examine the log protein solubility versus cosolvent molarity in water/protein/ polyethylene glycol (PEG) mixtures [42]. It was shown that there were almost linear behaviors for PEG 1000 and PEG

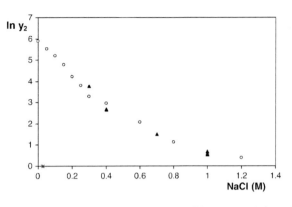

Fig. 1. Logarithm of lysozyme solubility y_2 [mg/ml] in aqueous solutions of sodium chloride at pH=4.5. (○) and (▲) are the experimental data from [15] and [18], respectively.

4000 but a nonlinear one for PEG 6000. The same nonlinear behaviors are expected to occur for higher molecular weights of PEG.

4. Discussion

Connections between the solubility of proteins in aqueous solutions and measurable quantities such the osmotic second virial coefficient and the preferential binding parameter are useful, because they can help to understand the aqueous protein solutions and to select the best conditions for protein crystallization. Such connections can be also used to predict the protein solubility on the basis of the experimental osmotic second virial coefficient or the preferential binding parameter.

Different scientific groups have demonstrated that there are direct connections between the solubility of proteins in aqueous solutions and both the osmotic second virial coefficient and the preferential binding parameter. The osmotic second virial coefficient B_{22} is a measure of the intermolecular protein–protein interaction and the preferential binding parameter $\Gamma_{23}^{(m)}$ is an indicator of the redistributions of water and cosolvent in the vicinity of a protein surface, hence a measure of their interactions with the solvent and cosolvent. The preferential binding parameter $\Gamma_{23}^{(m)}$ is defined at infinite dilution of a protein and is expected to be correlated to the protein solubility when the assumption of infinite protein dilution is valid [41,42]. For finite concentrations, both the osmotic second virial coefficient B_{22} and the preferential binding parameter $\Gamma_{23}^{(m)}$ (perhaps not defined at infinite dilution) are expected to affect the solubility of proteins in aqueous solutions. The osmotic second virial coefficient accounts for the interactions between proteins whereas the preferential binding parameter accounts for the interactions between the protein and the constituents of a mixed solvent.

5. Conclusion

The osmotic second virial coefficient and the preferential binding parameter, quantities which can be determined

Table 2
Comparison between the experimental slopes of the solubility vs. cosolvent mole fraction curve $\left(\frac{\partial \ln y_2}{\partial x_3}\right)$ with the results predicted by Eq. (9)

System	Experiment		Prediction	
	pH, reference[A]	Slope	pH, reference[B]	Slope
Water+ lysozyme+NaCl	pH=3.3 [18]	−0.32	pH=3.0 [62]	−0.37
Water+ lysozyme+NaCl	pH=4.3 [18]	−0.37	pH=4.5 [59]	−0.34
Water+ lysozyme+NaCl	pH=6.5 [18]	−0.50	pH=7.0 [62]	−0.32
Water+ lysozyme+NaAcO	pH=4.5 [20]	−0.24	pH=4.68–4.71 [59]	−0.45
Water+ lysozyme+MgCl$_2$	pH=4.1 [15]	−0.07	pH=4.5 [63]	−0.10
Water+human serum albumin+PEG 4000[C]	pH=3.8 [74]	−0.16	pH=7.0 [75][C]	−0.21
	pH=4.5 [74]	−0.23		
	pH=5.2 [74]	−0.21		
Water+bovine serum albumin+PEG 6000[D]	pH=4.0 [78]	−0.29	pH=7.0 [75]	−0.19
	pH=5.1 [78]	−0.27		
	pH=5.8 [78]	−0.26		

[A] References regarding the experimental values of solubility data.
[B] References regarding the experimental values of the preferential binding parameter.
[C] Experimental data for the preferential binding parameter in water+bovine serum albumin+PEG 4000 were used.
[D] More data regarding slopes for water+protein+PEG mixtures can be found in [42].

experimentally, can be used to correlate and predict the solubility of proteins in mixed aqueous solvents.

Appendix A

The aim of this Appendix is to provide a derivation of Eq. (8) on the basis of the Kirkwood–Buff theory of solutions [80]. The authors of the present paper derived the following expressions for $\Gamma_{23}^{(m)}$ [81]

$$\Gamma_{23}^{(m)} = \frac{c_3}{c_1} + c_3(G_{23} - G_{12} + G_{11} - G_{13}) \tag{A-1}$$

and $\Gamma_{23}^{(c)}$ [82]

$$\Gamma_{23}^{(c)} = c_3(G_{23} - G_{13}) \tag{A-2}$$

where G_{ij} are the Kirkwood–Buff integrals which are given by [80]

$$G_{ij} = \int_0^\infty (g_{ij} - 1) 4\pi r^2 dr \tag{A-3}$$

g_{ij} is the radial distribution function between species i and j, and r is the distance between the centers of molecules i and j.

Analytical expressions for the Kirkwood–Buff integrals G_{12} and G_{23} at infinite dilution of the protein are available in the literature [82]:

$$G_{12} = kTk_T - \frac{J_{21}V_3 c_3 + J_{11}V_2^\infty c_1}{(c_1 + c_1 J_{11} + c_3)} \\ - \frac{V_3 c_3(c_1 + c_3)(V_1 - V_3) + V_2^\infty(c_1 + c_3)}{(c_1 + c_1 J_{11} + c_3)} \tag{A-4}$$

and

$$G_{23} = kTk_T - \frac{J_{21}V_1 c_1 - J_{11}c_1 V_2^\infty}{(c_1 + c_1 J_{11} + c_3)} \\ + \frac{c_1 V_1(c_1 + c_3)(V_1 - V_3) - V_2^\infty(c_1 + c_3)}{(c_1 + c_1 J_{11} + c_3)} \tag{A-5}$$

where, $J_{11} = \lim_{x_2 \to 0} \left(\frac{\partial \ln \gamma_1}{\partial x_1}\right)$, $J_{21} = \lim_{x_2 \to 0} \left(\frac{\partial \ln \gamma_2}{\partial x_1}\right)$, x_i is the mole fraction of component i, γ_i is the activity coefficient of component i in a mole fraction scale and k_T is the isothermal compressibility of the mixture. Expressions for the Kirkwood–Buff integrals G_{11} and G_{13} at infinite dilution of a protein are also well-known from the literature (see for example [83,84]):

$$G_{11} = kTk_T - \frac{(c_1 + c_3)^2 V_1 V_3}{(c_1 + c_1 J_{11} + c_3)} + \frac{(c_1 + c_3)(V_3 - V_1) - J_{11}}{(c_1 + c_1 J_{11} + c_3)} \tag{A-6}$$

and

$$G_{13} = kTk_T - \frac{(c_1 + c_3)^2 V_1 V_3}{(c_1 + c_1 J_{11} + c_3)} \tag{A-7}$$

The insertion of Eqs. (A-4)–(A-7) into Eq. (A-1) leads after some algebra to Eq. (8) in the text.

References

[1] Cohn EJ. Physiol Rev 1925;5:349.
[2] Cohn EJ, Edsall JT. Proteins, amino acids and peptides. Reinhold; 1943.
[3] Green AA, Hughens WL. Methods Enzymol 1955;1:67.
[4] Fox SW, Foster JF. Introduction to protein chemistry. New York: Wiley; 1957.
[5] Arakawa T, Timasheff SN. Methods Enzymol 1985;114:49.
[6] Schein CH. Biotechnology (NY) 1990;8:308.
[7] Riès-Kautt M, Ducruix A. In: Ducruix A, Giégé R, editors. Crystallization of nucleic acids and proteins: a practical approach. New York: Oxford University Press; 1999.
[8] Tayyab S, Qamar S, Islam M. Med Sci Res 1993;21:805.
[9] Pace CN, Trevino S, Prabhakaran E, Scholtz JM. Philos Trans R Soc Lond B Biol Sci 2004;359:1225.
[10] Green AA. J Biol Chem 1931;93:495.
[11] Grönwall A. C R Trav Lab Carlsberg 1942;24:185.
[12] Ataka M, Tanaka S. Biopolymers 1986;25:337.
[13] Howard SB, Twigg PJ, Baird JK, Meehan EJ. J Cryst Growth 1986;90:94.
[14] Mikol V, Giege R. J Cryst Growth 1989;97:324.
[15] Ries-Kautt MM, Ducruix AF. J Biol Chem 1989;264:745.
[16] Cacioppo E, Pusey ML. J Cryst Growth 1991;114:286.
[17] Judge RA, Johns MR, White ET. J Chem Eng Data 1996;41:422.
[18] Retailleau P, Ries-Kautt MM, Ducruix AF. Biophys J 1997;72:2156.
[19] Forsythe RL, Judge RA, Pusey ML. J Chem Eng Data 1999;44:637.
[20] Retailleau P, Ducruix AF, Ries-Kautt M. Acta Crystallogr D Biol Crystallogr 2002;58:1576.
[21] Edmond E, Ogston AG. Biochem J 1968;109:569.
[22] Ogston AG. J Phys Chem 1970;74:668.
[23] Melander W, Horvath C. Arch Biochem Biophys 1977;183:200.
[24] Mahadevan H, Hall CK. AIChE J 1990;36:1517.
[25] Mahadevan H, Hall CK. AIChE J 1992;38:573.
[26] Mahadevan H, Hall CK. Fluid Phase Equilib 1992;78:297.
[27] Yu M, Arons JD, Smit JAM. J Chem Technol Biotechnol 1994;60:413.
[28] Agena SM, Bogle IDL, Pessoas FLP. Biotechnol Bioeng 1997;55:65.
[29] Jenkins WT. Protein Sci 1998;7:376.
[30] Agena SM, Pusey ML, Bogle IDL. Biotechnol Bioeng 1999;64:144.
[31] George A, Wilson WW. Acta Crystallogr D Biol Crystallogr 1994;50:361.
[32] George A, Chiang Y, Guo B, Arabshahi A, Cai Z, Wilson WW. Methods Enzymol 1997;276:100.
[33] Haas C, Drenth J, Wilson WW. J Phys Chem B 1999;103:2808.
[34] Guo B, Kao S, McDonald H, Asanov A, Combs LL, Wilson WW. J Cryst Growth 1999;196:424.
[35] Ruppert S, Sandler SI, Lenhoff AM. Biotechnol Prog 2001;17:182.
[36] Curtis RA, Blanch HW, Prausnitz JM. J Phys Chem B 2001;105:2445.
[37] Curtis RA, Ulrich J, Montaser A, Prausnitz JM, Blanch HW. Biotechnol Bioeng 2002;79:367.
[38] Pan X, Glatz CE. Cryst Growth Des 2003;3:203.
[39] Liu W, Bratko D, Prausnitz JM, Blanch HW. Biophys Chem 2004;107:289.
[40] Bolen DW. Methods 2004;34:312.
[41] Shulgin IL, Ruckenstein E. Biophys Chem 2005;118:128.
[42] Shulgin IL, Ruckenstein E. Biophys Chem 2006;120:188.
[43] Hill TL. Thermodynamics for Chemists and Biologists. Reading, Mass.: Addison-Wesley Pub. Co.; 1968
[44] McMillan WG, Mayer JE. J Chem Phys 1945;13:276.
[45] Velev OD, Kaler EW, Lenhoff AM. Biophys J 1998;75:2682.
[46] Neal BL, Asthagiri D, Velev OD, Lenhoff AM, Kaler EW. J Cryst Growth 1999;196:377.
[47] Bonnete F, Finet S, Tardieu A. J Cryst Growth 1999;196:403.
[48] Tardieu A, Le Verge A, Malfois M, Bonnete F, Finet S, Ries-Kautt M, et al. J Cryst Growth 1999;196:193.
[49] Tombs MP, Peacocke AR. The osmotic pressure of biological macromolecules. Oxford: Clarendon Press; 1974.
[50] Sauter C, Ng JD, Lorber B, Keith G, Brion P, Hosseini MW, et al. J Cryst Growth 1999;196:365.
[51] Casassa EF, Eisenberg H. Adv Protein Chem 1964;19:287.
[52] Wyman J. Adv Protein Chem 1964;19:223.

[53] Kuntz ID, Kauzmann W. Adv Protein Chem 1974;28:239.
[54] Timasheff SN. Annu Rev Biophys Biomol Struct 1993;22:67.
[55] Timasheff SN. In: Somero GN, Osmond CB, Bolis CL, editors. Water and life: comparative analysis of water relationships at the organismic, cellular, and molecular levels. Berlin: Springer-Verlag; 1992.
[56] Timasheff SN. Adv Protein Chem 1998;51:355.
[57] Zhang WT, Capp MW, Bond JP, Anderson CF, Record MT. Biochemistry 1996;35:10506.
[58] Courtenay ES, Capp MW, Anderson CF, Record MT. Biochemistry 2000;39:4455.
[59] Arakawa T, Timasheff SN. Biochemistry 1982;21:6545.
[60] Arakawa T, Timasheff SN. Biochemistry 1984;23:5912.
[61] Timasheff SN, Arakawa T. J Cryst Growth 1988;90:39.
[62] Arakawa T, Timasheff SN. Biochemistry 1987;26:5147.
[63] Arakawa T, Bhat R, Timasheff SN. Biochemistry 1990;29:1914.
[64] Record MT, Zhang WT, Anderson CF. Adv Protein Chem 1998;51:281.
[65] Felitsky DJ, Record MT. Biochemistry 2004;43:9276.
[66] Anderson CF, Courtenay ES, Record MT. J Phys Chem B 2002;106:418.
[67] Courtenay ES, Capp MW, Record MT. Protein Sci 2001;10:2485.
[68] Eisenberg H. Biological Macromolecules and Polyelectrolytes in Solution. Oxford: Clarendon Press; 1976.
[69] Baynes BM, Trout BL. J Phys Chem B 2003;107:14058.
[70] Lin TY, Timasheff SN. Biochemistry 1994;33:12695.
[71] Kulkarni AM, Zukoski CF. Langmuir 2002;18:3090.
[72] Gekko K, Timasheff SN. Biochemistry 1981;20:4677.
[73] Treece JM, Sheinson RS, McMeekin TL. Arch Biochem Biophys 1964;108:99.
[74] Atha DH, Ingham KS. J Biol Chem 1981;256:12108.
[75] Bhat R, Timasheff SN. Protein Sci 1992;1:1133.
[76] Middaugh CR, Tisel WA, Haire RN, Rosenberg A. J Biol Chem 1979;254:367.
[77] Arakawa T, Timasheff SN. Biochemistry 1985;24:6756.
[78] Juckes IRM. Biochim Biophys Acta 1971;229:535.
[79] Lee JC, Lee Y. J Biol Chem 1981;256:625.
[80] Kirkwood JG, Buff FP. J Chem Phys 1951;19:774.
[81] Shulgin IL, Ruckenstein E. Biophys J 2006;90:704.
[82] Shulgin IL, Ruckenstein E. J Chem Phys 2005;123:054909.
[83] Matteoli E, Lepori l. J Chem Phys 1984;80:2856.
[84] Shulgin I, Ruckenstein E. J Phys Chem B 1999;103:2496.

Local Composition in the Vicinity of a Protein Molecule in an Aqueous Mixed Solvent

Ivan L. Shulgin[†] and Eli Ruckenstein*

Department of Chemical and Biological Engineering, State University of New York at Buffalo, Amherst, New York 14260

Received: September 27, 2006; In Final Form: January 29, 2007

This paper is focused on the composition of a cosolvent in the vicinity of a protein surface (local composition) and its dependence on various factors. First, the Kirkwood–Buff theory of solution is used to obtain analytical expressions that connect the excess or deficit number of cosolvent and water molecules in the vicinity of a protein surface with experimentally measurable quantities such as the bulk concentration of the mixed solvent, the preferential binding parameter, and the molar volumes of water and cosolvent. Using these expressions, relations between the preferential binding parameter (at a molal concentration scale) and the above excesses (or deficits) are established. In addition, the obtained expressions are used to examine the effect of the nonideality of the water + cosolvent mixtures and of the molar volume of the cosolvent on the excess (or deficit) number of cosolvent molecules in the vicinity of the protein surface. It is shown that at least for the mixed solvents considered (water + urea and water + glucose) the nonideality of the mixed solvent is not an important factor in the local compositions around a protein molecule and that the main contribution is provided by the nonidealities of the protein–water and protein–cosolvent mixtures. Special attention is paid to urea as cosolvent, because urea is one of only a few compounds with a concentration at the protein surface larger than its concentration in the bulk. The composition dependence of the excess of urea around a protein molecule is calculated for the water + lysozyme + urea mixture at pH = 7.0 and 2.0. At pH = 7.0, the excess of urea becomes almost composition independent at high urea concentrations. Such independence could be explained by assuming that urea totally replaces water in some areas of the protein surface, whereas on the remaining areas of the protein surface both water and urea are present with concentration comparable to those in the bulk. The Schellman exchange model was used to relate the preferential binding parameter in water + lysozyme + urea mixtures to the urea concentration.

1. Introduction

Addition of an inorganic salt or of a small organic molecule (both will be called cosolvents in this paper) to an aqueous protein solution changes the microstructure and composition (the local composition) in the vicinity of the protein surface. It was suggested that the cosolvents can be subdivided into three groups:[1]

"1) the cosolvent is present at the protein surface in excess over its concentration in the bulk (this is what constitutes binding); 2) the water is present in excess at the protein surface; this means that the protein has a higher affinity for water than for the cosolvent (this situation is referred to as preferential hydration, or preferential exclusion of the cosolvent); 3) the protein is indifferent to the nature of the molecules (water or cosolvent) with which it comes in contact, so that no solvent concentration perturbation occurs at the protein surface"

The majority of cosolvents belong to the second group. They are inorganic salts, glycerol, sucrose, or similar compounds. There are only a few compounds that belong to the first group: urea, some derivatives of urea, guanidine hydrochloride, etc.

It should be mentioned that the compounds from the second group can stabilize at high concentrations the protein structure and preserve its enzymatic activity.[2–7] In contrast, the addition of a compound from the first group can cause protein denaturation.[2–7] It is remarkable that on the basis of their effect on protein solubility the cosolvents can also be subdivided into the same groups. The cosolvents of the first group increase the protein solubility compared to the solubility in water when small amounts are added to water.[8–9] The addition of a small amount of a compound from the second group decreases the solubility, and these compounds are well-known salting-out agents.[10–14]

There is no common understanding regarding the effect at the atomic level of various cosolvents on the stability of a protein in aqueous solutions. Kauzmann[4] pointed out the importance of hydrophobic interactions (the tendency of the nonpolar groups of two proteins to adhere to one another in an aqueous environment) in stabilizing the folded configuration in many native proteins. Another class of theories connects the effect of various cosolvents on the structure of water (structure-breaker or structure-maker) in the vicinity of a protein surface. The cosolvents belonging to the first group, such as urea and guanidine hydrochloride, are of particular interest because the understanding of their behavior in aqueous protein mixtures can provide insight into the mechanism of protein denaturation, processes in cells, etc.[15–37] Many explanations were also suggested for the effects of the above cosolvents on protein stability and properties. They range from energetic factors, such as the formation of H-bonds with the polypeptide backbone, which disrupts the internal hydrogen bond network of the

---

* Corresponding author. E-mail: feaeliru@acsu.buffalo.edu. Fax: (716) 645-3822. Phone: (716) 645-2911/ext 2214.
[†] E-mail: ishulgin@eng.buffalo.edu.

Local Composition of a Protein Molecule in Solvent

proteins, to the ability of these compounds to increase the solubility of both hydrophobic and hydrophilic moieties of the protein.

In this paper, the various factors that affect its behavior, such as the size of a cosolvent molecule, the nonideality of water + cosolvent mixture, and the interactions between water and cosolvent with the protein, will be analyzed from a theoretical point of view. Experimental data regarding the preferential binding parameter and the protein partial molar volume at its infinite dilution will be used. The emphasis will be on urea as cosolvent.

2. Theory

2.1. Preferential Binding Parameter.

The preferential binding parameter[5-7,38-39] can be defined in various concentration scales (component 1 is water, component 2 is a protein, and component 3 is a cosolvent):

(1) in molal concentrations

$$\Gamma_{23}^{(m)} \equiv \lim_{m_2 \to 0} (\partial m_3 / \partial m_2)_{T,P,\mu_3} \quad (1)$$

where m_i is the molality of component i, T is the absolute temperature, P is the pressure, and μ_i is the chemical potential of component i. A somewhat different quantity, namely $\lim_{m_2 \to 0} (\partial m_3 / \partial m_2)_{T,\mu_1,\mu_3}$, is determined experimentally. However, as demonstrated a long time ago,[40] both quantities are practically equal.

(2) in molar concentrations

$$\Gamma_{23}^{(c)} \equiv \lim_{c_2 \to 0} (\partial c_3 / \partial c_2)_{T,P,\mu_3} \quad (2)$$

where c_i is the molar concentration of component i. One should notice that $\Gamma_{23}^{(m)}$ and $\Gamma_{23}^{(c)}$ are defined at infinite dilution of the protein.

The preferential binding parameter $\Gamma_{23}^{(m)}$ was determined experimentally by sedimentation,[38] dialysis equilibrium,[39] vapor pressure osmometry,[26,41] etc. for numerous systems.[5-7,32,38-39,41-48]

$\Gamma_{23}^{(m)}$ provides information about the interactions between a protein and the components of a mixed solvent. $\Gamma_{23}^{(m)} < 0$ means that the protein is preferentially hydrated in the presence of a cosolvent from the second group of cosolvents defined above.[5,7,38-39,41-45] As already mentioned, these cosolvents stabilize at high concentrations the protein structure and preserve its enzymatic activity. $\Gamma_{23}^{(m)} > 0$ means that the protein is preferentially solvated by the cosolvent.[5,7,32,38-39,41-45] This occurs for instance for urea and other cosolvents belonging to the first group.[1]

As shown previously by the authors, the preferential binding parameters $\Gamma_{23}^{(m)}$ and $\Gamma_{23}^{(c)}$ can be expressed in terms of the Kirkwood−Buff integrals for ternary mixtures as follows:[49]

$$\Gamma_{23}^{(m)} = \frac{c_3^0}{c_1^0} + c_3^0(G_{23} - G_{12} + G_{11} - G_{13}) \quad (3)$$

and[50]

$$\Gamma_{23}^{(c)} = c_3^0(G_{23} - G_{13}) \quad (4)$$

where G_{ij} are the Kirkwood−Buff integrals (see Appendix for details) and c_1^0 and c_3^0 are the molar concentrations of water and cosolvent in the protein-free mixed solvent.

One can also write the following expression for the partial molar volume of a protein at infinite dilution in a mixed solvent (V_2^∞) in terms of the Kirkwood−Buff theory of solutions[51]

$$V_2^\infty = -c_1^0 V_1 G_{12} - c_3^0 V_3 G_{23} + kTk_T \cong \\ -c_1^0 V_1 G_{12} - c_3^0 V_3 G_{23} \quad (5)$$

where V_i is the partial molar volume of component i, k is the Boltzmann constant, and k_T is the isothermal compressibility of the protein-free mixed solvent.

Equations 3 and 5 allow one to calculate the Kirkwood−Buff integrals G_{12} and G_{23} using experimental data regarding the preferential binding parameters $\Gamma_{23}^{(m)}$ and the partial molar volume of a protein at infinite dilution in a mixed solvent V_2^{∞}.[49-50,52] The Kirkwood−Buff integrals G_{11} and G_{13} can be evaluated on the basis of the properties of protein-free mixed solvent water + cosolvent. It should be mentioned that recently the Kirkwood−Buff theory was used to analyze the effects of various cosolvents on the properties of aqueous protein solutions.[53-55]

2.2. Excess (Deficit) Number of Water and Cosolvent Molecules around a Protein Molecule.

The excesses (or deficits) number of water and cosolvent molecules around a protein molecule at infinite dilution are of great importance for the quantitative characterization of hydration in aqueous protein solutions. These excesses (or deficits) are results of the competition between the cosolvent and water molecules around a protein molecule or, in other words, how the mixed solvent has been altered in the vicinity of the protein surface in comparison with the bulk mixed solvent (far from the protein surface). The excess (deficit) number of water and cosolvent molecules around a protein molecule at infinite dilution can be calculated using the expressions[56]

$$\Delta n_{12} = c_1^0 G_{12} + c_1^0 (V_2^\infty - RTk_T) \quad (6)$$

and

$$\Delta n_{32} = c_3^0 G_{23} + c_3^0 (V_2^\infty - RTk_T) \quad (7)$$

where Δn_{12} and Δn_{32} are the excesses (deficits) of water and cosolvent molecules in the vicinity of a protein molecule in comparison with their bulk values and R is the universal gas constant. Because of the presence of the central protein molecule, there is a volume inaccessible to the water and cosolvent molecules. This effect, which was ignored in the traditional definition of the excesses, is accounted for in eqs 6 and 7. The detailed derivation of the above expressions was provided previously by the authors.[56]

The quantities G_{12} and G_{23}, and Δn_{12} and Δn_{32}, respectively, can be calculated using experimental data regarding the preferential binding parameters $\Gamma_{23}^{(m)}$ and the partial molar volume of a protein at infinite dilution in a mixed solvent V_2^∞.

More explicit expressions for Δn_{12} and Δn_{32} can be obtained by inserting the expressions for the Kirkwood−Buff integrals G_{12} and G_{23} in ternary mixtures (see Appendix) into eqs 6 and 7. One thus obtains

$$\Delta n_{12} = -\frac{c_1^0 c_3^0 V_3 (c_1^0 + c_3^0)(V_1 - V_3) + c_1^0 c_3^0 J_{21} V_3}{(c_1^0 + c_1^0 J_{11} + c_3^0)} \quad (8)$$

$$\Delta n_{32} = \frac{c_1^0 c_3^0 V_1(c_1^0 + c_3^0)(V_1 - V_3) + c_1^0 c_3^0 J_{21} V_1}{(c_1^0 + c_1^0 J_{11} + c_3^0)} \quad (9)$$

where $J_{11} = \lim_{x_2 \to 0}(\partial \ln\gamma_1 / \partial x_1)_{x_2}$, $J_{21} = \lim_{x_2 \to 0}(\partial \ln\gamma_2 / \partial x_1)_{x_2}$, x_i is the mole fraction of component i, and γ_i is the activity coefficient of component i in a mole fraction scale.

Equations 8 and 9 show that only the derivative J_{21} is related to the infinitely dilute protein; all other quantities in eqs 8 and 9 (c_1^0, c_3^0, V_1, V_3, and J_{11}) are provided by the protein-free mixed solvent water + cosolvent mixture.

Let us consider some particular forms of eqs 8 and 9.

2.2.1. Ideal Mixed Solvent Approximation. In this case, $J_{11} = 0$, $V_1 = V_1^0$, and $V_3 = V_3^0$ (V_1^0 and V_3^0 being the molar volumes of the pure water and cosolvent), and eqs 8 and 9 acquire the forms

$$\Delta n_{12} = -c_1^0 c_3^0 V_3^0 (V_1^0 - V_3^0) - \frac{c_1^0 c_3^0 J_{21} V_3^0}{(c_1^0 + c_3^0)} \quad (10)$$

and

$$\Delta n_{32} = c_1^0 c_3^0 V_1^0 (V_1^0 - V_3^0) + \frac{c_1^0 c_3^0 J_{21} V_1^0}{(c_1^0 + c_3^0)} \quad (11)$$

This approximation implies that the nonidealities between the protein and the constituents of the mixed solvent are much stronger than those between the constituents of the mixed solvent. In other words, in this case, the main contribution to the nonideality of the very dilute mixture protein + mixed solvent stems from the nonideality of the protein with the mixed solvent and not from the nonideality of the mixed solvent itself. This means that the activity coefficients and their derivatives with respect to the concentrations in the pairs protein−water and protein−cosolvent are much larger than for the pair water−cosolvent.

2.2.2. Ideal Ternary Solution. In this idealized case, $J_{11} = 0$, $J_{21} = 0$, $V_1 = V_1^0$, and $V_3 = V_3^0$, and eqs 8 and 9 become

$$\Delta n_{12} = -c_1^0 c_3^0 V_3^0 (V_1^0 - V_3^0) \quad (12)$$

and

$$\Delta n_{32} = c_1^0 c_3^0 V_1^0 (V_1^0 - V_3^0) \quad (13)$$

In this case, the excesses (or deficits) occur because of differences in the sizes of the cosolvents.

2.3. Relation between the Preferential Binding Parameter $\Gamma_{23}^{(m)}$ and the Excesses (Deficits) Number of Water and Cosolvent Molecules around a Protein Molecule. It is well-known from numerous experimental measurements that $\Gamma_{23}^{(m)} < 0$ means preferential hydration of a protein in the presence of a cosolvent and $\Gamma_{23}^{(m)} > 0$ means that the protein is preferentially solvated by the cosolvent.[5,7,32,38−39,41−45] A relation between $\Gamma_{23}^{(m)}$ and Δn_{12} or Δn_{32} can be obtained as follows.

First, eq 5 can be rewritten as

$$V_1 \Delta n_{12} + V_3 \Delta n_{32} = 0 \quad (15)$$

The combination of this equation with eqs 3, 6, and 7 leads to the following expressions

$$\Gamma_{23}^{(m)} = -\frac{\Delta n_{12}}{c_1^0 V_3} + \frac{c_3^0}{c_1^0} + c_3^0(G_{11} - G_{13}) \quad (16)$$

or

$$\Gamma_{23}^{(m)} = \frac{\Delta n_{32}}{c_1^0 V_1} + \frac{c_3^0}{c_1^0} + c_3^0(G_{11} - G_{13}) \quad (17)$$

The right-hand sides of eqs 16 and 17 contain three terms. For usual cosolvents, the absolute value of $-(\Delta n_{12})/(c_1^0 V_3) = (\Delta n_{32})/(c_1^0 V_1)$ is much larger than $(c_3^0)/(c_1^0) + c_3^0(G_{11} - G_{13})$. For example, for the water (1) + lysozyme (2) + urea (3) mixture (pH 7.0, 20 °C)[32,49] and $c_3^0 = 1.5$ mol/L, $(\Delta n_{12})/(c_1^0 V_3) = 1.03$, $(c_3^0)/(c_1^0) = 0.03$, and $c_3^0(G_{11} - G_{13}) = 0.04$, whereas, for $c_3^0 = 3.0$ mol/L, $(\Delta n_{12})/(c_1^0 V_3) = 10.1$, $(c_3^0)/(c_1^0) = 0.06$, and $c_3^0(G_{11} - G_{13}) = 0.08$. Comparable results can be obtained for other mixtures. One can, therefore, conclude that, for $\Gamma_{23}^{(m)} < 0$, $\Delta n_{12} > 0$ ($\Delta n_{32} < 0$) and, for $\Gamma_{23}^{(m)} > 0$, $\Delta n_{12} < 0$ ($\Delta n_{32} > 0$). However, in the unlikely case when the water (1) + protein (2) + cosolvent (3) mixture is close to an ideal ternary mixture, all the terms in eqs 16 and 17 play a role.

The preferential binding parameter at a molar scale $\Gamma_{23}^{(c)}$ does not correlate with Δn_{12} or Δn_{32} as the preferential binding parameter $\Gamma_{23}^{(m)}$ does. $\Gamma_{23}^{(c)}$ can have the same sign as $\Gamma_{23}^{(m)}$ but can also have the opposite sign.[5,57] A thermodynamic relation between $\Gamma_{23}^{(m)}$ and $\Gamma_{23}^{(c)}$ can be found elsewhere.[5,50]

3. Calculation of Δn_{12} (Δn_{32}) and Their Dependence on Various Factors

The excesses (or deficits) Δn_{12} or Δn_{32} were calculated with eqs 16 and 17 for (1) water (1) + lysozyme (2) + urea (3) (pH 7.0, 20 °C), (2) water (1) + lysozyme (2) + urea (3) (pH 2.0, 20 °C), and (3) water (1) + lysozyme (2) + glucose (3) (pH 3.0, 20 °C) mixtures. Whereas in the first mixture lysozyme is in its native state, in the second (at pH 2.0), it undergoes a transition to a denatured state between 2.5 and 5.0 M urea.[32] The water + lysozyme + glucose mixture was also considered because, in contrast to urea, glucose is a protein stabilizer.[58]

The results of the calculations are plotted in Figure 1 where only Δn_{12} is presented. All experimental data required to perform the calculations were taken from the literature.[32,58] In these calculations, the term $(G_{11} - G_{13})$ was approximated by $(V_3 - V_1)$, which, as demonstrated before,[52] constitutes a good approximation. The excesses (or deficits) Δn_{12} or Δn_{32} for the water (1) + lysozyme (2) + urea (3) mixtures were previously calculated by a different method involving a reference state.[50] The two calculations are in agreement. One can see from Figure 1 that, whereas the lysozyme is preferentially hydrated in the presence of glucose, urea preferentially pushes out water from the vicinity of the protein at both pH = 7 and 2. The results obtained are in agreement with the literature, which classifies urea as belonging to the first group of cosolvents,[1] in which water is preferentially excluded from the protein surface.[5,7,32,38−39,41−45] In contrast, the proteins are preferentially hydrated in the presence of glucose, which belongs to the second group of cosolvents.[5,7,32,38−39,41−45] Figure 1 shows that Δn_{12} for the water (1) + lysozyme (2) + urea (3) (pH 7.0) mixture becomes almost composition independent for $c_3^0 > 2$ mol/L. Such a behavior can be explained by a mechanism in which urea totally replaces water in some areas of the protein surface (probably because of urea−peptide H-bonds), whereas on the remaining part of the protein surface both water and urea are present with concentrations comparable to those in the bulk. It should be emphasized that, at pH = 7.0, the lysozyme is in its native state at all compositions, whereas at pH = 2.0 the

5 Aqueous solutions of biomolecules

Local Composition of a Protein Molecule in Solvent

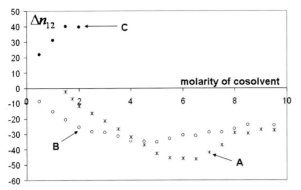

Figure 1. Excesses (or deficits) of water molecules in the vicinity of infinitely dilute lysozyme for various cosolvents. Δn_{12} is expressed in mol/mol: (A) Water (1) + lysozyme (2) + urea (3) (pH 2.0, 20 °C). (B) Water (1) + lysozyme (2) + urea (3) (pH 7.0, 20 °C). (C) Water (1) + lysozyme (2) + glucose (3) (pH 3.0, 20 °C).

TABLE 1: Δn_{12} in Ternary Ideal Mixture for Various Molar Volumes of the Cosolvent

c_3^0 [mol/L]	Δn_{12} [mol/mol]		
	$V_3^0 = 60$ [cm³/mol]	$V_3^0 = 114$ [cm³/mol]	$V_3^0 = 1000$ [cm³/mol]
0.5	0.07	0.29	25.73
1	0.12	0.54	47.90
1.5	0.17	0.76	67.60
2.0	0.22	0.94	84.36

lysozyme is in its native state only for $c_3^0 < 2$ mol/L and becomes denatured for $c_3^0 > 5$ mol/L. The lysozyme is a mixture of both native and denatured states in the composition range 2 mol/L $< c_3^0 <$ 5 mol/L. The composition dependence of the preferential binding parameter $\Gamma_{23}^{(m)}$ for water + lysozyme + urea at pH = 2.0 was examined previously.[32]

Now, we will try to examine the contributions to these excesses and deficits of various factors.

3.1. Contribution to the Excesses (or Deficits) Δn_{12} or Δn_{32} Due to Different Volumes of Water and Cosolvent Based on Ideal Ternary Mixture. The excesses (or deficits) in an ideal ternary mixture were calculated using eqs 12 and 13, and the results are listed in Table 1. The calculations were carried out for ideal mixtures with molar volumes of the pure components as in the mixture water + lysozyme + cosolvent. Table 1 shows that the infinitely dilute component 2 is usually preferentially hydrated because the cosolvents considered have molar volumes larger than that of water. When the molar volume of the cosolvent is small, Δn_{12} is also small. For example, for $V_3^0 = 114$ [cm³/mol] (this volume corresponds to the partial molar volume of glucose in an aqueous solution), Δn_{12} is small. However, when the size (molar volume) of the cosolvent is large (e.g., polyethylene glycol), Δn_{12} is large and its contribution to the total preferential hydration is important. Such a behavior was suggested in an unpublished opinion of Kauzmann (quoted by Timasheff[32]) that

> "the bulkiness of the cosolvent molecules creates around a protein molecule a zone that is impenetrable to the cosolvent, the thickness of which is determined by the distance of closest approach between protein and ligand molecules. This region can be penetrated by the smaller water molecules. Hence, it is enriched in water relative to the bulk solvent".

Whereas the ternary mixture used as an example cannot be considered an ideal mixture, it is interesting to note that such a

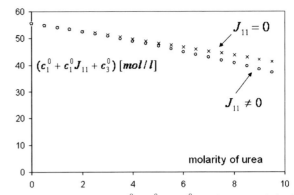

Figure 2. Dependence of $(c_1^0 + c_1^0 J_{11} + c_3^0)$ on the urea molarity in water + urea mixture for two cases: (1) $J_{11} = 0$ and (2) $J_{11} \neq 0$.

TABLE 2: Values of J_{21} for Water (1) + Lysozyme (2) + Urea (3) Mixture

c_3^0 [mol/L]	J_{21}			
	pH = 2.0		pH = 7.0	
	$J_{11} = 0$	$J_{11} \neq 0$	$J_{11} = 0$	$J_{11} \neq 0$
0.5			403.6	403.5
1.0			360.5	360.2
1.5	37.0	37.0	319.9	319.2
2.0	137.2	136.7	300.3	299.3
2.5	157.1	156.2	268.5	267.0
3.0	172.0	170.1	229.0	227.2
3.5	184.3	182.2	217.4	214.9
4.0	194.7	191.8	209.6	206.5
4.5	204.7	200.8	191.1	187.4
5.0	213.1	208.0	175.8	171.6
5.5	209.9	203.7	151.9	147.4
6.0	197.3	190.3	132.0	127.2
6.5	186.9	178.9	125.0	119.7
7.0	158.3	150.3	109.2	103.7
7.5	132.7	124.9	104.5	98.4
8.0	100.6	93.7	91.5	85.4
8.5	97.2	89.6	79.7	73.5
9.0	85.8	78.2	85.8	78.2
9.5	83.7	75.3	75.5	68.0

TABLE 3: Values of J_{21} for Water (1) + Lysozyme (2) + Glucose (3) Mixture

c_3^0 [mol/L]	J_{21}	
	$J_{11} = 0$	$J_{11} \neq 0$
0.5	−384.67	−384.6
1.0	−276.2	−275.9
1.5	−236.2	−235.6
2.0	−175.5	−174.7

description reflects correctly, at least qualitatively, the effect of the molar volume of the cosolvent belonging to the second kind of cosolvents (see Introduction).

3.2. Contribution to the Excesses (or Deficits) Δn_{12} or Δn_{32} by the Nonideality of the Mixed Solvent. The effect of the nonideality of the mixed solvent on Δn_{12} can be extracted from eq 8 (or eq 9) and stems from $J_{11} = \lim_{x_2 \to 0} ((\partial \ln \gamma_1)/(\partial x_1))_{x_2}$—the derivative of the activity coefficient of water in the protein-free mixed solvent (in the molar fraction scale) with respect to the mole fraction of water. Of course, $J_{11} = 0$ for an ideal mixed solvent, and $J_{11} \neq 0$ for a nonideal (real) mixed solvent. To evaluate how much the nonideality of the solvent mixture affects Δn_{12}, we compared $(c_1^0 + c_1^0 J_{11} + c_3^0)$ in eq 8 for both $J_{11} = 0$ and $J_{11} \neq 0$ (Figure 2). J_{11} was calculated from accurate data regarding the activity coefficient of water in the binary mixture

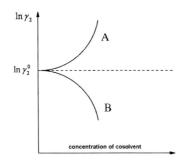

Figure 3. Illustration for the effect of the addition of a small amount of a cosolvent to an aqueous protein solution on the protein activity coefficient γ_2 (γ_2^0, protein activity coefficient at infinite dilution of the protein): (A) Cosolvents belonging to the second group (inorganic salts, glucose, glycerol, and similar substances). (B) Cosolvents belonging to the first group (urea, guanidine hydrochloride etc.).

water + urea.[59] One can see from Figure 2 that $(c_1^0 + c_1^0 J_{11} + c_3^0)$ has almost the same values for both $J_{11} = 0$ and $J_{11} \neq 0$ even for high urea molarities. Equation 8 was used to calculate $J_{21} = \lim_{x_2 \to 0}((\partial \ln \gamma_2)/(\partial x_1))_{x_2}$ for both $J_{11} = 0$ and $J_{11} \neq 0$. The calculations were carried out for (1) water (1) + lysozyme (2) + urea (3) (pH 7.0, 20 °C), (2) water (1) + lysozyme (2) + urea (3) (pH 2.0, 20 °C), and (3) water (1) + lysozyme (2) + glucose (3) (pH 3.0, 20 °C) mixtures, and the results are listed in Tables 2 and 3. Tables 2 and 3 demonstrate that $J_{11} = 0$ and $J_{11} \neq 0$ lead to almost the same results not only for Δn_{12} (or Δn_{32}), but also for J_{21}. Even at high urea concentrations ($c_3^0 > 7.0$ M), the difference between J_{21} calculated with $J_{11} = 0$ and $J_{11} \neq 0$ does not exceed 10% (see Table 2). Therefore, our calculations show that the nonideality of the urea−water solvent does not constitute an important factor in the calculation of the local compositions around a protein molecule. A similar conclusion is valid for the water−glucose mixed solvent (see Table 3).

Tables 2 and 3 also demonstrate that J_{21} constitutes a criterion for the cosolvent behavior. For cosolvents belonging to the first group (urea, guanidine hydrochloride, etc.; see Introduction), $J_{21} > 0$ and $J_{23} < 0$ (because $J_{23} = \lim_{x_2 \to 0}((\partial \ln \gamma_2)/(\partial x_3))_{x_2} = -J_{21}$). In contrast for cosolvents belonging to the second group (inorganic salts, glucose, glycerol, etc.) $J_{21} < 0$ and $J_{23} > 0$. Consequently, the addition of urea increases $\ln \gamma_2$, whereas the addition of glucose decreases $\ln \gamma_2$ (Figure 3). Hence, J_{21} (or J_{23}) can be used as a criterium for the behavior of a protein in a mixed solvent.

4. Calculation of J_{21} (J_{23}) and the Excesses (or Deficits) Δn_{12} (Δn_{32}) Using Various Theories

4.1. Flory−Huggins Equation for Real Solutions. The Flory−Huggins equation for real solutions[60−62] will be used to derive an expression for J_{21} (J_{23}) for water (1) + protein (2) + cosolvent (3) mixtures. The activity coefficient of component 2 in the above ternary mixture can be written in the following form[60]

$$\ln \gamma_2 = \ln(\varphi_2/x_2) + 1 - \varphi_2 - \varphi_1 V_2^\infty/V_1^0 - \varphi_3 V_2^\infty/V_3^0 - \chi_{13}\varphi_1\varphi_3 V_2^\infty/V_1^0 + (\chi_{12}\varphi_1 V_2^\infty/V_1^0 + \chi_{32}\varphi_3 V_2^\infty/V_3^0)(\varphi_1 + \varphi_3)$$
(18)

where φ_i is the volume fraction of component i ($\varphi_i = x_i V_i^0/(x_1 V_1^0 + x_2 V_2^0 + x_3 V_3^0)$, and we assumed that $V_2^0 = V_2^\infty$) and χ_{ij} is the Flory−Huggins interaction parameter between molecules i and

j. Differentiation yields for $x_3 \to 0$ the expression

$$\lim_{x_3 \to 0} J_{21} = \frac{\chi_{12}V_2^\infty V_3^0 - \chi_{32}V_1^0 V_2^\infty + \chi_{13}V_2^\infty V_3^0 + (V_2^\infty - V_1^0)(V_1^0 - V_3^0)}{(V_1^0)^2}$$
(19)

Because the molar volume of water is much smaller than the molar volume of the protein ($V_2^\infty \gg V_1^0$), eq 19 reduces to $J_{21} \geq 0$ when

$$\chi_{12}V_3^0 - \chi_{32}V_1^0 + \chi_{13}V_3^0 + V_1^0 - V_3^0 \geq 0 \quad (20)$$

and $J_{21} \leq 0$ when

$$\chi_{12}V_3^0 - \chi_{32}V_1^0 + \chi_{13}V_3^0 + V_1^0 - V_3^0 \leq 0 \quad (21)$$

Equations 20 and 21 provide general conditions for cases B (eq 20) and A (eq 21) of Figure 3. These inequalities show that the sign of J_{21} (J_{23}) depends on the Flory−Huggins interaction parameters χ_{ij} and the molar volumes of water and cosolvent.

For an ideal mixed solvent, $\chi_{13} = 0$, and eqs 20 and 21 can be written in the more simple forms $J_{21} \geq 0$ when

$$V_3^0(\chi_{12} - 1) \geq V_1^0(\chi_{32} - 1) \quad (22)$$

and $J_{21} \leq 0$ when

$$V_3^0(\chi_{12} - 1) \leq V_1^0(\chi_{32} - 1) \quad (23)$$

Usually, the Flory−Huggins interaction parameters χ_{ij} are positive quantities smaller than unity.[60−61,63−64] There are, however, cases in which χ_{ij} has negative values. Because the interaction of a protein with water or urea is exothermic,[38,65] the above parameters are expected to be negative. Indeed,[60] according to the van Laar expression for the heat of mixing in a two component system, the Flory−Huggins interaction parameter χ_{ij} is proportional to the heat of mixing in the binary system $i-j$. Furthermore, as demonstrated in a previous section, for the water + lysozyme + urea mixture $J_{21} \geq 0$, and eq 22 in which $V_3^0/V_1^0 \approx 2.5$ leads to

$$\chi_{32} \leq 2.5\chi_{12} - 1.5 \quad (24)$$

It should be noted that urea is a solid at 20 °C and that we have used for V_3^0 the partial molar volume of urea in an urea + water mixture at infinite dilution of urea.

From eq 24, one can conclude that urea is preferentially adsorbed by the protein because its energy of interaction with the protein is more negative than the energy of interaction of water with the protein.

4.2. Binding Theory. The binding theory[66−67] is based on the equilibrium between a ligand in solution L and a ligand on the protein P surface L(P)

$$L + P \leftrightarrow L(P) \quad (25)$$

Schellman[68−70] considered that this classical binding equilibrium is incomplete and suggested to replace eq 25 with the exchange equilibrium

$$L + P \cdot N(H_2O) \leftrightarrow L(P) + N(H_2O) \quad (26)$$

In Schellman's model,[68−70] the protein surface is subdivided into N sites that can be occupied by water or cosolvent

Local Composition of a Protein Molecule in Solvent

molecules, and there are no unoccupied sites. According to this model, the preferential binding parameter $\Gamma_{23}^{(m)}$ can be written as the sum

$$\Gamma_{23}^{(m)} = \sum_{i=1}^{N} \Gamma_{23}^{(m)}(i) \quad (27)$$

where

$$\Gamma_{23}^{(m)}(i) = \frac{(K_i - 1)x_3^0}{1 + (K_i - 1)x_3^0} \quad (28)$$

and

$$K_i = \overline{K}_i \gamma_3^0 / \gamma_1^0 \quad (29)$$

In the above expressions, $\Gamma_{23}^{(m)}(i)$ is the preferential binding parameter for site i, γ_1^0 and γ_3^0 are the activity coefficients of water and cosolvent (in a molar fraction scale) in a protein-free mixed solvent, and $\overline{K}_i$ is the equilibrium constant for the exchange equilibrium (eq 26) on site i. For an ideal mixed solvent, $K_i = \overline{K}_i$. In the Schellman model, it is necessary to know how the protein surface is subdivided into various kinds of sites. Such information is not available for real protein solutions. However, when some simplifications are made, the Schellman model can provide some information regarding the effect of various cosolvents on the protein stability.[67–71]

The simplest approximation is to consider that all the sites are identical and independent, an approximation that was applied by Timasheff to several water + protein + cosolvent mixtures.[67] In this case

$$\Gamma_{23}^{(m)} = \frac{N(K-1)x_3^0}{1 + (K-1)x_3^0} = \frac{N(K-1)m_3}{m_1 + Km_3} \quad (30)$$

where m_1 and m_3 are the molalities of water and cosolvent in the protein-free mixed solvent. The constant $K > 1$ for the first group of cosolvents (urea, some derivatives of urea, guanidine hydrochloride, etc.) and $K < 1$ for the second group of cosolvents (inorganic salts, glycerol, sucrose, or similar compounds). Following a suggestion of Kuntz,[72] Timasheff approximated the total number of sites N as the number of water molecules of hydration of the protein ($N = 290$ for water + lysozyme + urea). We have used this estimation and eq 30 to correlate the preferential binding parameter $\Gamma_{23}^{(m)}$ with m_3 for the water (1) + lysozyme (2) + urea (3) mixture (pH 7.0, 20 °C). The constant K was considered as an adjustable parameter, and the best fit was found for $K = 1.05$. The results of the calculations and the experimental values of the preferential binding parameter $\Gamma_{23}^{(m)}$ are plotted in Figure 4, which shows that eq 30 does not correlate satisfactorily the composition dependence of the preferential binding parameter $\Gamma_{23}^{(m)}$. This is not surprising because the protein surface is not uniform. Indeed, there are polar, nonpolar, and charged sites on the protein surface. An estimation of the various surface areas of a protein is available in the literature.[30] According to this estimation,[30] 57% of the accessible protein surface of an average native protein is nonpolar, 13% is polar due to the peptide backbone and can form hydrogen bonds, 19% is charged, and 11% is polar due to other polar groups. Consequently, there are at least four kind of sites on the accessible protein surface. For the sake of simplicity, it will be considered that the accessible protein surface consists of two kinds of sites; one kind is polar and can

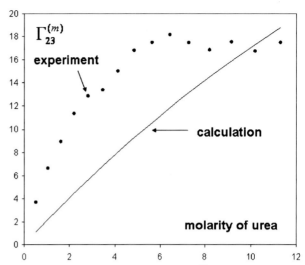

Figure 4. Comparison between experimental[32] and calculated preferential binding parameter $\Gamma_{23}^{(m)}$ for water (1) + lysozyme (2) + urea (3) mixture (pH 7.0, 20 °C).

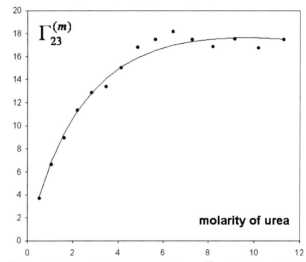

Figure 5. Comparison between experimental[32] and calculated (with eq 31) preferential binding parameter $\Gamma_{23}^{(m)}$ for water (1) + lysozyme (2) + urea (3) mixture at pH 7.0 and 20 °C. The solid line represent the results of calculations, and (•) are experimental points.

form hydrogen bonds with water and urea, and the other kind does not participate in hydrogen bonding.

In this case, eqs 27 and 28 lead to

$$\Gamma_{23}^{(m)} = \frac{N_\alpha(K_\alpha - 1)x_3^0}{1 + (K_\alpha - 1)x_3^0} + \frac{N_\beta(K_\beta - 1)x_3^0}{1 + (K_\beta - 1)x_3^0} = \frac{N_\alpha(K_\alpha - 1)m_3}{m_1 + K_\alpha m_3} + \frac{N_\beta(K_\beta - 1)m_3}{m_1 + K_\beta m_3} \quad (31)$$

where N_α and N_β represent the numbers of the two kinds of sites, $N_\alpha + N_\beta = N$, $N_\alpha = 0.13N$, $N_\beta = 0.87N$. Again, the constants K_α and K_β will be considered as adjustable parameters. We found that the best fit provides $K_\alpha = 15.29$ and $K_\beta = 0.79$. The results are plotted in Figure 5, which shows that eq 31 accurately represents the composition dependence of the preferential binding parameter $\Gamma_{23}^{(m)}$. Equation 31 can also be used

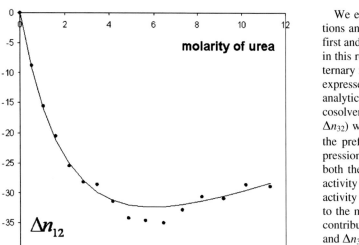

Figure 6. Δn_{12} for water (1) + lysozyme (2) + urea (3) mixture at pH 7.0 and 20 °C calculated with eq 16 and experimental data[32] on $\Gamma_{23}^{(m)}$ (•) and with $\Gamma_{23}^{(m)}$ provided by eq 31 (solid line).

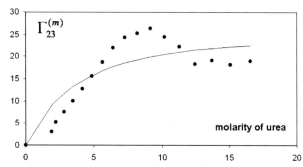

Figure 7. Comparison between experimental[32] and calculated (using eq 31) preferential binding parameter $\Gamma_{23}^{(m)}$ for water (1) + lysozyme (2) + urea (3) mixture at pH 2.0 and 20 °C. The solid line is the result of calculations, and (•) are the experimental points.

to calculate the deficit of water in the vicinity of the protein surface Δn_{12} via eq 16 (as before, the term $(G_{11} - G_{13})$ is approximated as $(V_3 - V_1)$). The results are plotted in Figure 6.

However, eq 31 is not accurate for the water (1) + lysozyme (2) + urea (3) mixture at pH 2.0 and 20 °C (see Figure 7).

5. Discussion and Conclusion

In this paper, water (1) + protein (2) + cosolvent (3) mixtures have been examined with the focus on the local composition of water and cosolvent in the vicinity of the protein surface. Cosolvents can affect the local composition of water/cosolvent in the vicinity of the protein surface in various ways:

(1) a small group of cosolvents (urea, some derivatives of urea, guanidine hydrochloride, etc.) preferentially expel water from the protein surface (at least in dilute solutions), and the local composition of the cosolvent is higher in the vicinity of the protein surface than in the bulk, and

(2) usual cosolvents (inorganic salts, glycerol, sucrose, or similar compounds) are preferentially expelled from the protein surface, the protein is preferentially hydrated, and the local water composition is higher that that in the bulk far from the infinitely dilute protein molecule.

We examined various factors that affect the local compositions and tried to explain why some compounds belong to the first and other to the second group of cosolvents. The main tool in this research was the Kirkwood–Buff fluctuation theory for ternary mixture. First, the preferential binding parameters were expressed in terms of the Kirkwood–Buff integrals,[49–50] and analytical expressions for the excesses (or deficits) of water and cosolvent around an infinitely dilute protein molecule (Δn_{12} and Δn_{32}) were derived, which showed how they are connected to the preferential binding parameters $\Gamma_{23}^{(m)}$ and $\Gamma_{23}^{(c)}$. These expressions (eqs 8 and 9) revealed that Δn_{12} and Δn_{32} depend on both the properties of the protein-free mixed solvent and the activity coefficient of the protein, namely the derivative of the activity coefficient of the protein at infinite dilution with respect to the mixed solvent composition (J_{21}). It was shown that the contribution of the nonideality of the mixed solvent J_{11} to Δn_{12} and Δn_{32} is usually small for the cosolvents considered and that J_{21} constitutes a most important factor.

Three mixtures [(1) water (1) + lysozyme (2) + urea (3) (pH 7.0, 20 °C), (2) water (1) + lysozyme (2) + urea (3) (pH 2.0, 20 °C), (3) water (1) + lysozyme (2) + glucose (3) (pH 3.0, 20 °C)] were selected for numerical estimations (urea belonging to the first group of cosolvents and glucose to the second group).

The composition dependence of Δn_{12} (Δn_{32}) and J_{21} for these mixtures were calculated from experimental data regarding the preferential binding parameter and the protein partial molar volume at infinite dilution. It was found that J_{21} provides a criterium for the cosolvent behavior. For cosolvents belonging to the first group, $J_{21} > 0$. In contrast, for cosolvents belonging to the second group, $J_{21} < 0$. Consequently, the addition of urea decreases $\ln \gamma_2$, whereas the addition of, e.g., glucose, increases $\ln \gamma_2$ (Figure 3). In order to estimate J_{21}, the Flory–Huggins equation for ternary solutions was used and an analytical expression for J_{21} was derived. Using this expression, the inequalities $J_{21} > 0$ and $J_{21} < 0$ were recast in terms of the Flory–Huggins interaction parameters and molar volumes of water and cosolvent molecules.

The composition dependence of Δn_{12} (Δn_{32}) for the water (1) + lysozyme (2) + urea (3) mixture (Figure 1) indicates that Δn_{12} (Δn_{32}) is almost composition independent for $c_3^0 > 2$ mol/L (especially for pH = 7.0). It is worth noting that at pH = 7.0 the lysozyme is in its native state at all urea concentrations, whereas at pH = 2.0 the lysozyme is in its native state only for $c_3^0 < 2$ mol/L and becomes denatured for $c_3^0 > 5$ mol/L. The lysozyme is a mixture of native and denatured states in the composition range 2 mol/L $< c_3^0 <$ 5 mol/L urea. In order to explain such a dependence, the Schellman exchange model[68–70] was used. First, we considered a single kind of site and concluded that such a model cannot be applied to the water + lysozyme + urea mixture. Second, we considered that there are two kinds of sites on the protein surface: one kind of site can form H-bonds with water and urea, whereas the second kind of site does not form such bonds. Using such an approximation, the Schellman exchange model could describe the composition dependencies of $\Gamma_{23}^{(m)}$ and Δn_{12} (Δn_{32}) for the water (1) + lysozyme (2) + urea (3) mixture at pH 7.0 and 20 °C on the concentration of urea. However, the results for pH = 2.0 and 20 °C were not too accurate because at this pH the lysozyme is a mixture of native and denatured molecules whose composition depends on the urea concentration.

The large difference between the equilibrium constants (K_α = 15.29 and K_β = 0.79) indicates an enormous difference in the concentrations of urea in the two kinds of sites.

Appendix

The Kirkwood–Buff integrals are defined as[73]

$$G_{\alpha\beta} = \int_0^\infty (g_{\alpha\beta} - 1) 4\pi r^2 \, dr \qquad (A1)$$

where $g_{\alpha\beta}$ is the radial distribution function between species α and β and r is the distance between the centers of molecules α and β.

The Kirkwood–Buff integrals in an n-component mixture can be obtained from the following relation[73]

$$G_{\alpha\beta} = \frac{v|A|_{\alpha\beta}}{\langle N_\alpha \rangle \langle N_\beta \rangle |A|} - \frac{\delta_{\alpha\beta}}{c_\beta} \qquad (A2)$$

where $\langle N_\alpha \rangle$ and $\langle N_\beta \rangle$ are the average numbers of molecules of α and β, respectively, in the volume v, $\delta_{\alpha\beta}$ is the Kronecker symbol ($\delta_{\alpha\beta} = 1$ for $\alpha = \beta$ and $\delta_{\alpha\beta} = 0$ for $\alpha \neq \beta$), c_β is the bulk molecular concentration of species β ($c_\beta = (\langle N_\beta \rangle)/v$), $|A|_{\alpha\beta}$ represents the cofactor of $A_{\alpha\beta}$ in the determinant $|A|$, and $A_{\alpha\beta}$ is given by[73]

$$A_{\alpha\beta} = \frac{1}{kT}\left(\left(\frac{\partial \mu_\alpha}{\partial N_\beta}\right)_{T,P,N_\gamma} + \frac{\nu_\alpha \nu_\beta}{k_T v}\right) \qquad (A3)$$

In eq A3, k is the Boltzmann constant, μ_α is the chemical potential per molecule of species α, ν_α and ν_β are the partial molar volumes per molecule of species α and β, respectively, and k_T is the isothermal compressibility. The derivative $\mu_{\alpha\beta} = ((\partial \mu_\alpha)/(\partial N_\beta))_{T,P,N_\gamma}$ is taken under isothermal–isobaric conditions for $N_\gamma = $ const with $\gamma \neq \beta$.

Equation A2 can be recast in the following form:

$$G_{\alpha\beta} = \frac{|A|_{\alpha\beta}}{v c_\alpha c_\beta |A|} - \frac{\delta_{\alpha\beta}}{c_\beta} \qquad (A4)$$

$G_{\alpha\beta}$ in eqs A2 and A4 is expressed as volume per molecule. These equations are valid for any n-component system. Expressions for the Kirkwood–Buff integrals in ternary mixtures can be obtained from eq A4.[50] In particular,[50] one can obtain the following expressions for G_{12} and G_{23} for an infinitely dilute solute (component 2)

$$G_{12} = kTk_T - \frac{J_{21}V_3 c_3^0 + J_{11}V_2^\infty c_1^0}{(c_1^0 + c_1^0 J_{11} + c_3^0)} - \frac{V_3 c_3^0 (c_1^0 + c_3^0)(V_1 - V_3) + V_2^\infty (c_1^0 + c_3^0)}{(c_1^0 + c_1^0 J_{11} + c_3^0)} \qquad (A5)$$

$$G_{23} = kTk_T + \frac{J_{21}V_1 c_1^0 - J_{11} c_1^0 V_2^\infty}{(c_1^0 + c_1^0 J_{11} + c_3^0)} + \frac{c_1^0 V_1 (c_1^0 + c_3^0)(V_1 - V_3) - V_2^\infty (c_1^0 + c_3^0)}{(c_1^0 + c_1^0 J_{11} + c_3^0)} \qquad (A6)$$

The Kirkwood–Buff integrals G_{11} and G_{13} can be calculated from the characteristics of the protein-free mixed solvent (binary mixture 1–3) using the expressions[74–76]

$$G_{11} = kTk_T - \frac{(c_1^0 + c_3^0)^2 V_1 V_3}{(c_1^0 + c_1^0 J_{11} + c_3^0)} + \frac{(c_1^0 + c_3^0)(V_3 - V_1) - J_{11}}{(c_1^0 + c_1^0 J_{11} + c_3^0)} \qquad (A7)$$

and

$$G_{13} = kTk_T - \frac{(c_1^0 + c_3^0)^2 V_1 V_3}{(c_1^0 + c_1^0 J_{11} + c_3^0)} \qquad (A8)$$

References and Notes

(1) Timasheff, S. N. A physicochemical basis for the selection of osmolytes by nature. In *Water and life : comparative analysis of water relationships at the organismic, cellular, and molecular levels*; Somero, G. N., Osmond, C. B., Bolis, C. L., Eds.; Springer-Verlag: New York, 1992; pp 70–84.

(2) Mahler, H. R.; Cordes, E. H. *Biological chemistry*; Harper & Row: New York, 1966.

(3) Kauzmann, W. Denaturation of proteins and enzymes. In *The mechanism of enzyme action*; McElroy, W. D., Glass, B., Eds.; Johns Hopkins Press: Baltimore, 1954; pp 71–110.

(4) Kauzmann, W. *Adv. Protein Chem.* **1959**, *14*, 1–63.

(5) Casassa, E. F.; Eisenberg, H. *Adv. Protein Chem.* **1964**, *19*, 287–393.

(6) Wyman, J. *Adv. Protein Chem.* **1964**, *19*, 223–286.

(7) Timasheff, S. N. *Annu. Rev. Biophys. Biomol. Struct.* **1993**, *22*, 67–97.

(8) Robinson, D. R.; Jencks, W. P. *J. Am. Chem. Soc.* **1965**, *87*, 2462–2470.

(9) Pace, C. N.; Trevino, S.; Prabhakaran, E.; Scholtz, J. M. *Philos. Trans. R. Soc. London, Ser. B* **2004**, *359*, 1225–1234.

(10) Cohn, E. J.; Edsall, J. T. *Proteins, Amino Acids and Peptides*; Reinhold: New York, 1943.

(11) Green, A. A.; Hughens, W. L. *Methods Enzymol.* **1955**, *1*, 67–90.

(12) Arakawa, T.; Timasheff, S. N. *Methods Enzymol.* **1985**, *114*, 49–74.

(13) Schein, C. H. *Biotechnology* **1990**, *8*, 308–315.

(14) *Crystallization of nucleic acids and proteins: a practical approach*; Riès-Kautt, M., Ducruix, A., Eds.; Oxford University Press: New York, 1999.

(15) von Hippel, P. H.; Wong, K.-Y. *Science* **1964**, *145*, 577–580.

(16) von Hippel, P.; Schleich, T. In *Structure and Stability of Macromolecules in Solution*; Timasheff, S., Fasman, G., Eds.; Marcel Dekker: New York, 1969; Vol. 11.

(17) Greene, R. F.; Pace, C. N. *J. Biol. Chem.* **1974**, *249*, 5388–5393.

(18) Prakash, V.; Loucheux, C.; Scheufele, S.; Gorbunoff, M. J.; Timasheff, S. N. *Arch. Biochem. Biophys.* **1981**, *210*, 455–464.

(19) Yancey, P. H.; Clark, M. E.; Hand, S. C.; Bowlus, R. D.; Somero, G. N. *Science* **1982**, *217*, 1214–1222.

(20) Pace, C. N. *Methods Enzymol.* **1986**, *131*, 266–280.

(21) Nilsson, S.; Piculell, L.; Malmsten, M. *J. Phys. Chem.* **1990**, *94*, 5149–5154.

(22) Muller, N. *J. Phys. Chem.* **1990**, *94*, 3856–3859.

(23) Pace, C. N.; Laurents, D. V.; Thomson, J. A. *Biochemistry* **1990**, *29*, 2564–2572.

(24) Lin, T.-Y.; Timasheff, S. N. *Biochemistry* **1994**, *33*, 12695–12701.

(25) Liepinsch, E.; Otting, G. *J. Am. Chem. Soc.* **1994**, *116*, 9670–9674.

(26) Zhang, W. T.; Capp, M. W.; Bond, J. P.; Anderson, C. F.; Record, M. T. *Biochemistry* **1996**, *35*, 10506–10516.

(27) Schellman, J. A.; Gassner, N. C. *Biophys. Chem.* **1996**, *59*, 259–275.

(28) Wang, A.; Bolen, D. W. *Biochemistry* **1997**, *36*, 9101–9108.

(29) Zou, Q.; Habermann-Rottinghaus, S. M.; Murphy, K. P. *Proteins* **1998**, *31*, 107–115.

(30) Courtenay, E. S.; Capp, M. W.; Record, M. T. *Protein Sci.* **2001**, *10*, 2485–2497.

(31) Konermann, L. Protein unfolding and denaturants. In *Encyclopedia of life sciences*; Nature Publishing Group: New York, 2002.

(32) Timasheff, S. N.; Xie, G. *Biophys. Chem.* **2003**, *105*, 421–448.

(33) Bennion, B. J.; Daggett, V. *Proc. Natl. Acad. Sci. U.S.A.* **2003**, *100*, 5142–5147.

(34) Hong, J.; Capp, M. W.; Anderson, C. F.; Record, M. T. *Biophys. Chem.* **2003**, *105*, 517–532.

(35) Felitsky, D. J.; Record, M. T. *Biochemistry* **2004**, *43*, 9276–9288.

(36) Hong, J.; Capp, M. W.; Anderson, C. F.; Soecker, R. M.; Felitsky, D. J.; Anderson, M. W.; Record, M. T. *Biochemistry* **2004**, *43*, 14744–14758.

(37) Mountain, R. D.; Thirumalai, D. *J. Phys. Chem. B* **2004**, *108*, 6826–6831.

(38) Kuntz, I. D.; Kauzmann, W. *Adv. Protein. Chem.* **1974**, *28*, 239–345.

(39) Timasheff, S. N. *Adv. Prot. Chem.* **1998**, *51*, 355–432.

(40) Stigter, D. *J. Phys. Chem.* **1960**, *64*, 842–846.

(41) Courtenay, E. S.; Capp, M. W.; Anderson, C. F.; Record, M. T. *Biochemistry* **2000**, *39*, 4455−4471.
(42) Arakawa, T.; Timasheff, S. N. *Biochemistry* **1982**, *21*, 6545−6552.
(43) Arakawa, T.; Timasheff, S. N. *Biochemistry* **1984**, *23*, 5912−5923.
(44) Timasheff, S. N.; Arakawa, T. *J. Cryst. Growth* **1988**, *90*, 39−46.
(45) Arakawa, T.; Timasheff, S. N. *Biochemistry* **1987**, *26*, 5147−5153.
(46) Arakawa, T.; Bhat, R.; Timasheff, S. N. *Biochemistry* **1990**, *29*, 1914−1923.
(47) Record, M. T. J.; Zhang, W.; Anderson, C. F. *Adv. Protein Chem.* **1998**, *51*, 281−353.
(48) Anderson, C. F.; Courtenay, E. S.; Record, M. T. *J. Phys. Chem. B* **2002**, *106*, 418−433.
(49) Shulgin, I. L.; Ruckenstein, E. *Biophys. J.* **2006**, *90*, 704−707.
(50) Shulgin, I. L.; Ruckenstein, E. *J. Chem. Phys.* **2005**, *123*, 054909.
(51) Ben-Naim, A. *Statistical Thermodynamics for Chemists and Biochemists*; Plenum Press: New York, 1992.
(52) Shulgin, I. L.; Ruckenstein, E. *Biophys. Chem.* **2006**, *120*, 188−198.
(53) Shimizu, S. *Proc. Natl. Acad. Sci. U.S.A.* **2004**, *101*, 1195−1199.
(54) Smith, P. E. *J. Phys. Chem. B* **2004**, *108*, 16271−16278.
(55) Schurr, J. M.; Rangel, D. P.; Aragon, S. R. *Biophys. J.* **2005**, *89*, 2258−2276.
(56) Shulgin, I. L.; Ruckenstein, E. *J. Phys. Chem. B* **2006**, *110*, 12707−12713.
(57) Noelken, M. E.; Timasheff, S. N. *J. Biol. Chem.* **1967**, *242*, 5080−5085.
(58) Arakawa, T.; Timasheff, S. N. *Biochemistry* **1982**, *21*, 6536−6544.
(59) Miyawaki, O.; Saito, A.; Matsuo, T.; Nakamura, K. *Biosci. Biotechnol. Biochem.* **1997**, *61*, 466−469.
(60) Flory, P. *Principles of Polymer Chemistry*; Cornell University Press: Ithaca, NY, 1953.
(61) Prausnitz, J. M.; Lichtenthaler, R. N.; Gomes de Azevedo, E. *Molecular thermodynamics of fluid−Phase equilibria,* 2nd ed.; Prentice-Hall: Englewood Cliffs, NJ, 1986.
(62) Scott, R. L. *J. Chem. Phys.* **1949**, *17*, 268−279.
(63) Huggins, M. L. *Physical chemistry of high polymers*, Wiley: New York, 1958.
(64) Chao, K. C.; Greenkorn, R. A. *Thermodynamics of fluids: an introduction to equilibrium theory*; Marcel Dekker: New York, 1975.
(65) Paz Andrade, M. I.; Jones, M. N.; Skinner, H. A. *Eur. J. Biochem.* **1976**, *66*, 127−131.
(66) Tanford, C. *Physical chemistry of macromolecules*; Wiley: New York, 1961.
(67) Timasheff, S. N. *Biophys. Chem.* **2002**, *101−102*, 99−111.
(68) Schellman, J. A. *Biopolymers* **1987**, *26*, 549−559.
(69) Schellman, J. A. *Annu. Rev. Biophys. Biophys. Chem.* **1987**, *16*, 115−137.
(70) Schellman, J. A. *Biophys. Chem.* **1990**, *37*, 121−140.
(71) Schellman, J. A. *Biophys. J.* **2003**, *85*, 108−125.
(72) Kuntz, I. D. *J. Am. Chem. Soc.* **1971**, *93*, 514−516.
(73) Kirkwood, J. G.; Buff, F. P. *J. Chem. Phys.* **1951**, *19*, 774−777.
(74) Matteoli, E.; Lepori, L. *J. Chem. Phys.* **1984**, *80*, 2856−2863.
(75) Shulgin, I.; Ruckenstein, E. *J. Phys. Chem. B* **1999**, *103*, 2496−2503.
(76) Ruckenstein, E.; Shulgin, I. *J. Phys. Chem. B* **1999**, *103*, 10266−10271.

Local Composition in Solvent + Polymer or Biopolymer Systems

Ivan L. Shulgin and Eli Ruckenstein*

Department of Chemical & Biological Engineering, State University of New York at Buffalo, Amherst, New York 14260

Received: August 13, 2007; In Final Form: November 30, 2007

The focus of this paper is on the application of the Kirkwood−Buff (KB) fluctuation theory to the analysis of the local composition in systems composed of a low molecular weight solvent and a high molecular weight polymer or protein. A key quantity in the calculation of the local composition is the excess (or deficit) of any species i around a central molecule j in a binary mixture. A new expression derived by the authors (*J. Phys. Chem. B* **2006**, *110*, 12707) for the excess (deficit) is used in the present paper. First, the literature regarding the local composition in such systems is reviewed. It is shown that the frequently used Zimm cluster integral provides incorrect results because it is based on an incorrect expression for the excess (or deficit). In the present paper, our new expression is applied to solvent + macromolecule systems to predict the local composition around both a solvent and a macromolecule central molecule. Five systems (toluene + polystyrene, water + collagen, water + serum albumin, water + hydroxypropyl cellulose, and water + Pluronic P105) were selected for this purpose. The results revealed that for water + collagen and water + serum albumin mixtures, the solvent was in deficit around a central solvent molecule and that for the other three mixtures, the opposite was true. In contrast, the solvent was always in excess around the macromolecule for all mixtures investigated. In the dilute range of the solvent, the excesses are due mainly to the different solvent and macromolecule sizes. However, in the dilute range of the macromolecule, the intermolecular interactions between solvent and macromolecule are mainly responsible for the excess. The obtained results shed some light on protein hydration.

1. Introduction

Systems composed of low molecular weight solvents and high molecular weight polymers or proteins, etc., are prone to various types of molecular clustering. We will consider a cluster as a micropart of a system in which the concentration differs from the bulk concentration. The first kind of clustering is the solvent clustering on high-weight polymers when the solvent as a vapor is adsorbed on the polymer.[1−3] Examples are the clustering of benzene on rubber,[1] toluene on polystyrene,[1] water on cellulose,[2] etc. Another kind of aggregation is the adsorption of water on a protein that leads to an excess (compared to the bulk) of the concentration of water in the vicinity of the protein surface.[4−5] Finally, one more kind of clustering is the aggregation of polymer molecules in water or aqueous solutions. As an example, one should mention the self-assembled aggregation of polyether block copolymers in water and water + cosolvent mixtures under particular conditions (above a certain concentration (cmc) and temperature (cmt), which depend on the cosolvent type and amount).[6] All the above-mentioned molecular aggregations in systems containing polymers, proteins, etc., are of industrial significance.[1−6] In addition, the understanding of the molecular origin of such phenomena is of importance in the theoretical understanding of the above systems.

The molecular clustering in systems composed of low molecular weight solvents and high molecular weight polymers, proteins, etc., has been investigated both experimentally and theoretically. The present paper is focused on the theoretical investigation of clustering on the basis of the fluctuation theory of Kirkwood and Buff (KB).[7] In a previous paper,[8] we applied the KB theory to the local composition in binary systems composed of two low molecular weight components. In the present paper, the systems are composed of low molecular weight solvents and high molecular weight polymers, proteins, etc.

2. Theoretical Background

2.1. The Zimm Cluster Integral. Zimm[9] was the first to apply the KB theory to the solvent clustering in binary solvent (1) + polymer (protein) (2) mixtures. On the basis of the KB theory, he was the first to derive the following expression for the KB integral (KBI),

$$G_{11} = kTk_T + \frac{c_2 v_2}{c_1^2 \mu_{11}^{(c)}} - \frac{1}{c_1} \quad (1)$$

where T is the absolute temperature; v_α is the partial molar volume per molecule of species α; k is the Boltzmann constant; k_T is the isothermal compressibility; c_α is the bulk molar concentration of component α; $\mu_{11}^{(c)} = (1/kT)(\partial\mu_1/\partial c_1)_{T,P} = (\partial \ln a_1/\partial c_1)_{T,P}$, μ_1 being the chemical potential per molecule 1; P is the pressure; a_1 is the activity of component 1; and G_{ij} is the Kirkwood−Buff integral (KBI),[7] which is defined as

$$G_{ij} = \int_0^\infty (g_{ij} - 1) 4\pi r^2 \, dr \quad (2)$$

In the above equation, g_{ij} is the radial distribution function between species i and j, and r is the distance between the centers of molecules i and j. Equation 1 is equivalent to the usual

* Corresponding author. Phone: (716) 645-2911, ext. 2214. Fax: (716) 645-3822. E-mail: feaeliru@acsu.buffalo.edu.

expression employed[10] to calculate the KBI, which involves the derivative of the chemical potential with respect to the mole fraction (see the Supporting Information for details).

Because far from critical conditions the contribution of the compressibility term (kTk_T) is very small,[2,9–10] eq 1 acquires the form

$$G_{11} \approx \frac{c_2 v_2}{c_1^2}\left(\frac{\partial c_1}{\partial \ln a_1}\right)_{P,T} - \frac{1}{c_1} \quad (3)$$

Equation 3 can be rewritten in the following frequently used form[9]

$$G_{11} \approx -v_1 \varphi_2 \left(\frac{\partial(a_1/\varphi_1)}{\partial a_1}\right)_{P,T} - v_1 \quad (4)$$

where $\varphi_i = c_i v_i$ is the volume fraction of component i in the mixture.

Further, Zimm[9] and Zimm and Lundberg[1] introduced the notion of "cluster integral" G_{11}/v_1 to characterize the solvent clustering in the systems solvent (1)–polymer (protein) (2). They called G_{11}/v_1 a "cluster integral", because[1] "the quantity $\varphi_1 G_{11}/v_1$ is the mean number of type 1 molecules in excess of the mean concentration of type 1 molecules in the neighborhood of a given type 1 molecule; thus, it measures the clustering tendency of the type 1 molecules".

Using eqs 3 and 4, one can write the following expression for the cluster integral G_{11}/v_1:

$$G_{11}/v_1 \approx \frac{\varphi_2}{\varphi_1 c_1}\left(\frac{\partial c_1}{\partial \ln a_1}\right)_{P,T} - \frac{1}{\varphi_1} = -\varphi_2\left(\frac{\partial(a_1/\varphi_1)}{\partial a_1}\right)_{P,T} - 1 \quad (5)$$

The average number of solvent molecules in a cluster or the mean size of the cluster was considered to be given by[2,11]

$$\frac{\varphi_1 G_{11}}{v_1} + 1 = -\varphi_1 \varphi_2 \left(\frac{\partial(a_1/\varphi_1)}{\partial \ln a_1}\right)_{P,T} +$$

$$\varphi_2 = \varphi_2 \left(\frac{\partial \ln \varphi_1}{\partial \ln a_1}\right)_{P,T} \quad (6)$$

Hence, the average number of solvent molecules in a cluster was considered as the excess (deficit) (compared to the bulk) of solvent molecules plus one (the central molecule).

The Zimm[9] (or Zimm and Lundberg[1]) theory of solvent clustering became a popular tool for the estimation of solvent clustering in solvent (1)-polymer (protein) (2) systems.[2,11–34] It has been used to estimate the solvent clustering in numerous systems.[1,2,11–34] However, several authors noted some inconsistencies in the application of the above theory to the solvent clustering. For instance, Brown[19] expressed doubts regarding the interpretation of $\varphi_1 G_{11}/v_1$ as the excess solvent molecules around a central solvent molecule. Klyuev and Grebennikov[31] noted that the average number of solvent molecules in a cluster calculated with eq 6 can be less than one, even though the central molecule is already included in the cluster. The authors of the present paper recently found[8,35] that $\varphi_1 G_{11}/v_1$ does not represent the excess (or deficit) of molecules 1 around a central molecule 1 compared to the bulk. The interpretation of $\varphi_1 G_{11}/v_1 = c_1 G_{11}$ as the excess (or deficit) was used in all publications based on the Zimm and Lundberg theory of solvent clustering and in all the papers that employed the KB theory of solutions (for more details, see refs 8 and 35).

2.2. Expression for the Excess (or Deficit) in Solvent + Polymer (Protein) Systems. Recently, we demonstrated that the quantity $c_i G_{ij}$ does not represent the excess (or deficit) of molecules i around a central molecule j compared to the bulk. It was shown that the excess (or deficit) molecules i around a central molecule j is given by the expression[35]

$$\Delta n_{ij} = c_i G_{ij} + c_i(v_j - kTk_T) \quad (7)$$

In contrast to the traditional expression ($\Delta N_{ij} = c_i G_{ij}$) for the excess, the new expression takes into account that owing to the central molecule j, there is a volume that is not accessible to molecules i. The difference between the above two expressions for the excess (or deficit) is particularly large for central molecules that possess large volumes.[8,35]

Far from critical conditions, one can neglect the compressibility term (kTk_T) as compared to v_j. Consequently, one can write the following expression for the excess (or deficit) number of molecules i around a central molecule j:

$$\Delta n_{ij} \approx c_i G_{ij} + c_i v_j \quad (8)$$

The excess (deficit) of a solvent around a central solvent molecule $\varphi_1 G_{11}/v_1$ provided by the Zimm cluster integral is related to the true excess (or deficit) Δn_{11} via the expression

$$\Delta n_{11} = (\varphi_1 G_{11}/v_1) + \varphi_1 \quad (9)$$

Therefore, φ_1 should be added to the Zimm excess to get the true excess, Δn_{11}.

Can the above equations provide information about the clustering? Let us consider the clustering of molecules 1 around a central molecule 1 in a volume, V^{corr}, of radius R in which the concentration differs from that in the bulk. This volume V^{corr} is usually called correlation volume. The total number, n_{11}, of molecules 1 in this volume (which can be identified as the size of cluster) can be calculated using the expression[36–37]

$$n_{11} = c_1 \int_0^R g_{11} 4\pi r^2 \, dr \quad i,j = 1,2 \quad (10)$$

which can be rewritten as

$$n_{11} = c_1 \int_0^R (g_{11} - 1) 4\pi r^2 \, dr + c_1 \int_0^R 4\pi r^2 \, dr \quad (11)$$

As soon as R becomes large enough for g_{11} to become unity, eq 11 can be rewritten as[36]

$$n_{11} = c_1 \int_0^\infty (g_{11} - 1) 4\pi r^2 \, dr + c_1 \int_0^R 4\pi r^2 \, dr = c_1 G_{11} + \frac{c_1 4\pi R^3}{3} = c_1 G_{11} + c_1 V^{corr} \quad (12)$$

The number of species 1 in the cluster is given by $n_{11} + 1$. Comparison of this number with the size of the Zimm cluster (eq 6) reveals that $n_{11} + 1$ is larger than $\varphi_1 G_{11}/v_1 + 1$ by $c_1 V^{corr}$. Thus, eq 12 provides a cluster size that includes all molecules 1 and not only the excess (or deficit). It also provides a clue as to why the average size of a cluster calculated using the expression ($\varphi_1 G_{11}/v_1 + 1$) was often very small.[11,19,26–27,31] However, the correlation volume is not usually known, even though it can be determined experimentally by small-angle X-ray scattering, small-angle neutron scattering (SANS), light-scattering (LS), etc. Although it is not yet possible to calculate the cluster size, one can, however, calculate the excesses, Δn_{ij},

5 Aqueous solutions of biomolecules

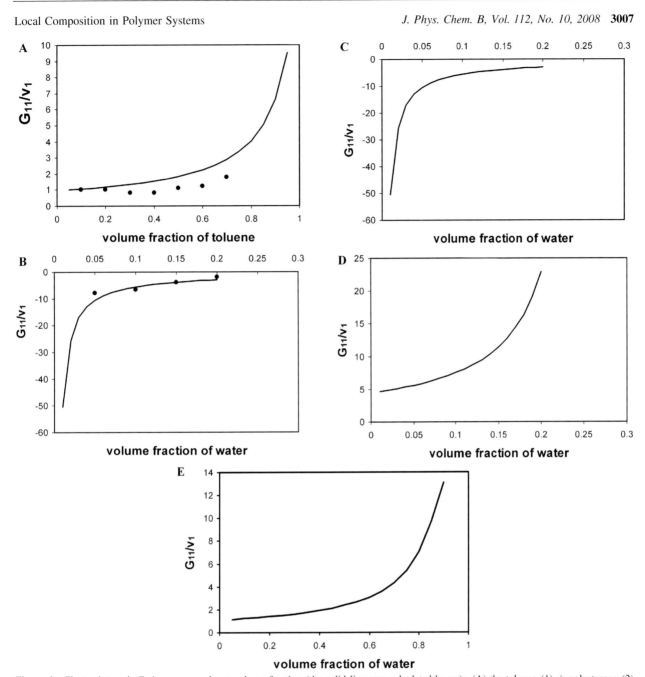

Figure 1. Cluster integral, G_{11}/v_1, versus solvent volume fraction (the solid line was calculated by us): (A) the toluene (1) + polystyrene (2) system (●, ref 1), (B) the water (1) + collagen (2) system (●, ref 12), (C) the water (1) + serum albumin (2) system, (D) the water (1) + hydroxypropyl cellulose (2) system, and (E) the water (1) + Pluronic P105 (2) system.

around central molecules, and these excesses will be considered as measures of the clustering.

In the next section, various solvent−polymer (protein) mixtures will be examined. The excesses Δn_{ij} will be used as measures of the clustering. Let us emphasize that our model implies that the systems considered behave like binary solutions. This behavior is debatable at low volume fractions of the solvent because then the macromolecules may acquire a gel-like structure imbibed with the solvent.

3. Excesses (or Deficits) in Various Solvent−Polymer (Protein) Mixtures

3.1. Toluene (1) + Polystyrene (2). The cluster integral G_{11}/v_1 for this system was calculated by Zimm and Lundberg[1] and later by Lundberg.[12] To the best of our knowledge, these papers contain the first calculation of the Kirkwood−Buff integrals (KBIs). The accuracy of the calculation of the KBIs mainly depends on the accuracy of the evaluation of the derivative $\mu_{11}^{(c)} = (1/kT)(\partial \mu_1/\partial c_1)_{T,P} = (\partial \ln a_1/\partial c_1)_{T,P}$ (see eq 3). Our calculations were carried out using a molecular weigh of polystyrene of 247 800 and the experimental[38] activity a_1 at $T = 323.15$ K. The activities were represented by the Flory−Huggins equation,[39] and the derivative $(\partial \ln a_1/\partial c_1)_{T,P}$ was calculated analytically. The molar volume, V, of toluene (1) + polystyrene (2) was calculated as $V = x_1 v_1^0 + x_2 v_2^S$, where v_1^0 is the molar volume of pure toluene, v_2^S is the molar volume of polystyrene in the toluene (1) + polystyrene (2) mixture calculated from literature data,[40] and x_i is the molar fraction of component i. The results of the calculations of the cluster

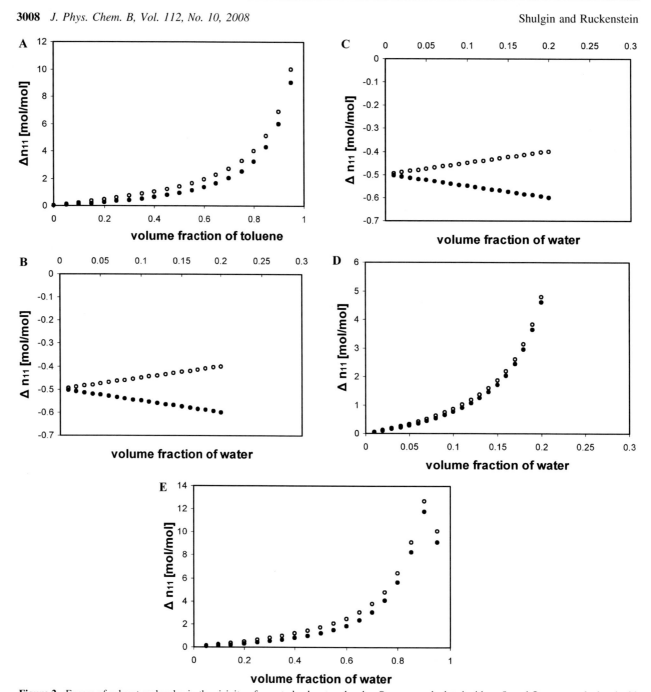

Figure 2. Excess of solvent molecules in the vicinity of a central solvent molecule: ○, excess calculated with eq 8; and ●, excess calculated with the Zimm expression ($\Delta N_{ij} = c_i G_{ij}$). (A) The toluene (1) + polystyrene (2) system, (B) the water (1) + collagen (2) system, (C) the water (1) + serum albumin (2) system, (D) the water (1) + hydroxypropyl cellulose (2) system, and (E) the water (1) + Pluronic P105 (2) system.

integrals G_{11}/v_1 are presented in Figure 1A. The other KBIs (G_{12} and G_{22}) have been calculated from G_{11} using the expressions[35] $v_1 \Delta n_{11} = -v_2 \Delta n_{21}$ and $v_1 \Delta n_{12} = -v_2 \Delta n_{22}$ with eq 8 for Δn_{ij}.

The obtained KBIs were used to calculate the excesses (or deficits) from eq 8, and the Zimm excesses (or deficits) ($\Delta N_{ij} = c_i G_{ij}$). The results of these calculations are plotted in Figures 2A and 3A. They demonstrate that toluene is in excess around both (toluene and polystyrene) central molecules. The excesses (or deficits) from eq 8 and the Zimm excesses (or deficits) ($\Delta N_{ij} = c_i G_{ij}$) will be compared in the Discussion and Conclusion Section of the article.

3.2. Water (1) + Collagen (2). The cluster integral G_{11}/v_1 for this system was calculated by Zimm and Lundberg[1] and Lundberg.[12] Bull's data at $T = 298.15$ K have been used for the activity of water in this system.[4] Starkweather[11] has determined the activity of water in the concentration range $0 \leq \varphi_1 \leq 0.2$ and noted that it depends on the volume fraction as $a_1 = 12\varphi_1^2$, the expression that was used in our calculations. The molecular weight and the partial specific volume of collagen were taken from ref 41. The results of the calculations are presented in Figures 1B, 2B, and 3B. In contrast to the toluene + polystyrene mixture, the solvent (water) is in deficit around a central water molecule but in excess around a protein molecule.

3.3. Water (1) + Serum Albumin (2). Again, Bull's data at $T = 298.15$ K have been used for the activity of water. Starkweather[11] has determined the activity of water in the

5 Aqueous solutions of biomolecules

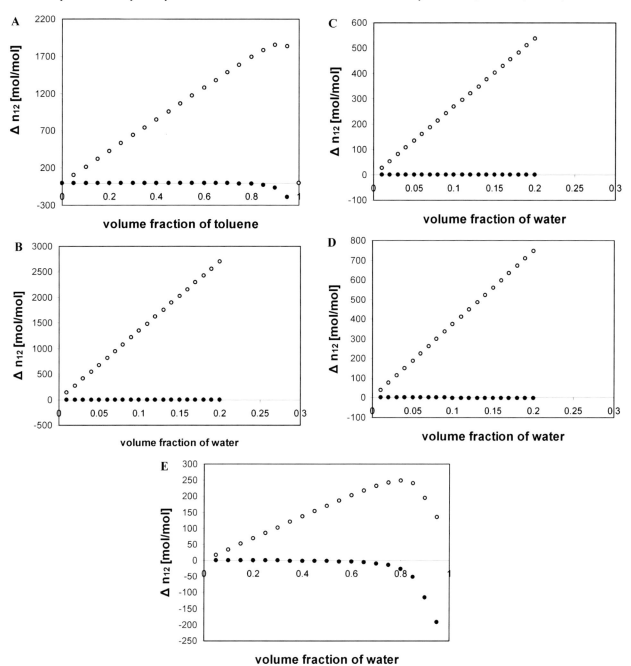

Figure 3. Excess of solvent molecules in the vicinity of a central solute molecule: ○, excess calculated with eq 8; ●, excess calculated with the Zimm expression ($\Delta N_{ij} = c_i G_{ij}$). (A) The toluene (1) + polystyrene (2) system, (B) the water (1) + collagen (2) system, (C) the water (1) + serum albumin (2) system, (D) the water (1) + hydroxypropyl cellulose (2) system, and (E) the water (1) + Pluronic P105 (2) system.

TABLE 1: Results of Calculations Regarding the Excess (or Deficits) in the Systems Investigated

system	composition range	$\Delta n_{11} > 0$	$\Delta n_{21} > 0$
toluene (1) + polystyrene (2)	$0 \leq \varphi_1 \leq 1$	yes	yes
water (1) + collagen (2)	$0 \leq \varphi_1 \leq 0.2$	no	yes
water (1) + serum albumin (2)	$0 \leq \varphi_1 \leq 0.2$	no	yes
water (1) + hydroxypropyl cellulose (2)	$0 \leq \varphi_1 \leq 0.2$	yes	yes
water (1) + Pluronic P105 (2)	$0 \leq \varphi_1 \leq 1$	yes	yes

concentration range $0 \leq \varphi_1 \leq 0.2$ and noted that it depends on the volume fraction as $a_1 = 29\varphi_1^2$, the dependence that was used in our calculations. The molecular weight and the partial specific volume of serum albumin were taken from ref 41. The results are plotted in Figures 1C, 2C, and 3C. The results obtained for the excess (or deficit) of water molecules around both central molecules (water and serum albumin) are comparable to those obtained for the water + collagen mixture.

3.4. Water (1) + Hydroxypropyl Cellulose (2). There are several papers[11,14,28] in which the KB theory of solutions was applied to water + cellulose (or cellulose derivatives) mixtures. The water + hydroxypropyl cellulose mixture was selected in this paper because there are accurate data regarding the activity of water.[42] The data in the range $0 \leq \varphi_1 \leq 0.2$ have been represented by the expression

$$\ln(a_1/\varphi_1) = t_1\varphi_2 + t_2\varphi_2^2 \quad (13)$$

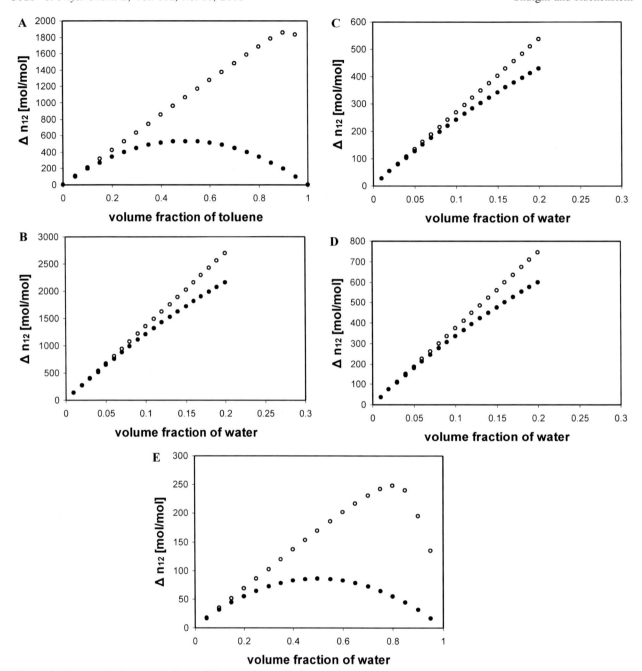

Figure 4. The contribution to Δn_{12} due to different volumes of solvent and solute: (A) toluene + polystyrene, (B) water + collagen, (C) water + serum albumin, (D) water + hydroxypropyl cellulose, and (E) water + Pluronic P105. ○, Δn_{12} calculated for a real system using eq 8 (see Figure 3) and ●, Δn_{12} calculated for an ideal system using eq 14.

where t_1 and t_2 are adjustable parameters ($t_1 = 2.925$ and $t_2 = -0.397$).

Equation 13 was used to calculate the derivative $\mu_{11}^{(c)}$, and the partial specific volume of hydroxypropyl cellulose was taken as that of the hydroxyethyl cellulose, which is provided in ref 43. The results are plotted in Figures 1D, 2D, and 3D.

3.5. Water (1) + Pluronic P105 (2). Pluronic P105 is[6] a polyether block copolymer ($EO_{37}PO_{58}EO_{37}$, where EO and PO denote ethylene oxide and propylene oxide segments, respectively). At elevated temperatures[6] (above 40−60 °C), this block copolymer self-assembles in water and in water + cosolvent as micelles. At low temperatures and concentrations, the block copolymer molecules are present in solution as independent polymer chains.[6] We carried out the calculation at the "low" temperature of 24 °C because activity data were available only at that temperature.[44] The molecular weight and the partial specific volume of Pluronic P105 were also taken from ref 44. The activity in the water + Pluronic P105 mixture was represented by the Flory−Huggins equation[44] and the derivative $(\partial \ln a_1/\partial c_1)_{T,P}$ was calculated analytically. The calculated excesses (or deficits) around both central molecules, water and Pluronic P105, are plotted in Figures 1E, 2E, and 3E.

4. Discussion and Conclusion

In this paper, the KB theory of solutions was applied to binary mixtures containing a low molecular weight solvent and a high molecular weight polymer, protein, etc.

Local Composition in Polymer Systems

We used our new expression[35] for the excess (or deficit) to examine the local composition in binary solvent (1) + polymer (protein) (2) mixtures. Several mixtures ((I) toluene (1) + polystyrene (2), (II) water (1) + collagen (2), (III) water (1) + serum albumin (2), (IV) water (1) + hydroxypropyl cellulose (2), and (V) water (1) + Pluronic P105 (2)) were considered. The excess (or deficit) of solvent molecules around a central solvent molecule (Δn_{11}) and around a central solute molecule (Δn_{12}) were calculated (see Figures 2 and 3), and the results are summarized in Table 1. In addition, the calculated excesses (or deficits) were compared with the Zimm excesses (or deficits) $\Delta N_{ij} = c_i G_{ij}$ in Figures 2 and 3. The comparison shows that Δn_{11} and ΔN_{11} are different in magnitude but have the same sign. Such a result is expected, because for the systems considered here, $|G_{11}| \gg v_1$, and for this reason, Δn_{11} and ΔN_{11} have comparable values. In contrast, the comparison between Δn_{12} and ΔN_{12} reveals striking difference between the two excesses (deficits). Whereas for all systems considered $\Delta n_{12} > 0$ and large, $\Delta N_{12} \leq 0$ and is smaller in absolute value than Δn_{12}. It is worth noting that $\Delta N_{12} \leq 0$ contradicts the experimental results,[41] which reveal that the proteins preferentially uptake water in aqueous solutions.

Let us emphasize the physical significance of the obtained results regarding the excesses (deficits). $\Delta n_{11} > 0$ means preferential hydration (or solvation in the case of toluene + polystyrene mixture) of the solvent molecules, and $\Delta n_{12} > 0$ means that the polymer (protein) molecules are preferentially hydrated (or solvated for the toluene + polystyrene mixture).

The signs and magnitudes of Δn_{11} and Δn_{12} depend on two factors that are of "enthalpic" and "entropic" nature. The former is due to differences in the intermolecular interactions: (1) solvent−solvent and solvent−solute for Δn_{11} and (2) solute−solute and solvent−solute for Δn_{12}. As suggested by Kauzmann (as quoted by Timasheff),[45] the latter is due to the different sizes of the solute and solvent molecules. According to Kauzmann, the smaller molecules can more easily penetrate in the vicinity of a central molecule, and for this reason, the vicinity of a central molecule is enriched in the smaller molecules.

One can see from Table 1 that $\Delta n_{11} > 0$ for three mixtures, but not for water + collagen and water + serum albumin mixtures. The very strong H-bonding[46] between two water molecules and the very small size of the water molecule suggest that $\Delta n_{11} > 0$. The opposite inequalities ($\Delta n_{11} < 0$ in water + collagen and water + serum albumin mixtures) can be attributed to the much stronger H-bonding of the water molecules to some functional groups of the protein (collagen and serum albumin) than to the water molecules.

In contrast to Δn_{11}, $\Delta n_{12} > 0$ for all mixtures investigated, and hence, the polymers or proteins are preferentially hydrated (solvated in the case of toluene + polystyrene mixture). Let us examine separately the contributions to Δn_{12} provided by the "entropic" and "enthalpic" factors. The contribution to Δn_{12} provided by the different sizes of the solvent and polymer (protein) molecules will be evaluated from the excess in an ideal mixture of components possessing the same volumes as the analyzed mixture. In this case,[35] one can write the following expression for Δn_{12},

$$\Delta n_{12}^{id} = \frac{x_1 x_2 v_2 (v_2 - v_1)}{(x_1 v_1 + x_2 v_2)^2} \quad (14)$$

where x_i is the bulk mole fraction of component i.

Δn_{12} and Δn_{12}^{id} are compared in Figure 4. One can see that the contribution of Δn_{12}^{id} is important in the dilute range of the

TABLE 2: Values of $100\Delta n_{12}^{id}/\Delta n_{12}$ (%) for $\varphi_1 = 0.2$

system	$100\Delta n_{12}^{id}/\Delta n_{12}$ (%)
toluene (1) + polystyrene (2)	88.0
water (1) + collagen (2)	88.0
water (1) + serum albumin (2)	88.0
water (1) + hydroxypropyl cellulose (2)	88.1
water (1) + Pluronic P105 (2)	88.2

solvent ($0 \leq \varphi_1 \leq 0.2$). The values of $100\Delta n_{12}^{id}/\Delta n_{12}$ for $\varphi_1 = 0.2$ are listed in Table 2. Figures 4A and E show that the "enthalpic" contribution to Δn_{12} becomes dominant at high φ_1 ($\varphi_1 \geq 0.4 - 0.5$).

Protein hydration has been the focus of attention for more than a century.[47] Many theories have been suggested to explain protein hydration.[41,48] These theories can be subdivided into three groups:[48] (1) those based on the physical adsorption of water vapor on the protein surface, (2) those based on the stoichiometric binding of water molecules to specific functional groups of the protein, and (3) models that have considered the water−protein system as a simple aqueous solution. Our results regarding the excess number of water molecules in the vicinity of a protein molecule indicate that at low humidity ($\varphi_1 \leq 0.2-0.3$), the excess of water in the vicinity of a protein molecule is due mainly to the difference in the sizes of the water and protein molecules.

Pauling[5] connected the hydration of proteins to the adsorption of water molecules on the polar groups of the former. He assumed that each polar group adsorbs one water molecule. Later, Kuntz and Kauzmann[41] criticized this approach because it provided only one-fourth of the hydration level found experimentally.

Our simple analysis indicates that the difference in the sizes of water and protein molecules may constitute an important factor in protein hydration. For water (1) + serum albumin (2) mixture, Figure 4C provides at $\varphi_1 = 0.2$, $\Delta n_{12} \approx 540$ and $\Delta n_{12}^{id} \approx 430$. Therefore, almost the entire increase in the hydration number compared to the bulk is due to the difference in the sizes of water and protein molecules and a smaller fraction (110 molecules) is probably due to the binding of the water molecules to the specific functional groups of serum albumin.

Supplementary Material.

The purpose of this part is to derive expressions for the Kirkwood-Buff integrals (KBIs) in binary systems. For binary mixtures, Kirkwood and Buff [7] obtained the following expressions for the partial molar volumes, the isothermal compressibility and the derivatives of the chemical potential with respect to concentrations

$$v_1 = \frac{1+(G_{22}-G_{12})c_2}{c_1+c_2+c_1c_2(G_{11}+G_{22}-2G_{12})} \tag{A-1}$$

$$v_2 = \frac{1+(G_{11}-G_{12})c_1}{c_1+c_2+c_1c_2(G_{11}+G_{22}-2G_{12})} \tag{A-2}$$

$$kTk_T = \frac{1+c_1G_{11}+c_2G_{22}+c_1c_2(G_{11}G_{22}-G_{12}^2)}{c_1+c_2+c_1c_2(G_{11}+G_{22}-2G_{12})} \tag{A-3}$$

$$\left(\frac{\partial \mu_1}{\partial x_1}\right)_{T,P} = \frac{kT}{x_1(1+x_1c_2(G_{11}+G_{22}-2G_{12}))} \tag{A-4}$$

$$\left(\frac{\partial \mu_1}{\partial c_1}\right)_{T,P} = \frac{kT}{c_1(1+c_1(G_{11}-G_{12}))} \tag{A-5}$$

where P is the pressure, T is the absolute temperature, v_α is the partial molar volume per molecule of species α, μ_1 is the chemical potential per molecule of component 1, k is the Boltzmann constant, k_T is the isothermal compressibility, c_α is the bulk molecular concentration of component α and x_α is the mole fraction of component α. Because $c_1v_1+c_2v_2=1$, one can solve Eqs. (A-1, A-3 and A-5) to obtain the following expressions for the KBIs

5 Aqueous solutions of biomolecules

$$G_{11} = kTk_T - \frac{1}{c_1} + \frac{c_2 v_2}{c_1^2 \mu_{11}^{(c)}} \tag{A-6}$$

$$G_{22} = kTk_T - \frac{1}{c_2} + \frac{v_1^2}{c_2 v_2 \mu_{11}^{(c)}} \tag{A-7}$$

and

$$G_{12} = kTk_T - \frac{v_1}{c_1 \mu_{11}^{(c)}} \tag{A-8}$$

where $\mu_{11}^{(c)} = \frac{1}{kT}\left(\frac{\partial \mu_1}{\partial c_1}\right)_{T,P} = \left(\frac{\partial \ln a_1}{\partial c_1}\right)_{T,P}$, a_1 is the activity of component 1 and k is Boltzmann's constant. Eq. (A-6) was derived by Zimm [9] (there is a misprint in Zimm's paper (eq. (10)) where the sign is (-) instead of (+) before the third term in the left hand side of eq. (A-6).

The KBIs can be also expressed in terms of $\mu_{11}^{(x)} = \frac{1}{kT}\left(\frac{\partial \mu_1}{\partial x_1}\right)_{T,P} = \left(\frac{\partial \ln a_1}{\partial x_1}\right)_{T,P}$.

Eqs. (A-1, A-3 and A-4) provides the following expressions for the KBIs

$$G_{11} = kTk_T - \frac{1}{c_1} + \frac{c_2 v_2^2 (c_1 + c_2)^2}{c_1^2 \mu_{11}^{(x)}} \tag{A-9}$$

$$G_{22} = kTk_T - \frac{1}{c_2} + \frac{v_1^2 (c_1 + c_2)^2}{c_2 \mu_{11}^{(x)}} \tag{A-10}$$

and

$$G_{12} = kTk_T - \frac{v_1 v_2 (c_1 + c_2)^2}{c_1 \mu_{11}^{(x)}} \tag{A-11}$$

These expressions are the same as those obtained by Matteoli and Lepori [10].

References and Notes

(1) Zimm, B. H.; Lundberg, J. L. *J. Phys. Chem.* **1956**, *60*, 425−428.
(2) Starkweather, H. W. *Polym. Lett.* **1963**, *1*, 133−138.
(3) *Water in Polymers*; Rowland, S. P., Ed.; ACS Symposium Series 127, American Chemical Society: Washington, DC, 1980.
(4) Bull, H. B. *J. Am. Chem. Soc.* **1944**, *66*, 1499−1507.
(5) Pauling, L. *J. Am. Chem. Soc.* **1945**, *67*, 555−557.
(6) Alexandridis, P.; Yang, L. *Macromolecules* **2000**, *33*, 5574−5587.
(7) Kirkwood, J. G.; Buff, F. P. *J. Chem. Phys.* **1951**, *19*, 774−777.
(8) Shulgin, I. L.; Ruckenstein, E. *Phys. Chem. Chem. Phys.* **2008**, *10*, 1097−1105.
(9) Zimm, B. H. *J. Chem. Phys.* **1953**, *21*, 934−935.
(10) Matteoli, E.; Lepori, L. *J. Chem. Phys.* **1984**, *80*, 2856−2863.
(11) Starkweather, H. W. *Macromolecules* **1975**, *8*, 476−479.
(12) Lundberg, J. L. *J. Macromol. Sci. Phys.* **1969**, *B3*, 693−710.
(13) Williams, J. L.; Hopfenberg, H. B.; Stannett, V. *J. Macromol. Sci. Phys.* **1969**, *B3*, 711−725.
(14) Orofino, T. A.; Hopfenberg, H. B.; Stannett, V. *J. Macromol. Sci. Phys.* **1969**, *B3*, 777−788.
(15) Lundberg, J. L. *J. Pure Appl. Chem.* **1972**, *31*, 261−281.
(16) Starkweather, H. W. Clustering of Solvents in Adsorbed Polymers; In *Structure Solubility Relationship in Polymers*; Harris, F. W., R. B. Seymour, R. B., Eds.; Academic Press: New York, 1977.
(17) Coelho, U.; Miltz, J. J.; Gilbert, S. G. *Macromolecules* **1979**, *12*, 284−287.
(18) Starkweather, H. W. In *Water in Polymers*; ACS Symposium Series 127; American Chemical Society: Washington, DC, 1980; p 433.
(19) Brown, G. L. In *Water in Polymers*; ACS Symposium Series 127; American Chemical Society: Washington, DC, 1980; p 441.
(20) Saeki, S.; Holste, J. C.; Bonner, D. C. *J. Polym. Sci. Polym. Phys. Ed.* **1982**, *20*, 805−814.
(21) Misra, A.; David, D. J.; Snelgrove, J. A.; Matie, G. *J. Appl. Polym. Sci.* **1986**, *31*, 2387−2398.
(22) Saeki, S.; Tsubokawa, M.; Yamaguchi, T. *Macromolecules* **1987**, *20*, 2930−2934.
(23) Saeki, S.; Tsubokawa, M.; Yamaguchi, T. *Macromolecules* **1988**, *21*, 2210−2213.
(24) Horta, A. *Macromolecules* **1992**, *25*, 5651−5654.
(25) Horta, A. *Macromolecules* **1992**, *25*, 5655−5658.
(26) Vancso, G. J.; Tan, Z. *Can. J. Chem.* **1995**, *73*, 1855−1861.
(27) Tan, Z.; Vancso, G. J. *Macromolecules* **1997**, *30*, 4665−4673.
(28) Beck, M. I.; Tomka, I. *J. Macromol. Sci. Phys.* **1997**, *B36*, 19−39.
(29) Benczedi, D.; Tomka, I.; Escher, F. *Macromolecules* **1998**, *31*, 3055−3061.
(30) Benczedi, D.; Tomka, I.; Escher, F. *Macromolecules* **1998**, *31*, 3062−3074.
(31) Klyuev, L. E.; Grebennikov, S. F. *Zh. Fiz. Khim.* **1999**, *73*, 1700−1702.
(32) Sabzi, F.; Boushehri, A. *Eur. Polym. J.* **2005**, *41*, 974−983.
(33) Sabzi, F.; Boushehri, A. *Eur. Polym. J.* **2005**, *41*, 2067−2087.
(34) Kilburn, D.; Claude, J.; Schweizer, T.; Alam, A.; Ubbink, J. *Biomacromolecules* **2005**, *6*, 864−879.
(35) Shulgin, I. L.; Ruckenstein, E. *J. Phys. Chem. B* **2006**, *110*, 12707−12713.
(36) Ben-Naim, A. Statistical Thermodynamics for Chemists and Biochemists; Plenum Press: New York, 1992.
(37) Hill, T. L. Statistical Mechanics: Principles and Selected Applications; McGraw-Hill: New York, 1956.
(38) Wong, H. C.; Campbell, S. C.; Bhethanabotla, V. R. *Fluid Phase Equilib.* **1997**, *139*, 371−389.
(39) Prausnitz, J. M.; Lichtenthaler, R. N.; Gomes de Azevedo, E. Molecular Thermodynamics of Fluid-Phase Equilibria, 2nd ed.; Prentice-Hall: Englewood Cliffs, NJ, 1986.
(40) Boyer, R. F.; Spencer, R. S. *J. Polym. Sci.* **1948**, *3*, 97−127.
(41) Kuntz, I. D.; Kauzmann, W. *Adv. Protein Chem.* **1974**, *28*, 239−345.
(42) Aspler, J. S.; Gray, D. G. *Macromolecules* **1979**, *12*, 562−566.
(43) Andersson, G. "The Partial Specific Volume of Polymers at High Pressures"; In *The Physics and Chemistry of High Pressures*; Papers read at the symposium held at Olympia, London, June 27−29, 1962; Society of Chemical Industry: London, 1963.
(44) Gu, Z.; Alexandridis, P. *Macromolecules* **2004**, *37*, 912−924.
(45) Timasheff, S. N.; Xie, G. *Biophys. Chem.* **2003**, *105*, 421−448.
(46) Eisenberg, D.; Kauzmann, W. *The Structure and Properties of Water*, Oxford University Press: Oxford, 1969.
(47) Timasheff, S. N. *Biochemistry* **2002**, *41*, 13473−13482.
(48) Shamblin, S. L.; Hancock, B. C.; Zografi, G. *Eur. J. Pharm. Biopharm.* **1998**, *45*, 239−247.

Various Contributions to the Osmotic Second Virial Coefficient in Protein−Water−Cosolvent Solutions[†]

Ivan L. Shulgin and Eli Ruckenstein*

Department of Chemical & Biological Engineering, State University of New York at Buffalo Amherst, New York 14260

Received: April 11, 2008; Revised Manuscript Received: June 10, 2008

An analysis of the cosolvent concentration dependence of the osmotic second virial coefficient (OSVC) in water−protein−cosolvent mixtures is developed. The Kirkwood−Buff fluctuation theory for ternary mixtures is used as the main theoretical tool. On its basis, the OSVC is expressed in terms of the thermodynamic properties of infinitely dilute (with respect to the protein) water−protein−cosolvent mixtures. These properties can be divided into two groups: (1) those of infinitely dilute protein solutions (such as the partial molar volume of a protein at infinite dilution and the derivatives of the protein activity coefficient with respect to the protein and water molar fractions) and (2) those of the protein-free water−cosolvent mixture (such as its concentrations, the isothermal compressibility, the partial molar volumes, and the derivative of the water activity coefficient with respect to the water molar fraction). Expressions are derived for the OSVC of ideal mixtures and for a mixture in which only the binary mixed solvent is ideal. The latter expression contains three contributions: (1) one due to the protein−solvent interactions $B_2^{(p-s)}$, which is connected to the preferential binding parameter, (2) another one due to protein/protein interactions ($B_2^{(p-p)}$), and (3) a third one representing an ideal mixture contribution ($B_2^{(id)}$). The cosolvent composition dependencies of these three contributions were examined for several water−protein−cosolvent mixtures using experimental data regarding the OSVC and the preferential binding parameter. For the water−lysozyme−arginine mixture, it was found that OSVC exhibits the behavior of an ideal mixture and that $B_2^{(id)}$ provides the main contribution to the OSVC. For the other mixtures considered (water−HmMalDH−NaCl, water−HmMalDH−$(NH_4)_2SO_4$, and water−lysozyme−NaCl mixtures), it was found that the contribution of the protein−solvent interactions $B_2^{(p-s)}$ is responsible for the composition dependence of the OSVC on the cosolvent concentration, whereas the two remaining contributions ($B_2^{(p-p)}$ and $B_2^{(id)}$) are almost composition independent.

1. Introduction

For more than 60 years it is known that B_2, the osmotic second virial coefficient (OSVC), constitutes an important thermodynamic characteristic of protein−protein interactions in protein solutions.[1−5] In a binary mixture solvent (1)−protein (2), B_2 is connected to the osmotic pressure (π) via the virial equation

$$\frac{\pi}{kT} = \rho_2 + B_2\rho_2^2 + B_3\rho_2^3 + \quad (1)$$

where ρ_2 is the number density of protein molecules; k is the Boltzmann constant; T is the absolute temperature; and B_3 is the osmotic third virial coefficient.

A similar expression can be written for multiple solvents. Our primary interest is on systems containing a protein (2) in a mixed solvent: water (1)−cosolvent (3). An inorganic salt or a small organic molecule is considered as a cosolvent.

B_2 is related to the radial distribution function g_{22} of a pair of protein molecules via the expression[1,2]

$$B_2 = -\frac{1}{2}\int_0^\infty (g_{22} - 1)4\pi r^2 dr \quad (2)$$

where r is the distance between the centers of the protein molecules and g_{22} is to be evaluated for an infinitely dilute protein solution.[2] Equation 1 can be converted into the following expression, when the composition of the protein is expressed as mass concentration C_2

$$\frac{\pi}{C_2RT} = \frac{1}{M_2} + B_{22}C_2 + B_{222}C_2^2 + \quad (3)$$

where B_{22}, B_{222},... are the second, third,... osmotic virial coefficients in the new concentration scale; R is the universal gas constant; and M_2 is the protein molecular weight. B_2 and B_{22} are related as follows

$$B_{22} = \frac{B_2}{M_2^2}N_A \quad (4)$$

where N_A is the Avogadro number.

In the last 10−15 years, a surge in the interest on both experimental and theoretical features of the OSVC of a protein (2) in mixed water (1)−cosolvent (3) mixtures has taken place. This surge was caused by the connection between OSVC and protein crystallization as well as its solubility in water and in aqueous mixed solvents.

First, George and Wilson[6] found empirically that protein crystallization can be correlated with B_{22}. They found that good crystals could be obtained when the osmotic second virial coefficient (B_{22}) has a value in the window (crystallization slot) between -2×10^{-4} and -8×10^{-4} mL mol/g². For $B_{22} > -2 \times 10^{-4}$, the protein−protein interactions are not strong enough for crystallization to occur, and for $B_{22} < -8 \times 10^{-4}$, the

---

[†] Part of the "Janos H. Fendler Memorial Issue".
* Corresponding author. E-mail: feaeliru@acsu.buffalo.edu. Fax: (716) 645-3822. Phone: (716) 645-2911ext. 2214.

protein−protein interactions are so strong that amorphous precipitates are formed because the process is too rapid for the protein to acquire a crystalline structure. George and Wilson's[6] findings constitute a useful screening criterion for protein crystallization because OSVC can be relatively easily obtained using various experimental techniques such as static light scattering,[7−10] small-angle X-ray,[11,12] neutron [7,13] scattering, membrane osmometry,[14−16] ultracentrifugation,[17] size-exclusion chromatography,[18] and self-interaction chromatography.[19−21]

Second, it was suggested that the solubility of a protein in aqueous mixed solvents can be correlated with the osmotic second virial coefficient.[9,22,23] However, whereas the connection of B_{22} to protein crystallization was widely accepted, the connection of B_{22} to the solubility of proteins requires additional investigation.[22,24]

The theoretical investigation of OSVC was mainly based on the following expression[1]

$$B_2 = -\frac{1}{2}\int_0^\infty [e^{-W_{22}/kT} - 1]4\pi r^2 dr \quad (5)$$

where W_{22} is the potential of mean force between the protein molecules. The main difficulty in the use of eq 5 is that the potential of mean force is unknown and a simplified expression has to be used. A detailed description of the work in this direction is available in refs 8 and 25−28.

In the present paper, a different approach based on the Kirkwood−Buff theory of solution[5] will be developed. First, an expression for the OSVC in ternary water (1)−protein (2)−cosolvent (3) mixtures will be obtained in terms of the partial molar volumes and the derivatives of the activity coefficients of the protein and water with respect to their concentrations. Further, the obtained expression will be subdivided into three components which reflect protein−protein, protein−water and cosolvent interactions, as well as an "ideal mixture" contribution. Finally, the OSVC of several water (1)−protein (2)−cosolvent (3) mixtures will be analyzed in order to compare the above contributions and their cosolvent composition dependencies.

2. Theory

2.1. Expression for B_2. The Kirkwood−Buff integral (KBI)[5] in n-component mixtures is provided by

$$G_{\alpha\beta} = \int_0^\infty (g_{\alpha\beta} - 1)4\pi r^2 dr \quad (6)$$

where $g_{\alpha\beta}$ is the radial distribution function (RDF) between species α and β and r is the distance between the centers of molecules α and β. For the ternary mixture water (1)−protein (2)−cosolvent (3), one can write (using eqs 2 and 6) that

$$B_2 = -\frac{1}{2}\lim_{x_2 \to 0} G_{22} = -\frac{1}{2}G_{22}^0 \quad (7)$$

where x_2 is the mole fraction of the protein in the ternary mixture water (1)−protein (2)−cosolvent (3). One of the attractive features of the Kirkwood−Buff theory of solutions is that the KBIs defined in terms of the RDFs can be expressed through measurable thermodynamic quantities, such as the derivatives of the chemical potentials with respect to concentrations, the isothermal compressibility, and the partial molar volumes. Such expressions provide relations between the microstructure of a solution (through RDF) and measurable thermodynamic quantities.

Expressions for the KBIs could be obtained for both binary[29] and ternary mixtures.[30] While an explicit expression for binary mixtures could be obtained relatively easily, explicit expressions for ternary and multicomponent mixtures required a long derivation.[30,31]

The expression for B_2 (G_{22}^0) in a ternary mixture has the following form (see Appendix 1 for the derivation)

$$B_2 = -\frac{1}{2}G_{22}^0 = -\frac{1}{2}\Bigg[kTk_T - 2V_2^\infty - \frac{J_{22}}{c_1 + c_3} + \frac{c_1 c_3 J_{21}^2}{(c_1+c_3)^2(c_1+c_3+c_1 J_{11})} + J_{21}\bigg(\frac{c_3^2 + c_1 c_3 + c_1^2 J_{11}}{(c_1+c_3)^2(c_1+c_3+c_1 J_{11})} + \frac{c_1 V_1 - c_3 V_3}{c_1+c_3+c_1 J_{11}}\bigg) + \frac{(c_1+c_3)(V_3 + c_1 V_1(V_1 - V_3))}{c_1 + c_3 + c_1 J_{11}} + \frac{c_1 J_{11}}{(c_1+c_3)(c_1+c_3+c_1 J_{11})}\Bigg] \quad (8)$$

where c_i ($i = 1, 3$) is the molar concentration of component i in a protein-free mixed solvent 1−3; V_i is the partial molar volume of component i in a protein-free mixed solvent 1−3; V_2^∞ is the partial molar volume of the protein at infinite dilution; k_T is the isothermal compressibility of the mixed solvent;

$$J_{11} = \lim_{x_2 \to 0}\left(\frac{\partial \ln \gamma_1}{\partial x_1}\right)_{x_2}$$

$$J_{21} = \lim_{x_2 \to 0}\left(\frac{\partial \ln \gamma_2}{\partial x_1}\right)_{x_2}$$

$$J_{22} = \lim_{x_2 \to 0}\left(\frac{\partial \ln \gamma_2}{\partial x_2}\right)_{x_1}$$

and γ_i is the activity coefficient of component i in a mole fraction scale.

Equation 8 for B_2 does not involve any approximations. Using eq 8, one can calculate B_2 from the properties of protein-free mixed solvents, such as the concentrations c_i ($i = 1, 3$), the partial molar volumes of the components V_i ($i = 1, 3$), the isothermal compressibility k_T, and J_{11} as well as those of infinitely dilute protein mixtures such as the partial molar volume of a protein at infinite dilution V_2^∞ and the derivatives J_{21} and J_{22}.

Let us examine various limiting cases of ternary mixtures.

2.1.1. Ideal Mixture Approximation [Superscript "(ideal)"]. In this case, all the activity coefficients are equal to unity, and all partial molar volumes are equal to the molar volumes of the pure components,[32] and eq 8 becomes

$$B_2^{(\text{ideal})} = -\frac{1}{2}[kTk_T^{(\text{id})} - 2V_2^0 + V_3^0 + c_1 V_1^0(V_1^0 - V_3^0)] \quad (9)$$

where V_i^0 is the molar volume of the pure component i.

2.12. Ideal Mixed Solvent Approximation[31,33,34] *(Superscript "is")*. In this case, the protein-free mixed solvent 1−3 behaves as an ideal mixture. This approximation implies that the main contribution to the nonideality of a very dilute protein + mixed solvent mixture stems from the nonideality due to the interactions of the protein with the mixed solvent and not from the nonideality of the mixed solvent itself. In this case, $J_{11} = 0$, and eq 8 acquires the form

Contributions to Osmotic Second Virial Coefficient

$$B_2^{(is)} = -\frac{1}{2}\left[kTk_T^{(id)} - 2V_2^{\infty} - \frac{J_{22}}{c_1+c_3} + \frac{c_1c_3J_{21}^2}{(c_1+c_3)^3} + J_{21}\left(\frac{c_3}{(c_1+c_3)^2} + \frac{c_1V_1^0 - c_3V_3^0}{c_1+c_3}\right) + V_3^0 + c_1V_1^0(V_1^0 - V_3^0)\right] \quad (10)$$

As we recently noted,[35] the ideal mixed solvent approximation is very accurate for water (1)−protein (2)−cosolvent (3) mixtures. In this paper, we will use this approximation which implies that $B_2 \approx B_2^{(is)}$. Compared to eq 8, eq 10 is simpler because it does not involve the activity coefficients of the components of the mixed solvent.

The above expression for the OSVC ($B_2^{(is)}$) can be subdivided into three contributions: (1) activity coefficient-free (or ideal mixture contribution) ($B_2^{(id)}$), (2) contribution due to protein/solvent interactions ($B_2^{(p-s)}$), (3) residual contribution or contribution due to direct protein/protein interactions ($B_2^{(p-p)}$).

The part of expression 10 free of activity coefficients (or ideal mixture contribution) is given by

$$B_2^{(id)} = -\frac{1}{2}\left[kTk_T^{(id)} - 2V_2^{\infty} + V_3^0 + c_1V_1^0(V_1^0 - V_3^0)\right] \quad (11A)$$

The difference between the ternary ideal mixture approximation (eq 9) and eq 11A consists in the molar volume of the protein: the partial molar volume of the protein at infinite dilution (V_2^{∞}) in eq 11A and the hypothetical molar volume of a pure protein (V_2^0) in eq 9.

The protein−solvent interactions depend[31] on J_{21}, and their contribution to the OSVC ($B_2^{(is)}$) is provided by

$$B_2^{(p-s)} = -\frac{1}{2}\left[\frac{c_1c_3J_{21}^2}{(c_1+c_3)^3} + J_{21}\left(\frac{c_3}{(c_1+c_3)^2} + \frac{c_1V_1^0 - c_3V_3^0}{c_1+c_3}\right)\right] \quad (11B)$$

It will be shown in the next section that J_{21} is directly related to the preferential binding parameter which is a result of the interactions between protein and water (or cosolvent).

The third contribution is the residual part of the OSVC. It depends on

$$J_{22} = \lim_{x_2 \to 0}\left(\frac{\partial \ln \gamma_2}{\partial x_2}\right)_{x_1}$$

and therefore can be attributed to the direct protein−protein interaction. This contribution to OSVC is provided by

$$B_2^{(p-p)} = \frac{J_{22}}{2(c_1+c_3)} \quad (11C)$$

Consequently

$$B_2^{(is)} = B_2^{(id)} + B_2^{(p-s)} + B_2^{(p-p)} \quad (12)$$

2.2. Relation between the Preferential Binding Parameter and OSVC. The effect of a cosolvent on the behavior of a protein is determined by the preferential binding parameter,[36−39] which, in molal concentrations, can be defined as follows[36]

$$\Gamma_{23}^{(m)} \equiv \lim_{m_2 \to 0}(\partial m_3/\partial m_2)_{T,P,\mu_3} \quad (13)$$

where m_i is the molality of component i; P is the pressure; and μ_i is the chemical potential of component i.

$\Gamma_{23}^{(m)}$ provides information about the concentrations of water and cosolvent on the protein surface. A negative value of $\Gamma_{23}^{(m)}$ indicates preferential hydration or preferential exclusion of the cosolvent, and a positive value of $\Gamma_{23}^{(m)}$ indicates preferential cosolvation on the protein surface. In the framework of the ideal mixed solvent approximation, $\Gamma_{23}^{(m)}$ can be expressed as follows (see Appendix 2 for the details of its derivation)

$$\Gamma_{23}^{(m)} = \frac{c_3J_{21}}{c_1+c_3} + \frac{c_3}{c_1} \quad (14)$$

Equation 14 shows that J_{21} can be calculated from experimental data regarding $\Gamma_{23}^{(m)}$. In the dilute cosolvent region, the preferential binding parameter $\Gamma_{23}^{(m)}$ is a linear function of cosolvent concentration,[31,39−41] and J_{21} can be considered to be a constant provided by the slope of $\Gamma_{23}^{(m)}$ versus c_3.

Equation 14 provides a relation between OSVC and the preferential binding parameter $\Gamma_{23}^{(m)}$, the two being related through J_{21}. Therefore, $\Gamma_{23}^{(m)}$ allows one to calculate the contribution to OSVC that is connected with the protein−solvent interactions $B_2^{(p-s)}$. The ideal mixture contribution ($B_2^{(id)}$) and the contribution due to protein/protein interactions ($B_2^{(p-p)}$) are not connected with $\Gamma_{23}^{(m)}$ and hence with the protein−solvent interactions.

3. Numerical Estimations for Various Systems

3.1. Calculation Procedure. Experimental data regarding the OSVC and the preferential binding parameter for several water (1)−protein (2)−cosolvent (3) mixtures will be used to estimate the above three contributions ($B_2^{(id)}$, $B_2^{(p-s)}$, and $B_2^{(p-p)}$) to OSVC. The ideal mixture contribution $B_2^{(id)}$ (eq 11A) can be calculated from the properties of the protein-free mixed solvent. The isothermal compressibility ($-0.5kTk_T^{(id)}$) contribution was neglected. We used the partial molar volumes of the components of the protein-free mixed solvent (V_1 and V_3) instead of those of pure components because they were available in the literature. The protein−solvent interaction contribution $B_2^{(p-s)}$ (eq 11B) can be calculated from the properties of the protein-free mixed solvent and data regarding the preferential binding parameter $\Gamma_{23}^{(m)}$. The direct protein/protein interactions $B_2^{(p-p)}$ (eq 11C) can be calculated using eq 12, experimental data regarding the OSVC, and the calculated values of $B_2^{(id)}$ and $B_2^{(p-s)}$. When data regarding the preferential binding parameter $\Gamma_{23}^{(m)}$ are not available, the experimental OSVC can be correlated using eq 10 with two adjustable parameters J_{21} and J_{22}.

3.2. Experimental Data and System Selection. We selected the water (1)−protein (2)−cosolvent (3) mixtures for which accurate data for both $\Gamma_{23}^{(m)}$ and OSVC were available. In addition, experimental information regarding the partial molar volume of a protein at infinite dilution (V_2^{∞}) is required. There are only a few systems for which accurate data for both $\Gamma_{23}^{(m)}$ and OSVC are available. For most systems, only incomplete data could be found. For example, for the popular water (1)−lysozyme (2)−NaCl (3) mixture, there are numerous experimental data regarding the OSVC, but accurate data for $\Gamma_{23}^{(m)}$ in the dilute region are missing. In contrast, for the water (1)−lysozyme (2)−urea (3) mixture, accurate data for $\Gamma_{23}^{(m)}$ are available,[42] but experimental information about OSVC is missing. We found only a few mixtures for which both $\Gamma_{23}^{(m)}$ and OSVC are available. In addition, we carried out calculations for one mixture (water−lysozyme−NaCl) with incomplete information. The selected mixtures are listed in Table 1.

3.3. Various Systems. 3.3.1. Water (1)−Malate Dehydrogenase (Hm MalDH) (2)−NaCl (3). For water (1)−Hm MalDH (2)−NaCl (3), experimental data for both $\Gamma_{23}^{(m)}$ and OSVC are available.[43,44] The partial molar volumes of the components of the protein-free mixed solvent (V_1 and V_3) were calculated from the densities of the water−NaCl

TABLE 1: Information About the Water (1)−Protein (2)−Cosolvent (3) Mixtures Selected for Calculations

		experimental data		
number	system	OSVC conditions, refs	$\Gamma_{23}^{(\mu q)}$ conditions, refs	V_2^{∞} conditions, refs
1	water (1)−Hm MalDH[a] (2)−NaCl (3)	pH = 8.2, ref 43	pH = 8.2, ref 44	pH = 8.2, ref 43
2	water (1)−Hm MalDH (2)−(NH$_4$)$_2$SO$_4$ (3)	pH = 8.2, ref 43	pH = 8.2, ref 44	pH = 8.2, ref 43
3	water (1)−lysozyme (2)−arginine (3)	pH = 4.5 (in the presence of 2 (w/v) % of NaCl), ref 45	pH = 5.7, refs 46 and 47	pH = 5.7, ref 46
4	water (1)−lysozyme (2)−NaCl (3)	pH = 4.5, refs 19, 28, 48	no data available for the dilute range	pH = 4.5, ref 49[b]

[a] Malate dehydrogenase. [b] V_2^{∞} was taken equal to the partial molar volume of the protein in a 1 M NaCl solution.

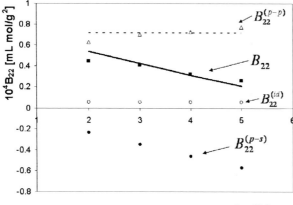

Figure 1. Dependence of the second virial coefficient on the concentration of salt in a water (1)−Hm MalDH (2)−NaCl (3) mixture. ■, experimental data; solid line, values predicted using eq 10 with j_{22} (see Table 2) as an adjustable parameter; ●, protein−solvent interaction contribution $B_2^{(p-s)}$ calculated using eq 11B; ○, ideal mixture contribution $B_2^{(id)}$ (eq 11A); △, the protein−protein interaction contribution $B_2^{(p-p)}$ calculated using eq 12 from experimental data regarding the OSVC and $B_2^{(id)}$ and $B_2^{(p-s)}$; broken line (- - -), protein−protein interaction contribution $B_2^{(p-p)}$ predicted by eq 11C with j_{22} as an adjustable parameter from Table 2. The values of B_2, $B_2^{(id)}$, $B_2^{(p-s)}$, and $B_2^{(p-p)}$ were converted in quantities with subscripts (22) using eq 4, which are expressed in [mL mol/g^2] units.

TABLE 2: Values of the Derivatives J_{21} and J_{22} used in Calculations

number	system	J_{21}	J_{22}
1	water (1)−Hm MalDH (2)−NaCl (3)	−1199[a]	141500[b]
2	water (1)−Hm MalDH (2)−(NH$_4$)$_2$SO$_4$ (3)	−2854[a]	283498[b]
3	water (1)−lysozyme (2)−arginine (3)	−171[a]	269[b]
4	water (1)−lysozyme (2)−NaCl (3)	−1257[b]	3863[b]

[a] Obtained from experimental $\Gamma_{23}^{(\mu q)}$ data (see Table 1). [b] Obtained from experimental OSVC data (see Table 1).

solutions.[50] The results of the calculations for this mixture are presented in Figure 1 and Table 2.

3.3.2. Water (1)−Hm MalDH (2)−(NH$_4$)$_2$SO$_4$ (3). For water (1)−Hm MalDH (2)−(NH$_4$)$_2$SO$_4$ (3), experimental data for both $\Gamma_{23}^{(\mu q)}$ and OSVC are available.[43,44] The partial molar volumes of the components of the protein-free mixed solvent (V_1 and V_3) were calculated from the densities of the water−(NH$_4$)$_2$SO$_4$ solutions.[50] The results of the calculations for this mixture are presented in Figure 2 and Table 2.

3.3.3. Water (1)−Lysozyme (2)−Arginine (3). Experimental data regarding $\Gamma_{23}^{(\mu q)}$ and OSVC for the water (1)−lysozyme (2)−arginine (3) mixture[45−47] were obtained under different conditions. The preferential binding parameter was measured[46] at pH = 5.7, whereas OSVC was measured[45] at pH = 4.5 and in the presence of various amounts of NaCl (2 or 5 (w/v) %).

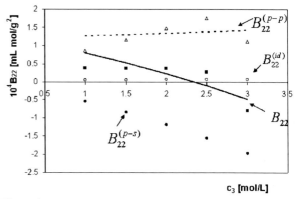

Figure 2. Dependence of the second virial coefficient on the salt concentration for the water (1)−Hm MalDH (2)−NH$_2$SO$_4$ (3) mixture. See Figure 1 for details.

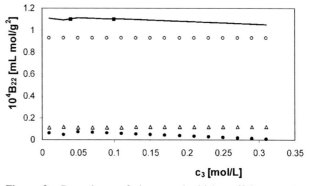

Figure 3. Dependence of the second virial coefficient on the concentration of cosolvent in the water (1)−lysozyme (2)−arginine (3) mixture. ■, experimental data; solid line, values predicted using eq 10 with j_{22} (see Table 2) as an adjustable parameter; ●, protein−solvent interaction contribution $B_2^{(p-s)}$ calculated using eq 11B; ○, ideal mixture contribution $B_2^{(id)}$ (eq 11A); △, protein−protein interaction contribution $B_2^{(p-p)}$ predicted by eq 11C with j_{22} as an adjustable parameter from Table 2. See details in Figure 1.

The data for 2 (w/v) % were employed. Let us note that $\Gamma_{23}^{(\mu q)}$ was measured without addition of NaCl, whereas the OSVC have been measured with addition of a certain amount of NaCl. The partial molar volumes of the components of the protein-free mixed solvent (V_1 and V_3) were taken from ref 46. The results of the calculations for this mixture are presented in Figure 3 and Table 2.

3.3.4. Water (1)−Lysozyme (2)−NaCl (3). Data regarding the preferential binding parameter for these mixtures are not available in the dilute range. For this reason, the experimental OSVC[19,28,48] were correlated using eq 10 with two adjustable parameters, J_{21} and J_{22}. The partial molar volumes of the components of the protein-free mixed solvent (V_1 and V_3) were

Contributions to Osmotic Second Virial Coefficient

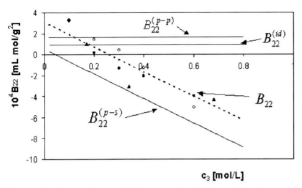

Figure 4. Dependence of the second virial coefficient on the concentration of NaCl in a water (1)−lysozyme (2)−NaCl (3) mixture. B_{22} (broken line) was calculated using eq 10 with both j_{21} and j_{22} used as adjustable parameters found by fitting the experimental data (●, ref 48; ○, ref 19; and ▲, ref 28). $B_2^{(id)}$, ideal mixture contribution predicted using eq 11A; $B_2^{(p-s)}$, protein−solvent interaction contribution calculated using eq 11B; $B_2^{(p-p)}$, protein−protein interaction contribution calculated using eq 11C. See Figure 1 for details.

calculated from the densities of water−NaCl solutions.[50] The results of the calculations for water (1)−lysozyme (2)−NaCl (3) mixtures are presented in Figure 4 and Table 2.

4. Results

Calculations were carried out for four mixtures: (1) water−Hm MalDH−NaCl, (2) water−Hm MalDH−$(NH_4)_2SO_4$, (3) water−lysozyme−arginine, and (4) water−lysozyme−NaCl mixtures. The composition dependence of OSVC in water−lysozyme−arginine mixtures is weak and close to that of an ideal mixture ($B_2^{(id)} \gg B_2^{(p-s)}$ and $B_2^{(p-p)}$). In contrast, OSVC has a strong dependence on the cosolvent composition for the water−Hm MalDH−NaCl, water−Hm MalDH−$(NH_4)_2SO_4$, and water−lysozyme−NaCl mixtures. Our calculations demonstrate that the three contributions ($B_2^{(id)}$, $B_2^{(p-s)}$, and $B_2^{(p-p)}$) to the OSVC depend differently on the cosolvent composition. While the ideal mixture contribution ($B_2^{(id)}$) and the contribution due to the direct protein/protein interactions ($B_2^{(p-p)}$) have very weak composition dependencies, the contribution due to the protein/solvents interactions ($B_2^{(p-s)}$) depends on composition in a way similar to the OSVC. In the dilute range of the cosolvent, $B_2^{(p-s)}$ and OSVC have almost the same slope. However, for the mixtures considered, the composition dependence of the OSVC is provided, at least in the dilute range, by the contribution due to the protein/solvent interactions ($B_2^{(p-s)}$) and can be evaluated from the preferential binding parameter ($\Gamma_{23}^{(\mu)}$).

The above observations about the various contributions to OSVC can provide some insight regarding the relation between the protein solubility and both $\Gamma_{23}^{(\mu)}$ and OSVC.[9,21−24,51−55] The relation between the solubility of protein crystals and the preferential binding parameter was noted[51,52] and was explained theoretically.[53−55] In particular, criteria for the occurrence of salting-out and salting-in were established on the basis of the sign of $\Gamma_{23}^{(\mu)}$, and in addition, equations relating the solubility of protein crystals to $\Gamma_{23}^{(\mu)}$ were established.[53−55] The correlation between OSVC and the protein solubility is more complex, since as shown by eq 10, OSVC depends not only on $\Gamma_{23}^{(\mu)}$ (through J_{21}), which is connected to the solubility, but also on J_{22} which is not related to $\Gamma_{23}^{(\mu)}$. This result allowed us to conclude that the solubility of the protein cannot always be correlated with the OSVC. There are some experimental observations in this direction concerning the solubility of lysozyme in water−alcohol mixtures. Indeed, when an alcohol is added to an aqueous lysozyme solution, the osmotic second virial coefficient increases,[56] whereas the alcohols are well-known protein precipitants; i.e., the aqueous lysozyme solubility decreases with the addition of an alcohol.[51,57,58]

5. Discussion and Conclusion

The purpose of this paper is to shed additional light on the cosolvent concentration dependence of OSVC in water−protein−cosolvent mixtures. The Kirkwood−Buff theory of solutions was used to derive an expression which connects OSVC to the thermodynamic properties of water−protein−cosolvent mixtures. These properties can be subdivided into two groups: (1) those due to a protein-free water−cosolvent mixture, such as concentrations, isothermal compressibility, partial molar volumes, and the derivative of the water activity coefficient with respect to the water molar fraction and (2) those of infinitely dilute (with respect to the protein) water−protein−cosolvent mixtures, such as the partial molar volume of the protein at infinite dilution (V_2^∞) and the derivatives of the protein activity coefficient with respect to the protein and water molar fractions (J_{21} and J_{22}). It was found that the derived expression for OSVC contains three contributions: (1) ideal mixture contribution ($B_2^{(id)}$), (2) contribution due to the protein/solvent interaction ($B_2^{(p-s)}$), (3) contribution due to the direct protein/protein interaction ($B_2^{(p-p)}$). The calculations were carried out for several systems for which $\Gamma_{23}^{(\mu)}$ and OSVC (or only OSVC) were available. The results revealed the dominant role of the protein/solvent interaction in the cosolvent composition dependence of OSVC. It is worth noting that the contribution due to the direct protein/protein interaction ($B_2^{(p-p)}$) is almost composition independent in the dilute range of the cosolvent.

The approach employed in the present paper did not require the knowledge of any intermolecular potentials, and this has some advantage compared to the traditional treatment of the OSVC in water−protein−cosolvent mixtures used in the literature. However, the suggested approach has several limitations. While eq 8 for the OSVC is a rigorous one, the calculations were carried out in the dilute range of the cosolvent where $\Gamma_{23}^{(\mu)}$ has a linear dependence on composition; in addition, the derivatives (J_{21} and J_{22}) are assumed to be constant in that composition range. It is also worth noting that the present treatment is based on the Kirkwood−Buff theory of ternary mixtures. However, the experimental results regarding the preferential binding parameter $\Gamma_{23}^{(\mu)}$ and the OSVC were obtained for mixtures which involve in addition a buffer, and the effect of the buffer is taken into account only indirectly via the preferential binding parameter $\Gamma_{23}^{(\mu)}$ and the OSVC.

Appendix 1: Derivation of Analytical Expressions for the Kirkwood−Buff Integrals (KBIs) in Ternary Mixtures

Generally speaking, analytical expressions for the KBIs in an n-component mixture are provided by the following relation[5]

$$G_{\alpha\beta} = \frac{v|A|_{\alpha\beta}}{\langle N_\alpha \rangle \langle N_\beta \rangle |A|} - \frac{\delta_{\alpha\beta}}{c_\beta} \quad (A1\text{-}1)$$

where $\langle N_\alpha \rangle$ and $\langle N_\beta \rangle$ are the average numbers of α and β molecules, respectively, in the volume v; $\delta_{\alpha\beta}$ is the Kronecker symbol ($\delta_{\alpha\beta} = 1$ for $\alpha = \beta$ and $\delta_{\alpha\beta} = 0$ for $\alpha \neq \beta$); c_β is the bulk molecular concentration of species β ($c_\beta = \langle N_\beta \rangle / v$); $|A|_{\alpha\beta}$ is the cofactor of $A_{\alpha\beta}$ in the determinant $|A|$; and $A_{\alpha\beta}$ is given by[5]

$$A_{\alpha\beta} = \frac{1}{kT}\left(\left(\frac{\partial \mu_\alpha}{\partial N_\beta}\right)_{T,P,N_\gamma} + \frac{v_\alpha v_\beta}{k_T v}\right) \quad \text{(A1-2)}$$

In eq A1-2, k is the Boltzmann constant; T is the absolute temperature; P is the pressure; μ_α is the chemical potential per molecule of species α; v_α and v_β are the partial molar volumes per molecule of species α and β, respectively; and k_T is the isothermal compressibility.

However, the derivation of analytical expressions for the KBIs in ternary and multicomponent mixtures is algebraically extensive and requires the use of an algebraic software, such as Mathematica or Maple.[31] A procedure was developed by us before[31] and used to obtain expressions for G_{12} and G_{23} in water (1)−protein (2)−cosolvent (3) for infinite protein dilution. The expressions have the forms

$$G_{12}^0 = kTk_T - \frac{J_{21}V_3 c_3 + J_{11}V_2^\infty c_1}{(c_1 + c_1 J_{11} + c_3)} - \frac{V_3 c_3(c_1 + c_3)(V_1 - V_3) + V_2^\infty(c_1 + c_3)}{(c_1 + c_1 J_{11} + c_3)} \quad \text{(A1-3)}$$

and

$$G_{23}^0 = kTk_T + \frac{J_{21}V_1 c_1 - J_{11}c_1 V_2^\infty}{(c_1 + c_1 J_{11} + c_3)} + \frac{c_1 V_1(c_1 + c_3)(V_1 - V_3) - V_2^\infty(c_1 + c_3)}{(c_1 + c_1 J_{11} + c_3)} \quad \text{(A1-4)}$$

Using the same technique, one can obtain an expression for G_{22} at infinite protein dilution

$$G_{22}^0 = kTk_T - 2V_2^\infty - \frac{J_{22}}{c_1 + c_3} + \frac{c_1 c_3 J_{21}^2}{(c_1 + c_3)^2(c_1 + c_3 + c_1 J_{11})} + J_{21}\left(\frac{c_3^2 + c_1 c_3 + c_1^2 J_{11}}{(c_1 + c_3)^2(c_1 + c_3 + c_1 J_{11})} + \frac{c_1 V_1 - c_3 V_3}{c_1 + c_3 + c_1 J_{11}}\right) + \frac{(c_1 + c_3)(V_3 + c_1 V_1(V_1 - V_3))}{c_1 + c_3 + c_1 J_{11}} + \frac{c_1 J_{11}}{(c_1 + c_3)(c_1 + c_3 + c_1 J_{11})} \quad \text{(A1-5)}$$

Expressions A1-3 and A1-4 are further used in Appendix 2.

Appendix 2: Relation between the Preferential Binding Parameter $\Gamma_{23}^{(m)}$ and the Derivatives of Activity Coefficient with Respect to the Mole Fractions

The preferential binding parameter $\Gamma_{23}^{(m)}$ can be expressed via the Kirkwood−Buff theory of solution as follows[59]

$$\Gamma_{23}^{(m)} = \frac{c_3}{c_1} + c_3(G_{23}^0 - G_{12}^0 + G_{11}^0 - G_{13}^0) \quad \text{(A2-1)}$$

In eq A2-1, all KBIs should be evaluated at infinite protein dilution. Expressions for G_{12}^0 and G_{23}^0 are provided in Appendix 1. Expressions for G_{11}^0 and G_{13}^0 are available in ref 60. They are given by

$$G_{11}^0 = kTk_T - \frac{1}{c_1} + \frac{c_3 V_3^2(c_1 + c_3)^2}{c_1(c_1 + c_3 + c_1 J_{11})} \quad \text{(A2-2)}$$

and

$$G_{13}^0 = kTk_T - \frac{V_1 V_3(c_1 + c_3)^2}{c_1 + c_3 + c_1 J_{11}} \quad \text{(A2-3)}$$

Inserting eqs A1-3, A1-4, A2-2, and A2-3 into eq A2-1, one obtains

$$\Gamma_{23}^{(m)} = \frac{c_3(J_{21} - J_{11})}{c_1 + c_3 + c_1 J_{11}} + \frac{c_3}{c_1} \quad \text{(A2-4)}$$

Equation A2-4 leads to eq 14 of the text for the ideal mixed solvent approximation because $J_{11} = 0$ in this case.

References and Notes

(1) McMillan, W.; Mayer, J. *J. Chem. Phys.* **1945**, *13*, 276−305.
(2) Zimm, B. H. *J. Chem. Phys.* **1946**, *14*, 164−179.
(3) Scatchard, G. *J. Am. Chem. Soc.* **1946**, *68*, 2315−2319.
(4) Scatchard, G.; Batchelder, A. C.; Brown, A. *J. Am. Chem. Soc.* **1946**, *68*, 2320−2329.
(5) Kirkwood, J. G.; Buff, F. P. *J. Chem. Phys.* **1951**, *19*, 774−777.
(6) George, A.; Wilson, W. W. *Acta Crystallogr. D* **1994**, *50*, 361−365.
(7) Velev, O. D.; Kaler, E. W.; Lenhoff, A. M. *Biophys. J.* **1998**, *75*, 2682−2697.
(8) Curtis, R. A.; Montaser, A.; Prausnitz, J. M.; Blanch, H. W. *Biotechnol. Bioeng.* **1998**, *58*, 11−21.
(9) Guo, B.; Kao, S.; McDonald, H.; Asanov, A.; Combs, L. L.; Wilson, W. W. *J. Cryst. Growth* **1999**, *196*, 424−433.
(10) Piazza, R.; Pierno, M. *J. Phys.: Condens. Matter* **2000**, *12*, A443−A449.
(11) Bonneté, F.; Finet, S.; Tardieu, A. *J. Cryst. Growth* **1999**, *196*, 403−414.
(12) Vivarès, D.; Bonneté, F. *Acta Crystallogr. D* **2002**, *58*, 472−479.
(13) Gripon, C.; Legrand, L.; Rosenman, I.; Vidal, O.; Robert, C.; Boué, F. *J. Cryst. Growth* **1997**, *178*, 575−584.
(14) Vilker, V. L.; Colton, C. K.; Smith, K. A. *J. Colloid Interface Sci.* **1981**, *79*, 548−566.
(15) Haynes, C. A.; Tamura, K.; Korfer, H. R.; Blanch, H. W.; Prausnitz, J. M. *J. Chem. Phys.* **1992**, *96*, 905−912.
(16) Wu, J. Z.; Prausnitz, J. M. *Fluid Phase Equilib.* **1999**, *155*, 139−154.
(17) Behlke, J.; Ristau, O. *Biophys. Chem.* **1999**, *76*, 13−23.
(18) Bloustine, J.; Berejnov, V.; and Fraden, S. *Biophys. J.* **2003**, *85*, 2619−2623.
(19) Tessier, P. M.; Lenhoff, A. M.; Sandler, S. I. *Biophys. J.* **2002**, *82*, 1620−1631.
(20) Dumetz, A. C.; Snellinger-O'Brien, A. M.; Kaler, E. W.; Lenhoff, A. M. *Protein Sci.* **2007**, *16*, 1867−1877.
(21) Valente, J. J.; Payne, R. W.; Manning, M. C.; Wilson, W. W.; Henry, C. S. *Curr. Pharm. Biotechnol* **2005**, *6*, 427−436.
(22) Haas, C.; Drenth, J.; Wilson, W. W. *J. Phys. Chem. B* **1999**, *103*, 2808−2811.
(23) Ruppert, S.; Sandler, S. I.; Lenhoff, A. M. *Biotechnol. Prog.* **2001**, *17*, 182−187.
(24) Ruckenstein, E.; Shulgin, I. L. *Adv. Colloid Interface Sci.* **2006**, *123−126*, 97−103.
(25) Coen, C. J.; Blanch, H. W.; Prausnitz, J. M. *AIChE J.* **1995**, *41*, 996−1004.
(26) Neal, B. L.; Asthagiri, D.; Lenhoff, A. M. *Biophys. J.* **1998**, *75*, 2469−2477.
(27) Neal, B. L.; Asthagiri, D.; Velev, O. D.; Lenho, A. M.; Kaler, E. W. *J. Cryst. Growth* **1999**, *196*, 377−387.
(28) Curtis, R. A.; Ulrich, J.; Montaser, A.; Prausnitz, J. M.; Blanch, H. W. *Biotechnol. Bioeng.* **2002**, *79*, 367−380.
(29) Matteoli, E.; Lepori, L. *J. Chem. Phys.* **1984**, *80*, 2856−2863.
(30) Ruckenstein, E.; Shulgin, I. *Fluid Phase Equilib.* **2001**, *180*, 281−297.
(31) Shulgin, I. L.; Ruckenstein, E. *J. Chem. Phys.* **2005**, *123*, 054909.
(32) Prausnitz, J. M.; Lichtenthaler, R. N.; Gomes de Azevedo, E. *Molecular thermodynamics of fluid - Phase equilibria*, 2nd ed.; Prentice - Hall: Englewood Cliffs, NJ, 1986.
(33) Ruckenstein, E.; Shulgin, I. *Int. J. Pharmaceut.* **2003**, *258*, 193−201.
(34) Ruckenstein, E.; Shulgin, I. *Environ. Sci. Technol.* **2005**, *39*, 1623−1631.
(35) Shulgin, I. L.; Ruckenstein, E. *J. Phys. Chem. B* **2007**, *111*, 3990−3998.
(36) Casassa, E. F.; Eisenberg, H. *Adv. Protein Chem.* **1964**, *19*, 287−393.
(37) Timasheff, S. N. *Annu. Rev. Biophys. Biomol. Struct.* **1993**, *22*, 67−97.

(38) Timasheff, S. N. A physicochemical basis for the selection of osmolytes by nature. In *Water and life: comparative analysis of water relationships at the organismic, cellular, and molecular levels*; Somero, G. N., Osmond, C. B., Bolis, C. L., Eds.; Springer-Verlag: Berlin; New York, c1992; pp 70−84.

(39) Record, M. T. J.; Zhang, W.; Anderson, C. F. *Adv. Protein Chem.* **1998**, *51*, 281−353.

(40) Courtenay, E. S.; Capp, M. W.; Anderson, C. F.; Record, M. T. *Biochemistry* **2000**, *39*, 4455−4471.

(41) Baynes, B. M.; Trout, B. L. *J. Phys. Chem. B* **2003**, *107*, 14058−14067.

(42) Timasheff, S. N.; Xie, G. *Biophys. Chem.* **2003**, *105*, 421−448.

(43) Costenaro, L.; Zaccai, G.; Ebel, C. *Biochemistry* **2002**, *41*, 13245−13252.

(44) Ebel, C.; Costenaro, L.; Pascu, M.; Faou, P.; Kernel, B.; Proust-De Martin, F.; Zaccai, G. *Biochemistry* **2002**, *41*, 13234−13244.

(45) Valente, J. J.; Payne, R. W.; Manning, M. C.; Wilson, W. W.; Henry, C. S. *Curr. Pharm. Biotechnol.* **2005**, *6*, 427−436.

(46) Kita, Y.; Arakawa, T.; Lin, T. Y.; Timasheff, S. N. *Biochemistry* **1994**, *33*, 15178−15189.

(47) Arakawa, T.; Ejima, D.; Tsumoto, K.; Obeyama, N.; Tanaka, Y.; Kita, Y.; Timasheff, S. N. *Biophys. Chem.* **2007**, *127*, 1−8.

(48) Rosenbaum, D. F.; Zukoski, C. F. *J. Cryst. Growth* **1996**, *169*, 752−758.

(49) Arakawa, T.; Timasheff, S. N. *Biochemistry* **1984**, *23*, 5912−5923.

(50) *Handbook of Chemistry and Physics*, 77th ed.; Lide, D. R., Editor in Chief; 1996−1997.

(51) Arakawa, T.; Timasheff, S. N. *Methods Enzymol.* **1985**, *114*, 49−74.

(52) Arakawa, T.; Bhat, R.; Timasheff, S. N. *Biochemistry* **1990**, *29*, 1914−1923.

(53) Shulgin, I. L.; Ruckenstein, E. *Biophys. Chem.* **2005**, *118*, 128−134.

(54) Shulgin, I. L.; Ruckenstein, E. *Biophys. Chem.* **2006**, *120*, 188−198.

(55) Shulgin, I. L.; Ruckenstein, E. *Fluid Phase Equilib.* **2007**, *260*, 126−134.

(56) Liu, W.; Bratko, D.; Prausnitz, J. M.; Blanch, H. W. *Biophys. Chem.* **2004**, *107*, 289−298.

(57) Cohn, E. J.; Edsall, J. T. *Proteins, Amino Acids and Peptides*; Reinhold: New York; 1943.

(58) Sauter, C.; Ng, J. D.; Lorber, B.; Keith, G.; Brion, P.; Hosseini, M. W.; Lehn, J. M.; Giege, R. *J. Cryst. Growth* **1999**, *196*, 365−376.

(59) Shulgin, I. L.; Ruckenstein, E. *Biophys. J.* **2006**, *90*, 704−707.

(60) Shulgin, I. L.; Ruckenstein, E. *Phys. Chem. Chem. Phys.* **2008**, *10*, 1097−1105.

JP803149T

Chapter 6

Water and dilute aqueous solutions

6.1 Simple computer experiments with ordinary ice.
6.2 Cooperativity in ordinary ice and breaking of hydrogen bonds.
6.3 The structure of dilute clusters of methane and water by ab initio quantum mechanical calculations.
6.4 Treatment of dilute clusters of methanol and water by ab initio quantum mechanical calculations.

Introduction to Chapter 6

Chapter 6 is devoted to pure water and dilute aqueous solutions. Some of the previous chapters (particularly chapters 4 and 5) were also concerned with aqueous solutions; however, in this chapter the emphasis is on the structure of water.

Why is water so unique and why are its properties so different from those of the "normal" liquids? These questions have been asked by numerous researchers, and so far, there are no absolute answers. However, there is one point on which almost all researchers agree: the network of hydrogen bonds in liquid water and ice (a water molecule can form up to four H-bonds) is the key to the understanding of this "mystery". Therefore, the main emphasis of Chapter 6 is on the H-bond network in water and dilute aqueous solutions.

The approach used in this chapter differs somewhat from those of the other chapters, because it is entirely based on computational methods, such as the ab initio quantum mechanical methods (6.3–6.4) or combinatorial computational methods (6.1–6.2).

Simple computer experiments (which employ 6–8 million water molecules) in which various fractions of H-bonds in ordinary ice are allowed to break are presented (6.1–6.2). The results of our calculations show that the small fraction of broken H-bonds (13–20%), which is usually considered enough for melting, is not sufficient to break up the network of H-bonds into separate clusters. Consequently, liquid water can be considered to be a deformed network with some ruptured H-bonds. The cooperative effect, first suggested by Frank and Wen, was examined by combining an ab initio quantum mechanical method with a combinatorial one (6.2). In agreement with the results obtained in (6.1), it is shown that 62–63% of H bonds must be broken in order to disintegrate a "piece" of ice (containing 8 million water molecules) into disconnected clusters.

The next two papers (6.3 and 6.4) deal with the application of an ab initio quantum mechanical method (the Møller-Plesset perturbation theory) to large binary clusters formed by water with methane or methanol. The molecules of methane or methanol were selected because they represent two extreme types of molecules: 1) methane, an entirely hydrophobic molecule and 2) methanol, which has both hydrophobic and hydrophilic parts and, in addition, can form H-bonds with water. These calculations allow one to analyze the changes in the H-bond network of water in the vicinity of both molecules when they are inserted into pure water. These two cases might be helpful in understanding much more complex molecules such as proteins.

J. Phys. Chem. B **2006**, *110*, 21381−21385

Simple Computer Experiments with Ordinary Ice

Ivan L. Shulgin[†] and Eli Ruckenstein*

Department of Chemical and Biological Engineering, State University of New York at Buffalo, Amherst, New York 14260

Received: July 18, 2006; In Final Form: August 25, 2006

Simple computer experiments in which various fractions of hydrogen bonds (H-bonds) in ice are allowed to break are presented in this paper. First, up to six million water molecules were used to build an artificial piece of ordinary hexagonal ice in the form of a cube, a monolayer, a bilayer, a trilayer, and thicker layers. Then, certain percentages of H-bonds were broken, and the obtained structures were examined. It was found that a large percentage of H-bonds must be broken in order to completely fragment the network of ice into clusters. For a cubic piece of ice, which can be considered bulk ordinary ice, this percentage is equal to 61% H-bonds, a figure also predicted as the threshold of the percolation theory for ice. If, as usually assumed, 13−20% of H-bonds are broken during melting (estimates based on the comparison between the heats of melting and sublimation of ice), the H-bond network of ice is not fragmented and the overwhelming majority of water molecules (>99%) belong to a new, distorted but unbroken network. The percentage of broken H-bonds required for full fragmentation of layers increases with the number of layers and reaches the bulk value of ice for 5−8 layers. This value is consistent with the literature observation that films of water thicker than 20−30 Å have properties close to those of the bulk structure.

1. Introduction

By considering that the main difference between liquid water and ice consists of the percentage of hydrogen bonds (H-bonds) of the latter being broken, we determined, by suitable computer experiments, the fractions of water molecules which are present as clusters and as a continuous network as a function of the percentage of broken H-bonds. The calculations have been carried out for both bulk and multilayer ice.

It is well-known [1,2] that water molecules in ordinary ice (Ih) have a tetrahedral structure. Every molecule is located in the center of a regular tetrahedron as a central molecule, has four nearest neighbors at the corners of this regular tetrahedron, and is linked through an H-bond with each of its nearest neighbors. This configuration leads to many (nonplanar) cycles that contain an even number of water molecules, the smallest cycle being hexagonal,[1−2] thus the reason this ice is called hexagonal (Ih).

Cold liquid water (liquid water at 0 °C) is a very structured liquid which possesses numerous features resembling the ordinary ice (Ih). Indeed,[1−6] (i) the number of nearest neighbors is 4.4 (4 in ice), (ii) the water molecules in cold water have only small deviations from the tetrahedral coordination of ice, (iii) the length of an H-bond (r_{OO} = 2.82 Å, r_{OO} being the distance between the centers of the oxygen atoms of two H-bonded water molecules) is only a little longer than in ice (r_{OO} = 2.76 Å), and (iv) the average number of H-bonds per molecule is 3.6 (in ice it is 4). This picture of liquid water was recently challenged by Wernet et al.[7] who investigated the first coordination sphere in liquid water by X-ray absorption spectroscopy and X-ray Raman scattering. According to them, most molecules in liquid water have only two H− bonds; in one of them the molecule acts as a strong acceptor and in the other as a strong donor.[7] This means that at room temperature in liquid water there are more than 80% broken H-bonds than in ice (Ih).[7−8] This opinion has been critically discussed in the literature.[8−9]

By comparing the heats of melting and sublimation one finds that only 13% of H-bonds in ice are broken upon melting [2]. A similar result (19%) was recently suggested on the basis of a heuristic density-functional method.[10] Many other estimates of the percentage of broken H-bonds are available in the literature.[3,11] These estimates, based both on experimental results obtained by various techniques and on theoretical models, provided values ranging from 2 to 72%,[3,11] which are dependent on the definition used for an H-bond in liquid water.[12−13]

In this paper, computer experiments have been performed in which various percentages of H-bonds in ice (Ih) were considered broken and the structures of the "liquids" thus obtained are presented as a function of them. A relatively large number, 6×10^6 molecules of water, were used to form a "piece" of ice (Ih). First, this piece of ice was created in the form of a cube, which can be considered as a model for bulk ice. Then, the same number of water molecules was used to "construct" an artificial "monolayer", "bilayer", "trilayer", and so on. Second, certain percentages of H-bonds in ice were randomly broken, and the structures thus obtained were examined. Finally, the obtained results were compared with the available models of liquid water and experimental results.

2. Methodology and Program Code

Experiment [1,2] has shown that every molecule of water in ice is located in the center of a tetrahedron and has four hydrogen bonds with its neighbors, which are located at the vertexes of the tetrahedron. This gives rise to a structure which can be represented as multiple interconnected sheets parallel to the *xy*-plane (see Figure 1 for an illustration of a single sheet). Each sheet consists of a grid of nonplanar hexagons with

---

* To whom correspondence should be addressed. E-mail: feaeliru@acsu.buffalo.edu. Fax: (716) 645-3822. Phone: (716) 645-2911 ext. 2214.
† E-mail: ishulgin@eng.buffalo.edu.

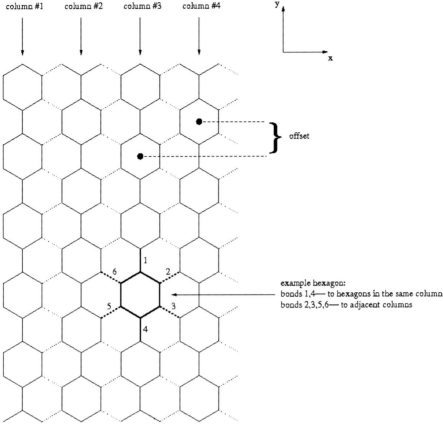

Figure 1. Illustration of the ice structure discussed in the paper (only one layer is shown for clarity).

vertexes in the negative and positive y direction, each containing six water molecules. The sheets are made up of parallel columns of hexagons in the y direction, with adjacent columns offset in the y direction by half the distance between the centers of adjacent hexagons. Each hexagon is connected with six neighboring hexagons in the sheet—one above and one below in the same column, as well as two in the left neighboring column and two in the right. The sheets also possess connections between them, as described below. Although the program uses rows of regularly spaced hexagons, its topological structure is as that of real ice, the regularity being simply a coding tool. Each water molecule in ice has four bonds: Two bonds are with neighboring molecules in the hexagon.[1-2] Another bond connects its hexagon to a neighboring hexagon in the same sheet; since each hexagon contains six molecules, it is connected to six other hexagons in the same sheet. Finally, the fourth bond is between sheets. If the molecules in a given hexagon are numbered around the perimeter from one through six, the odd-numbered molecules would bond to the sheet above, while the even-numbered ones would have bonds with the sheet below.

The program has four inputs—three for the dimensions of the lattice and one for the percentage of broken bonds. The first three inputs are integers specifying the width, length, and height of the ice structure. The height provides the number of sheets of molecules created; the width and length give the number of columns of hexagons and the length of each column. Thus, the total number of molecules created is six times the product of the width, length, and height. When the program receives the command-line inputs, it creates the lattice. The sheets are created individually by a loop that runs a number of iterations that is equal to the number of sheets. Then, the proper number of hexagons is created which are connected as described above.

Finally, the sheets are connected as already explained. Obviously, the molecules located on the surface of the lattice end up with less than four bonds.

Once the lattice is created, the bonds are randomly destroyed in a proportion specified at the command line. There is a linked list of sheets in the structure, and each sheet involves a linked list of bonds; thus, all bonds are processed using a double nested loop. For each bond, a random number between 0 and 1 is generated and compared to the specified probability (which is also between 0 and 1); if the random number is smaller, the bond is deleted. When a bond is deleted, it acquires the property "deleted", which is used later in the program.

Finally, the number and size of the resulting fragments must be found. If the probability used is very small, it is possible that no molecule or group of molecules becomes separated from the original structure and the original lattice is preserved. For calculating the size of a given piece, each molecule has the method "pull", which removes the molecule and any molecules attached to it from the lattice. Using the "pull" method, a depth-first search is run to find the number of molecules in each separate piece.[14]

3. Calculations

The hardware limitations of the program mainly rest with the creation of the lattice in the memory and the calculation of the cluster sizes. The program was designed on a personal machine with 512 MB RAM which could handle lattices of up to one million molecules in its memory, albeit very slowly. The lattices used for the final results, containing up to 6×10^6 molecules, had to be created on a supercomputer with up to 5 GB RAM. Thus, this aspect of the program is dependent directly

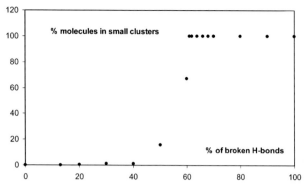

Figure 2. Percentage of molecules in small clusters (compared to the total of 6×10^6 molecules of water) as a function of percentage of broken H-bonds.

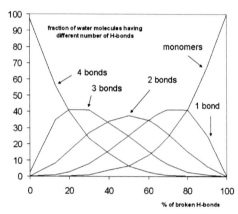

Figure 3. Percentages of water molecules having 4, 3, 2, 1, and 0 H-bonds as functions of the percentage of broken H-bonds.

TABLE 1: Average Size of Clusters as a Function of Percentage of Broken H-bonds

percentage of broken H-bonds (%)	average size of clusters expressed as number of water molecules in the cluster
61	3.9
64	3.4
66	3.0
68	2.7
70	2.5
80	1.7

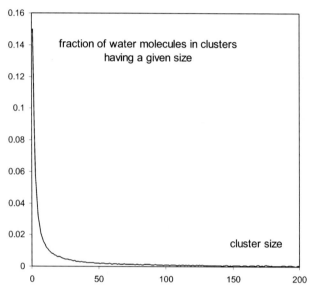

Figure 4. Distribution of the fraction of clusters when 61% of the H-bonds in ice are broken.

on the amount of RAM available and is indefinitely expandable. The second shortcoming, regarding the cluster sizes, does not have a simple solution. The depth-first search algorithm used for finding the sizes of the clusters ran into limitations in terms of the stack size on the machines used. Thus, only clusters smaller than 3000 molecules could be extracted. The solution to the problem would be a more efficient data structure for storing the molecules and their bonds, which allows for a more efficient breadth-first search.

The supercomputer used for calculations was the Young machine at the Center for Computational Research at the University of Buffalo. For the program to be executed using several processors, a computer-specific script was adopted.

4. Results of Computations

4.1. Bulk Ice. The results of the calculations have shown that a cubic piece of ice containing about 6×10^6 molecules of water cannot be broken up completely into small clusters when less than 61% of H-bonds are broken. This means that when 13−20% of H-bonds (usual estimates from the comparison between the heats of sublimation and melting of ice) are broken, the above piece of ice is still an unbroken entity and all the contained water molecules are connected in a network of H-bonds. The fraction of water molecules in clusters as a function of the percentage of broken H-bonds is presented in Figure 2. The fractions of water molecules having 4, 3, 2, 1, and 0 H-bonds calculated by the present simulations are presented in Figure 3. Figure 3 shows that there is a maximum for the fraction of water molecules having two H-bonds at $p = 0.5$, (where $p = 10^{-2} \times$ percentage of broken H-bonds), a maximum for the fraction of water molecules having one H-bond at $p = 0.75$, and a maximum for the fraction of water molecules having three H-bonds at $p = 0.25$. Let us note that the fractions of water molecules having 4, 3, 2, 1, and 0 H-bonds can be also calculated using probability theory (see refs 15−16) and that the results obtained via the latter theory can serve as a test for the correctness of our simulations. If H-bonds are broken randomly with a probability p, then the fraction of water molecules with four H-bonds is given by $(1-p)^4$, with three H-bonds by $C_4^1 \cdot p \cdot (1-p)^3$, with two H-bonds by $C_4^2 \cdot p^2 \cdot (1-p)^2$, with one H-bond by $C_4^1 \cdot p^3 \cdot (1-p)$, and with zero bonds (water monomers) by p^4, where C_i^j is the number of combinations of i objects taken j at a time. The results based on probability theory coincide with those provided by simulations and presented in Figure 3.

According to the results obtained by computer experiments, if more than 61% H-bonds are broken, the piece of ice will be completely fragmented into clusters. The average size of such clusters is provided in Table 1. The frequency of the clusters having various sizes is presented in Figure 4.

4.2. Several Layers of Ice. The same calculations were carried out for several (1−13) layers of ice. Each layer consisted of a sheet of hexagons as described in section 2 "Methodology and Program Code". As expected, in this case smaller percentages of H-bonds have to be broken for clusters to be generated. The results are presented in Figure 5 which shows that whereas for bulk ice (a cubic piece) 61% percent of broken H-bonds are required to generate separate clusters, this figure becomes 38% for a monolayer, 48% for a bilayer, and 61% for 5−8 layers of ice.

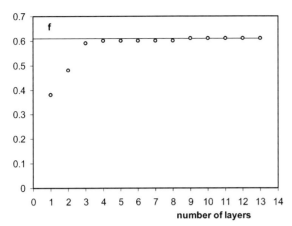

Figure 5. Fraction of broken H-bonds (f) required for full fragmentation in clusters of several layers of ice. The solid line represents the fraction of broken H-bond (0.61) required for full fragmentation of the bulk ice (a cubic piece).

5. Discussion of Results and Comparison with Available Models and Experimental Information

5.1. First, the results of computer experiments with a cubic piece of ice (Figures 2–4 and Table 1) will be examined.

A large number of water molecules (up to 6×10^6 water molecules) was considered in order to minimize the effect of water molecules located on the surface. Indeed, the number of molecules on the surface is small (approximately equal to $6((6 \times 10^6)^{1/3})^2 \approx 0.2 \times 10^6$, hence approximately 3.3% of the total number of water molecules). In addition, the numbers of water molecules having different numbers of H-bonds (Figure 3) coincide with those calculated using the probability theory.

Of course, the present simple computer experiment cannot be considered as an exact counterpart of the ice melting because[2] liquid water cannot be topologically related to ice by the simple breaking of the H-bonds of the latter. Indeed, our computer experiments deal only with energetic characteristics of the ice melting and do not account for changes in entropy and volume. However, the obtained results can be helpful in analyzing various liquid water models.

The number of models that describe the structure and properties of liquid water is enormous. They can be subdivided into two groups:[12] the uniform continuum models and the cluster or mixture models. The main difference between these two classes of models is their treatment of the H-bond network in liquid water; whereas the former assumes that a full network of H-bonds exists in liquid water, in the latter the network is considered broken at melting and that the liquid water is a mixture of various aggregates or clusters. The uniform continuum models stemmed from the classical publications of Bernal and Fowler,[17] Pople,[18] and Bernal.[19] Among the cluster or mixture models, reviewed in refs 2–6 and 12, one should mention the models of Samoilov,[20] Pauling,[21] Frank and Quist,[22] and Nemethy and Scheraga.[23]

If (as usually considered) 13% of H-bonds are broken during melting, the H-bond network of ice cannot be fragmented during melting and the overwhelming majority of water molecules (>99%) form a new distorted but unbroken network. This conclusion is not consistent with the cluster or mixture models, but it is compatible with the uniform continuum models. Arguments against the cluster or mixture models were presented by a number of authors.[2,15–16,24]

Our calculations lead to the conclusion that the H-bond network in ice is so dense that 13–20% of the H-bonds broken cannot generate a mixture of disconnected clusters. They also show that the network of H-bonds in ice can be completely fragmented when ∼61 % of the H-bonds are broken. Let us note that this value (∼61 %) corresponds exactly to the threshold in the percolation theory for ice (Ih).[15–16]

Consequently in the "liquid" obtained after breaking 13–20% of the H-bonds (i) all water molecules belong to one network and are connected to each other by a weakened network of H-bonds, (ii) there are some large cavities formed as a result of the rupture of H-bonds, and (iii) these cavities cannot be stable and are prone to entropy-driven structural transformations. These transformations contribute to the entropic and volumetric changes associated with ice melting.

5.2. The several layers of ice considered in this paper represent artificial constructions of ice in a vacuum in which water molecules on the surface have interactions (H-bonds) with other molecules in the layer and in neighboring layers. These layers represent slices of ice and are very different from bulk ice. For example, 99.9% of water molecules in the monolayer have three H-bonds, about 50% in the bilayer have three H-bonds, and about 50% have four H-bonds in the bilayer.

The calculations regarding several layers of ice have shown that the percentage of broken H-bonds needed for full fragmentation of layers increases with an increasing number of layers and reaches the value for bulk ice at 5–8 layers.

Thin films of ice and water are of interest in the nanosciences.[25–30] Usually these films are considered to be confined between two solid surfaces (walls). It was found that films thicker than 20–30 Å have properties close to those of the bulk.[25–30] These observations are consistent with our results that the percentage of broken H-bonds required for full fragmentation of layers reaches the bulk value for 5–8 layers.

6. Conclusion

Simple computer experiments were performed to examine the ordinary ice (Ih) in which a fraction of hydrogen bonds have been broken. A large number (6×10^6) of water molecules were considered. The sample of ice was first constructed in a cubic form, and then the same amount of water molecules was used to build up mono-, bi-, tri-, etc. layers of ice.

The results of our calculations indicated that in all cases (the cubic sample, mono-, bi-, tri-, and more layers) the small amount of 13–20% of broken H-bonds, usually considered enough for melting, is not sufficient to break up the network of H-bonds into separate clusters. The so-called cluster or mixture models are not consistent with the results of the present simulations. From our results one can conclude that liquid water can be considered to consist of a deformed network with some H-bonds ruptured. In the case of bulk ice more than 61% of the H-bonds has to be broken for its complete fragmentation into clusters to occur. The same result was obtained via percolation theory.[15–16]

The calculations, carried out for several layers of ice, indicated that the percentage of ruptured H-bonds required for full fragmentation of the layers increases with increasing number of layers and reaches the bulk value for 5–8 layers.

Acknowledgment. The authors are indebted to Leonid Shulgin (student from the Department of Physics, Princeton University) for writing the code.

References and Notes

(1) Petrenko, V. F.; Whitworth, R. W. *Physics of Ice*; Oxford University Press: Oxford, 1999.

(2) Stillinger, F. H. *Science* **1980**, *209*, 451.
(3) Eisenberg, D.; Kauzmann, W. *The Structure and Properties of Water*; Oxford University Press: Oxford, 1969.
(4) Narten, A. H.; Levy, H. A. *Science* **1969**, *165*, 447.
(5) Frank, H. S. *Science* **1970**, *169*, 635.
(6) Narten, A. H.; Levy, H. A. In *Water: A Comprehensive Treatise*; Franks, F., Ed.; Plenum: New York, 1972; Vol. 1.
(7) Wernet, P.; Nordlund, D.; Bergmann, U.; Cavalleri, M.; Odelius, M.; Ogasawara, H.; Naslund, L. A.; Hirsch, T. K.; Ojamae, L.; Glatzel, P.; et al. *Science* **2004**, *304*, 995.
(8) Smith, J. D.; Cappa, C. D.; Wilson, K. R.; Messer, B. M.; Cohen, R. C.; Saykally, R. J. *Science* **2004**, *306*, 851.
(9) Smith, J. D.; Cappa, C. D.; Wilson, K. R.; Cohen, R. C.; Geissler, P. L.; Saykally, R. J. *Proc. Natl. Acad. Sci. U.S.A.* **2005**, *102*, 14171.
(10) Hetényi, B.; De Angelis, F.; Giannozzi, P.; Car, R. *J. Chem. Phys.* **2004**, *120*, 8632.
(11) Falk, M.; Ford, T. A. *Can. J. Chem.* **1966**, *44*, 1699.
(12) Jeffrey, G. A. *An Introduction to Hydrogen Bonding*; Oxford University Press: New York, 1997.
(13) Scheiner, S. *Hydrogen Bonding: A Theoretical Perspective*; Oxford University Press: New York, 1997.
(14) Sedgewick, R. *Algorithms in Java*; Addison-Wesley: Boston, 2003.
(15) Geiger, A.; Stillinger, F. H.; Rahman, A. *J. Chem. Phys.* **1979**, *70*, 4185.
(16) Stanley, H. E.; Teixeira, J. *Chem. Phys.* **1980**, *73*, 3404.
(17) Bernal, J. D.; Fowler, R. H. *J. Chem. Phys.* **1933**, *1*, 515.
(18) Pople, J. A. *Proc. R. Soc. London A* **1951**, *205*, 163.
(19) Bernal, J. D. *Proc. R. Soc. London* **1964**, *A280*, 299.
(20) Samoilov, O. Y. *Zh. Fiz. Khim.* **1946**, *20*, 1411.
(21) Pauling, L. *The Nature of the Chemical Bond*, 3rd ed.; Cornell University Press: Ithaca, NY, 1960; Chapter 12.
(22) Frank, H. S.; Quist, A. S. *J. Chem. Phys.* **1961**, *34*, 604.
(23) Nemethy, G.; Scheraga, H. A. *J. Chem. Phys.* **1962**, *36*, 3382.
(24) Kauzmann, W. *Coll. Int. C.N R.S.* **1976**, *246*, 63.
(25) Israelachvili, J. N. *J. Colloid Interface Sci.* **1986**, *110*, 263.
(26) Bellissent-Funel, M. C.; Lal, J.; Bosio, L. *J. Chem. Phys.* **1993**, *98*, 4246.
(27) Bellissent-Funel, M. C.; Chen, S. H.; Zanotti, J. M. *Phys. Rev. E* **1995**, *51*, 4558.
(28) Raviv, U.; Laurat, P.; Klein, J. *Nature* **2001**, *413*, 51.
(29) Zangi, R. *J. Phys.: Condens. Matter* **2004**, *16*, S5371.
(30) Raviv, U.; Perkin, S.; Laurat, P.; Klein, J. *Langmuir* **2004**, *20*, 5322.

Cooperativity in Ordinary Ice and Breaking of Hydrogen Bonds

Eli Ruckenstein,*,† Ivan L. Shulgin,†,‡ and Leonid I. Shulgin§,£

Department of Chemical and Biological Engineering, State University of New York at Buffalo, Amherst, New York 14260, and Department of Physics, Princeton University, Princeton, New Jersey 08544

Received: February 26, 2007; In Final Form: April 3, 2007

The total interaction energy between two H-bonded water molecules in a condensed phase is composed of a binding energy between them and an energy due to a cooperative effect. An approximate simple expression is suggested for the dependence of the interaction energy between two H-bonded water molecules on the number of neighboring water molecules with which they are H-bonded. Using this expression, the probabilities of breaking a H bond with various numbers of H-bonded neighbors are estimated. These probabilities are used in computer simulations of the breaking of specified fractions of H bonds in an ordinary (hexagonal) ice. A large "piece" of hexagonal ice (up to 8 millions molecules) is built up, and various percentages of H bonds are considered broken. It is shown that 62–63% of H bonds must be broken in order to disintegrate the "piece" of ice into disconnected clusters. This value is only a little larger than the percolation threshold (61%) predicted both by the percolation theory for tetrahedral ice and by simulations in which all H bonds were considered equally probable to be broken. When the percentage of broken bonds is smaller than 62–63%, there is a network of H-bonded molecules which contains the overwhelming majority of water molecules. This result contradicts some models of water which consider that water consists of a mixture of water clusters of various sizes. The distribution of water molecules with unequal probabilities for breaking is compared with the simulation involving equal probabilities for breaking. It was found that in the former case, there is an enhanced number of water monomers without H bonds, that the numbers of 2- and 3-bonded molecules are smaller, and the number of 4-bonded molecules is larger than in the latter case.

1. Introduction

The many-body (or cooperative) effect in intermolecular interactions plays an important role in the modern view of condensed matter.[1] Hydrogen bonding in water constitutes one such system. This cooperativity explains some of the anomalies of water and aqueous systems.[1,2] For example, the cooperativity is responsible for the contraction of H bonds in ordinary ice and liquid water compared to the gaseous dimer.[3,4] Indeed, the length of a H bond (r_{oo} distance) in the gaseous dimer is about 2.98 Å, in liquid water it is about 2.85 Å, and in ordinary ice it is about 2.74 Å. The approaches based on pair additive interactions cannot properly describe the properties of ice, water, and aqueous solutions[1–6] because they ignore the cooperativity.

Frank and Wen were probably the first to emphasize the importance of cooperativity in hydrogen bonding in water.[7] They noted[7] that "the formation of hydrogen bonds in water is predominantly a cooperative phenomenon, so that, in most cases, when one bond forms several (perhaps 'many') will form, and when one bond breaks, then, typically, a whole cluster will 'dissolve'. This gives a picture of flickering clusters, of various sizes and shapes, jumping to attention, so to speak, and then relaxing 'at ease'". This simple intuitive idea has been very popular among scientists interested in water and aqueous solutions. For example, the well-known model of water[8] of Nemethy and Scheraga is based on "flickering clusters". The Frank and Wen image of cooperativity was further clarified by Eisenberg and Kauzmann.[2] The strength of a H bond between two water molecules is expected to be affected by the cooperativity and to depend on the number of neighboring water molecules with which the two interacting water molecules are H-bonded. The energy of a H bond in a dimer (zero H-bonded neighbors) differs from that in ice (six neighbors H-bonded to a pair of H-bonded water molecules) and from that in liquid water where the number of H-bonded neighbors can vary between zero and six.

One of the critical points in discussing cooperativity in ice, water, and aqueous mixtures is the evaluation of the H-bonding energy of a pair of water molecules. For ordinary ice (I_h), this energy has been estimated in various ways[2,10–13] which, however, provided quite different results. These estimates were based on the experimentally determined sublimation energy of ice (E_{subl}).[2,8,11–13] Because every water molecule in ordinary ice has four nearest neighbors with which it is linked through H bonds,[2,14] one can estimate the energy of a H bond between two neighboring water molecules in ice using the expression[2]

$$E_{H_2O-H_2O} = \frac{E_{subl} - E_{other}}{2} \quad (1)$$

where E_{other} represents the intermolecular energy associated with interactions other than H bonds (such as the van der Waals interaction). However, due to the ambiguity of the estimates of the "other interactions", the estimates[2] of $E_{H_2O-H_2O}$ vary between 17.8 and 32.2 kJ/mol. A more accurate estimation of the energy of a H bond in ordinary ice is based on the ice lattice energy[2] ($\Delta E_{lattice}$), which at 0 K, has the value[12,13]

---

* To whom correspondence should be addressed. E-mail: feaeliru@acsu.buffalo.edu. Fax: (716) 645-3822. Phone: (716) 645-2911 ext. 2214.
† State University of New York at Buffalo.
§ Princeton University.
‡ E-mail: ishulgin@eng.buffalo.edu.
£ E-mail: lshulgin@princeton.edu.

Cooperativity in Ordinary Ice and Breaking of H Bonds

$$\Delta E_{\text{lattice}} = 58.95 \text{ kJ/mol} \quad (2)$$

and provides the following energy for a H bond in ordinary ice

$$E_{\text{H}_2\text{O}-\text{H}_2\text{O}} = \frac{58.95 \text{ kJ/mol}}{2} = 29.48 \text{ kJ/mol} \quad (3)$$

The latter value will be used in what follows for the energy of a H bond in ordinary ice.

A powerful impetus in the calculations of the interactions between water molecules in a condensed phase has been provided by quantum mechanical ab initio methods. Indeed, the advent of powerful computers provided the opportunity to use quantum mechanical ab initio methods for large clusters of water (containing up to several dozen water molecules).[2,3,15–22] In addition, quantum mechanical ab initio methods are the only ones which allow one to calculate separately the energy contributions from the interactions between two, three, and more molecules and therefore allow one to make a direct estimate of the cooperative effect.

The total interaction energy (ΔE_{int}) of a cluster of n water molecules (n can be 2, 3, 4, ...) can be decomposed as follows[3] (for details, see Appendix A)

$$\Delta E_{\text{int}} = \Delta E_{\text{two-body}} + \Delta E_{\text{many-body}} \quad (4)$$

where $\Delta E_{\text{two-body}}$ is the pair interaction energy which involves the interaction energies between all pairs of molecules in the cluster, and $\Delta E_{\text{many-body}}$ is the interaction energy between various combinations of 3, 4, ..., n molecules (see eq A1 in Appendix A). According to Appendix A, $\Delta E_{\text{two-body}}$ includes contributions from all possible pairs in the clusters, which can be H-bonded and non-H-bonded. Consequently, $\Delta E_{\text{two-body}}$ can be separated into two parts, a contribution from H-bonded pairs ($\Delta E_{\text{H-bonds}}$) and a contribution from non-H-bonded pairs ($\Delta E_{\text{non-H-bonded}}$)

$$\Delta E_{\text{two-body}} = \Delta E_{\text{H-bonds}} + \Delta E_{\text{non-H-bonded}} \quad (5)$$

By combining eqs 4 and 5, one obtains the following expression for the total interaction energy of a cluster containing n water molecules

$$\Delta E_{\text{int}} = \Delta E_{\text{H-bonds}} + \Delta E_{\text{non-H-bonded}} + \Delta E_{\text{many-body}} \quad (6)$$

Quantum mechanical ab initio calculations of small (usually $n < 10$) water clusters[1,3,4,6,16–22] have shown that the many-body contribution to the total interaction energy represents more that 20% for clusters larger than the pentamer, with the main contribution (larger than 90%) of $\Delta E_{\text{many-body}}$ arising from the ternary interactions (ΔE_{ijk}). However, it is not yet known how large $\Delta E_{\text{many-body}}$ is for clusters consisting of a dozen or hundred molecules. In addition, the above-mentioned calculations have been carried out with artificially constructed clusters in the form of chains and rings, which are not present in ordinary ice. However, the analyses of six-member rings[16] and tetrahedral structures[17] provided some information of interest. For instance, it was found that for six-member rings,[16] $\Delta E_{\text{two-body}}$ represents 80% of the total interaction energy and $\Delta E_{\text{many-body}}$ 20%. The $\Delta E_{\text{two-body}}$ consists of 87.5% $\Delta E_{\text{H-bonds}}$ and 12.5% $\Delta E_{\text{non-H-bonded}}$. The $\Delta E_{\text{two-body}}$ in a cluster of five water molecules (they form a tetrahedron with one molecule in the center)[17] constitutes 95% of the total interaction energy, whereas $\Delta E_{\text{many-body}}$ is 5%. The $\Delta E_{\text{two-body}}$ consists of 92.2% $\Delta E_{\text{H-bonds}}$ and 7.8% $\Delta E_{\text{non-H-bonded}}$. The above two examples show that $\Delta E_{\text{many-body}}$ and ΔE_{vdW} provide non-negligible contributions.

In most cases, E_{other} (see eq 1) is due to the van der Waals interaction, and Pauling[11] suggested that it represents about 20% of the sublimation enthalpy of ordinary ice. By comparing eqs 1 and 6, one can conclude that the energy of a H bond calculated from the energy of sublimation includes a contribution from many-body interactions. This contribution probably constitutes the main reason for the difference between the H-bond strengths in ice, liquid water, and gaseous dimers.

Consequently, the cooperative effect represents one factor in the counting of broken H bonds during the melting of ordinary ice. In our previous paper,[23] simple computer experiments were carried out in which various fractions of H bonds were allowed to break in ice. In that paper, the H bonds were considered to have equal breaking probabilities. It was found that a large fraction of H bonds must be broken to completely disintegrate the network of ice into clusters. For a cubic piece of ice, this percentage was 61%, the value also predicted as the threshold of the percolation theory for ice. It is usually assumed that 13–20% of H bonds are broken during melting. These estimates are based on a comparison between the heats of melting and sublimation of ice. Through melting, the H-bond network of ice is not disintegrated, and the overwhelming majority of water molecules (>99%) belongs to a new distorted but unbroken network.

In the present paper, a treatment which accounts for the cooperativity of H bonds will be presented. First, an approximate simple expression will be suggested for the energy of a H bond between two water molecules as a function of the number of H bonds which the above two molecules make with the neighboring water molecules. This expression will be used to evaluate the probabilities of breaking the H bonds with various numbers of H-bonded neighbors. Further, the calculated probabilities will be used to simulate the breaking of various fractions of H bonds of a large "piece" of hexagonal ice. Finally, the results will be discussed and compared with available information.

2. H-Bond Energy Between Two Water Molecules with Various Numbers of Additionally Bound Water Molecules

The energy of a H bond is expected to depend on the number of its H-bonded neighbors with which the two water molecules are H-bonded, which can be 6 (hexagonal ice), 5, 4, 3, 2, 1, and 0 (gaseous dimer) (Figure 1). Therefore, the energy of a H bond in the various cases presented in Figure 1 should depend on the number of H-bonded neighbors.[7,9,24–26] Such a dependence was examined by Symons et al. for H-bonded methanol molecules.[24–26] They compared the energy (α) of the H bond in the methanol dimer A−B with those in the linear chain C−A−B−D (where A, B, C, and D are methanol molecules and − stands for a H bond), with the energies α_1 for the pair C−A and α_2 for the pair A−B (absolute values of the energies were considered). Their experimental results[24,25] have shown that $\alpha_1 > \alpha$ and $\alpha_2 >> \alpha$. Similar results are expected to occur for ice and water, but because of a different structure (tetrahedral) of the H bonds in ice and in water, the dependence of the H-bond energy on the number of neighboring water molecules with which the two water molecules are H-bonded can be very different.[7,26,27]

To calculate the probability of breaking a H bond with various numbers of H-bonded water neighbors, one must first obtain an expression for the energy of breaking a H bond. The total energy of breaking a H bond includes, in addition to the breaking energy between two water molecules, a cooperative energy due to the H bonding of the two water molecules with neighboring water molecules.

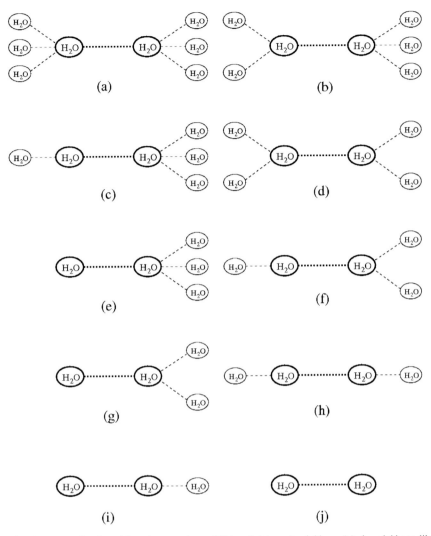

Figure 1. Two H-bonded (···) water molecules with various numbers of H-bonded (- - - -) neighbors; (a) six neighbors, like in hexagonal ice, (b) five neighbors, (c) and (d) four neighbors, (e) and (f) three neighbors, (g) and (h) two neighbors, (i) one neighbor, and (j) no neighbors, like in the gaseous dimer.

The total energy of a H bond will be written in the simple approximate form

$$E^{(N)}_{\text{H-bond}} = E^{(6)}_{\text{H-bond}} + \Delta(6 - N) \qquad (7)$$

where N is the number of broken bonds ($N = 0, 1, 2, 3, 4, 5,$ or 6), $E^{(N)}_{\text{H-bond}}$ and $E^{(6)}_{\text{H-bond}}$ are the breaking energies of the H bonds between two water molecules with N broken neighboring bonds and with all six broken H bonds (gas dimer), respectively, and Δ is a quantity which accounts for the cooperative effect.

There is no cooperative effect for a gaseous dimer (with no H-bonded neighbors), but there is cooperativity in ice which affects the total energy of the H bonds. Because both energies $E^{(0)}_{\text{H-bond}}$ and $E^{(6)}_{\text{H-bond}}$ are known, the parameter Δ can be evaluated. Of course, eq 7 is approximate because (i) it takes into account only the effect of H-bonded neighbors from the first "layer" on the energy of a pair of H-bonded water molecules, (ii) it does not account for differences in the energies of the H bonds of "isomers" with the same number of H-bonded neighbors (see, e.g., Figure 1 c and d, e and f, and g and h), and (iii) it does not account for differences in the energies of H bonds with donor or acceptor neighbors (e.g., a neighbor in Figure 1h can be both donor and acceptor, and in both cases, eq 7 provides the same energy for a H bond).

In essence, $E^{(N)}_{\text{H-bond}}$ is the sum of the energy $E^{(6)}_{\text{H-bond}}$ between two water molecules plus a cooperative effect contribution $\Delta(6 - N)$. The results of our ab initio quantum mechanical calculations listed in Table 1 show that the number of water molecules H-bonded with the two water molecules does not appreciably affect the energy $E^{(6)}_{\text{H-bond}}$ (some details regarding our calculations are presented in Appendix B).

A more rigorous expression for $E^{(N)}_{\text{H-bond}}$ is provided by a nonlinear expression which also accounts for the approximations listed above (i, ii, and iii) involved in eq 7. However, for the time being, there is not enough information to determine all of the parameters in such an expression, and we have to use eq 7 as a first-order linear approximation.

Further, it will be assumed that the probability to break a H bond with N broken neighboring bonds ($p^{(N)}$) is provided by the expression

$$p^{(N)} = w \exp\{-E^{(N)}_{\text{H-bond}}/RT\} \qquad (8)$$

where w is a constant independent of N, R is the universal gas constant, and T is the absolute temperature.

TABLE 1: The Average Pair Interaction Energy and Length of Two H-Bonded Water Molecule in a Dimer and Various Clusters

cluster	number of H-bonded neighbors	average binding energy between two H-bonded water molecules (kJ/mol)	average length of the H bonds (r_{o-o}) (Å)	reference
dimer	0	-19.7 ± 1.5	2.91	18, 28
clusters containing 20 and 25 water molecules	6 (Figure 1a)	-17.0	2.80	29[a]
	5 (Figure 1b)	-17.8	2.82	29
	4 (Figure 1c and d)	-18.9	2.81	29

[a] Details of the calculations regarding large water clusters are briefly summarized in Appendix B.

TABLE 2: Experimental Data ($E^{(0)}_{H-bond}$ and $E^{(6)}_{H-bond}$) Used to Calculate Δ by Eq 7

$E^{(0)}_{H-bond}$ (kJ/mol)[12,13]	$E^{(6)}_{H-bond}$ (kJ/mol)[18,28]	Δ (kJ/mol)
29.5	19.7	1.6

Using eqs 7 and 8, one can write the following relation between the probabilities

$$p^{(0)}/p^{(1)}/p^{(2)}/p^{(3)}/p^{(4)}/p^{(5)}/p^{(6)} = 1/\exp(\Delta/RT)/\exp(2\Delta/RT)/\exp(3\Delta/RT)/\exp(4\Delta/RT)/\exp(5\Delta/RT)/\exp(6\Delta/RT) \quad (9)$$

The experimental data for $E^{(0)}_{H-bond}$ and $E^{(6)}_{H-bond}$ and the value of Δ are listed in Table 2. Using these values, one can rewrite eq 9 as follows

$$p^{(0)}/p^{(1)}/p^{(2)}/p^{(3)}/p^{(4)}/p^{(5)}/p^{(6)} = 1/2.1/4.3/8.8/18.1/37.3/77.0 \quad (10)$$

The above expression shows that the probability to rupture a H bond with 6 H-bonded neighbors is 77 times lower than that to rupture a H bond in a dimer. This observation explains why no measurable dimer concentration was detected in liquid water. It also indicates that there are small probabilities for the existence of H bonds with four and five broken neighboring H bonds and, hence, that simple linear chains and cycles in which the water molecules possess at most two H bonds have low probabilities to be present in liquid water.

3. Algorithm, Code, and Calculations

A cubic piece of ice was built up as described in our previous paper.[23] As is well-known, hexagonal ice can be represented as multiple interconnected sheets parallel to the xy plane (see Figure 1 of our previous paper[23]), each sheet consisting of a grid of nonplanar hexagons containing six water molecules with vertices in the negative and positive y direction. The sheets are made up of parallel columns of hexagons in the y direction, with adjacent columns offset in the y direction by half of the distance between the centers of adjacent hexagons. Each hexagon is connected with six neighboring hexagons in the sheet, one above and one below in the same column, as well as two in the left neighboring column and two in the right neighboring column. The sheets also possess connections between them, as described below. Although the program uses rows of regularly spaced hexagons, its topological structure is as that of real ice, the regularity representing simply a coding tool. Each water molecule in ice has four bonds, two with the neighboring molecules in the hexagon, another connecting its hexagon to a neighboring hexagon in the same sheet (since each hexagon contains six molecules, it is connected to six other hexagons in the same sheet), and a fourth between sheets. If the molecules in a given hexagon are numbered around the perimeter from one through six, the odd-numbered molecules are bonded to the sheet above, while the even-numbered ones are bonded to the sheet below.

The program involves three groups of inputs, one for the dimensions of the lattice, another one for the total number of bonds to be broken, and finally a third one for the probabilities of bond breaking ($p^{(0)}$, $p^{(1)}$, $p^{(2)}$, $p^{(3)}$, $p^{(4)}$, $p^{(5)}$, and $p^{(6)}$). The first group consists of three integers, specifying the width, the length, and the height of the ice structure. The height is merely the number of sheets of molecules created; the width and length are the number of columns of hexagons and the length of each column, respectively. Consequently, the total number of molecules created is 6 × width × height × length. When the program receives the command-line inputs, it creates the lattice. The sheets are created individually by a loop that runs a (height) number of iterations. Each sheet is characterized by the parameters height and length; then, the proper number of hexagons is created, and they are connected accordingly. Finally, the sheets are connected as explained above. Obviously, some molecules end up with less than four bonds since they are located on the bottom and top sheets; therefore, they cannot be linked down and up, respectively. Also, the molecules on the border of each sheet have less than four bonds because there are fewer hexagons to link to the same sheet.

The second and third groups of inputs deal with the breaking of the bonds. The second group is a decimal M specifying the total fraction of bonds that will be broken (i.e., $M = 0.13$ for 13% of all bonds). The third group consists of a set of probabilities for a bond to be broken, different (increasing) decimals for bonds with six neighboring bonds (the maximum), five, four, three, two, one, and zero. Only first-neighbor bonds affect the probabilities; second-order effects are ignored.

Bonds are broken in a multistage process. First, a small fraction (this fraction can also be selected as equal to zero) of bonds are broken at the "bulk" probability to "seed" the melting process. Once that is accomplished, a "breaking" loop is executed repeatedly until the fraction M of the bonds has been broken. In each iteration of the "breaking" loop, the program looks at each bond in the lattice in order. It calculates the number of neighboring bonds that still exists. This number is originally six for all bonds on the inside of the lattice (less near the boundaries, as explained above) but can decrease to anywhere between five and zero as nearby bonds disappear. On the basis of the number of neighbors, a probability of breaking is retrieved for this bond (from the third group of inputs). Say this probability is 0.02 (2%), then a random decimal between 0 and 1 is generated. If it is less than the probability, the bond is deleted (on average, this process deletes the bond with a 2% probability). Then, the loop moves on through the lattice until all bonds have been visited. After each iteration of the "breaking" loop, the program calculates the total fraction of bonds in the lattice that have been broken. If that fraction is less than the number specified in the second input group, it repeats the loop. Otherwise, the program moves on to the final stage.

Finally, the number and size of the resulting fragments must be found. If the total fraction of bonds broken is very small, it is possible that no molecule or group of molecules is broken off from the original structure. To distinguish the separated fragments in the code, a loop over all molecules is used. At first, all molecules are flagged as "unexamined". When an unexamined molecule is reached, it is added to a queue and set as "examined". Then, the program removes the molecule from the queue and looks at its neighbors. If they are unexamined, they are, in turn, placed into the queue and set to examined. This process continues recursively until all of the molecules in the same fragment as the original molecule have been, at some point, placed in the queue and flagged as examined. Then, the number of molecules thus reached is counted and reported as the size of the fragment. The program then loops through the lattice until it finds another unexamined molecule; clearly, it must be part of a new fragment. A new queue is created, and all of the molecules in the new fragment are labeled and tallied. The process continues until all of the molecules in the lattice have been examined.

The hardware limitations of the program mainly rest with the creation of the lattice in memory. The lattices used for the final results, up to 8 million water molecules in size (in the form of a cube with length, width, and height of 110, containing 7986000 molecules), were created on a supercomputer with 2 GB of RAM. Thus, this aspect of the program depends directly on the amount of RAM available and is indefinitely expandable. The supercomputer used for calculations was the U2 machine at the Center for Computational Research at the University of Buffalo. In order to run the program on a parallel-processor supercomputer, we had to use a computer-specific script.

4. Results and Discussion

For this computer experiment, a large (about 8 million water molecules) "piece" was built up, as described above. (In our previous calculations,[23] a piece containing about 6 million water molecules was used.) Various fractions (from 5 to 95%) of the H bonds were allowed to break in a process that takes into account different probabilities of rupture of various types of H bonds. After that, the resulting piece of "ice" was examined.

4.1. Relation between the Number of Broken Bonds and Structure. Percolation Threshold.
Figure 2 presents the fraction of water molecules in small clusters as a function of the fractions of broken H bonds. The calculations show that the small amount of 13–20% of broken H bonds, usually considered to occur in melting, is not sufficient to disintegrate the network of H bonds into separate clusters and that the overwhelming majority of water molecules (>99%) belongs to a new distorted but unbroken network. This result was also obtained by us before[23] when we assumed equal probability of rupture of H bonds and also by others a long time ago.[30,31] It may be used as a test for any models of the water structure. For instance, the so-called cluster or mixture models[2] are not consistent with the above conclusion.

Figure 2 also shows that when 62–63% of H bonds are broken, the piece of ice is disintegrated into small separate clusters, and the network of H bonds is completely broken down. This result is slightly different from that[23] (60–61%) obtained by assuming equal probability of rupture of all H bonds. Let us note[30,31] that 60–61% is also the threshold provided by the percolation theory for the tetrahedral structure.

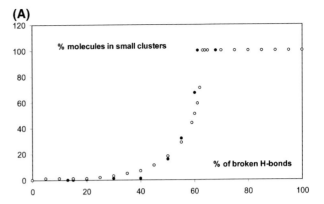

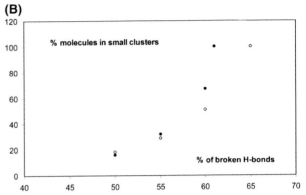

Figure 2. Percentage of molecules in small clusters (compared to the total of about 8×10^6 molecules of water) as a function of the percentage of broken H bonds; ○ present calculations, ● results from our equal probability previous calculations.[23] In B, the results are plotted for the percentage of broken H bonds between 40 and 70%.

4.2. H-Bond Statistics.
If H bonds are broken randomly with equal probability (p) for each H bond, then the fractions of water molecules with four intact H bonds is given[30,31] by $(1-p)^4$, with three intact H bonds by $C_4^1 \cdot p \cdot (1-p)^3$, with two intact H bonds by $C_4^2 \cdot p^2 \cdot (1-p)^2$, with one intact H bond by $C_4^1 \cdot p^3 \cdot (1-p)$, and with zero bonds (water monomers) by p^4, where C_i^j is the number of combinations of i objects taken j at a time. However, when the cooperative effect is taken into account, the H bonds do not rupture with equal probability. The breaking of one H bond is described mathematically by eqs 7–10. Figure 3 compares the fractions of water molecules having various numbers of H bonds in the two cases. One can observe that the numbers of 2- and 3-bonded molecules are smaller and the numbers of 4-bonded molecules and monomers are larger for the unequal than the equal probabilities of rupture. It should be mentioned that several models of the water structure involve a large fraction of monomer water molecules.[11,32] Our calculation show that when 20% of H bonds in ice are broken, 0.5% of the monomer is formed; when 25% of H bonds in ice are broken, 1% of the monomer is generated. It is worth noting that the equal probability assumption leads to 0.16% monomer when 20% of the H bonds in ice are broken and to 0.39% monomer when 25% of the H bonds in ice are broken. Due to their capability to fill the "holes" in the ice structure, the monomers can play an important role. For instance, Frank[33] suggested that "the structure of cold water seems likely to consist, for the most part, of hydrogen-bonded, four-coordinated, framework regions, with interstitial monomers occupying some fraction of the cavities that the framework encloses. The precise geometry of the framework has not been specified, but some evidence

Cooperativity in Ordinary Ice and Breaking of H Bonds

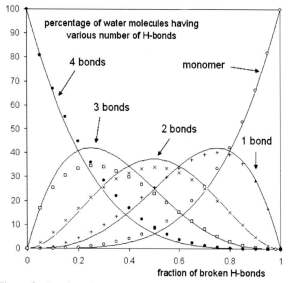

Figure 3. Fraction of water molecules having various numbers of H bonds. The solid lines are for equal probability of rupture of H bonds. Results of the present calculation: ● fraction of water molecules having four H bonds, □ fraction of water molecules having three H bonds, × fraction of water molecules having two H bonds, + fraction of water molecules having one H bond, ○ fraction of water molecules having no H bonds.

suggests that it is rather regular at low temperatures and becomes more random as the water gets warmer.

There are important differences between the literature models[11,32] and our results. In our case, (i) the number of monomers is smaller than that in the Pauling model (where they are present in clathrate-like cages),[11] and (ii) they "coexist" with a disturbed but still infinite, not disintegrated network of water molecules. In contrast, the models in refs 11 and 32 do not involve a network but only a distribution of clusters.

4.3. How Many H Bonds are Ruptured during Melting?
Various estimates for the fraction of H bonds ruptured during melting are available, which are reviewed in references 2 and 34. A convenient method is the comparison of the heats of melting and sublimation. Pauling found that about 15% of H bonds are ruptured upon melting,[11] and Stillinger found a value of about 13%.[14] These calculations involve the enthalpy of sublimation of ice ($\Delta H_{subl} = 51.6$ kJ/mol at 273.15 K) and the enthalpy of fusion of ice ($\Delta H_{fusion} = 6.01$ kJ/mol at 273.15 K),[2] which constitutes about 11.8% of the enthalpy of sublimation of ice. However, these simple estimates imply several assumptions; (1) the enthalpies of sublimation and fusion are due only to the H-bond rupture, and (2) the energies of a single H bond of the hexagonal ice and cold water (at 273.15 K) are the same. It is clear that assumption 1 constitutes only a rough approximation (see eq 1) because those enthalpies also include (see eq 6) other interactions (such as the van der Waals interactions). Eisenberg and Kauzmann[2] employed the ice lattice energy at 0 K ($\Delta E_{lattice}$) to estimate the H-bond energy (see eq 3), and this led to 10.2% (($100 \times 6.01)/58.95 = 10.2\%$) for the fraction of H bonds ruptured during melting.

The present procedure allows one to take into account the cooperative effect in the breaking of H bonds during melting. Figure 4 presents the contributions of various types of H bonds (see Figure 1) to the "pool" of ruptured H bonds. One can note that more than 95% of the ruptured H bonds have six or five H-bonded neighbors. The results from Figure 4 can be combined with eq 7 to calculate the energy required to rupture a certain fraction of H bonds. A comparison of the calculated energy

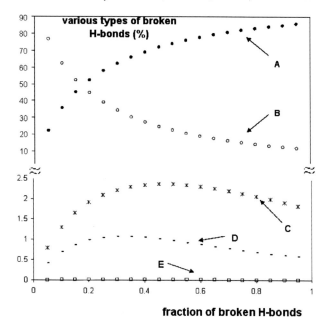

Figure 4. The contributions (%) from various types of broken H bonds to the total number of broken H bonds. (A) six H-bonded neighbors, (B) five H-bonded neighbors, (C) four H-bonded neighbors, (D) three H-bonded neighbors, and (E) two H-bonded neighbors (see also the caption to Figure 1).

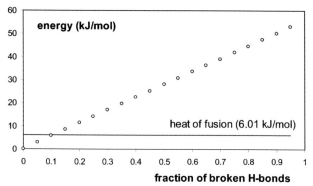

Figure 5. The dependence of the energy required to rupture a certain fraction of H bonds on the fraction ruptured. The solid line represents the enthalpy of fusion of hexagonal ice.

with the heat of fusion allows one to calculate the fraction of broken H bonds upon melting. The results of such calculations are presented in Figure 5. One can see that the fraction of H bonds ruptured upon melting is between 11 and 12%, a value only a little larger than the 10.2% calculated by comparing with the heat of vaporization.

4.4. Average Cluster Size. As already mentioned above, 62−63% of H bonds must be ruptured to achieve the full fragmentation of the hexagonal ice structure into small clusters. When the number of broken H bonds is less than 62−63%, most of the water molecules belong to an unbroken network (see Figure 2). The average cluster sizes when the number of broken H bonds is larger than 62−63% are listed in Table 3.

5. Conclusion

The present paper is focused on the role of the cooperativity (or many-body interactions) on the structure of ordinary ice and liquid water. For this purpose, the energy of a H bond was expressed as a function of the number of H-bonded neighbors (see eq 7). Then, the probabilities of breaking various types of

TABLE 3: Average Size of Clusters versus the Percentage of Broken H Bonds

percentage of broken H bonds (%)	average size of clusters expressed as the number of water molecules in the cluster
64	3.3 (3.4)[a]
65	3.2
70	2.4 (2.5)[a]
75	2.0
80	1.6 (1.7)[a]
85	1.4
90	1.2
95	1.1

[a] The values in parentheses are for the equal probability of breaking of H bonds.[23]

H bonds (see Figure 1) were calculated to conclude that there are large differences in the probabilities of breaking various types of H bonds. For example, the probability to rupture the H bond of a water dimer, which is H-bonded to zero water neighbors (gaseous dimer), is 77 times larger than that of a water dimer which is H-bonded to six water neighbors because of the cooperative effect in the latter.

The predicted probabilities were used to simulate the rupture of H bonds in an ordinary (hexagonal) ice. For this purpose, a large "piece" (up to 8 million water molecules in the form of a cube) of ordinary ice was built up, and various fractions (from 5 to 95%) of H bonds were allowed to rupture.

It was found that 62−63% of H bonds must be broken to disintegrate the above piece of ice into small clusters. This value is a little larger than that (61%) predicted for equal probability of rupture for all H bonds.

In addition, it was found that the obtained structure contained an enhanced (compared with the case of equal probability of rupture[23] of H bonds) number of water monomers. This result supports the viewpoint[11,32,33] that the water monomers play a role in the liquid water structure.

Our calculation showed that the percentage of ruptured H bonds during melting is 11−12%.

Acknowledgment. The authors are indebted to the Center for Computational Research (CCR) of the University at Buffalo for the use of its facilities.

Appendix A

The total interaction energy of a cluster of n water molecules (n can be 2, 3, ...) can be decomposed as follows[3]

$$\Delta E_{int} = \sum_{i=1}^{n-1} \sum_{j>i}^{n} \Delta E_{ij} \quad \text{"two-body"}$$
$$+ \sum_{i=1}^{n-2} \sum_{j>i}^{n-1} \sum_{k>j}^{n} \Delta E_{ijk} \quad \text{"three-body"}$$
$$+ \sum_{i=1}^{n-3} \sum_{j>i}^{n-2} \sum_{k>j}^{n-1} \sum_{l>k}^{n} \Delta E_{ijkl} \quad \text{"four-body"}$$
$$\cdots$$
$$+ ..., \quad \text{"}n\text{-body"} \quad (A1)$$

where $\Delta E_{m-\text{body}}$ is the m-body interaction energy of the cluster ($m = 2, 3, ..., n$) and ΔE_{ij}, ΔE_{ijk}, and so forth in eq A1 are defined as

$$\Delta E_{ij} = E_{ij} - E_i - E_j$$
$$\Delta E_{ijk} = E_{ijk} - [E_i + E_j + E_k] - [\Delta E_{ij} + \Delta E_{ik} + \Delta E_{jk}]$$
$$\cdots$$
$$\Delta E_{N-\text{body}} = E_{N-\text{body}} - [E_i + E_j + E_k] - [\Delta E_{ij} + \Delta E_{ik} + \Delta E_{jk}] - ... \quad (A2)$$

where E_i is the energy of an isolated molecule i, E_{ij} is the energy of the complex consisting of molecules i and j, E_{ijk} is the energy of a complex consisting of molecules i, j, and k, and so forth. Equation 4 for the total interaction energy of a cluster of n water molecules can be rewritten in the form

$$\Delta E_{int} = \Delta E_{\text{two-body}} + \Delta E_{\text{many-body}} \quad (A3)$$

where $\Delta E_{\text{two-body}}$ is the pair interaction energy which includes the interaction energy between all pairs of molecules in the cluster of n molecules and $\Delta E_{\text{many-body}}$ is the interaction energy between various combinations of 3, 4, ..., n molecules. $\Delta E_{\text{two-body}}$ can be separated into two parts, one contribution from H-bonded pairs in the cluster and another contribution from non-H-bonded pairs in the cluster

$$\Delta E_{\text{two-body}} = \Delta E_{\text{H-bonds}} + \Delta E_{\text{non-H-bonded}} \quad (A4)$$

Appendix B

The second-order Møller-Plesset perturbation theory (MP2) with a large 6-311++G (3d, 2p) basis set was employed to optimize the geometries of the clusters containing 20 and 25 water molecules.[29] The computations were carried out as in our previous publications.[35,36]

The computational procedure has allowed one (1) to find the optimal geometries for the clusters considered and (2) to determine the distances and the interaction energies between pairs of H-bonded water molecules.

The computational procedure consists of three steps. (i) An initial cluster configuration was constructed using the Cerius² 4.2 software. The configuration thus built was processed using the Cerius² CLEAN function, and the obtained structure was considered as the initial guess. (ii) The cluster geometry was obtained by optimizing the guess with respect to all coordinates using the MP2 method with a 6-311++G (3d, 2p) basis set. For large clusters, it is difficult to reach the global minimum because there are many local minima. To avoid the effect of the initially selected guess, the minimization procedure was carried out for a large number of initial guesses (five−ten). (iii) All H-bonded pairs in the optimized clusters were selected and ascribed to various groups shown in Figure 1. (iv) The interaction energies between H-bonded molecules α and β ($E_{\alpha\beta}^{int}$) were calculated using the supermolecular approach[37,38]

$$E_{\alpha\beta}^{int} = E_{\alpha\beta}\{\alpha\beta\} - E_{\alpha}\{\alpha\beta\} - E_{\beta}\{\alpha\beta\} \quad (B1)$$

where $E_{\alpha\beta}\{\alpha\beta\}$ is the total energy of an $\alpha\beta$ pair obtained with an $\{\alpha\beta\}$ basis set and $E_{\alpha}\{\alpha\beta\}$ and $E_{\beta}\{\alpha\beta\}$ are, respectively, the energies of the individual α and β molecules obtained also with the $\{\alpha\beta\}$ basis set. (v) The average interaction energies and lengths between the H-bonded water molecules were calculated (Table 1) for various H-bonded water molecules (see Figure 1).

References and Notes

(1) Elrodt, M. J.; Saykally, R. J. *Chem. Rev.* **1994**, *94*, 1975.

(2) Eisenberg, D.; Kauzmann, W. *The Structure and Properties of Water*; Oxford University Press: Oxford, U.K., 1969.
(3) Ojamäe, L.; Hermansson, K. *J. Phys. Chem.* **1994**, *98*, 4271.
(4) Ludwig, R. *Angew. Chem., Int. Ed.* **2001**, *40*, 1809.
(5) Barnes, P.; Finney, J. L.; Nicholas, N. D.; Quinn, J. E. *Nature* **1979**, *282*, 459.
(6) Xantheas, S. S. *Chem. Phys.* **2000**, *258*, 225.
(7) Frank, H. S.; Wen, W. Y. *Discuss. Faraday Soc.* **1957**, *24*, 133.
(8) Nemethy, G.; Scheraga, H. A. *J. Chem. Phys.* **1962**, *36*, 3382.
(9) Perram, J. W.; Levine, S. *Mol. Phys.* **1971**, *21*, 701.
(10) Bernal, J. D.; Fowler, R. H. *J. Chem. Phys.* **1933**, *1*, 515.
(11) Pauling, L. *The Nature of Chemical Bond*, 3rd ed.; Cornell University Press: Ithaca, NY, 1960.
(12) Petrenko, V. F.; Whitworth, R. W. *Physics of Ice*; Oxford University Press: Oxford, U.K. 1999.
(13) Whalley, E. In *The Hydrogen Bond*; Schuster, P., Zundel, G., Sandorfy, C., Eds.; North-Holland: Amsterdam, The Netherlands, 1976; Vol. III, p 1425.
(14) Stillinger, F. H. *Science* **1980**, *209*, 451.
(15) Wales, D. J. In *Encyclopedia of Computational Chemistry*; Schleyer, P. v. R., Allinger, N. L., Clark, T., Gasteiger, J., Kollman, P. A., Schaefer, H. F., III, Schreiner, P. R., Eds.; Wiley: New York, 1998.
(16) Hermansson, K. *J. Chem. Phys.* **1988**, *89*, 2149.
(17) White, J. C.; Davidson, E. R. *J. Chem. Phys.* **1990**, *93*, 8029.
(18) Xantheas, S. S.; Dunning, T. H., Jr. *J. Chem. Phys.* **1993**, *99*, 8774.
(19) Xantheas, S. S. *J. Chem. Phys.* **1995**, *102*, 4505.
(20) Milet, A.; Moszynski, R.; Wormer, P. E. S.; van der Avoird, A. *J. Phys. Chem. A* **1999**, *103*, 6811.
(21) Dunning, T. H. *J. Phys. Chem. A* **2000**, *104*, 9062.
(22) Lee, H. S.; Tuckerman, M. E. *J. Chem. Phys.* **2006**, *125*, 154507.
(23) Shulgin, I. L.; Ruckenstein, E. *J. Phys. Chem. B* **2006**, *110*, 21381.
(24) Symons, M. C. R.; Shippey, T. A.; Rastogi, P. P. *J. Chem. Soc., Faraday Trans. 1* **1980**, *76*, 2251.
(25) Symons, M. C. R.; Thomas, V. K. *J. Chem. Soc., Faraday Trans. 1* **1981**, *77*, 1883.
(26) Symons, M. C. R. *Philos. Trans. R. Soc. London, Ser. A* **2001**, *359*, 1631.
(27) Frank, H. S. *Proc. R. Soc. London, Ser. A* **1957**, *247*, 481.
(28) Szalewicz, K.; Cole, S. J.; Kolos, W.; Bartlett, R. J. *J. Chem. Phys.* **1988**, *89*, 3662.
(29) Ruckenstein, E.; Shulgin, I. L. **2007**, unpublished results.
(30) Geiger, A.; Stillinger, F. H.; Rahman, A. *J. Chem. Phys.* **1979**, *70*, 4185.
(31) Stanley, H. E.; Teixeira, J. *J. Chem. Phys.* **1980**, *73*, 3404.
(32) Samoilov, O. Y. *Zh. Fiz. Khim.* **1946**, *20*, 1411.
(33) Frank, H. S. *Science* **1970**, *169*, 635.
(34) Falk, M.; Ford, T. A. *Can. J. Chem.* **1966**, *44*, 1699.
(35) Ruckenstein, E. E.; Shulgin, I. L.; Tilson, J. L. *J. Phys. Chem. B* **2003**, *107*, 2289.
(36) Ruckenstein, E. E.; Shulgin, I. L.; Tilson, J. L. *J. Phys. Chem. B* **2005**, *109*, 807.
(37) Chalasinski, G.; Gutowski, M. *Chem. Rev.* **1988**, *88*, 943.
(38) Jeziorski, B.; Szalewicz, K. In *Encyclopedia of Computational Chemistry*; Schleyer, P. v. R., Ed.; Wiley: New York, 1998; Vol. 2, p 1376.

J. Phys. Chem. A **2003**, *107*, 2289−2295

The Structure of Dilute Clusters of Methane and Water by ab Initio Quantum Mechanical Calculations

Eli Ruckenstein,*,[†] Ivan L. Shulgin,[†] and Jeffrey L. Tilson[‡]

Department of Chemical Engineering and Center for Computational Research, State University of New York at Buffalo, Amherst, New York 14260

Received: October 21, 2002; In Final Form: January 16, 2003

Ab initio quantum mechanical methods have been used to examine clusters formed of molecules of methane and water. The clusters contained one molecule of one component (methane or water) and several (10, 8, 6, 4, and 1) molecules of the other component. The Møller−Plesset perturbation theory (MP2 method) was used in the calculations. The cluster geometries were obtained via optimization and the interaction energies between the nearest neighbors were calculated for the geometries obtained in the first step. It is shown that the interaction energies and intermolecular distances between the molecules of methane and water are quite different in the clusters $CH_4 \cdots (H_2O)_{10}$ and $H_2O \cdots (CH_4)_{10}$. They are also different from those in the water/methane dimer. The structure of the cluster $CH_4 \cdots (H_2O)_{10}$ is highly affected by the hydrogen bonding among the water molecules, and the methane molecule is located inside a cage formed of water molecules. In contrast, the molecules of methane and water are randomly distributed in the cluster $H_2O \cdots (CH_4)_{10}$. The average methane/water intermolecular distance in the cluster $CH_4 \cdots (H_2O)_{10}$ provided by the quantum mechanical calculations is in agreement with the experimental and simulation results regarding the position of the first maximum in the radial distribution function $g_{oc} = g_{oc}(r_{oc})$ in dilute mixtures of methane in water, where r_{oc} is the distance between the C atom of methane and the O atom of water. It is shown that the water molecules in the vicinity of a central methane molecule can be subdivided into two groups, A and B. Molecules of type A are touching nearest neighbors of the central methane molecule. They are located on a sphere with a radius corresponding to the first maximum in the radial distribution function $g_{oc} = g_{oc}(r_{oc})$ and are tangentially oriented toward the central methane molecule. The layer of A water molecules is somewhat denser than bulk water. The molecules of type B are also located in the first hydration layer of a central methane molecule (up to a distance given by the position of the first minimum of the radial distribution function $g_{oc} = g_{oc}(r_{oc})$), but are not touching nearest neighbors. They are distributed more randomly than the molecules of type A, because they are less affected by the hydrophobic core of the solute.

1. Introduction

The interactions in mixtures of nonpolar substances, such as noble gases and hydrocarbons, with water constitute the simplest manifestation of the hydrophobic effect. A large number of publications (many thousands) have been devoted to this topic and information about the hydrophobic effect was summarized in books and recent reviews.[1−7] The hydrophobic effect is germane to chemistry (gas solubility in water, phase separation, and self-assembling in aqueous mixtures), biology (protein folding and micellization), and even geology (undersea deposits of methane hydrates). Two manifestations of the hydrophobic effect can be considered: the interaction of one molecule of a nonpolar solute with the surrounding water molecules (hydrophobic hydration) and the interactions of nonpolar molecules among themselves in a water environment (hydrophobic interactions).[1−7] While Kauzmann was the first to introduce the notion of hydrophobic interactions in the 1950s,[8,9] some fundamentals of hydrophobic hydration were established earlier in the 1930s and 1940s in the publications of Butler and those of Uhlig and Eley.[10−13] In their papers, they tried to explain the poor solubilities of nonpolar molecules in water (as a rule they are smaller by 1−3 orders of magnitude than those in organic substances) by dividing the dissolution process into two steps: (1) the creation of a "cavity" in the bulk water and (2) the insertion of the nonpolar molecule into the cavity. This scheme became classic and was used to explain the behavior of various thermodynamic functions that characterize the dissolution of nonpolar substances in water under ambient conditions: the free energy change ΔG^d is positive (unfavorable), the enthalpic change ΔH^d is negative (favorable), the entropic change ΔS^d is negative (unfavorable with a larger absolute value of $T\Delta S^d$ than ΔH^d), and the change in the isobaric heat capacity Δc_p^d is large and positive.[14] Reliable values of these thermodynamic functions are now available for numerous substances.[14−16] Frank and Evans[17] provided an additional insight in the understanding of the hydrophobic hydration by suggesting that during the second step the layers of water around the solute molecule become more ordered. The formation of the more ordered structures (icebergs) around a molecule of a nonpolar solute was in their opinion the cause of the great loss of entropy in the process of dissolution. This idea dominated the field for several decades and more detailed theories were

---

* Corresponding author. E-mail: feaeliru@acsu.buffalo.edu. Fax: (716) 645-3822. Phone: (716) 645-2911/ext. 2214.
[†] Department of Chemical Engineering.
[‡] Center for Computational Research.

10.1021/jp0222671 CCC: $25.00 © 2003 American Chemical Society
Published on Web 03/08/2003

developed to provide quantitative explanations for the behavior of the above thermodynamic functions and for the temperature and pressure dependencies of the solubility in water.[1-7] The concept of iceberg led to the conclusion that the decrease in entropy caused by the organization of the water molecules is responsible for the low solubility of hydrocarbons in water. In reality, the change in entropy due to ordering is compensated by the change in enthalpy caused by the interactions between the hydrocarbon molecule and water.[18-22] Shinoda[18,19] concluded that the formation of a cavity constitutes the main effect, while Ruckenstein[20-22] has shown, on the basis of a simple thermodynamic approach, that while the formation of a cavity provides the largest contribution, the "iceberg" formation also plays a role.

A different interpretation of the hydrophobic effect was suggested by Lucas and Lee.[23-24] They suggested that the poor solubility of nonpolar compounds in water is due to an excluded volume effect, which is amplified, in the case of liquid water, by the small size of the water molecules, and that the entire hydrophobic effect is a result of their small size. The combination of this idea[25-28] with Muller's two-state water structure[29] provided reasonable results regarding the hydrophobic hydration. The more recent application[30-32] of information theory to the treatment of the hydrophobic effect was used to explain[5] (a) the temperature dependence of the hydrophobic hydration, (b) the water/hydrogen isotope effect, etc.

During the last 2 decades, the availability of powerful computers and the wide use of modern experimental methods, especially X-ray and neutron scattering, allowed one to obtain valuable information about the nanostructure of mixtures containing hydrophobic solutes. Moreover, one can observe a certain redirection in the research of the hydrophobic effect. While in the past the main goal was to obtain reliable data concerning the thermodynamics of the hydrophobic hydration and to interpret them using different models, the goal now is to obtain information about the nanostructure of water around a hydrophobic solute and to find out how this nanostructure differs structurally and energetically from that of bulk water. While the existence of several layers of water molecules around a hydrophobic solute which are affected by the solute is beyond doubt (it was demonstrated experimentally[33-35]), the characteristics of this "perturbed" water are not yet well-known. Several questions arise regarding them and the difference from bulk water: (1) how many water molecules are involved or how many water molecules are affected by the presence of a hydrophobic solute? (2) is its structure more ordered than that of bulk water? (3) what is the local density of this "perturbed" water? and so on.

It is clear that these questions can be answered if information about the local structure and intermolecular interactions in the layers of the "perturbed" water can be obtained.

An important step in understanding the local structure around a nonpolar solute in water was made by Jorgensen et al.[36] Using Monte Carlo simulations based on an intermolecular potential, which contained Lennard-Jones and Coulomb contributions, they determined the number of water molecules in the first hydration layer (located between the first maximum and the first minimum of the radial distribution function) around a nonpolar solute in water. This number (20.3 for methane, 23 for ethane, etc.) was surprisingly large compared with the coordination numbers in cold water and ice (4.4 and 4, respectively). These results provided evidence that major changes occur in the water structure around a nonpolar solute and that the perturbed structure is similar to that of the water−methane clathrates,[37] which involve 20−24 water molecules that form a clathrate cage around a methane molecule. The conclusions of Jorgensen et al.[36] were verified both experimentally[34,35] (ref 35 provided a value of 16 for the number of water molecules in the first hydration layer) and by molecular simulations.[38-40] Similar results regarding the number of water molecules in the first hydration layer were obtained for infinitely dilute aqueous solutions of noble gases,[38,41] oxygen,[42] etc. As expected the number of water molecules in the first hydration layer depends on the size of the nonpolar solute: this number is about 20 for methane, 17 for oxygen, 19 for argon, 22 for krypton, 23 for xenon, 23 for ethane, 27 for n-propane, and 30 for n-butane.[36,38-42]

The local density of water around a nonpolar solute was found to be somewhat larger than the bulk density under ambient conditions,[42-43] but lower[42] for $T > 311$ K and approaching[42] the bulk density of water at sufficiently high temperatures. Another important characteristic of the aforementioned "perturbed" water is the number of hydrogen bonds (H-bonds) per water molecule. Molecular dynamics[40] and Monte Carlo[44] simulations indicated that the number of H-bonds per water molecule in the first hydration layer was slightly smaller than that in bulk water. It was found that the number of water molecules in the first hydration layer that possess four H-bonds was slightly lower and those with 1, 2, and 3 H-bonds slightly larger when compared to bulk water. However, as noted by Meng and Kollman[45] the water molecules in the first hydration layer have the same average number of H-bonds as the bulk water molecules. These results appear to favor the opinion[4] that "water does not undergo a major structural change in the presence of an apolar solute but maintains its original structure by accommodating the apolar solute in its original hydrogen bond network. The unique property of water is that it can dissolve an apolar solute of limited size without sacrificing a significant number of hydrogen bonds".[4] There is good agreement between the X-ray, neutron scattering and molecular simulations regarding the radial distribution functions g in dilute mixtures of nonpolar species and water. In the particular case of methane,[34-36,39-40,44-46] the position of the first maximum in the dependence $g_{oc} = g_{oc}(r_{oc})$, where r_{oc} is the distance between a C atom of methane and an O atom of water, was found to be at about 3.5−3.7 Å. The first minimum was found at 5.1−5.7 Å and the second maximum[39] at about 6.3 Å. Neutron diffraction scattering[34,41] indicated that the second maximum was very shallow. This means that one or at most two adjoining layers of water are affected by the presence of a nonpolar solute (methane). The water molecules in these adjoining layers have peculiar properties, the nearest to the nonpolar molecule being tangentially oriented toward its surface,[34] due to the "hydrophobic wall" (or hard core) effect of the solute. Their H-bonds are slightly shorter,[44] and the average number of their nearest neighbors slightly smaller than in bulk water.[44,46-47]

Another approach to investigate the hydrophobic effect is the ab initio quantum mechanical technique.[48,49] It is based on first principles (the Schrödinger equation), and this constitutes its main advantage compared to molecular dynamics and Monte Carlo approaches, which are based on classical potentials. At the present time, the ab initio quantum mechanical methods have limitations connected to the complexity and size of the molecular clusters considered.[48,50,51] Nevertheless, these methods have been often used to accurately predict the structure and energy of a system of two molecules (dimers),[50-52] such as the system methane/water.[49,53-57] However, the structure and energy of a

Clusters of Methane and Water

dimer are different from those in a condensed phase. Let us consider pure water as an example. The water dimer was investigated using various quantum mechanical ab initio methods, and reliable information about its structure and interaction energy is available.[50] They are different from those in condensed mixtures, where the effect of the nearest neighbors is an important factor. For pure water, it was clearly demonstrated[58] how the equilibrium intermolecular distance depends on the number of water molecules involved in the ab initio calculations. As already emphasized,[49] for the methane/water mixture, "the system of final interest is not $CH_4 \cdots H_2O$ ..., but $CH_4 \cdots (H_2O)_n$".[49] Sandler and co-workers have used quantum mechanical ab initio calculations for a group of several molecules to simulate the condensed mixtures and calculated the intermolecular interaction energies between a solute and the solvent molecules.[59-61] They employed the Hartree−Fock self-consistent field approximation[50,51] to calculate the intermolecular interaction energies for aqueous solutions of alcohols. The obtained energies were used to calculate the Wilson and UNIQUAC parameters and then to (successfully) predict the activity coefficients. Recently,[62] we used a quantum mechanical ab initio method [the Møller−Plesset perturbation theory[50,51] (MP2 method)] to compute the intermolecular energies for the $CF_4 + CCl_4$ dimer and used the results to (accurately) predict the solubility of solid CCl_4 in supercritical CF_4.

The Møller−Plesset perturbation theory will be employed in this paper to investigate the mixture methane/water. We selected this mixture because it is an ideal candidate for investigating the hydrophobic hydration. The mixture methane/water has also importance in understanding the structure and intermolecular interactions of the methane hydrates, though the specifics of these hydrates will not be addressed in this paper. These hydrates constitute[63] a major potential fuel reserve.

The main goal of the present paper is to obtain information about the intermolecular interactions and distances between several molecules of water (10, 8, 6, 4, and 1) and a single molecule of methane and vice versa using quantum mechanical ab initio methods. In addition, the interactions between the nearest neighbors water molecules in the vicinity of the hydrophobic solute will be calculated and compared to those of the bulk water phase.

The paper is organized as follows: in the next section, the quantum mechanical ab initio method employed will be presented. This will be followed by the results obtained for dilute mixtures of methane and water. Further, these results will be compared with the available information obtained experimentally and by simulations. Finally, they will be used to examine the hydrophobic hydration and shed light on the structure and other features of the water molecules in the vicinity of a hydrophobic solute.

2. Methodology of Calculations

It would be ideal to use for these calculations molecular clusters containing a single molecule of a solute and many (dozens or even hundreds of molecules) of a solvent. Unfortunately, at the present time, the ab initio methods based on the Møller−Plesset perturbation theory have computational limitations regarding the size of the cluster.[64] Therefore, we will have to compromise between a "dilute solution" and a relatively small number of solvent molecules. The largest investigated molecular clusters will contain a single molecule of methane (water) surrounded by 10 molecules of water (methane). To verify whether this cluster (1:10) is sufficiently large to capture the essential physics of the interactions, the same procedure will

TABLE 1: Bond Lengths in Methane and Water Molecules in the Optimized Clusters

	bond length [Å]	
component	exptl[67−68]	calcd
water (r_{OH})	0.9571	0.986
methane (r_{CH})	1.089	1.096

be carried out with smaller clusters (1:8, 1:6, 1:4, and 1:1) and with a larger one (1:11), and the trends will be analyzed. For the $CH_4 \cdots H_2O$ pair the dispersion interactions are vitally important[49] (this statement is valid for all mixtures involving weak interactions[50,65]). Therefore, the second-order Møller−Plesset perturbation theory, which partially accounts for dispersion interactions, constitutes a suitable though not ideal approximation. The cluster geometries will be obtained by optimizing each of them with respect to all coordinates, using the MP2 method with a compact 6-31G basis set. This basis set makes tractable the numerous geometry optimizations required in this work. The convergence to an energy minimum was confirmed by calculating the vibrational frequencies. There is another important feature concerning the quantum mechanical ab initio calculations for clusters containing several molecules, namely, the effect of the initial configuration. Indeed, the equilibrium structure of a weakly interacting cluster, for example, 1:10, can be affected by the initial guess of the configuration which can lead to a local minimum. To minimize the errors associated with the initial guesses, we carried out the minimization for every cluster composition several times (at least eight times) starting from different initial configurations.

After generating optimized clusters, the intermolecular interaction energies between pair molecules α and β ($E^{int}_{\alpha\beta}$) in the cluster were calculated using the supermolecular approach[59,60,66]

$$E^{int}_{\alpha\beta} = E_{\alpha\beta}\{\alpha\beta\} - E_{\alpha}\{\alpha\beta\} - E_{\beta}\{\alpha\beta\} \quad (1)$$

where $E_{\alpha\beta}\{\alpha\beta\}$ is the total energy of an αβ pair with the $\{\alpha\beta\}$ basis set, and $E_{\alpha}\{\alpha\beta\}$ and $E_{\beta}\{\alpha\beta\}$ are the energies of α and β molecules with the $\{\alpha\beta\}$ basis set, respectively, calculated by the ghost atoms method.[64] This method partially accounts for the basis set superposition error (BSSE). The energies ($E_{\alpha\beta}\{\alpha\beta\}$, $E_{\alpha}\{\alpha\beta\}$, and $E_{\beta}\{\alpha\beta\}$) were computed with a much better basis set than that employed for the cluster geometry optimization. Specifically all MP2 pair energies were calculated with the triple-ζ 6-311++G(3d,2p) basis set. This basis set includes polarization and diffuse functions. All of the ab initio computations were performed using the Gaussian 94 program on the IBM SP at the Center for Computational Research (CCR), at the University at Buffalo.

3. Results of the ab Initio Computations

3.1. The Dilute Mixture of Methane in Water. The calculated bond lengths in methane and water molecules of the optimized clusters (1:10) are listed in Table 1 together with data from literature. The arithmetic average distance and interaction energy between a methane molecule and the nearest touching water molecules in the cluster $CH_4 \cdots (H_2O)_n$ are listed in Table 2.

One of the typical minimized clusters 1 (methane):10 (waters) is presented in Figure 1a,b. They show that the methane molecule is enclosed in a cavity formed by water molecules. The two spheres centered on a methane molecule, with radii of 3.6 and 5.35 Å, correspond to the first maximum and the first minimum in the radial distribution function $g_{oc} = g_{oc}(r_{oc})$ in dilute mixtures of methane in water. It is worth noting that

TABLE 2: Arithmetic Average Distance and Interaction Energy^a between a Methane Molecule and Touching nearest Neighbors Water Molecules (Type A Water Molecules) in the Clusters $CH_4\cdots(H_2O)_n$

cluster $CH_4\cdots(H_2O)_n$	r_{CO} [Å]	$E^{int}_{CH_4-H_2O}$ [KJ/mol]	data from literature
$n = 1$ (dimer)	3.69	−1.06	(1) r_{CO} =3.5 Å (exptl value[34] of the position of the first peak in the radial distribution function $g_{oc} = g_{oc}(r_{oc})$ in dilute methane−water mixtures)
$n = 4$	3.73	−1.02	(2) r_{CO} = 4.0 Å (in solid methane hydrate[69])
$n = 6$	3.77	−0.93	(3) (a)r_{CO} =3.6 Å, (b) r_{CO} =3.73 Å (the position of the first peak in the radial distribution function $g_{oc} = g_{oc}(r_{oc})$ in dilute methane−water mixtures found by (a) Monte Carlo[44] and (b) by molecular dynamics[70] simulations)
$n = 8$	3.80	−0.79	
$n = 10$	3.74	−0.75	

^{a} The interaction energies were calculated between a central methane molecule and all the water molecules located not further than 4.1 Å from the central methane molecule (type A water molecules) as arithmetic averages. The values listed for the distances are also arithmetic averages.

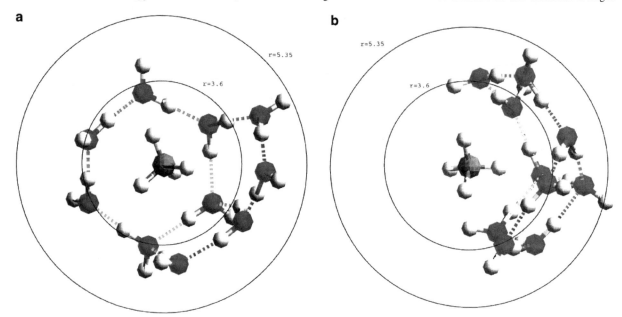

Figure 1. Optimized methane (1):water (10) cluster. (a) The front view. (b) The view from the right. The two circles in Figure 1 correspond to the first maximum (3.6 Å) and first minimum (5.35 Å) of the radial distribution function $g_{oc} = g_{oc}(r_{oc})$.

Jorgensen et al.[36] used the distance of 5.35 Å in their Monte Carlo simulations as a cutoff distance for the definition of the number of water molecules in the first hydration layer around a methane molecule in water. One can see from Figure 1a,b that the water molecules between the spheres with radii 3.6 and 5.35 Å include not only touching nearest neighbors but also water molecules from a second sublayer which are non touching nearest neighbors. Consequently, the space between the first maximum and first minimum in the radial distribution function $g_{oc} = g_{oc}(r_{oc})$ around a central molecule of methane is filled with water molecules of type A (or first sublayer), which are touching nearest neighbors, and water molecules of type B (or second sublayer), which are nontouching nearest neighbors of the methane molecule. The water molecules of types A and B are quite different because their distances from the central methane molecule, their orientations toward it, the number of H-bonds per water molecule, the energy of interaction with the central molecule, etc. (see also Discussion), are different.

The distances between the C atom of methane and O and H atoms of the water molecules of type A (C−O and C−H) are almost equal to each other (see Table 3) and are tangentially oriented toward the surface of the methane molecule, as was also found experimentally.[34] However, the B water molecules have a different orientation (see Table 3). Another important characteristic of the water molecules in the vicinity of a hydrophobic molecule, besides their interaction with the latter, is the interactions between themselves. The intermolecular distances between the water molecules in the vicinity of a methane molecule are listed in Table 4. Following the suggestion of ref 40 we define the water molecules which are located not

TABLE 3: Orientation of the Average Water Molecules in the First and Second Sublayers Surrounding the Methane Molecule

	the average distance between the carbon atom of methane and the oxygen and hydrogen atoms of the water molecules in the cluster $CH_4\cdots(H_2O)_{10}$ [Å]		
layer	r_{OC}	$r_{CH(1)}$	$r_{CH(2)}$
$r_{OC} \leq 4.1$ Å (type A)	3.704	3.769	3.771
4.1 Å $< r_{OC} \leq 5.6$ Å (type B)	4.570	4.339	5.138

TABLE 4: The Average Intermolecular Distances between nearest Water Molecules in the Vicinity of a Methane Molecule

pair of water molecules	r_{OO} [Å]	data from literature[71,72]
both are of type A	2.69	r_{OO} =2.84 Å in liquid water at 4 °C
one is of type A and the other of type B	2.73	and r_{OO} =2.759 Å in ice at 223 K

Clusters of Methane and Water

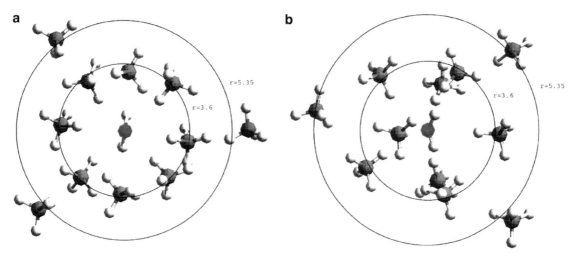

Figure 2. Optimized water (1):methane(10) cluster. (a) The front view. (b) The view from the right. The two circles in Figure 2 correspond to the first maximum (3.6 Å) and first minimum (5.35 Å) of the radial distribution function $g_{oc} = g_{oc}(r_{oc})$.

TABLE 5: The Average Distance and Interaction Energy between a Water Molecule and nearest Methane Molecules in the Clusters $H_2O\cdots(CH_4)_n$

cluster $H_2O\cdots(CH_4)_n$	r_{OC} [Å]	$E^{int}_{CH_4-H_2O}$ [KJ/mol]
$n = 8$	3.70	−1.79
$n = 10$	3.69	−1.83

further than 3.5 Å from a central water molecule as its nearest neighbors. For comparison, the same parameters for pure water are also listed in Table 4.

3.2. The Dilute Mixture of Water in Methane. When one molecule of water is surrounded by methane molecules, the molecule of water behaves like a regular nonpolar molecule (see Figure 2, where one of the typical minimized clusters 1 (water):10 (methane) is presented). The average intermolecular distance and interaction energy between a water molecule and the nearest neighbors methane molecules in the clusters $H_2O\cdots(CH_4)_{10}$ are listed in Table 5.

4. Discussion

4.1. Comparison between Two Clusters: $CH_4\cdots(H_2O)_n$ and $H_2O\cdots(CH_4)_n$. The results listed in Tables 2 and 5 show that the interaction energies between a water and a methane molecule in the two clusters $CH_4\cdots(H_2O)_n$ and $H_2O\cdots(CH_4)_n$ are very different not only one from another but also from that in the $H_2O\cdots CH_4$ dimer. While the intermolecular distances r_{oc} are not very different from one another and from the intermolecular distance r_{oc} in the methane/water dimer, the interaction energies between methane and water depend on the cluster type. It reflects the fact that the two kinds of clusters represent two different physical systems. Indeed, the two extreme cases at mole fractions $x_1 \rightarrow 0$ and $x_2 \rightarrow 0$ are very different. When one molecule of methane is located in water, the molecules of water are subjected to hydrogen bonding which will affect the interaction between H_2O and CH_4.[73] When one molecule of water is located in CH_4, the water molecule is no longer subjected to hydrogen bonding, but the interactions with the other methane molecules interfere with the interaction between H_2O and CH_4.

The single molecule of methane is located inside a cage formed by water molecules which are bound through hydrogen bonds (Figure 1). In contrast, in the second case (a single water molecule and 10 molecules of methane, Figure 2), the molecules of methane do not form a cage around a water molecule, but simply surround it. This generates a difference between the water/methane intermolecular interaction energies in the two cases and clearly indicates that the intermolecular interaction energy between the molecules of water and methane depends on the composition, which must be taken into account when calculating phase equilibria. However, usually, the models based on lattice theories ignore this dependence and use the pair intermolecular energy in the condensed mixtures as a composition independent quantity.

4.2. Molecules of Water in the Vicinity of a Methane Molecule. First, one should clearly emphasize the difference between (1) the number of water molecules in the first hydration layer around a methane molecule and (2) the coordination number of a methane molecule in an infinitely dilute aqueous solution. Jorgensen et al.[36] defined the number of water molecules in the first hydration layer around a methane molecule as the water molecules located between the spheres with radii 3.6 and 5.35 Å. Hence, Jorgensen's first hydration layer contains both A and B species. However, the coordination number in a liquid is usually defined[74] as the number of nearest touching neighbors and corresponds to A type molecules.

Let us consider two spheres with radii 3.6 and 5.35 Å in pure liquid water. Under ambient conditions and assuming the density equal to that of bulk water one can easily compute that there are 6.5 water molecules inside the first sphere and 24.5 water molecules inside the second, and hence, that there are 18 water molecules between the two spheres. If this number is compared to that of Jorgensen et al.[36] (20.3), one can conclude that the water layer around a central methane molecule is slightly denser than the bulk water. Our calculations regarding the intermolecular distance between neighboring water molecules in the vicinity of a central methane molecule (Table 4) is in agreement with this observation. It is not possible to calculate accurately the number of water molecules of types A and B because the number of water molecules considered in the calculations (10) is smaller than their number between two spheres of radii 3.6 and 5.35 Å in pure liquid water. However, a simple evaluation can be made by taking into account that a central molecule of methane replaces 6.5 molecules of water inside the first sphere. Because each of these molecules of water has 4.4 nearest neighbors in pure water, by subtracting the nearest neighbors that are present among them, one obtains that the methane molecule has 15−16 water molecules of type A. Our computa-

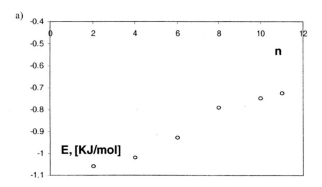

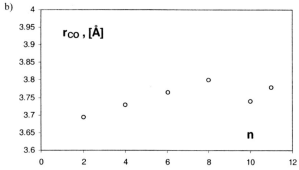

Figure 3. The dependence of the interaction energy (a) and average distance (b) between a methane molecule and water molecules of type A in the clusters $CH_4\cdots(H_2O)_n$ on the number of water molecules in the cluster n.

tions (Table 4) showed that the A water molecules have hydrogen bonds slightly shorter between them, than with the water molecules of type B. These results are in agreement with the Monte Carlo simulations.[44] Table 3 provides information about the orientation of the water molecules in the vicinity of a central methane molecule. One can see that the water molecules in the first sublayer (type A) are oriented almost tangentially toward to methane molecule; this peculiar orientation was also found experimentally.[34] However, the water molecules in the second sublayer (type B) are oriented more randomly. We agree with the previous authors[40,44] that the tangential orientation is due to the "hydrophobic wall" effect. Consequently, the water molecules of type A are quite different not only from the "bulk" water molecules, but also from the water molecules of type B. If one returns to the Frank and Evans[17] "iceberg", one can state that the iceberg is not uniform, but contains at least two types of water molecules. A simplified model of water like the Mercedes-Benz (MB) model[75] probably can be used to simulate the first sublayer of water (type A molecules) around the central methane molecule.

4.3. Cluster Size. The critical problem is how many water molecules should be used to accurately represent the hydration layers around nonpolar molecules. As already mentioned, it would be ideal to use in such calculations molecular clusters containing a single molecule of solute and hundreds of solvent molecules. Unfortunately, the ab initio methods based on MP2 are computationally expensive and this limits the size of clusters that can be attempted.[64] To fully understand whether the cluster 1:10 is sufficiently large to capture the essential physics of the interactions, we carried out additional calculations for clusters containing a single molecule of methane surrounded by 11 molecules of water (1:11). A comparison of the average distance and interaction energy between a methane molecule and the water molecules of type A in the cluster $CH_4\cdots(H_2O)_{11}$ with those for smaller clusters (see Table 2) is presented in Figure 3. This figure shows that the arithmetic average distance and interaction energy between a methane molecule and the nearest water molecules (type A) is almost the same in the clusters $CH_4\cdots(H_2O)_{10}$ and $CH_4\cdots(H_2O)_{11}$. One can, therefore, conclude that the cluster $CH_4\cdots(H_2O)_{10}$ is large enough for a correct estimation of the average distance and interaction energy between a methane molecule and the A water molecules in dilute solutions of methane in water. A more detailed analysis of the clusters $CH_4\cdots(H_2O)_{11}$ leads to the same conclusions regarding the water molecules in the vicinity of a central methane molecule for the clusters $CH_4\cdots(H_2O)_{10}$ (see section 4.2). However, the clusters ($CH_4\cdots(H_2O)_{10}$ and $CH_4\cdots(H_2O)_{11}$) are not large enough to accurately represent the characteristics of the B water molecules and of the water molecules in the second, third, and so on, hydration layers.

4.4. Influence of Temperature. As is well-known, the ab initio quantum mechanical methods provide results valid at 0 K and zero pressure. While the interactions depend on temperature and pressure,[76,77] this dependence is expected to be weak up to normal conditions. Indeed, it was shown[76,77] that for pure water the length of the hydrogen bond changed by at most four parts per thousand when the temperature varied by 100 K.

5. Conclusion

In this paper, the Møller–Plesset perturbation theory was applied to clusters formed by one molecule of methane and several molecules of water, or one molecule of water and several molecules of methane. The goal was to determine the intermolecular distances and interaction energies between a water molecule and a methane molecule in the clusters $CH_4\cdots(H_2O)_n$ and $H_2O\cdots(CH_4)_n$ and to compare the obtained results with available experimental data.

It was found that the intermolecular distances and interaction energies between a water molecule and a methane molecule are quite different in the clusters $CH_4\cdots(H_2O)_{10}$ and $H_2O\cdots(CH_4)_{10}$. The average intermolecular distance between a central methane molecule and the touching nearest neighbor water molecules is in agreement with the experimental value regarding the position of the first minimum in the radial distribution function $g_{oc} = g_{oc}(r_{oc})$. It was shown that the water molecules in the vicinity ($r_{oc} \leq 5.35$ Å) of a central methane molecule can be subdivided into two groups. A first group of water molecules (type A) in the immediate vicinity ($r_{oc} < 4.1$ Å) of the central methane molecule, which are touching nearest neighbors and a second group of water molecules (type B) in the second sublayer (4.1 Å $< r_{oc} \leq 5.6$ Å). The molecules of type A are tangentially oriented toward the central methane molecule and have shorter hydrogen bonds; the central methane molecule interacts with the water molecules (type A) through the so-called "hydrophobic wall" effect.

Acknowledgment. The authors are indebted to the Center for Computational Research (CCR) of the State University of New York at Buffalo for the use of its facilities.

References and Notes

(1) Tanford, C. *The Hydrophobic Effect: Formation of Micelles and Biological Membranes*, 2nd ed.; Wiley: New York, 1980.
(2) Ben-Naim, A. *Hydrophobic Interactions*; Plenum: New York, 1980.
(3) Dill, K. A. *Biochemistry* **1990**, *29*, 7133.
(4) Blokzijl, W.; Engberts, J. B. F. N. *Angew. Chem., Int. Ed.* **1993**, *32*, 1545.
(5) Hummer, G.; Garde, S.; Garcia, A. E.; Pratt, L. R. *Chem. Phys.* **2000**, *258*, 349.
(6) Schmid, R. *Monatsh. Chem.* **2001**, *132*, 1295.
(7) Ludwig, R. *Angew. Chem. Int. Ed.* **2001**, *40*, 1809.

(8) Kauzmann, W. Denaturation of Proteins and Enzymes. In *The Mechanism of Enzyme Action*; McElroy, W. D., Glass, B., Eds.; Johns Hopkins Press: Baltimore, 1954; p 71.
(9) Kauzmann, W. *Adv. Protein Chem.* **1959**, *14*, 1.
(10) Butler, J. A. V. *Trans. Faraday Soc.* **1937**, *33*, 229.
(11) Uhlig, H. H. *J. Phys. Chem.* **1937**, *41*, 1215.
(12) Eley, D. D. *Trans. Faraday Soc.* **1939**, *35*, 1281.
(13) Eley, D. D. *Trans. Faraday Soc.* **1944**, *40*, 184.
(14) Abraham, M. H. *J. Am. Chem. Soc.* **1982**, *104*, 2085.
(15) Abraham, M. H. *J. Chem. Soc., Faraday Trans. 1*, **1984**, *80*, 153.
(16) Abraham, M. H.; Matteoli, E. *J. Chem. Soc., Faraday Trans. 1*, **1988**, *84*, 1985.
(17) Frank, H. S.; Evans, M. W. *J. Chem. Phys.* **1945**, *13*, 507.
(18) Shinoda, K. *Principles of Solution and Solubility*; M. Decker: New York, 1977.
(19) Shinoda, K. *J. Phys. Chem.* **1977**, *81*, 1300.
(20) Ruckenstein, E. A Simple Interpretation of Hydrophobic Interactions and Critical Concentrations in Micellar Solutions. In *Progress in Microemulsions*; Martelucci, S., Chester, A. H., Eds.; Plenum Press: New York, 1989; p 31.
(21) Ruckenstein, E. *Colloids Surf.* **1992**, *65*, 95.
(22) Ruckenstein, E. *J. Dispersion Sci. Technol.* **1998**, *19*, 329.
(23) Lucas, M. J. *J. Phys. Chem.* **1976**, *80*, 359.
(24) Lee, B. *Biopolymers* **1985**, *24*, 813.
(25) Lee, B.; Graziano, G. J. *J. Am. Chem. Soc.* **1996**, *118*, 5163.
(26) Graziano, G. *J. Phys. Soc. Jpn.* **2000**, *69*, 3720.
(27) Graziano, G. *J. Phys. Chem. B* **2000**, *104*, 9249.
(28) Graziano, G.; Lee, B. *J. Phys. Chem. B* **2001**, *105*, 10367.
(29) Muller, N. *J. Solution Chem.* **1988**, *17*, 661.
(30) Hummer, G.; Garde, S.; Garcia, A. E.; Pohorille, A.; Pratt, L. R. *Proc. Natl. Acad. Sci. U.S.A.* **1996**, *93*, 8951.
(31) Hummer, G.; Garde, S.; Garcia, A. E.; Paulaitis, M. E.; Pratt, L. R. *J. Phys. Chem. B* **1998**, *102*, 10469.
(32) Garde, S.; Hummer, G.; Garcia, A. E.; Paulaitis, M. E.; Pratt, L. R. *Phys. Rev. Lett.* **1996**, *77*, 4966.
(33) Haselmeier, R.; Holz, M.; Marbach, W.; Weingartner, H. *J. Phys. Chem.* **1995**, *99*, 2243.
(34) De Jong, P. H. K.; Wilson, J. E.; Neilson, G. W.; Buckingham, A. D. *Mol. Phys.* **1997**, *91*, 99.
(35) Koh, C. A.; Wisbey, R. P.; Wu, X. P.; Westacott, R. E.; Soper, A. K. *J. Chem. Phys.* **2000**, *113*, 6390.
(36) Jorgensen, W. L.; Gao, J.; Ravimohan, C. *J. Phys. Chem.* **1985**, *89*, 3470.
(37) Glew, D. N. *J. Phys. Chem.* **1962**, *66*, 605.
(38) Lue, L.; Blankschtein, D. *J. Phys. Chem.* **1992**, *96*, 8582.
(39) Guillot, B.; Guissani, Y. *J. Chem. Phys.* **1993**, *99*, 8075.
(40) Chau, P. L.; Forester, T. R.; Smith, W. *Mol. Phys.* **1996**, *89*, 1033.
(41) Broadbent, R. D.; Neilson, G. W. *J. Chem. Phys.* **1994**, *100*, 7543.
(42) Fois, E.; Gamba, A.; Redaelli, C. *J. Chem. Phys.* **1999**, *110*, 1025.
(43) Matubayasi, N.; Levy, R. M. *J. Phys. Chem.* **1996**, *100*, 2681.
(44) Chau, P. L.; Mancera, R. L. *Mol. Phys.* **1999**, *96*, 109.
(45) Meng, E. C.; Kollman, P. A. *J. Phys. Chem.* **1996**, *100*, 11460.
(46) Mancera, R. L.; Buckingham, A. D.; Skipper, N. T. *J. Chem. Soc., Faraday Trans.* **1997**, *93*, 2263.
(47) Ikeguchi, M.; Shimizu, S.; Nakamura, S.; Shimizu, K. *J. Phys. Chem. B* **1998**, *102*, 5891.
(48) Clementi, E. *Computational Aspects for Large Chemical Systems*; Springer-Verlag: Berlin, 1980.
(49) Bolis, G.; Clementi, E.; Scheraga, H. A.; Tosi, C.; Wertz, D. H. *J. Am. Chem. Soc.* **1983**, *105*, 355−360.
(50) Levine, I. N. *Quantum Chemistry*, 4th ed.; Prentice Hall: Englewood Cliffs, NJ,
(51) Szabo, A.; Ostlund, N. S. *Modern Quantum Chemistry. Introduction to Advanced Electronic Structure Theory*; Dover Publication: New York, 1996.
(52) Frisch, M. J.; Delbene, J. E.; Binkley, J. S.; Schaefer, H. F. *J. Chem. Phys.* **1986**, *84*, 2279.
(53) Owicki, J. C.; H. A. Scheraga, H. A. *J. Am. Chem. Soc.* **1977**, *99*, 7413.
(54) Swaminathan, S.; Harrison, S. W.; Beveridge, D. L. *J. Am. Chem. Soc.* **1978**, *100*, 5705.
(55) Novoa, J. J.; Tarron, B.; Whangbo, M.-H.; Williams, J. M. *J. Chem. Phys.* **1991**, *95*, 5179.
(56) Szczesniak, M. M.; Chalasinski, G.; Cybulski, S. M.; Cieplak, P. *J. Chem. Phys.* **1991**, *95*, 5179.
(57) Cao, Z. T.; Tester, J. W.; Trout, B. L. *J. Chem. Phys.* **2001**, *115*, 2550.
(58) Liu, K.; Cruzan, J. D.; Saykally, R. J. *Science* **1996**, *271*, 929.
(59) Sum, A. K.; Sandler, S. I. *Ind. Eng. Chem. Res.* **1999**, *38*, 2849.
(60) Sum, A. K.; Sandler, S. I. *Fluid Phase Equilib.* **1999**, *160*, 375.
(61) Lin, S. T.; Sandler, S. I. *AIChE J.* **1999**, *45*, 2606.
(62) Ruckenstein, E.; Shulgin, I. *Ind. Eng. Chem. Res.* **2001**, *40*, 2544.
(63) Appenzeller, T. *Science* **1991**, *252*, 1790.
(64) Frisch, M. J.; Trucks, G. W.; Schlegel, H. B.; Gill, P. M. W.; Johnson, B. G.; Robb, M. A.; Cheeseman, J. R.; Keith, T.; Petersson, G. A.; Montgomery, J. A.; Raghavachari, K.; Al-Laham, M. A.; Zakrzewski, V. G.; Ortiz, J. V.; Foresman, J. B.; Cioslowski, J.; Stefanov, B. B.; Nanayakkara, A.; Challacombe, M.; Peng, C. Y.; Ayala, P. Y.; Chen, W.; Wong, M. W.; Andres, J. L.; Replogle, E. S.; Gomperts, R.; Martin, R. L.; Fox, D. J.; Binkley, J. S.; Defrees, D. J.; Baker, J.; Stewart, J. P.; Head-Gordon, M.; Gonzalez, C.; Pople, J. A. *Gaussian94*, revision C.2; Gaussian, Inc.: Pittsburgh, PA, 1995.
(65) Woon, D. E. *Chem. Phys. Lett.* **1993**, *204*, 29.
(66) Chalasinski, G.; Gutowski, M. *Chem. Rev.* **1988**, *88*, 943.
(67) Eisenberg, D.; Kauzmann, W. *The Structure and Properties of Water*; Oxford University Press: New York, 1969.
(68) Allen, F. H.; Kennard, O.; Watson, D. G.; Brammer, L.; Orpen, A. G.; Taylor, R. *J. Chem. Soc., Perkin Trans.* **1987**, *2*, 2, S1.
(69) McMullen, R. K.; Jeffrey, G. A. *J. Chem. Phys.* **1965**, *42*, 2725.
(70) Skipper, N. T. *Chem. Phys. Lett.* **1993**, *207*, 424.
(71) Page, R. H.; Frey, J. G.; Shen, Y. R.; Lee, Y. T. *Chem. Phys. Lett.* **1984**, *106*, 373.
(72) Huisken, F.; Kaloudis, M.; Kulcke, A.; Voelkel, D. *Infrared Phys. Technol.* **1995**, *36*, 171.
(73) Ruckenstein, E.; Shulgin, I. *Ind. Eng. Chem. Res.* **1999**, *38*, 4092.
(74) Prausnitz, J. M.; Lichtenhaler, R. N.; Gomes de Azevedo, E. *Molecular Thermodynamics of Fluid-Phase Equilibria*, 2nd ed.; Prentice Hall: Englewood Cliffs, NJ, 1986.
(75) Silverstein, K. A. T.; Dill, K. A.; Haymet, A. D. *J. Fluid Phase Equilib.* **1998**, *151*, 83.
(76) Dougherty, R. C. *J. Chem. Phys.* **1998**, *109*, 7372.
(77) Dougherty, R. C.; Howard, L. N. *J. Chem. Phys.* **1998**, *109*, 7379.

Treatment of Dilute Clusters of Methanol and Water by ab Initio Quantum Mechanical Calculations

Eli Ruckenstein,*[,†] Ivan L. Shulgin,[†,§] and Jeffrey L. Tilson[‡,ǁ]

Department of Chemical and Biological Engineering and Center for Computational Research, University at Buffalo, State University of New York, Amherst, New York 14260

Received: August 2, 2004; In Final Form: November 17, 2004

Large molecular clusters can be considered as intermediate states between gas and condensed phases, and information about them can help us understand condensed phases. In this paper, ab initio quantum mechanical methods have been used to examine clusters formed of methanol and water molecules. The main goal was to obtain information about the intermolecular interactions and the structure of methanol/water clusters at the molecular level. The large clusters ($CH_4O\cdots(H_2O)_{12}$ and $H_2O\cdots(CH_4O)_{10}$) containing one molecule of one component (methanol or water) and many (12, 10) molecules of the other component were considered. Møller–Plesset perturbation theory (MP2) was used in the calculations. Several representative cluster geometries were optimized, and nearest-neighbor interaction energies were calculated for the geometries obtained in the first step. The results of the calculations were compared to the available experimental information regarding the liquid methanol/water mixtures and to the molecular dynamics and Monte Carlo simulations, and good agreement was found. For the $CH_4O\cdots(H_2O)_{12}$ cluster, it was shown that the molecules of water can be subdivided into two classes: (i) H bonded to the central methanol molecule and (ii) not H bonded to the central methanol molecule. As expected, these two classes exhibited striking energy differences. Although they are located almost the same distance from the carbon atom of the central methanol molecule, they possess very different intermolecular interaction energies with the central molecule. The H bonding constitutes a dominant factor in the hydration of methanol in dilute aqueous solutions. For the $H_2O\cdots(CH_4O)_{10}$ cluster, it was shown that the central molecule of water has almost three H bonds with the methanol molecules; this result differs from those in the literature that concluded that the average number of H bonds between a central water molecule and methanol molecules in dilute solutions of water in methanol is about two, with the water molecules being incorporated into the chains of methanol. In contrast, the present predictions revealed that the central water molecule is not incorporated into a chain of methanol molecules, but it can be the center of several (2–3) chains of methanol molecules. The molecules of methanol, which are not H bonded to the central water molecule, have characteristics similar to those of the methane molecules around a central water molecule in the $H_2O\cdots(CH_4)_{10}$ cluster. The ab initio quantum mechanical methods employed in this paper have provided detailed information about the H bonds in the clusters investigated. In particular, they provided full information about two types of H bonds between water and methanol molecules (in which the water or the methanol molecule is the proton donor), including information about their energies and lengths. The average numbers of the two types of H bonds in the $CH_4O\cdots(H_2O)_{12}$ and $H_2O\cdots(CH_4O)_{10}$ clusters have been calculated. Such information could hardly be obtained with the simulation methods.

1. Introduction

Alcohol/water systems have attracted the attention of many scientists and technologists for a number of reasons: (i) The low cost of the lower members of the aliphatic alcohols and their miscibility with water make the alcohol/water mixtures useful as industrial solvents for a variety of chemical reactions and for small- and large-scale separation processes. In particular, the aqueous solutions of alcohols are often employed in the extraction and manipulation of labile materials such as proteins.[1] (ii) They have unusual thermodynamic properties that depend in a complex way on composition, pressure, and temperature.[1–8]

(iii) They constitute a model for the investigation of the hydrophobic effect. Although the interaction of a nonpolar solute molecule such as methane (or other hydrocarbons, noble gases, etc.) with the surrounding water molecules represents the simplest manifestation of the hydrophobic hydration, the interactions with molecules of a dual nature, such as alcohols, involve not only the hydrophobic hydration of the nonpolar moiety of the molecule but also the hydrophilic interactions between the polar groups and the water molecules. (iv) Because alcohol molecules have a dual nature, details regarding the aqueous solvation of alcohols can be used to improve our understanding of aqueous solutions of much more complex amphiphilic molecules, such as proteins, drugs, and biomolecules. (v) Pure water and alcohol generate different H-bond networks in the liquid state; it is therefore natural to ask how these networks reorganize in water/alcohol mixtures and how

---

* Corresponding author. E-mail: feaeliru@acsu.buffalo.edu. Fax: (716) 645-3822. Phone: (716) 645–2911/ext. 2214.
† Department of Chemical Engineering.
‡ Center for Computational Research.
§ E-mail: ishulgin@eng.buffalo.edu.
ǁ E-mail: jtilson@eng.buffalo.edu.

the hydrophobic hydration of the nonpolar part and the H bonds formed between water and methanol cooperate.

The macroscopic properties of alcohol/water systems were carefully investigated, and excellent reviews[1-2,4-5] and books[6-8] are available. In contrast, the structural and energetic features on the nanometer level have not been as well investigated.

During the last two decades, the wide use of modern experimental tools, especially X-ray, neutron scattering, and modern spectroscopic methods,[9-17] allowed one to obtain valuable information about the nanostructure of aqueous mixtures containing alcohols or various hydrophobic solutes. The availability of powerful computers combined with refined methods of molecular simulations, such as molecular dynamics and Monte Carlo, were actively used to investigate the nanolevel scale of aqueous solutions.[18-25]

The present paper is devoted to the application of ab initio quantum mechanical investigation to dilute clusters of methanol and water. The ab initio quantum mechanical methods are based on the Born–Oppenheimer approximation to the Schrödinger equation and do not involve the traditional model interaction potentials that are employed in molecular dynamics and Monte Carlo simulations. The ab initio quantum mechanical methods have been used frequently to determine the geometry and energy of small molecular clusters such as dimers, and the obtained results were usually used to fit various intermolecular pair potentials. More recently, the ab initio quantum mechanical methods have been applied to large molecular clusters[26-30] formed of the same molecules or of molecules of two different kinds. Large molecular clusters can be considered as intermediate states between gas and condensed phases and can be helpful in the understanding of some properties of the latter phase, particularly the local organization of the molecules and the interactions between them.

The aim of the present paper is to use ab initio quantum mechanical methods, such as Møller–Plesset perturbation theory[31,32] (MP2 method), to examine large clusters formed of one molecule of methanol (or water) and up to 10–12 molecules of water (or methanol). Methanol was selected because it is one of the simplest amphiphile-like molecules. Furthermore, the results will be compared to those obtained for dilute clusters of methane and water.[33]

The paper is organized as follows: In the next section, the literature results regarding some features of the nanostructure of pure water or methanol and the water/methanol binary clusters will be summarized. Then, the quantum mechanical ab initio method that was employed will be described. This will be followed by the presentation of the results that were obtained for the dilute clusters of methanol and water. Furthermore, the results will be compared to the available information regarding the liquid methanol/water mixtures that were obtained experimentally and by simulations. Finally, they will be used to shed some light on the structure and other features of water molecules in the vicinity of an amphiphilic solute.

2. Nanometer Features of Water and Methanol and their Mixture

There is no single theory that can provide explanations for all of the properties of the most mysterious substance: water. However, much information is available about the properties of water and about the organization of molecules in liquid water at the molecular level. Cold liquid water (liquid water at 0° C) is a very structured liquid with many features resembling the nanostructure of ice. Indeed[34-38] (i) the number of nearest neighbors is 4.4 (4 in ice); (ii) the water molecules in cold water have tetrahedral coordination as in ice, with only a small deviation; (iii) the length of a H bond ($r_{oo} = 2.82$ Å) is only a little longer than that in ice ($r_{oo} = 2.76$ Å); and (iv) the average number of H bonds per molecule is 3.6 (in ice it is 4). However, there are many subtle characteristics in which liquid water is very different from ice. For instance, the fraction of four H-bonded molecules in water is about 55%, whereas in ice almost all of the molecules have four H bonds.

Methanol molecules form in the liquid-state chains of hydrogen-bonded molecules.[2,39-40] The average number of H bonds per methanol molecule in the liquid state is about 1.8, whereas a methanol molecule can form three H bonds: two as acceptors and one as a donor. The average distance between two H-bonded methanol molecules is 2.8 Å. As for water, there is a similarity with the methanol in the solid state, where the molecules form infinite chains with two H bonds per molecule.[41,42]

The nanostructure and energetic features of the liquid water/methanol mixture were investigated both experimentally and by molecular simulations. Neutron diffraction data[13] of a water-rich region (mole fraction of water 0.9) revealed that a hydration shell of water molecules is located at a distance of about 3.7 Å from the carbon atom of a methanol molecule. Although the water molecules in this shell generated a disordered cage, they retained roughly the tetrahedral local coordination of pure water. The water molecules in the above hydration shell were not greatly affected by the presence of methanol molecules.[13] This observation is in disagreement with the famous hypothesis of Frank and Evans[43] that an ordered structure (iceberg) is formed around a nonpolar solute in water. This iceberg structure was frequently[1-2,4-5] used to explain the large loss of entropy during the process of dissolution. A similar observation for the water-rich region was made by using the depolarized Rayleigh light-scattering technique[15] and by coupling neutron diffraction with hydrogen/deuterium isotope substitution.[17] Although no reorganization of the water surrounding the nonpolar groups was detected, a compression of the second-neighbor water–water contact distance was observed, which might constitute a structural feature of the hydrophobic hydration. The structure of methanol/water clusters and its dependence on the methanol mole fraction was investigated by mass spectrometry using clusters isolated from submicrometer droplets by adiabatic expansion in vacuum and by X-ray diffraction of bulk binary solutions.[44] It was found that in the water-rich range the water molecules had a tetrahedral orientation, the length of a H bond at a mole fraction of water of 0.9 being 2.82 Å, and the average distance between the carbon of a methanol molecule and the oxygen of the nearest touching water molecule (not H bonded to methanol) being 3.40 Å. However, in the methanol-rich region, chain clusters of methanol molecules became predominant, the length of a H bond at a mole fraction of methanol of 0.9 being 2.80 Å, and the average distance between the oxygen of a water molecule and the carbon of the nearest touching methanol molecule (not H bonded to methanol) being 3.38 Å.

The above experimental results provided many features regarding the local microscopic structure of methanol/water mixtures. However, for the time being, the experiment could not provide some subtleties regarding the local structure. For example, the hydration picture in the water-rich region was expected to be different around the hydrophobic moiety of methanol (methyl group) and around its hydrophilic hydroxyl group, but the experiment could not provide the details of the difference.

Meaningful results regarding the structural and energetic characteristics of methanol/water mixtures were obtained by

TABLE 1: Some Experimental and Computational Results Regarding the Local Structure of Water/Methanol Mixtures ($T = 298.15$ K)

	water-rich range	methanol-rich range	reference	comments
number of nearest neighbors that satisfy the condition ($r_{C_M O_W} \leq 3.5$ Å)	3.1^a		47	experimental data
	3.4^a, 3.24^a	2.0^b, 1.9^b	44	
	3.4^c		18	molecular simulation
	$1.62^{d,e}$		22	
	2.9^a	2.6^b	46	
	2.6^f	3^g	48	
	2.5^a	2.51^h	49	
	3.4^i		23	
number of nearest neighbors that satisfy the condition (3.5 Å $\leq r_{C_M-O_W} \leq 5.5$ Å)^j	10.7^a		47	experimental data
	10^a		13	
	17.6^k		17	
	20^c		18	molecular simulation
	11.3^d		22	
	16^f	8^g	48	
	12.4^i		23	
	$\sim 13^l$		45	
number of water/methanol H bonds with the central molecule that satisfy the condition ($r_{C_M-O_W} \leq 3.5$ Å)	2.3^c		18	molecular simulation
	2.4^l		45	
average length of H bonds, $r_{O_M-O_W}$ (Å)	2.83^a		47	experimental data
	2.84^a, 2.82^a	2.76^b, 2.80^b	44	
	2.8^c		18	molecular simulation
	2.85^a	2.85^b	46	
average distance from a central molecule to the nearest neighbors, $r_{C_M-O_W}$ (Å)	3.7^a		47	experimental data
	3.7^a		13	
	3.4^a	3.38^b	44	
	3.7^i		23	molecular simulation
	3.7^l		45	

^{a} The mole fraction of methanol is 0.1. ^{b} The mole fraction of methanol is 0.9. ^{c} The mole fraction of methanol is 0.008. ^{d} The mole fraction of methanol is 0.002. ^{e} $r_{C_M-O_W} \leq 3.3$ Å. ^{f} The mole fraction of methanol is 0.125. ^{g} The mole fraction of methanol is 0.875. ^{h} The mole fraction of methanol is 0.75. ^{i} The mole fraction of methanol is 0.003. ^{j} The nearest neighbors listed in the previous part of the Table ($r_{C_M-O_W} \leq 3.5$ Å) are excluded. ^{k} The mole fraction of methanol is 0.05. ^{l} The mole fractions of methanol are 0.003 and 0.015.

molecular simulation. Two important papers[18,19] regarding the Monte Carlo simulations of dilute solutions of methanol in water were published about 20 years ago, and they provided some conflicting results. Okazaki et al.[19] concluded that by introducing one methanol molecule into water the potential energy and the structure of water had the tendency to be stabilized as a whole. This stabilization was attributed to the structural stabilization around the methyl group and to the strong H bonding in the hydrophilic region that acts cooperatively with the structural stabilization in the hydrophobic region. In contrast, according to Jorgensen and Madura,[18] the main feature of the hydration in the water-rich region is the favorable solute−solvent hydrogen bonding. They found that the first shell around the carbon of a methanol molecule (from 0 to 3.5 Å) contained 3.4 water molecules, which formed 2.3 hydrogen bonds with the methanol molecule, and an average of 2.9 water−water hydrogen bonds per water molecule. Consequently, they formed a total of 3.6 hydrogen bonds per water molecule, which was exactly the same as that for a water molecule in pure water. However, the water molecules from the second shell (from 3.5 to 4.5 Å) had a slightly lower average number of hydrogen bonds (3.39) than the bulk water (3.57). Although Okazaki et al.[19] found an iceberg-like structure of water molecules around a methanol molecule, Jorgensen and Madura[18] did not observe a large distortion of the water molecules around the methyl group. Results supporting the findings of Jorgensen and Madura[18] were recently obtained by Fidler and Rodger[24] via molecular dynamics simulations. They found[24] that the structure of water around the hydrophobic moiety of alcohol was essentially the same as that found in bulk water; in particular, there was no evidence of the presence of a clathrate-like cage around the hydrophobic moiety of the alcohol. Some change in the water structure was found in the vicinity of the hydroxyl group of the alcohol, with a hydrogen-bonding network closer to tetrahedral in the solvation shell than in bulk water. The Monte Carlo investigation of Hernandez-Cobos and Ortega-Blake[23] and the molecular dynamics results of Meng and Kollman[22] for dilute solutions of methanol in water also supported the results of Jorgensen and Madura.[18] The recently published density functional theory (DFT) based on molecular dynamics simulation[45] found that the "speculations that the normal water structure is significantly affected by the hydrophobic alkyl group are groundless". However, much less information is available regarding the structural and energetic characteristics in the methanol-rich region. The molecular dynamics simulation of Palinkas, Hawlicka, and Heinzinger[46] for dilute solutions of water in methanol showed that when very little water was added to pure methanol (methanol-rich region) the water molecules associated with methanol were incorporated into the chains of the latter.

Some experimental and simulation results regarding the structural characteristics of methanol/water mixtures are listed in Table 1.

In contrast to the experimental methods, the molecular simulation techniques, such as molecular dynamics and Monte Carlo methods, allowed one to obtain some details about the molecular arrangements on the nanometer scale. However, the simulation techniques are very sensitive to the model potentials that are employed.

So far, ab initio quantum mechanical techniques were applied to the methanol/water dimer and the methanol/water/water trimer. It is well known that the methanol/water dimer can adopt two possible configurations depending on whether water (WdM) or methanol (MdW) acts as the hydrogen-bond donor. Because there is no large energetic difference between the two dimers, it was not easy to select the more stable dimer. Nevertheless, it was recently established that the dimer in which the water molecule is the proton donor (WdM) is more stable.[50,51] However, one should point out that a dimer or a trimer cannot represent a real patch of a dilute condensed phase because they cannot represent, for instance, the cooperative effect of many molecules. To achieve this goal, one must consider a much larger cluster.

3. Methodology of Calculations

Ab initio quantum mechanical methods were recently applied to the analysis of large clusters formed of one solute molecule and several molecules of solvent for water/methane mixtures.[33] It was shown[33] that they can provide information regarding the interaction energies and intermolecular distances between the molecules of methane and water. The obtained results were compared to the available experimental and molecular simulations regarding condensed mixtures, and agreement was found. A similar methodology of calculations will be used in the present paper as well.

Two types of clusters will be considered: (1) clusters with 1 methanol and 12 water molecules and (2) clusters with 1 water and 10 methanol molecules. Such clusters represent a computational compromise between the current capabilities of the modern ab initio methods and computer power on one hand and a feasible representation of dilute binary condensed mixtures on the other hand.

Second-order Møller−Plesset perturbation theory (MP2) is the quantum mechanical approach selected for the calculations because the sizes of the clusters that were employed were too large to use more accurate methods. In addition, MP2 provides accurate results regarding the calculation of the interaction energies for both H-bonded pairs[52] and van der Waals interacting pairs.[53]

The computational procedure presented below has the following objectives: (1) to find an optimal geometry for the clusters considered and (2) to determine the distances and the interaction energies between a central "solute" molecule and its nearest neighbors ("solvent" molecules).

The computational procedure consisted of three steps: (i) An initial cluster configuration was constructed using the Cerius^2 4.2 software. The solute molecule was placed in the center and was randomly surrounded by the molecules of the solvent. The configuration that was built was processed using the Cerius^2 CLEAN function, and the obtained structure was considered as the initial guess. (ii) The cluster geometry was obtained by optimizing the guess with respect to all coordinates using the MP2 method with a 6-31G basis set. This basis set makes the numerous geometry optimizations required tractable. For large clusters, it is difficult to reach the global minimum because the minimum reached can be a local one. To avoid the effect of the initially selected guess, we carried out the minimization procedure for a large number of initial guesses (12−16). In addition, vibrational frequencies were used to ensure that the optimized geometries were located at real minima. (iii) All of the pairwise intermolecular interaction energies were calculated for all of the optimized geometries. All of the interaction energies between molecules α and β ($E_{\alpha\beta}^{int}$) were calculated using the supermolecular approach[26,27,54]

$$E_{\alpha\beta}^{int} = E_{\alpha\beta}\{\alpha\beta\} - E_{\alpha}\{\alpha\beta\} - E_{\beta}\{\alpha\beta\} \qquad (1)$$

where $E_{\alpha\beta}\{\alpha\beta\}$ is the total energy of an $\alpha\beta$ pair obtained with an $\{\alpha\beta\}$ basis set and $E_{\alpha}\{\alpha\beta\}$ and $E_{\beta}\{\alpha\beta\}$ are the energies of the individual α and β molecules, respectively, also obtained with the $\{\alpha\beta\}$ basis set. The basis set superposition error (BSSE)[55] was partially accounted for by using the function counterpoise method (FCP).[56]

In contrast to the geometry optimization, which was carried out with the smaller 6-31G basis set, a larger 6-311++G (3d, 2p) basis set was employed to calculate the energies because, at least for small clusters,[57−59] the geometry is less sensitive, whereas the energies are very sensitive to the size of the basis set used.

We were tempted to use the same basis set for the cluster geometry optimization as that used for the calculations of the interaction energies between molecules (6-311++G (3d, 2p)). However, the present computer capabilities have not allowed us to perform such calculations in a reasonable amount of time.

4. Results of the ab Initio Computations

4.1. Dilute Mixture of Methanol in Water. Sixteen initial guesses, each containing 1 molecule of methanol and 12 molecules of water, were optimized in the present paper. The optimized clusters were treated as follows: (1) The geometries of the clusters were used to calculate the distances between the carbon and oxygen atoms of the central methanol molecule and the oxygen atoms of the water molecules. (2) The interaction energies between the central methanol molecule and the surrounding water molecules were calculated using eq 1.

Experimental data and simulation results for dilute solutions of methanol in water[13,23] indicated that the radial distribution function $g_{C_MO_W}$ has the first maximum at a distance of about 3.7 Å and the first minimum at about 5.2−5.3 Å from the central methanol molecule. The water molecules located in the layer between 3.7 and 5.3 Å constitute the first solvation shell. According to recent data,[17] there are about 18 water molecules around a central methanol molecule in the first solvation shell. These molecules can be roughly subdivided into two groups:[33] (1) touching nearest neighbors and (2) nontouching nearest neighbors. The molecules of the first group are in contact with the central methanol molecule. In our paper regarding dilute clusters of methane in water,[33] we used (somewhat arbitrarily) a distance of 4.1 Å from the central methane molecule to separate these two groups of molecules from each other. These two groups of molecules have very different interaction energies with the central methane molecule. Besides, the molecules of the first group (touching nearest neighbors) are tangentially oriented toward the central methane molecule, and this sublayer is somewhat denser than the bulk water. We will use the same separation of water molecules around the central methanol molecule in the first solvation shell. A similar subdivision was used by Rossky and Karplus in a paper regarding dipeptide hydration.[60]

For each of the water molecules belonging to the first group, the distances between the O atom of water and the carbon atom of the methanol molecule and the intermolecular interaction energy were calculated. The results of these calculations are listed in Tables 2−4. These Tables contain the average distances and interaction energies as double arithmetic averages. First,

TABLE 2: Arithmetic Averages of the Distances and Interaction Energies[a] between a Central Solute Molecule and Touching Nearest-Neighbor Solvent Molecules[b] in the $CH_4O\cdots(H_2O)_{12}$ and $H_2O\cdots(CH_4O)_{10}$ Clusters

cluster	type of solvent molecules in the cluster	$r_{C_MO_W}$, Å	$E^{int}_{CH_4O-H_2O}$, kJ/mol
$CH_4O\cdots(H_2O)_{12}$	not H bonded with the central methanol molecule	3.45	−2.7
	H bonded with the central methanol molecule	3.61	−15.79
$H_2O\cdots(CH_4O)_{10}$	not H bonded with the central water molecule	3.65	−1.8
	H bonded with the central water molecule	3.62	−17.29

[a] The interaction energies were calculated between a central solute molecule and all of the solvent molecules located not further than 4.1 Å from the central solute molecule. [b] The solvent molecules located not further than 4.1 Å from the central solute molecule were considered to be touching nearest neighbors of a central solute molecule.

TABLE 3: Average Orientation of Water Molecules[a] Not H Bonded with a Central Methanol Molecule toward the Central Methanol Molecule

	average distance between the carbon atom of methanol and the oxygen and hydrogen atoms of the water molecules in the $CH_4O\cdots(H_2O)_{12}$ cluster, Å		
layer	$r_{C_MO_W}$	$r_{C_MH_W(1)}$	$r_{C_MH_W(2)}$
$r_{C_MO_W} \leq 4.1$ Å	3.45	3.85	3.77

[a] The water molecules are not located further than 4.1 Å from the central methanol molecule.

calculations were made for all of the water molecules of a cluster belonging to one of the groups, and second, for all of the 16 clusters investigated. One of the typical minimized clusters $(CH_4O\cdots(H_2O)_{12})$ is presented in Figure 1a and b. The molecules of water in the vicinity of a central methanol molecule can be subdivided into two classes: (1) not H bonded with the central methanol molecule and (2) H bonded with it. (A hydrogen bond is defined here as suggested by Jorgensen and Madura.[18] Namely, any pair of molecules with an interaction energy of −9.5 (kJ/mole) or less is considered to be hydrogen bonded.) Although the average distances between the central methanol molecule and these two types of water molecules are almost the same, the interaction energies are enormously different (see Tables 2 and 4).

The arithmetic averages of the distances and of the interaction energies between a central methanol molecule and the water molecules belonging to each of the above two classes in the $CH_4O\cdots(H_2O)_{12}$ cluster are listed in Table 2. The orientation of the non-H-bonded water molecules with respect to the central methanol molecule is presented in Table 3.

Details regarding the second class of water molecules (H bonded with the central methanol molecule) in the $CH_4O\cdots(H_2O)_{12}$ cluster are listed in Table 4. A comparison with the literature data regarding the liquid methanol/water mixtures (Table 1) reveals that (1) the average number of water molecules having H bonds with a central methanol molecule is about 2.8, whereas molecular simulation[18,45] predicted 2.3−2.4 and (2) the obtained lengths of the H bonds are somewhat shorter than those obtained by molecular simulation[18,45] (see Discussion).

4.2. Dilute Mixture of Water in Methanol. Twelve initial guesses, each containing 1 molecule of water and 10 molecules of methanol, were optimized and then analyzed in the same way as the $CH_4O\cdots(H_2O)_{12}$ clusters in the preceding section. Namely, (1) the geometries of each cluster were determined, and the distances between the oxygen atom of water and the carbon and oxygen atoms of the methanols were calculated and (2) the interaction energies between a central water molecule and the surrounding methanol molecules were calculated using eq 1. The results of these calculations are presented in Tables 2 and 4. One of the typical minimized clusters $(H_2O\cdots(CH_4O)_{10})$ is presented in Figure 2a and b. As in the previous case $(CH_4O\cdots(H_2O)_{12}$ cluster), the molecules of methanol in the vicinity of a central water molecule can be subdivided into two different classes: (1) not H bonded with water and (2) H bonded with it.

The arithmetic averages of the distances and of the interaction energies between a central water molecule and the methanol molecules belonging to each of the above classes in the $H_2O\cdots(CH_4O)_{10}$ cluster are listed in Table 2. It is interesting to note that the methanol molecules not H bonded to a central water molecule have characteristics very similar to those of methane molecules around a central water molecule in the $H_2O\cdots(CH_4)_{10}$ cluster.[33] Indeed, in the latter cluster, the average distance and interaction energy between a central water molecule and the nearest touching methane molecules ($r_{C_{CH_4}O_{H_2O}} = 3.69$ Å and $E^{int}_{CH_4-H_2O} = -1.83$ kJ/mol) are very close to the corresponding values in Table 2 ($r_{C_MO_W} = 3.65$Å and $E^{int}_{H_2O-CH_4O} = -1.79$ kJ/mol). However, the average distance between a central water molecule and the nearest touching methanol molecules from Table 2 is somewhat different from the distance (3.38 Å, see Table 1) obtained experimentally[44] for a dilute solution of water in methanol.

Let us consider in more detail the second class of methanol molecules in the $H_2O\cdots(CH_4O)_{10}$ cluster, which are H bonded to a central water molecule (see Table 4). Comparing Tables 1 and 4, one can conclude that the lengths of H bonds in the $H_2O\cdots(CH_4O)_{10}$ clusters are in agreement with the experimental lengths;[44] the simulations[46] provided somewhat longer H bonds. Our results regarding the average number of H bonds between a central water molecule and methanol molecules in the $H_2O\cdots(CH_4O)_{10}$ cluster differ from those predicted in the literature.[46] According to the literature,[44,46] the average number of H bonds between a central water molecule and methanol molecules in a dilute solution of water in methanol is about 2, and the molecular dynamics simulation of Palinkas, Hawlicka, and Heinzinger[46] showed that the water molecules are associated with methanol, being incorporated into the chains of the latter. Our results (Table 4) provide for the average number of H bonds between a central water molecule and methanol molecules a value of about 3. This means that a central water molecule

TABLE 4: Methanol/Water Hydrogen Bonds in the $CH_4O\cdots(H_2O)_{12}$ and $H_2O\cdots(CH_4O)_{10}$ Clusters

cluster	average number of H bonds between a central solute molecule and the solvent molecules		average length of the H bonds ($r_{O_M-O_W}$), Å		average energy of the H bond, kJ/mol	
	WdM[a]	MdW[b]	WdM[a]	MdW[b]	WdM[a]	MdW[b]
$CH_4O\cdots(H_2O)_{12}$	1.75	0.94	2.74	2.67	−18.34	−11.04
$H_2O\cdots(CH_4O)_{10}$	1.75	1.08	2.77	2.72	−18.50	−14.75

[a] The water molecule is the proton donor. [b] The methanol molecule is the proton donor.

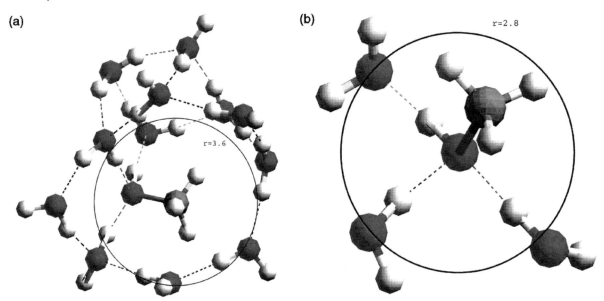

Figure 1. Optimized methanol (1)/water (12) cluster: (a) full cluster and (b) fragment containing only the central methanol molecule and the water molecules H bonded to the central one. (- - -) denotes H bonds. The circles (in a, $r = 3.6$ Å centered on a carbon atom of the methanol and in b, $r = 2.8$ Å centered on an oxygen atom of the methanol) are included for illustration.

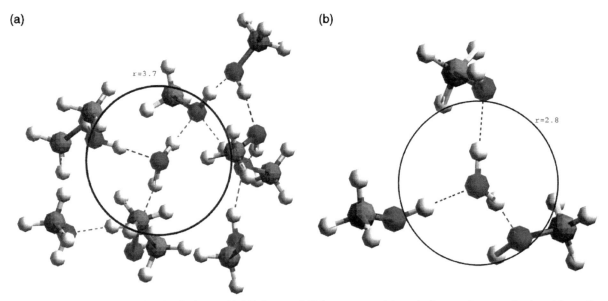

Figure 2. Optimized water (1)/methanol (10) cluster: (a) full cluster and (b) fragment containing only the central water molecule and the methanol molecules H bonded to the central one. (- - -) denotes H bonds. The circles (in a, $r = 3.7$ Å centered on an oxygen atom of the water, and in b, $r = 2.8$ Å centered on an oxygen atom of the water) are included for illustration.

cannot be incorporated into the middle of a chain of methanol molecules but can be, for instance, the center of several (2−3) chains of methanol molecules.

5. Discussion

Sixteen different $CH_4O\cdots(H_2O)_{12}$ clusters and twelve different $H_2O\cdots(CH_4O)_{10}$ clusters were optimized, and the results listed in Tables 2−4 represent double arithmetic averages per cluster and all of the clusters of the calculated properties. Table 5 provides the mean percentage deviation of the average H-bond lengths and energies (Table 2) from the values provided by the optimized clusters. Table 5 shows that the number of initial cluster configurations considered (16 and 12) is large enough to represent accurately the distances and the interaction energies between a central solute molecule and its nearest neighbors (solvent molecules).

The results listed in Tables 2−4 show that the interaction energies and the intermolecular distances between the molecules of water and methanol are quite different in the $CH_4O\cdots(H_2O)_{12}$ and $H_2O\cdots(CH_4O)_{10}$ clusters. This difference reflects the fact that the two kinds of clusters represent two different physical systems. Indeed, the two extreme cases at mole fractions of $x_1 \rightarrow 0$ and $x_2 \rightarrow 0$ are very different because the solute molecules have different solvent environments in the two clusters. This generates a difference between the water/methanol intermolecular interaction energies in the two cases and clearly indicates that the intermolecular interaction energy between the molecules of water and methanol depends on composition. This

TABLE 5: Mean Percentage Deviation of the Average H-Bond Length and Energy (Table 2) from the Values Obtained for All of the Optimized Clusters

cluster composition	number of optimized clusters	deviation (%)[a] length of the H bonds ($r_{O_M-O_W}$)	energy of the H bonds
$CH_4O\cdots(H_2O)_{12}$	16	3.0	9.8
$H_2O\cdots(CH_4O)_{10}$	12	3.3	11.8

[a] Deviation (%) is the mean percentage deviation defined as

$$\frac{100\sum_{i=1}^{N}|\frac{x_i-x}{x}|}{N}$$

where x_i is either the length of the H bonds or the energy of the H bonds in the optimized cluster i, x is the arithmetic average value of these quantities (see Table 2), and N is the number of optimized clusters (here, 16 and 12).

TABLE 6: Average Intermolecular Distances between H-bonded Water Molecules in the Vicinity[a] of a Solute Molecule

cluster	r_{OO}, (Å) b	r_{OO}, (Å) c	data from the literature[35,37] regarding the length of H bonds in liquid water
$CH_4O\cdots(H_2O)_{12}$	2.74	2.74	$r_{OO} = 2.82$ (Å) in liquid water at 4 °C and $r_{OO} = 2.84$ (Å) in liquid water at 20 °C
$CH_4\cdots(H_2O)_{10}$[d]	2.73		

[a] The pairs of water molecules were selected such that at least one water molecule was located not further than 4.5 Å from the central solute molecule. [b] At least one water molecule is H bonded to the central solute molecule. [c] Neither water molecule is H bonded to the central solute molecule. [d] Data for the $CH_4\cdots(H_2O)_{10}$ cluster were calculated on the basis of previous results.[33]

fact should be taken into account when the intermolecular interaction energies are employed to calculate the thermodynamic properties at different compositions of a binary mixture.[61]

The molecules of solvent in both clusters ($CH_4O\cdots(H_2O)_{12}$ and $H_2O\cdots(CH_4O)_{10}$) can be subdivided into two very different classes: (1) H bonded with a central solute molecule and (2) not H bonded with a central solute molecule. As expected, these two classes possess striking energy differences. Indeed, although they are located at about the same distance from the central solute molecule ($r_{C_MO_W}$ in Table 2), their intermolecular interaction energies with the central molecule are extremely different. One can see from Table 2 that the H-bond energy between a central water molecule and a neighboring methanol molecule is almost 10 times larger (in magnitude) than when the water/methanol pair is not H bonded.

Let us examine the properties of the water molecules around a solute molecule by discussing how these molecules differ from those in pure water, how they are oriented toward the surface of the solute, and the properties of the H bonds between the water molecules (their lengths and energies). Such characteristics of the water molecules in the vicinity of a solute molecule are important not only for small molecules, such as alcohols and hydrocarbons, but also for "large" molecules such as biomolecules.

First, let us consider the lengths of the H bonds between the water molecules in the vicinity of a central methanol molecule in the $CH_4O\cdots(H_2O)_{12}$ clusters (Table 6). One can see from Table 6 that the lengths of the H bonds between the water molecules in the vicinity of a central methanol molecule are almost the same as those in the $CH_4\cdots(H_2O)_{10}$[33] cluster. However, the H bonds in the above clusters are shorter than those in pure water (see Table 6). This observation can be explained if one supposes that the layer of water molecules around a solute molecule (methane or methanol) is a little denser than that in pure water. This increase in density is probably caused by the so-called hydrophobic wall effect, which occurs when the water molecules have a hydrophobic surface on one side and cannot form four H bonds. Because the average number of H bonds per molecule of water in pure water is 3.6, the water molecules in the vicinity of a solute molecule should have a smaller number of H bonds, and therefore their characteristics should be different from those in pure water.

Second, the orientation of the water molecules (not H bonded with a central solute molecule) toward the hydrophobic surface of the solute plays a role in the understanding of the hydration of molecules of dual nature, such as methanol. Although for pure hydrophobic solutes, such as hydrocarbons and noble gases, the water molecules in the vicinity of a solute are tangentially oriented toward the surface of the solute (a fact also observed experimentally[62]), such an orientation is not obvious for the molecules of dual nature. Indeed, we found that the average orientation of the water molecules in the vicinity of a central methanol molecule is not fully tangential (Table 3). However, it is clear that not all of the water molecules in the vicinity of a central methanol molecule are under "the same conditions". The water molecules located on the C_M-O_M line beyond the CH_3 group can be tangentially oriented toward the surface of the CH_3 group; however, it is hardly possible for the water molecules located in the vicinity of the hydrophobic/hydrophilic interface of a methanol molecule to be tangentially oriented toward the methanol molecule.

As already noted, the methanol/water dimer can adopt two possible configurations (WdM or MdW), depending on whether the water or methanol molecule acts as the hydrogen-bond donor. It was recently established that the dimer in which the water molecule is the proton donor (WdM) is more stable.[50,51] However, in a condensed phase, which is represented here approximately by the $CH_4O\cdots(H_2O)_{12}$ and $H_2O\cdots(CH_4O)_{10}$ clusters, both types of H-bond configurations are present (see Table 4). Let us compare some of the characteristics of dimer WdM or MdW in the gas phase with the different H-bond configurations in the clusters investigated (Table 7). One can see from Table 7 that the lengths of the H bonds (O−O distances) in both kinds of H-bond configurations (WdM and MdW) in the $CH_4O\cdots(H_2O)_{12}$ and $H_2O\cdots(CH_4O)_{10}$ clusters are shorter and their interaction energies are smaller (in magnitude) than those of the gas-phase dimers. These results can be explained by the mutual steric hindrances between the solvent molecules that cannot be oriented in their optimal positions as they are for the gas-phase dimers. Table 7 also reveals that the average number of WdM configurations is twice as large as the average number of MdW configurations. For the $CH_4O\cdots(H_2O)_{12}$ clusters, this result can be explained by observing that the methanol molecule can donate only a single H bond but can accept two. However, for the $H_2O\cdots(CH_4O)_{10}$ clusters, the number of WdM and MdW configurations is determined by energetic and steric factors.

Although the average number of H bonds per molecule in cold water is 3.6 (in ice it is 4), in the $H_2O\cdots(CH_4O)_{10}$ clusters, the average number of H bonds per molecule of water is 2.7. Therefore, a molecule of water has lost about 1 H bond when compared to cold water. However, the average number of H bonds per molecule in liquid methanol is about 2, and we found that the average number of H bonds per molecule of methanol

TABLE 7: Comparison between the Two Types of Water/Methanol Hydrogen Bonds in the $CH_4O\cdots(H_2O)_{12}$ and $H_2O\cdots(CH_4O)_{10}$ Clusters

H-bond configuration	O—O distance, Å			H-bond energy, kJ/mole			average number of H bonds per solute molecule	
	a	b	c	a	b	c	a	b
WdM	2.74	2.77	2.85	−18.34	−18.50	−22.8	1.75	1.75
MdW	2.67	2.72	2.90	−11.04	−14.75	−19.2	0.94	1.08

[a] $CH_4O\cdots(H_2O)_{12}$. [b] $H_2O\cdots(CH_4O)_{10}$. [c] Water/methanol dimer.[50,51]

in the $CH_4O\cdots(H_2O)_{12}$ clusters is about 2.8; a molecule of methanol acquires about 0.8 additional H bond compared to pure liquid methanol, probably because the water molecules are smaller.

Two additional issues should be examined at least in passing: (1) the effect of cluster size and (2) the effect of temperature. (1) In our previous publication,[33] we considered clusters of 1 methane and several (1, 4, 6, 8, 10 and 12) water molecules, and the conclusion was that the clusters with 10–12 water molecules provided accurate results for the "average" distances and interaction energies between a methane molecule and the nearest water molecules in the $CH_4\cdots(H_2O)_n$ cluster. Of course, a full picture of the hydration phenomenon of small solutes can be obtained only if larger clusters (1 molecule of solute and 24 or 36 molecules of water) are considered. In particular, larger clusters are required to understand the structure and intermolecular interactions in the second and probably third hydration layers. The same conclusion is likely to be valid for the $CH_4O\cdots(H_2O)_{12}$ and $H_2O\cdots(CH_4O)_{10}$ clusters considered in the present paper. The clusters considered in the present paper are helpful in understanding the interactions between a central solute molecule and its nearest neighbors (solvent molecules). (2) The ab initio quantum mechanical methods provide results at 0 K and zero pressure. Although the interactions depend on temperature and pressure, this dependence is expected to be weak up to normal conditions.[63,64] Indeed, it was shown[63,64] for pure water that the length of the hydrogen bonds has changed by only 4 parts per 1000 at most when the temperature varied by 100 K. Furthermore, a comparison between the local structure of supercooled water and liquid water under ambient conditions indicated that the number of nearest neighbors and the position of the maximum on the radial distribution function $g_{oo} = g_{oo}(r_{oo})$ (where r_{oo} is the distance between the oxygen atoms of two water molecules) are only slightly different.[65–69] In addition, a simple procedure to account for the effect of temperature was suggested.[63,64] One should note that the results of the ab initio quantum mechanical method, such as those obtained in the present paper, cannot provide information about the temperature effect on the hydrophobic hydration. In contrast, the molecular simulation methods can provide such information but involve model interaction potentials.

6. Conclusions

In this paper, the Møller–Plesset perturbation theory was applied to clusters formed by 1 molecule of methanol and 12 molecules of water or 1 molecule of water and 10 molecules of methanol. The goal was to determine the intermolecular distances and interaction energies between water and methanol molecules in the $CH_4O\cdots(H_2O)_{12}$ and $H_2O\cdots(CH_4O)_{10}$ clusters to compare the obtained results with the available experimental data and to shed some light on the nanostructure and molecular interactions in dilute solutions of methanol and water.

It was found that the solvent molecules in both clusters ($CH_4O\cdots(H_2O)_{12}$ and $H_2O\cdots(CH_4O)_{10}$) can be subdivided into two classes: (1) not H bonded with a central solute molecule and (2) H bonded with a central solute molecule. Although they are located at almost the same distances ($r_{C_MO_W}$) from the central solute molecule, these two classes possess striking differences regarding the interaction energies with a solute.

The solvent molecules, which are not H bonded with a central solute molecule, do not exhibit any peculiar features different from those of pure solvents (water or methanol). However, the H bonds in the clusters investigated demonstrated the presence of salient features, which seem to be important in the understanding of the molecular interactions in dilute mixtures formed by water and methanol.

In general, the ab initio quantum mechanical method that was employed in the present paper provided useful information about the hydrogen bonds in the systems investigated. In particular, it gives full information about two types of H bonds (WdM and MdW) between water and methanol molecules, including information about their energies and lengths; it provided a relationship between the numbers of the two types of H bonds in the $CH_4O\cdots(H_2O)_{12}$ and $H_2O\cdots(CH_4O)_{10}$ clusters. Such unique information could hardly be obtained by other methods.

Acknowledgment. We are indebted to the Center for Computational Research (CCR) of the University at Buffalo for the use of its facilities and to Professor H. F. King (Department of Chemistry, State University of New York at Buffalo) for helpful discussions.

References and Notes

(1) Franks, F.; Desnoyers, J. E. *Water Sci. Rev.* **1985**, *1*, 171.
(2) Franks, F.; Ives, D. *J. Rev. Chem. Soc.* **1966**, *20*, 1.
(3) Ott, J. B. *J. Chem. Thermodyn.* **1990**, *2*, 1129.
(4) Franks, F. In *Water: A Comprehensive Treatise*; Franks, F., Ed.; Plenum: New York, 1973; Vol. 2.
(5) Franks, F.; Reid, D. S. In *Water: A Comprehensive Treatise*; Franks, F., Ed.; Plenum: New York, 1973; Vol. 2.
(6) Tanford, C. *The Hydrophobic Effect: Formation of Micelles and Biological Membranes*, 2nd ed.; Wiley: New York, 1980.
(7) Ben-Naim, A. *Hydrophobic Interactions*; Plenum: New York, 1980.
(8) Belousov, V. P.; Panov, M. Y. *Thermodynamics of Aqueous Solutions of Non-Electrolytes* (in Russian); Khimiya: Leningrad, Russia, 1983.
(9) Nishikawa, K.; Kodera, Y.; Iijima, T. *J. Phys. Chem.* **1987**, *91*, 3694.
(10) Nishikawa, K.; Hayashi, H.; Iijima, T. *J. Phys. Chem.* **1989**, *93*, 6559.
(11) Hayashi, H.; Nishikawa, K.; Iijima, T. *J. Phys. Chem.* **1990**, *94*, 8334.
(12) Hayashi, H.; Udagawa, Y. *Bull. Chem. Soc. Jpn.* **1992**, *65*, 155.
(13) Soper, A. K.; Finney, J. L. *Phys. Rev. Lett.* **1993**, *71*, 4346.
(14) Turner, J.; Soper, A. K. *J. Chem. Phys.* **1994**, *101*, 6116.
(15) Micali, N.; Trusso, S.; Vasi, C.; Blaudez, D.; Mallamace, F. *Phys. Rev. E* **1996**, *54*, 1720.
(16) Dixit, S.; Poon, W. C. K.; Crain, J. *J. Phys.: Condens. Matter* **2000**, *12*, L323.
(17) Dixit, S.; Soper, A. K.; Finney, J. L.; Crain, J. *Europhys. Lett.* **2002**, *59*, 377.
(18) Jorgensen, W. L.; Madura, J. D. *J. Am. Chem. Soc.* **1983**, *105*, 1407.
(19) Okazaki, S.; Nakanishi, K.; Touhara, H. *J. Chem. Phys.* **1983**, *78*, 454.
(20) Nakanishi, K.; Ikari, S.; Okazaki, H.; Touhara, J. *Chem. Phys.* **1984**, *80*, 1656.

(21) Nakanishi, K. *Chem. Soc. Rev.* **1993**, *22*, 177.
(22) Meng, E. C.; Kollman, P. A. *J. Phys. Chem.* **1996**, *100*, 11460.
(23) Hernandez-Cobos, J.; Ortega-Blake, I. *J. Chem. Phys.* **1995**, *103*, 9261.
(24) Fidler, J.; Rodger, P. M. *J. Phys. Chem. B* **1999**, *103*, 7695.
(25) Kiselev, M.; Ivlev, D. *J. Mol. Liq.* **2004**, *110*, 193.
(26) Sum, A. K.; Sandler, S. I. *Ind. Eng. Chem. Res.* **1999**, *38*, 2849.
(27) Sum, A. K.; Sandler, S. I. *Fluid Phase Equilib.* **1999**, *160*, 375.
(28) Maheshwary, S.; Patel, N.; Sathyamurthy, N.; Kulkarni, A. D.; Gadre, S. R. *J. Phys. Chem. A* **2001**, *105*, 10525.
(29) Weinhold, F. *J. Chem. Phys.* **1998**, *109*, 367.
(30) Ludwig, R.; Weinhold, F. *Phys. Chem. Chem. Phys.* **2000**, *2*, 1613.
(31) Levine, I. N. *Quantum Chemistry*, 4th ed.; Prentice Hall: Englewood Cliffs, NJ,
(32) Szabo, A.; Ostlund, N. S. *Modern Quantum Chemistry: Introduction to Advanced Electronic Structure Theory*; Dover Publication: New York, 1996.
(33) Ruckenstein, E.; Shulgin, I. L.; Tilson, J. L. *J. Phys. Chem. A* **2003**, *107*, 2289.
(34) Eisenberg, D.; Kauzmann, W. *The Structure and Properties of Water*; Oxford University Press: Oxford, U.K., 1969.
(35) Narten, A. H.; Levy, H. A. *Science* **1969**, *165*, 447.
(36) Frank, H. S. *Science* **1970**, *169*, 635.
(37) Narten, A. H.; Levy, H. A. In *Water: A Comprehensive Treatise*; Franks, F., Ed.; Plenum: New York, 1972; Vol. 1.
(38) Stillinger, F. H. *Science* **1980**, *209*, 451.
(39) Narten, A. H.; Habenschuss, A. *J. Chem. Phys.* **1984**, *80*, 3387.
(40) Jorgensen, W. L. *J. Am. Chem. Soc.* **1980**, *102*, 543.
(41) Tauer, K.; Lipscomb, W. N. *Acta Crystallogr.* **1952**, *5*, 606.
(42) Nagayoshi, K.; Kitaura, K.; Koseki, S. et al. *Chem. Phys. Lett.* **2003**, *369*, 597.
(43) Frank, H. S.; Evans, M. W. *J. Chem. Phys.* **1945**, *13*, 507.
(44) Takamuku, T.; Yamaguchi, T.; Asato, M.; Matsumoto, M.; Nishi, N. *Z. Naturforsch., A: Phys. Sci.* **2000**, *55*, 513.
(45) van Erp, T. S.; Meijer, E. J. *Chem. Phys. Lett.* **2001**, *333*, 290.

(46) Palinkas, G.; Hawlicka, E.; Heinzinger, K. *Chem. Phys.* **1991**, *158*, 65.
(47) Bako, I.; Palinkas, G.; Heinzinger, K. *Z. Naturforsch., A: Phys. Sci.* **1994**, *49*, 967.
(48) Ferrario, M. Haughney, M.; McDonald, I. R.; Klein, M. L. *J. Chem. Phys.* **1990**, *93*, 5156.
(49) Freitas, L. C. G. *J. Mol. Struct.: THEOCHEM* **1993**, *101*, 151.
(50) González, L.; Mó, O.; Yáñez, M. *J. Chem. Phys.* **1998**, *109*, 139.
(51) Kirschner K. N.; Woods, R. J. *J. Phys. Chem. A* **2001**, *105*, 4150.
(52) Dunning, T. H. *J. Phys. Chem. A* **2000**, *104*, 9062.
(53) Woon, D. E. *Chem. Phys. Lett.* **1993**, *204*, 29.
(54) (a) Chalasinski, G.; Gutowski, M. *Chem. Rev.* **1988**, *88*, 943. (b) Jeziorski, B.; Szalewicz, K. In *Encyclopedia of Computational Chemistry*; Schleyer, P. v. R., Ed.; Wiley: New York, 1998; Vol. 2, p 1376.
(55) Liu, B.; McLean, A. D. *J. Chem. Phys.* **1973**, *59*, 4557.
(56) Boys, S. F.; Bernardi, F. *Mol. Phys.* **1970**, *19*, 553.
(57) Tsuzuki, S.; Uchimaru, T.; Matsumura, K.; Mikami, M.; Tanabe, K. *J. Chem. Phys.* **1999**, *110*, 11906.
(58) Frisch, M. J.; Delbene, J. E.; Binkley, J. S.; Schaefer, H. F. *J. Chem. Phys.* **1986**, *84*, 2279.
(59) Gonzalez, L.; Mo, O.; Yanez, M. *J. Chem. Phys.* **1998**, *109*, 139.
(60) Rossky, P. J.; Karplus, M. *J. Am. Chem. Soc.* **1979**, *101*, 1913.
(61) Ruckenstein, E.; Shulgin, I. *Ind. Eng. Chem. Res.* **1999**, *38*, 4092.
(62) De Jong, P. H. K.; Wilson, J. E.; Neilson, G. W.; Buckingham, A. D. *Mol. Phys.* **1997**, *91*, 99.
(63) Dougherty, R. C. *J. Chem. Phys.* **1998**, *109*, 7372.
(64) Dougherty, R. C.; Howard, L. N. *J. Chem. Phys.* **1998**, *109*, 7379.
(65) Narten, A. H.; Thiessen, W. E.; Blum, L. *Science* **1982**, *217*, 1033.
(66) Kimura, N.; Yoneda, Y. *Phys. Lett. A* **1982**, *92*, 297.
(67) Bosio, L.; Chen, S. H.; Teixeira, J. *Phys. Rev. A* **1983**, *27*, 1468.
(68) Corban, R.; Zeidler, M. D. *Ber. Bunsen-Ges. Phys. Chem.* **1992**, *96*, 1463.
(69) Botti, A.; Bruni, F.; Isopo, A.; Ricci, M. A.; Soper, A. K. *J. Chem. Phys.* **2002**, *117*, 6196.